ENVIRONMENTAL CONSIDERATIONS
IN ENERGY PRODUCTION

EDITED BY JOHN R. CRAYNON

Published by the
Society for Mining, Metallurgy, and Exploration, Inc.

Society for Mining, Metallurgy, and Exploration, Inc. (SME)
12999 E. Adam Aircraft Circle
Englewood, Colorado, USA 80112
(303) 948-4200 / (800) 763-3132
www.smenet.org

SME advances the worldwide mining and minerals community through information exchange and professional development. SME is the world's largest association of mining and minerals professionals.

Electronic edition published 2013.

Cover photo credit: Leigh-Anne Krometis

ISBN 978-0-87335-380-9
Ebook 978-0-87335-381-6

Library of Congress Cataloging-in-Publication Data

Cataloging-in-publication data has been applied for.

Contents

Preface vii

1.1 **WATERSHEDS**

Water Quality and Appalachian Mining: Analysis of Watershed-Based Conclusions 2

A Value of Information Study Examining the Effectiveness of Spatial Datasets for Modeling Landscape to Water Quality Relationships 3

Water Quality Permitting Under Managed Discharge 4

Scenario Analysis and the Watershed Futures Planner: Predicting Future Aquatic Conditions in an Intensively Mined Appalachian Watershed 5

The Mine Land Stewardship Initiative 20

2.1 **ENVIRONMENTAL CONSIDERATIONS IN COAL PREPARATION AND WASTE MANAGEMENT**

Application of Electronic Sorting Technology for Reducing Environmental Impacts of Coal Processing 22

Stabilization of Fine Coal Waste as an Alternative Disposal Technique 32

Application of a Pilot-Scale Plate Filter Press in Dewatering Coal Slurries 42

Diesel Range Organics In Coal Preparation 51

Minimizing the Environmental Impacts of Coal Waste Disposal 60

2.2 **COMMUNITY WELL-BEING**

Decision Tools for Integrating Sustainability Considerations in Large Capital Projects 78

The Mining Industry and Community Poverty Across Appalachia 86

Edgework and Boundary Crossings: Assessing Foundations for Public Ecology in the Appalachian Region 99

Determinants of Local Community Acceptance of Energy Resources and Materials Mining 111

Ethics and Community Well-Being 121

3.1 **COAL, GAS, AND ENERGY PRODUCTION**

Appalachian Coal: Past, Present, and Future 124

Environmental Considerations and Operational Parameters of Underground Coal Gasification 125

The Marcellus Shale Gas Play and the Gas Migration/Stray Gas Problem 137

Experiences in Reservoir Model Calibration for Coal Bed Methane Production in Deep Coal Seams in Russell County, Virginia 140

Sustainable Degasification Strategies: A Case Study in the Qinshui Basin of China 153

3.2 PUBLIC HEALTH AND ENERGY PRODUCTION

Evaluation of Research Addressing Chronic Health in Coal-Dependent Communities in Central Appalachia . . . 172
Comparison of Mortality Rates Between Appalachian Coal Mining Counties with Non-Coal Mining . . . 189
Population and Exposure Characteristics of Coal Mining and Non-Coal Mining Counties in West Virginia . . . 202
Inventory of Bacterial Impairments in Central Appalachia . . . 214
Identification of Toxic Agents and Potential Exposure Routes to Appalachian Coal Mining Communities . . . 229

4.1 TOTAL DISSOLVED SOLIDS AND ENVIRONMENTAL IMPACTS 1

Comparison of Groundwater Seepage Modeling in Conventional and Geomorphic Valley Fill Design . . . 246
Predicting Total Dissolved Solids Release from Overburden in Appalachian Coal Fields . . . 255
Natural Attenuation of TDS in Mountaintop Mines . . . 265
Predicting TDS Release from Appalachian Mine Spoils . . . 275
Long-Term Trends of Electrical Conductivity in Streams Flowing Out of Headwater Valley Fills in Virginia . . . 286

4.2 ENVIRONMENTAL CONSIDERATIONS IN UNDERGROUND MINING

Event Tree Failure Mode Development for Mining Near Bodies of Water . . . 288
Applications of Subsurface Subsidence Model to Study Longwall Subsidence Influences on Overburden Hydrological System . . . 297
Addressing the Effect of Sloping Terrain on Ground Movements Due to Underground Mining . . . 308
Application of Subsidence Prediction Methodologies for Sizing Barrier Pillars for Stream Protection in Appalachia . . . 319
Preliminary Assessment of Factors Affecting Springs Undermined by Longwall Mines in Pennsylvania . . . 336

5.1 STREAM BIOTA IMPACTS

Methods for Evaluating the Effects of a Simulated Mine Effluent with Elevated Ionic Concentration to Field Collected Benthic Macroinvertebrates . . . 348
Exploring Benthic Impairment and Total Dissolved Solids in the Dumps Creek Watershed . . . 362
Benthic Macroinvertebrate Response to Total Dissolved Solids in Coal Mine Streams . . . 371
How Well Does Electrical Conductivity Predict Stream Impairment? . . . 372
Evaluation of Current Measures of Aquatic Biological Integrity in the Central Appalachian Coalfields: Efficacy and Implications . . . 381

5.2 LANDFORMS AND GEOMORPHIC RECLAMATION

Sustainable Landform Construction on Coal Mining Operations That Require Valley Fills . . . 396
Methodology for Geomorphic Landform Design of Valley Fills in Appalachia Surface Mine Reclamation . . . 397

Investigating Regional Design Parameters in the Design of West Virginia Valley Fills to Support Application of Geomorphic Landform Design Principles 405
Slope Stability Analysis of Geomorphic Landform Profiles Versus Approximate Original Contour Applied to Valley Fill Designs........ 415

6.1 **SELENIUM IN THE ENVIRONMENT**
Subsurface Sequestration of Leached Waste-Rock Selenium in Reactive Ferrihydrite Beds..... 426
Selenium Concentrations and Solid-State Chemical Speciation in Black Shale Associated with a West Virginia Coal Seam 438
Selenium Treatment for Mine Water Discharge Compliance 443
Strategies for Remediation of Selenium-Laden Waters 450
Analysis of Low-Level Selenium in Coal Mine Discharges 451

6.2 **STREAM RESTORATION**
Natural Streams: Biological Effects of Natural Channel Design for Aquatic Habitat Restoration........ 454
Hydrologic Assessment of the Guy Cove Stream Restoration Project........ 463
Guy Cove Stream Restoration Project: Recreating a Headwater Stream System on a Head-of-Hollow Fill........ 464
Leaf Litter Breakdown as a Functional Assessment Tool for Appalachian Coal Mine Stream Restorations........ 465
Substrate and Seeding Treatment Effects on Native Tree Establishment and Herbaceous Cover on an Appalachian Coal Mine: First Growing Season Results 467

7.1 **TOTAL DISSOLVED SOLIDS AND ENVIRONMENTAL IMPACTS 2**
Conductivity Sensor Accuracy and Temporal Stability: A Laboratory and Field Study........ 470
Screening-Level Specific Conductivity Testing of Drill Core Strata 471
Field Data and Regulatory Aquatic Life Criteria 472
Chronic Toxicity Testing in Mining Influenced Streams of West Virginia 474

7.2 **ENVIRONMENT, COAL MINING, AND SUSTAINABILITY**
Surface Coal Mine Blasting Optimization and Mitigation of Environmental Impacts: Mine to Fill 476
Production, Cost, and Environmental Impact of Loading Equipment in Surface Coal Mining in Appalachia........ 490
NOx Emission of Equipment and Blasting Agents in Surface Coal Mining 501
Micro-Hydroelectric Power Generation Utilizing Mine Drainage........ 512
Minepool Geothermal in Pennsylvania 520
Sustainability in Surface Coal Mine Operations: The Importance of Good Geotechnical Design 529
Ethics in Energy Sustainability 530

Index........ 531

Preface

The Environmental Considerations in Energy Production symposium provides an avenue for researchers to present their reviews and findings for a wide variety of topics on how coal mining, oil and gas production, electricity generation, and other related activities impact the environment and the communities in which we live and work.

This proceedings volume contains not only complete papers, but also abstracts for presentations where the presenters do not feel that a full paper is currently warranted. These abstracts are included to provide a resource to readers who may be interested in contacting those individuals. It is hoped this book will give you a flavor of the breadth and depth of this symposium and the topics that come under the umbrella of such a meeting. The papers and presentations range from reviews of work previously completed and discussions of preliminary investigations to thorough reports of research and recommended changes in how things are done.

The full papers in this volume (not the abstracts) were subjected to a double-blind peer-review process, wherein neither the reviewers nor authors were identified. Each paper was reviewed by at least two experts in the fields of work related to the topic. After the reviewer comments were received, authors were provided with those remarks to make changes to their papers. Authors were encouraged to accept reviewer comments but were also given the opportunity to disagree with the reviewers and provide reasons why particular comments were not accepted. As the editor, I had the final say as to whether the author's rationale was adequate or whether further changes were necessary.

I extend my thanks to Brad Kelley and Rita Caldwell of the Virginia Center for Coal and Energy Research for their critical assistance in editing and compiling the papers in this volume. The SME staff was also of immense and essential assistance in pulling this work together. Michael Karmis provided wise and experienced oversight of this effort, and I am honored to learn from and work with him.

To the nearly 100 brave souls, both longtime professional colleagues and respected new acquaintances who provided peer reviews, I give my humble gratitude. Your dedication and time showed in the thoughtful and thorough reviews you provided. Through your efforts, we together continue to advance science and engineering to address real and sometimes difficult challenges.

And finally, to the authors and presenters of the works contained herein, I give my hearty congratulations for a job well done. It has been a privilege to work with each of you and to read these papers and abstracts. I am encouraged that, as a society, we can successfully address the issues we face in energy production as a result of your work.

John R. Craynon
Editor

1.1 Watersheds

Water Quality and Appalachian Mining: Analysis of Watershed-Based Conclusions 2

A Value of Information Study Examining the Effectiveness of Spatial Datasets for Modeling Landscape to Water Quality Relationships 3

Water Quality Permitting Under Managed Discharge 4

Scenario Analysis and the Watershed Futures Planner: Predicting Future Aquatic Conditions in an Intensively Mined Appalachian Watershed 5

The Mine Land Stewardship Initiative 20

Water Quality and Appalachian Mining: Analysis of Watershed-Based Conclusions

Steven Gardner
ECSI, LLC, Lexington, Kentucky

ABSTRACT

Surface coal mining operations in central Appalachia are coming under increasing scrutiny and regulation as operators attempt to obtain permits for pollutant discharge or hollow fills. Well-publicized studies, such as those by Gregory Pond and others, have raised questions about the long-term impact of Appalachian mining on aquatic life in mountain streams. Recent summary work such as that by Palmer and others and the EPA's evaluation of mountaintop mining draw broad conclusions about mining's impact on water quality, but the conclusions are based on studies with varied methodologies. The variation is particularly evident when the studies attempt to quantify the extent of mining and other changes in land use.

State regulatory authorities, the U. S. Army Corps of Engineers, and the U. S. Environmental Protection Agency are making critical decisions on water discharge permits issued under Section 402 of the Clean Water Act and dredge and fill permits issued under Section 404 of the Clean Water Act. In connection with evaluation of these permits, EPA and the Corps entered into a Memorandum of Understanding to conduct "Enhanced Surface Coal Mining Pending Permit Coordination Procedures." The EPA also issued an executive order, "Detailed Guidance: Improving EPA Review of Appalachian Surface Coal Mining Operations under the Clean Water Act, National Environmental Policy Act and the Environmental Justice." These recent actions by the Corps and by the EPA have been challenged in the federal courts because of their alleged failure to follow the notice and comment process in adopting rules and regulations.

This paper is a metastudy of the primary research on which the EPA and others base their evaluation of mining's impact. The paper will evaluate the methodology and assumptions of those studies with particular emphasis on how the extent of mining is characterized. For example, the paper will consider whether the mining at issue was conducted before the adoption of the Surface Mining Control and Reclamation Act versus more modern, highly regulated mining. It will also evaluate the extent to which other land use patterns have been included for consideration. Finally, the studies will be evaluated regarding whether or how they included consideration of the distinctive geological, geomorphological, and biogeographical characteristics of the affected watersheds. The analysis in the proposed paper will provide valuable insights and a framework for meaningful public comment when the regulatory actions under consideration are presented for the notice and comment.

ACKNOWLEDGMENT

The author thanks W. Blaine Early III and Douglas Mynear for their contributions to this paper.

A Value of Information Study Examining the Effectiveness of Spatial Datasets for Modeling Landscape to Water Quality Relationships

Michael P. Strager
Division of Resource Management, West Virginia University, Morgantown, West Virginia

J. Todd Petty
West Virginia University, Morgantown, West Virginia

Paul Ziemkiewicz
West Virginia Water Research Institute, West Virginia University, Morgantown, West Virginia

ABSTRACT

A critical part of building landscape to water quality relationships involves the timely and accurate mapping of landscape features at a scale, resolution, and detail necessary to meet the modeling needs. There are many imagery choices available to classify and map landscape features. Each tends to differ temporally, by positional accuracy, by the attributes that can be derived, and by cost. As part of our ongoing work to develop and maintain a regional spatial database for the ARIES project, our goal was to compare the utility of different imagery and their potentially derived cover types in the statistical modeling for stream conditions. We wanted to answer the question of which imagery dataset and classes are most effective in statistically modeling stream conditions. We assessed the imagery inputs in terms of classes that can be obtained, cost to obtain end product, time to acquire, final thematic map accuracy, goodness of model fits, and the significance of input variable coefficients in a boosted regression tree model. The importance of asking the most appropriate questions was critical to match the geospatial data collection to describing the landscape to receiving water relationship. Statistical models provided a basis to evaluate the analyzed spatial data to both water quality and benthic macro invertebrate endpoints. The decision analysis approach called "value of information" provided the theoretical framework for analyzing the problem and matching a desired set of research questions to the needed inputs and classification approach. The study provides a decision tree matrix for how best to assemble spatial data to match the research questions being asked regarding the landscape to receiving water interaction.

ACKOWLEDGMENT

This study was sponsored by the Appalachian Research Initiative for Environmental Science (ARIES). The views, opinions and recommendations expressed herein are solely those of the authors and do not imply any endorsement by ARIES employees, other ARIES-affiliated researchers or industrial members. Information about ARIES can be found at http://www.energy.vt.edu/ARIES.

The authors thank Aaron E. Maxwell for his contribution to this paper.

Water Quality Permitting Under Managed Discharge

P.F. Ziemkiewicz
West Virginia Water Research Institute, West Virginia University, Morgantown, West Virginia

ABSTRACT

The Federal Clean Water Act (CWA) recognizes pollutants as either toxic or non-toxic. Non-toxic pollutants, while not hazardous to humans, may impair designated stream uses such as: fishery, public water supply and recreation. Impairment is defined through numeric standards or narrative water quality standards. Narrative standards such as the West Virginia Stream Condition Index (WVSCI) rely on the composition of the stream's insect population to indicate stream health. These stream condition indices are then assigned a critical value below which a stream is deemed impaired. Regulatory agencies are faced with developing discharge limits for regulated discharges such that streams will not be impaired. This normally involves estimating the effects of a proposed discharge on stream water quality then an estimation of how it will affect numeric standards. Often the effect is estimated by determining the effect of maximum discharge load at minimum stream flow. The CWA, however, only regulates point source discharges from active operations. As a result in order to bring a stream into compliance with narrative standards, further restriction of point sources is often the only option exercised by the regulatory agencies. This approach often imposes significant burdens on the holders of 'point source' discharges and, since other development activities and abandoned mines are often significant causes of impairment it often fails to improve the stream's condition. A process is needed that yields greater benefits to the stream. The challenge is finding a way to ensure regulatory compliance accountability and measurable environmental benefits.

This paper discusses a case study in which a voluntary, managed program has maintained stream water quality at predetermined levels through modulation of point source discharge rates. Implications for other water quality challenges are discussed.

ACKNOWLEDGMENT

The author thanks R.G. McLusky for his contribution to this paper.

Scenario Analysis and the Watershed Futures Planner: Predicting Future Aquatic Conditions in an Intensively Mined Appalachian Watershed

J. Todd Petty
Division of Forestry, Natural Resources, and Design, West Virginia University, Morgantown, West Virginia

Michael P. Strager
Division of Resource Management, West Virginia University, Morgantown, West Virginia

Eric M. Merriam
Division of Forestry, Natural Resources, and Design, West Virginia University, Morgantown, West Virginia

Paul F. Ziemkiewicz
West Virginia Water Research Institute, West Virginia University, Morgantown, West Virginia

ABSTRACT

Scenario analysis is a quantitative process for evaluating potential future conditions of aquatic resources under various development/mitigation scenarios. We have constructed an analytical software package entitled the Watershed Futures Planner (WFP), which integrates data management, statistical and process modeling, results summary and visualization, and decision support into a comprehensive analytical tool. The objectives of this study were to describe our overall scenario analysis approach and apply this approach to questions of mine development and watershed restoration within the Coal River, an intensively mined watershed in the mountaintop removal mining region of West Virginia. Application of theWFP in the Coal River watershed produced the following insights: (1) Additional surface mining in the absence of restorative action will likely produce measurable impacts to water quality and biological condition; (2) The greatest benefits to water quality in the future would most likely come from managing the effects of deep mine effluents. These benefits would be overwhelmed, however, in the absence of improved surface mine reclamation if all currently permitted surface mines were mined out; (3) The greatest benefits to biological conditions would most likely come from managing the effects of residential development. In fact, we predict that substantial benefits to biological conditions could be achieved through mitigation of development impacts even if all of the currently permitted mines are mined out. We conclude that potential impacts from future surface mining are manageable, but only when coupled with strategic mitigation of pre-existing impacts caused by other stressors. Application of our scenario analysis approach throughout the central Appalachian region may be used to identify additional opportunities for watershed scale improvements to aquatic conditions in intensively mined watersheds.

INTRODUCTION

Changes in land cover can affect water quality through alteration of flows and introduction of contaminants. These changes have consequences for aquatic resources at a watershed scale (Allan 2004). For example, clearing of forests and large-scale surface mining in small headwater catchments can influence water quality and biological conditions downstream (Lindberg et al. 2011, Merriam et al. 2011). Likewise, residential development activities can dramatically alter stream flows (Kepner et al. 2012), water quality (Utz et al. 2011), and invertebrate assemblages (King et al. 2011) throughout a watershed.

Because landscape change is a wide-ranging problem throughout the US, analytical tools are needed that allow managers to make informed decisions regarding watershed development, conservation, and restoration. Kepner et al. (2012) argue that in order to contribute to the sustainable management of water resources in actively developing watersheds, scientists must more accurately describe and predict how land use decisions today might impact future aquatic ecosystem conditions.

However, complexities of ecosystem functioning and lack of scientific information at relevant spatial scales make it difficult for aquatic resource managers to understand and predict the potential consequences of present-day land use decisions on future conditions of water resources (Kepner et al. 2012). For example, despite several recent studies examining the direct effects of surface mining on downstream ecosystems, our understanding of how mining interacts with other land use activities to affect aquatic conditions remains weak (but see Petty et al. 2010, Merriam et al. 2011). This is an important problem because individual stressors rarely occur in isolation, and consequently, quantitative measures of stressor interactions are needed to improve our ability to predict future watershed conditions in actively developing watersheds (Seitz et al. 2011, Kepner et al. 2012).

Alternative future scenario analysis (henceforth "scenario analysis") is a quantitative process for evaluating potential future conditions of aquatic resources under a range of watershed development scenarios (Mahmoud et al. 2011, Kepner et al. 2012). Scenario analysis can improve the utility of scientific information and increase the transparency of the decision making process. Consequently, scenario analysis has become recognized as a useful tool for solving highly contentious water resource problems; it can provide a better understanding of how and why an alternative future may evolve and the necessary steps that decision makers are required to take in order to adapt to such a future (Kepner et al. 2012).

The mountaintop removal valley fill (MTR-VF) coal mining province of central Appalachia represents a region in which management decisions surrounding land use and aquatic resources are acute and proper action is imperative. Recent research has identified additive effects of multiple surface mines on downstream aquatic resources (Bernhardt et al. 2012, Lindberg et al. 2012). Furthermore, our own research has identified context dependent effects of coal mining on water quality and biological conditions, where impacts of surface mines often depend on the precise spatial location of mines and the presence and extent of pre-existing stressors, such as other surface mines, deep mines, and residential development (Petty et al. 2010, Merriam et al. 2011, Merriam et al., unpublished manuscript).

In this paper, we demonstrate how scenario analysis can be used to better understand how multiple stressors interact to influence aquatic conditions within an intensively mined central Appalachian watershed. Specifically, the objectives of this study are to: (1) describe our overall scenario analysis approach, including: data sources, statistical modeling, GIS modeling and data management, and final scenario assessment and application; and (2) conduct a scenario analysis of mine development and mitigation within the Coal River watershed in south-central West Virginia. We demonstrate how strategic mitigation decisions, when coupled with mine development decisions, can potentially result in improved watershed scale conditions despite expanding impacts from new mine development.

ALTERNATIVE FUTURES SCENARIO ANALYSIS

Scenario Analysis

Formal scenario analysis is a method of using data and models to predict the effects of alternative management actions. Phases of scenario analysis include: (1) Scenario Definition; (2) Scenario Construction; (3) Scenario Analysis; and (4) Scenario Assessment (Liu et al. 2008; Kepner et al. 2012). During scenario development,

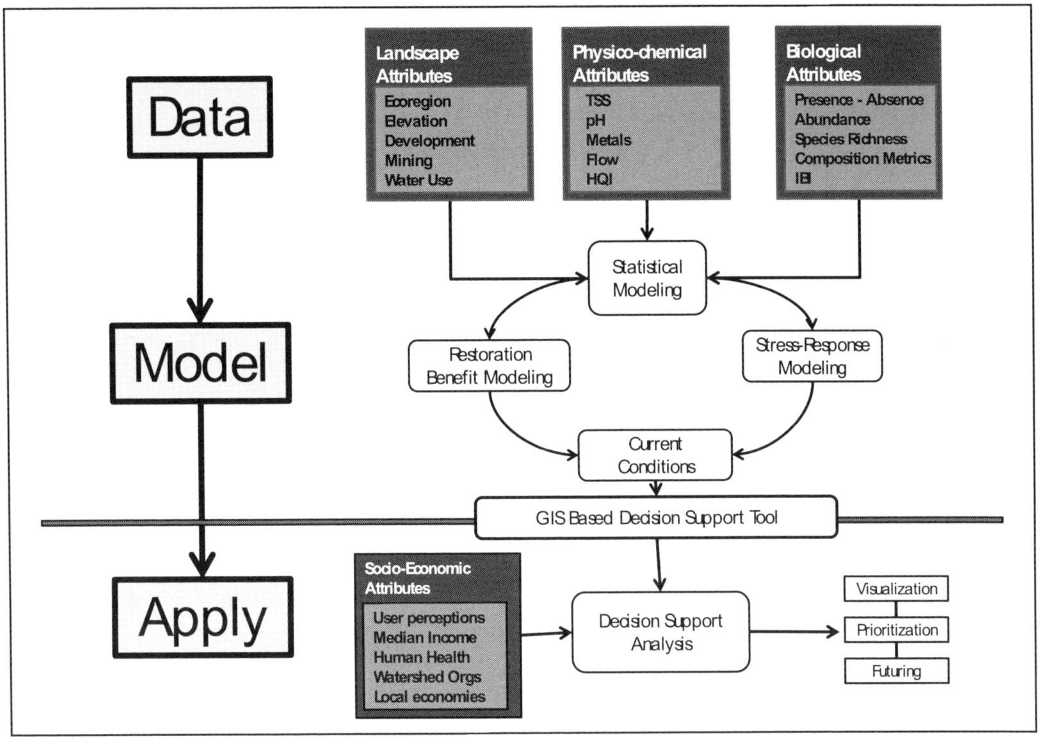

Figure 1. Flow diagram of the overall watershed modeling and scenario analysis process. The Watershed Futures Planner (WFP) is a decision support tool that integrates data, modeling algorithms, mapping tools, and scenario analysis within a GIS-based spatially explicit framework.

analysts determine which driving forces to evaluate, the environmental endpoints to assess, and characteristics that will be used to differentiate among scenarios. Scenario construction consists of the actual modeling of environmental response to alternative landscape changes, and outcomes of scenario projections are compared during scenario analysis. Finally, during scenario assessment, alternative scenario outcomes are presented to stakeholders for discussion and final decision making. It is during this phase that socioeconomic and institutional constraints must be considered. The purpose of scenario analysis is to identify priority areas within a watershed for restoration and conservation, as well as to highlight the potential value of development techniques that reduce impacts to aquatic resources.

Watershed Futures Planner

The Watershed Futures Planner (WFP) is a comprehensive modeling system that our research team has developed over the past ten years as a tool for conducting multi-scale scenario analyses in highly impacted or rapidly developing watersheds (Figure 1) (Petty and Thorne 2005, Strager et al. 2009, 2010, Petty et al. 2010, Merriam et al. 2011). The WFP integrates data along with empirical and process-based models to predict changes in aquatic ecosystem conditions over time that may result from alternative watershed development/restoration scenarios. Outcomes from the WFP are then used by stakeholders to make informed, objective, science-based decisions for watershed restoration and conservation (Figure 1).

The WFP contains a set of analysis tools to enhance decision making for local, state, federal agencies, and stakeholders. The modeling platform integrates spatial data with watershed networks and statistical approaches (Figure 2). Instrumental to the integration is Microsoft .NET architecture with C# utilizing ESRI ArcObjects libraries. The WFP was developed as an independent third party set of software tools that resides on the industry standard GIS proprietary software ESRI ArcGIS Desktop 10.x. The underlying mapping tools in ESRI GIS software with National Hydrography Data allow cumulative watershed calculations. R statistical software was used for performing statistical modeling (e.g., boosted regression trees; BRT) procedures on the GIS watershed data. R software is always evolving and provides a free to use but strong development platform. For the statistical modeling BRT provides a combination of two different algorithms of regression trees (models that relate a response to their predictors by recursive binary splits) and boosting (an adaptive method for combining many simple models to give improved predictive performance). R Scripting language is used to run the BRT models with map libraries for spatial calculations. The integration of the C# .NET watershed models with the statistical package R was performed using an independent execution library rather than functional integration to preserve the integrity of two different languages which helps for future maintenance and updates. This provides limitless options depending on available landscape variables and applicability of boosted regression on the study areas. This modeling platform allows a unique combination of GIS and statistical tools that produce BRT results. The software identifies a set of columns read from a csv file that corresponds with spatial data (GIS Map Attribute table) using Unique ID field to populate resulting data. This data then can be modified using the GIS interface for symbology and other geoprocessing tools.

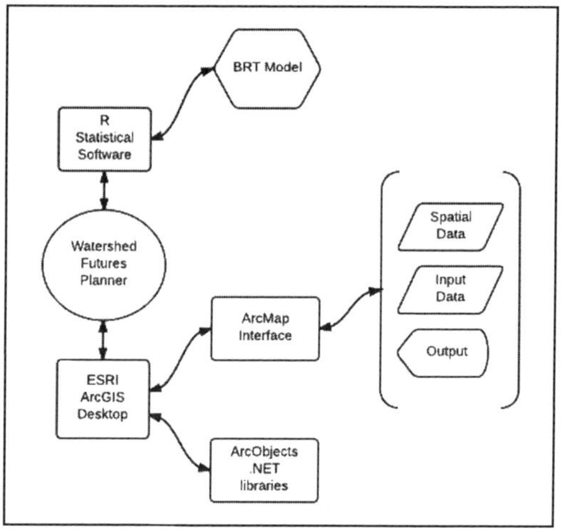

Figure 2. Watershed futures planner software structure

Data Structure and Modeling Approach

Recent advances in remote sensing and geographic information system (GIS) technologies have vastly improved our ability to construct and link high resolution land cover and use data to in-stream conditions (Carlisle et al. 2009). Raster-based land cover and use datasets are typically used to characterize and quantify landscape characteristics upstream of the site of interest. For example, numerous studies have linked degraded aquatic conditions to increases in residential/urban (Utz et al. 2010; King et al. 2011), agricultural (Cuffney et al. 2000; King et al. 2005), and mining-related (Petty et al. 2010; Merriam et al. 2011; Lindberg et al. 2012) land cover. However, ancillary GIS data are often necessary to elucidate cause-effect relationships between land use and altered in-stream condition. Additional anthropogenic features, such as residential and commercial structures (Merriam et al. 2011) and dams (Jones et al. 2001), have been shown to be valuable when modeling and predicting in-stream conditions.

Empirical modeling approaches have proven very effective at modeling in-stream conditions based on statistical relationships with land cover and use data. Studies have related landscape-based indicators to altered in-stream physical and chemical characteristics (Utz et al. 2011), macroinvertebrate (King et al. 2011; Merriam et al. 2011) and fish community structure (Utz et al. 2010), and ecosystem functions (Clapcott et al. 2010).

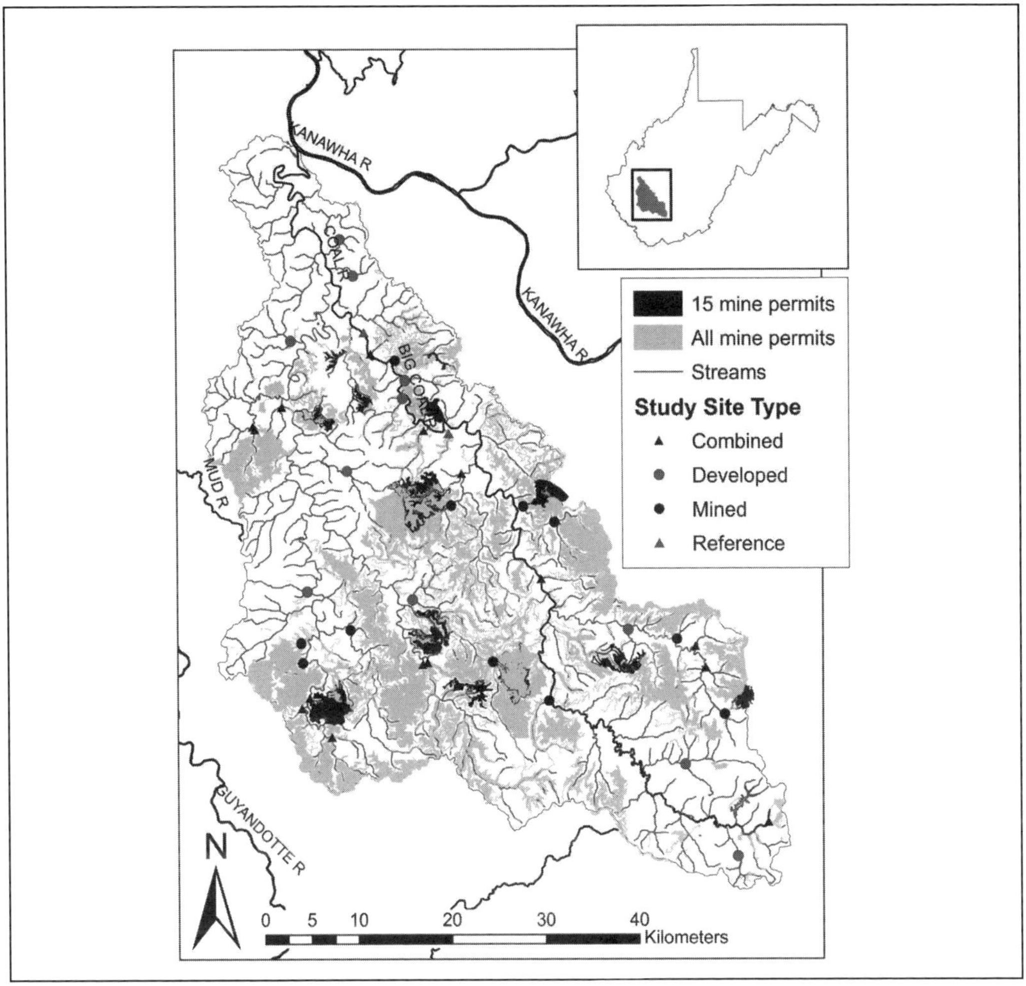

Figure 3. Map of the Coal River watershed illustrated with 1:100,000-scale stream segments, study sites, and surface mining permits. All mine permits and 15 mine permits correspond to alternative future mining scenarios. Stream flow is to the northwest.

Despite these successes, geospatial data have presented statistical problems that complicate traditional linear modeling approaches (i.e., skewed distributions, complex covariance structures, interactions, and nonlinear effects; King et al. 2005). Machine learning techniques, specifically classification and regression tree-based methods (e.g., boosted regression trees; BRT), overcome many of these statistical issues and have been increasingly used to model ecological data sets (Elith et al. 2008; Carlisle et al. 2009). BRT is an extension of classification and regression techniques that attempts to improve model structure and predictive performance by iteratively fitting many simple models from subsets of data and then combining them to offer better estimations of the response (Elith et al. 2008). Moreover, because BRT is a tree-based method, it can handle complicated data (i.e., skewed distributions, non-linearity, and multiple data types) and model complex interactions (Elith et al. 2008; Carlisle et al. 2009).

COAL RIVER SCENARIO ANALYSIS

Study Area

The Coal River is a headwater 8-digit Hydrological Unit Code (HUC) watershed located in south-central West Virginia (WV; Figure 3). The Coal River drains approximately 2307 km^2 and flows northwest until its confluence with the Kanawha River. The study basin is approximately 80% forested. Developed and mining-related cover classes represent the dominant land uses within the study region. Steep topography confines development to narrow floodplains, while coal mining is focused along ridgelines and headwater catchments. Surface mining land use comprises 9% of the total watershed area, and there are >425 underground mine NPDES permits. The low S content of the coal and calcareous nature of surrounding strata result in alkaline mine drainage typified by increased pH, alkalinity, ionic strength (SO_4^{2-}, Ca^{2+} $Mg^{2+}$and HCO^{3-}), and trace elements (Mn and Se; Pond et al. 2008; Lindberg et al. 2011).

Methods

We quantified local (i.e., for individual stream segments) and cumulative landscape (i.e., all upstream land area) characteristics for all 1:24,000 segment-level watersheds (SLWs) within the Coal River basin (Strager et al. 2009; Merriam et al. 2011). We used spatial analysis functions in ArcView 3.3 (Environmental Systems Research Institute, Redlands, California) in conjunction with flow tables developed by the West Virginia Resources Analysis Center to calculate cumulative measures of all landscape attributes (Strager et al. 2009).

We used 2009 and 2010 National Agriculture Imagery Program (NAIP) orthophotography to map land cover for the study area. NAIP orthophotography was obtained from the Aerial Photography Field Office of the Farm Service Agency and has a 1 meter pixel resolution at a scale of 1:10,000. We used the Feature Analyst software (VLS 2004) to differentiate between open water, forest, grassland/pasture, and barren. We then used the mining-permit boundaries layer developed by the West Virginia Department of Environmental Protection (WVDEP) to differentiate mining-related open water (i.e., slurry impoundments), barren (i.e., active mine lands) and grasslands (i.e., reclaimed mine lands) from non-mining open water, barren (i.e., residential impervious surfaces) and grasslands. We manually digitized valley fill faces and kept their areal extent separate from other reclaimed mine lands. We used national pollution discharge elimination system (NPDES) data obtained from the WVDEP to calculate the cumulative density of all NPDES permits (#/km^2). We also calculated underground mine NPDES permit density to represent deep mine influence. We used the 2003 Statewide Addressing and Mapping Board structures layer in conjunction with information on public service districts to calculate the density of serviced (i.e., connected to public sewage system) and un-serviced residential and commercial structures (#/km^2). We calculated road density (km/km^2) using data created by US Census Bureau, Geography Division. We also calculated the cumulative percentage of coal belonging to the Allegheny and Kanawha formations. Geology data were created by the WV Geological and Economic Survey.

We selected 40 segment-level watersheds as study sites within the Coal River watershed (Figure 3) across a range of influence from residential development and mining (Figure 4). Eleven sites made up a distinct residential gradient (structure density: 5–76 structures/km^2) and had <2% surface mining. Eleven sites were part of a mining gradient (surface mining: 7 to 64%) and a structure density <2 structures/km^2 (Figure 4). We included 2 reference sites with minimal mining and development (<2% surface mining and <2 structures/km^2) (Figure 4) to represent the best possible conditions within the study area and anchor each gradient. The remaining 16 sites, referred to as combined sites, were affected by a combination of residential development and coal mining (Figure 4). Drainage areas ranged from 1–36 km^2, with similar averages among mined, developed, and combined sites (12.5, 12.9, and 11.3 km^2, respectively). All sites were selected to

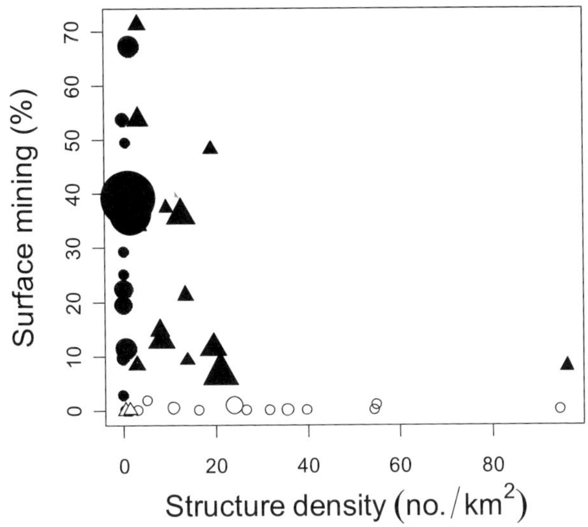

Figure 4. Magnitude of mining and residential development for study sites. Closed circles = mined; open circles = developed; closed triangles = combined; open triangles = reference. Symbol size = # deep mine NPDES permits, with increasing size corresponding to an increasing number of permits.

be independent of one another and not linked by flow.

We visually assessed physical-habitat quality at each site with US Environmental Protection Agency (EPA) rapid visual habitat assessments (RVHA; Barbour et al. 1999). Habitat was assessed during normal baseflow conditions from July 15 to August 15, 2011. We delineated reach lengths as 40× mean stream width, with minimum and maximum reach lengths of 150 and 300 m, respectively. We took water samples during the summer of 2010 (August 9–27). Instantaneous measures of temperature (°C), pH, specific conductance (µS/cm), and dissolved O_2 (mg/L) were obtained with a YSI 650 equipped with a 600XL sonde (Yellow Springs Instruments, Yellow Springs, Ohio) calibrated prior to each sampling date. We collected a 250-mL filtered sample from each site with a Nalgene filtration unit (mixed cellulose ester membrane filter, 0.45-µm pore size). The filtered sample was fixed with nitric acid to a pH <2 and analyzed for Al, Ba, Ca, Cd, Cr, Fe, Mg, Mn, Na, Ni, Se, Zn, and K. We also collected 2 unfiltered samples from each site. The first was fixed with sulfuric acid to a pH <2 and analyzed for NO_3 and NO_2, and total P. The second was analyzed for total and bicarbonate alkalinity, Cl, SO_4, and total dissolved solids. We stored samples at 4°C until analyses were completed at Research Environmental & Industrial Consultants, Inc., Beaver, WV. EPA standard methods were used for all analyses. We took duplicate samples from 10% (n = 4) of sites. One field blank was obtained on each sampling date.

We sampled benthic macroinvertebrate communities during the summer of 2010 (August 9–27) following procedures established by WV Department of Environmental Protection (WVDEP) Watershed Assessment Program and the USEPA Rapid Bioassessment Protocols for wadeable streams (Barbour et al. 1999; WVDEP 2009). We obtained kick samples (net dimensions 335 × 508 mm with 500-µm mesh) from 4 riffles at each site. We combined organisms and debris from the 4 kick samples into a single composite sample for each site and preserved samples in 95% ethanol. In the laboratory, macroinvertebrates from combined kick samples were subsampled using the 200-count method described by WVDEP (2009). Individuals were enumerated and identified to genus with keys in Merritt and Cummins (2008). We used two benthic macroinvertebrate multimetric indices (MMI) to assess ecological condition at each site. The West Virginia Stream Condition Index (WVSCI; Gerritsen et al. 2000) is a family-level index applied statewide within a single index period. The Genus Level Index of Most Probable Stream Status (GLIMPSS; Pond et al. 2012) is a genus-level index calibrated by region and season. We used GLIMPSS (CF), a version of GLIMPSS not requiring genus-level identification of individuals within the taxa Chironomidae and Oligochaeta. Both WVSCI and GLIMPSS are scored on a scale of 0–100, with scores <68 and <54 being categorized as impaired, respectively. Duplicate samples were obtained and identified from 10% (n = 4) of sites.

Boosted Regression Tree Models

We constructed boosted regression tree (BRT) models to predict current physical (RVHA), chemical (conductivity, TDS, and Se) and biological (WVSCI and GLIMPSS) conditions at the watershed scale. Optimal models were identified by altering tree complexity (number of splits within each tree and degree of interaction; *tc*) and learning rate (contribution of each tree to the growing model; *lr*). A tree complexity of 1 (no interaction among predictor variables) was chosen for all models because predictive performance did not improve with increased depth. Bag fraction (proportion of data randomly selected for each tree) was set at 0.75. We removed redundant variables (r > 0.90) and variables with minimal variation among study sites. We constructed global models with all remaining variables and quantified their relative influences using an out of bag procedure. We assessed predictive performance by calculating mean model deviance (and standard error) and cross- validated (CV) predictive deviance (and standard error) from 10 folds of the data. We then simplified global models using scree plots of predictor relative influences. Simplified models were retained if their CV predictive deviance was better than or equal to that of the global model. We created partial dependency plots to show effects of the 3 most important variables for select models. BRT models were fitted using code provided by Elith et al. (2008).

Alternative Future Scenarios

We used final BRT models to predict current chemical (specific conductance) and biological (GLIMPSS) conditions for all SLWs. We then selected actual surface mine permits from the WVDEP permit boundary dataset to developed 2 future mining scenarios. The first scenario consisted of 15 permits distributed throughout the study basin (Figure 3). The second scenario consisted of all permits (Figure 3). We updated the current landscape for each scenario by converting all land area within the permit boundaries to reclaimed mine land, removing all intersecting vector features, and re-accumulating landscape attributes for all SLWs within the study basin. We also simulated management opportunities under each scenario. We simulated advanced materials handling techniques (i.e., decreased downstream impacts from mines of equivalent size) for new mines by iteratively decreasing the effect size of future surface mines under each scenario by 25, 50, 75, and 100%. In other words, a 100 acre mine constructed with advanced reclamation techniques would be equivalent to a 75, 50, 25, and 0 acre mine constructed with current techniques. We also simulated mitigation of deep mine and development-related impacts by iteratively decreasing their effect size. We compared the length of stream exceeding accepted criteria for specific conductance (500 µS/cm; Pond et al. 2008) and GLIMPSS (54; Pond et al. 2012) for each scenario.

Results

Landscape features associated with residential development were the dominant predictors of RVHA (Table 1, Figure 5). Conversely, surface mining-related land use was the dominant predictor of Se (Table 1). Selenium exhibited sharp increases at low levels of surface mining activity (Figure 5). Specific conductance was strongly influenced by surface mining, deep mining, and development-related land-use (Table 1, Figure 5).

Table 1. Boosted regression tree results for RVHA, specific conductance (SpC), selenium, GLIMPSS, and WVSCI*

	RVHA	ln SpC	ln Selenium	GLIMPSS	WVSCI
# Trees	2150	3300	3750	2800	9850
Learning rate	0.001	0.005	0.001	0.005	0.001
Total Deviance	145.75	0.73	1.77	179.75	134.45
Residual Deviance	99.91	0.06	0.35	24.38	26.80
Variance Explained	31%	92%	80%	86%	80%
CV Deviance (se)	141.11 (18.89)	0.18 (0.06)	0.76 (0.16)	82.47 (14.64)	66.10 (11.29)
CV Deviance Explained	3%	75%	57%	54%	51%
Surface mining land cover variables					
Cumulative reclaimed mine land (%)	.	31.7	19.8	.	.
Cumulative% barren mine land (%)	.	9.1	26.1	.	.
Cumulative% valley fills (%)	.	12.6	43.7	.	.
Underground mining landscape variables					
Cumulative underground mine NPDES permit density (#/km^2)	.	13.0	.	.	.
Human development land cover and use variables					
Cumulative serviced structure density (#/km^2)	11.4	.	.	9.4	.
Cumulative unserviced structure density (#/km^2)	.	.	.	.	.
Cumulative road density (km/km^2)	.	.	.	5.6	14.4
Cumulative grassland and pasture (%)	38.6	14.8	.	21.2	20.0
Cumulative NPDES permit density (#/km^2)	.	.	.	8.2	18.0
Local serviced structure density (#/km^2)	24.0	.	.	4.6	.
Local grassland and pasture (%)	.	.	.	10.8	.
Local barren and developed land cover (%)	.	.	.	9.2	14.8
Natural landscape variables					
Cumulative forest (%)	26.0	18.8	.	31.0	32.8
Local forest (%)	.	.	5.1	.	.
Cumulative Allegheny group coal outcrop (%)	.	.	5.3	.	.

* Relative influences (%) of variables are presented. See methods for description of model statistics. NPDES = national pollution discharge elimination system; ln = natural log transformation.

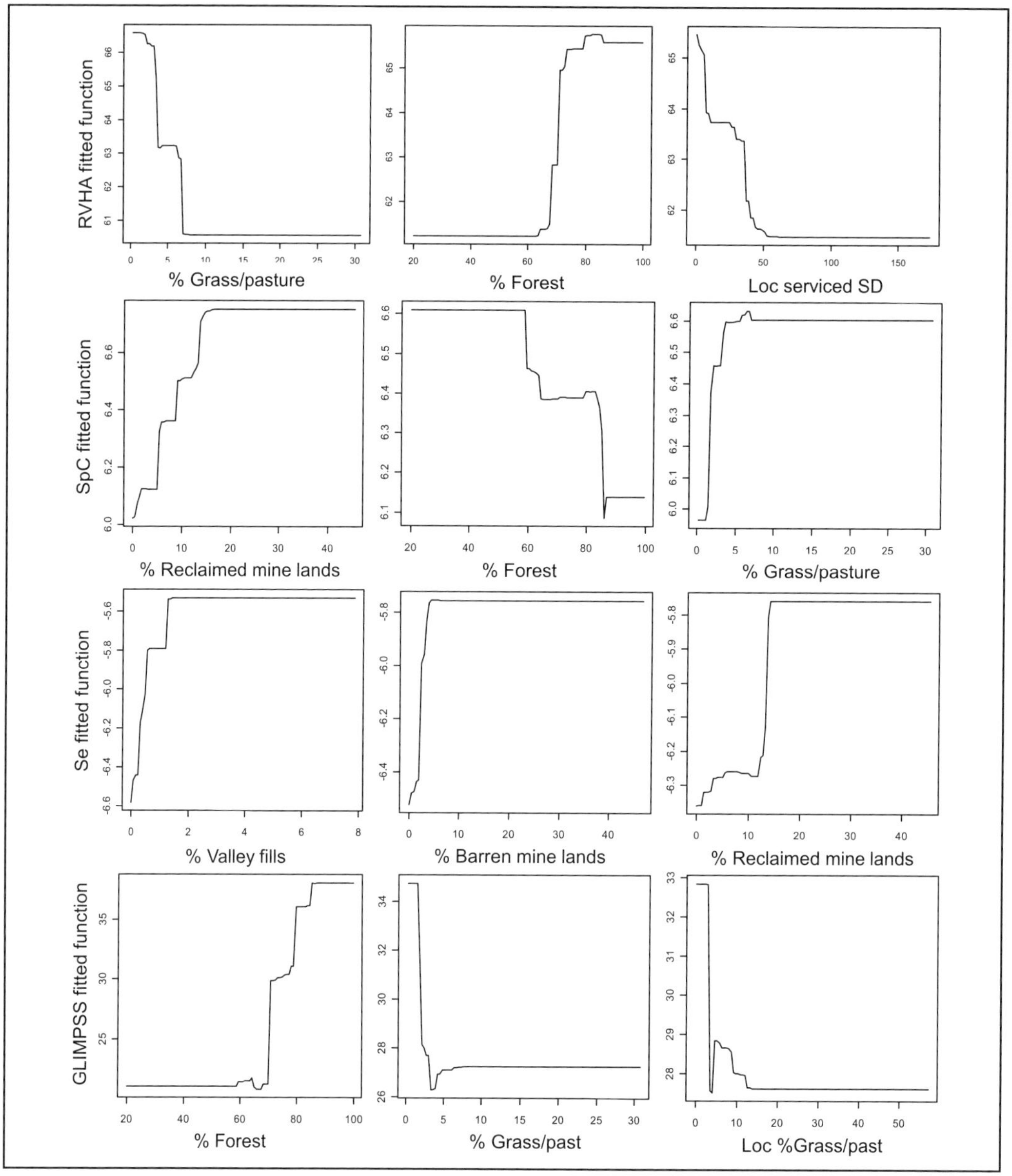

Figure 5. Fitted functional response curves for RVHA, specific conductance, selenium, and GLIMPSS to the three most influential variables in each respective BRT model. Fitted functions represent the marginal response to a given predictor variable, holding all others constant.

Landscape features associated with human development were the strongest predictors of WVSCI and GLIMPSS (Table 1 and Figure 5). BRT explained >80% of the total variation in chemical and biological data (Table 1). Cross validation indicated that models predicting chemical endpoints (57–75% of CV predictive deviance explained) were more certain than those predicting biological condition (51–54%) and habitat quality (3%).

BRT models predicted 1125 km (33%) and 3379 km (98%) of stream as currently exceeding

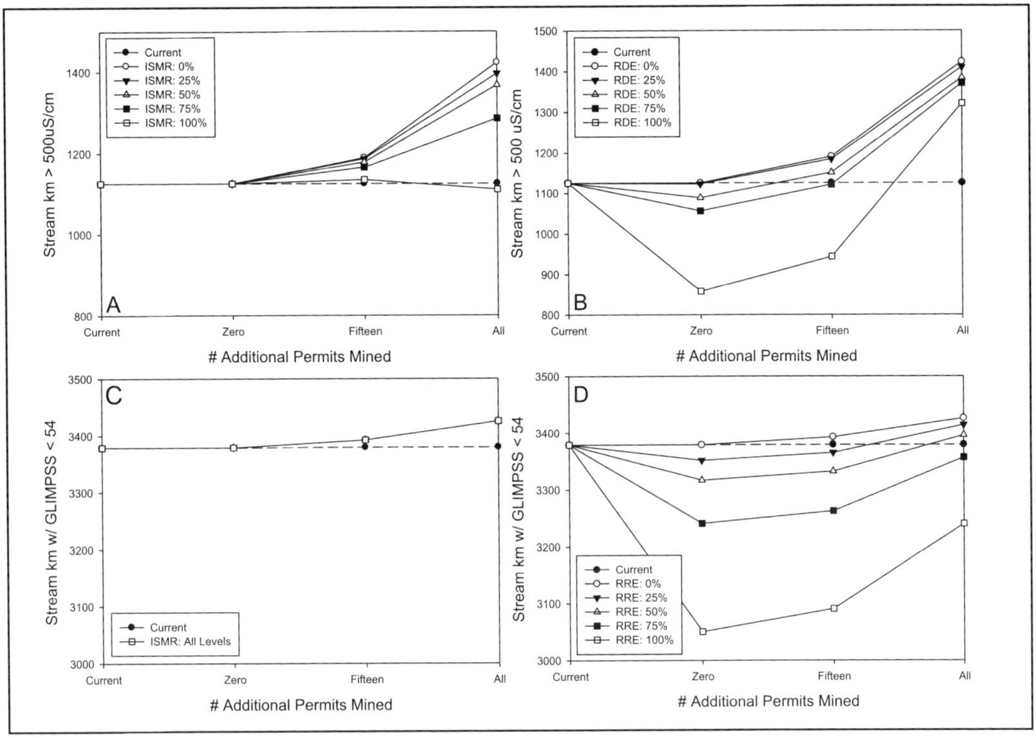

Figure 6. Predicted specific conductance (A and B) and GLIMPSS (C and D) in response to three alternative future mining scenarios (Zero = zero additional surface mining; Fifteen = 15 permit future scenario; All = all permit future scenario) and mitigation activities (ISMR = Increased surface mine reclamation; RDE = reduced deep mine effluents; RRE = Reduced residential effect). Conditions under each scenario are shown relative to conditions under current surface mining, deep mining, and residential land use (Current). Total stream length is 3449 km.

established criteria for specific conductance and GLIMPSS, respectively (Figure 6A, C). In the absence of any mitigation activity, additional mining under the 15 permit scenario resulted in an additional 66 (2%) and 7 km (0.2%) of stream exceeding specific conductance and GLIMPSS criteria, respectively. When all permits were simulated as mined, an additional 299 (9%) and 18 km (0.5%) of stream exceeded specific conductance and GLIMPSS criteria, respectively (Figure 6A, C). Improved surface mine reclamation had no effect on GLIMPSS (Figure 6C). Improved surface mine reclamation practices decreased the length chemically impaired streams under the 15 (0.3–5% of stream length) and all permit (2–22%) scenarios as compared to equal mining under current practices; however, improvement above current conditions did not occur at the watershed-scale (Figure 6A). All levels of decreased deep mine effluent in the current landscape resulted in watershed-scale improvements in the length of stream exceeding 500 µS/cm (0.1–8% of total watershed stream length; Figure 6B). In the 15 mine future scenario, decreasing the effect of deep mining by 75 and 100% resulted in watershed-scale improvements in specific conductance (0.1 and 5% of total watershed stream length, respectively; Figure 6B, Figure 7). The number of streams exceeding 500 µS/cm increased when all permits were mined, regardless of the level of reduced deep mining (Figure 6B). All simulated reductions in residential effect size resulted in watershed-scale improvements in GLIMPSS under both the current (0.8–10% of total watershed stream length) and 15 permit scenario (0.4–8% of total

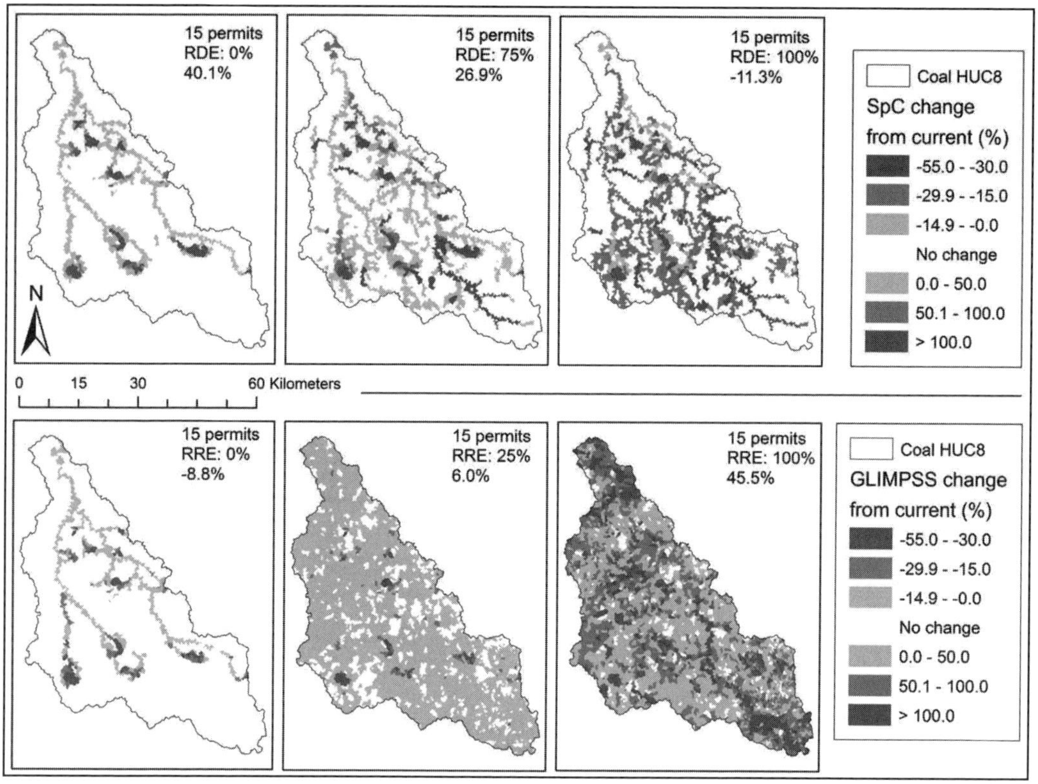

Figure 7. Predicted specific conductance under current landscape conditions (A) and a simulated 100% reduction in deep mine effects (RDME; B) for all mapped stream segments. Predicted GLIMPSS under current landscape conditions (C) and a simulated 100% reduction in residential effect (RRE; D) for all mapped stream segments.

watershed stream length) landscapes (Figure 6D, Figure 7). We predicted watershed-scale improvements in GLIMPSS following simulated mine out of all permits with a 75 (0.6% total watershed stream length) and 100% (4% total watershed stream length) reduction in development-related impacts (Figure 6D).

CONCLUSIONS

Scenario analysis coupled with predictive modeling provides an effective way to facilitate the decision making process on highly contentious issues, such as MTR-VF mining. When conducted properly, scenario analysis is transparent and science-based and provides an objective platform for constructive discussion (Kepner et al. 2012). Scenario analysis also provides a straightforward way for scientists to transfer complex information to stakeholders from a variety of backgrounds. However, the value of scenario analysis often is limited by the quality of underlying datasets and models. If data are flawed, then there may be considerable uncertainty in the modeling output. If modeling approaches are weak, then modeling outputs may be unreliable. In many applications, data on aquatic conditions and landscape characteristics may be highly confounded. For example, when aquatic conditions are affected by multiple stressors and the multiple stressors co-vary within the dataset, then it may be impossible to determine the effects of each stressor independently (King et al. 2005).

In this paper, we present the results of a scenario analysis designed to facilitate decisions within the MTR-VF mining region of West Virginia. There are several aspects of this research that we believe make it a valuable addition to the current debates regarding how best to manage mining impacts. First, stream sampling was designed to minimize the confounding effects of multiple stressors. We know from previous research that surface mining is only one of several possible sources of stress in this region. Other known stressors include deep mining effluents, urban runoff from residential development, and untreated wastewater (Petty et al. 2010, Merriam et al. 2011). We attempted to avoid a confounded dataset by sampling along independent stressor axes as well as within streams affected by multiple stressors simultaneously. Second, models linking landscape characteristics to aquatic conditions were based on landscape data derived from time appropriate aerial photography. Most previous studies are based on stream data collected across many years and landscape data derived from aerial photos that may come several years before or several years after stream sampling. We avoided this problem by analyzing aerial photography from 2009–2010 and linking it to stream data collected in summers of 2010 and 2011. Third, Boosted Regression Trees (BRT) represents the state-of-the-art approach for linking landscape attributes to in-stream conditions. BRT allows for co-varying predictor variables, incorporates interactive effects of multiple stressors, and produces highly precise estimates of stream conditions at the stream segment scale (Elith et al. 2008, Carlisle et al. 2009). Finally, integration of landscape change and BRT modeling within a GIS platform (i.e., the Watershed Futures Planner) enables us to run multiple complex scenarios and produce easily interpreted, visual, and spatially explicit results.

So, what are we able to learn from the scenario analysis of mining in the Coal River watershed? (1) There currently is an extremely high degree of water quality and biological impacts to streams resulting from pre-existing surface mining, deep mining, and residential development. Nearly 1200 km of streams have a conductivity that exceeds a 500 µS/cm benchmark, and nearly 3400 km of streams have a biological condition that is below a critical threshold. (2) Additional surface mining in the absence of restorative action will likely produce measurable additional impacts to water quality and biological condition. Expected response to mining is especially pronounced for specific conductance. Expected negative response of biological condition is relatively low, but only because current conditions are so poor. (3) The greatest benefits to water quality in the future would come from managing the effects of deep mine effluents. These benefits would be overwhelmed, however, in the absence of improved surface mine reclamation if all currently permitted surface mines were mined out. (4) The greatest benefits to biological conditions would come from managing the effects of residential development. In fact, we predict that substantial benefits to biological conditions could be achieved through mitigation of development impacts even if all of the currently permitted mines are mined out. However, full mitigation of development impacts is unlikely. Consequently, improved conditions in the future will require both mitigation and minimization of future mining impacts. (5) Potential impacts from future surface mining are manageable, but only when coupled with strategic mitigation of pre-existing impacts caused by other stressors.

There were several limitations with our analysis that should be addressed. First, we were unable to externally validate the models underlying our scenario analysis. However, internal cross validation procedures underlying the BRT process suggested our models are successfully predicting in-stream chemical and biological conditions. Although these results are specific to the Coal River watershed, it is a continued goal of our research program to construct and validate landscape-based models of in-stream conditions for the entire MTR-VF region. Moreover, the results of our models are consistent with other studies relating land use changes to altered specific conductance (Merriam et al. 2011), selenium (Lindberg et al. 2011), and biological condition (Merriam et al. 2011) in complex mined landscapes. Another limitation of our analysis was the treatment of simulated mitigation and reclamation activities. Iteratively and uniformly decreasing the effect size of surface mining, deep mining, and residential development does not reflect specific practices associated with potential mitigation of anthropogenic activities, such as providing centralized wastewater services to currently under serviced residences. Thus, it is unknown whether actual mitigation and reclamation efforts would result in stream responses similar to those predicted in our analysis. Despite the broad treatment of mitigation and restoration, our scenario analysis successfully identified management needs within the Coal River watershed and provides a framework for studies in other watersheds.

Scenario analysis represents an important step in the management process, but it is not a conclusive step. In order to make progress on recovery in the Coal River watershed, more specific restoration and mitigation plans that target deep mine effluents and stream impacts from poor residential development will be needed. In addition, the Coal River is but a single watershed within a highly impacted region. Scenario analyses for other watersheds will be needed in order to meet water quality management goals for the region.

ACKOWLEDGMENTS

This study was sponsored by the Appalachian Research Initiative for Environmental Science (ARIES). The views, opinions and recommendations expressed herein are solely those of the authors and do not imply any endorsement by ARIES employees, other ARIES-affiliated researchers or industrial members. Information about ARIES can be found at http://www.energy.vt.edu/ARIES.

We thank the numerous field and laboratory helpers that were involved in various aspects of this study, especially Eric Miller and Donna Hartman. Funding for this study was also provided by the U.S. Environmental Protection Agency Region III. Although a portion of the research described in this article has been funded by the U.S. Environmental Protection Agency, it has not been subjected to the agency's required peer and policy review and therefore does not necessarily reflect the views of the agency, and no official endorsement should be inferred.

REFERENCES

Allan, J.D. 2004. Landscapes and riverscapes: the influence of land use on stream ecosystems. *Annu. Rev. Ecol. Evol. Syst.* 35:257–284.

Barbour, M.T., Gerritsen, J., Snyder, B.D., and Stribling, J.B. 1999. *Rapid bioassessment protocols for use in streams and wadeable rivers: periphyton, benthic macroinvertebrates, and fish. 2nd edition.* EPA/841-B-99-022. US Environmental Protection Agency, Washington DC, USA.

Bernhardt, E.S., Lutz, B.D., King, R.S., Fay, J.P., Carter, C.E., Helton, A.M., Campagna, D., and Amos, J. 2012. How many mountains can we mine? Assessing the regional degradation of central Appalachian rivers by surface coal mining. *Environ. Sci. Technol.* 46:8115–8122.

Carlisle, D.M., Falcone, J., and Meador, M.R. 2009. Predicting the biological condition of streams: use of geospatial indicators of natural and anthropogenic characteristics of watersheds. *Environ. Monit. Assess.* 151:143–160.

Clapcott, J.E., Young, R.G., Goodwin, E.O., and Leathwick Jr, J.R. 2010. Exploring the response of functional indicators of stream health to land-use gradients. *Freshw. Biol.* 55:2181–2199.

Cuffney, T.F., Meador, M.R., Porter, S.D., and Gurtz, M.E. 2000. Response of physical, chemical, and biological indicators of water quality to a gradient of agricultural land use in the Yakima River Basin, Washington. *Environ. Monit. Assess.* 64:259–270.

Elith, J., Leathwick, R., and Hastie, T. 2008. A working guide to boosted regression trees. *J. Anim. Ecol.* 77:802–813.

Gerritsen, J., Burton, J., and Barbour, M.T. 2000. *A stream condition index for West Virginia wadeable streams.* Tetra Tech, Inc., Owings Mills, Maryland.

Jones, K.B., Ritters, K.H., Wickham, J.D., Tankersley Jr, R.D., O'Neill, R.V., Chaloud, D.J., Smith, E.R., and Neale, A.C. 1997. *An ecological assessment of the United States Mid-Atlantic Region: a landscape atlas.* U.S. Environmental Protection Agency. EPA/600/R-97/130.

Kepner, W.G., Ramsey, M.M., Brown, E.S., Jarchow, M.E., Dickinson, K.J.M., and Mark, A.F. 2012. Hydrologic futures: using scenario analysis to evaluate impacts of forecasted land use change on hydrologic services. *Ecosphere* 3:1–25.

King, R.S., Baker, M.E., Whigham, D.F., Weller, D.E., Jordan, T.E., Kazyak, P.F., and Hurd, M.K. 2005. Spatial considerations for linking watershed land cover to ecological indicators in streams. *Ecol. Appl.* 15:137–153.

King, R.S., Baker, M.E., Kazyak, P.F., and Weller, D.E. 2011. How novel is too novel? Stream community thresholds at exceptionally low levels of catchment urbanization. *Ecol. Appl.* 21:1659–1678.

Lindberg, T.T., Bernhardt, E.S., Bier, R., Helton, A.M., Merola, R.B., Vengosh, A., and Di Giulio, R.T. 2011. Cumulative impacts of mountaintop mining on an Appalachian watershed. *Proc. Natl. Acad. Sci. U.S.A.* 108:20929–20934.

Liu, Y., Mahmoud, M., Hartmann, H., Stewart, S., Wagener, T., Semmens, D., Stewart, R., Gupta, H., Dominguez, D., Hulse, D., Letcher, R., Rashleigh, B., Smith, C., Street, R., Ticehurst, T., Twery, M., van Delden, H., and White, D. 2008. Formal scenario development for environmental impact assessment studies. In *Environmental modeling, software, and decision support.* Edited by A. Jakerman, A. Voinov, A Rizzolli, and S. Chen. Elsevier Science, New Your, New York, USA.

Mahmoud., M.I., Gupta, H.V., and Rajogopal, S. 2011. Scenario development for water resources planning and watershed management: methodology and semi-arid region case study. *Ecological Modeling & Software* 26:873–885.

Merriam, E.R., Petty, J.T., Merovich, G.T., Fulton, J.B., and Strager, M.P. 2011. Additive effects of mining and residential development on stream conditions in a central Appalachian watershed. *J. North Am. Benthol. Soc.* 30:399–418.

Merritt, R.W., and Cummins, K.W. 2008. *An introduction to the aquatic insects of North America 4th edition*. Kendall/Hunt Publishing Co., Dubuque, Iowa.

Petty, J.T. and Thorne, D. 2005. An ecologically based approach for identifying restoration priorities in acid impacted watersheds. *Restoration Ecology* 13:348–357.

Petty, J.T., Fulton, J.B., Strager, M.P., Merovich, G.T., Stiles, J.M., and Ziemkiewicz, P.F. 2010. Landscape indicators and thresholds of stream ecological impairment in an intensively mined Appalachian watershed. *J. North Am. Benthol. Soc.* 29:1292–1309.

Pond, G.J., Bailey, J.E., Lowman, B.M., and Whitman, M.J. 2012. Calibration and validation of a regionally and seasonally stratified macroinvertebrate index for West Virginia wadeable streams. *Environ. Monit. Assess.* DOI:10.1007/s10661-012-2648-3.

Pond, G.J., Passmore, M.E., Borsuk, F.A., Reynolds, L., and Rose, C.J. 2008. Downstream effects of mountaintop coal mining: comparing biological conditions using family- and genus-level macroinvertebrate bioassessment tools. *J. North Am. Benthol. Soc.* 27:717–737.

Seitz, N.E., Westbrook, C.J., and Noble, B.F. 2011. Bringing science into river systems cumulative effects assessment practice. *Environ. Impact Assess. Rev.* 31:172–179.

Strager, M.P., Fletcher, J.J., Strager, J.M., Yuill, C.B., Eli, R.N., Petty, J.T., and Lamont, S.J. 2010. Watershed analysis in GIS: the Watershed Characterization and Modeling System Software application. *Comput. Geosci.* 36:970–976.

Strager, M.P., Petty, J.T., Strager, J.M., and Barker-Fulton, J. 2009. A spatially explicit framework for quantifying downstream hydrologic conditions. *J. Envron. Man.* 90:1854– 1861.

Utz, R.M., Hilderbrand, R.H., and Reasly, R.L. 2010. Regional differences in patterns of fish species loss with changing land use. *Biol. Cons.* 143:688–699.

Utz, R.M., Eshleman, K.N., and Hilderbrand, R.H. 2011. Variation in physicochemical responses to urbanization in streams between two Mid-Atlantic physiographic regions. *Ecol. Appl.* 21:402–415.

WVDEP (West Virginia Department of Environmental Protection). 2009. *Standard operating procedures.* West Virginia Department of Environmental Protection, Charleston, WV.

The Mine Land Stewardship Initiative

David Ledford
Appalachian Wildlife Foundation, Corbin, Kentucky

ABSTRACT

The Appalachian Wildlife Foundation (AWF), in partnership with coal mining companies, mineral and landowners, electric utility companies, wildlife conservation organizations, and coal communities is developing the Mine Land Stewardship Initiative (MLSI). The MLSI is a program of voluntary performance standards for wildlife and ecosystem conservation within the context of mining for coal and the restoration of mined lands. The ongoing debate over the mining and use of coal among regulators, environmental groups, elected officials, coal operators, and the public is very heated and contentious at times. The seemingly continuous re-write of mining regulations by federal and state regulators, and the routine legal challenges to mining activity have created an enormous amount of uncertainty in the coal and utility industries. Concurrently, electric utility companies and banks are questioning coal suppliers about ecological performance and are looking for performance baselines with which to assess coal operators. We believe that, over time, implementation of the voluntary performance standards developed through the MLSI can answer much of the criticisms aimed at coal mining as they relate to ecological issues. Having continual improvement as a driving force, voluntary programs can be innovative and reward the high performers as opposed to regulation which sets performance minimums and has to accommodate the lowest common denominator. A general overview of the MLSI will be presented. Conservation and land use planning is a major objective of the MLSI, and this will be discussed specifically.

2.1

Environmental Considerations in Coal Preparation and Waste Management

Application of Electronic Sorting Technology for Reducing Environmental Impacts of Coal Processing 22

Stabilization of Fine Coal Waste as an Alternative Disposal Technique 32

Application of a Pilot-Scale Plate Filter Press in Dewatering Coal Slurries 42

Diesel Range Organics In Coal Preparation 51

Minimizing the Environmental Impacts of Coal Waste Disposal 60

Application of Electronic Sorting Technology for Reducing Environmental Impacts of Coal Processing

Michael Kiser
Department of Mining & Minerals Engineering, Virginia Tech, Blacksburg, Virginia

Robert Bratton
Department of Mining & Minerals Engineering, Virginia Tech, Blacksburg, Virginia

Gerald Luttrell
Department of Mining & Minerals Engineering, Virginia Tech, Blacksburg, Virginia

Charles Roos
Mineral Separation Technologies, Inc., Coal Center, Pennsylvania

ABSTRACT

Coal preparation plants serve as an important link in the electrical power supply chain by providing high quality fuel for coal-fired utilities, industrial boilers and metallurgical coking operations. The plants utilize a wide array of solid-solid and solid-liquid separation processes to remove inorganic material that contains unwanted impurities such as ash and sulfur from run-of-mine coals. Essentially all of the metallurgical coal and more than one third of the power station coal produced in the U.S. is prepared for market by these facilities. Unfortunately, these facilities also produce a substantial amount of coarse waste rock and fine slurry as byproducts that must be disposed of at the mining sites. To minimize this problem, new processing technologies need to be developed that can improve how waste is managed, handled and disposed. In this article, a new dry separation process has been evaluated that can remove significant amounts of waste rock as close to the working face as possible. This capability would permit the storage of waste in underground and in-pit operations that would significantly reduce environmental issues associated with releases from surface waste storage facilities.

INTRODUCTION

Coal preparation plants are commonly used to upgrade the quality of run-of-mine coals from surface and underground mining operations. These plants are necessary since freshly mined coals contain a heterogeneous mixture of both organic (carbonaceous) and inorganic (mineral) matter. The inorganic matter includes noncombustible materials such as shale, slate and clay. These impurities reduce the coal heating value, leave behind an undesirable ash residue, and increase the cost of transporting coal to market. The presence of unwanted surface moisture also reduces the coal heating value and can lead to handling and freezing issues for consumers. Moreover, as the first step in the power cycle, coal preparation plants improve the environmental acceptability of coal by removing impurities that may be transformed into harmful gaseous or particulate pollutants when burned (Akers, 1996; Couch, 2000). These pollutants typically include particulates (fly ash) and sulfur dioxide (SO_2), as well as air toxics such as mercury. The presence of mineral impurities can also adversely affect the suitability of coal for high-end uses such as the manufacture of metallurgical coke or

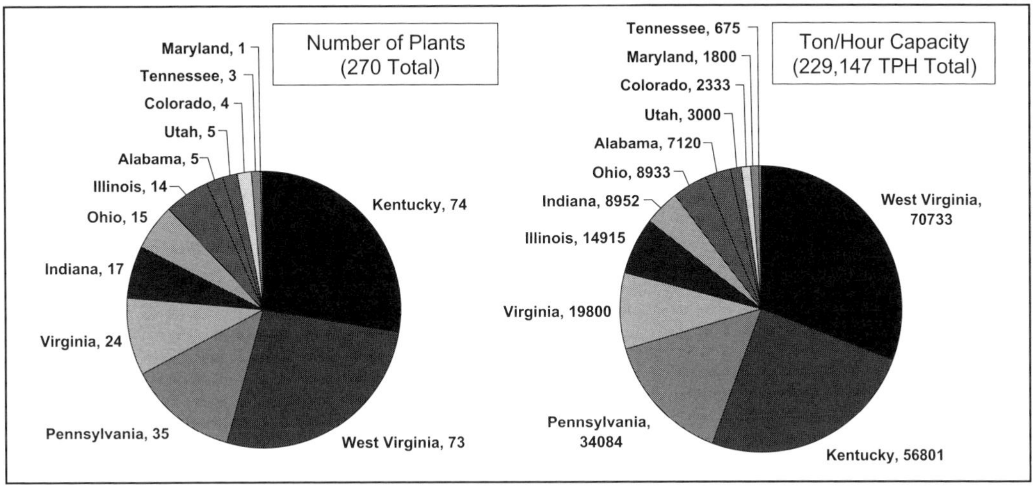

Figure 1. Distribution of coal preparation plants in the United States (Fiscor, 2007)

generation of petrochemicals and synthetic fuels. Coal preparation plants also reduce greenhouse gas emissions by increasing the thermal efficiency of coal-fired boilers (Couch, 1995). Therefore, essentially all coal supply agreements impose strict limitations on the purity of purchased coal.

According to the annual census of coal preparation plants conducted by Coal Age (Fiscor, 2007), the United States operates 270 coal preparation plants in 12 states. This number is relatively small by comparison to the worldwide fleet which is estimated to be 2,283 plants (Kimpnich, 2003). As shown in Figure 1, the preparation plants are largely concentrated in the eastern coalfields. The states of West Virginia and Kentucky have the largest number of coal processing facilities with 74 and 73 plants, respectively. When combined with the plants in Pennsylvania (35) and Virginia (24), these four states represent more than three-quarters of all preparation facilities in the U.S. Due to gaps in the survey data reported by Coal Age, the total capacity of these plants cannot be exactly calculated. However, the data that is available suggests that the average plant has a capacity of 849 tons per hour, with about 20% of the fleet exceeding 1,000 tons per hour and 20% under 500 tons per hour. When weighed by the number of plants, the state-by-state averages suggest that the total feed capacity for the fleet is in the range of 229,147 tons per hour. This production level equates to a total maximum capacity of just over 2 billion tons of feed annually (assuming around-the-clock operation). An estimate compiled from company production records suggest that about half this tonnage would report as clean coal. Based on this value, the theoretical maximum production of clean coal from the current fleet of plants is estimated to be approximately 1 billion tons annually. More than 85% of this capacity is available in the combined states of Kentucky, West Virginia, Pennsylvania, Virginia and Illinois. However, the actual production capability is probably much smaller than this upper limit due to constraints associated with plant availability. An availability correction accounts for losses in production due to mechanical failures, weather problems, power outages, or other shutdowns attributed to the operation of the preparation facility. While this factor is highly site specific, an average availability of 85% is realistically attainable based on production records provided by many of the larger producers. This correction reduces the total clean coal capacity for the U.S. fleet to about 850 million tons per year. Another problem is that preparation plants are typically located as mine mouth facilities in order to keep haulage costs as low as possible. As such, many of the older plants are in locations which have been depleted of large blocks of coal reserves, thereby making full utilization of the plant capacity unrealistic. A survey of industry suggests that one-third to one-half of the current fleet is impacted by this limitation. Therefore, based on these factors, the maximum production capability of the existing fleet of coal preparation plants is estimated to be only in the order of 430–570 million tons of clean coal annually.

One of the downsides of coal preparation is that the separation processes generate large volumes of waste rock and slurry. The percentage of clean coal tonnage generated from each ton of run-of-mine coal is commonly referred to as plant yield. The average yield produced by coal preparation plants has steadily declined over the years due to a depletion of lower ash feeds and increased mining mechanization. For example, Figure 2 shows the yield of clean coal currently obtained from a random survey of several major plants operating in the eastern U.S. It is not uncommon for eastern operations to experience yields under 30–35%, thereby producing only one ton of clean coal from three or more tons of mined product. An estimate complied from production records supplied by coal producers suggest that the average yield from coal preparation plants is now less than 50% (i.e., 49.8%+3.5%) for the U.S. This situation is expected to worsen as eastern reserves become thinner and more challenging to mine. A study reported by Weisenfluh et al. (1998) indicated that nearly 52% of the remaining eastern Kentucky coalfield resources are located in coal seams that are 14–28 inches in thickness, while 31% are in 28–42 inch thick seams. Likewise, a study of the Virginia coalfields found that 30% of the total reserve base (203 million tons) exists in seams with a thickness less than 28 inches (Sites and Hostettler, 1991). Consequently, ever increasing amounts of rock from out-of-seam dilution are being mined and passed to preparation plants for removal and disposal.

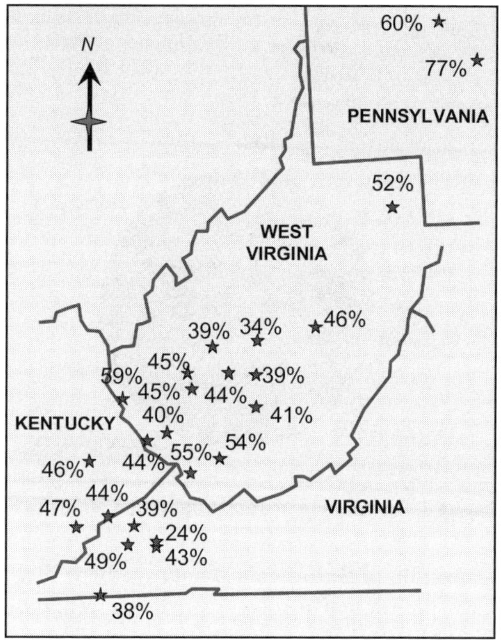

Figure 2. Clean coal yields for a random sampling of eastern preparation plants (Luttrell, 2008)

Waste Disposal

One of the downsides of coal cleaning operations is that these on-site facilities produce large volumes of waste that must be discarded into refuse piles or impoundments. Refuse piles are designed to receive coarse particles of waste rock that can be easily dewatered. The waste contains solid and liquid components that may present long-term disposal problems depending on the sizes, types and quantities of minerals present and the conditions under which the wastes are stored (e.g., dry vs. wet, loose vs. compacted, etc.). The factors play a key role in establishing the structural integrity (e.g., slope stability, surface water runoff, sediment containment, seepage, etc.) and chemical nature (e.g., acid generation, metal dissolution, etc.) of the wastes. Fine wastes have historically been discarded into earthen impoundments for permanent disposal. The waste, which is difficult to dewater, is normally pumped from the preparation plant thickener to the impoundment as slurry containing approximately 30% solids. According to the National Research Council (Ore et al., 2002), the coal industry discards 70–90 million tons of fine wastes each year into existing impoundments. In addition, it has been estimated that more than 2.5 billion tons of fine coal waste has been discarded in the past into existing and abandoned impoundments. Impoundments can pose a safety and environmental risk if not properly constructed, monitored and maintained. Potential problems include structural failures, seepage/piping, overtopping, acid drainage, and black water discharges. Since the well-known Buffalo Creek dam failure in 1972, strict engineering standards have been mandated by government agencies to regulate the design and operation of impoundments. No failures of impoundment dams or overspills have occurred since this legislation was enacted. However, several breakthroughs of slurry into old mine workings beneath impoundments have occurred (www.coalimpoundment.org). A number of small accidental releases of black water have also been reported at various plant sites.

Several alternatives to impoundments have been employed by the coal industry in an attempt to avoid any future potential for environmental damage. These alternatives have been examined extensively by Gardner et al. (2003). For example, some mines use new modes of slurry disposal such as slurry cells and underground injection wells. Slurry cells have been used successfully in some cases (albeit at higher cost), but limitations associated with maintaining less than 20 acre-ft of settling area make this alternative difficult to apply in all cases. Likewise, the use of injection wells has raised public concerns about groundwater contamination and well water quality (Breen, 2007; Wilcox, 2007). To overcome these problems, various types of mechanical solid-liquid separators have been investigated as a means of more fully dewatering the fine solids prior to disposal. Notable examples include paste thickeners and different types of filters (pressure, vacuum, belt press, plate-and-frame, etc.). Unfortunately, these systems each have specific problems of their own such as high costs, large chemical demand, poor performance, high energy consumption and safety concerns. Consequently, it is not surprising that the Committee on Coal Waste Impoundments that recently examined the issue of slurry disposal concluded that "...although there are alternatives to disposing of coal waste in impoundments, no specific alternative can be recommended in all cases."

Water Demand

Conventional coal preparation facilities require large amounts of water for coal upgrading and slurry transportation. As should be expected, the water demand varies with the relative proportions of water shipped with the clean coal product and discarded with the coarse refuse and fine slurry streams. While plant designers typically report water usage based on plant feed tonnage, a value of more interest to the public is the amount of water consumed per unit of clean coal product shipped to market. Water demand reported on this basis has risen in recent decades due to a continued decline in clean coal yield resulting from the processing of poorer quality reserves. To illustrate this trend, the water demands associated with the clean coal product, coarse refuse and fine slurry waste were calculated and plotted in Figure 3. Figure 3(a) shows that the amount of moisture reporting with the clean coal product is independent of plant yield. For typical product moistures of 6–8%, approximately 15–21 gallons of water are transported to consumers with each ton of clean coal shipped. In addition, the coarse and fine waste streams also consume water when placed into disposal areas. Figure 3(b) shows that the water demand for the coarse refuse, which typically leaves the plant at 7–10% moisture, varies considerably depending on the clean coal yield of the plant. For a 50% clean coal yield, the coarse refuse would consume 21–27 gallons of additional water per ton of clean coal generated. The largest amount of water used by the plant, however, is consumed during the disposal of the fine slurry waste. For conventional operations, the waste slurry is typically passed from the plant thickener/clarifier to the impoundment at about 30% solids. As shown in Figure 3(c), this solids content corresponds to a water demand of about 42 gallons per ton of clean coal produced. In total, the cumulative amount of water consumed for a 50% yield plant would be in the order of 80–85 gallons per ton of clean coal shipped to market. This demand drops to 55–65 gallons of water per ton of clean coal for plants treating feeds with clean coal yields in the 65% range and increases to 115–120 gallons per ton of clean coal

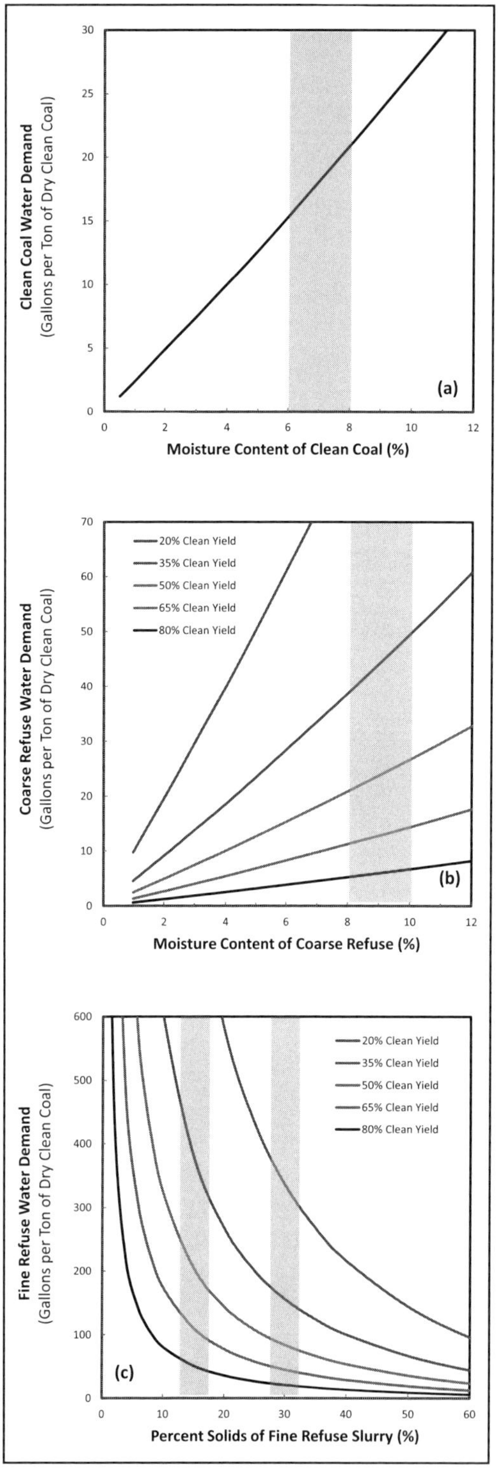

Figure 3. Plant water demand requirements

for plants treating feeds with clean coal yields in the 35% range. Therefore, as a rule-of-thumb, a coal preparation facility would be expected to consume about 1–2 GPM of water per TPH of clean coal generated.

It is also interesting to compare how different waste disposal technologies impact the water consumption. For example, underground slurry injection systems normally dispose of the fine refuse at a lower solids content of about 15% solids. Consequently, this approach to slurry disposal results in a much higher water consumption of 102 gallons of water per unit of clean coal generated (1.7 GPM water per TPH clean coal). On the other hand, if belt presses are used, then the solids content of the fine refuse could be in the range of about 70% solids. For this low-water technology, the plant disposal system would consume less than 7.7 gallons of water per ton of clean coal (0.13 GPM water per TPH clean coal). Not surprisingly, the capital, operating and maintenance costs for these alternative waste disposal systems are inversely proportional to their water demand (i.e., slurry injection is usually the least costly and belt presses are the most costly).

DRY COAL SORTING TECHNOLOGY

Coal preparation offers may attractive benefits including lower transportation costs, improved utilization properties and reduced emissions of particulate and gaseous pollutants. However, the industry also faces several challenges associated with increased solid waste disposal requirements and higher demands for process water (Meenan, 2005). To address these concerns, researchers at Virginia Tech have been working with equipment manufacturers to develop the next generation of coal preparation technology. One particularly promising process is electronic sorting. Electronic ore sorters were first introduced to the minerals processing industry in the late 1940s. Since that time, faster microprocessors, improved sensors and lower equipment costs have allowed this unique technology to evolve and become commercially attractive for a variety of applications. Recent estimates indicate that nearly 300 industrial-scale sorters are now used worldwide in the minerals industries for ore concentration. Electronic sorters utilize specially-designed sensors to evaluate the quality of feed particles that are spread across the surface of a moving conveyor belt. High-speed microprocessors use the sensor data to control pneumatic actuators located at the end of the conveyor. The pneumatic actuators are sequenced so that particles meeting the target quality are diverted into the product stream. This system offers many benefits for coarse coal upgrading including mechanical simplicity, high capacity, low cost and minimal environmental impacts. Moreover, the compact size and low unit cost of sorter technology improves the viability of separating rock from run-of-mine coal as close to the working face as possible utilizing a system that is integrated within the production process so that the surface disposal of wastes and water demands could be minimized.

EXPERIMENTAL

Several series of dry coal cleaning tests were conducted using an electronic dual-energy X-ray coal sorter developed specifically for the coal industry as a cooperative effort with Mineral Separation Technologies, Inc. The separator, which is being marketed under the tradename DriJet™, subjects particles that travel along a high speed conveyor to hundreds of sequential X-ray scans (Figure 4). The X-rays are transmitted through the solid particles in proportion to their atomic number. This phenomenon makes it possible to distinguish coal (organic carbonaceous matter) composed mostly of carbon having a low atomic number from rock (inorganic mineral matter) composed of various elements such as Si and Al with higher atomic numbers (Figure 5). The resolution and speed of the proprietary scanner and associated electronics is of sufficient quality so that a compositional profile of each particle can be reconstructed in fractions of a second. Once identified, microbursts of compressed air from a horizontal array of pneumatically actuated jets divert unwanted particles of rock into the reject stream, while coal particles follow their normal trajectory into the clean coal product stream.

The pilot-scale test work was conducted using a refuse coal sample from an abandoned mine site located in southwestern Virginia. Due to production constraints, the feed sample was sized into a 2 × ¼ inch fraction using a pilot-scale screening system. The sized feed was then manually fed to the separator through a feed hopper onto a feed belt. The feed belt discharged onto a scanning belt that moved at a speed of about 3 m/sec (i.e., just under 10 ft/sec). This configuration allowed the feed particles to be evenly spread into a thin layer of particles for the x-ray scanning system. The resultant clean coal and reject products were each collected and placed into separate containers

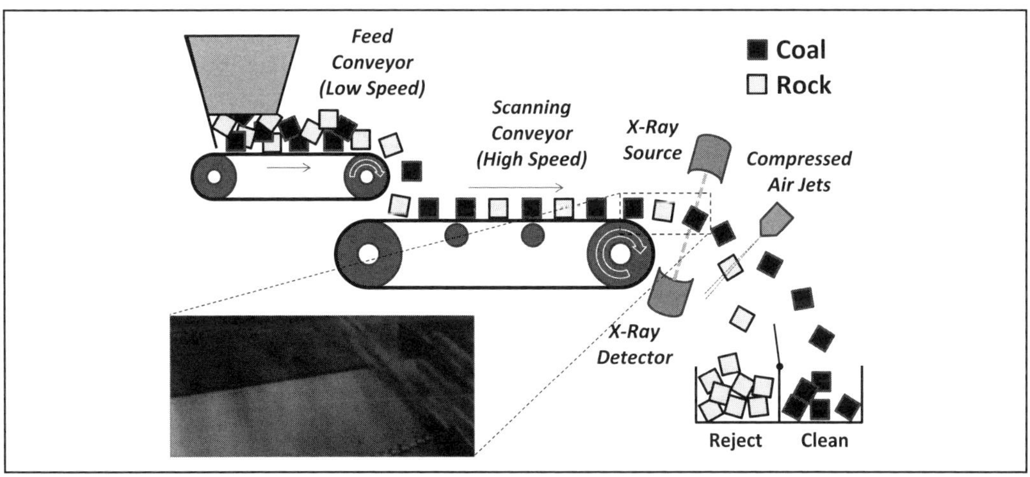

Figure 4. Schematic of the X-ray sorting process

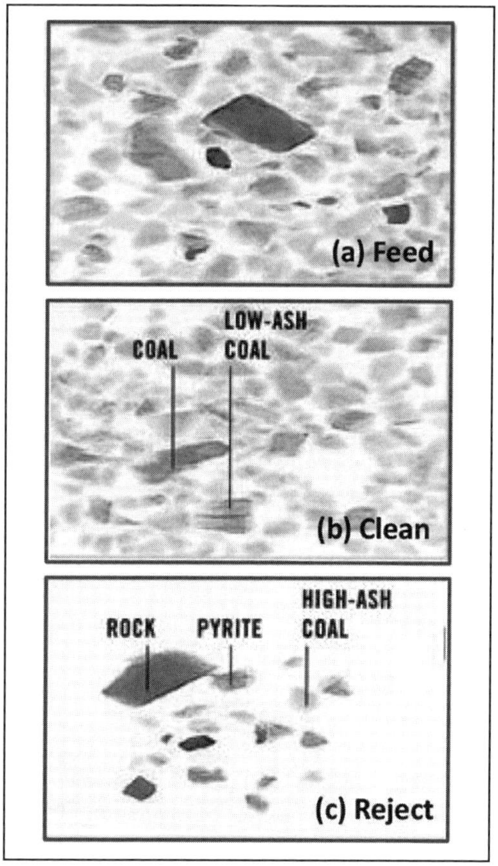

Figure 5. X-ray images of run-of-mine feed, clean coal, and rejected material

for transport to a coal analysis laboratory. In the laboratory, each product was sized into 2 × 1, 1 × ½, ½ × ¼ and minus ¼ inch fractions. Each of these size fractions was then crushed, split into smaller representative lots, and then subjected to ash analysis.

RESULTS AND DISCUSSION

Figure 6 shows the size-by-size recovery and rejection values obtained using the pilot-scale coal sorter. The recovery (R) represents the percentage of combustible matter present in the feed that reports to clean coal. Mathematically, this performance indicator was calculated using:

$$R = Y \times \left(\frac{100 - Ac}{100 - Af}\right) \quad (1)$$

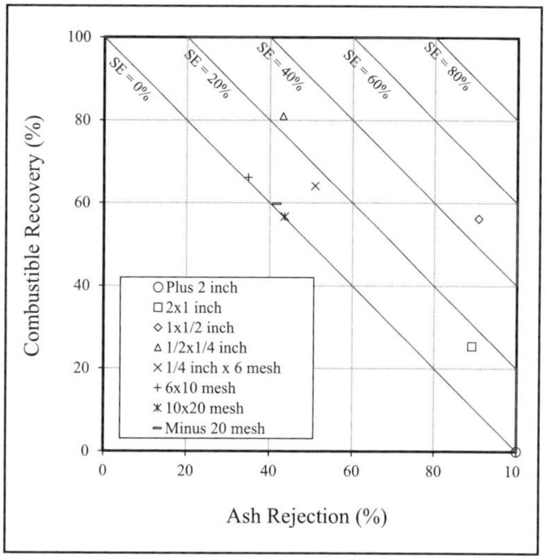

Figure 6. Size-by-size separation performance for upgrading waste coal using sorter technology

where Y is the clean coal yield (%), Ac is the clean coal ash content (%) and Af is the feed ash content (%). The rejection (J), which represents the percentage of ash present in the feed that reports to the reject stream, was calculated using:

$$J = (100 - Y) \times \left(\frac{Ar}{Af}\right) \quad (2)$$

where Ar is the ash content (%) of the solids reporting to the reject stream. The diagonal red lines in Figure 6 represent different levels of separation efficiency (SE) as mathematically described using:

$$SE = R - (100 - J) \quad (3)$$

An ideal separation, which is never achieved in actual practice, would have a separation efficiency of 100%, which corresponds to 100% recovery of combustibles and 100% rejection of ash. A completely nonselective process, such as a material splitter, would have a separation efficiency of 0% and would be represented by the diagonal line along which the recovery and rejection add to 100% at any point.

The test data provided in Figure 6 shows that the separator performed well in rejecting material in the large size classes. The corresponding data table, which is provided in Table 1, shows that these coarser fractions in the waste coal sample were mostly composed of high-ash particles with ash contents ranging from a low of 49.8% to a high of 89.2%. The separator provided higher recoveries of particles in the middle size classes comprised of the ½ × ¼ inch and ¼ inch × 6 mesh particles, but at a lower overall rejection of unwanted ash. All of the remaining fine material less than 6 mesh tended to fall along the 0% separation efficiency line, which indicates that the sorter did not selectively place material of this size into one stream or the other (i.e., this material is simply being non-selectively split between the two streams). It is believed that the overall product quality might be improved if all material larger than 1 inch was removed prior to being treated or if the feed had been divided into two size classes (2 × 1 inch and 1 × ¼ inch) and treated separately. Unfortunately, no such tests were conducted due to time constraints imposed by this publication.

One interesting correlation was discovered while analyzing the data. As shown in Figure 7, a linear relationship exists between feed ash content and combustible recovery achieved by the sorting device. This trend holds true for the size range the machine is designed to treat (i.e., 2 × ¼ inch), but begins to scatter below this particle size range. This finding supports the manufacturer's recommendations that only particles coarser than approximately ¼ inch are likely to be significantly upgraded using the current configuration of the coal sorter technology. This limitation is not surprising considering the requirement that a single layer of particles needs to be presented to the X-ray scanner. In fact, the limitation imposed by this constraint on particle presentation makes it possible to estimate the theoretical maximum production that can be attained using the new sorter technology. The effective spatial volume (Q) moving through the scanner can be calculated using:

$$Q = W \times Dp \times V \times \beta \qquad (4)$$

where W is the width of the scanner belt, Dp is the particle diameter (bed height), V is the belt velocity and β is the particle packing efficiency. For spherical mono-sized particles placed back to back along the conveyor, β cannot exceed a value of $\pi/6$ (i.e., ratio of sphere-to-cube volume). From these expressions, the maximum mass flow rate (M) that can be passed as a single layer of particles through the separator is given by:

$$M = \rho \times Q \qquad (5)$$

where ρ is the composite density of particles passing through the separator. Typically, this density value would correspond to specific gravity values of 1.5 to 1.9 SG for ash levels observed in typical eastern feed coals. A plot of the theoretical capacity as a function of particle diameter for different feed densities is provided in Figure 8. The plot

Table 1. Size-by-size ash contents for the waste coal fed to the sorter technology

Passing Size	Retained Size	Ash (%)
—	2 inch	89.22
2 in.	1 inch	81.97
1 in.	½ inch	72.85
½ in.	¼ inch	64.19
¼ in.	6 mesh	58.08
6 mesh	10 mesh	49.84
10 mesh	20 mesh	51.60
20 mesh		71.19
	Total	70.58

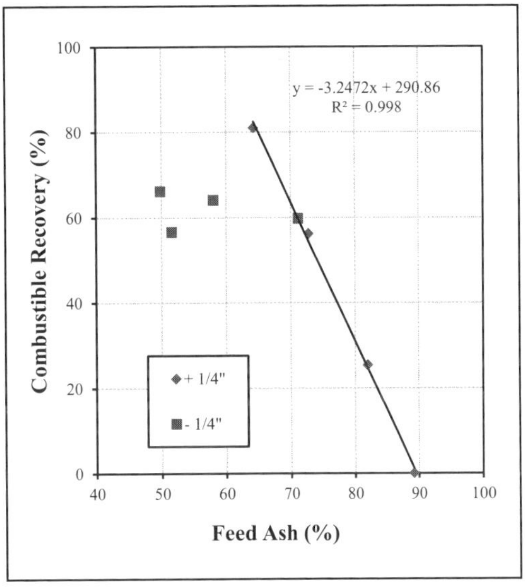

Figure 7. Linear correlation between feed ash and combustible recovery for coarser particles

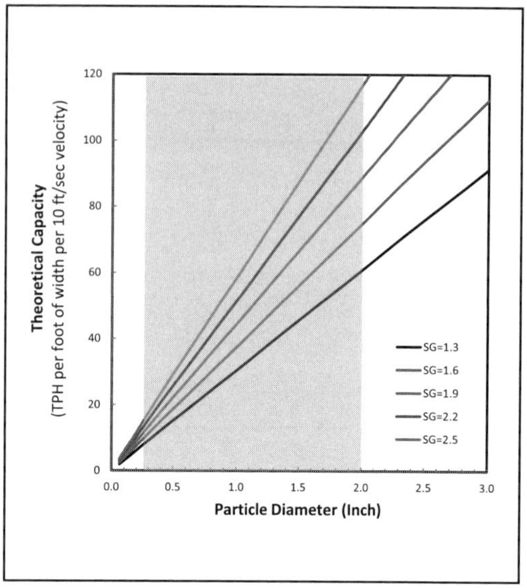

Figure 8. Theoretical maximum sorter capacity for mono-sized particles of different densities

was generated using a belt velocity of 10 ft/sec. As should be expected, the theoretical maximum throughput capacity for larger 2-inch particles would fall in the range of 60 to 120 TPH per foot of belt width, depending on the feed particle density. In contrast, the theoretical capacity would fall sharply to 7–14 TPH per foot of belt width for particles smaller than ¼ inch in diameter. Therefore, these finer particles can be easily missed when intermixed with coarser particles that are being separated at much higher production rates.

SUMMARY

Several series of experimental test runs were conducted to evaluate the potential of an electronic coal sorter for upgrading of waste coal from an eastern U.S. refuse pile. The test data indicate that this novel sorting technology can effectively remove unwanted mineral matter impurities contained in coarse (2 × ¼ inch) coal feeds. Due to inherently low capital and operating costs, this unique technology has the potential to serve as a viable coal cleaning alternative for sites that are water constrained or that have too low tonnage to justify a full-scale coal preparation facility (e.g., highwall miner applications, small contract mines with long truck haulage routes, etc.). More importantly, as a dry process, this method of separation avoids issues related to water usage and waste disposal that typically occur using traditional water-based separation processes. The compact footprint of this process may also allow the technology to be integrated into mining production units in underground mines, thereby reducing the demand for transporting and disposing wastes in dedicated surface refuse areas.

REFERENCES

Akers, D.J., 1996. "Coal Cleaning Controls HAP Emissions," Power Engineering, June.

Breen, T., 2007. "Lawmakers Want Study of Coal Slurry in Drinking Water," Associated Press, The Daily Mail, January 9, 2007.

Couch, G.R., 1995. Power From Coal—Where To Remove Impurities?, IEA Coal Research, London IEACR/82, 87 pp.

Couch, G.R., 2000. "Opportunities for Coal Preparation to Lower Emissions," International Energy Agency (IEA), CCCr30, London, 46 pp.

Fiscor, S., 2007. "U.S. Preparation Plant Census 2007," Coal Age, September, pp. 38–46.

Gardner, J.S., Houston, K.E. and Campoli, A. (2003) Alternatives analysis for coal slurry impoundments. SME Annual Meeting, Feb. 24–26, Cincinnati, OH, Preprint 03-032, 1–5.

Luttrell, G.H., 2008. "Coal Preparation," Chapter 4, in Meeting Projected Coal Production Demands in the USA: Upstream Issues, Challenges, and Strategies, M. Karmis (Chair), National Commission on Energy Policy, Published by Virginia Center for Coal and Energy Research (VCCER) of Virginia Tech, December, 2008, pp. 106–143.

Meenen, G., 2005. "Implications of New Dewatering Technologies for the Coal Industry," Invited Lecture, Center for Advanced Separation Technologies (CAST) Workshop, July 26–28, 2005, Virginia Tech, Blacksburg, Virginia, 14 pp.

Orr, F.M. (Committee Chair), 2002. "Coal Waste Impoundments: Risks, Responses and Alternatives," Committee on Coal Waste Impoundments, National Research Council, National Academy Press, Washington, D.C., 230 pp.

Sites, R.S. and Hostettler, K.K. (1991) Available coal resource study of Appalachia 7.5-minute quadrangle Virginia-Kentucky, Virginia Division of Mineral Resources Publication 118, 1–51.

Weisenfluh, G.A., Andrews, W.A. and Hiett, J.K. (1998) Availability of coal resources for the development of coal. Kentucky Geological Survey, Report, U.S. Department of Interior Grant 14-08-0001-A0896, 1–32.

Wilcox, L., 2007. "Coal Industry Official Disputes Research About Health of People Living Near Mines," The Exponent Telegram, Clarksburg Publishing Company, Clarksburg, West Virginia, October 30, 2007.

Stabilization of Fine Coal Waste as an Alternative Disposal Technique

R.Q. Honaker
University of Kentucky, Lexington, Kentucky

Z. Duan
University of Kentucky, Lexington, Kentucky

D. Cobb
Armstrong Coal, Madisonville, Kentucky

S. Baird
Armstrong Coal, Madisonville, Kentucky

ABSTRACT

The permitting of new fine coal waste impoundments and slurry cell systems has become a nearly improbable venture. When faced with limited or no waste slurry storage areas, the common practice is to dewater the thickener underflow to a solids concentration greater than 55% by weight or greater and co-dispose with coarse reject. However, the combined reject may be difficult to handle and sometimes fails to meet regulatory requirements. A potential solution is stabilization of the reject by the addition of Portland cement. The result is a combined refuse that meets compaction requirements and is impermeable thereby eliminating mine acid formation and leaching of trace elements. Conceptual proof of the technique as a valid option is provided by details of a mining operation utilizing co-disposal with cement addition as well as the findings of a laboratory investigation.

INTRODUCTION

Increased mechanization in the coal mining industry has been accompanied by decreased selectivity and increased proportion of waste materials in the raw coal. As a result, large volumes of waste have been generated from the coal washing process. The simplest approach to dispose fine coal tailing is to impound the tailing slurry by constructing dams by solid mining waste (Snyder et al. 1977). The low cost and convenience have made impoundments the most popular tailings disposal method (Almes 1979; Williams 1991).

Unfortunately, several environmental and safety issues have made the process of obtaining new impoundment permits from the U.S. Environment Protection Agency (EPA) difficult. Moreover, due to the ultrafine particle size of fine coal tailings, the exposed surface area presents potential for the leaching of trace elements associated with the mineral matter contained within the raw coal.

As an alternative to impoundments, industry is choosing to dewater the fine reject and, in most cases, co-dispose with coarse waste material using a stacking method. Belt filter presses have been the most commonly used technology for dewatering fine waste to levels between 55% and 75% solids by weight. Relatively high throughput capacity values, while achieving effective dewatering on clay-rich material, are cited as advantages of belt filter presses. However, chemical consumption is high, with costs in the range of $0.30 to $0.50 per ton of plant feed. Furthermore, the product moisture may not be sufficiently high to provide enough stability to allow refuse storage using stacking techniques.

Recent commercial applications of plate-and-frame filter presses have provided excellent results when dewatering thickener underflow. Filter cake containing as high as 80% solids by weight is reportedly being produced from plate-and-frame units installed to dewater fine coal waste at preparation plants located in southern West Virginia, U.S. Similar findings have been reported in the literature (Xie et al. 2010; Verma et al. 2010; Forster 1977). Using only the chemicals necessary for the initial solids concentration in a conventional thickener, the plate-and-frame filter provides a cake that is sufficiently stable to meet EPA requirements for compaction. Results reported in this essay support this fact. However, plate-and-frame filters require large capital to install and their semi-continuous operation requires the use of two units. Moreover, the dewatering process does not effectively decrease the exposed particle surface area. When subjected to rain events, the filter cake decomposes back into slurry form. As such, leaching potential of trace elements is not permanently controlled.

The concept of waste solidification could provide a potential solution for the aforementioned issues associated with belt press and plate-and-frame filters. Waste solidification was originally applied to radioactive waste disposal in the 1970s, which involved the deployment of chemical stabilization agents to transform the waste from a liquid phase to solid form with adequate physical strength to allow disposal in a landfill. In fact, waste solidification is commonly employed in the extraction of metallic deposits using the cut-and-fill mining method. The primary factor influencing solidification performance is the type and amount of chemical agents employed. The primary chemical is cement, which is known widely for its hydration reaction with water to generate high strength within a short time. When the solids concentration of the waste is low (i.e., <50% by weight), sodium silicate can be added, which solidifies the water into a 'quick gel' that can be further solidified by the addition of cement. Hydraulic lime is also effective in water adsorption.

Proper selection of the agents and their concentrations is based on the physical strength of the stabilized product, which can be stored in a safe and environmentally friendly manner. The EPA considers a stabilized/solidified material with strength of 0.35 MPa (50 psi) to have a satisfactory unconfined compressive strength to provide a stable foundation for materials placed upon it, including construction equipment and impermeable caps and cover material (Malviya et al. 2006; USEPA). However, according to other researchers, such as Conner (1990), a minimum bearing strength of 0.1 MPa (14 psi) is acceptable for landfill purposes. For wheeled vehicles and other more stringent requirements, 0.24 MPa (35 psi) bearing pressure is sufficient.

As previously mentioned, waste stabilization provides another very significant environmental benefit. The stabilized product has reduced leaching rates due to fixation of the heavy metals within the solidified matrix or by chemical reaction between stabilization agents and waste elements (Chen et al. 2009; Paria et al. 2006; Pinarli et al. 2005; Dermatas et al. 2003; Mulligan et al. 2001; Johnston et al. 1997; Conner et al. 1990, 1998; Myers 1986).

Due to compaction issues with the co-disposal of coarse and fine reject, a coal producer located in western Kentucky utilizes a waste stabilization process that employs the use of Portland cement. This publication provides a detailed description of the process along with the results of a detailed laboratory investigation that focused on process optimization.

COAL PREPARATION AND WASTE STABILIZATION PLANT

The Armstrong Dock and Preparation Facility located in Centertown, Kentucky, is operated by Armstrong Coal Company. The plant has a throughput capacity of 1,200 tons/hr and cleans coal from a blend of the Kentucky No. 12, No. 13 and No. 14 seams with a mass yield to the product of around 70%. Coal from the Kentucky

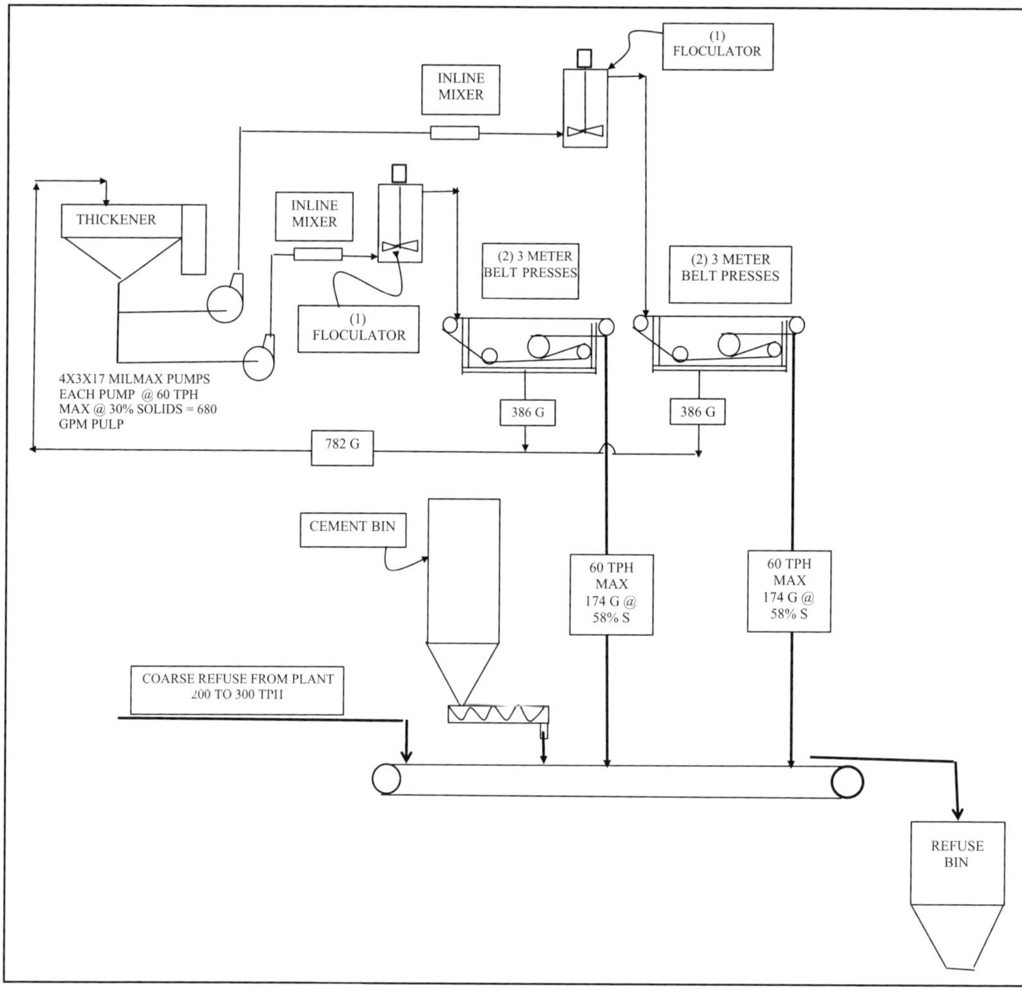

Figure 1. Flow sheet of the Armstrong refuse treatment process

No. 9 and No. 11 seams are treated separately with typical mass yield values of 60% and 68%, respectively. The plant utilizes dense medium cyclones and spiral concentrators to clean the particle size fractions coarser than 0.15 mm.

The amount of coarse refuse varies with the plant feed source but typically ranges between 200 and 300 tons/hr. The mass flow rate of fine refuse is relatively constant at 120 ton/hr due to the capacity constraints of the belt filter presses. As a result of this constraint, the plant feed rate must be reduced below the 1,200 tons/hr rated capacity if the amount of clays or fines become unmanageably high. There are times when the plant feed rate is reduced to 900 tons/hr to handle an elevated amount of fines.

The coarse and fine waste materials are co-disposed using a series of lifts. The fine waste is dewatered in four Phoenix Model WX-3.OG6 belt filters. The chemical cost to run the belt presses averages \$0.55 per ton of plant feed, whereas the other chemical costs at the plant are around \$0.045 per ton.

When the co-disposal practice was initiated, the material did not reach the required compaction levels. Hydrated lime was added in an effort to enhance the stability of the fine waste, which worked to a degree but did not provide the

Figure 2. Components of the refuse treatment process (a) cement bin and conveyor and (b) dozer handling of the combined refuse

desired impact on compaction. Portland cement was tried and found to provide the desired effect. By adding cement at a rate of three tons/hr, which is around 2.5% of the fine refuse, the co-disposed reject material can be handled in a matter of days and compacted to the required level in as little as two weeks. The cement cost is $0.27 per ton of plant feed.

The waste treatment process involves splitting the thickener underflow using two variable speed pumps with the output of each pump having volumetric and mass solids flow rates of 680 gallons/min and 60 tons/hr, respectively. As shown in Figure 1, the output of each pump is fed through an in-line mixer and flocculator prior to entering to belt filter presses. The plant coarse refuse is placed on a conveyor belt in the plant and run under a bin where the Portland cement is added (Figure 2a). The belt containing the coarse refuse and cement then enters the belt filter facility where the belt filter product is placed on the belt. The combined coarse and fine refuse with the cement is dumped through a pug mill and the final mixture subsequently placed on another conveyor that transports the material to the waste disposal area. Dozers are used to spread the material (Figure 2b).

A sample of the combined refuse was collected from the placement area and tested to determine the unconfined compressive strength (UCS). The measurement was conducted according to ASTM standard D1633-84: Standard Test Methods for Compressive Strength of Molded Soil-Cement Cylinders (USEPA). The sample was molded into a 25.4 mm (2-inch) diameter by 50.8 mm (4-inch) cylinders and cured for 1, 2, 4, 5 and 7 days. The samples were sealed in molds for one day and subsequently cured outside the mold in open air.

As shown in Figure 3, unconfined compressive strength of the combined refuse is around 20 psi after seven days of curing using one percent cement concentration, which is substantially lower than the EPA requirement of 50 psi. Nonetheless, the strength is sufficient for grading and compaction using dozers and for trucks to travel across after setting for a few days.

Proposed Plate-and-Frame Filter Installation

The recent commercial application of plate-and-frame press filters for dewatering fine waste has shown the ability to achieve 80% solid concentrations in the filter cake and excellent stability when placed with coarse refuse, which is a fact supported by the laboratory test data presented in this publication. The performances have been achieved with no additional chemical costs beyond those associated with conventional thickening of the plant fine waste. As a result, a technical and financial costs analysis was performed to compare belt filter presses and cement addition system with plate-and-frame presses. As shown in Tables 1 and 2, the amount of capital needed to purchase plate-and-frame units to treat the 120 tons/hr of fines is slightly greater than $7.5 million as compared to nearly $2 million for the belt filter presses. Despite the high capital cost, the plate-and-frame filters are a significantly better economic option due to the reduced annually operating costs, which are largely associated with cationic flocculation and cement additions.

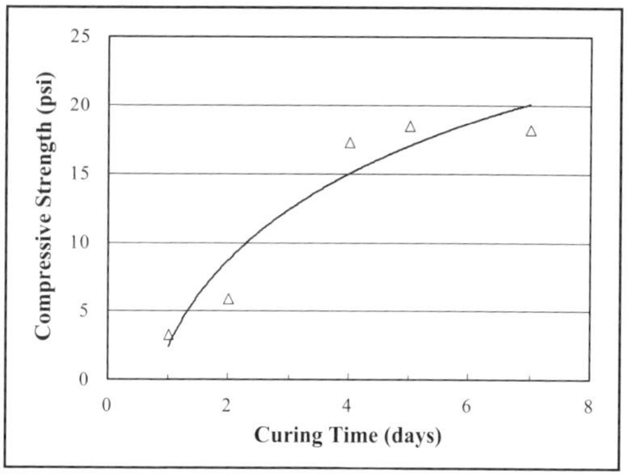

Figure 3. Unconfined compressive strength of the co-disposed reject material with cement

LABORATORY EXPERIMENTAL STUDY

Thickener Underflow Samples

A representative sample of the thickener underflow was collected in three 55-gallon drums. Upon arrival at the laboratory, the sample was split into representative sample lots of five gallons each. A sample was collected from one of the five gallon buckets and analyzed for the particle size distribution, ash and sulfur contents. The total solids content of the slurry was 35.09% by weight. A proximate analysis showed the sample contained 51.17% ash and 4.95% total sulfur on a dry basis. The particle size distribution is provided in Table 3.

Stabilization Tests

In each test, around 900 grams of thickener underflow slurry with the desired solids content

Table 1. Capital and operating costs of plate-and-frame filters for fine coal waste dewatering and placement

	Plate-and-Frame Filter			
Year	Capital Cost	Operating Cost	Total	Cumulative
2012	$7,586,400	$215,600	$7,802,000	$7,802,000
2013		$215,600	$215,600	$8,017,600
2014		$215,600	$215,600	$8,233,200
2015		$215,600	$215,600	$8,448,800
2016		$215,600	$215,600	$8,664,400
2017		$215,600	$215,600	$8,880,000
		NPV @ 10%	$7,714,021	

Table 2. Capital and operating costs belt press and 1% cement application installation for fine coal waste dewatering and placement

	Belt Press Filter & Cement Addition			
Year	Capital Cost	Operating Cost	Total	Cumulative
2012	$1,925,000	$4,724,268	$6,649,268	$6,649,268
2013		$4,724,268	$4,724,268	$11,373,536
2014		$4,724,268	$4,724,268	$16,097,804
2015		$4,724,268	$4,724,268	$20,822,072
2016		$4,724,268	$4,724,268	$25,546,340
2017		$4,724,268	$4,724,268	$30,270,608
		NPV @ 10%	$19,658,693	

Table 3. Particle size distribution of the thickener underflow

Particle Size Fraction (mm)	Weight (%)
+0.425	2.23
0.425 × 0.212	14.13
0.212 × 0.150	4.94
0.150 × 0.075	11.49
0.075 × 0.045	9.70
–0.045	57.21
Total	100.00

was mixed thoroughly with the Portland cement for about five minutes, and then injected into three 50-mm diameter by 100-mm plastic cylindrical molds to cure. Typical curing times were 1, 7, 14, and 28 days (EPA/625/6-89/022). Treated tailings were sealed in molds for one day and then removed to finish curing in open air. Specimens were cured in a storage chamber with around 50% relative humidity and 18 °C temperature. After the desired curing time, the specimens were subjected to an unconfined compressive strength (UCS) test. The measurement was conducted according to ASTM standard D1633-84: Standard Test Methods for Compressive Strength of Molded Soil-Cement Cylinders (USEPA). Results of the unconfined compressive strength of three cured specimens were obtained and averaged.

Tests were conducted over a range of cement and solid concentrations to determine the parameter effects on the compressive strength. The solid concentration values of 50%, 60% and 75% by weight were evaluated by filtering the thickener underflow slurry by pressure filtration and then adding water if necessary to achieve the desired solids concentration. It is noted that the fine tailings slurry with solid content of 50% and 60% by weight was in liquid form and fluid enough to be mixed with an overhead propeller mixer at 2,000 rpm.

In addition, since co-disposal with coarse reject was utilized at the Armstrong plant, the fine-to-coarse reject ratio was varied. The evaluation was achieved by adding limestone having a size of about about 6mm (0.25-inch) as a pseudo-coarse refuse to the thickener underflow at fine-to-coarse mass ratio of 8:2, 7:3 and 6:4, respectively. For the tests involving a solids concentration of 75% solids by weight, the waste is no longer in slurry form. As such, a ball mill was applied to mix the simulated coarse waste into the fine waste of 75% solids.

LABORATORY RESULTS AND DISCUSSION

The Armstrong plant currently adds cement at a concentration of around one percent of the total dry weight of the combined coarse and fine coal refuse. The same cement dosage was maintained in the initial tests while evaluating the effect of solids concentration and the fine-to-coarse coal ratio. The coarse reject acts as a bulking agent and reduces the volume of waste that needs to be treated with cement. The thickener underflow is filtered to achieve samples of 50%, 60% and 75% solid contents. Coarse material was added to achieve fine-to-coarse ratios of 8:2, 7:3 and 6:4 by weight. The mixed waste compounds were cured in molds for 24 hours and then the molds were removed and cured in open air for one more day.

The results shown in Figure 4 indicate a very significant impact by solid concentration. After only two days of curing, the samples containing 75% solids by weight obtained an unconfined compressive strength of nearly 55 psi with fine reject by itself, whereas a value of almost 90 psi were obtained with fine-to-coarse ratios of 7:3 and 6:4. For the samples with 50% and 60% solids contents, compressive strengths of 20 psi and less were obtained. The results indicate that there is a critical solids concentration between 60% and 75%, below which a stabilized refuse will require significantly greater amounts of cement. It should be noted that the results agree with industrial reports indicating plate-and-frame filter presses are providing filter cake with solid concentrations of around 80% and excellent stability when placed in a waste disposal area where the material is stacked.

The addition of coarse reject to the fine reject improves the overall strength of the combined refuse at all solid concentrations. The unconfined compressive strength nearly doubles under all

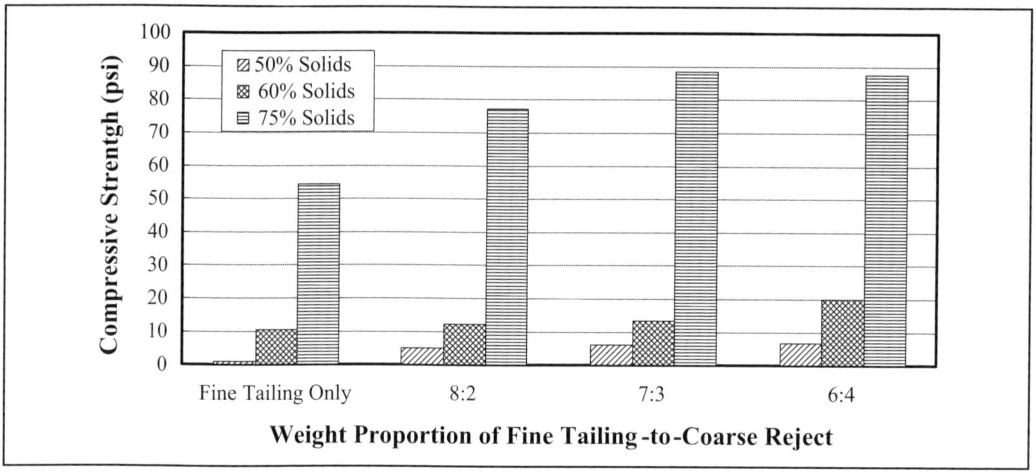

Figure 4. Mixing of fine tailings and coarse reject with multiple mixing ratios; 2-day curing and 1% cement

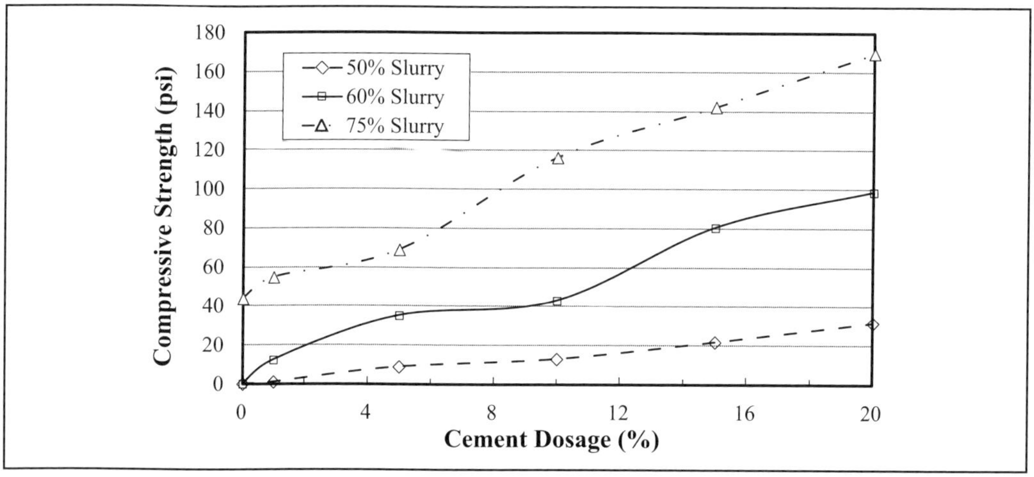

Figure 5. Cement dosage impact on compressive strength after two curing days over a range of solid concentrations

three solid concentrations as the coarse-to-fine ratio was adjusted from no coarse reject addition to a ratio of six parts fines to four parts coarse.

An evaluation performed with the thickener underflow material without the coarse material found that the material having a 75% solids concentration provided natural unconfined strength without the need for cement addition. Figure 4 shows that material formed into cylinders and cured with no cement had an unconfined compressive strength of around 40 psi. The dewatered thickener underflow containing 50% and 60% solids by weight with no cement had no compressive strength. Adding cement at a one percent concentration provided a two-day strength of 12 psi for 60% solids content material, whereas high cement additions were required for the 50% solids material (Figure 5).

Significantly greater compressive strengths were realized after seven days of curing as shown in Figure 6. With no cement, the strength of the material formed from a 75% solid content sample obtained a value greater than 150 psi. Cement addition over the range tested doubled the overall

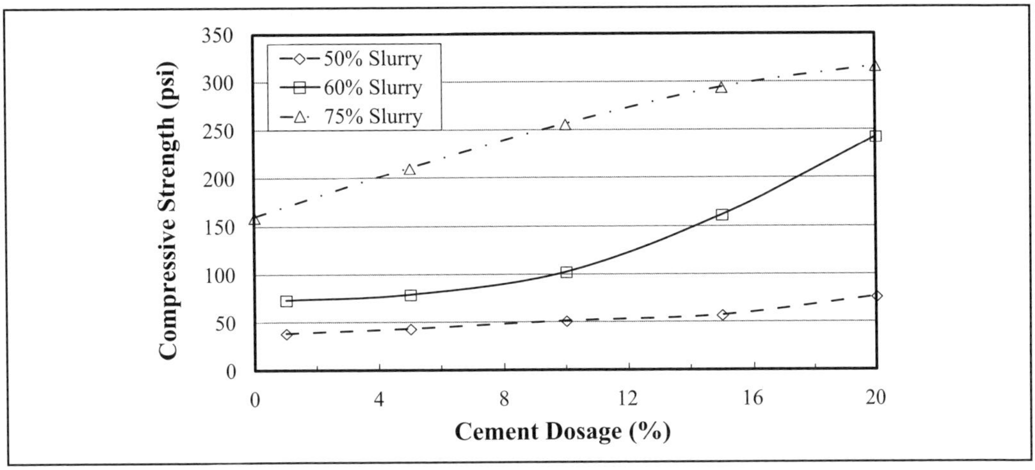

Figure 6. Cement dosage impact on compressive strength after seven curing days over a range of solid concentrations.

strength. However, the natural strength of the material explains the successes being reported by industrial applications of the plate-and-frame filter press.

However, if dewatering of the fine reject results in solid concentrations of 60% or lower or weather conditions decreases the solids content of the reject to 60% or lower, the addition of cement can provide a viable solution, as shown in Figure 6. Cement additions of one percent by weight provided compressive strengths above the EPA requirements of 50 psi for the 60% solids material and just below the level for fine waste with 50% solids.

Tests have been conducted with fine reject samples treated with cement and submersed in water for a number of days. Samples that were not treated with cement immediately returned to slurry form. However, samples treated with five percent or more cement retained their shape during submersion. The fine reject treated with one percent concrete became soft and broke up into smaller particle agglomerates. In addition, the pH of the water in which the cement treated specimens were submersed was around 9.0, whereas the pH was slightly acidic for samples not treated with cement. The results indicate potentially favorable environmental impacts from treating the waste with cement.

CONCLUSIONS

The preferred method for disposing of fine coal refuse is placement in slurry impoundments. However, the practice may be discontinued in the future due to regulatory issues. Operations that have been unable to obtain permits for slurry impoundments have already adopted alternative methods, which typically involve dewatering the fine refuse using belt filter presses and, more recently, plate-and-frame filters.

The Armstrong Dock and Preparation Facility employs belt filter presses to dewater the fine reject. The belt press product is combined with the coarse reject and disposed by stacking the combined reject. However, compaction issues required stabilization of the waste using Portland cement, which has proven to be very effective. Hydrated lime was tested and found to be ineffective and more costly on a per ton basis than Portland cement. The amount of cement added is approximately 2.7% by weight of the fine reject material. The total chemical cost of operating the belt presses and cement addition system is $0.55 per ton of plant feed.

Laboratory tests revealed that the addition of cement at a level of one percent by weight is sufficient for the fine coal waste at the Armstrong operation after filtration to a solids concentration of 60% by weight. The one percent cement-treated material produced unconfined compressive strengths exceeding the EPA requirements of 50 psi needed for stable foundations that can support construction equipment, impermeable caps and cover material. The stabilized material also limits the exposure of particle surface area, which should provide positive environmental impacts.

An economical comparison between a plate-and-frame application and the current belt press and cement treatment process at the Armstrong Facility indicates significant financial gains for the plate-and-frame system. The technical benefits of the plate-and-frame is the production of filter cake containing 80% solids without the need of chemicals beyond those typically applied in the conventional thickening process. Laboratory tests conducted on 75% percent solid material showed significant natural compressive strength without the need for cement. After seven curing days, the unconfined compressive strength of fine reject containing 75% solids was slightly greater than 150 psi. As such, the plate-and-frame filter provides a product that is stackable and able to support construction equipment.

Lab results showed that solid concentrations of 60% or lower required the use of cement to obtain the required compressive strength. If the plate-and-frame material is subjected to high rainfall, the material may return to slurry form and thus require the cement addition to meet the desired stability.

ACKNOWLEDGMENTS

This study was sponsored by the Appalachian Research Initiative for Environmental Science (ARIES). The views, opinions and recommendations expressed herein are solely those of the authors and do not imply any endorsement by ARIES employees, other ARIES-affiliated researchers or industrial members. Information about ARIES can be found at http://www.energy.vt.edu/ARIES.

Financial support for the project was provided by the U.S. Department of Energy through the Center for Advanced Separation Technologies (CAST). The authors acknowledge Edward Zawadzki and Fred Murrell of Carbon Resources of Florida, Inc., for their generous financial and technical support. Technical assistance from Mr. Ed Thompson who is the Laboratory Supervisor for the University of Kentucky Mining Engineering Department is deeply appreciated.

REFERENCES

Alam, N., Ozdemir, O., Hampton, M. and Nguyen, A. 2011. Dewatering of coal plant tailings: Flocculation followed by filtration. *Fuel* 90 (1): 26–35.

Chen, Q.Y., Tyrer, M., Hills, C.D., Yang, X.M. and Carey, P. 2009. Immobilisation of heavy metal in cement-based solidification/stabilization: a review. *Waste Management* 29(1): 390–403.

Conner, J.R. and Hoeffner, S.L. 1998. A critical review of stabilization/solidification technology. *Critical Reviews in Environmental Science and Technology* 28(4): 397–462.

Conner, J.R. 1990. *Chemical Fixation and Solidification of Hazardous Wastes*, New York, John Wiley and Sons.

Dermatas, D., Dadachov, M., Mirabito, M. and Meng, X. 2003. Strength development of solidified/stabilized organic waste and optimum treatment design. *Journal of the Air and Waste Management Association* 53(11): 1363–1372.

Forster, H.W. 1977. *Pressure Filtration Technology in the Coal Industry*. Final report to the Institute for Mining and Minerals Research (IMMR) 32-M4-77: 15–18.

Johnston, A.G., Davies, M.C.R. and Williams, K.P. 1997. Cement stabilization as an aid to the disposal of coal preparation plant tailings. *International Journal of Surface Mining & Reclamation* 11(4):169–174.

Malviya, R. and Chaudhary, R. 2006. Factors affecting hazardous waste solidification/stabilization: a review. *Journal of Hazardous Materials* 137(1): 267–276.

Mulligan, C.N., Yong, R.N. and Gibbs, B.F. 2001. Remediation technologies for metal-contaminated soils and groundwater: an evaluation. *Engineering Geology* 60(1-4): 193–207.

Myers, T.E., 1985. Simple procedure for acceptance testing of freshly prepared solidified waste. *Hazardous and Industrial Solid Waste Testing: Fourth Symposium*. Arlington, VA, USA.

Paria, S. and Yuet, P.K. 2006. Solidification-stabilization of organic and inorganic contaminants using portland cement: a literature review. *Environmental Reviews* 14(4): 217–255.

Pinarli, V., Karaca, G., Salihoglu, G. and Salihoglu, N.K. 2005. Stabilization and solidification of waste phosphate sludge using portland cement and fly ash as cement substitute. *Journal of Environmental Science and Health–Part A Toxic/Hazardous Substances and Environmental Engineering* 40(9):1763–1774.

Snyder, G.A., Zuhl, F.A. and Burch, E.F. 1977. Solidification of fine coal refuse. *Mining Congress Journal* 63(12):43–46.

Verma, S. and Klima, M.S. 2010. Evaluation of a pilot-scale, plate-and-frame filter press for dewatering thickener underflow slurries from bituminous coal-cleaning plants. *International Journal of Coal Preparation and Utilization* 30(1):1–19.

Williams, D.J. 1991. *Developments in Coal Mine Tailings Disposal and Rehabilitation. Queensland Coal Symposium*, Brisbane, Australia: Australian Institute of Mining & Metallurgy.

Xie, G., Wu, L. and Yu H. 2010. The selection of fine coal flotation and dewatering technology in a coal preparation plant. *Proceedings of the 16th International Coal Preparation Congress*, Edited by R.Q. Honaker. Littleton, CO:SME.

Application of a Pilot-Scale Plate Filter Press in Dewatering Coal Slurries

M.S. Klima
John and Willie Leone Family Department of Energy and Mineral Engineering, Penn State University, University Park, Pennsylvania

B.J. Arnold
PrepTech, Inc., Apollo, Pennsylvania

J. Munjack
PrepTech, Inc., Apollo, Pennsylvania

B. Barry
John and Willie Leone Family Department of Energy and Mineral Engineering, Penn State University, University Park, Pennsylvania

ABSTRACT

A plate filter press provides a potential option for dewatering ultrafine clean coal and refuse slurries, which could benefit plant operations through reduced product moistures, improved handle-ability of the dewatered products, and increased amount of water that is available for reuse in the cleaning facility. In this study, a series of tests was carried out with a pilot-scale plate filter press to evaluate its effectiveness in dewatering various coal slurries. The filter press consisted of a single set of 0.4 m × 0.4 m plates and was equipped with a hydraulic system, which operated the filter plates and feed pumps. Primary dewatering was carried out at a slurry feed pressure of approximately 860 kPa (125 psi) and additional dewatering was done using an air-blow system that provided compressed air of approximately 620 kPa (90 psi). The feed slurries were obtained from four bituminous coal cleaning facilities and included an effluent from a screen-bowl centrifuge dewatering a flotation product, a clean coal froth product, and three thickener underflow slurries. Filter cake moistures ranged from 20% for the clean coal product to approximately 30% for the thickener underflow material, though in one case, cake formation did not occur. Solids recoveries around 99% were obtained for all tests.

INTRODUCTION

Improved dewatering of both the clean coal and refuse material can provide various benefits to the plant operator. For example, when producing a clean coal product that must meet a specified heating value, a reduction in the product moisture content could be offset by increasing the amount of higher ash value material in the product so that the desired heating value is the same. This approach would increase the overall clean coal yield (Mohanty et al. 2012). In other cases, ultrafine material that was discarded because it could not be dewatered sufficiently could be recovered as a product, also increasing the clean coal yield. Likewise, refuse slurry could be dewatered to produce material that could be mixed with the coarser refuse for disposal. This would eliminate the need for a slurry impoundment and its potential environmental liabilities. Moreover, improved dewatering increases the amount of water that can be reused in the coal cleaning facility. The result is that less makeup water is needed and possible environmental effects associated with the treatment of additional water is reduced.

One technique with the potential for improving product dewatering is pressure filtration.

Pressure filtration involves separating solids from the liquid under pressures higher than atmospheric. The slurry is presented to a filter medium, which retains the dewatered solids (filter cake) while passing the liquid (filtrate). The pressure difference across the across the filter cake forces the filtrate from the slurry. The pressure is typically produced by pumping the slurry into the filter chamber. Typically, higher pressures will produce drier filter cakes; though for compressible filter cakes, higher pressures can decrease cake permeability, resulting in a lower filtration rate (Tiller 1975). In general, pressure filtration is used when dewatering slurries containing large percentages of ultrafine particles typically at feed solids concentrations greater than 10% solids (by weight). In the case of coal slurries, product moistures less than 25% are not unusual using a plate filter press (Patwardhan et al. 2006; Verma and Klima 2010a, 2010b).

Plate filter press operation is a batch process characterized by a total cycle time, which includes loading of the filter cavity with slurry, a ramping pressure period in which cake formation and filtrate production begin, constant pressure operation, and discharge of the cake. Prior to cake discharge, additional processing is often done to increase moisture removal. This may include the use of flexible membrane units that squeeze moisture from the cake, but this is more frequently accomplished through post-filtration air blowing. However, dewatering by air blowing may be limited by high capillary forces encountered with small pore sizes (Buscall and White 1987). Typically, air blowing is effective for cakes with permeabilities of 10^{-11} to 10^{-15} m^2 (Carleton and Salway 1993). The air must be applied at sufficient pressure so that the cake entry pressure is exceeded, at which point pore liquid is removed. Liquid in the pores first exits the cake through plug flow and then the remaining particle surface moisture is removed by the air. Evaporation effects are also present but desaturation by evaporation is often minimal unless given extensive blowing time, which is not typical in industry (Carleton and Salway 1993). After a period of blowing, an equilibrium or residual saturation is reached at which time the driving pressure differential and capillary forces of the cake are balanced so that an irreducible amount of moisture exists within the pores. The residual saturation is dependent on porosity as much as cake pore structure (Hosten and San 2002).

Large commercial plate filter press units may contain more than 100 plates, with plate areas over 2 m^2. Feed pressures can range from 700 kPa to well over 1,500 kPa. Moreover, a plate filter press minimizes or eliminates the need for additional flocculant as might be required for other pressure filtration devices (Prat 2012). In this study, the performance of a pilot-scale plate filter press was evaluated for dewatering bituminous coal slurries. This included slurries containing different amounts of –38 μm material and with a range of ash values and feed solids concentrations.

EXPERIMENTAL APPROACH

Sample Collection and Characterization

Slurry samples were obtained from four bituminous coal cleaning facilities located in the United States and Canada. Plant 1 samples were collected by plant personnel in 55-gallon drums from the effluent of a screen-bowl centrifuge, which had been thickened in a clean-coal thickener. The centrifuge was treating the clean coal product from a froth flotation circuit. Samples of the clean coal froth product were obtained from Plant 2 in 5-gallon buckets (Sample 2a) as well as from the thickener underflow (Sample 2b). Thickener underflow samples from the other two facilities were collected by plant personnel in 5-gallon buckets (Plant 3) and 55-gallon drums (Plant 4). At the laboratory, the material in each drum was mixed and then split into 5-gallon buckets to facilitate transferring the slurry to the filter press feed tank. The mixing/splitting approach ensured that the material in each bucket was representative of the entire drum. Each plant sample was characterized for solids concentration, size distribution, and ash values.

Filter Press Testing

Tests were carried out using a pilot-scale, plate filter press manufactured by Tecnicas Hidraulicas.

The filter press (Figure 1) consists of a 100-L feed tank with a variable-speed mixer, a set of two (0.4 m × 0.4 m) filter plates equipped with a hydraulic system for opening and closing the plates and powering diaphragm feed pumps, and a filtrate collection tank. Each plate is covered with a polypropylene cloth having a pore size of 0.051 mm. The filtrate collection tank is equipped with a conductivity probe, which can be raised or lowered to control the filtration time. The filtration time is the time during which the slurry is fed continuously to the filter press. The filtration cycle will stop if the probe is exposed before the maximum filtration time is reached, which would occur when the filtrate discharge rate decreases sufficiently. When the filtration cycle ends, compressed air at approximately 620 kPa is used to provide additional dewatering by blowing the air through the filter cake voids. The air-blow time can be adjusted as required. More detail on these operations can be found elsewhere (Verma 2009; Prat 2012).

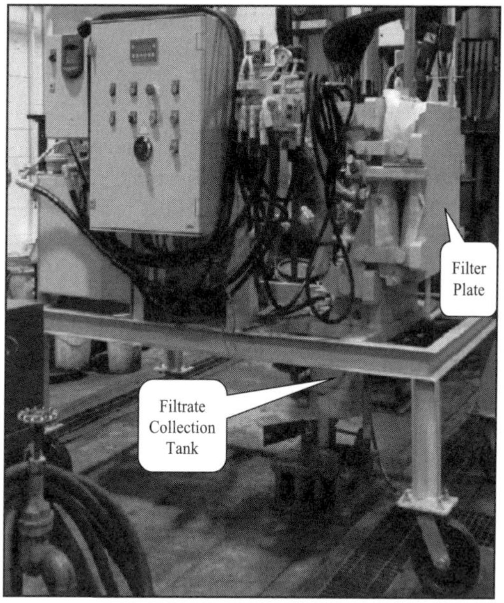

Figure 1. Pilot-scale plate filter press

For each test, the appropriate slurry sample was added to the feed tank in approximately 19-L batches. The operating variables, including maximum filter time and air-blow time, were set prior to the start of each test. For all tests, the conductivity probe height was lowered to the minimum height to maximize the filtration time. The filter plates were closed, and the feed pumps began filling the filter chamber. The filtrate began discharging into the collection tank. During the filtration cycle, the feed pressure was approximately 860 kPa. As the solids loading in the filter chamber increased, the filtrate rate decreased. The filter cycle ended either when the filter time was reached or when the filtrate rate decreased such that the conductivity probe was exposed in the collection tank. At that point, the air-blow cycle started and continued until the maximum air-blow time was reached. Prior to cake discharge, a valve was opened to drain slurry from the feed line to avoid diluting the filter cake during cake discharge. When the filter plates opened, the cake discharged by gravity, completing the cycle.

For each test, the filtrate was collected and weighed over timed intervals during the filter and air-blow cycles. The filter cake was collected and weighed. Samples of the filter cake were taken from several locations in the cake and were used for moisture determination.

RESULTS AND DISCUSSION

Sample Analyses

Table 1 summarizes the results from the size and ash analyses for the feed material. Plant 1 material (centrifuge effluent) had a solids concentration of 9.7% (by weight) and an ash value of 11.6% on a dry basis. Over 90% of the material was finer than 38 µm, which had an ash value of 11.7%. Plant 2a material (froth product) had a solids concentration of 17.5% and an ash value of 7.1%. Nearly 52% of the material was finer than 38 µm, which had an ash value of 10.6%. Plant 2b material (thickener underflow) had a solids concentration of 10% and an ash value of 45.6%. Approximately 84% of the solids were finer than 38 µm, which had an ash value of 45.7%. Plant 3 material (thickener underflow) had a solids concentration of 12.4% and an ash value of 54.3%. For this material, 70% of the material was finer than 38 µm, which had an ash value of 66%. Plant 4 material (thickener underflow) had a solids concentration of 22.8% and an ash

Table 1. Summary of the feed analyses and test results for the plant samples

Size Interval, µm	Plant 1		Plant 2 (Sample A)		Plant 2 (Sample B)		Plant 3		Plant 4	
	Wt.,%	Ash,%	Wt.,%	Ash,%	Wt.,%	Ash,%	Wt.,%	Ash,%	Wt.,%	Ash,%
+150	1.19	51.9	4.81	4.0	6.56	54.8	8.79	46.7	11.15	64.5
–150 + 75	1.93	7.35	22.67	2.8	3.39	58.3	9.26	20.0	5.95	62.0
–74 + 38	6.30	3.75	20.64	3.7	6.14	26.7	11.72	18.8	3.76	55.7
–38	*90.58*	*11.7*	*51.88*	*10.6*	*83.91*	*45.7*	*70.23*	*66.2*	*79.14*	*34.5*
Total	100	11.6	100	7.1	100	45.6	100	54.3	100	40.3
Feed solids concentration,%	9.7		17.5		10.0		12.4		22.8	
Filter cake moisture,%	25.3		20.4		29.5		NA*		29.3	
Unit capacity, kg/hr/m^2	146		245		49.6		NA*		161	

*No filter cake was formed.

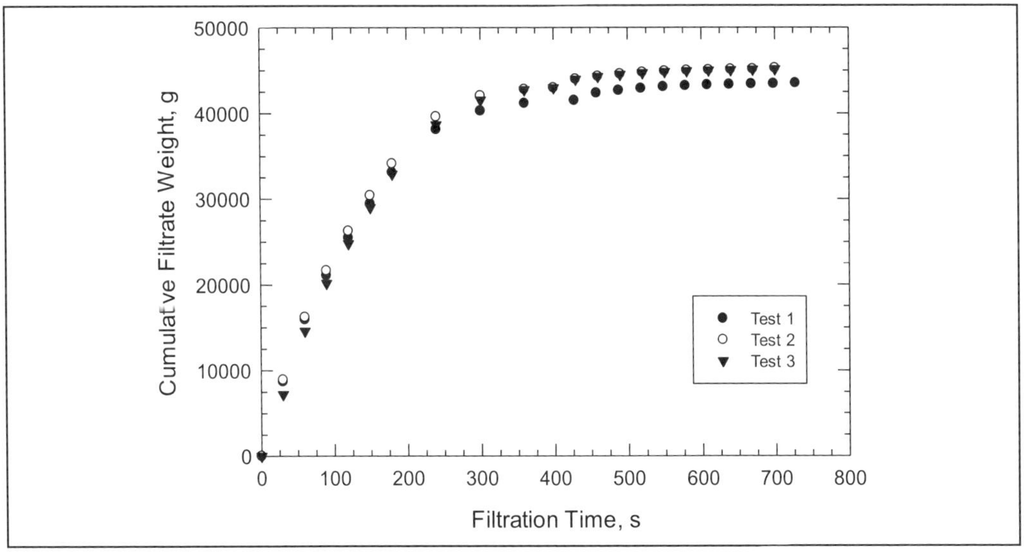

Figure 2. Reproducibility testing for Plant 1 material

value of 40.3%. Approximately 79% of the material was finer than 38 µm, though the ash value was 34.5%, which was lower than the other two underflow samples.

Filter Press Testing

Plant 1

The screen-bowl effluent is currently being thickened and recycled in the plant, though this is being done inefficiently at best. Increased recovery of this material is desired as it has a relatively low ash value and could be sold as a coking coal if the moisture could be reduced sufficiently. Several tests were conducted with this material to evaluate different aspects of the filtration process. The normal filtrate and air-blow filtrate samples were collected in timed increments and then weighed. This process was repeated for two additional runs to evaluate the reproducibility of the process. These results are shown in Figure 2. Very similar results were obtained for all three tests. As expected, the rate of filtration was the greatest over the first 300 seconds and then decreased at the longer times. After 450 seconds, there was a slight increase in the filtration rate when the air-blow cycle started. Similar results were observed in previous testing using both anthracite and bituminous coal slurries (Verma and Klima 2010a, 2010b).

Figure 3 shows a typical filter cake for this material. The cake was approximately 50 mm thick, which was the depth of the filter cavity. The cake discharged easily from the filter cloth by gravity when the filter plates opened. At the start of the filter cycle, the filtrate was slightly turbid. However, the clarity quickly improved as the solids collected on the cloth, with the cake acting as the filter medium.

Figure 3. Filter cake using Plant 1 material

Using the data from Test 1, an overall solids and water balance was performed. These results are shown in Table 2. The reconstituted solids concentration and ash values compare well with the values from the feed sample analysis given in Table 1. Approximately 99% of the total solids were recovered in the filter cake. This value is similar to those obtained in previous testing with this unit (Verma and Klima 2010a, 2010b). Analysis of the filtrate solids indicated that 89% of the material was finer than 25 µm and had an ash value of 13.5%, which was similar to the ash value of the feed solids.

Another filtration test was conducted, and the results were used to determine the change in cake moisture with time (Figure 4). As with the other tests, the recovered filtrate increased with increasing filter time, while cake moisture decreased (and the solids concentration increased). For example, the cake moisture decreased from 90% (10% feed solids concentration) to about 40% after the filter cycle ended at 450 seconds. During the air-blow cycle, additional filtrate was removed, reducing the final cake moisture to 25%.

The filter unit capacity was estimated by Equation 1.

Table 2. Solids and water balance for Plant 1 material

	Filter Cake	Filtrate (Filtration Cycle)	Filtrate (Air Blow Cycle)	Reconstituted Feed
Weight of Water, g	1626.6	41491	2011	45128.6
Weight of Solids, g	4757.4	55.3	1.3	4814.0
Distribution of Solids.%	98.82	1.15	0.03	100.00
Solids Concentration, wt.%	74.52	0.13	0.03	9.64*
Ash%, dry basis	11.19	12.75	65.01	11.22*

Size Interval, µm	Wt.,%	Ash,%
+150	0.81	39.55
+150–75	1.85	5.64
+75–38	6.89	3.86
+38–30	8.70	3.12
–30	81.75	12.78
Total	100.	11.41

*Values compare well with feed sample analysis.

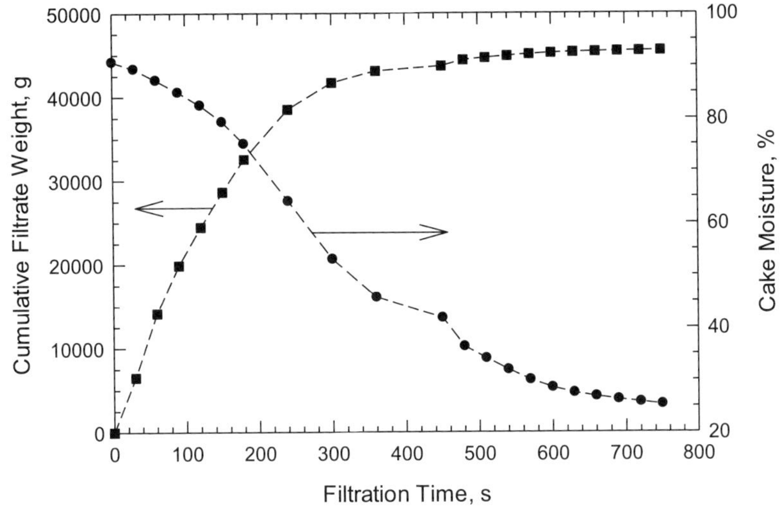

Figure 4. Variation of filtrate weight and cake moisture with filtration time for Plant 1 material

$$\text{Unit Capacity} = \frac{\text{Final Dry Cake Weight (kg)}}{\text{Cycle Time (hr)} \times \text{Filter Plate Area (m}^2\text{)}} \quad (1)$$

The cycle time is the sum of the filter and air-blow times, which was 750 seconds for this test. This gave a unit capacity of 146 kg/hr/m^2. Higher unit capacities would be associated with higher filter cake moistures. For example, in a previous study using the same test unit, it was found that the unit capacity increased from 60 to 210 kg/hr/m^2 with an increase in filter cake moisture from 17% to 26% (Verma and Klima 2010a). In this case, anthracite slurry from a thickener underflow was processed. This material was 61% finer than 38 µm and had an ash value of almost 68%.

Plant 2

For the froth product material (Sample 2a), the filter cycle lasted for 176 seconds at which time the air-blow cycle began. This occurred because the flow of filtrate was very slow and essentially no material was being pumped into the filter cavity. The air-blow cycle continued for an additional 300 seconds. The filter cake discharged from the filter cloth with no difficulty and was very dry to the touch. There was less than a 1% loss of solids to the filtrate over the entire filter cycle. The final cake moisture was 20.4%, and the calculated unit capacity was 245 kg/hr/m^2. The high unit capacity was not surprising considering that this was a clean coal product and would be expected to dewater well.

For the thickener underflow material (Sample 2b), a very long filtration time was used to maximize water removal. After 40 minutes, the filter cycle stopped when the conductivity probe was exposed. The air-blow cycle continued for an additional seven minutes. Analysis of the filtrate indicated that that approximately 99% of the feed solids were recovered in the filter cake. After the air-blow cycle, the plates opened and the cake discharged by gravity with no difficulty.

Figure 5 shows the cake moisture as a function of air-blow time. Before the air-blow cycle began, the cake moisture was about 38%. At this point, the filter cycle ended due to the very low filtrate flow. However, during the air-blow cycle, additional filtrate was removed. After only 60 seconds, the cake moisture was reduced over four percentage points and nearly six percentage points after 120 seconds. An additional two percentage points of moisture were removed over the remainder of the air-blow cycle, and a final

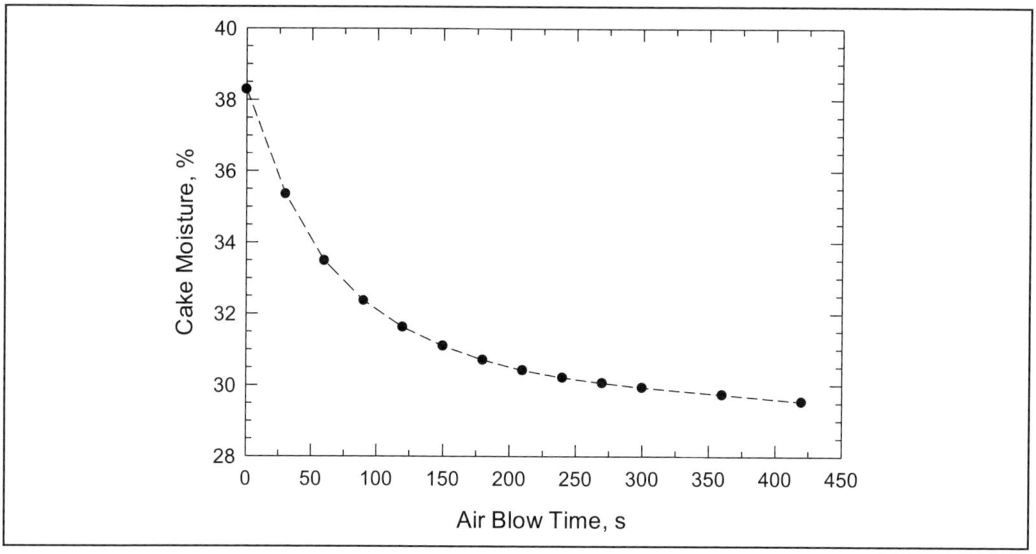

Figure 5. Variation of filter cake moisture with air blow time for Plant 2 (Sample 2b) material

moisture content of 29% was obtained. The unit capacity was 49.6 kg/hr/m^2 due to the relatively long cycle time.

Plant 3

For this material, no cake was formed during filtration. As the slurry was pumped into the filter chamber, the filter cloth became coated with a fine layer of solids, blinding the cloth and preventing filtration. This test was repeated and the same result occurred. Although only a thin film of solids was recovered, this material was dried, and the solids concentration was approximately 20% in both cases compared to 12.4% solids in the feed. In previous testing with this unit, filter cakes were formed in all cases (Verma and Klima 2010a, 2010b). It was likely that the presence of ultrafine clays in the Plant 3 material led to the filter cloth blinding, preventing filter cake formation.

Plant 4

A filter time of 720 seconds and an air-blow time of 120 seconds were used for this test. The recovered filtrate and cake moisture as a function of filtration time are shown in Figure 6. As was seen for the other tests, the filtration rate was the highest initially and then decreased at the longer times. As more filtrate was removed, the cake moisture decreased from 80% moisture to a final value of 29.3% after a total time of 840 seconds. Based on these results, the unit capacity was estimated to be 163 kg/hr/m^2, with an overall solids recovery of 98.5%.

Estimation of Filter Cake Moisture

In an attempt to estimate the moisture contents for the various filter cakes, a simple regression model was developed based on the percentage of –38 µm material in the feed and the corresponding ash value for the four tests in which a filter cake was produced. Another data point was also used, which was obtained from previous testing using the same filter press (Verma and Klima 2010b). The regression equation is given by

$$M = 0.174A + 0.115B + 9.694 \quad (2)$$

where M is the filter cake moisture (%); A is the % of –38 µm in the feed material; and B is the ash value (%) of the –38 µm material. Figure 7 compares the actual and calculated moistures for the five tests. The fit was relatively good considering the variation in feed materials that were used. The adjusted r-squared value was 0.775. It has been shown that for some dewatering devices (i.e., screens, centrifuges, and vacuum filters) the

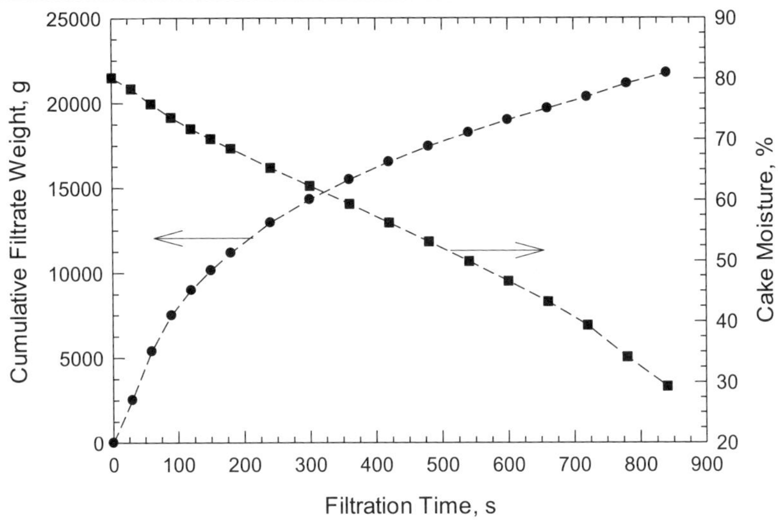

Figure 6. Variation of filtrate weight and cake moisture with filtration time for Plant 4 material

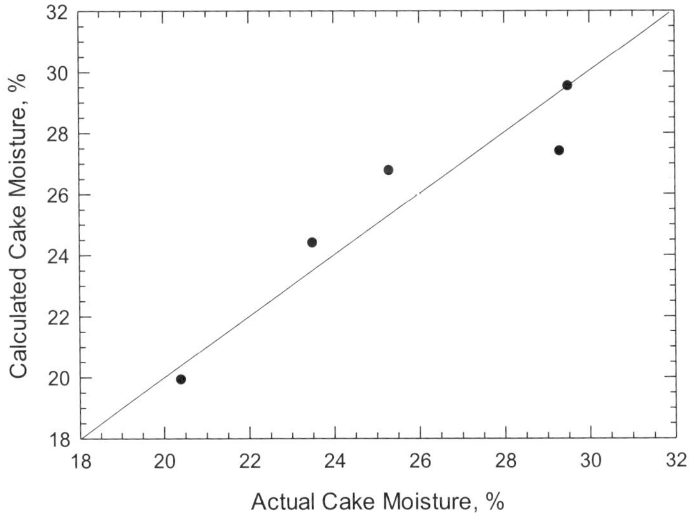

Figure 7. Comparison of the actual and calculated cake moisture contents

product moisture could be related to the percentage of –75 µm material in the feed (Arnold 1999). Further work will be needed to develop better relationships between final cake moisture and properties of the material to be dewatered, taking into account, perhaps, particle shape and a measure for overall average particle size.

SUMMARY AND CONCLUSIONS

Five bituminous coal slurries were evaluated using a pilot-scale plate filter press. One sample was obtained from the effluent of a screen-bowl centrifuge, which was dewatering clean coal from a froth flotation circuit. Starting with a slurry solids concentration of 10%, a filter cake with approximately 42% moisture was obtained after a filter time of 400 seconds. Following an additional 300 seconds of air blowing, a final cake moisture of approximately 25% was obtained. Based on the dry cake weight and cycle time, a unit capacity of 146 kg/hr/m^2 was determined. The froth flotation product was dewatered to a final moisture content of 20.4% with a unit capacity of 245 kg/hr/m^2.

The other three coal slurries were obtained from thickener underflow streams at different processing facilities. For two of the samples (Plants 2 and 4), the final cake moistures were approximately 29%, with unit capacities of 49.6 and 163 kg/hr/m^2, respectively. On the other hand, a suitable cake could not be formed with the Plant 3 material. The type of clays in this material likely blinded the filter cloth, preventing cake build up and subsequent filtration. Overall it appears that a plate filter press provides an option for dewatering various clean coal and/or refuse slurries. This would benefit plant operations through reduced product moistures, improved handleability of the dewatered products, and an increased amount of water available for reuse in the cleaning facility. Moreover, additional chemical treatment would likely only be needed for materials that would blind the filter cloth.

REFERENCES

Arnold, B.J., 1999. Simulation of Dewatering Devices for Predicting the Moisture Content of Coals. *Coal Preparation,* 20:35–54.

Buscall, R., and White, L.R., 1987. The Consolidation of Concentrated Suspensions. Part 1.-The Theory of Sedimentation. *J. Chem. Soc. Faraday Trans. I* 83 (3):873–891.

Carleton, A.J., and Salway, A.G., 1993. Dewatering of Cakes. *Filtration & Separation.* 30 (7):641–646.

Hosten, C., and San, O., 2002. Reassessment of Correlations for the Dewatering Characteristics of Filter Cakes. *Minerals Engineering,* 15 (5):347–353.

Mohanty, M.K., Akbari, H., and Luttell, G.H. 2012. Fine Coal Drying and Plant Profitability. In *Challenges in Fine Coal Processing, Dewatering, and Disposal,* edited by M.S. Klima, B.J. Arnold, and P.J. Bethell, SME, Littleton, CO, pp. 329–344.

Patwardhan, A., Chugh, Y.P., Arnold, B.J., and Terblanche, A.N., 2006. Dewatering Ultrafine Clean Coal in a T.H. Filter Press, *Coal Preparation,* 26:33–54.

Prat, G., 2012. Dewatering Fine Coal and Tailings with a Filter Press, In *Challenges in Fine Coal Processing, Dewatering, and Disposal,* edited by M.S. Klima, B.J. Arnold, and P.J. Bethell, SME, Littleton, CO, pp. 279–292.

Tiller, F.M., 1975. *Theory and Practice of Solid-Liquid Separation,* 2nd ed., University of Houston, 247–287.

Verma, S., and Klima, M.S., 2010a. Evaluation of a Pilot-Scale, Plate-and-Frame Filter Press for Dewatering Fine Anthracite Refuse, In *Proceedings of the XVI International Coal Preparation Congress,* Edited by R. Honaker, SME, Littleton, CO, pp. 516–524.

Verma, S., and Klima, M.S., 2010b. Evaluation of a Pilot-Scale, Plate-and-Frame Filter Press for Dewatering Thickener Underflow Slurries from Bituminous Coal Cleaning Plants, *International Journal of Coal Preparation and Utilization,* 30:1–19.

Verma, S., 2009. An Evaluation of a Pilot-Scale Plate-and-Frame Filter Press for Dewatering Fine Coal Refuse and Silica Sand, M.S. Thesis, The Pennsylvania State University, University Park.

Diesel Range Organics In Coal Preparation

J.P. Morris
Department of Mining and Minerals Engineering, Virginia Tech, Blacksburg, Virginia

E.A. Sarver
Department of Mining and Minerals Engineering, Virginia Tech, Blacksburg, Virginia

G.H. Luttrell
Department of Mining and Minerals Engineering, Virginia Tech, Blacksburg, Virginia

ABSTRACT

Recent laboratory testing has suggested that partitioning of petro-diesel collector reagents in coal flotation circuits is not perfectly ideal. In this paper, we investigate the persistence of diesel range organic (DRO) compound residuals in process waters under a number of physio-chemical conditions. Additionally, we investigate desorption of DROs from coal surfaces exposed to fresh water. In both cases, we also examine the behavior of individual PAH compounds. Results are discussed in the context of potential environmental transport and fate of DRO compounds in water from coal preparation plants. While our results indicate that DRO concentrations in process waters are expected to be at sub-ppm levels under normal operating conditions, we note that "green" collectors are available for coal flotation.

INTRODUCTION

The primary function of coal preparation is removal of mineral matter (i.e., ash), which detracts from the coal value. Preparation plants typically have multiple circuits for processing materials of different particle sizes. In the fine circuits (i.e., –100 mesh particles), froth flotation is often used to separate coal from ash. In 2012, about 30% of the coal preparation plants in the US (i.e., 82 of 289) were reported to operate fine coal flotation circuits, and virtually all of these (i.e., 80 of 82) are located in the Central Appalachian basin (Fiscor 2012). For reference, this basin also accounted for about 85% of the total preparation plants (i.e., 248 of 289).

Froth flotation of coal works on the basic principle that coal particles are relatively hydrophobic and lightweight, such that they easily float to the top of the flotation columns or cells and can be recovered from a froth that forms there; ash mineral particles, on the other hand, are relatively hydrophilic and dense, such that they sink and are rejected to the tailings stream. Coal particles are naturally hydrophobic to some degree, depending on their specific chemical properties (e.g., surface oxidation, liberation from mineral matter), but "collector" reagents that increase their hydrophobicity are often used to aid the flotation process (Wills 2006). Globally, petro-diesel (termed "diesel" in this paper) is the most commonly used collector given its relatively low cost and proven performance (Laskowski 2001). However, collector dosages are quite variable (e.g., diesel may be dosed in the range of 0–5 lb/ton) due to flotation circuit parameters and feed quality; dosage at a single plant may be adjusted frequently.

While collectors work by sorbing to the coal surface (Kondrat'ev 2009), recent studies have indicated that the partitioning of diesel between coal and process water is not perfectly ideal (Morris et al. 2012). In bench-scale experiments, under a variety of test conditions mixing water, coal and diesel, low-level diesel range organic (DRO) compound residuals were consistently measured in process water. Results suggested that the DRO residuals may primarily consist of the water soluble fraction (WSF) of the diesel, which exists at sub-ppm levels. At present, DROs are not monitored in impoundments or discharges, however, given increasing concerns over diesel contamination of water resources by leaking underground storage tanks, highway and blacktop runoff, and events such as the recent Deepwater Horizon oil spill in the Gulf of Mexico (Sementelli and Simons 1997; Lloyd and Cackette 2011; Osofsky et al. 2011), it is important to understand the potential transport mechanisms and fate of DROs from coal processing. It should be noted that alternative "green" collector reagents such as bio-diesel and pine oil products are already being utilized for fine coal flotation in some instances. These collectors are significantly more expensive than diesel but may be required in special circumstances, such as safeguarding groundwater resources when coal waste slurry is intended to be disposed via underground injection (WV-DEP 2009).

Environmental Fate and Transport of Diesel Compounds in Water

Diesel has a mixed composition of roughly C_{10} to C_{19} hydrocarbons, which, as for many petroleum products, varies based on the crude oil source and refining process(es) (ATSDR 1999). This variability makes it difficult to uniquely classify the chemical and environmental characteristics of diesel, and also to accurately measure individual components (ATSDR 1995). Instead, diesel is usually described in terms of major compound categories: total saturated hydrocarbons (e.g., alkanes and cycloalkenes) and total aromatic hydrocarbons (e.g., naphthalenes, acenaphthenes, and acenaphthylenes). Saturated hydrocarbons typically account for about 90% (by weight) of total diesel, while aromatics account for about 10% (Wang et al. 2003). Any remaining components (e.g., waxes and resins) usually make up a very small fraction of the total diesel and are not often quantified.

In addition to broad categorical classification, the nature of diesel and other petroleum products is sometimes characterized by measuring a series of ideal (i.e., n-alkanes) and/or priority compounds (i.e., specific aromatics). Typical analytical techniques include GC-FID, GC-MS and fluorometry (Wang et al. 2003). It is important to note that while specific compounds may be targeted by these techniques, the large number of individual compounds present in diesel (e.g., branched alkanes and functionalized aromatics) makes it impossible to quantify each and every one (ATSDR 1995). Moreover, the diesel composition may change dramatically in the environment as it weathers (e.g., via volatilization, bio- or photo-degradation). Such complexities make predicting environmental implications of diesel releases quite challenging, but offer unique opportunities for source tracking in some cases (Wang et al. 1996).

A basic understanding of environmental transport mechanisms and the fate of diesel can be gleaned from properties of the compounds in the broad categories mentioned above. Generally speaking, the saturated hydrocarbons are relatively volatile (Fingas 1994; Fingas 1995), insoluble in water, and photo- and bio-degradable (Olson et al. 1999; Marquez-Rocha et al. 2001; Cohen et al. 2002; Kakkar et al. 2011). Thus, at relatively low levels, this group of compounds does not present major concerns for water resources. Aromatics in the diesel range are also typically volatile and insoluble in water, but are much more persistent in the environment because they resist degradation (Olson et al. 1999). Further, some monocyclic and polycyclic aromatic hydrocarbons (MAHs and PAHs) have been classified as possible or probable human carcinogens (ATSDR 2009), and have been linked to acute or chronic toxicity in aquatic organisms (Schein et al. 2008). Thus, if significant concentrations and exposure pathways exist, these compounds may present ecological and human health hazards. Indeed, the US EPA has developed a list of 16 priority PAHs, although only one compound (i.e., benzo-[a]-pyrene) is regulated by a maximum contaminant level in drinking water (EPA 2011). Total DRO is not federally regulated (EPA 2003).

PAHs exist naturally in fossil fuels and their derivatives, and can also be formed via incomplete combustion of these and other sources of organic carbon (e.g., wood and other plant matter) (SCF 2002). Common contributors to DRO and PAHs in water include oil spills and leaks (e.g., from underground storage tanks), urban run-off, and municipal and industrial effluents. PAHs in the atmosphere, which originate from sources such as volcanic eruptions, forest fires, combustion of fossil fuels, waste incineration, coke and asphalt production, oil refining, and aluminum production, can also deposit into surface waters (Manoli and Samara 1999). In diesel, the most prevalent PAH groups are typically those with the lowest molecular weights (e.g., naphthalenes, fluorenes, and phenanthrenes); these PAHs have weights in the same range as saturate compounds in diesel (e.g., C_{10}–C_{19} alkanes). The fate of individual PAHs is dependent on their physicochemical properties (e.g., molecular weight, structure, water solubility, and vapor pressure) (Manoli and Samara 1999; Wick et al. 2011), but it can be generally stated that light-weight PAHs tend to be more soluble and volatile than their heavier counterparts. As molecular weight increases, water solubility and vapor pressure decrease and the compound becomes more difficult to degrade (Wick et al. 2011).

With regard to diesel reagents used in fine coal processing, it is important reiterate that under normal operating conditions, nearly all but the soluble DROs are expected to be effectively sorbed to the coal surface based on prior work (Morris et al. 2012). Thus, only those compounds that do not sorb, desorb or that are soluble can potentially move with process water. While the sum of EPA-listed PAHs (and their simple alkylates) do not generally make up a significant fraction of total diesel (i.e., less than a few percent) (EPA 1990), the current attention on this broad category of compounds warrants investigation to ensure that coal preparation activities do not present cause for concern. Here, our objective was to gain a better fundamental understanding of potential fates and transport mechanisms of diesel, including PAH compounds, from fine coal processing.

EXPERIMENTAL METHODS

Partitioning studies were carried out to obtain preliminary data on the likely fate of residual DROs, including PAH compounds from fine coal flotation. In one set of tests, the potential for diesel removal from process water by heating, stirring, and aeration was analyzed; in another test, desorption of diesel from coal surfaces was investigated. A sample of raw coal was obtained from the Hernshaw coal seam in Kanawha, West Virginia, which was determined to be 44.6% ash. The coal was prepared for testing only by sizing. It was crushed using a laboratory jaw crusher followed by a roll crusher, and then screened to obtain material in the range of 74–149 microns (i.e., 100 × 200 mesh). A sample of diesel was purchased from a local fuel station, and stored in an amber, airtight glass jar. Basic characterization of the diesel via GC-FID (as described in EPA 2003) showed that it consisted of 87.4% saturates and 11.5% aromatics.

For each test, a slurry sample (5% solids by weight) was prepared by adding the dry, sized coal to deionized water (DI); then diesel was added to achieve an effective dosage of 10 lb diesel/ton coal (Figure 1). Slurries were mixed using a hand blender for two minutes, and then the coal particles were separated from the water by filtration (through 25 μm paper) using a vacuum pump.

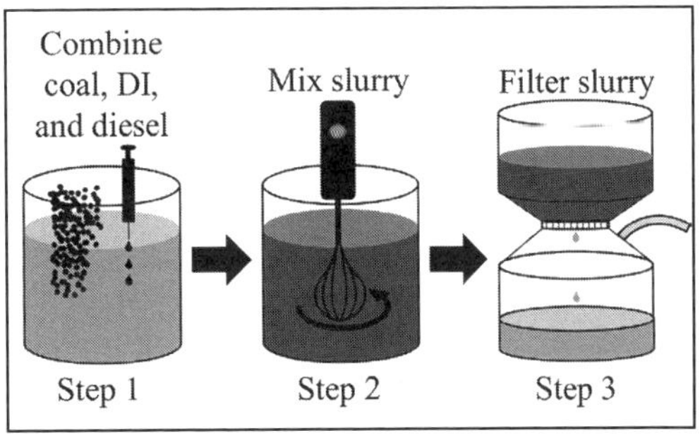

Figure 1. Sample preparation

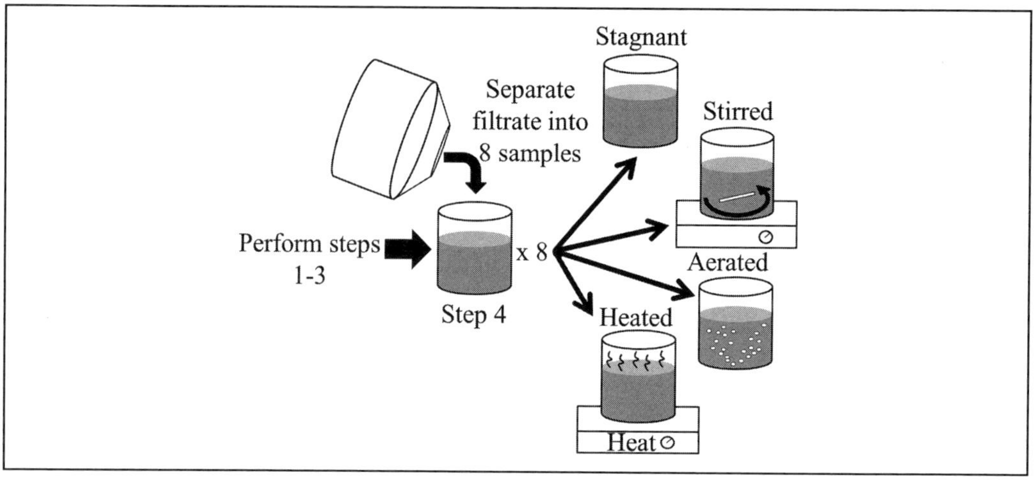

Figure 2. DRO removal—experimental methods

DRO Removal

For the diesel removal tests, a total of 2 L of filtrate was prepared, with 150 mL of this being immediately stored in an airtight amber jar for analysis (i.e., t = 0). The rest of the filtrate was split into 8 equal subsamples for individual test conditions (Figure 2). The test conditions were: stagnant (open jar in ambient room temperature and pressure), gentle stirring, gentle aeration and heating (40°C); and for each, one sample was conditioned for 2 hours and another was condition for 4 hours. At the end of those time increments, the subsamples were stored for analysis.

Diesel Desorption

For the diesel desorption tests, slurry was also prepared and filtered. Following collection and storage of the initial filtrate, coal particles remaining on the filter paper were carefully removed (Figure 3) and mixed into a new volume of deionized water (again at 5% solids)—this time without the addition of diesel. The slurry was then filtered, and the filtrate was again collected for analysis. This process was repeated multiple times to investigate the trend in DRO and PAH concentrations for successive rinses. In the first test (Test A) a total of four rinse cycles were completed, one immediately after the next. In the next test (Test B), a total of six rinse cycles were completed; however, for the last two cycles, the coal particles were allowed to air dry for 24 hours.

Analytical Methods

Residual diesel in each sample was analyzed using an Agilent 5890 gas chromatograph equipped with a flame ionization detector (GC-FID); we followed EPA Method 3150 for quantifying diesel range organics (DRO) in water samples. Samples were also analyzed for a group of target PAHs using a Thermo Trace GC equipped with a Thermo DSQ II mass spectrometer (GC-MS); we followed EPA Method 3535A: Solid-Phase Extraction (SPE). The target PAHs were: benz(a)anthracene, benzo(a)pyrene, benzo(b)fluoranthene, benzo(k)fluoranthene, chrysene, dibenz(a,h)anthracene, indeno(1,2,3-cd)pyrene, acenaphthene, acenaphthylene, anthracene, benzo(ghi)perylene, fluoranthene, fluorine, naphthalene, phenanthrene, and pyrene.

RESULTS

Results of all tests confirmed that a small fraction of diesel tends to partition to the process water—and this is typically limited to the soluble fraction. The converse side of this observation is therefore that the coal particles have a very high capacity for sorption of insoluble diesel.

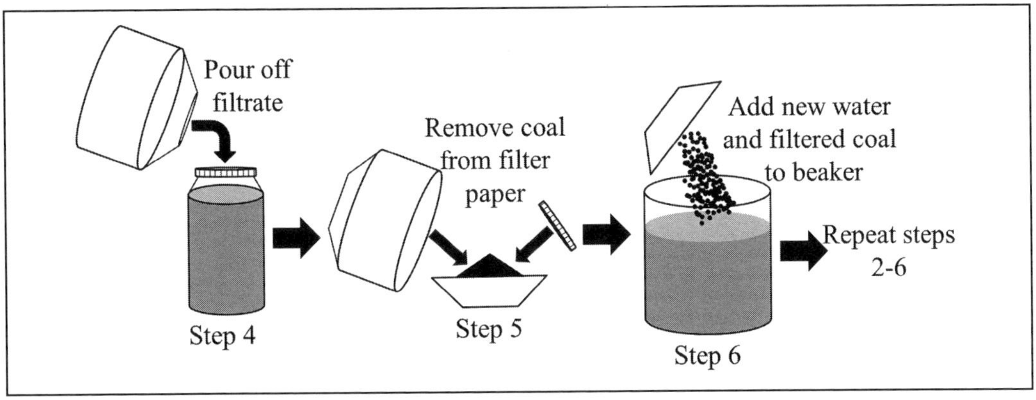

Figure 3. Diesel desorption—experimental methods

Table 1. DRO and PAH removal results

Test Condition	Time (hr)	PAHs (ppb) Naphthalene	Acenaphthene	Fluorene	Anthracene	Fluoranthene	Pyrene	DRO (ppm)
Initial	0	60	6.5	6.8	2.7	0.14	0.18	0.43
Stagnant	2	56	7.1	7.6	2.8	NF	0.27	0.59
	4	34	5.1	5.6	NF	NF	0.25	0.53
Stirring	2	2.5	0.99	1.6	1.5	NF	0.16	0.30
	4	0.51	0.15	0.37	0.81	NF	NF	0.27
Aeration	2	0.98	0.10	NF	NF	NF	NF	0.27
	4	0.69	0.41	NF	NF	NF	NF	0.29
Heating	2	17	2.9	3.5	2.0	NF	0.13	0.48
	4	12	2.0	2.7	1.6	NF	0.12	0.38

NF indicates the compound was not found.

Potential for Diesel Removal from Process Water

Table 1 presents results for DRO and PAH residuals in the tests designed to investigate potential for diesel removal from process water. Figure 4 also graphically displays the results for DRO and naphthalene. The rate of DRO removal was significantly higher for the stirred and aerated conditions than for the heated and stagnant conditions, with DRO concentrations leveling off just below 0.3 ppm. This concentration corresponds to the solubility limit of diesel in water at ambient temperature and pressure. For the heated and stagnant conditions, DRO actually increased after two hours (more significantly for the stagnant sample), and then began decreasing after four hours. Over the test period, the heated sample did lose significant volume due to evaporation, and it is also possible that the observed increase in DRO in this sample is related to the time required for equilibration (i.e., as water evaporates, DRO does not immediately come out of solution and vaporize). This would also explain the slower DRO removal in the stagnant condition, wherein there was no additional driving force to quickly remove all insoluble diesel from the water (vs. the stirred and aerated conditions). In an additional experiment to test this hypothesis, the stagnant and aerated conditions were repeated and samples were collected and measured after 24 hours; results confirmed that the total DRO was similar for both conditions. After 48 hours, total DRO was again similar between the two conditions.

PAH removal trends appeared to be similar to those of the total DRO for all tests, in that (1) the concentrations decreased over time, (2) the rate of decrease was fastest for the stirred and aerated

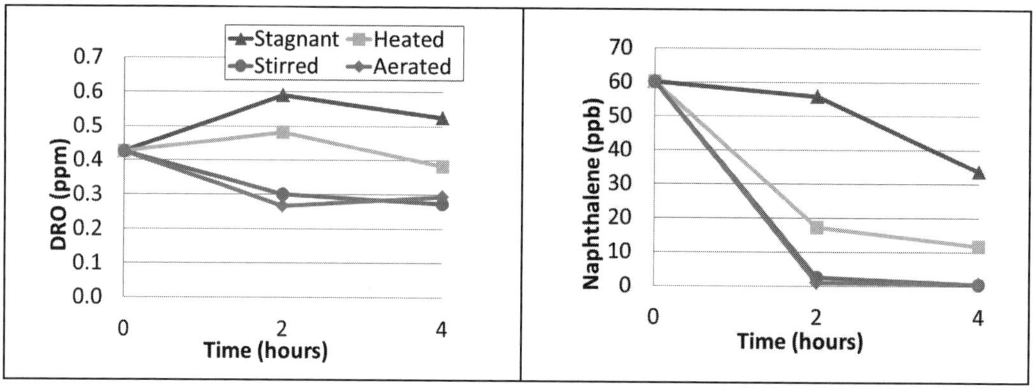

Figure 4. DRO and naphthalene concentration as a function of time under stagnant, heated, stirred, and aerated conditions

conditions and slowest for the stagnant and heated conditions. However, it should be noted that the PAHs tended to be removed from the process water much more quickly overall than the total DRO. This is illustrated in Figure 4 by comparing DRO to naphthalene, which had the highest measured concentrations of all target PAHs, followed by acenapthene and fluorene. For the stirred and aerated conditions, while the DRO does decline (to about 70% of its initial concentration), the naphthalene is removed down to sub-ppb levels (to about 1% of its initial concentration). The other PAHs behaved similarly. Thus, it appears that the relative abundance of PAHs in the WSF of diesel may be much smaller than in the total diesel.

DRO Desorption

Previous work has demonstrated that the sorption capacity of raw coal for diesel is very high (e.g., all but the soluble portion of diesel sorbs to the coal)—even beyond practical operating conditions for fine coal flotation (Monsalve 2010; Morris et al. 2012). The goal of the present testing was to determine if the sorbed diesel might be easily removed again from the coal, for example during dewatering or other instances where the coal product contacts water, or in instances where ultrafine coal with sorbed diesel reports to tailings impoundments. Figure 5 shows the total DRO concentrations measured in filtrate samples following successive rinsing of diesel-contacted coal

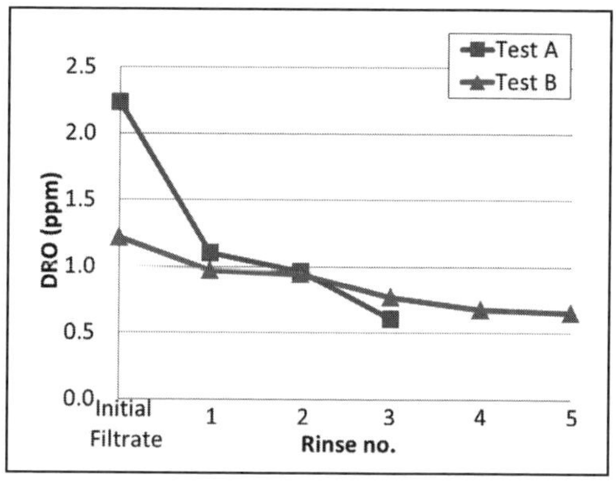

Figure 5. DRO content after each rinse cycle for coal that has previously been in contact with diesel

for two separate tests (i.e., A and B). The declining trends indicate that, initially, the clean water is likely rinsing free diesel from the coal surfaces (i.e., diesel that was not actually sorbed to the coal, but rather caught between particles); however, with subsequent rinses, diesel is desorbing from the coal. With more rinsing cycles, it is expected that the diesel would eventually desorb from the coal to satisfy the solubility limit in the fresh water.

For test B, the target PAHs were also measured (Figure 6). It should be noted that the initial naphthalene concentration measured in this test was much lower than that in the tests presented in Table 1 (by nearly an order of magnitude). An explanation for this is not clear. However, the significant decline in naphthalene shown in

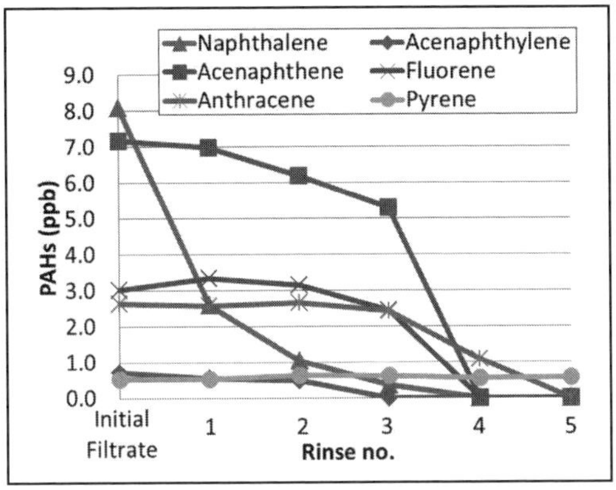

Figure 6. PAH content after each wash cycle for coal that has previously been in contact with diesel

Figure 6 following the first rinse cycle is consistent with the DRO observations and the idea that naphthalene is not very abundant in the WSF of diesel. Acenaphthene, fluorene and anthracene also exhibit similar behaviors, although these compounds require more rinse cycles before significant decline is observed. Pyrene and acenaphthylene are present in only very small concentrations and remain relatively constant.

DISCUSSION

The results presented here provide some fundamental insights to the probable behavior of diesel compounds in fine coal flotation circuits. It is clear that some low-level diesel will likely report with process water to tailings impoundments or other waste repositories. In open-air systems, insoluble DRO may be gradually removed from process water, and removal may be accelerated in systems that promote aeration, mixing or heating; and these conditions which are commonly encountered in both impoundments (or slurry cells) and preparation plants where recycled flotation-process water may be utilized. Under normal operating conditions, a system is expected to continuously equilibrate such that only the WSF of diesel persists to any appreciable degree, and thus high concentrations of DROs are not expected accumulate. In a typical impoundment, the WSF will likely also be subject to photo- and bio-degradation.

We have also shown that diesel-contacted coal may release diesel to fresh water. In terms of environmental fate, this may be most important for clean coal stocks that are wetted, or in the instance of ultrafine coal that ends up in an impoundment. In the latter case, it is expected that diesel will desorb from the coal particles only until equilibrium with respect to diesel solubility in the impoundment is reached.

In regards to behavior of PAHs, it has been demonstrated here that some of these compounds may be expected to quickly leave an impoundment via volatilization, which may be encouraging—at least for water quality. Naphthalene was the most abundant of any of the PAHs targeted in this work, and at even the highest measured concentration of approximately 60 ppb, this is still lower than EPA lifetime advisory limits of 100 ppb for drinking water (EPA 2006).

It should be noted that the PAH contents measured in process waters in this study do appear to suggest some concentration effects (vs. the pure diesel), meaning that all components of the diesel do not remain in constant proportions. For instance, naphthalene in the pure diesel was measured to be approximately 0.1%, while in the initial filtrate sample used in the diesel removal tests the naphthalene was calculated to be about 13% (i.e., 60.1 ppb naphthalene in 0.43 ppm total DRO). This may suggest that coal selectively sorbs saturate compounds over aromatics. However, according to the observed behavior of PAHs under conditions of stirring, aeration, stagnation and heating, this concentration effect might be quickly reduced or even reversed.

Another important point of discussion in regards to PAHs in diesel is that of the relative abundance of target vs. actual compounds. While ideal PAHs (i.e., simple aromatics of fused benzene rings without functional groups) do exist in diesel, it is well established that alkylated PAHs are typically present at much higher concentrations (Irwin et al. 1997). The simplest alkylated PAHs are formed when the parent PAH compound is functionalized by addition of one or more methyl groups, which may occur during the digenesis of fossil fuels from organic sediments. Naphthalene can accommodate up to four methyl groups, and there are 22 individual compounds in the class of methylated naphthalenes (Abraham et al. 2005). Since quantification of specific compounds is highly complex, oftentimes only the parent compound (i.e., the EPA priory compound) such as naphthalene is measured to characterize fuel products or environmental samples. Additional testing of the diesel used in this work has revealed that di-and tri-methylated naphthalene compounds are present at about 3x the concentration of pure naphthalene; while mono- and tetra-compounds are present at more similar concentrations as the pure naphthalene. Considering this, it may be prudent to target the more abundant compounds in future testing—although, if the behavior of the alkylated compounds are similar to that of their parent, environmental implications may not differ significantly.

CONCLUSIONS

Diesel is a common collector reagent in fine coal flotation circuits in both the US and abroad. Despite coal's very high sorption capacity for this reagent, sub-ppm levels of DROs dominated by the WSF of diesel are expected to be present in process waters. In a typical tailings impoundment or open-air portions of the flotation circuit, DROs may be removed relatively quickly via volatilization and/or degradation. The most prevalent PAHs may also volatilize quickly, particularly where mixing, aeration or heating of the water occurs. DROs may desorb from diesel-contacted coal; however, in the context of settled ultrafine coal in an impoundment, this process should be limited diesel solubility in the impoundment water. Relative concentrations of both total DRO and PAHs targeted in this study do not appear to present significant concerns for water quality under normal operating conditions—but variances from such conditions should clearly be avoided. "Green" reagents such as bio-diesel and pine oil products, both of which should not contain PAHs, are being considered as alternative collectors for fine coal flotation; however, these have not yet gained widespread use throughout the industry.

ACKNOWLEDGMENTS

This study was sponsored by the Appalachian Research Initiative for Environmental Science (ARIES). The views, opinions and recommendations expressed herein are solely those of the authors and do not imply any endorsement by ARIES employees, other ARIES-affiliated researchers or industrial members. Information about ARIES can be found at http://www.energy.vt.edu/ARIES.

We also thank Jody Smiley for her invaluable mentoring and assistance with analytical results.

REFERENCES

Abraham, M., Autenrieth, R. and Dimitriou-Christidis, P. (2005). "The estimation of physicochemical properties of methyl and other alkyl naphthalenes." *The Royal Society of Chemistry* 2005(7): 5.

ATSDR (1995) "Fuel Oils–Chemical and Physical Information." 105–109.

ATSDR (1999). Total Petroleum Hydrocarbons–Identity and Analysis of Total Petroleum Hydrocarbons. Atlanta, Georgia, Agency for Toxic Substances and Disease Registry: 315.

ATSDR (2009). Toxicity of Polycyclic Aromatic Hydrocarbons (PAHs): 68.

Cohen, M., Williams, J. and Yamasaki, H. (2002). "Biodegradation of Diesel Fuel by an Azolla-Derived Bacterial Consortium." *Environmental Science and Health* A37(9): 14.

EPA (1990). Documentation of 7-PAH and 16-PAH National Emission Estimates.

EPA (2003) "Nonhalogenated Organics using GC/FID." 1–37.

EPA. (2006). "Drinking Water Health Advisories 2006 Edition." from http://water.epa.gov/action/advisories/drinking/dwstandards.cfm.

EPA (2011). 2011 Edition of the Drinking Water Standards and Health Advisories. Washington, DC, Office of Water: 18.

Fingas, M. (1994). "Evaporation of Oil Spills." *ASCE*: 8.

Fingas, M. (1995). "A literature review of the physics and predictive modelling of oil spill evaporation." *Hazardous Materials* 42(1995): 19.

Fiscor, S. (2012). "U.S. Prep Plant Census 2012." *Coal Age* 117(10): 14.

Irwin, R., VanMouwerik, M., Stevens, L., Seese, M. and Basham, W. (1997) "Environmental Contaminants Encyclopedia Entry on Alkyl PAHs (Alkyl Homologs of PAHs)." 18.

Kakkar, P., Saxena, R., Rathee, N. and Joshi, M. (2011). "Water soluble fraction of diesel fuel induced histopathological alterations in the liver of Channa punctatus." *Toxicology International* 18(1): 3.

Kondrat'ev, S. (2009). "Physically Sorbed Collectors in Froth Flotation and their Activity. Part II." *Journal of Mining Science* 45(2): 9.

Laskowski, J.S. (2001). *Developments in Mineral Processing*. Amsterdam, The Netherlands, Elsevier Science B.V.

Lloyd, A. and Cackette, T. (2011). "Diesel Engines: Environmental Impact and Control." *Journal of the Air and Waste Management Association* 51(6): 38.

Manoli, E. and Samara, C. (1999). "Polycyclic aromatic hydrocarbons in natural waters: sources, occurrence and analysis." *Trends in Analytical Chemistry* 18(6): 12.

Marquez-Rocha, F., Hernandez-Rodri, V. and Ma, L. (2001). "Biodegradation of Diesel Oil in Soil by a Microbial Consortium." *Water, Air, & Soil Pollution* 128(3–4): 8.

Monsalve, C. (2010). Process for the Containment of Oil Spills. *United States Patent Office*. P.A. Publication. United States. US 2011/0203996 A1: 8.

Morris, J., Sarver, E., Luttrell, G. and Novak, J. (2012). Reagents in Coal Preparation: Where do they go? *Annual Conference of Metallurgists 2012*. Niagara Falls, Canadian Institute of Mining, Metallurgy and Petroleum (CIM): 12.

Olson, J., Mills, G., Herbert, B. and Morris, P. (1999). "Biodegradation Rates of Separated Diesel Components." *Environmental Toxicology and Chemistry* 18(11): 6.

Osofsky, H., Baxter-Kauf, K., Hammer, B., Mailander, A., Mares, B., Pikovsky, A., Whitney, A. and Wilson, L. (2011). "Environmental Justice and the BP Deepwater Horizon Oil Spill." *New York University Environmental Law Journal* 20(1): 81.

SCF (2002). Polycyclic Aromatic Hydrocarbons–Occurrence in foods, dietary exposure and health effects. Brussels, Belgium, European Commission.

Schein, A., Scott, J., Mos, L. and Hodson, P. (2008). "Oil Dispersion Increases the Apparent Bioavailability and Toxicity of Diesel to Rainbow Trout (Oncorhynchus Mykiss)." *Environmental Toxicology and Chemistry* 28(3): 8.

Sementelli, A. and Simons, R. (1997). "Regulation of Leaking Underground Storage Tanks: Policy Enforcement and Unintended Consequences." *Economic Development Quarterly* 11(3): 13.

Wang, Z., Hollebone, B., Fingas, M., Fieldhouse, B., Sigouin, L., Landriault, M., Smith, P., Noonan, J. and Thouin, G. (2003). Characteristics of Spilled Oils, Fuels, and Petroleum Products: 1. Composition and Properties of Selected Oils.

Wang, Z., Hollebone, B., Fingas, M., Landriault, M., Sigouin, L. and Feng, Y. (1996). *Using Systematic and Comparative GC/MS and GC/FID Data to Identify the Source on Contaminated Birds*. Nineteeth Arctic Marine Oilspill Program, Ottawa, Canada.

Wick, A., Haus, N., Sukkariyah, B., Haering, K. and Daniels, W. (2011). Remediation of PAH-Contaminated Soils and Sediments: A Literature Review. Blacksburg, VA, Virginia Polytechnic Institute and State University: 102.

Wills, B. (2006). *Wills' Mineral Processing Technology*. Burlington, MA, Elsevier Ltd.

WV-DEP (2009). An Evaluation of the Underground Injection of Coal Slurry in West Virginia. *Phase I: Environmental Investigation*.

Minimizing the Environmental Impacts of Coal Waste Disposal

Mohammad Rezaee
Department of Mining Engineering, University of Kentucky, Lexington, Kentucky

Frank E. Huggins
Department of Chemical and Materials Engineering, University of Kentucky, Lexington, Kentucky

Rick Q. Honaker
Department of Mining Engineering, University of Kentucky, Lexington, Kentucky

ABSTRACT

To assess strategies aimed at minimizing the impact of disposal of coal waste materials on the environment, two long-term leaching experiments of up to five months duration were performed using waste materials from two plants cleaning high and low sulfur bituminous coal. The tests evaluated the mobility of major trace elements under different disposal scenarios: (i) a static leaching test designed to simulate the quiescent conditions encountered by coal waste material stored under water in a stable impoundment, and (ii) a dynamic test to simulate waste materials exposed to the atmosphere, either in variable wet/dry storage conditions, or in unusual circumstances like those resulting from breaching of an impoundment containment wall. The results indicate that the mobility of most elements is enhanced under highly alkaline or acidic conditions with a few being mobilized under both conditions, suggesting that the minimization of element mobility requires the pH value of the medium to be maintained around neutral. In this paper, the effects of a strategy of co-disposing coarse and fine refuse to control pH are presented as a method to maintain near neutral pH values and thus minimize trace element mobility. The co-disposal strategy was found to reduce the mobility of trace elements in the static leaching tests whereas the opposite was found from dynamic experiments. The results indicate that such controlled storage under water could retard acid generation and the mobility of trace elements.

INTRODUCTION

Trace elements generally refer to elements present in a natural material at concentrations below 1,000 ppm. Most of the trace elements of environmental concern in coal (e.g., As, Cr, Se, etc.) are mineral-associated and therefore tend to be enriched in refuse from coal cleaning operation. As a result, the behavior of trace elements in coal cleaning waste streams must be well understood in disposal situations, particularly with the aim of protecting ground water quality. Illite and pyrite are generally the two major coal minerals that contain most of the trace elements in eastern U.S. bituminous coals. The clay mineral, illite, typically contains most of the lithophilic trace elements such as vanadium and chromium, whereas the sulfide mineral, pyrite, typically controls the behavior of chalcophilic elements such as arsenic, selenium, mercury and lead. A number of other elemets, such as manganese, nickel and zinc, may have a mixed association with both of these minerals (Huggins et al., 2009). Therefore, these two minerals typically host most of the trace elements that are subject to federal and state regulations and statutes, such as the Clean Air Act Amendments (U.S. Congress 1990), and may be of concern as a result of possible mobilization from coal waste impoundments. The behavior of illite and pyrite in coal cleaning processes determines to a large extent how much and with what tailing fraction trace elements are distributed in coal wastes.

The oxidative weathering of pyrite has the potential to degrade water quality, yielding high acidity and mobilization of metals in concentrations that exceed water quality standards (Moses 1990). The weathering of pyrite can yield highly acidic waters even when initiated in waters of near-neutral pH. "On the other hand, pyrite weathering does not always produce very high levels of acidity or metals because of pH buffering by natural chemical or biological processes, such as carbonate mineral dissolution" (Toran, 1987; Nicholson et al. 1988) or bacterial reduction (Herlihy et al., 1988; Moses, 1990). Also Huggins et al. (2011) in their study of leaching of rejects of Illinois based coal concluded that the dissolution of first Na and then Ca, which give rise to the formation of insoluble sulfate minerals, and other alkaline earth elements is an effective buffer that maintains the pH of the leachates close to neutral and minimizes mobilization of iron and trace elements associated with pyrite (Evangelou and Zhang, 1995; Moses and Herman, 1991, Huggins et al., 2011).

Such observations suggest that leaching of trace elements can be repressed by enhancing the buffering capacities of the waste streams. Clearly, one strategy to enhance such capability would be to add alkali or alkaline-earth buffering capability like lime or fly ash to the impoundment, (Jastrow et al., 1981, Stewart et al., 1997, Hassett et al., 2005), but it is likely to add significantly to the cost of the disposal and storage of the cleaning rejects and related combustion wastes as well as increase the mass of material in the impoundment. A better cost strategy, which is explored in this study, would be to modify coal cleaning processes and/or storage practices to maximize the buffering capability of the waste materials and thereby repress the leachability of trace elements from coal waste streams.

EXPERIMENTAL

Coal Samples

Samples of raw coal, clean coal, and different coal waste streams (coarse refuse, thickener underflow, spiral refuse, high frequency screen over) from two different processing plants treating two different coal seams (Coalburg seam, WV and 5-B central seam, IN) were collected. 5-B Central seam is characterized as high-sulfur bituminous while the Coalburg seam is an oxidized bituminous coal seam of low sulfur content. The following analyses were carried out on each major coal fraction:

- Ultimate, proximate, forms of sulfur, heating value, size analysis;
- X-ray fluorescence (XRF) of major elements in 500°C ash from each fraction;
- Trace element analyses by inductively coupled plasma mass spectrometry (ICP-MS);
- Mineral matter by X-ray diffraction (XRD) using Rietveld method.
- Fe speciation and oxidation state determinations by Mössbauer spectroscopy;
- Speciation of sulfur and key trace elements by X-ray absorption fine structure (XAFS) spectroscopy.

Leaching Tests

The standard ASTM method D 5744 was designed specifically to meet regulatory requirements for mining wastes and ores. As shown in Figure 1, the experimental setup involved holding the waste streams samples in humidity cells made of PVC. The samples rested on 22-µm polypropylene felt filter media to allow air and water to flow through while retaining the 150 µm (100 mesh) samples. The dynamic tests were carried out on two parallel sets of representative coal seam samples with each weighing 200 g. The humidity cells were setup to receive an upward flow of air through the bottom valve. Dry air was obtained from the laboratory compressed air line routed through a desiccant, while the humid air was generated by passing air through water held at 30°C in the humidifier. The water temperature of the humidifier was kept constant by immersing it in a water bath with a thermostat controlled temperature setting.

The samples in the humidity cells were subjected to a weekly cycle involving alternating periods of dry air (3 days), humid air (3 days), followed by leaching with 200 ml of de-ionized water on the seventh day. Each flooded cell was

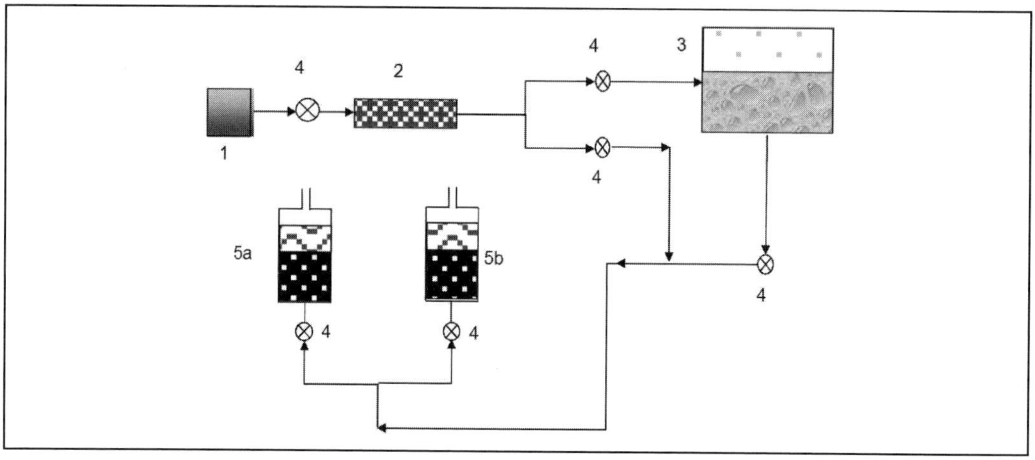

Figure 1. Experimental setup for the dynamic leaching test based on ASTM method D 5744–96. Numbered items indicate: (1) compressed air source, (2) desiccant chamber, (3) humidifier, (4) air-flow control valves, and (5) humidity series cells containing 200 g of (a) high sulfur content waste samples, and (b) low sulfur content waste samples (Huggins et al., 2012).

then allowed to sit for a period of 2 hour before draining the leachant into plastic bottles. The two-hour leach time commenced after all of the leachant had been placed in the cell, thus ensuring that the samples were saturated and completely immersed by the leachant water. The valve at the bottom of the humidity cell was opened and the leachant allowed to drain into the plastic bottles. The volume, pH, and conductivity of the collected leachant solutions were then recorded. At the conclusion of the 7th day, the cycle was repeated. After a number of weekly cycles, the collected leachant waters were combined, based on the pH change, and submitted for Eh, sulfate concentration, and major and trace elements composition analysis using ICP-MS. The 0th cycle corresponds to the initial aqueous leaching of the rejects conducted prior to the first cycle of exposure to dry and humid air.

In the dynamic column leaching test weekly leaching with deionized water, or low ionic strength water, promote the removal of leachable mineral dissolution product generated from the previous week's weathering cycle. The purpose of the three-day dry air part of the weekly cycle is to evaporate some of the water that remains in the pores of the sample after the weakly leach without totally drying out the sample. Therefore, the sample saturation is reduced and air contact is enhanced. During the three-day dry-air period of the weekly cycle, the oxygen diffusion rate through the sample may be enhanced several orders of magnitude as compared to its diffusion rate under more saturated conditions of the leach, which helps promote the oxidation of such constituents as iron sulfide. This evaporation also increased pore water cation/anion concentrations and may cause increased acidity (e.g., by increasing the hydrogen ion concentration generated from previously oxidized iron sulfide). Increased acid generation will increase the dissolution of additional sample constituents. As evaporation continues, the remaining water might become over-saturated in regard to some mineral phases, and therefore causing them to precipitate. Some precipitated mineral are potential sources of acidity when re-dissolved (like melanterite, $FeSO_4{\cdot}7H_2O$; and jarosite, $K_2Fe_6(OH)_{12}(SO_4)_4$). Three-day wet (saturated)-air portion of the cycle helps maintain a relatively constant mass of pore water in the sample, and promote some diffusion of weathering products (e.g, re-dissolved precipitation products) in the residual pore water without totally saturating the sample and adversely affecting oxygen diffusion (ASTM D5744, Huggins et al., 2012).

Table 1. Ultimate, forms of sulfur analysis of coal samples

	Ultimate Analysis of Coal Samples						Forms of Sulfur			
Samples	%Ash	%Moisture	%C	%H	%N	%O	%Total S	%Sulfate	%Organic	%Pyritic
High Sulfur Feed	19.4	8.3	53.6	4.9	0.6	14.1	7.4	0.3	3.5	3.6
High Sulfur Coarse Refuse	54.5	1.6	17.9	1.6	0.4	<0.01	29.3	1.1	7.6	20.7
High Sulfur Spiral Refuse	66.8	1.2	16.4	1.6	0.4	<0.01	20.1	0.6	4.1	15.3
High Sulfur Thickener Underflow	41.7	1.8	45.2	3.1	0.6	<0.01	9.6	0.6	2.1	6.9
High Sulfur Clean coal	10.4	8.5	64.8	5.7	0.8	14.7	3.6	0.4	1.2	2.0
Coalburg Coarse Refuse	62.3	1.1	29.3	2.6	0.5	4.9	0.3	<0.01	0.2	0.2
Coalburg Thickener Underflow	55.5	1.4	35.2	2.7	0.5	5.6	0.5	0.0	<0.01	0.5
Coalburg High Freq. Screen Over	72.5	0.9	19.5	2.0	0.5	5.0	0.5	0.0	0.4	0.1

Table 2. Semi-quantitative mineralogy of coal fractions (5-B Central seam, IN)

	%Mineral (as wt% of mineral matter)				
Coal Fractions	Pyrite	Quartz	Illiite	Kaolinite	Calcite
Spiral Refuse	28	17	15	21	17
Thickener underflow	23	18	36	23	<5
Coarse Reject	43	20	19	18	<5
Feed Coal	25	16	16	43	<5
Clean Coal	27	23	27	22	<5

RESULTS AND DISCUSSION

Coal Analyses

Ultimate and forms of sulfur analyses of coal samples from the two different sources are summarized in Table 1.

Table 1 indicates that the Coalburg, WV, samples have relatively low sulfur content, while 5-B Central seam, IN, samples have high sulfur content. Among the reject fractions of 5-B Central seam, coarse refuse has the highest total sulfur and pyritic sulfur, and thickener underflow has the lowest total and pyritic sulfur, indicating coarse refuse generates lower pH in contact with water and since has higher pyritic sulfur, more trace elements associated with pyrite would be released.

The mineralogical information of high sulfur coal samples obtained from XRD is shown in Table 2.

Table 2 indicates that there is significant variation in the mineralogy of the three coal refuse fractions. This variation arises because of the different size distributions and ways in which the individual minerals occur physically in the coal, resulting in different behavior they exhibit in the coal cleaning processes. The spiral refuse is the only fraction to contain significant calcite ($CaCO_3$); this result is consistent with the ash chemistry analyses (Table 3), which show that the spiral refuse sample has by far the highest CaO percentage. Conversely, the coarse reject fraction is much enriched in pyrite compared to the other refuse fractions and the corresponding ash chemistry also shows the highest iron oxide content.

Weight percentages (ash basis) of major elements obtained from ash chemistry analysis of coal samples are shown in Table 3. Higher amount of Al, iron and silicon than other elements can be recognized in this table. It is also the same for calcium in spiral refuse.

More advanced characterization methods were also performed on these materials, including speciation analyses by 57Fe Mossbauer and X-ray absorption fine structure (XAFS) spectroscopies. Based on the Mossbauer data, almost all of the iron in the coal and refuse fractions was present in the form of pyrite (FeS2). Preliminary XAFS results have shown that arsenic is quite significant and largely associated with pyrite, whereas selenium is barely detectable in the coal. Both iron Mossbauer and arsenic XAFS spectra indicate that slight oxidation of the pyrite has already commenced, particularly in the finer refuse fractions.

Table 3. Major element analyses (ash basis) of 500°C ash prepared from the coal samples

Oxide*	Clean Coal	Feed	Course Refuse	Spiral Refuse	Thickener Underflow
Al_2O_3	18.27	15.76	10.41	13.02	16.76
BaO	0.04	0.03	0.02	0.03	0.07
CaO	0.62	1.66	0.45	14.71	5.41
Fe_2O_3	35.23	40.99	61.48	35.68	26.99
MgO	0.83	0.75	0.58	0.72	1.09
P_2O_5	0.21	0.19	0.15	0.2	0.24
K_2O	2.37	2.05	1.38	1.56	2.28
SiO_2	40.7	36.25	24.41	28.99	44.13
Na_2O	0.43	0.36	0.24	0.23	0.35
SrO	0.05	0.04	0.02	0.03	0.03
SO_3	0.32	1.17	0.45	4.41	1.91
TiO_2	0.93	0.75	0.43	0.45	0.73

*by X-ray fluorescence (XRF)

Leaching Characteristics of Waste Samples

In this study, static and kinetic leaching tests were initiated on the individual samples of the different coal refuse streams. Frequent measurements of pH, Eh, and conductivity have been made on the supernatant liquids in the static test and similar measurements are also being conducted on the leach water collected weekly from the kinetic tests. Variation of pH with time for dynamic and static leaching tests is shown in Figures 2 and 3. The values of pH and Eh of the supernatant liquids in both tests showed a wide range for the different fractions, varying from neutral to relatively acidic (pH ~3 for the coarse high sulfur coal refuse), and each remained fairly constant for the entire leaching period (19 weeks). The variation of pH in both static and dynamic tests for refuse samples of the two different coal seams also is shown in Figures 4 and 5 separately. The difference in the pH values is due to the difference in mineralogy of the different fractions. Among high pyritic coal refuse samples, the coarse refuse has the highest content of pyritic sulfur, and in contact with water generates a lower pH than the other fractions. The spiral refuse and thickener underflow fractions have higher amounts of alkali and alkaline earth elements which buffer the pH of the solution, especially in static leaching test where the ions stay in the solution of the entire period of leaching test.

The pH of the supernatant liquids in contact with various coal refuse fractions of oxidized Coalburg seam in the leaching tests is close to neutral. The pH of the supernatant liquids in the static leaching test is slightly lower than those of the dynamic leaching test. Our hypothesis is that since these samples are oxidized coal, in contact with water they release humic acid and therefore slightly decrease the pH of the solution (Pawlik et al., 1977).

The electrical conductivity of a solution provides an indirect measure of the amount of dissolved ions in that solution (Rhoades, 1982; Stewart et al., 2001). The variation of conductivity with time for the dynamic leaching test is shown in Figure 6.

Figure 6 shows that the conductivity of supernatant liquids in contact with different refuse fractions of the Coalburg seam is much lower than those of the high pyritic coal seam. The reason is likely due to the higher pyritic content of the 5-B central coal seam, IN, and therefore more sulfate and trace elements associated with pyrite are released, resulting in increasing the conductivity of the leachate solution.

In Figure 7, the conductivity of the combined leachate of the dynamic leaching tests versus static tests, for the different coal waste stream is depicted.

The correlation of conductivity and released ions in the dynamic column leaching test for the

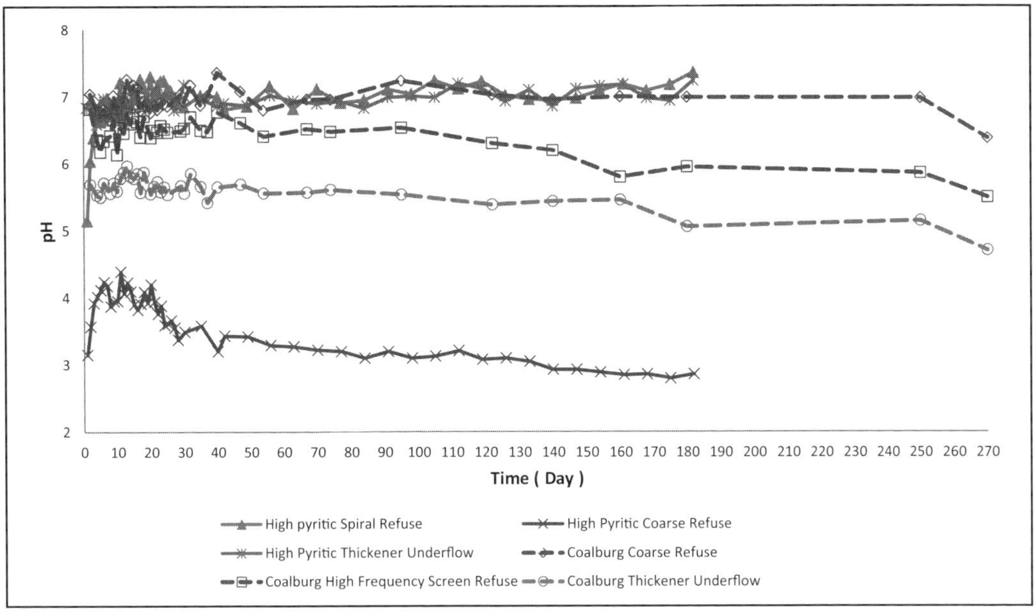

Figure 2. Plots of the variation over time of the pH of supernatant liquids in contact with various coal refuse fractions in the static leaching test. Note the wide range of pH exhibited by the different fractions.

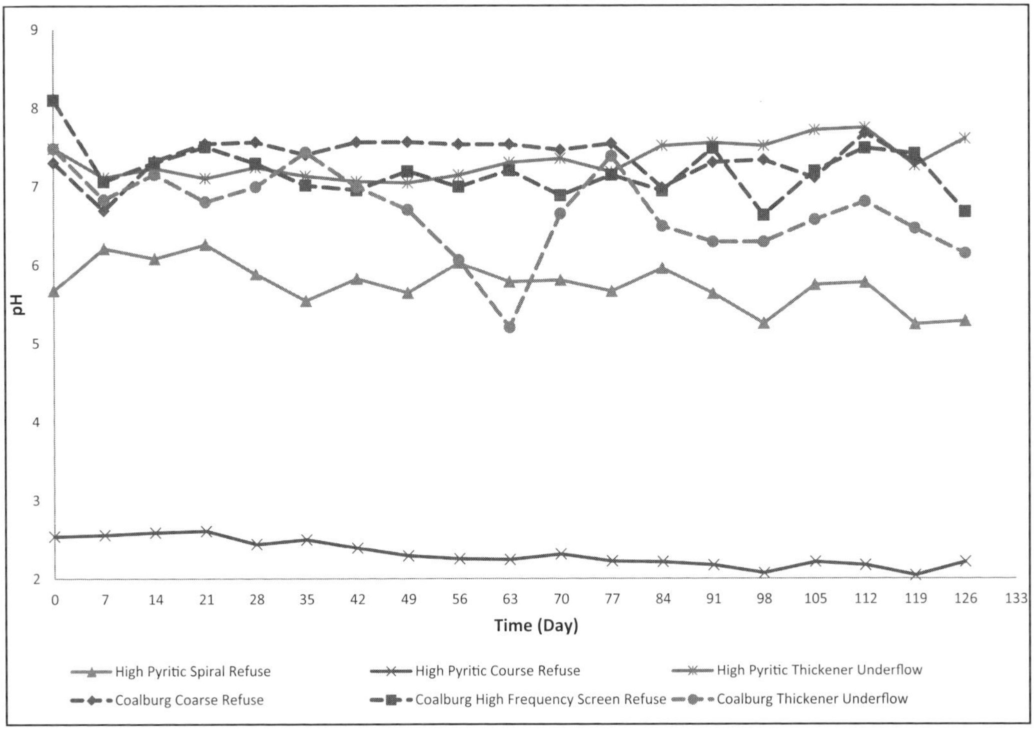

Figure 3. Plots of the variation over time of the pH of supernatant liquids in contact with various coal refuse fractions in the dynamic leaching test. Note the wide range of pH exhibited by the different fractions.

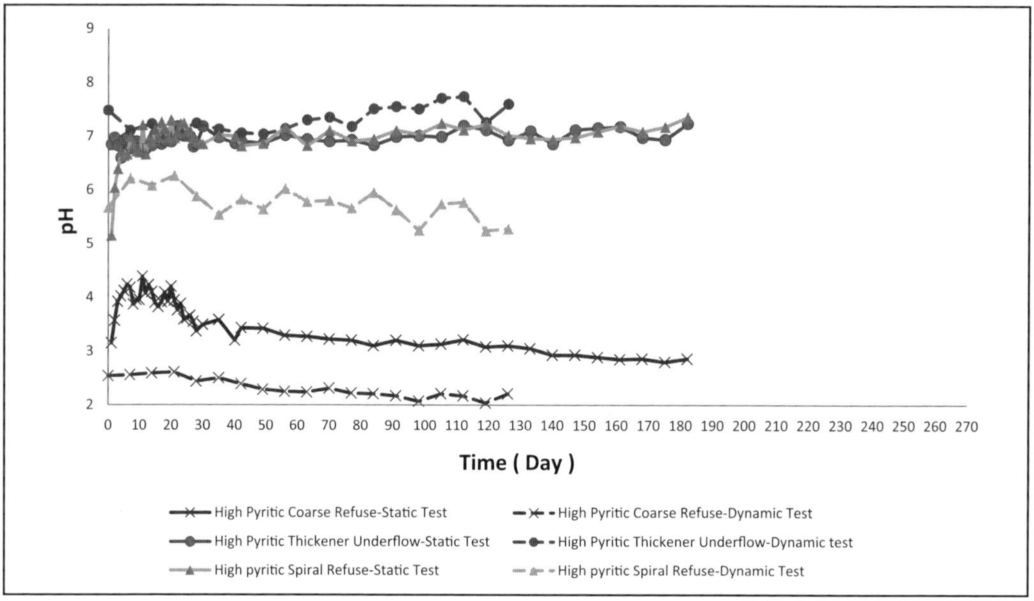

Figure 4. Plots of the variation over time of the pH of supernatant liquids in contact with various coal refuse fractions of high pyritic coal seam, IN, in the both static and dynamic leaching test. Note the wide range of pH exhibited by the different fractions.

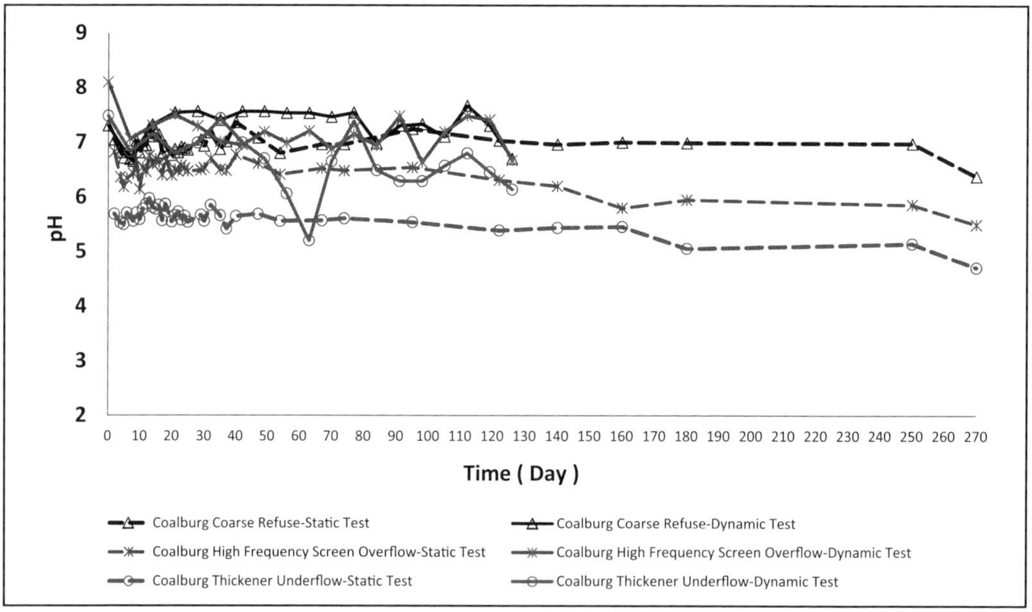

Figure 5. Plots of the variation over time of the pH of supernatant liquids in contact with various coal refuse fractions of Coalburg seam, WV, in the both static and dynamic leaching test

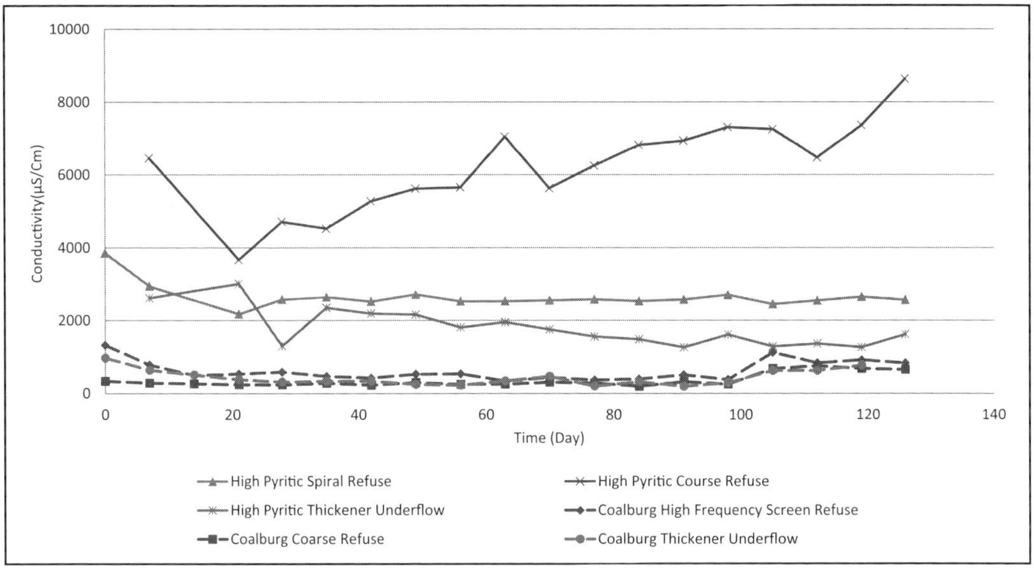

Figure 6. Plots of the variation over time of the conductivity of supernatant liquids in contact with various coal refuse fractions in the dynamic leaching test

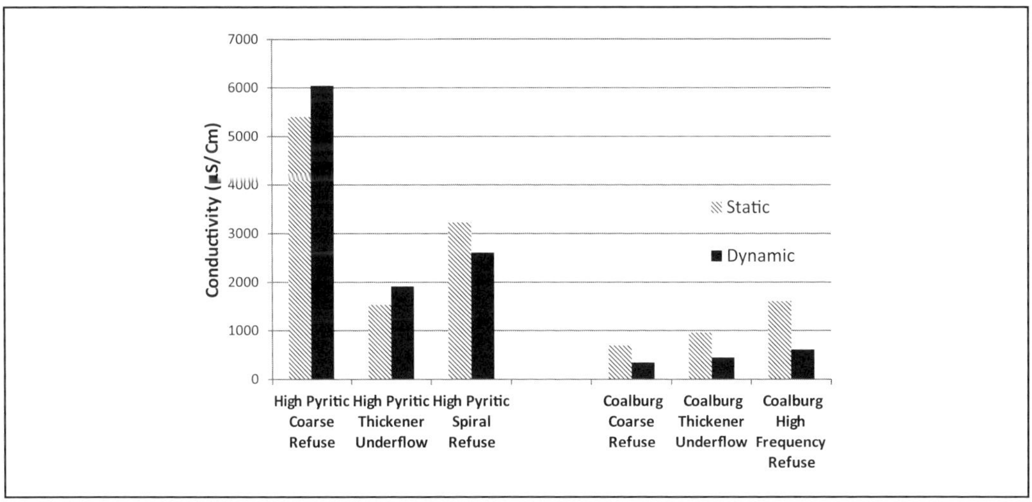

Figure 7. Plots of the conductivity of supernatant liquids in contact with various coal refuse fractions in the static and dynamic leaching tests

high pyritic thickener underflow, spiral refuse and coarse refuse has been determined to be 0.95, 0.85, and 0.33, respectively. The high correlation between conductivity and sulfate for thickener underflow and spiral refuse implies that sulfate anions are the predominant soluble anions being extracted in the dynamic leaching test for these samples (Huggins et al., 2012; Gray, 1996). The lower correlation value for the coarse refuse sample would appear to indicate a more complicated situation at lower pH values.

The concentration of thirty elements in leachates was determined by inductively coupled plasma optical emission spectrometry (ICP-OES). In Table 4, the ICP data of 19 weeks cyclic dynamic leaching test are listed, even for those

Table 4. Element concentration (mg/L) data from ICP measurements on leachates from the kinetic leaching test on high pyritic waste material

	Coarse Refuse						Thickener Underflow						Spiral Refuse						
Elements\Cycle(s)	1_2	3_4	5_7	8_11	12_15	16_19	1_2	3_4	5_7	8_11	12_15	16_19	1	2	3_4	5_7	8_11	12_15	16_19
Aluminum	56.5	18	24	32.5	44.1	55.4	<MDL	<MDL	<MDL	<MDL	<MDL	<MDL	<MDL	<MDL	<MDL	<MDL	0.07	<MDL	0.16
Antimony	0.1	0.05	0.06	0.07	0.07	0.08	<MDL	<MDL	<MDL	<MDL	<MDL	<MDL	<MDL	<MDL	<MDL	<MDL	<MDL	<MDL	<MDL
Arsenic	1.55	0.32	1.12	3.21	5.17	6.3	<MDL	<MDL	<MDL	<MDL	<MDL	<MDL	<MDL	<MDL	<MDL	<MDL	<MDL	<MDL	<MDL
Barium	0.02	0.02	0.02	0.02	0.02	0.02	0.04	0.03	0.02	0.025	0.03	0.04	0.025	0.02	0.02	0.02	0.02	0.02	0.02
Beryllium	0.07	0.02	0.01	0.009	0.006	0.006	<MDL	<MDL	<MDL	<MDL	<MDL	<MDL	0.001	<MDL	<MDL	<MDL	0.001	0.001	0.002
Boron	4.77	1.99	2.55	2.51	3.97	5.81	1.35	1.85	1.08	0.67	0.71	0.78	0.91	0.62	0.33	0.33	0.21	0.39	0.27
Cadmium	0.08	0.02	0.04	0.03	0.07	0.08	0.001	<MDL	<MDL	<MDL	<MDL	<MDL	0.02	0.01	0.005	0.008	0.004	0.005	0.005
Calcium	257	135	88.7	70.9	57.5	50.1	400	686	642	508	357	337	540	605	514	673	679	666	662
Chromium	0.81	0.32	0.42	0.45	0.4	0.41	<MDL	<MDL	<MDL	<MDL	<MDL	<MDL	0.04	<MDL	<MDL	<MDL	<MDL	<MDL	<MDL
Cobalt	2	0.39	0.18	0.11	0.09	0.09	1.46	0.28	0.15	0.04	0.03	0.03	3.79	1.21	0.41	0.35	0.13	0.09	0.07
Copper	0.79	0.26	0.27	0.27	0.37	1.14	0.02	0.02	0.01	0.006	0.02	0.02	0.03	<MDL	0.009	0.04	0.05	0.07	0.07
Gold	<MDL	<MDL	<MDL	<MDL	<MDL	<MDL	<MDL	<MDL	<MDL	<MDL	<MDL	<MDL	<MDL	<MDL	<MDL	<MDL	<MDL	<MDL	<MDL
Iron	2570	1070	1550	1900	1960	2130	<MDL	<MDL	<MDL	<MDL	<MDL	0.06	204	16.3	3.42	32.5	39.8	46.5	38.3
Lead	0.09	<MDL	<MDL	0.02	0.06	0.09	<MDL	<MDL	<MDL	<MDL	<MDL	<MDL	<MDL	<MDL	<MDL	<MDL	<MDL	<MDL	<MDL
Lithium	0.3	0.08	0.06	0.09	0.14	0.17	0.11	0.06	0.02	0.01	0.02	0.02	0.25	0.08	0.03	0.03	0.02	0.03	0.04
Magnesium	108	40.7	26.7	24.2	36.8	41.5	105	80	22.7	7.75	6.87	7.58	106	82.1	24.9	17.7	10.2	11.1	13.9
Manganese	31.4	14.5	10.4	6.89	4.09	3.57	7.29	6.73	4.04	1.47	1.5	1.73	22.6	18.8	9.39	12.2	8.71	9.9	12
Nickel	6.77	1.29	0.79	0.62	0.64	0.58	2.31	0.44	0.24	0.08	0.09	0.09	7.12	2.45	0.81	0.93	0.54	0.45	0.42
Phosphorus	2.15	0.44	1.54	5.33	8.8	9.38	0.1	0.03	0.02	0.02	0.02	0.03	0.04	0.02	<MDL	<MDL	0.01	0.02	0.02
Potassium	7.9	3.66	4.6	5.17	33.7	12.1	16.3	14.3	7.33	4.06	7.69	5.84	13.4	6.3	3.36	3.66	3.57	10.3	6.93
Selenium	<MDL	<MDL	<MDL	<MDL	<MDL	<MDL	0.16	0.12	0.05	0.025	0.015	<MDL	0.09	0.19	0.07	0.06	0.04	0.03	0.04
Silicon	4.56	3.18	3.02	5.04	24.1	19.7	1.62	2.28	1.89	1.9	9.14	9.44	2.65	2.83	2.09	2.91	3.06	11.7	8.37
Silver	<MDL	<MDL	<MDL	<MDL	0.004	0.007	<MDL	<MDL	<MDL	0.005	0.005	<MDL	<MDL	<MDL	<MDL	<MDL	<MDL	0.004	0.005
Sodium	149	10.7	5.97	7.28	23.1	29.3	111	34	7.99	3.51	3.92	5.25	230	80.8	11.1	5.67	4.08	4.57	4.81
Strontium	1.56	0.37	0.21	0.23	0.29	0.32	1.52	1.57	1.07	0.73	0.52	0.48	1.72	0.97	0.59	0.74	0.84	0.86	0.96
Sulfur	2240	891	1190	1400	1480	1670	576	898	600	397	286	253	927	747	526	712	736	669	612
Thalium	0.08	<MDL	<MDL	<MDL	<MDL	<MDL	<MDL	<MDL	<MDL	<MDL	<MDL	<MDL	<MDL	<MDL	<MDL	<MDL	<MDL	<MDL	<MDL
Tin	0.02	<MDL	<MDL	<MDL	<MDL	0.013	<MDL	<MDL	<MDL	<MDL	<MDL	<MDL	<MDL	<MDL	<MDL	<MDL	<MDL	<MDL	<MDL
Vanadium	0.09	0.03	0.05	0.05	0.04	0.03	<MDL	<MDL	<MDL	<MDL	<MDL	<MDL	<MDL	<MDL	<MDL	<MDL	<MDL	<MDL	<MDL
Zinc	1.52	0.71	0.59	0.57	0.74	1.3	0.07	0.07	0.03	0.02	0.02	0.01	0.55	0.32	0.07	0.13	0.17	0.35	0.5
Sulfate	6600	2560	3730	1590	4730	5250	1440	1670	1470	1100	868	818	2500	1820	1240	1560	1460	1670	1900
Leachate volume (ml)	265	299	412	552	536	504	139	192	315	407	354	337	120	130	288	397	511	540	472

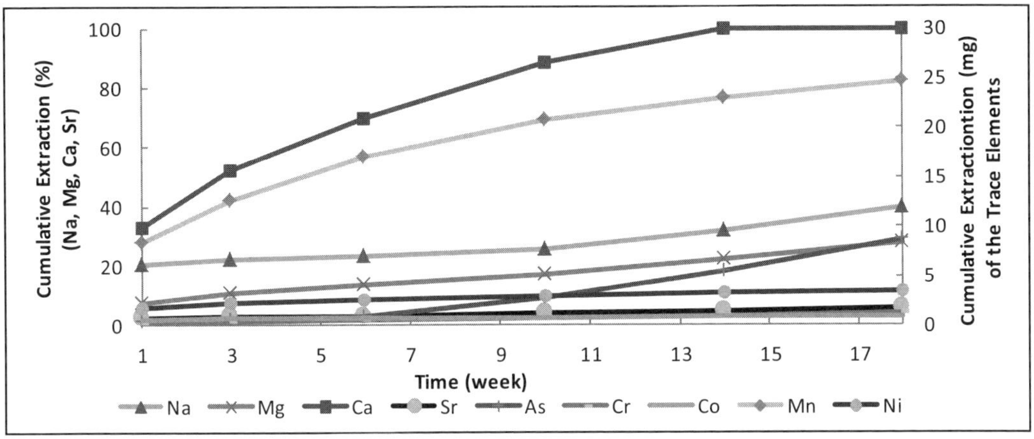

Figure 8. ICP-MS results for the coarse refuse of high sulfur content coal

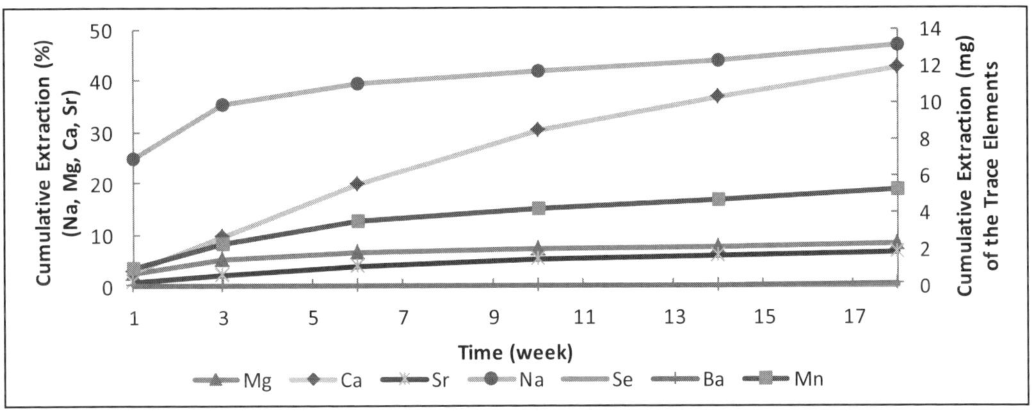

Figure 9. ICP-MS results for the thickener underflow of high sulfur content coal

having concentrations below minimum detection limit (MDL) in all samplings. The elemental mass leached can be easily derived by multiplying the concentration by the volume of leachant collected for each cycle. The coarse refuse also shows by far the highest leaching of iron and arsenic (q.v. Table 4) and this is due to the high content of pyrite in the fraction, the strong association of arsenic with pyrite, and the highly acidic solution that is generated when water is in contact with the coarse refuse. In contrast, the spiral refuse shows minimal or negligible releases of arsenic and iron, but significant release of calcium. For this fraction, dissolution of calcium is retarding acid generation and related leaching of elements from pyrite. These observations help explain the different pH response and leaching behavior that we have seen with the different refuse fractions and emphasize the importance of the role of the mineralogy in determining the leachability of elements from different coal refuse streams.

The extraction amount (mg) of the trace elements of concern in the CAAA over the period of time of the dynamic leaching test for different high pyritic waste samples are depicted in Figures 8 to 10. In Figures 8 to 10, Na, Mg, Ca, Sr release (%) are shown separately since these metal ions are relatively readily soluble in water and can buffer the pH of solution, while the larger metal ions in these groups, K^+ and Ba^{2+} were more resistant

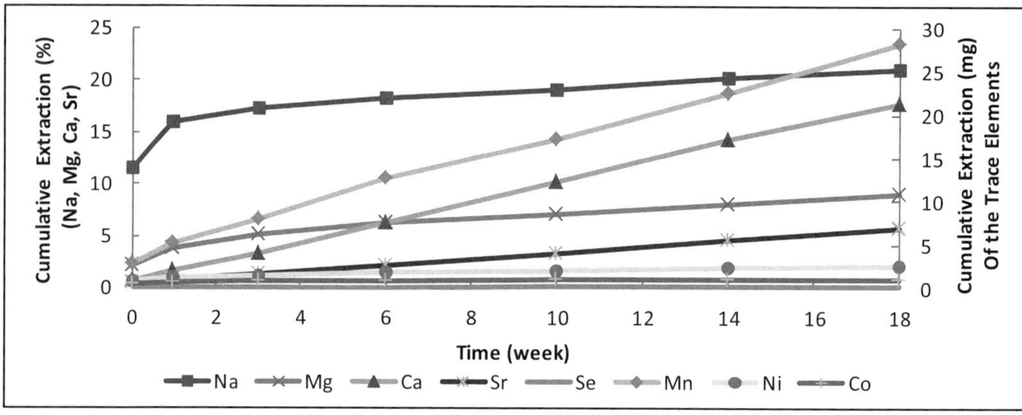

Figure 10. ICP-MS results for the spiral refuse of high sulfur content coal

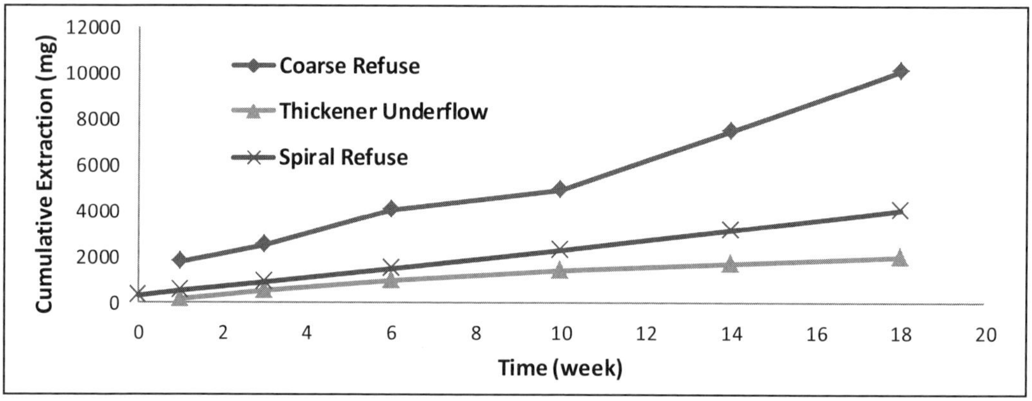

Figure 11. Cumulative extraction of sulfate during cyclic dynamic leaching test

to leaching. The trace elements that were not released or their amount of release was negligible are not plotted.

In Figure 11, the cumulative extraction (mg) of sulfate for the different waste samples during the dynamic leaching test has been depicted. The sulfate concentration and iron concentration are indicators of pyrite oxidation in each sample and the highest amount and rate for both ions and therefore for the pyrite oxidation is related to the coarse refuse.

The total extractions of trace elements from the high pyritic waste samples after 19 weeks of the dynamic leaching tests and 6 months of the static leaching tests are summarized in Table 5.

These data confirm that the dynamic leaching test is more aggressive than the static leaching test with respect to leaching of most elements from the waste samples. The maximum release of the trace elements is related to coarse refuse sample where it is subjected to the dynamic leaching test, where the high amount of pyrite is oxidized and most of the trace elements are released under the acidic conditions. In contrast, lower amount of release of the trace elements under neutral pH condition in spiral refuse and thickener underflow leaching tests. Hence, to minimize mobility of elements, a good strategy would be to keep the pH as close to neutral as possible.

The total extractions (mg) of trace elements from the waste samples of low pyritic Coalburg coal seam after 19 weeks of the dynamic leaching tests also are summarized in Table 6.

Table 5. Cumulative amounts (in mg) based on ICP measurements of elements in leachates over the 19 week long dynamic and 6 month long static leaching tests on 200 g of each waste sample

	Coarse Refuse		Spiral Refuse		Thickener Underflow	
Elements	Dynamic	Static	Dynamic	Static	Dynamic	Static
Aluminum	99.74	2.79	0.11	0.00	0.00	0.00
Antimony	0.18	0.00	0.00	0.00	0.00	0.00
Arsenic	8.69	0.01	0.00	0.00	0.00	0.00
Barium	0.05	0.00	0.05	0.00	0.05	0.00
Beryllium	0.04	0.01	0.00	0.00	0.00	0.00
Boron	9.35	0.43	0.86	0.07	1.67	0.13
Cadmium	0.14	0.00	0.02	0.00	0.00	0.00
Calcium	240.22	70.88	1577.74	62.94	836.25	40.26
Chromium	1.15	0.02	0.00	0.00	0.00	0.00
Cobalt	0.88	0.02	1.02	0.00	0.34	0.00
Copper	1.32	0.06	0.12	0.00	0.03	0.00
Gold	0.00	0.00	0.00	0.00	0.00	0.00
Iron	4812.46	95.17	104.01	0.00	0.02	0.00
Lead	0.11	0.01	0.00	0.00	0.00	0.00
Lithium	0.34	0.05	0.11	0.02	0.05	0.01
Magnesium	105.79	27.98	55.36	8.71	45.25	5.10
Manganese	24.74	9.12	28.16	0.30	5.29	0.10
Nickel	3.48	0.23	2.49	0.00	0.58	0.01
Phosphorus	13.72	0.00	0.03	0.00	0.05	0.00
Potassium	32.10	1.14	15.50	2.08	13.66	1.03
Selenium	0.00	0.00	0.14	0.00	0.08	0.00
Silicon	29.03	6.42	14.28	0.13	8.45	0.18
Silver	0.01	0.00	0.00	0.00	0.00	0.00
Sodium	76.31	36.96	50.37	28.00	29.06	7.26
Strontium	1.05	0.19	2.14	0.23	1.49	0.13
Sulfur	3758.05	187.44	1668.72	79.07	789.56	38.67
Thalium	0.02	0.01	0.00	0.00	0.00	0.00
Tin	0.01	0.00	0.00	0.00	0.00	0.00
Vanadium	0.12	0.00	0.00	0.00	0.00	0.00
Zinc	2.22	0.21	0.69	0.00	0.05	0.00
Sulfate	10110.16	578.16	4057.70	238.56	2014.49	122.65
TM	5463.27	251.71	1853.20	102.48	942.37	54.21

TM = total metals leached (Σ of all elements except sulfur).

The comparison between the cumulative amount (mg) of release of the trace elements of the dynamic leaching tests for the refuse samples of high pyritic 5-B central seam, IN, and low pyritic Coalburg seam, WV, shows that for trace elements especially those elements expected to be associated with pyrite or other sulfide minerals in the coal refuse (viz. chalcophilic elements such as Sb, As, Cd, Co, Fe, Pb, Mn, Ni, Cu, Se and Th) there are huge differences in their apparent mobility in the waste streams of processing of the two coal seams, and release of alkali and alkaline earth elements have enough buffer capability to keep the medium neural and prevent the release of the trace elements. Therefore, adding the processing units to the plants to separate pyrite from refuse streams, and then storing pyrite separately after applying some other strategies like solidification to keep pyrite from oxidation, may decrease not only the conductivity (Figure 7) but also the release of the trace elements.

Table 6. Cumulative amounts (in mg) based on ICP measurements of elements in leachates over the 19 week long dynamic on 200 g of refuse samples of the Coalburg seam

Element	Coarse Refuse	High Freq. Over	Thickener Underflow
Aluminum	0	0	0
Antimony	0	0	0
Arsenic	0	0	0
Barium	0.41	0.07	0.14
Beryllium	0	0	0
Boron	0.14	0.04	0.07
Cadmium	0	0	0
Calcium	152.3	100.22	35.48
Chromium	0	0	0
Cobalt	0.01	0.07	0.02
Copper	0.14	0.43	0.07
Gold	0	0	0
Iron	0	0.02	0.04
Lead	0.07	0	0.02
Lithium	0.2	0.11	0.03
Magnesium	57.33	44.39	17.91
Manganese	0.54	5.52	0.71
Nickel	0.05	0.07	0.02
Phosphorus	0.54	0.07	0.16
Potassium	58.15	27.35	17.56
Selenium	0.2	0.09	0.03
Silicon	40.28	7.91	11.77
Silver	0	0	0
Sodium	123.87	55.83	61.73
Strontium	0.81	0.81	0.59
Sulfur	141.47	140.13	77.39
Thalium	0	0	0
Tin	0	0	0
Vanadium	0	0	0
Zinc	0.47	0.43	0.1
Sulfate	494.14	425.98	233.03
TM	435.51	243.43	146.45

TM = total metals leached (Σ of all elements except sulfur).

Intermediate Leaching Experiment

Based on the stated strategy, in order to preliminary feasibility study of neutralizing the pH of coarse refuse stream and minimizing the release of trace elements of this stream by using buffer capability of the other waste streams, a set of static leaching tests were conducted over a range of thickener underflow-to-coarse refuse ratios. Test results shown in Figure 12 indicate that all ratios ≥2 provide the buffering action needed to maintain the pH near neutral.

Leaching Characteristics of Co-Disposed Coarse Refuse and Thickener Underflow

To evaluate the long term effect of co-disposal coarse refuse and thickener underflow strategy, longer static leaching (9 months) and dynamic leaching (5 months) tests were conducted on the co-disposed samples with the ratio of 1 and 2 thickener underflow to coarse refuse.

Variation of pH with time for dynamic and static leaching tests is shown in Figures 13. The pH variation of supernatant liquids in static leaching tests is consistent with the results of intermediate test, and the ratio of thickener underflow to coarse refuse should be at least equal to 2 in order to have enough buffering capability to neutralize the pH of the medium. However, in dynamic leaching tests, even with mixing ratio of 2, pH decreased to acidic conditions and trace elements were released. This is because in dynamic test condition pyrite is oxidized much faster than static test and also alkali and alkaline earth elements are leached out from the sample.

The comparison of the conductivity of combined leachant water from the samples over the entire period of dynamic leaching tests and the related values of static leaching tests are demonstrated in Figure 14, indicating the more effectiveness of static condition than dynamic condition to decrease the conductivity of high pyritic sulfur coarse refuse stream.

The total extractions (mg) of trace elements from the co-disposed samples after 19 weeks of the dynamic leaching tests and 9 months of the static leaching tests are summarized in Table 7.

The data indicates that the effectiveness of the test conditions, exist of enough buffering ions in the solution and the pH of the solution in mobility of the trace elements: Dynamic thickener underflow-to-coarse refuse: 1:1 >> Dynamic thickener underflow-to-coarse refuse: 2:1 >> Static thickener underflow-to-coarse refuse: 1:1 >> Static thickener underflow-to-coarse refuse: 2:1. Comparing the data of table 5 and 6 indicates that storage of co-disposed coarse refuse and

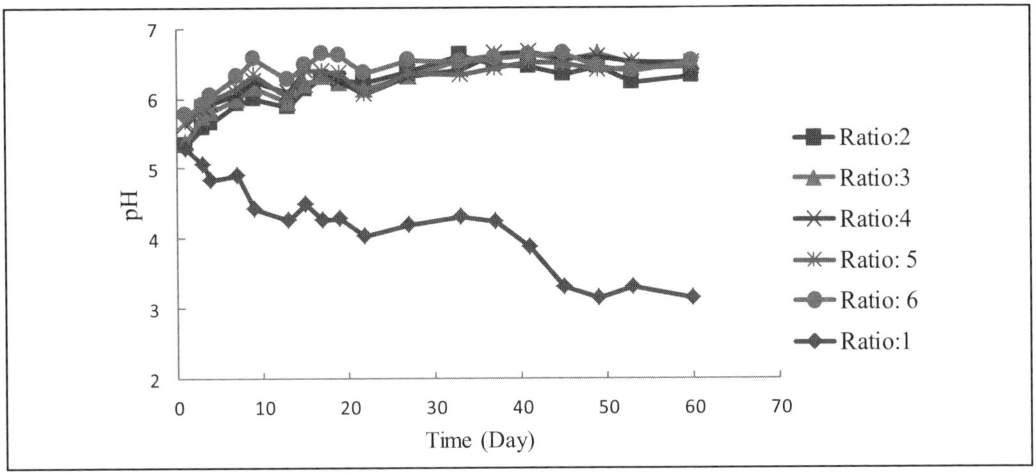

Figure 12. Variation of leachate pH over a range of thickener underflow-to-coarse coal refuse ratios for the high sulfur content coal

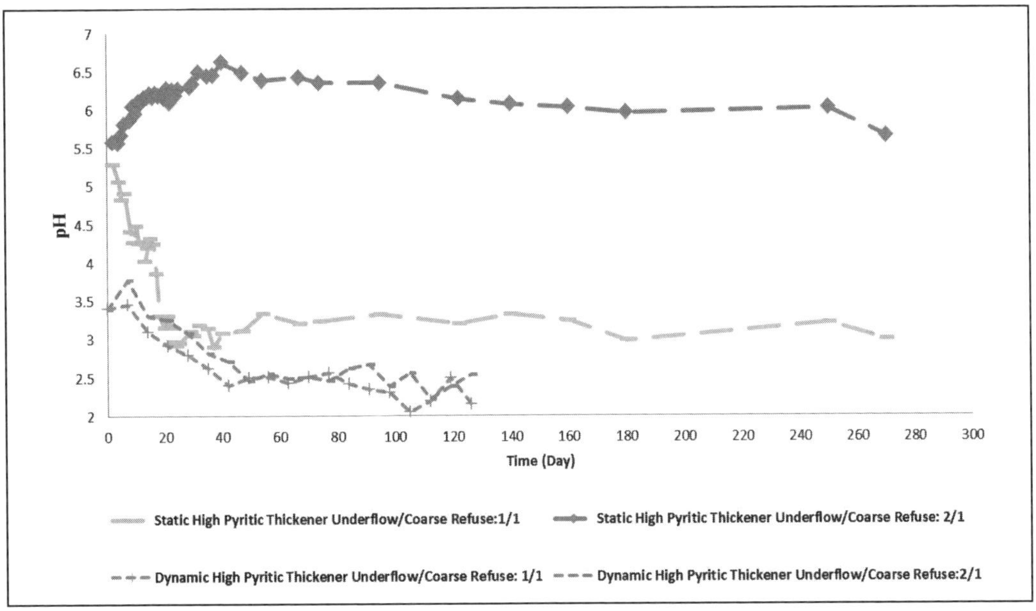

Figure 13. Plots of the variation over time of the pH of supernatant liquids in contact with co-disposed coarse refuse and thickener underflow of high pyritic coal samples, in the both static and dynamic leaching test

thickener underflow under water retards pyrite oxidation, acid generation and release of trace elements by neutralizing the pH using buffer capability of alkali and alkaline earth elements.

The other advantages of co-disposal have been well discussed by Williams (1995). Some of the advantages are: it eliminate the need for a coarse reject truck hauling system, the disposal operating costs can be reduced by up to 85%, it can provide up to 40% more water recovery from the waste than that of the separate disposal of coarse reject and wet tailings, it provides 40% a denser packing results due to the wide size distribution and therefore can save 40% storage volume compared with separate disposal.

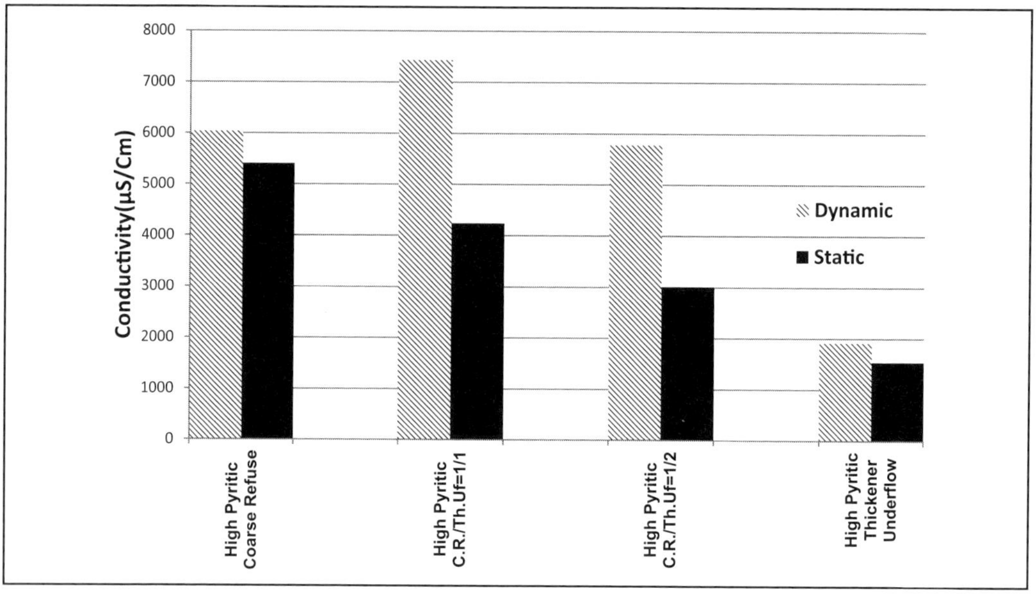

Figure 14. Plots of the conductivity of supernatant liquids in contact with various coal refuse and co-disposed coarse refuse and thickener underflow, in the static and dynamic leaching tests

CONCLUSIONS

To develop strategies to minimize the release of the trace elements in coal waste impoundments, coal waste samples (coarse refuse, thickener underflow, spiral refuse and high frequency screen overflow) from plants treating high and low sulfur coal content were subjected to two types of long term leaching experiments. These two leaching tests, static and dynamic, simulate cola waste disposal under different disposal scenarios, and provide the data regarding the amount, the rate and oxidation form of the elements released during the tests.

In contact with water, the different waste samples generated a wide range of the pH conditions in both types of leaching tests due to differences in their mineralogy differences. The elements data analysis by ICP-OES, suggested that most of the trace elements are released from coal waste material under either acidic or alkaline pH conditions, and just a few of them can be released in both conditions. So a good strategy to minimize the release of trace elements is keep the pH of medium as close as neutral and the best cost effective method for that is modifying coal cleaning processes and/or storage practices to maximize the buffering capability of the waste materials and thereby repress the leachability of heavy metals from coal reject streams.

Among the samples, the high pyritic coarse refuse sample had the highest amount of pyrite and in contact with water generated highly acidic condition and released much more trace elements in compare to the other streams. In contrast, high pyritic spiral refuse and thickener underflow samples had higher alkali and alkaline earth elements which neutralize the pH and repress acid generation and release of trace elements. On the other hand, low pyritic coal refuse samples showed much lower conductivity and very low mobility of trace elements especially those associated with pyrite.

The co-disposal coarse refuse and thickener underflow scenario was examined to evaluate the self-neutralizing potential of the waste streams, by using static and dynamic leaching tests. The results showed that storage of the co-disposed waste material with the ratio of thickener underflow to coarse refuse ≥2 under water retards the pyrite oxidation, acid generation and the release of the trace elements.

Table 7. Cumulative amounts (in mg) based on ICP measurements of elements in leachates over the 19 week long dynamic and 9 month long static leaching tests on 200 g of each codisposed sample

	Th.Uf/C.R=1/1		Th.Uf/C.R=1/2	
Elements	Dynamic	Static	Dynamic	Static
Aluminum	160.03	1.75	113.97	0.21
Antimony	0.16	<MDL	0.11	<MDL
Arsenic	2.84	0	0.58	<MDL
Barium	0.04	0	0.04	0
Beryllium	0.04	0	0.04	<MDL
Boron	8.18	0.31	5.92	0.19
Cadmium	0.09	0	0.09	<MDL
Calcium	1175.61	58.61	1164.2	68.79
Chromium	0.98	0	0.73	<MDL
Cobalt	1.65	0.01	1.76	0
Copper	3.31	0.03	2.34	0.01
Gold	<MDL	<MDL	<MDL	<MDL
Iron	4385.07	2.42	3000.65	1.46
Lead	0.11	0.01	0.09	<MDL
Lithium	0.45	0.04	0.33	0.02
Magnesium	129.85	32.3	103.51	16.82
Manganese	37.55	9.36	35.84	1.59
Nickel	6.01	0.03	5.61	0
Phosphorus	23.47	0	11.11	0.01
Potassium	17.28	1.76	10.37	1.15
Selenium	<MDL	0.01	<MDL	0
Silicon	38.22	4.3	38.73	0.42
Silver	0.01	0	0.01	0
Sodium	86.94	19.09	48.97	9.54
Strontium	1.27	0.35	1.47	0.27
Sulfur	4347.08	111.22	3323.42	82.13
Thalium	0.11	0.01	0.09	<MDL
Tin	0.04	<MDL	0.04	<MDL
Vanadium	0.25	<MDL	0.16	<MDL
Zinc	6.28	0.09	5.94	0.01
Sulfate	12739.5	348.54	9660.84	279.56
TM	6085.84	130.48	4552.7	100.49

In overall, based on the results of this study separation high density fraction especially pyrite from refuse streams and storing that separately after applying solidification, encapsulation or other methods to prevent pyrite from oxidation, and then applying codisposal scenario for lower density fraction is suggested to be an ideal solution regarding minimizing conductivity and release of trace elements from waste streams.

REFERENCES

ASTM D5744–D5796. 1996. *Standard test methods for accelerated weathering of solid materials using a modified humidity cell.* West Conshohocken, PA: ASTM International.

Evangelou, V.P., and Zhang, Y.L. 1995. A review: pyrite oxidation mechanisms and acid mine drainage prevention. Critical Reviews in Environmental Science and Technology. 25: 141–199.

Gray, N.F. 1996. Field assessment of acid mine drainage contamination in surface and groundwater. Environmental Geology. 27: 358–361.

Hassett, D.J., Pflughoeft-Hassett, D.F., and Heebink, L.V. 2005. Leaching of CCBs: observations from over 25 years of research. Fuel. 84: 1378–1383.

Herlihy, A.T., Mills, A.L., and Herman, J.S. 1988. Distribution of reduced inorganic sulfur compounds in lake sediments receiving acid mine drainage. Appl. Geochem. 3: 333–344.

Huggins, F.E., Seidu, L.B.A., Shah, N., Backus, J., Huffman, G.P., and Honaker, R.Q. 2012, Mobility of elements in long-term leaching tests on Illinois #6 coal rejects. [doi:10.1016/j.coal.2011.04.006] Int. J. Coal Geol. 94 (2012): 326–336.

Huggins, F.E., Seidu, L.B.A., Shah, N., Huffman, G.P., Honaker, R.Q., Kyger, J.R., Higgins, B.L., Robertson, J.D., Pal, S., Seehra, M.S. 2009. Elemental modes of occurrence in an Illinois #6 coal and fractions prepared by physical separation techniques at a coal preparation plant. International Journal of Coal Geology. 78: 65–76.

Jastrow, J.D., C.A. Zimmerman, A.J., Dvorak, and R.R. Hinchman. 1981. plant growth and trace metal uptakeon acidic coal refuse amended with lime and fly ash. J. Environ. Qual.10: 154–160.

Moses, C.O., Herman, J.S. 1991. Pyrite oxidation at circumneutral pH, Geochimica et Cosmochimica Acta. 55: 471–482.

Nicholson, R.V., Gillham, R.W., and Reardon, E.J. 1988. Pyrite oxidation in carbonate-buffered solution: 1. Experimental Kinetics, Geochim. Cosmochim. Acta. 52: 1077–1085.

Pawlik, M., Laskowski, J.S., and Liu, H., 1997. Effect of humic acid and coal surface properties on rhelogy of coal-water slurries, Coal Preparation. 18: 129–150.

Stewart, B.R., Daniels, W.L., and Jackson, M.L., 1997. Evaluation of leachate quality from codisposed coal fly ash and coal refuse. J. Environ. Qual. 26: 1417–1424.

Stewart, B.R., Daniels, W.L., and Jackson, M.L., 1997, Evaluation of leachate quality from codisposed coalfly ash and coal refuse. J. Environ. Qual. 26. 1417–1424.

Toran, L., 1987. Sulfate contamination in groundwater from a carbonate-hosted mine. J. Contam. Hydrol. 2: 1–29.

Williams, D.J., and Kuganathan, V. 1992. Co-disposal of coal mine tailing and coarse rejects. Proceeding of the third large open pit mining conference, Mackey. 429–432.

Williams, D., M. Gowan, and P. Keefer. 1995. Practical co-disposal deposition. Proceedings of the Seventh Australian Coal Preparation Conference J. Smitham, ed. Australian Coal Preparation Society.

2.2 Community Well-Being

Decision Tools for Integrating Sustainability Considerations in Large Capital Projects 78

The Mining Industry and Community Poverty Across Appalachia 86

Edgework and Boundary Crossings: Assessing Foundations for Public Ecology in the Appalachian Region 99

Determinants of Local Community Acceptance of Energy Resources and Materials Mining 111

Ethics and Community Well-Being 121

Decision Tools for Integrating Sustainability Considerations in Large Capital Projects

Andrea Ramage
CH2M HILL, Inc., Seattle, Washington

ABSTRACT

Companies in the extractives sector are advancing the concepts of sustainable development (SD), but for many companies, implementation of SD in large capital projects falls short of commitments made in corporate publications and policies. The missing link can be addressed with tools and processes for integrating SD considerations into decision-making. These elements help operationalize SD elements in capital projects in a way that supports organizational strategies, integrates with business decisions and operational activities, and addresses stakeholder relationships. This paper presents a case study on how to integrate sustainable development considerations within the corporate decision-making process, as applied to significant capital projects, through the use of decision tools and a supporting process. The paper reviews definitions and implementation arenas for SD, discusses challenges in successful SD implementation, introduces the SD decision tools, describes how the tools are applied, and enumerates the benefits of SD tool deployment.

INTRODUCTION

Since late 2010, CH2M HILL has been working with a confidential client who holds a long-term commitment to sustainable development (SD). The commitment is aligned with all three elements of the triple bottom line: people/society, ecosystems/environment, and financial/economic systems. These commitments are carried forward by the firm's executives, managers, and operational sites; they are evident in materials posted on the company's public website; and, performance against these commitments is reported through a number of publicly available documents, including the annual report. Consequently, our client wished to consider SD issues more effectively and consistently in their decision processes for large capital projects. The company engaged CH2M HILL to meet this goal by developing SD decision tools for project teams to use from project conception to construction.

DEFINITIONS

Definitions of Sustainable Development and Sustainability

In the context of this paper, "sustainable development" and "sustainability" are used interchangeably to denote the pursuit and achievement of environmentally responsible solutions and restorative outcomes, approaches to stakeholder and community relations that are transparent and fair while promoting safety and well-being, and technologies and processes that are efficient, less consumptive of natural resources, durable, and economically viable.

This perspective is aligned with the well-known definition of sustainable development put forward by the Brundtland Commission: "to meet the needs of the present without compromising the ability of future generations to meet their own needs" (World Commission on Environment and Development, 1987). It is also compatible with the ten principles of sustainable development that are promoted by the International Council on Mining and Metals (ICMM) (ICMM, 2012), and is aligned with ceres' twenty key expectations for corporations of this century, such as "substantive stakeholder dialogue," "engaging suppliers," and "design for sustainability" (ceres, 2010).

Three Implementation Arenas for Sustainable Development

SD is an exceedingly broad topic, akin to other broad realms such as "justice" and "agriculture." It is therefore helpful to compartmentalize sustainability within three implementation arenas more specifically applicable to organizations and capital projects, described below. Sustainability implementation can begin in any or all of these three arenas, but successful enterprise-wide implementation requires action in all of them. The focus of the SD decision tools discussed in this paper is on the third arena (project sustainability), which incorporates elements from the other two arenas (organizational and operational sustainability).

Organizational Sustainability

Organizational sustainability consists of policy frameworks, governance systems, management structure and processes, culture, leadership, education and training, strategy, corporate communications, and various systems that support endorsement and integration of sustainability within the organization. Organizational sustainability is typically associated with corporate functions such as board relations, finance and administration, human resources, procurement, fleet management, accounting, information technology, and corporate environmental management.

Organizational sustainability elements provide the support structure for sustainability achievements in the other two arenas: operational and project sustainability. In other words, organizational sustainability guides sustainability outcomes in the other two arenas.

Companies that are successful in this arena typically do the following:

- Establish a formal corporate Sustainability Policy.
- Develop a Sustainability Strategy aligned with business strategy, including prioritized issues, goals, strategies, and key performance indicators.
- Establish executive and managerial accountability for implementing the Sustainability Strategy, and provide adequate resources to implement the strategy.
- Publish sustainability performance regularly in a Sustainability Report, which may follow the Global Reporting Initiative (GRI) guidelines (Global Reporting Initiative, 2011).
- Integrate sustainability considerations into corporate decision-making, including strategy development, risk management, business function conduct (e.g., procurement, fleet management), and investment choices.

Operational Sustainability

Operational sustainability consists of the various processes and systems that consume natural resources (water, energy, materials), deploy human resources, and generate products and wastes. For extractives enterprises, the scale of operations is typically greater than the scale of organizational or project activities. Therefore, for these companies, operational sustainability represents the primary leverage point for improving sustainability performance. The opportunities to do so are related to resource consumption (land, water, energy, raw materials, and products) and waste production (solid waste, hazardous waste, air emissions, effluents), which affect environmental quality and social well-being, directly and indirectly.

Enterprises that are successful in this arena typically do the following:

- Establish site-specific sustainability goals and key performance indicators, linked to the corporate Sustainability Policy and Strategy.
- Establish site-specific managerial accountability (such as sustainability manager for the site), adequate staff, and funds for implementing the site's sustainability goals.
- Track and report sustainability performance regularly, which may be aggregated into the corporate Sustainability Report.

- Integrate sustainability considerations into site-specific decision-making regarding land management, water and energy consumption, waste minimization, process efficiency, productivity, regulatory compliance, community relations, and health and safety.

Project Sustainability

Project sustainability consists of planning, design, engineering, procurement, and construction activities needed to realize large capital projects. The functional departments typically involved in projects include finance, environmental permitting, design and engineering, and procurement as well as community relations, health and safety, and process management. Project sustainability is important because early-stage planning and design have the power to significantly shape the project concept. Therefore, these stages represent a key leverage point for integrating sustainability considerations into project decision-making.

Organizations that are successful in this arena do the following:

- Establish project-specific sustainability goals and key performance indicators, linked to the site-specific sustainability goals (and by inference, the corporate Sustainability Strategy).
- Establish project-specific managerial accountability (such as a sustainability manager for the project), adequate staff, and funds for implementing the project's sustainability goals.
- Establish an action plan for meeting project-specific sustainability goals.
- Track and report sustainability performance regularly, relative to the project's goals, key performance indicators, and action plan.
- Integrate sustainability considerations into project-specific decision-making in the planning, design/engineering, procurement, and construction phases.
- Document sustainability decisions in normal project documentation such as feasibility reports, engineering plans, and construction contract documents.
- Document lessons-learned about sustainable approaches (cost/benefit, design flaws or successes, construction experience, etc.).

The three arenas of enterprise sustainability are mutually supportive and necessary for achieving sustainability that is meaningful from business and sustainability perspectives. For example, without organizational sustainability to provide an overarching business case for sustainability, facility sites and projects would likely be less successful in initiating or carrying out sustainability initiatives. Without operational sustainability, the organization's key environmental and social impacts would not be adequately addressed, thereby weakening the credibility of sustainability efforts in the other two arenas. And, without project sustainability, companies lose a key leverage point with which to "bake in" sustainable approaches to new facilities and increase opportunities for operational sustainability. Therefore, all three are needed.

Challenges in Sustainability Implementation

Regardless of the arena in which an organization is integrating sustainability, the challenges to implementation are similar. Whether initiating or already engaged in sustainability efforts, the most common pitfalls include the following:

- Not understanding that successful implementation is a long-term, enterprise-wide undertaking that requires consistent and committed leadership, often for more than a decade.
- Not recognizing that sustainability implementation shares many characteristics and critical success factors with other organizational change efforts.
- Not making the linkage between sustainability initiatives and business strategy, and thereby failing to gain the support of key business leaders.
- Failing to acknowledge the many sustainability achievements already completed

by teams across the organization and in projects.

- Neglecting to communicate and translate corporate commitments to sustainability (and performance expectations) to operations staff at facility sites.
- Using sustainability performance data only as a report card, instead of essential data for making management decisions at appropriate levels.
- Not leveraging the storehouse of knowledge and experience of staff at facility sites and in projects to generate innovative and viable new solutions.
- Failing to communicate sustainability performance (challenges and achievements) to various stakeholders, including communities around sites, suppliers, consultants, regulators, employees, investors, and shareholders.

These challenges lead to many undesirable results, including frustrated managers, disenfranchised personnel, mediocre to poor sustainability results, loss of credibility with stakeholders, wasted money/time/resources, and lost opportunities. In contrast, an effective approach to sustainability implementation can lead to endorsement for sustainability at all levels of the organization, stellar sustainability results that contribute to business performance, high integrity ratings for the organization, enthusiastic employees, improved operational efficiency, enhanced social license to operate, innovation, and improved market competitiveness.

SD DECISION TOOLS

From 2010 to the present, CH2M HILL has been working with a confidential client to develop two SD decision tools and the supporting implementation process. The purpose of this undertaking is to help our client's project teams to integrate social, environmental, and economic considerations into all stages of their capital improvement projects. As presented above, the three implementation arenas for sustainable development and the challenges in sustainability implementation provided essential context for development of these tools. Specifically, the SD tools and the process for using them are designed to achieve success in project sustainability while addressing typical implementation challenges.

The two tools include the SD Checklist and the SD Action Plan. The following paragraphs describe the steps taken to develop the tools, the features of the tools, how the tools are used in practice, lessons learned from implementing the tools with several project teams around the world, and the challenges in and benefits of using the tools.

Development Process for SD Tools

CH2M HILL worked collaboratively with two senior managers from the client's organization to develop the SD tools. The tool development process occurred over the span of a year, starting late in 2010, and included the following steps.

Conduct a Literature Review

To develop the SD Checklist tool, CH2M HILL reviewed our client's SD-related values, policies, goals, targets, and commitments. From these documents and conversations, the CH2M HILL team created a comprehensive list of SD issues and topics that are most relevant to the client's projects. To create a robust list for consideration, the CH2M HILL team also identified SD-related topics used by companies in sectors outside the client's industry.

Establish Tool Design Parameters

Before the tools were designed, CH2M HILL worked with our client to identify the ideal characteristics of the SD tools. These characteristics included:

- Completeness—The SD tools should help project teams understand the full range of SD issues that can be associated with various projects; thus, the tool should be comprehensive in covering SD topics.
- Ease of Use—The tools should be easy to use, since the people using the tool may have limited knowledge about sustainability but be specialists in their own fields.

- Ability to Prioritize—The tools should help project teams identify priorities, so that financial and technical resources can be focused on SD issues with material significance to the project and the company.
- Ability to Stimulate Ideas—Lastly, the SD Checklist should help to stimulate new thinking for the purpose of improving project outcomes.

These characteristics formed the design criteria for the SD tools that were developed.

Select a Conceptual Model

The CH2M HILL team presented a variety of existing SD decision tools and discussed the pros and cons of each with our client. This step provided the starting framework from which a tailored SD tool could be developed for the client's project teams. Two conceptual frameworks that were considered included the International Standards Organization (ISO) 14001 standard for Environmental Management Systems (ISO, 2004) and a skeleton draft of the recently-launched envisionTM sustainable infrastructure rating tool developed by the Institute for Sustainable Infrastructure (ISI, 2012).

Prepare a Draft Checklist and Conduct Pilot Tests

In the next step, CH2M HILL developed a draft SD Checklist. This was tested in a desk-top exercise by filling out the Checklist based on information about a project in progress. CH2M HILL's observations from this pilot test and an in-depth discussion with our client led to several improvements. Subsequently, four project teams in different parts of the world conducted "live" pilot-tests of the tool. The project teams provided valuable feedback to further improve the SD Checklist tool. In addition, the pilot testing process revealed the need for a second tool, called the SD Action Plan. Pilot-testing was essentially completed by the end of 2011.

SD Tool Descriptions

The following paragraphs describe each of the SD Tools.

Description of SD Checklist Tool

The SD Checklist tool is constructed as a spreadsheet that covers six SD categories. These include water, energy and carbon management, environment and biodiversity, people and community, health and safety, and supply chain. Each SD category includes a set of predefined SD goals and a series of thought-provoking questions for project teams to consider. This structure enables teams to identify elements of their project that will affect SD goals, thereby either posing a risk to achieving SD goals or creating an opportunity to enhance achievement of SD goals. The SD Checklist prompts the project teams to record these SD risks and opportunities and prioritize them using a scoring methodology. The risk/opportunity scoring system was designed to align with the client's existing risk assessment methodology, thereby building on a framework with which project teams are already familiar. Once the Checklist is completed, only the priority SD risks and opportunities are carried forward into the next tool, the SD Action Plan.

The SD Checklist is meant to be a living document that is reviewed by the project team at every major project milestone, to incorporate new information and update the Checklist with any new risks or opportunities. It is most effective if applied in the early stages of a project, when the ideas generated through its use can influence decisions about project configuration or investments. If used in later stages, such as design and engineering, the Checklist can document team decisions to further SD goals in the project. However, applying the tool only at this late-stage function fails to capture the potential value that sustainable approaches can offer to capital projects.

Description of SD Action Plan

The SD Action Plan uses a spreadsheet format that builds on the structure established by the SD Checklist. It is populated with the priority SD risks and opportunities identified in the SD Checklist. The structure of the Action Plan then helps project teams develop and agree on tailored strategies for mitigating the SD risks or taking advantage of opportunities to improve performance with respect to the SD goals outlined in the SD Checklist. For each risk or opportunity, project teams also identify the discrete, assignable actions to be taken, parties responsible, and due dates. Space is left in the Action Plan for project managers to record progress or exceptions as they occur during the project.

The SD Action Plan works hand-in-hand with the SD Checklist. Like the Checklist, it is a working document intended for regular updating by the project team, to incorporate new information and address any new risks or opportunities arising from new information about the project. As such, it is an effective project management tool for keeping a team on track to mitigate SD risks and reap the value from SD opportunities.

How the SD Tools Are Used in Practice

In working with our client, the SD Checklist tool and the SD Action Plan were substantially complete and ready for more widespread use among the client's project teams by early 2012. Therefore, six of the client's large capital projects introduced the SD Tools into their project development process in 2012. The favored approach was to introduce SD concepts and use the SD Tools in a workshop setting. Workshops typically took place over two days and included approximately fifteen project team members who represented a variety of technical disciplines and functional responsibilities at the corporate, site, and project levels. This diversity of perspectives is one of the key strengths of the SD Tool process, in that it facilitates system-wide discussions and generation of new ideas.

Once the SD Tools were completed for the first time, each project team assigned a responsible party for ensuring that the tool was reviewed at regular intervals. These intervals vary by team; some meet every six weeks while others review their progress every three to six months. At every major milestone in the project, a more formal SD review meeting is held to ensure that the project is on track with respect to addressing SD risks and opportunities.

Once a project is completed, the ideal situation is that all SD risks and opportunities identified through use of the tools will have been mitigated or realized. If any such risks or opportunities remain, they are transferred into the risk management system under operations management. In this way, each identified SD risk or opportunity is dealt with.

Lessons Learned from SD Tool Development and Implementation

The process by which these SD tools were developed provided a number of valuable lessons. First, pilot-testing the SD tools was essential for arriving at user-friendly tool designs and creating a streamlined process for implementing the tools in and between workshops. Second, the project teams, as direct users of the tools, provided invaluable feedback that markedly improved the tools compared to their initial configuration. Third, the development process demonstrated that sufficient time (about a year) was needed to validate ideas, gather input from tool users, and integrate feedback. Finally, the initial design parameters outlined earlier in this paper were confirmed: the current tools meet the design parameters, and the project teams have not challenged them.

Similarly, project teams' experience with using the SD Tools also provided lessons learned. First, because SD is a familiar concept to our client's project teams, little effort was needed to educate tool users about SD as the underlying concept for the tools. This facilitated rapid uptake of the tool by each project team. Second, the workshops are an effective venue for discussing SD solutions in the context of large capital projects. We believe this is because large capital projects are complex systems that require multidisciplinary and cross-functional coordination, thereby facilitating the system-oriented solutions that are the hallmark of sustainable solutions. Therefore, workshops that include a diversity of participants can be highly generative environments for brainstorming, group learning and information sharing, discussing, clarifying, and most importantly—deciding. The third lesson learned from implementation to date is that project teams appreciate and benefit from the presence of senior leaders in the workshops; senior leaders add gravitas to the conversation, confirm the importance of the company's commitment to SD issues, and facilitate decision-making. Fourth, the SD tools are most effective when used early in the project life cycle, when creative thinking in the SD workshop can lead to improved approaches to project design, construction, and operations.

Challenges Faced During SD Tool Implementation

Our client's project teams faced challenges in SD tool implementation that project teams in other companies can also expect to encounter. During a team's initial use of the SD tools, one of the most common challenges is learning to shift from the familiar operational mindset (which focuses on cost, schedule, safety, and productivity impacts) to one that focuses on SD issues and opportunities. A good workshop facilitator who is expert in sustainability topics can help a project team overcome this hurdle. The result is a team that can think in both terms and stimulate new solutions.

For the duration of the project, another challenge has been establishing and maintaining a robust process that supports the use of the SD tools and thereby leverages value out of their use. The critical success factors for addressing this challenge include (1) having leadership support for use of the SD tools, at both corporate, project, and operational levels, (2) assigning a "champion" or owner of the SD tools, (3) conducting periodic reviews and updates of the tools to ensure that SD is factored into project decision-making, and (4) obtaining the project team's commitment to integrating SD considerations into their project.

Finally, another common challenge is dealing with lack of time due to aggressive project schedules or lack of human resources to do the work associated with the SD tools in addition to other project responsibilities. These constraints are realities for most project teams and must be addressed. In our client's case, these issues are in part addressed by asking consultants such as CH2M HILL to provide assistance.

Benefits of Using the SD Tools

The SD Checklist and the SD Action Plan offer many benefits to the project teams that choose to use them:

- The primary benefit of using the SD tools is that a company has a structured, consistent methodology for integrating SD considerations into the decision-making process for large capital projects. This establishes and reinforces a common SD language, set of goals, and action areas within the organization, thereby growing project teams' capacity to deliver SD solutions.
- The prioritization function of the SD Checklist helps to focus limited financial and human resources into areas that will provide the most value.
- Both the comprehensive nature of the tools and the diversity of experience present at SD review workshops facilitate a systems-oriented approach to problem solving for the project. This leads to improved coordination, less duplication of effort, and avoided costs.
- The process of using the tools builds credibility with stakeholders because the tools document a project team's thought process and achievements over time.
- The SD tools help track and report sustainability performance. As such, the SD tools can support a stakeholder engagement and public outreach efforts associated with the project. Stakeholders typically include senior management, government regulators, advocacy groups, and communities affected by the project.
- The "low-tech" spreadsheet framework of the tools, which allows project teams in all parts of the world to use the tools, regardless of technology sophistication levels.
- The unintended but beneficial consequence of applying the SD tools in multidisciplinary and cross-functional workshop settings is that valuable discussion and

coordination occur on other project concerns in addition to SD issues.

CONCLUSIONS

This paper has described CH2M HILL's work with a confidential client to develop and implement SD decision tools. The purpose of the tools is to help this client's project teams effectively and creatively consider SD issues in every stage of a large capital project's life cycle. The SD Checklist tool identifies priority SD risks and opportunities, and the SD Action Plan creates the strategies and actions to mitigate the risks and take advantage of the SD opportunities. Both tools are designed to integrate with existing decision and risk management processes and to be used by project staff and consultants. While some challenges can be expected in implementation, the SD tools provide many benefits, including meaningful action toward company commitments to sustainable development, improved project team capacity to deliver SD solutions, clear documentation of decisions made and SD achievements realized, and improved credibility with stakeholders.

REFERENCES

ceres (2010) "The 21st Century Corporation: The ceres Roadmap for sustainability." [online] Available at http://www.ceres.org/resources/reports/ceres-roadmap-to-sustainability-2010/view (accessed on 19 September 2012).

Global Reporting Initiative (2011) 'Sustainability Reporting Guidelines, G3.1' [online] Available at https://www.globalreporting.org/resourcelibrary/G3.1-Guidelines-Incl-Technical-Protocol.pdf (Accessed May 2012).

Institute for Sustainable Infrastructure (2012) "envisionTM Sustainability Rating System." [online] Available at http://www.sustainableinfrastructure.org/rating/index.cfm (accessed on 20 September 2012.

International Council on Mining and Metals (ICMM) (2012) '10 Principles' [online] Available at: http://www.icmm.com/our-work/sustainable-development-framework/10-principles (Accessed on 20 September 2012).

International Standards Organization (2004) "ISO 14001:2004, Environmental Management Systems—Requirements with Guidance for Use." [online] Available at http://www.iso.org/iso/catalogue_detail?csnumber=31807 (accessed 20 September 2012).

World Commission on Environment and Development (1987) *Our Common Future*, Oxford University Press, Oxford, p. 43.

The Mining Industry and Community Poverty Across Appalachia

Linda Lobao
Rural Sociology Program, School of Environment and Natural Resources, The Ohio State University, Columbus, Ohio

Mark Partridge
Department of Agricultural, Environmental and Development Economics, The Ohio State University, Columbus, Ohio

Lawrence A. Brown
Department of Geography, The Ohio State University, Columbus, Ohio

Michael Betz
Department of Agricultural, Environmental and Development Economics, The Ohio State University, Columbus, Ohio

Michael Farren
Department of Agricultural, Environmental and Development Economics, The Ohio State University, Columbus, Ohio

ABSTRACT

This study takes a new look at coal mining and poverty in Appalachia using interdisciplinary approaches from economics, sociology, and geography. Social scientists refer to a "natural resource curse" in which mining-dependent places tend to have poorer conditions relative to other areas. We evaluate this relationship by analyzing the present period and including the more modern influence of mountain-top-mining (MTM) with its broader environmental footprint. Independent variables include the share of employment in the coal mining industry and a gradient measure of mountain-top mining based on coal production. We also include a series of robust control-variables lagged in time to be causally prior to the dependent variables. Poverty rates for 2000 and 2010 are the dependent variables. Multivariate models were constructed for the Appalachian Regional Commission (ARC) counties (N=417) and for all U.S. counties (N=2, 611). These models explain from 88–94% of the variance in poverty rates. For ARC counties, the share of coal mining employment in 1990 is related to higher poverty in 2000 but net change in coal employment over the 1990–2000 decade has no statistically significant relationship. Outside of the ARC, the share of coal employment has no statistically significant relationship with poverty. Turning to the next decade, for the ARC counties, the share of coal employment in 2000 had no statistically significant relationship with subsequent (2010) poverty rates. For the ARC counties, coal employment also tends to be related to better local conditions relative to employment in oil/gas, as the latter sector is correlated with higher poverty in 2010. Within the ARC counties, MTM had no statistical relationship to poverty. Overall, the results suggest that coal's historically observed natural resource curse or positive relationship with poverty may be changing in the post-2000 period. To evaluate this unique relationship found in our study, future research should aim to model the potential nonrandom nature of mining location. Coal mining is determined by geology, national/global demand, and other nonrandom, local factors, including potential siting in business-friendly locales. Hence, future work should focus on the nonrandom nature of mining location such as through the use of instrumental variables to further appraise whether the effects of coal mining are truly changing.

INTRODUCTION

This study takes a new look at coal mining and poverty in Appalachia. We develop an interdisciplinary approach by building from research in economics, sociology, and geography. In doing so, we improve conceptually and methodologically on past research questioning the degree to which the coal mining industry influences poverty.

Economists, sociologists, geographers, and other social scientists have long studied poverty in Appalachia. Across the social sciences, researchers refer to a "natural resource curse" in which mining-dependent areas tend to have poorer conditions relative to other areas. The natural resource curse, however, has been analyzed mainly across nations, particularly the developing world and within the U.S. is sometimes applied to the early history of Appalachia. In this study, we integrate contemporary social science research on poverty with that pertaining to coal mining and the natural resource curse—to analyze the degree to which the coal mining industry today influences poverty.

We improve upon earlier studies in several ways. First, we analyze the impacts of coal mining employment in the present period until 2010 and we include the more modern influence of mountain-top-mining (MTM) with its broader environmental footprint. Second, we provide a nationally comparative study that not only analyzes Appalachia but compares the case of coal in Appalachia with that of other regions. Third, we add new independent, control variables pertaining to community context that include distance from urban areas, other employment including employment in oil and gas, and past history of poverty. These variables improve ability to control for potentially spurious determinants and also allow us to assess the relative performance of employment in coal compared to gas/oil in alleviating poverty. Finally, we improve upon measures of MTM used in past research (Hendryx, 2011; Partridge et al., 2013), as well as use improved measures of coal mining employment. To measure poverty, we use the official federal poverty rate measure, the proportion of an area's population whose pre-tax cash income falls below the federally-set threshold; this measure is the most commonly-used, well-established poverty measure in scholarly work and its importance is well-recognized by policymakers. Certainly researchers can construct other measures of poverty. Nevertheless, the poverty rate is a key measure in virtually all federal, state, and local assessments of well-being, and it is one of the three indicators used in the Appalachian Regional Commission (ARC) distress index. The standard federal threshold measure in 2011 was set at $11,484 for a family of one and $23,021 for a family of four (DeNavas-Walt et al. 2012:49).

We limit this study to a focus on poverty for several reasons. First, poverty has been historically one of the key indicators of well-being in Appalachia, as seen for example in President Johnson's War on Poverty in the 1960s and its use in the ARC index. Second, recent research continues to seek to disentangle the relationship between poverty and mining in Appalachia (Hendryx 2011; Partridge et al. 2013). Finally, we focus on a single indicator measure to better provide more depth of analysis over time and place. Nevertheless, it should be noted, that analysts have been concerned with mining and an array of other indicators of well-being.

Research on the natural resource curse by economists has typically focused on economic growth indicators and diversity of economic sectors (James and Aadland 2011). Partridge and Rickman (2006) show that poverty is an outcome of economic growth: thus poverty can be considered as a down-stream indicator of the conventional natural resource curse. Sociologists have stressed not only growth and diversity, but distribution (such as poverty), social disruption, and environmental degradation as indicators of well-being affected by natural resource extraction (Stedman et al. 2012). Our study is thus necessarily limited in its consideration of well-being and future work should build on a wider-range of indicators.

This paper first provides a brief overview of the conceptual and methodological issues important for social science assessment of the impacts of the mining industry on community well-being. The second section documents the steps to address some of these issues and provides an empirical demonstration. We present ten quantitative analyses that assess the extent to which poverty rates across communities are related to the mining industry.

SOCIAL SCIENCE CONCEPTUAL ISSUES IN ASSESSING IMPACTS OF MINING

Economists, sociologists, and geographers have long analyzed the impacts of different employment sectors on poverty and other indicators of community well-being (Brown et al. 2005; Lobao 1990; Lobao, Hooks, and Tickamyer 2007; Partridge and Rickman 2006). Drawing from this past research, we denote key issues that need to be considered in empirical research aimed at evaluating the influence of the mining industry.

(1) Community-specific issues pertaining to selection bias need to be considered. By selection bias, we refer to the degree to which past community conditions (such as the presence of coal deposits, previous socioeconomic conditions, and population demographic attributes) have influenced the location of the mining industry. Appalachian coal mining is concentrated in some of the nation's poorest counties. Some researchers suggest coal mining may be a cause of these very poor conditions (Hendryx 2011). Yet, coal mining elsewhere in the United States is not associated with such geographical clusters of poverty (e.g., coal mining in Wyoming and much of the Great Plains). Thus, it is reasonable to question whether coal mining is the ultimate "cause" of abject conditions in parts of Appalachia or whether other persistent factors cause these pockets of poverty. In the effort to sort out this causal relationship, researchers must deal with the issue of endogeneity: in this case, poor communities in the past may have adopted coal development but coal mining today does not causally determine poverty. For example, there are host of reasons that could explain why coal mining communities in Appalachia are poor. Foremost are their general remoteness from large cities and a workforce with inadequate education. In addition, poor communities may be prone to accept coal development as an avenue for new jobs. This relationship was suggested by Gould (1991)in a series of case studies of mining dependent areas. Here coal mining would not be the cause of deprived development, but rather an outcome of the process. Hence, robust research is necessary to uncover coal mining's ultimate impact. Still, it needs to be recognized that the issue of endogeneity (i.e., establishing causality) affects virtually all social science studies, ours included. It is an inherent problem for studies assessing the impacts of economic and social processes across communities that researchers must take steps to minimize.

(2) The time period of focus and changes in relationships over time need to be considered. Past community conditions likely carry forward to influence present-day levels of poverty net of mining itself. Extant studies on mining in Appalachia have addressed its impacts on poverty rates usually through cross-sectional and/or correlational-type analyses (e.g., Hendryx 2011). A more robust empirical approach is needed that takes into account past local conditions. Further, mining may not have consistent effects across time. As the industry has required more skilled, higher paid labor, its impacts on communities may have improved in recent years.

To address, issues (1) and (2) in our empirical analyses, we first describe the degree to which poverty and other conditions were different (a priori) in counties characterized by mining operations, including mountaintop mining. We use control variables lagged in time back to 1990 where possible to control for past conditions and to omit reverse causality. For example, we use lagged poverty rates to control for whether poverty initially was higher in coal mining counties. Other persistent economic conditions are measured by past job growth of the county. We also include the presence of other mining in gas/oil as proportion of county employment, to assess the comparative impacts of other energy industries. Also included as control variables are demographic attributes of race/ethnicity, age, education, and family structure and a detailed mix of population agglomeration variables pertaining to distance and size of the nearest metropolitan areas. Past research finds that poverty rates are positively associated with distance from metropolitan areas (all else equal) (Partridge and Rickman 2008a). Finally, to assess the degree to which the impacts of mining might vary over time, we examine two time periods, the respective decades of 1990–2000 and 2000–2010.

(3) Comparisons to other U.S. places are also needed. With regard to Appalachia as a whole, socioeconomic conditions in mining communities may not be much different from conditions in other communities. Conversely, rural Appalachian communities potentially have experienced more severe declines in sectors such as manufacturing relative to mining in recent decades; mining in fact may be a buffer to decline. As noted above, comparing relationships between Appalachia and other regions is important. Should coal mining be associated with higher poverty in Appalachia but not in other regions, this would suggest some unique attributes of Appalachia may be driving the results.

In summary, research is needed to: identify coal mining's impacts using methodological approaches that go beyond extant correlational designs; evaluate whether the effects of coal mining employment are shifting over time; and assess whether relationships found in Appalachia persist in other parts of the country. The analyses below move in these directions.

EMPIRICAL ANALYSES: COAL MINING EMPLOYMENT, MOUNTAIN-TOP MINING, AND POVERTY

Modeling Approach and Empirical Implementation

Our modeling approach follows that taken in a previous study of mining and poverty by Partridge, Betz, and Lobao (2013) and developed from other published poverty studies (Levernier, Partridge, and Rickman 2000; Partridge and Rickman 2005, 2008a). The aforementioned studies use a disequilibrium partial-adjustment model in order to evaluate poverty relationships across geographic space, using counties as the unit of analysis. The model assumes each county's poverty rate is a function of its characteristics such as job growth, and demographic, geographic attributes. These characteristics influence the county's poverty rate over time, though there may be a significant time-lag between changes in these characteristics and the actual poverty rate itself. For example, when college degree attainment-levels increase, it may take years before more educated workers are employed in better paying jobs and the poverty rate declines. Because of the sluggish adjustment to a new equilibrium poverty rate, the current poverty rate is assumed to be a function of the past poverty rate as well as current demographic, economic, and geographic attributes. For further details about the modeling approach, see Partridge, Betz and Lobao (2013). Below we provide a brief overview of the empirical implementation of the modeling approach.

The empirical models are estimated with ordinary least squares (OLS) and state-fixed effects. OLS models have shortcomings that we try to mitigate, one concern being endogeneity discussed above which raises concern about reverse causality. Generally, we time-lag the independent variables to reduce the concern with reverse causality.

Two samples are analyzed: (1) Appalachian counties as designated by the Appalachian Regional Commission (ARC counties) and (2) all other US counties. A finding that coal mining has a different effect in the rest of the US compared to the ARC region would suggest that coal mining may not be the cause of high poverty, but possibly something else uniquely related to Appalachia (e.g., regional institutions, culture or historic remoteness). Deaton and Niman (2012) find that coal mining is associated with high-levels of persistent poverty in the ARC region. Further, James and Aadland (2011), Papyrakis and Gerlagh (2007), and Kilkenny and Partridge (2009) find evidence of a "natural resources curse" whereby mining has a similar (generally negative) effect on economic well-being across the country (and the world).

The dependent variable is the total (individual) poverty rate, measured for two time periods (2000 and 2010). We measure it using data from the 2000 Census, where data refers to the previous year's poverty threshold (we refer to it as the 2000 poverty rate) and we measure the 2010 poverty rate from the U.S. Census Bureau SAIPE. We first estimate a base model with (non-mining) independent variables following Partridge, Betz, and Lobao (2013); then we add the *MINING* employment and mountain-top/intrusive surface mining *MTM* variables, the model taking the form:

$$\begin{aligned} POV_{it} = {} & \alpha POV_{it-10} + \beta MTM \\ & + \varphi MINING \\ & + \psi MTM * MINING \\ & + \lambda AVGNEIGHPOV_{it-10} \\ & + \gamma PROXIMITY_i + \theta CITYTYPE_i \\ & + \delta ECON_{it-10} + \phi DEMOG_{it-10} \\ & + \sigma_s + \varepsilon_{it} \end{aligned} \quad (1)$$

where POV_{it-10} is the ten-year lagged poverty rate measured in 1989 (from the 1990 Census) for the 2000 poverty rate model and measured in 1999 (from the 2000 Census) in the 2010 poverty rate model. Regression coefficients are represented by, α, β, φ, ψ, λ, γ, θ, δ, and ϕ. State fixed effects are represented by σ_s to capture factors associated with each state such as welfare policies, tax and regulatory policies, and cultural and historic factors such as those found in Central Appalachia. The error term is denoted by ε_{it}, which we assume is heteroskedastic. We correct for heteroskedasticity by estimating robust standard errors.

We measure mining through indicators of employment and mountain-top/other intrusive surface mining. For mining employment activities, we include three (place of work) employment shares for (1) coal mining; (2) oil and gas mining, and (3) other mining activities using four digit NAICS codes using data from EMSI consulting company.* We also include the ten-year percent change in the three mining sector's county employment shares (1990–00 and 2000–10), which should be exogenous because these goods are almost exclusively traded on national international markets. Thus, we account for lagged historic effects of mining and the contemporaneous effects of recent changes in mining employment—i.e., do new mining operations have different marginal effects than long-term legacy effects. By separating the three mining industries, we also can ascertain whether coal mining has different effects. Mountain-top/other intrusive surface mining (MTM) is measured by a continuous variable we developed from the United States Energy Information Association's FERC-423 data. This improves upon the standard presence/absence of MTM used in some studies

* EMSI uses other data sets and an algorithm to estimate values for data suppressed in the U.S. Department of Labor's *Quarterly Census of Employment and Wages*—see Dorfman et al. (2011) for more details. The county share of coal employment is calculated as follows. Total employment in coal mining employment (NAICS 2121) is used as our base value for coal mining employment. To this value we add an estimate of county employment in coal mining support activities. Our employment data is only detailed to the 4-digit level, which gives us information on total mining support employment (NAICS 2131), but not coal mining support employment (NAICS 21314), which is recorded as a 5-digit NAICS category. To derive a county estimate for employment in coal mining support we calculate the share of mining support employment attributable to coal mining support employment at the national level using publicly available data from the Bureau of Labor Statistics. We multiply the national share of mining support attributable to coal mining support by the county's employment in mining support to derive an approximate measure of county employment in coal mining support activities. We add this to our base value of coal mining employment and divide by total county employment to get the share of county employment attributable to coal mining and coal mining support activities. By contrast, Partridge et al. (2013) did not adjust for coal mining support activities, though this is a small share of coal employment.

by allowing for a gradient measure and allows us to capture the intensity of surface mining in more mountainous counties. This improves upon the rough measure of the simple existence of mountain-top mining based on satellite data (e.g., see Hendrxy, 2011; Partridge et al., 2013), including giving MTM a slightly larger geographical scope. The USEIA generated data has information on where the coal originates and whether underground or surface mining was used. The USEIA data does not distinguish mountain-top mining specifically from other surface mining so we consider them in aggregate. Although this also includes contour mining or contour auger mining, they are all intrusive and disruptive forms of mining. In order to develop a gradient measure, we then consider positive values of surface mining for counties that have a USDA ERS topography code>18, ("high-elevation mountainous counties"), assigning counties with a topography code<18 a value of zero. The measure we create is the total quantity of surface mining measured in tons of coal for high elevation counties. We also interact the MTM variable with the coal mining employment variable. This allows us to parse out whether the effects of the level and growth of coal mining are different in MTM counties.

Based on extant research, it is not clear *a priori* whether the mining industry has a positive or negative overall effect on poverty rates. For example, increased mining activities will likely provide jobs. Black et al. (2005) find that about 100 new coal mining jobs provided on net about 25 additional indirect jobs in Appalachian coal communities. However, in terms of attracting population and reducing poverty, it is not clear whether these economic benefits outweigh the negative environmental and health externalities often associated with mining.

Our empirical models also include key variables that social scientists identify as important control variables in regional poverty studies (Blank 2005; Glasmeier 2002; Partridge and Rickman 2006; Weber et al. 2005). We employ the same control variable measures as those used by Partridge, Lobao, and Betz (2013). Further detail on data sources and measures of the variables are provided in Partridge and Rickman (2005, 2008a). *AVGNEIGHPOV* is the average poverty rate of counties contiguous to county *i* measured in 1990 to account for spatial spillovers across counties. *PROXIMITY* is a vector of distance variables that reflect remoteness to increasingly larger cities. Proximity to urban agglomeration economies tends to provide better employment opportunities for workers near the poverty line, while communities themselves benefit from stronger commuting linkages, knowledge spillovers, and tighter input-output linkages with a larger set of suppliers and consumers nearer the urban core. Agglomeration measures include distance in kilometers to the nearest metropolitan area as well as to larger metro areas. *CITYTYPE* is a vector that includes the lagged county population and measures of city-size within counties. The *ECON* vector contains measures of local job growth from the U.S. Bureau of Economic Analysis data (BEA) based on growth over the five years directly prior to the poverty measurement and then five years prior to that. So for the 2000 poverty model, employment growth was measured over 1990–1995 and 1995–2000. Demographic composition differences are controlled with the *DEMOG* vector that includes age, race, education, immigration, and single-parent households. Demographic variables are also lagged ten years prior to the dependent variable (i.e., measured in 1990 or 2000) to mitigate concerns about endogenous relationships with the dependent variable.

Descriptive Statistics

Figures 1a and 1b show coal mining's share of total county employment for 1990 and 2000 in the Central Appalachian region. In cross-hatching, we display whether or not the county had a positive value for our mountaintop mining variable. These counties have both positive values for surface coal mining and a USDA ERS topography code greater than 18. Consistent with the long-term trend of declining economic importance of coal mining in the region over the past 60 years, the maps show fewer counties with coal employment shares above 16% of total employment in

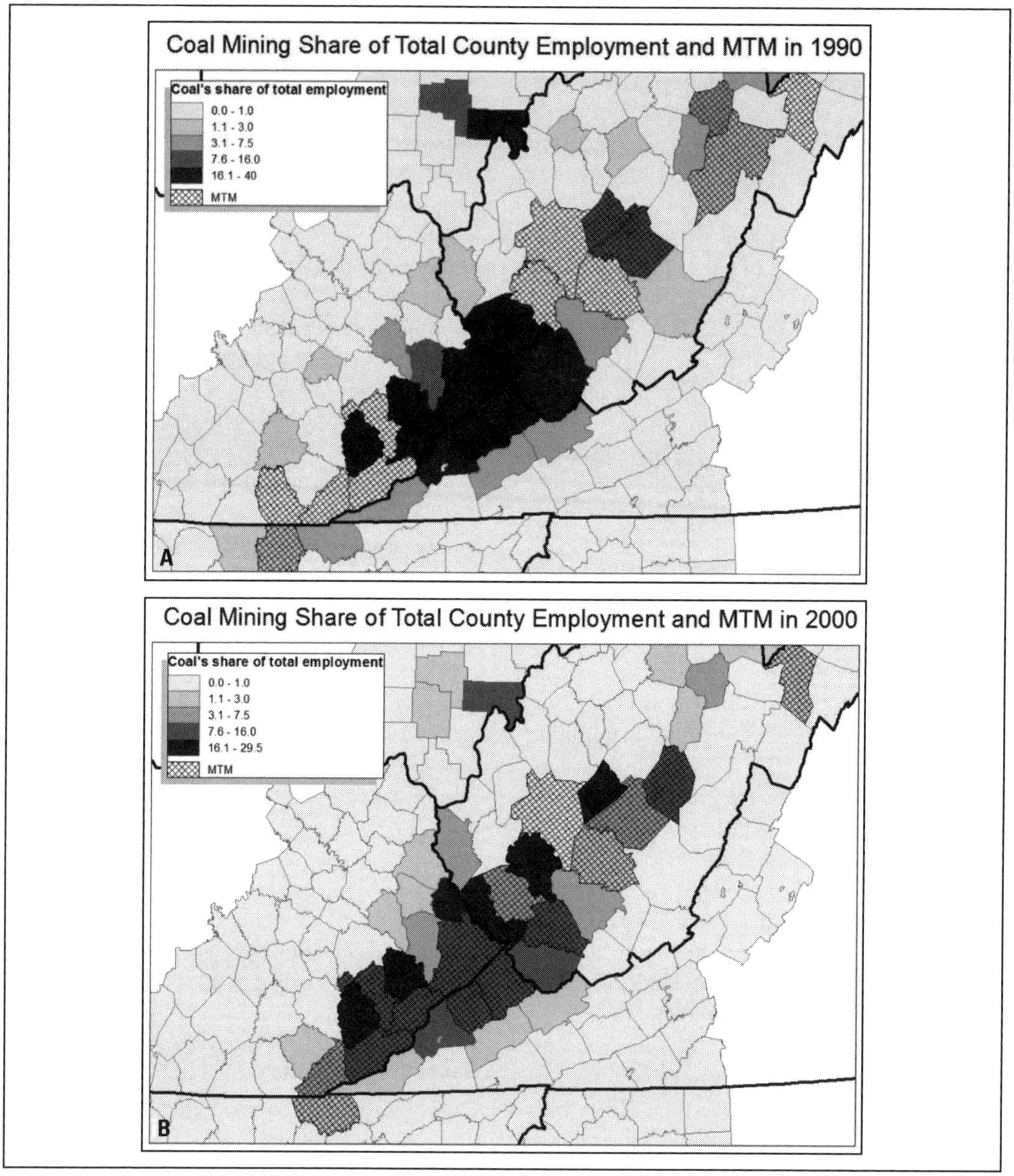

Figure 1. Central Appalachia

2000 than in 1990 as well as fewer counties for which employment in coal mining constitutes more than one-percent of total employment.

Table 1 contains selected descriptive statistics for the ARC, the rest of the U.S, and Central Appalachian MTM counties for 1990 and 2000. The lowest poverty rates occur in the rest of the U.S. with higher rates in the ARC region, and the highest rates in MTM counties. The latter counties have the greatest average distance to the nearest metropolitan area, suggesting predominantly remote, rural areas that would face significant economic barriers. The ARC region has slower employment growth than the rest of the U.S. and MTM counties have the slowest job growth.

Table 1. Selected descriptive statistics by region: 1990 and 2000

	US			ARC			MTM		
	1990	2000	2010	1990	2000	2010	1990	2000	2010
Poverty	16.34	13.78	16.37	19.06	16.38	19.52	30.04	26.34	25.06
	(7.79)	(6.38)	(6.17)	(7.90)	(6.39)	(5.38)	(7.25)	(6.31)	(5.41)
MTM	0.12	0.11		1.15	1.65		13.01	19.23	
	(3.85)	(4.01)		(6.54)	(10.97)		(18.83)	(33.15)	
Km to nearest MA	74.41			53.02			72.08		
	(62.10)			(31.85)			(26.33)		
Incremental distance MA of 250k	61.84			22.7			3.06		
	(103.27)			(31.02)			(9.62)		
Incremental distance MA of 500k	41.12			38.68			116.84		
	(69.19)			(50.85)			(60.12)		
Incremental distance MA of 1.5m	88.48			99.39			64.38		
	(121.57)			(98.25)			(72.18)		
Nearest metro population	455,918			213,224			193,341		
	(1,460,000)			(346,714)			(84,147)		
Employment growth 1990–95 (2000–05)	11.25	3.06		9.25	2.29		5.67	0.57	
	(18.62)	(10.94)		(9.41)	(9.67)		(9.27)	(7.40)	
Employment growth 1995–2000 (2005–2010)	9.43	1.26		7.58	−1.45		1.00	−0.38	
	(9.97)	(7.76)		(10.67)	(6.84)		(7.40)	(6.49)	
Percent high school graduate	34.35	34.39		34.55	37.4		30.01	34.44	
	(6.02)	(6.44)		(6.73)	(6.41)		(4.29)	(4.29)	
Percent some college	17.07	21.1		12.14	16.05		10.28	13.73	
	(4.33)	(4.10)		(2.94)	(3.12)		(2.03)	(2.35)	
Percent associates degree	5.55	5.85		4.06	4.8		2.69	3.63	
	(2.10)	(1.99)		(1.63)	(1.66)		(1.03)	(1.11)	
Percent college grad	13.89	16.97		10.48	12.96		7.63	9.23	
	(6.51)	(7.75)		(4.97)	(5.89)		(2.41)	(2.83)	
Percent female-headed HH with children	5.46	6.15		5.19	5.74		6.12	6.12	
	(2.35)	(2.43)		(1.58)	(1.65)		(1.02)	(0.94)	
Percent male-headed HH with children	1.37	2.12		1.3	1.98		1.4	1.9	
	(0.56)	(0.65)		(0.42)	(0.50)		(0.47)	(0.48)	
Percent under age 6	10.23	9.05		9.13	8.56		9.1	8.29	
	(1.44)	(1.47)		(1.08)	(1.10)		(0.93)	(0.61)	
Percent age 7–17	16.89	16.72		16.16	15.17		18.61	15.29	
	(2.40)	(2.03)		(1.80)	(1.45)		(1.30)	(1.19)	
Percent age 18–24	8.99	8.81		10.29	9.18		9.74	9.12	
	(3.43)	(3.33)		(3.23)	(3.27)		(1.02)	(0.69)	
Percent age 60–64	4.69	4.45		4.81	4.82		4.5	4.67	
	(1.01)	(0.98)		(0.73)	(0.76)		(0.53)	(0.49)	
Percent over age 65	15.06	14.87		14.38	14.6		12.59	13.54	
	(4.52)	(4.26)		(2.65)	(2.74)		(2.08)	(1.77)	
N	2,596	2,596	2,596	413	413	413	37	37	37

Notes: See the text or Partridge and Rickman (2005, 2008a, 2008b) for more details of variable definitions.

Table 2. Determinants of 2000 (1999) poverty rates

	(1) Base		(2) Levels and Change		(3) MTM
	US	ARC	US	ARC	ARC
Poverty rate 1990	0.53‡	0.59‡	0.53‡	0.59‡	0.58‡
	(25.95)	(12.96)	(25.97)	(12.89)	(12.70)
MTM					0.01
					(0.44)
MTM*share coal					–2.2e–03
					(–1.56)
MTM*change coal					–4.4e–05
					(–0.16)
Share coal	0.02	0.09‡	0.02	0.09‡	0.12‡
	(0.50)	(2.74)	(0.50)	(2.76)	(3.10)
Share oil and gas	0.04†	–0.02	0.04†	–0.01	–0.02
	(2.09)	(–0.33)	(2.03)	(–0.29)	(–0.40)
Share other mining	4.8e–03	–0.13	4.60E–03	–0.15	–0.15
	(0.29)	(–1.27)	(0.28)	(–1.37)	(–1.37)
Change in coal 1990–2000			–1.7e–05‡	2.6e–04	2.6e–04
			(–2.90)	(0.65)	(0.67)
Change in oil and gas 1990–2000			–4.3e–05	1.4e–04	1.5e–04
			(–0.34)	(0.43)	(0.44)
Change in other mining 1990–2000			1.1e–06	–1.2e–04	–1.3e–04
			(1.41)	(–0.33)	(–0.37)
N	2611	417	2611	417	417
R–sq	0.908	0.932	0.908	0.932	0.933

Notes: See the text or Partridge and Rickman (2005, 2008a, 2008b) for more details of variable definitions.
All standard errors are robust.
* p<.10, † p<.05, ‡ p<.0

Results from the Empirical Models

Table 2 displays the results for the determinants of 2000 county poverty rates for the ARC sample and the rest of the U.S. sample. Both samples include two base specifications, with an additional model for the ARC sample that includes MTM and the MTM interaction variables. To condense discuss and better focus on the mining results, the tables report only the results for those variables (net of other control variables above that are also included in the estimations).

The models in Table 2 perform quite well, explaining over 90% of the variation in the dependent variable. Consistent with theory, lagged poverty rates are positively and significantly related to current poverty rates. Persistence is greater in ARC counties compared to non-ARC counties, consistent with the long-term disadvantages felt in the region. For the base models (1), share of oil/gas is statistically significant for the U.S. 2000 model but coal is not. This relationship is interesting for it shows that coal may perform better for well-being than other types of mining. Conversely in ARC counties, coal mining's employment share is positive and statistically significant but the other mining shares are statistically insignificant. These results support the view that coal mining in Appalachia is associated with higher poverty, but the same does not apply elsewhere in the U.S.

Models 2 (Table 2) add the corresponding ten-year percent change in county shares of coal mining, oil and gas mining, and other mining to the model. For the U.S. sample, poverty has a negative relationship with change in mining employment shares. For the ACR, there is no statistical association between percent change in mining employment shares and poverty. These

Table 3. Determinants of 2010 poverty rates

	(1) Base		(2) Levels and Change		(3) MTM
	US	ARC	US	ARC	ARC
Poverty rate 2000	0.69‡	0.56‡	0.69‡	0.56‡	0.56‡
	(30.40)	(9.97)	(30.30)	(9.90)	(9.85)
MTM					4.3e–03
					(0.22)
MTM*share coal					–3.4e–04
					(–0.54)
MTM*change coal					–9.6e–05
					(–0.45)
Share coal	–0.14‡	–0.05	–0.14‡	–0.05	–0.05
	(–2.74)	(–1.30)	(–2.74)	(–1.30)	(–0.82)
Share oil and gas	–0.08‡	0.13†	–0.08‡	0.13†	0.12†
	(–3.03)	(2.20)	(–3.02)	(2.16)	(2.11)
Share other mining	–0.02	–0.18	–0.02	–0.19	–0.19
	(–1.09)	(–1.49)	(–1.09)	(–1.57)	(–1.57)
Change in coal 2000–2010			–1.4e–06†	–6.8e–07	–6.4e–07
			(–2.51)	(–0.56)	(–0.52)
Change in oil and gas 2000–2010			–7.9e–09	4.7e–06	2.9e–06
			(–0.12)	(0.04)	(0.02)
Change in other mining 2000–2010			–4.1e–06	–5.6e–05*	–5.6e–05*
			(–0.54)	(–1.78)	(–1.77)
N	2611	417	2611	417	417
R–sq	0.902	0.884	0.902	0.884	0.884

Notes: See the text or Partridge and Rickman (2005, 2008a, 2008b) for more details of variable definitions.
All standard errors are robust.
* p<.10, † p<.05, ‡ p<.0

results suggest that poverty's association with coal mining is more of a legacy effect than an association with increases in coal production over the 1990–2000 period.

Model 3 shows no statistically significant relationship between MTM and 2000 ARC poverty rates. We also interact the MTM indicator variable separately with coal's share of total employment and the percent change in coal's share of total employment. Again these terms are not statistically significant and suggest MTM has no additional impact on poverty. We point out, however, that our measure of MTM does not control for non-random selection of MTM sites. Companies interested in carrying out MTM may strategically pursue counties with certain demographic, geographic, or economic characteristics that are in turn associated with poverty rates. To deal with these selection issues, further analysis using more sophisticated statistical techniques, such as instrumental variables, is necessary.

With regard to other (control) variables, though we do not display results, we note that most have results similar to those of Partridge and Rickman (2008a, 2008b). In particular, these results highlight the statistically significant effects of distance and education. That is, poverty rates are positively related to distance to higher tiered MAs and to a smaller share of the adult (age 25 or over) having completed high school and college. The effects of education and distance are important because most mining operations are located in remote rural areas with less educated populations. Such variables are important to control for effects that may bias the MTM and other mining variables.

The regression results for the 2010 poverty model are reported in Table 3. Past poverty is again a significant contributor to current poverty, but unlike the 2000 poverty model, persistence is higher in the rest of the U.S. compared to the ARC region. It is significant that ARC counties are less tied to past poverty rates in the 2000s compared to the previous decade because that implies the region may be breaking its historical disadvantages. For the ARC counties, in both Models 1 and 2, the positive relationship between coal mining's share of total employment and poverty in the 2000 poverty model disappears in the 2010 poverty model as the relationship is no longer significant. Further, the coal employment share relationship is negative and statistically significant in the 2010 U.S. model. This suggests that the negative legacy of large coal operations may be dissipating over time. In Model 2, percent change in coal employment now has a negative and significant relationship with poverty in the rest of the U.S. Interestingly, the share of oil and gas mining now has a positive and significant relationship with poverty in the ARC model. Mountain-top mining further has no significant relationship with poverty, suggesting that at least in the short term, MTM has a similar relationship to poverty as coal mining in general. These results suggest the historic legacy of coal mining being associated with higher poverty in Appalachia may be dated.

It should be noted that distance and education both have significant effects in the 2010 poverty model, but the influence of remoteness has lessened. All levels of education still have a positive and significant relationship to poverty, but it seems the poverty-reducing threshold for education has moved from a high school diploma to an Associate's degree (in the U.S. and ARC region). This is consistent with the larger trend of decreasing employment in goods-producing sectors and increased demand for higher skilled labor.

CONCLUSIONS

In this study we examine a cross-disciplinary social science research question: does mining adversely affect population's well-being? Often framed as the "natural resource curse," this question has been studied globally across nations and conventionally assumed to apply to the case of coal mining regions in Appalachia. We report the results from a pilot-study aimed at improving past research on the coal industry. Our study: uses a robust set of control variables lagged in time to improve causal understanding; compares Appalachia with the rest of the nation for two time periods; and analyzes mountain-top mining as a gradient measure that better reflects its intensity. For ARC counties, we find that coal employment did have a positive relationship to poverty rates in the past (i.e., prior to 2000). However, over the last decade, the share of coal employment has no statistically significant relationship to poverty. Neither do we find mountain-top mining related to poverty in the ARC in either of the two time periods analyzed. It should be noted that the MTM variable is a gradient measure that measures the intensity of surface mining. It is different from the binary measure that we used in a previous study (Partridge et al. 2013) which produced different results. In the previous study, the binary measure was negatively related to poverty (and statistically significant) for 2010. These differences illustrate why careful replication across different research designs is needed to uncover the true effects of coal mining.

Further, we find the impact of coal mining in the ARC counties appears to diverge from its impact in other U.S. regions. Outside Appalachia, the share of coal employment is related to lower poverty in the contemporary (2010) period and increases in coal employment are related to lower poverty rates over time. This indicates processes unique to Appalachia beyond mining per se may be influencing the conventionally assumed link between poverty and mining.

Future social science research needs to sort out the causal impacts of coal employment and other coal mining indicators in more depth and detail. In our future work, we will build on the results of this study to develop improved methodological designs to better address cause-effect issues (i.e., endogeneity) which is an inherent concern of all empirical studies of the mining industry. Such designs include development of instrumental variable approaches.

Finally, future research should give greater focus to the coal industry's impact on other well-being indicators such as local economic growth and diversity and environmental conditions; and coal mining employment relative to other traditional industries in Appalachia such as manufacturing, which has experienced long-term decline, and to expanding employment in shale gas. We find some evidence that relative to gas/oil mining, coal mining has more positive effects on economic well-being in Appalachia. More attention is needed to comparing coal mining employment today to other sectors affecting the region's development.

ACKOWLEDGMENT

This study was sponsored by the Appalachian Research Initiative for Environmental Science (ARIES). The views, opinions and recommendations expressed herein are solely those of the authors and do not imply any endorsement by ARIES employees, other ARIES-affiliated researchers or industrial members. Information about ARIES can be found at http://www.energy.vt.edu/ARIES.

REFERENCES

Ahern, M.M., Hendryx, M., Conley, J., Fedorko, E. Ducatman, A.and Zullig, K G.2011. "The Association between Mountaintop Mining and Birth Defects among Live Births in Central Appalachia, 1996–2003." *Environmental Research.*11 (6): 838–846.

Black, D. McKinnish, T., and Sanders, S. 2005. "The economic Impact of the Coal Boom and Bust." *Economic Journal.* 115: 449–476.

Blank, Rebecca. 2005. "Poverty, policy, and place: how poverty and policies to alleviate poverty are shaped by local characteristics." *International Regional Science Review* 28 (4): 441–464.

Brown, L. Lee, S., Lobao, L. and Chung, S. 2005. "Continuity amidst restructuring the U.S. gender division of labor in geographic perspective." *International Regional Science Review.* 28 (3):271–301.

Deaton, B.J. and Niman, E. 2012. "An empirical examination of the relationship between mining employment and poverty in the Appalachian region." *Applied Economics*, 44: 303–312.

DeNavas-Walt, C., Proctor, B.D., and Smith, J. 2012. *Income, Poverty, and Health Insurance Coverage in the United States: 2011.* U.S. Census Bureau, Current Population Reports, P60–243. Washington, D.C.: U.S. Government Printing Office.

Dorfman, J., Partridge M.D., and Galloway, H. 2011. "Are high-tech employment and natural amenities linked: answers from a smoothed Bayesian spatial model." *Spatial Economic Analysis.* (6): 397–422.

Glasmeier, A.K. 2002. "One nation pulling apart: the basis of persistent poverty in the USA." *Progress in Human Geography* 26 (2): 155–173.

Ghose, M.K. 2007. "Generation and quantification of hazardous dusts from coal mining in the Indian context."*Environmental Monitoring and Assessment,* 130, 35–45.

Gould, K.A. 1991. "The sweet smell of money: economic dependency and local environmental political mobilization." *Society and Natural Resources: An International Journal.* 4 (2): 133–150.

Hendryx, M. 2011. "Poverty and mortality disparities in central Appalachia: mountaintop mining and environmental justice." *Journal of Health Disparities Research and Practice.* (4): 44–53.

James, A. and Aadland, D. 2011. "The curse of natural resources: an empirical investigation of U.S. counties."*Resource and Energy Economics* 33 (2): 440–453.

Kilkenny, M. and Partridge, M.D. 2009. "Export sectors and rural development."*American Journal of Agricultural Economics* 91 (4): 910–929.

Levernier, W., Partridge, M.D. and Rickman, Dan S. 2000. "The causes of regional variations in U.S. poverty: a cross-county analysis." *Journal of Regional Science*, 40, 473–498.

Lobao, L. *Locality and Inequality: Farm and Industrial Structure and Socioeconomic Conditions.* 1990. Albany: The State University of New York Press.

Lobao, L.M., Hooks, G. and Tickamyer, A.R. (editors) 2007. *The Sociology of Spatial Inequality.* Albany: The State University of New York Press.

McAuley, S.D. and M.D. Kozar. 2006. "Ground-water quality in un-mined areas and near reclaimed surface coal mines in the northern and central Appalachian coal regions, Pennsylvania and West Virginia," *Scientific Investigations Report, US Department of the Interior, US Geological Survey*: 2006–5059.

Papyrakis, E. and Gerlagh, R. 2007. "Resource abundance and economic growth in the U.S." *European Economic Review.* 51: 1011–1039.

Partridge, M.D. and Rickman, Dan S. 2005. "Persistent high-poverty in nonmetropolitan America: can economic development help?" *International Regional Science Review*, 28, 415–440.

Partridge, M.D. and Rickman, Dan S. 2006. *The Geography of American Poverty: Is There a Need for Place-Based Policies?* Kalmazoo, MI: Upjohn.

Partridge, M.D. and Rickman, Dan S. 2008a. "Distance from Urban Agglomeration Economies and Rural Poverty." *Journal of Regional Science*, 48, 285–310.

Partridge, M.D. and Rickman, Dan S. 2008b. "Place-Based Policy and Rural Poverty: Insights from the Urban Spatial Mismatch Literature." *Cambridge Journal of Regions, Economy and Society,* 1: 131–156.

Partridge, M. D, Betz, M. and Lobao, L. 2013 "Natural resource curse and poverty in Appalachian America" *American Journal of Agricultural Economics* 95 (2):449–456.

Skytruth. 2009. Mountaintop *Removal Mining. Part 1: Measuring the Extent of Mountaintop Removal in Appalachia.* Available from: http://blog.skytruth.org/2009/12/measuring-mountaintop-removal-mining-in.html (accessed December 2011).

Stedman, R.C., Patriquin, M.N. and Parkins, J.R. 2012. "Dependence, diversity, and the well-being of rural community: building on the Freudenberg legacy." *Journal of Environmental Studies Science.* 2:28–38.

US Department of Labor Mine Safety and Health Administration. 2010. *Effects of Blasting on Air Quality.* Available from: www.msha.gov/Illness_Prevention/healthtopics/blasting.HTMS (accessed December 2011).

Weber, B., L. Jensen, K. Miller, J. Mosely, and M. Fisher. 2005. "A critical review of rural poverty literature: is there truly a rural effect?" *International Regional Science Review* 28 (4):381–414.

Woods, B.R. and Gordon, J.S. 2011. "Mountaintop removal and job creation: exploring the relationship using spatial regression." *Annals of the Association of American Geographers,* 101 806–815.

Edgework and Boundary Crossings: Assessing Foundations for Public Ecology in the Appalachian Region

Mary Hufford
Virginia Tech, Blacksburg, Virginia

Betsy Taylor
Virginia Tech, Blacksburg, Virginia

ABSTRACT

Public ecology is one name for an interdisciplinary, multisectoral approach to the study and management of complex socio-ecological systems. Emerging around the world under a variety of names such as civic environmentalism, community-based natural resource management, collaborative ecosystem management, and cultural conservation, public ecology traverses institutionalized boundaries, framing expert-driven environmental research and decision-making. We describe "edgework" as a metadiscursive practice in the peer-reviewed literature that marks pathways emerging across five sets of "boundaries." Edgework and boundary crossings operate pragmatically to affect the epistemological and ontological shifts required for new multisectoral research communities engaging complex, multicausal, interdisciplinary objects of sustainability. The status of these shifts in relation to three objects of sustainability—water, forests, and carbon-based energy production—bears significantly on our study of foundations for public ecology in the Appalachian region.

INTRODUCTION

The past twenty years have witnessed the remarkable growth, worldwide, of interdisciplinary, multisectoral partnerships for the study and management of environmental resources. Such partnerships blur disciplinary boundaries as they build teams of experts with the diverse specializations needed to understand multicausal, complex, socioecological processes. Such partnerships blur entrenched boundaries between sectors (community, civil society, government, industry, and academia) when local knowledge of ordinary citizens makes valuable contributions to understanding and managing watersheds, forests, and other environmental resources. Finally, such partnerships blur *conceptual* boundaries that categorize objects of environmental stewardship according to whether their values are economic, social, or ecological.

The variety of names for these hybrid partnerships, approaches, and reconceptualized objects of environmental stewardship map emerging pathways across multiple boundaries: "civic environmentalism" (Agyeman and Angus 2003), "grassroots ecosystem management" (Weber 2000), "citizen science" (Puckett et al. 2012), "civic science" (Craynon et al. 2012), "collaborative adaptive management" (Hull 2008), "cultural conservation" (Howell 2002, Hufford 1994), "corporate social responsibility" (Hilson 2012), "social ecological systems" (Brondizio et al. 2009), "collaborative ecosystem governance" (Karkkainen 2003), and "adaptive ecosystem management" (Norton 2005) to name just a few. Such terms are linked by a common objective: the integration of social, economic, and ecological values in the research and management of environmental resources. This integration is a complex undertaking that entails learning to communicate across boundaries within and between academic disciplines and social sectors, and revising the processes linking knowledge making with the formation and application of public policy. Drawing on the insights of colleagues working across schools at Virginia Tech over the past decade, we are calling this complex undertaking "public ecology" (Craynon et al. 2012; Robertson and Hull 2003b; Luke 2005).

PUBLIC ECOLOGY

Public ecology names an interdisciplinary, multisectoral approach to the study and management of complex adaptive socio-ecological systems. John Craynon et al. (2012) locate public ecology at the "nexus of science, engineering, public policy and interest, citizen views and values, market forces and environmental protection statutes and regulations." The authors further define public ecology as "an open and participatory discourse… intended to ensure ecological systems continue to function as societies operate within and derive benefits from them."

This approach requires an alternative epistemological and ontological framework for a shared public process of knowledge making and for researching and managing objects of environmental stewardship. As Robertson and Hull (2003b) summarize it, public ecology:

- entails public dialogue and collaboration across disciplinary, institutional, professional, and cultural boundaries;
- integrates local knowledge with professional expertise in studying, monitoring, and designing for environmental health;
- measures environmental resources in terms of ecological and social as well as economic values;
- addresses the interdependencies of local human ecosystems with larger regional and global systems.

A growing number of theoretical and case studies suggest that a public ecology approach could be used to resolve thorny environmental issues, including conflicts over the production of energy from coal. As Craynon, Sarver, and Robertson ask in the title of their recently published article, "Can a Public Ecology Approach Help to Resolve the Mountaintop Mining Controversy?" (Craynon et al. 2012). The history of struggle between labor and industry (Corbin 1989; Eller 1982; Fisher 1993) and within and between coalfield communities and industry over environmental and health impacts of coal extraction has left a legacy of distrust that could seriously impede a public ecology approach (Craynon et al. 2012). At the same time, the social history of the region points to an abundance of the kind of social and political capacity (Couto 1999; Fisher and Smith 2012) foundational to the kinds of multisectoral partnerships envisioned for public ecology (Figure 1).

Our broad research questions are: What are the foundations for public ecology in the Appalachian Region? Where in the region are multidisciplinary, multisectoral approaches used to build knowledge about forests, water, and energy production in relation to complex socio-ecological systems? How are people working across sectors to sustain the region's historic ecological, social, and economic values to meet the needs of the present generation without compromising the ability of future generations to meet their needs?

To place our study in a global context, we are presently surveying the literature on interdisciplinary, multisectoral approaches to the research and management of complex socio-ecological systems around the world. We are specifically

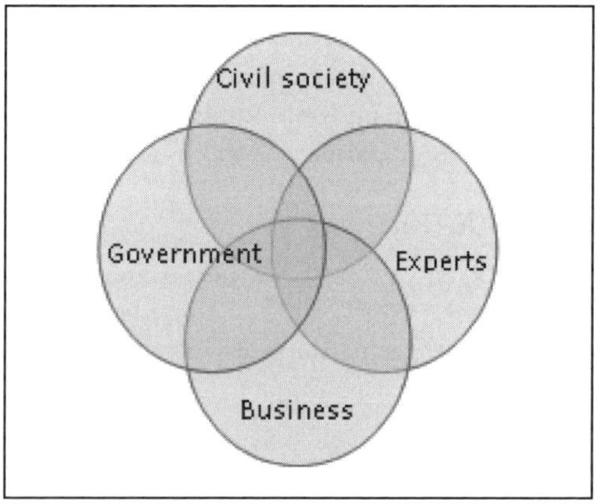

Figure 1. Multisectoral partnerships for sustainability

reviewing interdisciplinary, multisectoral efforts organized around the collaborative study and management of three "objects, water, forests, and energy production. Our broad questions are: Around the world, what are innovative models for public ecological practices in industrial development and planning that could be effective in the energy sector? How have these efforts been evaluated in peer reviewed studies? What are the shared attributes and best practices? What foundations for public ecology are being laid in the Appalachian region? We are limiting our review to peer-reviewed articles that describe and evaluate interdisciplinary, multisectoral approaches to three objects of sustainability: water, forests, and carbon-based energy production (coal mining).

In this paper we characterize two of the broader issues facing public ecologists: (1) the formation of collaborative multisectoral research communities; and (2) the constitution of an object of environmental research and stewardship, that is, a shared object of sustainability. Both of these issues require a good deal of attention to emerging ways of communicating across boundaries to shape a new, public, environmental discourse.

EDGEWORK AND BOUNDARY CROSSINGS

Considerable attention is given across the literature to innovative discourse practices and their effectiveness in communicating across boundaries (see Agrawal 2005, Hajer 2005, Hufford 1994, Huttunen 2009, Rodela 2013, Scott 1988). The public ecology literature is itself a site of numerous linguistic operations that identify and traverse disciplinary and sectoral boundaries. Introduced by Katherine Young in her study of the phenomenology of narrative communication (1987) and drawing on the work of Husserl, Schutz, James, Luckman and Goffman, "edgework" is our term for metadiscourse that reflexively shapes conversations across boundaries to reframe both the research community and its object of sustainability. This reframing is displayed in terms such as "integrated assessment" (German 2008), "collaborative management," "community-based participatory research," "community-based natural resource management," "multi-level collaboration," "civic science," and "environmental citizenship." In what follows, we identify edgework that integrates what the boundaries divide in emerging forms of public ecology around the world and in the Appalachian region.

The public ecology literature is especially attentive to boundaries that impede public communication, social learning, and democratic decision-making. While boundary maintenance, (called "boundary work" in the history of disciplines, see Klein 1996) is crucial to the design and conduct of scientific research, the object of environmental stewardship cannot be adequately known and managed without open lines of communication that reveal the multiplicity of ways in which environments provide for human communities. In the public ecological literature we see at least five sets of boundaries that are repeatedly framed and traversed:

- boundaries maintained between research and policy phases of environmental management;
- boundaries dividing social, ecological, and economic functions of environmental resources;
- boundaries preserved among academic disciplines and professions;
- boundaries distinguishing the public sector from government, industry, and the academy; and
- spatio-temporal boundaries defining local, regional, and global ecological scales.

We will focus here on the edgework indicated by the proliferation of hybrid terms, and the epistemological and ontological implications for emerging multisectoral research communities and the complex socio-ecological systems that form their objects of sustainability.

EPISTEMOLOGICAL AND ONTOLOGICAL IMPLICATIONS

Between Research and Public Policy: The Research Community

In the literature on public ecology one often finds the use of qualifiers for terms such as "science," "assessment," "management," "ecology," and

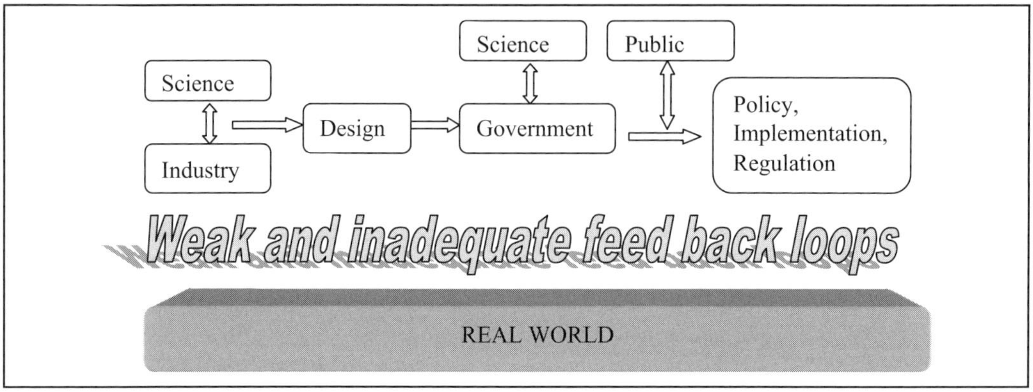

Figure 2. Serial model, separating scientific research phase from public engagement phase

"development." Drawing attention to what we have taken for granted about such terms, the qualifiers are meta-discursive, prompting reflection on our own assumptions and awareness of multiple points of view. Hence, the term "science," long an unmarked category, is now qualified by a contrast between, for example, "conventional" and "postnormal" science, or "traditional" and "civic" science. This distinction marks an ontological shift in the role for science and scientists in public life; and in the models for producing and circulating the knowledge that supports democratic deliberation and policy setting. Moving science from a "traditional' or "conventional" position outside of society to a "postnormal" or "civic" position within, this ontological shift alters relationships and identities throughout the system.

These insights are gradually changing official frameworks and processes for environmental decision-making, from what environmental philosopher Bryan Norton calls a "serial" process to a process characterized by feedback loops linking phases of research, policy making, and management (Norton 2005). The serial process prescribed by the U.S. EPA positioned science as an input from outside the system, yielding data and a model to be turned over to managers responsible for applying the science (Figure 2).

The environmental assessment process is initiated after the science and technological design have informed a "proposed action." This externalized location for the science and the linear, unidirectional flow of information through the regulatory channels expresses the ideology that science cannot be objective if the scientist engages policy questions. (For critiques of this position see Harding 1992; Hess 2007; Robertson and Hull 2003b; and Innocenti and Albrito). Yet, because the serial process does not accommodate feedback when the models fail to predict real world scenarios, the serial process itself is increasingly seen as an impediment to good science. (Examples of studies that focus on overcoming the problems faced by managers attempting to use "model-based tools" generated through the serial process include Biggs et al. 2010 and Borowsky and Hare 2007.)

Alternative models for multi-directional research and management processes express the growing recognition of academic researchers as citizens with specialized skill sets who can design and conduct research leading to information that supports democratic decision-making (Boyte 2009; Fischer 2000). Publicly engaged collaborative research models such as action research (Barton et al. 2009) and learning action networks (Craynon et al. 2012) are characterized by feedback loops that make the research, evaluation, and management phases of shared inquiry and action iterative, muiltidirectional, and ongoing (Figure 3).

Removing the boundary between science and society opens channels of communication that contribute to a collective knowledge making process that can never be seen as complete or finalized (Norton 2005; Rittel and Webber 1973). Civic scientists are specialized citizens who participate

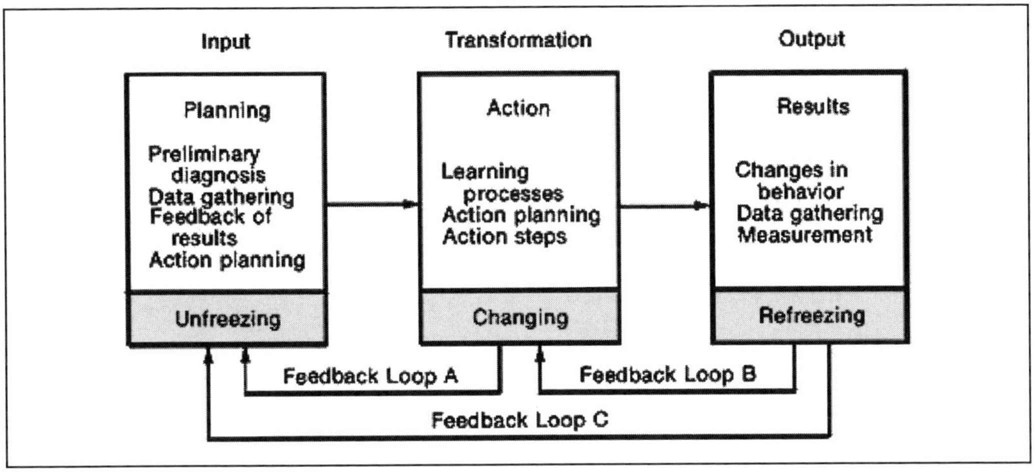

Figure 3. Systems model of action research, showing feedback loops integrating research and management (action) phases Source: Wikipedia Commons http://en.wikipedia.org/wiki/File:Systems_Model_of_Action-Research_Process.jpg downloaded 01/20/2013

in knowledge communities foundational to public ecology (Fischer 2000). Edgework helps to track emerging kinds of public space in which scientists and citizens constitute a world in common for which they co-exist and to which they contribute from different perspectives.

In the Appalachian region, the serial model is still used for the permitting of valley fills necessitated by mountaintop removal. The flow chart in Figure 4 depicts the model presently used for environmental research and decision-making when the object of sustainability is the production of carbon-based energy through mountaintop removal mining (Figure 4). In this model, the science has already been completed and applied to the design of technologies for mining and reclamation before the public decision-making process begins. The assumption that the science has been completed, embodied in the form of data, models, and existing technologies, contributes to a "one size fits all" approach to permitting.

The space entitled "Application Evaluated" provides for feedback on the real world impacts of this design through regulatory review and optional "hearings." In practice, hearings are not spaces for public dialogue; rather they comprise one-way communication events in which government employees record citizen comments for review by decision-makers. By sequestering public input away from the research and design process, the serial model makes it difficult to address impacts on "esthetics, recreation, fish values, land use, wildlife value, economics, food production, road damage prevention, safety, conservation, environmental concerns and historic value."

Federal policy instruments such as the Executive Order on Environmental Justice (E.O. 12898) are changing the relationship between scientists and the public, and between research and management, particularly as these bear on the sustainability of water and forests. There have been a number of community-based participatory research projects led by scientists in the Appalachian region with both federal and corporate funding, including citizen monitoring of forest and stream health, community-based biomonitoring of water and of body burden pollution, as well as numerous academic-community partnerships (Wills et al. 1997; Emmett et al. 2009; Haynes et al. 2011; Puckett et al. 2012; Hendryx et al. 2012; Cook and Taylor 2001). And a recent memorandum issued by the US EPA advises mining regulators to incorporate more public input and to be attentive to environmental justice issues in the course of mine permitting (USEPA 2011). Our study will look closely at peer-reviewed scientific articles describing and evaluating these and other

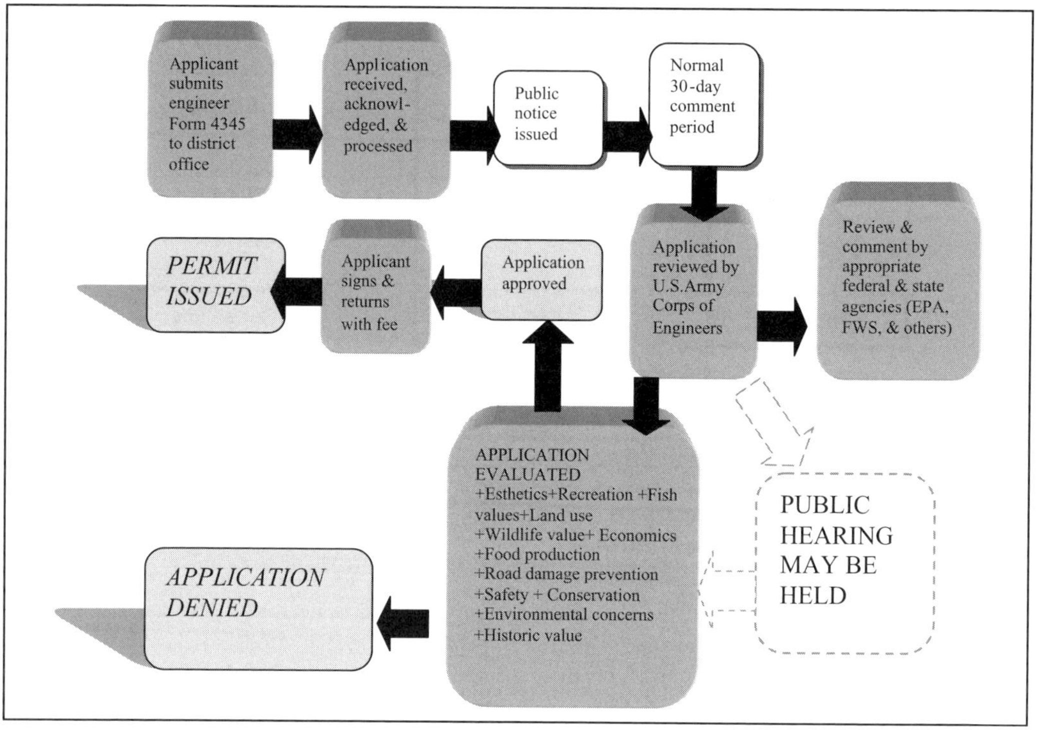

Figure 4. Flow chart depicting U.S. Army Corps of Engineers review process for Section 404 dredge-and-fill permits issued for mountaintop mining and valley fill projects. Based on graphic downloaded on 02-01-2013 http://water.usgs.gov/nwsum/WSP2425/legislation.html

emerging forms of interdisciplinary, multisectoral planning for sustainability related to forests, water, and energy production.

Between Social, Ecological, and Economic Functions at Local, Regional, and Global Scales: The Object of Sustainability

A number of environmental philosophers cite the need for what Robertson and Hull (2003b) describe as "a vision of nature and human society that encourages people to create healthy human ecosystems and sustainable communities at local, regional, and global scales." Arriving at such a vision entails the historically unprecedented project of building a culture of sustainability across disciplinary, sectoral, and scalar boundaries.

Bringing multiple perspectives into dialogue, ongoing conversations between scientists and the public have illuminated the extent to which categories that compartmentalize nature, society, and economy into separate spheres for research and management impede planning for sustainability. Integrating these spheres transforms the object of environmental protection from an inventory of static "resources" into a complex multifunctional system. Comprising social as well as ecological and economic values, this system is situated at a public nexus of interdisciplinary and multisectoral perspectives (Figure 5).

Terms for such systems, emerging in the public ecology literature, signal innovative experiments underway for managing water, forests, and energy production within interdisciplinary, multisectoral frameworks that integrate social, ecological, and economic values. Among many terms embodying the integration of these three "pillars of sustainability" we find "socio-ecological systems" (Brondizio et al. 2009), "industrial ecology (Hoffman 2003)," "community forestry" (Baker and Kusel 2003; Gibson et al. 2000), "corporate social responsibility" (Hilson 2012, Hutchins et al. 2007, Jenkins and Yakovleva

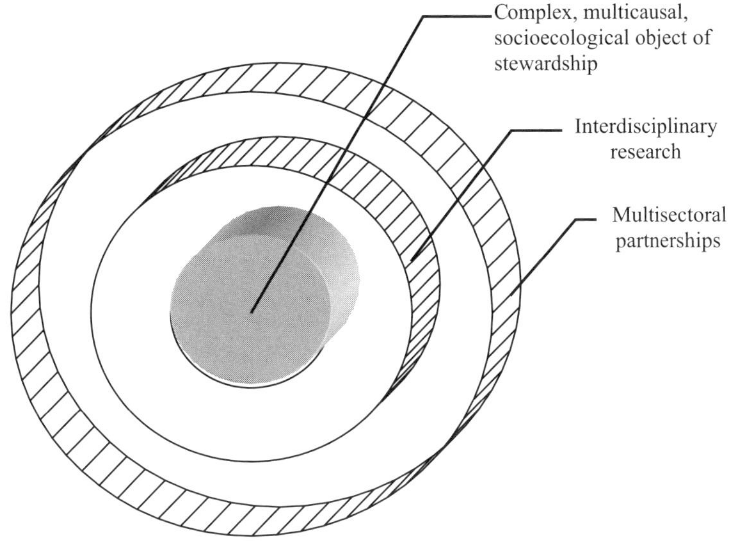

Figure 5. Sectors, disciplines, and the object of stewardship

2006), "corporate partnerships for sustainability" (Juniper and Moore 2002), "cultural conservation" (Hufford 1994, Howell 2002), "traditional ecological knowledge" (Huntington 2000), "mixed mesophytic community forest and watershed" (Hufford et al. 2008), and "commons" (Reid and Taylor 2010; Boyer 2007; Hufford 2002).

"The land community" was Aldo Leopold's term for what he arrived at as the object of adaptive management at the local level (Leopold 1948). Integrating economy, ecology, and society, Leopold's term complicates the opposition often drawn between economy and ecology. Norton argues that Leopold's vision not only transcended the polarization of ecological and economic values, it challenged positivist approaches restricting the assessment of risk to physical and economic impacts. Establishing standards for the physical survival of individual persons, species, and artifacts, such approaches fail to recognize negative impacts on cultural institutions that produce and reproduce a land ethic: a set of ideas and practices established over generations that enable communities to meet their needs in the present without compromising the needs of future generations to meet their needs (Norton, 2005: 72–73; King 1998).

Aldo Leopold's notion of the land community guided by a land ethic highlights the importance of ongoing collective cultural processes for sustainability over multiple generations. Terms like "landcare" (Kimmel et al. 2011; Kimmel et al. 2012), "commons" (Hufford 2002; Nonini 2007; Reid and Taylor 2012) and "common pool resource systems" (Armitage 2005; McKean 2000) capture the institutional arrangements subtending communities of land and people. While such arrangements may seem amorphous when viewed from academic, industrial, and government sectors, they have functioned over many generations to articulate and transmit land ethics governing the production and reproduction of communities of land and people (Reid and Taylor 2010). Such arrangements are strongest at the local level. (Efforts to integrate social and cultural into design professions and resource management have produced terms like "soft systems" (Checkland 2000), "intangible cultural resources" (King 2008), and "traditional cultural properties" (King 2003).)

Recognition that local knowledge and culture have a foundational role to play in sustainability at regional and global scales underpins growing interest in locally-centered models like community-based natural resource management (CBNRM) and grassroots ecosystem management (GREM) (Weber, 2000), as well as place-based approaches and the study and application of traditional ecological knowledge (TEK) (see Agrawal 1999; Armitage 2005; Huntington 2000; See also Taylor 2009). Such terms appear in connection with innovative efforts around the world to engage land communities in the collaborative study and integration of traditional management practices with scientific knowledge for long-term sustainability across scales. Implications explored in the literature include the possibility of developing international markets for global services provided by land communities such as the preservation of biodiversity and carbon sequestration (Jenkins et al. 2004).

In the U.S., a growing number of studies focus on the sustainability of local commons and common pool resource systems and related implications for public policy. The Executive Order on Environmental Justice (E.O. 12898) directs all federal agencies to consider the needs of communities for continued access to fish, game, and non-timber forest products when making decisions that could affect that access (US EPA 2004; US EPA 2011). Since then a number of studies of nontimber forest products have sought to clarify the implications of this policy outside of Native American contexts, including "forest practitioners" in the Appalachian region, whose harvesting is not only economically productive, but socially, culturally, and ecologically reproductive (Fortmann 1990; Goodman 2002; Jones et al. 2002, Chamberlain 1998). Nontimber forest products are yet another example of edgework—employing the marked category to reveal what is hidden by the unmarked term, and ultimately diversifying constituencies whose knowledge can contribute to sustainability.

One of the most challenging questions raised in the literature is that of how to promote communication across sectors and disciplines. The public ecology literature identifies and evaluates a number of theories and strategies for promoting communication as a means of reaching common ground among stakeholders, including social learning (Cundill and Rodela 2012; Rodela 2013; Schusler 2003; Ison et al. 2007; Craynon et al. 2012), participatory reason (Reid and Taylor, 2010), and storylines (Hendriks 2005). A question we have is whether the full complement of disciplines with relevant expertise have been engaged. Given the importance of cultural processes to the sustainability of land communities, efforts made to understand such processes could fruitfully engage the skills of ethnographers and public humanists. Such studies presently inform the integration of TEK with managerial ecology, relying on ethnomethodologists for translation. But in communities like those in the Appalachian region, where citizens and scientists share a common language, it is not translation that is needed so much as appreciation for the social, economic, and ecological functions embedded in land communities. Prompting reflection on local forms of management for multiple values, collaborative ethnographies can contribute to learning that integrates scientific and community perspectives into public culture for sustainability.

CONCLUSION

As an approach to sustainable development, public ecology is in its infancy. The simultaneous emergence, worldwide, of cognate approaches to sustainability signals that public ecology is part of an important epistemological and ontological shift that is underway, rewriting the interrelations of research and management, science and public policy, and the social, ecological, and economic values ascribed to environmental resources. Salient aspects of this shift are marked in the hybrid terms appearing over the past two decades in peer-reviewed environmental studies literature. These terms are metadiscursive and pragmatic, both naming and constituting new alliances across disciplines and sectors. Linking values systems operating at local, regional, and global scales, these terms also constitute new, public objects of sustainability. In effect, these are metadiscursive operations that illuminate emerging trailheads to public ecology, some of which appear in the Appalachian region. Awaiting discovery, through our systematic review of the peer-reviewed literature, are the theoretical and practical routes to best practices for public ecology, in the Appalachian region and around the world.

ACKOWLEDGMENT

This study was sponsored by the Appalachian Research Initiative for Environmental Science (ARIES). The views, opinions and recommendations expressed herein are solely those of the authors and do not imply any endorsement by ARIES employees, other ARIES-affiliated researchers or industrial members. Information about ARIES can be found at http://www.energy.vt.edu/ARIES.

REFERENCES

Agrawal, A. 2005. Environmentality: Technologies of Government and the Making of Subjects. Durham, NC: Duke University Press.

Agrawal, A. 1999. Enchantment and disenchantment: the role of community in natural resource conservation. World Development 27(4):629–649.

Agyeman, J. and Angus, B. 2003. The role of civic environmentalism in the pursuit of sustainable communities. Journal of Environmental Planning and Management. 46(3):345–363.

Armitage, D. 2005. Adaptive capacity and community-based natural resource management. Environmental Management. 35(6):703–715.

Baker, M. and Kusel, J. 2003. Community Forestry in the United States: Learning from the Past, Crafting the Future. Washington, D.C.: Island Press.

Barton, J., Stephens, J., and Haslett, Tim. 2009. Action research: its foundations in open systems thinking and relationship to the scientific method. Systems Practice and Action Research. 22:477–488.

Biggs, R., Westley, F.R., and Carpenter, S.R. 2010. Navigating the back loop: fostering social innovation and transformation in ecosystem management. Ecology and Society. 15(2):1–25.

Borowsky, I., and Hare, M. 2007. Exploring the gap between water managers and researchers: difficulties of model-based tools to support practical water management. Water Resources Management. 21(7):1049–1074.

Boyer, J.C. 2007. Reinventing the Appalachian commons. In The Global Idea of 'the Commons.' Ed. By D.M. Nonini. New York: Berghahn Books.

Boyte, H. 2009. Civic agency and the cult of the expert. Kettering Foundation.

Brondizio, E.S., Ostrom, E., and Young, O.R., 2009. Connectivity and the governance of multilevel social-ecological systems: the role of social capital. Annual Review of Environment and Resources. 34:253–278.

Chamberlain, J., Bush, R., and Hammet, A.L. 1998. Non-timber forest products. Forest Products Journal. 48:10–19.

Checkland, P. 2000. Soft-systems methodology: a thirty year retrospective. Systems Research and Behavioral Science. 17:S11-S58.

Cook, S. and Taylor, B. eds. 2001. Academics, Activism, and Place-Based Education in the Appalachian Coal Belt. Special issue of Practicing Anthropology. 23(2):1–32.

Corbin, D.A. 1989. Life, Work, and Rebellion in the Coal Fields: The Southern West Virginia Miners 1880–1922. Urbana, IL: University of Illinois Press.

Couto, R. 1999. Making Democracy Work Better: Mediating Structures, Social Capital, and the Democratic Prospect. Chapel Hill: University of North Carolina Press.

Craynon, J.R., Sarver, E., and Robertson, D.P. 2012. Could a public ecology approach help resolve the mountaintop mining controversy? Resources Policy 37.

Cundill, G., Cummings, G.S., Biggs, D., and Fabricius, C. 2012. Soft systems thinking and social learning for adaptive management. Conservation Biology. 26(1):13–20.

Cundill, G., and Rodela, R. 2012. A review of assertions about the processes and outcomes of social learning in natural resource management. Journal of Environmental Management. 113:7–14.

Eller, R.D. 1982. Miners, Millhands, and Mountaineers: Industrialization of the Appalachian South 1880–1930. Knoxville: University of Tennessee Press.

Emmett, E., Zhang, H., Shofer, F.S., Roday, N., Desai, C., Freeman, D., and Hufford, M. 2009. Development and application of a successful "community first" communication model for community-based environmental health research. Journal of Occupational and Environmental Medicine. 51 (2):146–156.

Fischer, F. 2000. Citizens, Experts, and the Environment: The Politics of Local Knowledge. Durham: Duke University Press.

Fisher, S.L., ed. 1993. Fighting Back in Appalachia: Traditions of Resistance and Change. Philadelphia: Temple University Press.

Fortmann, L. 1990. Locality and custom: non-aboriginal claims to customary usufructuary rights as a source of rural protest. Journal of Rural Studies. 6:195–208.

Gabrielson, T. 2008. Green citizenship: a review and critique. Citizenship Studies. 12 (4):429–446.

German, L., Mansoor, H., Alemu, G., Mazengia, W., Amede, T., and Stroud, A. 2007. Participatory integrated watershed management: evolution of concepts and methods in an ecoregional program of the eastern African highlands. Agricultural Systems. 94(2):189–204.

Gibson, C.C., McKean, M.A., and Ostrom, E. eds. 2000. People and Forests: Communities, Institutions, and Governance. Cambridge, MA: MIT Press.

Goffman, E. 1974. Frame Analysis: An Essay on the Organization of Experience. Boston: Northeastern University Press.

Goodman, E.C. 2002. Nontimber forest products customary claims. In Nontimber Forest Products in the United States. Edited by E.T. Jones, R.J. McLain, and J. Weigand. Lawrence, KS:University Press of Kansas.

Gray, B. 2004. Strong opposition: frame-based resistance to collaboration. Journal of Community and Applied Social Psychology. 14(3):166–176.

Hajer, M. and Versteeg, W. 2005. A decade of discourse analysis of environmental politics: achievements, challenges, perspectives. Journal of Environmental Policy and Planning. 7(3):175–184.

Harding, S. 1992. After the neutrality idea: science, politics, and "strong objectivity.' Social Research 59 (3): 567–87.

Hayles, N.K. 1999. How We Became Posthuman: Virtual Bodies in Cybernetics, Literature, and Informatics. Chicago: University of Chicago Press.

Haynes, E.N., Beidler, C., Wittberg R., Meloncon, L., Parin, M., Kopras, E.J., Succop, P., Dietrich, K.N., 2011. Developing a bidirectional academic-community partnership with an Appalachian-American community for environmental health research and risk communication. Environmental Health Perspectives. 119 (10): 1364–1372.

Hendriks, C.M. 2005. Participatory storylines and their influence on deliberative forums. Policy Sciences. 38(1):1–20.

Hendryx, M., Wolfe, L., Luo, J., and Webb, B. 2012. Self-reported cancer rates in two rural areas of West Virginia with and without mountaintop coal mining. Journal of Community Health.

Hess, D.J. 2007. Alternative Pathways in Science and Industry: Activism, Innovation, and the Environment in an Era of Globalization. Cambridge, MA: MIT Press.

Hilson, G. 2012. Corporate social responsibility in the extractive industries: experiences from developing countries. Resources Policy. 37(2):131–137.

Hoffman, A.J. 2003. Linking social systems analysis to the industrial ecology framework. Organization and Environment. 16 (1):66–85.

Hoffman, A.J. 2003. Linking social systems analysis to the industrial ecology framework. Organization and Environment. 16(1):66–85.

Howell, B., ed. 2002. Culture, Environment, and Conservation in the Appalachian South. Urbana, IL: University of Illinois Press.

Hufford, M. ed. 1994. Conserving Culture: A New Discourse on Heritage. Urbana, IL: University of Illinois Press.

Hufford, M. 2002. Reclaiming the commons: narratives of progress, preservation, and ginseng. In Culture, Environment, and Conservation in the Appalachian South. Urbana, IL: University of Illinois Press.

Hufford, M. 2003. Knowing ginseng: the social life of an Appalachian root. Cahiers de Litterature Orale. 53-54:265–91.

Hufford, M., Carroll, T., and Moonsammy, R. 2008. Ethnographic Overview and Assessment of the New River Gorge National River and the Gauley River National Recreation Area. Boston: National Park Service Ethnography Program and Philadelphia: Center for Folklore and Ethnography, University of Pennsylvania.

Hull, B.R. 2008. Adaptive management. In Encyclopedia of Environmental Ethics and Philosophy. Edited by J. Baird Callicott and R. Frodeman. New York, NY: MacMillan.

Huntington, H.P. 2000. Using traditional ecological knowledge in science: methods and applications. Ecological Applications. 10(5):1270–1274.

Hutchins, M.J., Walck, C.L., Sterk, D.P., and Campbell, G.A. 2007. Corporate social responsibility: a unifying discourse for the mining industry? Greener Management International. 52:17–30.

Huttunen, S. 2009. Ecological modernization and discourses on rural non-wood bioenergy production in Finland from 1980–2005. Journal of Rural Studies. 25(2): 239–247.

Innocenti, D. and Albrito, P. 2011. Reducing the risks posed by natural hazards and climate change: the need for a participatory dialogue between the scientific community and policy makers. Environmental Science and Polity. 14(7):730–733.

Ison, R., Roling, N., and Watson, D. 2007. Challenges to science and society in the sustainable management and use of water: investigating the role of social learning. Environmental Science and Policy. 10(6):499–511.

Jenkins, H., and Yakovleva, N. 2006. Corporate social responsibility in the mining industry: exploring trends in social and environmental disclosure. Journal of Cleaner Production. 14(3/4):271–284.

Jenkins, M., Scherr, S.J., and Inbar, M. 2004. Markets for biodiversity services: potential roles and challenges. Environment 46(6):32–42.

Jones, E.T., McLain, R.J., and Weigand, J. 2002. Nontimber Forest Products in the United States. Lawrence, KS: University Press of Kansas.

Juniper, C., and Moore, M. 2002. Synergies and best practices of corporate partnerships for sustainability. Corporate Environmental Strategy. 9(3):267–276.

Karkainen, B.C. 2003. Collaborative ecosystem governance: scale complexity, and dynamism. Virginia Environmental Law Journal. 21:189–243.

Kimmel, C., Hull, R.B., Stephenson, M., Robertson, D. and Cowgill, K. 2012. Building community capacity and social infrastructure through landcare: a case study of land grant engagement. Higher Education. 64(2):223–235.

King, T. 1998. How the archeologists stole culture: a gap in the environmental assessment process and how to fix it. Environmental Impact Assessment Review. 18:117–133.

King, T. 2003. Places That Count: Traditional Cultural Properties in Cultural Resource Management. New York: Altamira Press.

King, T. 2008. Cultural Resource Laws and Practice. New York: Altamira Press.

Klein, J.T. 1996. Crossing Boundaries: Knowledge, Disciplinarities, and Interdisciplinarities. Charlottesville, VA: University Press of Virginia.

Leopold, A. 1948. A Sand County Almanac and Sketches from Here and There. New York: Oxford University Press.

Luke, T.W. 2005. The death of environmentalism or the advent of public ecology? Organization and Environment 18(4):489–494.

McKean, M.A. 2000. Common property: what is it, what is it good for, and what makes it work? In People and Forests: Communities, Institutions, and Governance. Edited by C.C. Gibson, M.A. McKean, and E. Ostrom. Cambridge, MA: MIT Press.

Nonini, D.M. ed. 2007. The Global Idea of the Commons. New York: Bergahn Books.

Norton, B.G. 2005. Sustainability: A Philosophy of Adaptive Ecosystem Management. Chicago: University of Chicago Press.

Puckett, A., Fine, E., Hufford, M., Taylor, B., and Kingsolver, A. 2012. Who knows? Who tells? Creating a knowledge commons. In Transforming Places: Lessons in Movement Building from Appalachia. Edited by S. Fisher and B.E. Smith. Urbana, IL: University of Illinois Press.

Reid, H.G. and Taylor, B. 2010. Recovering the Commons: Democracy, Place, and Global Justice. Urbana, IL: University of Illinois Press.

Rittel, H. and Webber, M. 1973. Dilemmas in a general theory of planning. Policy Sciences. 4:155–169.

Robertson, D.P. and Hull, R.B. 2003a. Biocultural ecology: exploring the social construction of the southern Appalachian ecosystem. Natural Areas Journal. 23(2):180–189.

Robertson, D.P. and Hull, R.B. 2003b. Public ecology: an environmental science and policy for global society. Environmental Science and Policy. 6:399–410.

Rodela, R. 2013. The social learning discourse: trends, themes, and interdisciplinary influences on current research. Environmental Science and Policy. 25:157–166.

Scott, J.C. 1998. Seeing Like A State: How Certain Schemes to Improve the Human Condition Have Failed. New Haven, CT: Yale University Press.

Shuman, A. 1983. Dismantling local culture. Western Folklore. 52(2-4): 345–364.

Soja, E. 2011. Postmodern Geographies: The Reassertion of Space in Critical Social Theory. New York: Verso Books.

Stivers, C. 2009. The ontology of public space: grounding governance in social reality. American Behavioral Scientist. 52:1095–1108.

Taylor, B. 2009. "Place" as prepolitical grounds of democracy: an Appalachian case study in class conflict, forest politics, and civic networks. American Behavioral Scientist. 52:826–845.

US EPA. 2004. Mountaintop Mining and Valley Fill: Programmatic Environmental Impact Assessment.

US EPA. 2011. Improving EPA Review of Appalachian Surface Coal Mining Operations Under the Clean Water Act, National Environmental Policy Act, and the Environmental Justice Executive Order. Memorandum to Regional Administrators.

Weber, E.P. 2000. A new vanguard for the environment: grass-roots ecosystem management as a new environmental movement. Society and Natural Resources 13:237–259.

Wills, K., Loucks, O., Kalisz, P., Fritsch, A., Rees, A., and Davidson, W. 1997. Patterns of forest health: a report on citizen monitoring in the eastern mountains 1994–1997. Washington, D.C.: Trees for the Planet. Downloaded 02-01-2013 from http://memory.loc.gov/ammem/collections/tending/title_textP.html

Young, K.G. 1987. Taleworlds and Storyrealms: The Phenomenology of Narrative. Dordrecht: Martinus Nijhoff Publishers.

Determinants of Local Community Acceptance of Energy Resources and Materials Mining

Sisi Que
Missouri University of Science & Technology, Rolla, Missouri

Kwame Awuah-Offei
Missouri University of Science & Technology, Rolla, Missouri

ABSTRACT

The significance of local community acceptance to the success of a mine cannot be over-stated. Also, community acceptance has implications for project sustainability, as the development and operation of a mine can be viewed as a development intervention with concrete social, economic and environmental impacts. However, very little has been done to understand these relationships in a way that is useful for planning and design of mining operations. Ongoing research at Missouri University of Science & Technology seeks to provide further insight on the determinants (demographics and mine characteristics) of local community acceptance. This paper presents results of a comprehensive literature review and proposes a research framework based on discrete choice theory. The literature review shows that mining has significant economic and social impacts in addition to the well-established environmental impacts. Some attempts have been made, particularly in Australia, to use choice experiments and multinomial logit (MNL) models to understand the determinants of community acceptance. However, these attempts do not account for the most significant determinants. Also, the MNL model is limited since it imposes independence of irrelevant alternatives property on all individual choice. The proposed approach will address these short comings and provide a means to incorporate this knowledge into predictive modeling approaches. This will allow engineers and regulators to better evaluate community acceptance and attendant effects on sustainability.

INTRODUCTION

Local community acceptance of natural resource projects (mining, hydraulic fracturing, drilling etc.) are important for successful permitting and operation of these projects. The community's acceptance or opposition is critical to obtaining permits prior mining. In fact, in some jurisdictions (e.g., Peru) community acceptance has been included as a requirement for permitting. In the United States (US), public participation is required through the environmental impact assessment, although, the community's acceptance is not necessarily a requirement for granting a permit. A good illustration of the importance of community acceptance during permitting is Newmont Mining's Minas Conga copper and gold mine in Peru, which was halted in 2012 due to opposition from local residents worried about water pollution (Anon. 2012). During mining, local community acceptance still plays an important role. Safe and smooth operation, devoid of protests (both at mine sites and annual general meetings) and vandalism, which can lead to denial of mine access and shut-downs, depend on community acceptance. Community acceptance also affects the sustainability impacts (perceived and real) of a mine through its socioeconomic implications. A mine, after all, is a development intervention with sustainable development implications.

Sustainable development is defined as the ability of current generations to meet their needs without compromising the ability of future generations to meet their own needs (Bundtland, 1987). It is well established that this includes, in equal parts, social, economic and environmental impacts (Munashinge and Shearer, 1995). Mining has social, economic, and environmental impacts—both positive and negative. There is an inter-relationship between the support, or lack thereof, from the local community and the socioeconomic impacts of mining. Community acceptance affects mining's social sustainability impacts such as human rights and governance (e.g., depending on whether there is prolonged community opposition and government intervention). The mining project produces social, economic, and environmental impacts as well as affects several demographic aspects of the geographic area. At the same time, the community acceptance and sustainability impacts affect the success of the project.

There has been very little research, beyond qualitative linkages, done to understand the determinants of community acceptance. Discrete choice theory, based on the Nobel winning work by McFadden (1973), has been successfully used in econometrics and other disciplines to understand consumer behavior, choice processes, among others. In energy, choice theory has been used to evaluate community acceptance of renewable energy projects (Dimitropoulos and Kontoleon, 2009).This provides a framework to objectively and quantitatively understand the characteristics of mining projects and demographic factors that are determinants of community acceptance.

The goal of current research at Missouri University of Science & Technology is to provide further insight on the determinants of local community acceptance using discrete choice theory. The objectives of this paper are to present: (i) results of a comprehensive literature review; and (ii) a proposed research framework. This is done by reviewing peer-reviewed journal papers to analyze relevant choice models. The results of the analysis are used to propose a framework for future research on community acceptance of mining.

This work will be a significant contribution to the literature on sustainable mining. The research will provide needed insight for mining professionals and government regulators to facilitate meaningful community consultation. This work provides the basis for planned future work that will result in a framework to incorporate community acceptance and sustainability considerations in mine planning and design.

DETERMINANTS OF COMMUNITY ACCEPTANCE OF MINING

The key factors that determine (determinants) the level of community acceptance (say percentage of the population in favor of the mine) of a mining project can be classified into two groups: (i) characteristics and impacts of the mine; and (ii) demographic factors. Selecting the key determinants from all possible explanatory variables can be a large undertaking. A possible approach is a two-step process: (i) select possible determinants through thorough literature review; then (ii) use a focus group(s) to select the key determinants (Hoyos 2010).

Sustainable Development Impacts of Mining

Building and operating a mine or an oil and gas production field is a developmental intervention with significant sustainable development implications. Over its life cycle (exploration, development, production, and closure), a mine will have significant but varying social, economic, and environmental impacts. Demographic Factors summarizes the sustainability impacts of mining. The environmental impacts are well known among the general populace, although they may be misinformed or under informed on some aspects. However, the social and economic impacts are not that obvious to many people, even if they know of these linkages. These impacts (real or perceived) have an impact on local community acceptance (Table 1).

Table 1. Sustainable development impacts of mining

Impact	Examples
Environmental	
Material and resource consumption	• US surface coal mining consumes 7.8 to 9.4 kg Sb-eq./tonne of coal (Ditsele and Awuah-Offei, 2012)
Air emissions	
Water pollution	• >11,500 miles of streams have been disturbed/impacted by coal mining (OSMRE 1978–2008) • >23,000 km of streams are affected by acid mine drainage from mines (Kim et al., 1982; USFS, 1993).
Land use impacts	• >1.5 million acres of land have been disturbed/impacted by coal mining (OSMRE 1978–2008)
Noise	
Social	
Human rights	
Working conditions	
Health and safety	• Migration into mining towns (often higher male migration) can stress local services and lead to health problems, such as HIV-AIDS (Hunter, 2007; Petkova et al., 2009) • Influx of itinerants lead to crime and impacts on community identity (Petkova et al., 2009)
Cultural heritage	
Governance	
Socio-economic repercussions	
Economic	
Job creation	• US mining supports nearly 1.98 million jobs (NMA 2012) • In rural economies, mining often creates more jobs than the population can take (Petkova et al., 2009)
Taxes	• US mining generated $50 billion in federal, state and local taxes in 2010 (NMA 2012)
Downstream effects	• Petkova et al. (2009) reports that nearly all business respondents in a survey administered (in rural Australia) reported a generally positive impact on their business since the coal mining boom.

Demographic Factors

Various demographic factors have been shown to affect an individual or a household's likelihood to support (proposed) mining in their community. For instance, Ivanova et al. (2011) shows that gender, number of children, and age significantly ($p < 0.05$) affect community acceptance. Other factors include household income (Petkova et al., 2009), buffer from the mine (del Rio and Burguillo, 2008; Ivanova et al., 2011) and perceived or actual net gain from the project (del Rio and Burguillo, 2008).

It has long been posited that one of the key motivations for human decisions is utility maximization (say in consumer behavior). We postulate that utility maximization is the framework to understanding individual/household support, or lack thereof, of mining. Any demographic factor, which significantly affects the utility function, will be a determinant of community support.

Discussion

It is clear from the work cited in this section that: (i) mining (and other energy production activities) contributes to sustainable development (positively and negatively); and (ii) demographic factors do affect community acceptance. The linkage between environmental impacts and community acceptance has long been accepted. The example of the Minas Congas project, previously cited, is one example among many (Anon. 2012). However, the connection between the social and economic impacts have not been discussed that much. Yet, the literature supports this connection too. For example, Petkova et al. (2009) shows that employment opportunities, increase in local business opportunity, the influx of itinerants, and community identity were all important to local residents in determining community acceptance. We, therefore, draw the motivation from the literature for one of the key hypothesis of this research: *project sustainability is a key determinant of community acceptance.*

This hypothesis needs to be rigorously tested with further research. The other question that arises from this is: which of the sustainability impacts of a mining project are critical determinants of community acceptance? Are impacts on employment more important than environmental impacts? Are contributions to the local tax base and economic development more important than the social ills associated with migration? These questions need good objective answers that only further research, using decision theory, can provide.

DISCRETE CHOICE THEORY AND MODELING

Discrete choice analysis can be employed to describe the influence of decision makers' characteristics and attributes of alternatives on choices. The most popular discrete choice models are multinomial logit (MNL), multinomial probit (MNP), nested logit (NL), and mixed multinomial logit (ML) models. While NL is widely used in other fields, we do not discuss it in this paper since the mining acceptance problem is not a "nested" problem. Discrete choice theory and the other three models are discussed separately in the following.

Discrete Choice Theory

The basic theory of discrete choice modeling is random utility maximization (Marschak 1959). The individual decision maker's overall preference of a choice alternative is a function of the utility, which the alternative holds for the individual. This individual's utility (U_{ni}) for an alternative is separable into two components: (i) the component, which can be explained by the observed (by a researcher) variables; and (ii) the component, which can be explained by unobserved variables—often, deemed random.

$$U_{ni} = V_{ni} + \varepsilon_{ni} \tag{1}$$

U_{ni} = utility of alternative i to individual n
V_{ni} = observed component measured for alternative i of individual n
ε_{ni} = unobserved random component

It is postulated that an individual will prefer the choice alternative perceived to have the greatest utility. If individual n prefers the mining project or plan i of choice set J, the probability is:

$$P_{ni} = Prob\begin{pmatrix} U_{ni} \geq U_{nj}, \\ \forall\ i \neq j, \\ i\ and\ j\ \in J \end{pmatrix}$$

$$= Prob\begin{pmatrix} V_{ni} + \varepsilon_{ni} \geq V_{nj} + \varepsilon_{nj}, \\ \forall\ i \neq j, \\ i\ and\ j\ \in J \end{pmatrix}$$

$$= Prob\begin{pmatrix} \varepsilon_{ni} - \varepsilon_{nj} \geq V_{nj} - V_{ni}, \\ \forall\ i \neq j, \\ i\ and\ j\ \in J \end{pmatrix} \tag{2}$$

Multinomial Logit Model

The multinomial logit (MNL) model was first formulated by McFadden in the 1970s (McFadden 1974). The observed component V_{ni} is a linear function of X_{ni} (a vector of observed variables that relate to the alternative and/or the decision maker) and the random component (ε). The error terms, ε, are assumed to be independently and identically distributed (IID) with type 1 extreme value distribution.

$$U_{ni} = V_{ni} + \varepsilon_{ni} = \beta'_n\ X_{ni} + \varepsilon_{ni} \tag{3}$$

β_n = vector of coefficients of variables for person n representing that person's taste.

The cumulative distribution function (cdf) of ε is given by Equation 4.

$$F(\varepsilon_{ni}) = e^{-e^{-\varepsilon_{ni}}} \quad (4)$$

The probability of choice i to individual n is:

$$P_{ni} = \frac{\exp(V_{ni})}{\sum_{j=1}^{J} \exp(V_{nj})} \quad (5)$$

The multinomial logit (MNL) model is widely used in different fields (Cho, 1997; Walekhwa et al., 2009; Kwak and Clayton-Matthews, 2002). However, the MNL model has three main shortcomings. Firstly, the coefficient vector β_n is fixed in an MNL model. It means that different people with the same surveyed characteristics will make the same choice given the same choice set. Actually, even twins might make different choices. Thus, the fixed coefficient β_n is not reasonable.

Secondly, the MNL model has an independence of irrelevant alternatives (IIA) property, since the error terms, ε_n, are assumed to be independently and identically distributed (Train 2002). The probability ratio of alternatives i and k only depend on i and k in the MNL model and it is not relative to other alternatives (Equation 6).

$$\frac{P_{ni}}{P_{nk}} = \frac{\exp(V_{ni})/\sum_{j=1}^{J} \exp(V_{nj})}{\exp(V_{nk})/\sum_{j=1}^{J} \exp(V_{nj})} = \exp(V_{ni} - V_{nk}) \quad (6)$$

The IIA property means that there is no cross elasticity among the alternatives. If an attribute of one alternative j is changed, the changes of other alternatives' probabilities are not relative to the changed alternative j. Yet, this is not true in some choice situations. For example, assume there are three kinds of vehicles in a market: large gasoline cars, small gasoline cars, and small electric cars. Their current market shares are 66%, 33% and 1%, respectively. Also, assume that a government subsidy increases the market share of the small electric car from 1% to 10%. Using the MNL model, the market share of the other two cars would be predicted to drop while still maintaining the same ratio. The market share of large gasoline cars would drop from 66% to 60%, and that of small gasoline cars would drop from 33% to 30% (maintaining the 2:1 ratio). The ratio of the market share of these two vehicles have to be 2:1 since their utility rate is 2:1, and is not relative to any other alternative. However, this prediction is unrealistic. Since the electric car is small, subsidizing it can be expected to draw more from small gas cars than from large gas cars.

Thirdly, the MNL model has the potential to capture dynamics of repeated choice. But the repeated choice has to be independent over time since the error terms, ε_n, are assumed to be independently and identically distributed (IID) in the MNL model. Thus, the MNL model cannot handle repeated choice situations if they are correlated over time.

Multinomial Probit Model

The first binary probit model was derived by Thurstone (1927). Hausman and Wise (1978) and Daganzo (1979) employed and developed it for choice behavior. The utility equation is same as the MNL model (Equation 3), but the ε_{ni} are assumed to be normally distributed with mean 0 and covariance matrix Ω.

The probability density function (pdf) of ε_n is:

$$\Phi(\varepsilon_n) = \frac{1}{(2\pi)^{\frac{1}{2}} |\Omega|^{\frac{1}{2}}} \exp\left(-\frac{1}{2}\varepsilon'\Omega^{-1}\varepsilon_n\right) \quad (7)$$

ε_n'is a vector of ε_{ni}, and $\varepsilon_n' = (\varepsilon_{n1}, \varepsilon_{n2,\ldots,} \varepsilon_{nJ})$

For individual n, probability of choice i is:

$$P_{ni} = Prob\begin{pmatrix} \varepsilon_{ni} - \varepsilon_{nj} \geq V_{nj} - V_{ni}, \\ \forall\, i \neq j \end{pmatrix}$$

$$= \int I\begin{pmatrix} \varepsilon_{ni} - \varepsilon_{nj} \geq V_{nj} - V_{ni}, \\ \forall\, i \neq j \end{pmatrix} \Phi(\varepsilon_n)\, d\varepsilon_n \quad (8)$$

I(•) is an indicator function, it equals 1 when the expression inside the parenthesis is real and 0 otherwise.

Compelling progress was made by Haaijer et al. (1998). They accounted for random variation in the coefficients β over decision-makers, instead of being fixed as before. The coefficients β were assumed to be normally distributed in the population with mean b and covariance W. And the parameters β and W can be estimated by the MNP model.

Thus, the three limitations of the MNL model are all relaxed in the MNP model. Firstly, two people who have the same surveyed characteristics can make different choices since the coefficients are not fixed (they are normally distributed with covariance W). Secondly, MNP does not have the IIA property and represents any substitution pattern, because ε_n are assumed to be normally distributed with mean 0 and covariance matrix Ω. In the previous example, the large gasoline and small gasoline cars' market share would not have to maintain the 2:1 ratio after the small electric car's market share changed. Their market share will be relative to the change in small electric car's market share. Finally, the MNP model can handle repeated choice situations where choices are correlated over time by expanding the covariance matrix Ω of the errors ε_n. The details are not explained here since the current research does not include dynamic choice modeling of mining local community acceptance.

Mixed Multinomial Logit Model

The mixed, multinomial logit (ML) model was proposed by McFadden and Train (2000). In the ML model, the distribution of coefficents, $f(\beta)$, is not limited to the normal distribution like in the MNP model. The ML model can utilize any distribution for the random coefficients. The most popular distributions of the random parameters are uniform, triangular, normal and lognormal distributions.

Mixed logit probabilities are the integral of standard logit probabilities over the coefficents-distribution function, $f(\beta)$.

$$P_{nj} = \int L_{nj}(\beta) f(\beta) d\beta \tag{9}$$

$L_{ni}(\beta)$ is the logit probability evaluated at parameters β

$$L_{nj}(\beta) = \frac{\exp[V_{nj}(\beta)]}{\sum_{j=1}^{J} \exp[V_{nj}(\beta)]} \tag{10}$$

Then,

$$P_{nj} = \int \frac{\exp[V_{nj}(\beta)]}{\sum_{j=1}^{J} \exp[V_{nj}(\beta)]} f(\beta) d\beta \tag{11}$$

McFadden and Train (2000) show that any choice model can be approximated by the ML model with appropriate specification of the observed variables and distribution of coefficients. The MNP is a special case of the ML model where the coefficent distribution function, $f(\beta)$, is a normal distribution.

Discussion

The multinomial logit (MNL) model has been used to understand the decision-making process of local communities regarding preferred mineral project development choice (Ivanova, et al., 2007; Ivanova and Rolfe, 2011). However, their work did not aim to determine the effect of sustainability on community acceptance, which affected the choice of mining attributes included in the model. Also, they tracked very few attributes of mining (as indicated by the high alternative specific constant). The ML model is more suitable for mining local acceptance since it is not limited by fixed taste coefficients β_n and the IIA property.

Fixed taste coefficient is unsuitable for mining community acceptance modeling. In mining local community acceptance modeling, it is not possible to include all demographic factors (due to survey and data limitations). This means people/households with the same surveyed demographic factors will have the same characteristics. However, it is not reasonable to expect that all individuals/households with the same surveyed characteristics will make the same choice (they are still different in other ways). Thus, we have to appreciate the effect of unobserved variables. And

Table 2. Sample choice set

Choice Options	Acres of Wetlands Impacted by Mine	New Jobs Created
Option 1	0	500
Option 2	1,000	500
Option 3	1,000	1,000

the model needs to allow preferences to vary with the unobserved variables and randomly.

Additionally, the IIA property is not true for some mining local acceptance choice situations. Consider the choice set presented in Table 2, for example. Assume that 66%, 33% and 1%, respectively, choose options 1, 2, and 3. Assume also that after public education, more people have been convinced of the ability of the mining company to implement the proposed 3:1 wetland compensation plan (i.e., 3 acres of wetlands will be built elsewhere for every acre of wetlands impacted), leading to an increase in the percentage of people in favor of Option 3 to 10%. The MNL model (with the IIA property) will predict proportional decreases in Options 1 and 2 to 60% and 30%, respectively (same as the previous example). However, it is likely that more of those in favor of Option 1 (those influenced mostly by wetland concerns) will change their mind with this change than those who were in favor of Option 2 (wetlands were not an issue for those).

These two constraints (constant β_n and IIA property) are relaxed in the mix multinomial logit model. Also, the ML model allows each random coefficient to follow any distribution (instead of being restricted to the normal distribution, as in the MNP model), which should be realistic in a complex situation such as mining community acceptance. Additionally, the ML model has the potential to handle repeated choice situations, even if the repeated choice is correlated over time. The ML model will still be suitable for dynamic mining local acceptance model if we choose to do it later. Thus, we posit that the mix multinomial logit model should be the most appropriate one for mining local acceptance model.

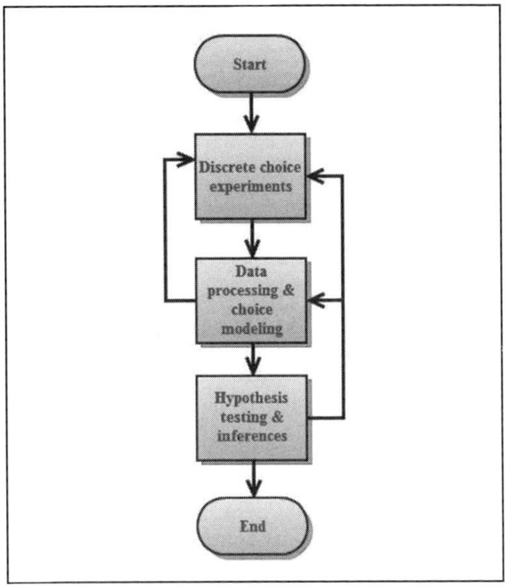

Figure 1. Proposed research framework

PROPOSED FRAMEWORK FOR FUTURE RESEARCH

The proposed framework for this work is shown in Figure 1. The approach is to design and administer discrete choice experiments to representative samples in various communities to acquire data for choice modeling. The data from these choice experiments will then be used to determine the mixed multinomial logit model parameters. The modeling results will be used then to test the research hypotheses. Since the surveys (choice experiments) will be done in the selected communities sequentially, feedback from the latter stages (choice modeling and hypothesis testing) will be incorporated into the earlier stages (Figure 1).

The first task (beyond literature review) is to conduct choice experiments, which will provide the data for discrete choice modeling. The research team will use established best practice to design and administer the questionnaires (Louviere et al., 2000; Hoyos, 2010). In order to test the research hypothesis (*project sustainability is a key determinant of community acceptance*), the set of variables included in the choice alternatives

Table 3. Mine development characteristics and levels used by Ivanova, et al. (2011)

Variable	Levels
Additional annual costs to household	1. $0 2. $250 3. $500 4. $1,000
Housing and rental prices	1. 25% increase 2. No change 3. 25% decrease
Level of water restrictions	1. Some for households, town parks and gardens are drier than now 2. None for households, town parks and gardens are drier than now 3. None for households, townparks and gardens are greener than now
Buffer for mine impacts close to town	1. Moderate impacts from noise, vibration and dust 2. Slight impacts from noise, vibration and dust 3. No additional impacts
Population in work camps	1. No more housing and 5,000 in work camps 2. 1,000 in housing and 4,000 in work camps 3. 4,000 in housing and 1,000 in work camps

need to include mining sustainability impacts (e.g., Table 1).

The challenge is to limit the number of variables to only the most important variables (Table 3 shows the variables used by Ivanova, et al. (2011)). The literature shows that the cognitive burden (the cognitive requirements of the choice decision) increases with the complexity of the choice decision (Caussade et al., 2005; Hoyos, 2010). This can lead to a gap between the cognitive ability of respondents and the cognitive burden of the decision they are asked to make. This can be avoided by using focus groups and pilot surveys (Hoyos 2010). The second challenge in the choice experiment design is determining the appropriate levels, which the included variables can be set, to generate the choice alternatives (Table 3). Expert opinion will be sought to determine appropriate levels. These will also be tested in the focus groups and pilot surveys (Louviere et al., 2000).

The second task is data processing and model fitting. Although, we postulate that the mixed multinomial logit model is best suited for the mining community acceptance problem, we intend to fit the MNL and MNP models to the data as well. Best practices for model goodness-of-fit will be followed (Louviere et al., 2000; Hoyos, 2010). The goodness of fit of estimated choice models is usually measured by the likelihood ratio index (LRI) or pseudo-rho squared. The LRI is zero if the model is no better than no model and 1 if it perfectly predicts each choice. A rule of thumb is to accept models with LRI greater than 0.2 (Hoyos, 2010).

Testing the research hypothesis is the final part of this research. The goodness-of-fit tests used for choice model fitting provide tests of statistical significance. The results provide the coefficients and the significance level of each factor (variable) in the model. To test the hypothesis that mining project sustainability is a key determinant, we will examine the significance (say, at $\alpha = 0.05$) of these variables in the model that are related to sustainability. By the same logic, we would evaluate the significance of demographic factors to test the hypothesis that demography is a key determinant of community acceptance.

Sustainability assessment and evaluation has evolved over the last decade. The practice has evolved from a top-down one to a bottom-up one. Community consultation is now seen as a crucial part of sustainability assessment. However, in most cases, the community is presented with a project to accept or reject. The community consultation process often leads to changes in the mine (project) plan. However, the changes can sometimes ignore major concerns and do not address "willingness-to-pay," a key criterion for consumer behavior, for instance. What is the community willing to give up for their preferred mine (project) plan? This approach requires a comprehensive discussion of sustainability (environmental, social, and economic impacts) as well as demographics. Discrete choice theory provides the framework to handle all these issues to facilitate useful and meaningful community consultation.

The current research effort will provide insights into these critical issues. These insights will be beneficial for both regulators and mining professionals. The proposed research framework will provide objective assessment of the factors affecting local community acceptance of resource (energy) projects. Further research, beyond what is proposed in this paper, will develop a framework that helps mine planning engineers to design mines in the context of sustainability (including the effect of community acceptance). The ultimate goal is to facilitate design in the context of sustainability (Fletcher and Dewberry, 2002).

CONCLUSIONS

Community acceptance is important for mining (energy) project sustainability and success. The literature shows a clear interaction between community acceptance and sustainability. Measurable demographic factors have been shown to affect community acceptance. Research using choice theory to understand mining community acceptance is minimal: only two examples (in the same communities by the same researchers) were found in this literature survey (Ivanova, et al., 2007 and 2011). These examples appear not to have tracked the most significant attributes of mining development options (high alternative specific constant or low adjusted R^2). This review also shows the multinomial logit model is not the most appropriate for mining community acceptance.

The current work proposes a framework based on choice experiments and mixed multinomial logit modeling. The work will include significant sustainability (environmental, social, and economic) attributes of mining in choice experiments to be administered in several communities. The authors propose to use focus groups and expert opinion to limit cognitive burden in the choice experiments, while including the most significant attributes and demographic factors. The results of the modeling will be used to test the hypothesis that project sustainability is a key determinant of community acceptance.

This work will provide significant insights for regulators and mining professionals and transform current community consultation practices. The authors recommend that future research (beyond what is covered in this paper) needs to concentrate on how to use results of choice experiments and modeling to make a shift from the current mine planning and design framework to one where sustainability is the context for the design.

REFERENCES

Anon. (2012), "Protests Shut Down Conga Project in Peru," Engineering & Mining Journal, Vol. 213(1), pp. 10.

Bundtland G, (1987), Our common future: the world commission on environment and development. Oxford: Oxford University Press.

Cho, C.-J., (1997). Joint Choice of Tenure and Dwelling Type: A Multinomial Logit Analysis for the City of Chongju. Urban Studies, 24(9), pp. 1459–1473. Available at: http://usj.sagepub.com/cgi/doi/10.1080/0042098975510.

Dimitropoulos, A., and Kontoleon, A. (2009). Assessing the determinants of local acceptability of wind-farm investment: A choice experiment in the Greek Aegean Islands. Energy Policy, 1842–1854.

Ditsele, O. and Awuah-Offei, K. (2012), "Effect of Mine Characteristics on Life Cycle Impacts of US Surface Coal Mining," Int. Journal of Life Cycle Assessment, Vol. 17(3), Ecomed Publishers, Landsberg, Germany, pp. 287–294, DOI: 10.1007/s11367-011-0354-y http://www.springer.com/alert/urltracking.do?id=Lb6af b6M97db76Sab094e3.

Hoyos, David (2010), "The state of the art of environmental valuation with discrete choice experiments," Ecological Economics, Vol. 69, pp. 1595–1603.

Hunter, M. (2007), "The Changing Political Economy of Sex in South Africa: The Significance of Unemployment and Inequalities to the Scale of the AIDS Pandemic," Social Science & Medicine, Vol. 64, pp. 689–700, doi:10.1016/j.socscimed.2006.09.015.

Ivanova, G. & Rolfe, J., 2011. Assessing development options in mining communities using stated preference techniques. Resources Policy, 36(3), pp. 255–264. Available at: http://linkinghub.elsevier.com/retrieve/pii/S0301420711000195 [Accessed November 15, 2012].

Kate Fletcher, Emma Dewberry, (2002) "Demi: a case study in design for sustainability," International Journal of Sustainability in Higher Education, Vol. 3 Iss: 1, pp. 38–47.

Kim, A.G., Heisey, B., Kleinmann, R. and Duel, M. (1982), Acid Mine Drainage: Control and Abatement Research, U.S. DOI, Bureau of Mines IC 8905.

Kwak, C. & Clayton-Matthews, A., 2002. Multinomial Logistic Regression. Nursing Research, 51(6), pp. 404–410. Available at: http://content.wkhealth.com/linkback/openurl?sid=WKPTLP:landingpage&an=00006199-200211000-00009.

Louviere, J.J., Hensher, D.A., Swait, J.D., 2000. Stated Choice Methods: Analysis and Applications. Cambridge University Press, New York.

Marschak, J., 1959. Binary choice contraints on random utility indications. Mathematical methods in the social sciences, Stanford Univeristy press, pp. 312–329.

McFadden, D. (1973). Conditional Logit Analysis of Qualitative Choice Behavior. In P. Zarembka, Frontiers in Econometrics (pp. 105–142). New York: Academic Press.

McFadden, D., 1974. Coditional logit analysis of qualitative choice behavior. In Frontiers in Econometrics. New York: Academic Press, pp. 105–142.

Mcfadden, D. & Train, K., 2000. Mixed MNL models for discrete response. Journal of applied econometrics, 15(5), pp. 447–470.

Munashinge, M. and Shearer, W. (1995), Defining and measuring sustainability. Washington: The World Bank.

National Mining Association (NMA) (2012), The Economic Contributions of U.S. Mining in 2010, Report by NMA, Available at http://nma.org/pdf/economic_contributions.pdf (Accessed Nov. 29, 2012).

Office of Surface Mining, Reclamation & Enforcement (1978–2008), Surface Mining Control and Reclamation Act of 1977 Annual Reports, http://www.osmre.gov/Reports/AnnualReport/AnnualReport.shtm (Accessed Nov. 29, 2012).

Train, K., 2002. Discrete Choice Methods with Simulation, Cambridge University.

US Forest Service (1993), Acid Mine Drainage From Mines on the National Forests, A Management Challenge, Program Aid 1505.

Walekhwa, P.N., Mugisha, J. & Drake, L., 2009. Biogas energy from family-sized digesters in Uganda: Critical factors and policy implications. *Energy Policy*, 37(7), pp. 2754–2762. Available at: http://linkinghub.elsevier.com/retrieve/pii/S030142150900161X [Accessed November 15, 2012].

Ethics and Community Well-Being

Peter H. Denton
Red River College, Winnipeg, Manitoba

ABSTRACT

There are metrics of "community well-being," whether measured in terms of economic externalities like GDP/GNP or by other indicators like the Gross Happiness Index. The World Happiness Report, presented to the UN Conference on Happiness on April 2, 2012, explored difficulties with such metrics at the same time as it emphasized the importance of "well-being" for global sustainability. This paper argues, however, that such metrics reflect the problems of unsustainable systems and not their resolution. Sustainability at all levels is to be found through a better understanding of the relationship between ethics and technology.

3.1
Coal, Gas, and Energy Production

Appalachian Coal: Past, Present, and Future 124

Environmental Considerations and Operational Parameters of Underground Coal Gasification 125

The Marcellus Shale Gas Play and the Gas Migration/Stray Gas Problem 137

Experiences in Reservoir Model Calibration for Coal Bed Methane Production in Deep Coal Seams in Russell County, Virginia 140

Sustainable Degasification Strategies: A Case Study in the Qinshui Basin of China 153

Appalachian Coal: Past, Present, and Future

J. Steven Gardner
ECSI, LLC, Lexington, Kentucky

ABSTRACT

Coal in Appalachia has gone through dramatic changes in the last few years and is likely to undergo even more changes in the future. The industry has been under an unprecedented, sustained attack by Federal agencies and NGOs on many fronts and has been affected by weak commercial markets and a changing energy landscape. As litigation issues are likely to continue and political positions become solidified, it is important to ask what is next for the industry. This presentation will take a brief look at the industry's past, before shifting focus to the current state of the coal industry, including the regulatory environment and the broader energy industry. It will also look ahead at new coal market opportunities that are developing domestically and abroad and at innovative operating approaches that may help the industry survive current challenges.

Environmental Considerations and Operational Parameters of Underground Coal Gasification

Zeshan Hyder
Department of Mining and Minerals Engineering, Virginia Tech, Blacksburg, Virginia

Michael Karmis
Virginia Center for Coal and Energy Research, Virginia Tech, Blacksburg, Virginia

ABSTRACT

Underground coal gasification (UCG) is a technology that provides an economic and environment friendly alternative for exploitation of coal reserves. It provides an excellent opportunity to utilize low grade, steeply dipping and deep-seated coal reserves through in situ gasification. Several factors make this method environmentally and economically superior compared to other coal exploitation methods. These factors include elimination of conventional mining, coal cleaning and transportation to and from the surface, reduced surface and water utilization, minimal waste production and waste handling infrastructure, and utilization of coalbed methane in the gasification process.

Despite economic, environmental and technical benefits, if not properly managed, this technology has the potential to create hazards of atmospheric emissions, ground water contamination, uncontrollable cavity growth and underground fires, subsidence, CO_2 management and human impacts like noise, dust, and increased traffic. Thus, in order to manipulate the full economic, environmental and technical potential of this technology, complete understanding and proper control over the environmental and operational parameters of UCG is imperative. This paper highlights some of the environmental aspects and operational parameters of UCG that are very important in realization of the full potential of this technology for exploitation of coal reserves in an environmentally friendly, technically sound and economic manner.

INTRODUCTION

Underground coal gasification (UCG) is a process involving in situ burning and conversion of coal into a gaseous product, commonly known as synthetic gas or syngas. This syngas is at a high temperature and pressure and can be used for several processes including but not limited to electricity, heat and power generation, feedstock for several chemical processes such as Fischer-Tropsch, synthetic natural gas and hydrogen production, iron reduction and chemical products like ethylene, polyolefin, methanol, petrol, acetic acid and formaldehyde (Liu et al. 2003; Courtney 2009; Oukaci 2009). In its simplest form, the process involves two wells, an injection and production well, drilled up to the coal seam. Then steam, air or oxygen is injected under pressure through the injection well to start the combustion process, and gases having high temperature and pressure are produced and collected at the production well.

The composition of syngas is largely H_2, CH_4, CO and CO_2, with a calorific value of 850–1,200 kcal/Nm^3 (~3.5–5 MJ/m^3) (Ghose and Paul 2007). However, the composition and calorific value of syngas varies according to site characteristics (coal ranks, strata type, depth and amount of moisture, etc.) and type of gases used for injection (air, oxygen or steam). The calorific value for air injected syngas typically ranges from

4.0–5.5 MJ/m^3 and approximately doubles when oxygen is injected instead of air (Walker 1999).

RENEWED INTEREST IN UCG

UCG is not a new concept and can be traced back to as early as 1868, when Sir William Siemens, a German scientist first presented the idea of burning ash-rich coal underground before the Fellows of the Chemical Society of London (Olness and Gregg 1977). The famous Russian scientist, Dmitry Mendeleyev mentioned the same idea some twenty years later in 1888, when he remarked that in the future coal would be burnt underground and supplied as gas via pipes for consumption. Mendeleyev exclusively suggested his depiction of UCG during a visit to Kizel plant in Ural Russia in 1899, where he espoused the idea of controlling the underground coal fires by drilling of injection and production wells, with a limited supply of air through one pipe and withdrawing syngas by suction though another pipe (Klimenko 2009). In the U.S., research and development about UCG started in the 1940s. The program accelerated in the 1970s due to the energy crisis, and more than 30 experiments were conducted between 1972 and 1989 (PWC 2008). However, groundwater contamination resulting from these pilot tests and reduced natural gas and oil prices in the 1990s hampered further development and commercialization of UCG projects in the U.S. (PWC 2008).

UCG is now attracting considerable global attention as a viable process to provide a "clean" and economic fuel from coal. This technology has the potential to exploit energy from low grade, deep seated, thin coal seams in an economic, environmentally friendly and sustainable manner. This technology can be applied to abandoned coal mines, remnants of already exploited reserves and deposits considered uneconomic and technically difficult for conventional mining methods.

The renewed interest in commercial development of UCG technology can be attributed to the following factors:

- the stringent environment and safety regulations specific to coal mining
- the unstable and politically motivated increases in petroleum prices
- an increasing negative public perception about coal mining
- rapidly increasing mining costs
- a decrease in economically minable resources
- the environmental and economic potential of UCG (Meany and Maynard 2009)
- the ability to recover energy from previously mined coal seams (Lamb 1977)
- the applicability of UCG to low heating value and deep and relatively thin coal seams, considered economically unminable through conventional mining methods (Lamb 1977)
- the lower capital cost due to elimination of coal mining, transportation and ash management facilities (Burton et al. 2006)

In addition to the aforementioned factors, another factor favoring the promotion of UCG is its ability to be used for various purposes. The product of UCG is a gas at high temperature and pressure. This is a combustible gas and can be utilized for electricity generation in a combined cycle gas turbine after minimal processing. After further processing and refining, it can be used for production of a wide variety of gases, liquid fuels and chemicals (PWC 2008). Following is the list of possible syngas uses:

- Electric power
- Hydrogen
- Ammonia
  - Urea
  - Fertilizers
- Substitute natural gas
- Methanol
  - Dimethyl Ether
  - Acetic acid
  - Formaldehyde
  - Gasoline
  - Olefins
- Other Chemicals
- Fischer-Tropsch liquids
- Gasoline
  - Diesel

- Jet fuel
- Liquefied Petroleum Gas (LPG)
- Waxes
- Lube oils
- Higher alcohols
- Chemicals
- Naphtha
- Petrochemicals
- Solvents

A typical procedure includes gas cleaning before its application for other uses. Generally, the gas cleaning plant is proposed to be onsite to improve economic viability of the project.

Despite economic, environmental and technical benefits, if not properly managed, this technology has the potential to create hazards of atmospheric emissions, ground water contamination, uncontrollable cavity growth and underground fires, subsidence, CO_2 management and human impacts like noise, dust, and increased traffic. Thus, in order to extract the full economic and environmental potential of this technology, complete understanding and proper control over the environmental and operational parameters of UCG is imperative. This paper highlights some of the environmental aspects and operational parameters of UCG that are very important in the realization of the full potential of this technology for exploitation of coal reserves in an environmentally friendly and economic manner.

ENVIRONMENTAL CONSIDERATIONS OF UCG

UCG has several economic and environmental benefits when compared to conventional coal mining, surface gasification processes and even coalbed methane drainage procedures (Meany and Maynard 2009). In the gasification process ash and heavy minerals remain underground and do not report to the surface (Fergusson 2009), thus resulting in decreased waste management cost and related infrastructure. Sulfur and nitrogen come out to the surface with product gas but conventional sour gas cleaning technologies can remove sulfur from the gas (Fergusson 2009). As reported by Meany and Maynard, ash content of syngas is one-seventh (1/7th) that of coal burnt on the surface, and various concentrations of toxic nitrogen oxides (NO_x) and sulfur dioxide (SO_2) are reduced by UCG (Meany and Maynard 2009).

Due to the elimination of mining, there is no overburden dump, no spoil dumps and no ash or tailing dams as most of the waste remain underground (Meany and Maynard 2009). The elimination of conventional mining significantly reduces the environmental problems associated with dirt handling and disposal, coal washing and fines disposal, coal stocking and transportation, thereby resulting in a smaller surface footprint (Creedy et al. 2001). In addition, Creedy et al. (2001) have highlighted following environmental benefits of UCG over coal mining:

- Reduction in dust, noise and visual impact on the surface
- Lower water consumption
- Lower risk of surface water pollution
- Reduced methane emission
- Elimination of dirt handling and disposal at mine
- Elimination of coal washing and fines disposal at mine site
- Elimination of ash handling and disposal at power station sites
- Elimination of coal stocking and transport
- Smaller surface footprint
- Elimination of mine waste recovery and significant surface hazard liabilities on abandonment

Similarly, the process creates minimal surface disruption, thus resulting in reduced land acquisition and rehabilitation requirements (Ghose and Paul 2007). Another important aspect of UCG is its synergy with carbon capture and sequestration (CCS). CO_2 is the main component of UCG syn gas and may be present in the range of 25–40%. Integration of UCG and CCS may result in a critical climate change mitigation technology capable of producing power from coal, and many studies suggest it as a low cost, above ground, low carbon form of coal power production (Redman et al. 2009). The technology to capture CO_2 from the syngas and store it in geological formations is readily available; however, research is in progress to store CO_2 in the voids and cavities created by UCG processes (Friedmann 2009; Roddy and Younger 2010).

The gasification process requires a certain amount of water to facilitate the chemical reaction (Ag Mohamed et al. 2011), which results in minimal mine water recovery. During the burning process, UCG not only consumes the coal in the strata but also the entrapped coalbed methane present in the strata. This gives an added advantage to UCG over other coal exploitation methods, where entrapped methane has to be drained either through a ventilation system or through venting into the atmosphere (EPA 2010). As reported by EPA, the methane emissions from natural gas systems were 6.6 Tg in 2010, and for coal mines the amount was 4.9 Tg (EPA. 1999a).

However, like all other technologies, UCG possesses some environmental risks, if the operations are not managed adequately. Major environmental concerns of this technology are ground water contamination and surface subsidence. UCG can contaminate nearby aquifers through gas leakage and leachate (Sury et al. 2004). The burning process results in organic pollutants like phenols, benzene, polycyclic aromatic hydrocarbons, and heterocyclic compounds and inorganic compounds like ammonia, cyanides, sulfates, and heavy metals (Humenick and Mattox 1978; Kapusta and Stanczyk 2011). The amount and nature of contaminants depends upon the type and rank of coal, e.g., hard coal, subbituminous or lignite. (Kapusta and Stanczyk 2011). These pollutants can move out of the cavity either in the aqueous phase or along with gas leakage. In order to avoid flow of contaminations from the cavity to the underground water table and to minimize the loss of product gases, the pressure of the burn cavity must be maintained below hydrostatic pressure, to ensure a small and continuous influx of water into the cavity to aid the burning process and to act as a barrier to contaminants outflow (Shafirovich and Varma 2009).

The cavities created by UCG resemble long wall panels and result in the unsupported rocks and strata overlying the cavity. This unsupported mass will gradually settle or subside and the effect can reach the surface depending upon the size of cavity, type of strata, depth of the coalbed and strength of surrounding rocks. The impacts of subsidence include damage to surface structures and facilities like roads, pipes and buildings, loss of agricultural land through formation of surface fissures, changes in ground slope and surface drainage and hydrological changes including changes in water quantity, quality and flow paths (Blodgett and Kuipers 2002). UCG-induced subsidence is expected to progress depending on the geometry of the cavity and depth—the greater the depth, the smaller the chances of subsidence, depending upon the mechanical properties of the rock and stress regime in the area (Creedy et al. 2001). However, most experimental work conducted for UCG did not report any significant surface subsidence, possibly because of the small size of active operations. For commercial large-scale projects, however, more focused research is required for assessing, managing and reducing the subsidence impacts of UCG (Friedmann 2009).

The syngas produced by UCG contains a component of vaporized or produced water that may contain residual hydrocarbons, benzenes, phenols and polycyclic hydrocarbons (Moorhouse et al. 2010). These water vapors need to be removed before combusting the gas in a power plant. If mixed with surface water streams and channels, this water has the potential to contaminate them. These water vapors are, however, fully treatable; industries have been treating these products for about 60 years (Moorhouse et al. 2010).

The atmospheric emissions from the UCG process include emissions during the process and emissions during the transport and use of the syngas. Combustion of product gas and transport to other location produces harmful pollutants; however, the actual UCG process itself does not contribute criteria pollutants to the atmosphere (Ag Mohamed et al. 2011). The main emissions from UCG include CH_4, CO_2, CO, H_2, S, organic N_2, H_2S and NH_3; however, the pollutants can be separated from the product gas using proven technologies like cyclones, bag-house filters and electrostatic precipitators (Creedy et al. 2001; Ray et al. 2010).

Potential environmental hazards of UCG can be mitigated effectively through careful site selection, appropriate operational controls, proper shut down process and effective environment monitoring (Sury et al. 2004). It is worth noting that only two of over 30 UCG trials in the U.S. resulted in clear evidence of environmental contamination (Burton et al. 2008).

OPERATIONAL PARAMETERS OF UCG

In addition to the environmental considerations of UCG, it is very important to have full understanding of and proper control over the operational and engineering parameters of this technology. The operational parameters such as start of burn, cavity development, pressure, volume, velocity and amount of gasifying agents, number of wells and type and layout of instruments and instrumentation wells are some of the factors that are essential in controlling the production of syngas, its quantity, quality, and the environmental impacts of this technology. Some of the important operational parameters of UCG are discussed in below to communicate their importance in the process and the understanding of their control.

Ignition

Ignition of coal seam generally starts after the wells have been drilled and the gasifier is ready to operate. Several methods can be used for the in situ ignition of coal seams but the most popular method is utilizing electrical resisting heating. In this method, two electrical lighters (generally 1 KW) are lowered into the well along with a thermocouple that measures the changes in the temperature (Thorsness et al. 1977). After the lighters are lowered into the well close to the coal seam, charcoal briquettes, wood chips or other flammable material is placed at the bottom of the hole until the lighters are covered with this material. At this moment, the controlled flow of air is directed at the lighters and charcoal. The charcoal and wooden chips ignite in a few minutes and in turn ignite the coal seam. This is demonstrated by changes in temperature indicated by the thermocouple. Once the ignition has started, the gasification process starts and the injection of gases and steam starts at the rate and pressure designed for the particular gasifier.

Another method used for initial ignition of the coal seam is through explosive material. In this method, propane, gasoline or lighting fluid is injected into the coal seam just prior to start of ignition and a light explosive charge is set off at the bottom of the hole. The flame generated by the explosion starts a fire that is sustained through the injection of oxygen or air. Wooden chips or other flammable material can also be dumped into the hole, if required.

Well Pattern

Generally, four types of wells are drilled to form the gasifiers. They are

- Injection wells
- Production wells
- Instrumentation/monitoring wells
- Dewatering wells

Injection wells are used for injecting the oxidizers, i.e., air, gas or steam, into the seam. They have compressors or compressing stations installed around them and inject high-pressure gases into the gasifiers. The production wells are used to extract the syngas form the gasifier. The injection and production wells are sometimes collectively referred to as process wells (Blinderman and Jones 2002). The instrumentation wells are scattered in the area as per the design of the cavity. They contain instruments for measuring pressures, temperature and flow rates and monitoring the flows in or out of the cavity. The dewatering wells are used to dewater the seam either prior to gasification or during the gasification process. These wells are also used to collect samples for measurement of water quality prior to, during and after the gasification process.

Well Spacing

The spacing and number of process wells in the gasifier depend upon coal thickness, coal rank, geology of the area (i.e., number of prominent fractures or folds in the area, permeability of the coal seam, gasification rate and economics of the project). The number of wells required for thicker seams is less than that required for thinner seams; thus, in thicker seams wells can be at wider spacing than those in thinner seams. The most common well spacing lies between 18 m (60 ft.) to 30 m (~100 ft.). The increase in well spacing decreases the drilling cost and price of the gas. A sensitivity analysis of well spacing and cost of gas, however, indicates that the cost benefits decrease rapidly after the well spacing is more than 150 feet (Boysen 1978). The reason is the contribution of drilling cost to the operating cost, which has reduced impact after this spacing. Monitoring wells are generally placed around 15 m and 30 m radius (Snoeberger 1977).

Wells Structure

The process wells are generally steel-cased and cemented to withstand the higher temperatures and pressures. These wells can be completed in the upper portion of the seam and perforated at the lower or bottom part of the seam (Fischer et al. 1977). The diameter of process wells is in the range of 14 to 20 inches (Boysen 1978). The instrumentation wells are generally not cemented or cased and are small diameter holes. They carry thermocouples and tubes for measuring temperatures and water levels and for gas sampling. The dewatering wells are steel-cased to the top of coal seam and are perforated within the coal seam and have steel pump sections that carry water pumps to collect water from sumps (Thorsness et al. 1977).

Instrumentation

Instruments are used to measure temperatures, air and gas pressures, water level, flow rates, and for gas sampling. Thermocouples are used to measure temperature variations within the coal seam and in the surrounding strata. A single instrument hole can carry several thermocouples at different depths. They are generally placed within the seam at varying depths (top, center and bottom of seam) and at 5, 10, 15 and 20 feet in the over and under burden to have complete temperature profile of the gasifier and surround rocks (Thorsness et al. 1977). The air and gas flow meters, transducers and gas chromatographs measure the flow rate, pressure and composition of gas. Submersible pumps are used in dewatering wells.

Well Linkages

The most important step during the creation of the in situ gasifier is the development of linkage between the injection and production well. This can be achieved through any of the following methods.

- Explosive fracturing
- Reverse combustion linkage
- Hydraulic fracturing
- In seam channels through drilling
- CRIP (Controlled Retraction Injection Point)

In the explosive fracturing technique, a high explosive is detonated at the bottom of the hole, either in one of the process wells or at a specially drilled well for this purpose. The idea behind this technique is that explosive fracturing will enhance the permeability of the coal, hence the linkage between the wells. An experimental blast consisting of two high explosive spherical shots at Hoe Creek, Wyoming, revealed that this technique creates three regions after explosions (Thorsness et al. 1977).

- The region where there is no change in the permeability and the permeability value remains as it was before explosions. This region is at a radial distance of about 50 feet from explosions depending upon type of explosive and geology of the area.
- A very high permeability area within 10 feet of explosive charges and
- An intermediate area where permeability is significantly enhanced after explosions

However, this technique is not very successful as in the areas very close to explosion; the permeability actually decreased due to plugging of fractures by fine coal particles resulting from the blast (Thorsness et al. 1977).

Reverses combustion linkage is a somewhat more successful method than explosive fracturing and has been tried in several experimental trials and pilots in Europe, the U.S. and other places (Zamzow 2010). In the reverse combustion linking, the fire is started at one well, Well A in Figure 1. The air is initially injected in the ignition well (Well A) to sustain the fire. Once the combustion of coal is started, the air injection is switched and injected to the other well, Well B in Figure 1. Now the air current moves towards the ignition front through the coal seam and the combustion moves from ignition well (Well A) to the injection well (Well B). The combustion zone is moving against the current of air towards oxygen source (Well B) and because of this movement against the air current, the process is called reverse combustion. Because of this movement of the combustion zone towards the injection well, a highly permeable localized zone is created in the seam. After the combustion zone reaches the injection well, the reverse combustion is completed (Fischer et al. 1977). At this point high volume, low-pressure air is injected into the injection well (Well B) and combustion zone now moves towards ignition well (Well A). At this point the air current and combustion zone move in the seam direction and this is called forward combustion. The position of the reverse combustion link at the bottom of the seam is very important, as in the forward combustion this undercut zone will result in the falling of fresh coal in the reaction zone, yielding high efficiency and producing a packed bed (Fischer et al. 1977).

Hydraulic fracturing is another technique used to improve the linkage between wells. In this process, the low-pressure water is first used to washout the borehole. The pressure is then increased gradually until the cracks are generated in the seam. After that, hydraulic fluids are pumped into the cracks. Finally, the area is flushed with water and then with air to remove any plugging (Olness 1981). Control in hydraulic fracturing is very important, as the process can increase the potential of gas losses significantly due to irregularity of fractures development.

In-seam channels are one of the better options to improve well linkages. In this method, an in-seam channel is created by drilling a horizontal borehole in the lower part of the coal seam, as shown in Figure 2. Directional drilling reduces the surface disruption and increases the amount of coal that can be accessed by a pair of wells (Ahner 2008).

The CRIP technology is a modern technique, giving excellent control in the cavity formation and linking the wells and is based on advancements in directional drilling. CRIP stands for Controlled Retraction of Injection Point. In this technique, the injection point is retracted occasionally to expose the fresh coal to the injecting gases. In this method, the link between the wells is established through a directionally drilled horizontal borehole between the injection and projection well. After a cavity has been formed in the coal at a specific location, the injection point is retracted either mechanically or by burning the portion of the injection tube, thus exposing the fresh coal to the stream of gases and developing a new cavity. Ten to 20 such movements can occur during the lifetime of a set of wells (Clean Coal Limited 2007).

Operating Pressures

The operating pressure required for gasification depends upon the type of coal, the permeability, depth of seam, type of oxidant used, amount of water and moisture in the coal and strata, and most importantly the hydrostatic pressure around the burn cavity. Experiments in the U.S. indicate that the operating pressure should never exceed the hydrostatic pressure of the cavity; otherwise the hazardous waste and pollutants can migrate out of the cavity and can result in the contamination of surrounding area and water resources (Ahern and Frazier 1982; Younger et al. 2010). The air pressure also depends upon the method used for linking the wells. For reverse combustion and hydraulic fracturing high pressure is required to move the ignition front whereas low pressure is required for CRIP and in-seam borehole. Gas pressures at the beginning of ignition are high to

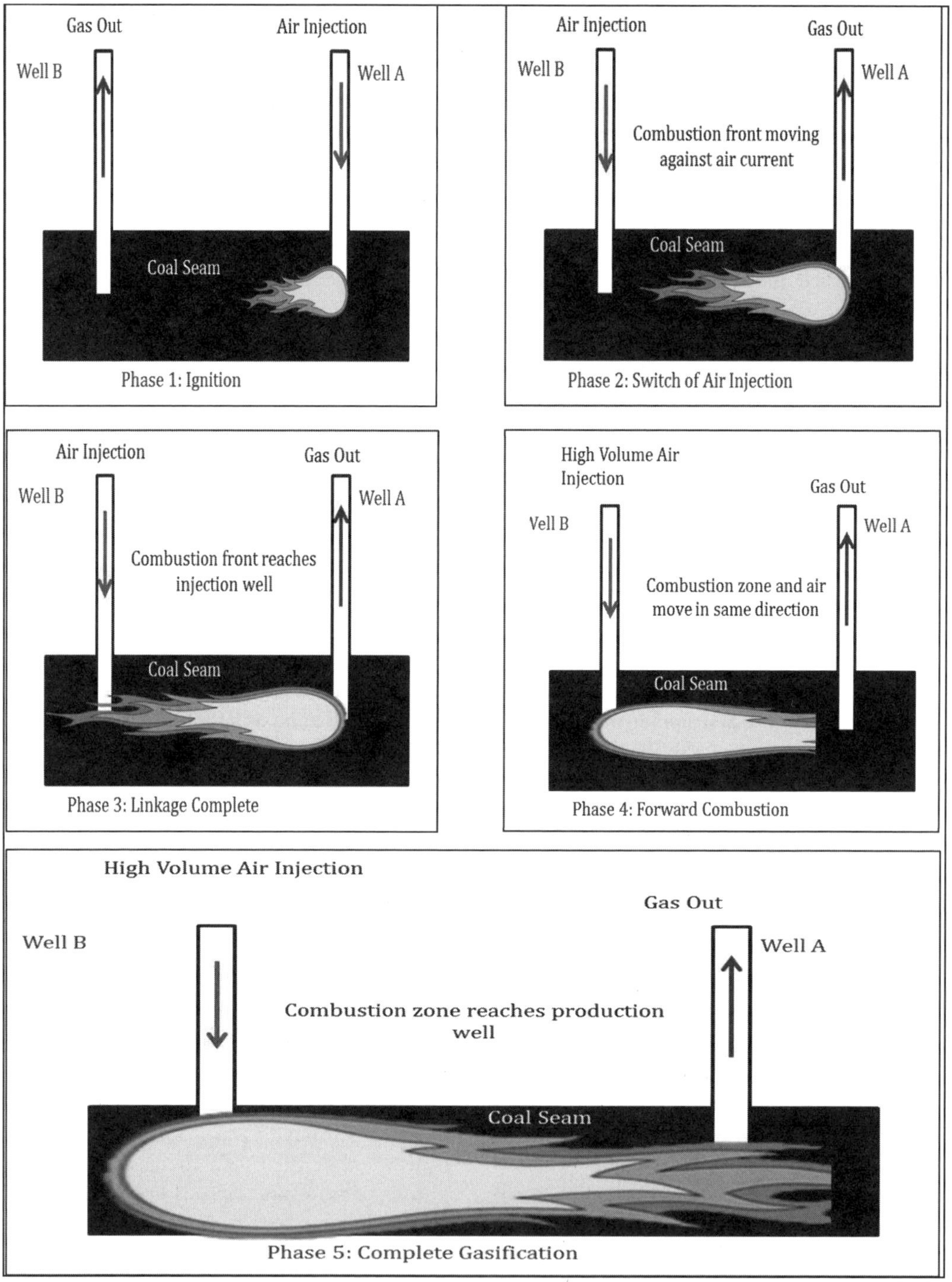

Figure 1. Reverse combustion linkage

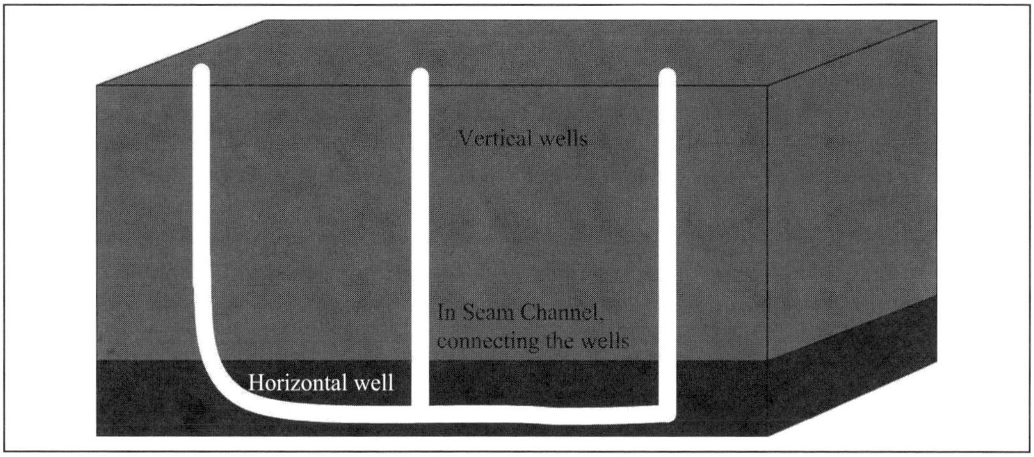

Figure 2. In-seam channel

promote combustion of coal, but once combustion is achieved, injection pressure is reduced and flow rate is increased. Typical injection pressures range between 60 to 120 psi and the flow rates vary from around 200 scfm (Standard Cubic Feet per Minute) in the beginning to around 4,500 scfm at various stages of gasification (Fischer et al. 1977).

Cavity Development

After ignition, the cavity starts to develop along the line of linkage between the wells. In case of reverse combustion linkage, the cavity develops in the form of a cylinder with the final radius almost equal to the distance between the process wells (Fischer et al. 1977). The initial development of the cavity is in the horizontal direction but later cavity development occurs in a vertical direction and burning takes place near the top of seam (Roehl et al. 1977). Thus, the initial linking path should be close to the bottom of the seam during reverse combustion to get maximum recovery in forward combustion. For directional drilling and CRIP, the cavity develops around the borehole and expands laterally.

Water Control

Water plays a crucial role in the gasification process. The presence of water affects the nature and quality of the product gas (Walters and Nuttall 1977). The presence of water is necessary for sustaining the reaction and water inflow of about 1 gpm is considered good for gasification (Snoeberger 1977). Water inflow depends upon the presence of the aquifer in the area and permeability of the seam. In the high permeability area, the water influx can be higher than the desired levels and can decrease the heating value of the gas, as vaporization of the water consumes part of thermal energy. Excessive water influx can be controlled by using the dewatering wells, controlling the cavity or gas pressure, gasifying the seam in an up dip direction and two-stage pyrolysis gasification (Gunn 1977; Vanderborgh et al. 1977). In the two-stage gasification pyrolysis process, CO_2 is recovered from the product gas, heated and injected into the coal seam prior to gasification in that portion. This heated CO_2 dries the coal seam, thus resulting in less water content and increased porosity of coal because of the water removal from pores (Vanderborgh et al. 1977).

Gas Cleaning

The level of cleaning required for the product gas depends upon the amount of sulfur compounds, moisture contents, particulate matter and gas composition. Sulfur and nitrogen report to the surface with the product gas but sulfur can be removed using sour gas cleaning technologies that are already available and quite common (Fergusson 2009). CO_2 removal from the syngas stream is quite easy and economical and technologies are currently available to separate CO_2 from syngas either onsite or at power plants in a cost-effective manner (Clean Air Task Force 2009). The gas stream exiting the production well is at elevated temperatures and gas cleaning is accomplished more easily if the gas is at lower temperatures (Vanderborgh et al. 1977). This is achieved through the introduction of a heat exchanger in the system before transporting gas to the gas cleaning facility. If the gasification is taking place at lower depths and the pressure of syngas is less than the operating pressures required for turbine, a compressor can be inducted into the process line to pressurize the gas at desired levels (Blinderman and Jones 2002). However, for gasification of deep-seated coal seams, the compressor is not required as product gas is already at elevated pressures. Cleaning of product gas from UCG is easier and relatively less expensive than that from the surface gasifiers because the more viscous coal tars from surface gasifiers plug the valves and other equipment, whereas tars from UCG have low viscosity similar to that of oils (Gunn 1977) and are easy to remove for gas cleaning.

CONCLUSIONS

UCG has distinctive environmental and economic advantages as technology capable of exploiting coal reserves; however, the application of UCG requires careful consideration related to its environmental impacts and proper control of operational parameters. This technology has several benefits that can help in the development of coal resources in an environmentally friendly and sustainable manner. The technology has the potential to mitigate environmental hazards typically associated with mining coal. However, it is imperative that potential impacts of this technology are also evaluated when developing sites for commercial utilization of UCG.

It is very important to analyze and understand the operational parameters of UCG in order to determine their significance and to evaluate the effective range of values for proper control of the process. Cavity pressures, gas and water flow rates and linkage between process wells are some of the most important parameters that help in the continuous flow of consistent quality product gas and depend upon strata conditions and coal quality parameters. Improvements in the well linking techniques have made it possible to have a constant control of cavity development, gas flow and quality of gas. The monitoring of contaminant movements and environmental impacts of cavity development and gas flows is essential to control the environmental impacts of UCG. With the advent of sophisticated equipment and reliable online systems, monitoring of the UCG process has improved and can be done remotely. Improved understanding of the process through experiments and pilot studies will help in the evolution of better process and operational controls.

REFERENCES

Ag Mohamed, A., Batto, S. F., Changmoon, Y., Chien, S. C., Choe, J. I., Cole, K. R., Wisely, A. M., 2011. *Viability of Underground Coal Gasification with Carbon Capture and Storage in Indiana*. Bloomington School of Public and Environmental Affairs, Indiana University.

Ahern, J. J. and Frazier, J. A., 1982. *Water Quality Changes at Underground Coal Gasification Sites—A Literature Review*. Water Resources Research Institute, University of Wyoming, Laramie, Wyoming.

Ahner, P., 2008. *Underground Coal Gasification (UCG) Consulting* from http://ucgconsulting.com/ucg-process-description.html.

Blinderman, M. S. and Jones, R. M., 2002. *The Chinchilla IGCC Project to Date: Underground Coal Gasification and Environment*. 2002 Gasification Technologies Conference. San Francisco, USA.

Blodgett, S. and Kuipers, J. R., 2002. *Underground Hard-rock Mining: Subsidence and Hydrologic Environmental Impacts*. Bozeman, MT, Center for Science in Public Participation: 50.

Boysen, J. E., 1978. *An Economic Sensitivity Study of UCG Based on Field Performance, Theory And Operational Experience*. Masters, University of Wyoming.

Burton, E., Friedmann, J. and Upadhye, R., 2006 *Best Practices in Underground Coal Gasification*. Draft Technical Report (Lawrence Livermore National Laboratory, U.S. Department of Energy contract No. W-7405-Eng-48.), 119. from http://www

.purdue.edu/discoverypark/energy/pdfs/cctr/BestPracticesinUCG-draft.pdf.

Burton, E., Friedmann, J. and Upadhye, R., 2008. *Effectively Managing UCG Environmental Issues.* Lawrence Livermore National Laboratory.

Clean Air Task Force, 2009. *Coal Without Carbon: An Investment Plan for Federal Action.* A Clean Air Task Force Report, Funded by the Doris Duke Charitable Foundation.

Clean Coal Limited, 2007. *The Linked Vertical Well Technique and CRIP.* Retrieved August, 2012, from http://www.cleancoalucg.com/index.php?option=com_content&view=article&id=87&Itemid=124.

Courtney, R., 2009. *Underground Coal Gasification.* UCG Workshop, Pittsburgh Coal Conference. Pittsburgh, PA.

Creedy, D. P., Garner, K., Holloway, S., Jones, N. and Ren, T. X., 2001. *Review of Underground Coal Gasification Technological Advancements.* Report No. COAL R211, DTI/Pub URN 01/1041 Under Department of Trade & Industry, UK.

EPA., 1999a. *U.S. Methane Emissions 1990–2020: Inventories, Projections, and Opportunities for Reductions.* Washington DC. U.S. Environmental Protection Agency, Office of Air and Radiation.

EPA., 2010. *Greenhouse Gas Emissions Reporting From the Petroleum and Natural Gas Industry, Background Technical Support Document.* U.S. Environmental Protection Agency Climate Change Division. Washington DC,.

Fergusson, K. J., 2009. *A Cleaner, Cheaper, Indigenous Fuel for Combined Cycle Plants.* Modern Power Systems, 29(8), 24–26.

Fischer, D. D., King, S. B. and Humphrey, A. E., 1977. *A Report on the Successful Development of Underground Coal Gasification at Hanna, Wyoming.* Fuel, 22(4), 49–63.

Friedmann, J., 2009. *Accelerating Development of Underground Coal Gasification: Priorities and Challenges for U.S. Research and Development.* Coal without Carbon: An investment plan for Federal Action, A Clean Air Task Force Report (Chapter 1), 1–16.

Ghose, M. K. and Paul, B., 2007. *Underground Coal Gasification: A Neglected Option.* International Journal of Environmental Studies, 64:777–783.

Gunn, R. D., 1977. *Problems Solved and Problems Not Solved In UCG.* Fuel, 22(4), 64–75.

Humenick, M. J. and Mattox, C. F., 1978. *Groundwater Pollutants from Underground Coal Gasification.* Water Research, 12(7), 463–469.

Kapusta, K. and Stanczyk, K., 2011. *Pollution of Water during Underground Coal Gasification of Hard Coal and Lignite.* Fuel, 90(5), 1927–1934.

Klimenko, A., 2009. *Early Ideas in Underground Coal Gasification and Their Evolution.* Energies, 2(2), 456–476.

Lamb, G. H., 1977. *Underground Coal Gasification.* Park Ridge, New Jersey, NOYES Data Corporation.

Liu, J., Mallet, C., Beath, A., Elsworth, D. and Brady, B., 2003. *A Coupled Flow-Transport-Deformation Model for Underground Coal Gasification.* GeoProc. Sweden.

Meany, R. A. and Maynard, A., 2009. *A Review of the Potential for Underground Coal Gasification and Gas to Liquids Applications in Pedirka basin, Onshore Northern Territory and Pela 77 Pedirka basin, Onshore South Australia* Mulready Consulting Services Pty Ltd.

Moorhouse, J., Huot, M. and McCulloch, M., 2010. *Underground Coal Gasification: Environmental Risks and Benefits.* The Pembina Institute. Alberta.

Olness, D. and Gregg, D. W., 1977. *The Historical Development of Underground Coal Gasification.* Lawrence Livermore Laboratory, University of California, Livermore.

Olness, D. U., 1981. *The Podmoskovnaya Underground Coal Gasification Station.* University of California, Livermore, California, Lawrence Livermore Laboratory.

Oukaci, R., 2009. *Syngas Uses.* 26th Annual International Pittsburgh Coal Conference, UCG Tutorial, Pittsburgh, PA.

PWC, 2008. *Industry Review and an Assessment of the Potential of UCG and UCG Value Added Products.* Linc Energy Limited, PriceWaterHouseCoopers.

Ray, S. K., Panigrahi, D. C. and Ghosh, A. K., 2010. *Cleaner Energy Production with Underground Coal Gasification–A Review.* The Institute of Engineers India: IE(I) Journal-MN, 91.

Redman, E., Fenerty, K. and Fowler, M., 2009. *Mobilizing Next Generation Coal Gasification Technology for Carbon Capture and Sequestration.* Coal without Carbon: An investment plan for Federal Action, A Clean Air Task Force (Chapter 2), 17–36.

Roddy, D. J. and Younger, P. L., 2010. *Underground Coal Gasification with CCS: A Pathway to Decarbonizing Industry.* Energy & Environmental Science, 3(4), 400–407.

Roehl, A. A., Brown, A. S. and Jensen, E. J., 1977. *Underground Coal Gasification Field Test in Alberta–1976.* Fuel, 22(4), 92–104.

Shafirovich, E. and Varma, A., 2009. *Underground Coal Gasification: A Brief Review of Current Status.* Industrial & Engineering Chemistry Research, 48(17), 7865–7875.

Snoeberger, D. F., 1977. *Field Hydrological Tests of Explosively Fractured Coal.* Fuel, 22(4), 22–34.

Sury, M., Kirton, J., White, M. and Hartwell, D., 2004. *Review of Environmental Issues of Underground Coal Gasification—Best Practice Guide.* DTI Cleaner Coal Technology Transfer, UK, Report No. COAL R273 DTI/Pub URN 04/1881.

Sury, M., White, M., Kirton, J. and Carr, P., 2004. *Review of Environmental Issues of Underground Coal Gasification.* DTI Cleaner Coal Technology Transfer, UK, Report No. COAL R272 DTI/Pub URN 04/1880.

Thorsness, C. B., Hill, R. W. and Stephens, D. R., 1977. *Preliminary Results from an In Situ Coal Gasification Experiment Using Explosive Fracturing.* Fuel, 22(4), 1–21.

Vanderborgh, N. E., Wewerka, E. M., Trela, W. J., Williams, J. M. J. and Elliott, G. R. B., 1977. *Combined CO_2-O_2 Underground Pyrolysis-Gasification for Southwestern Coals.* Fuel, 22(4), 35–48.

Walker, L., 1999. *Underground Coal Gasification: A Clean Coal Technology Ready for Development.* The Australian Coal Review,19–21.

Walters, E. A. and Nuttall, H. E., 1977. *Potential for Underground Coal Gasification in The SouthWest.* Fuel, 22(4).

Younger, P., González, G. and Amezaga, J. M., 2010. *Water Management Issues In The Underground Gasification of Coal and The Subsequent Use of Resultant Voids For Long-Term CO_2 Storage.* International Mine Water Association, Sydney.

Zamzow, K. L., 2010. *Underground Coal Gasification: History, Environmental Issues, and the Proposed Project at Beluga, Alaska.* Center for Science in Public Participation. Cook Inlet Region, Inc. (CIRI) UCG Project, Alaska.

The Marcellus Shale Gas Play and the Gas Migration/ Stray Gas Problem

Julie Vandenbossche
Swanson School of Engineering, University of Pittsburgh, Pittsburgh, Pennsylvania

Anthony Iannacchione
Swanson School of Engineering, University of Pittsburgh, Pittsburgh, Pennsylvania

John Brigham
Swanson School of Engineering, University of Pittsburgh, Pittsburgh, Pennsylvania

ABSTRACT

The Marcellus shale-gas play represents a significant energy resource for the northeastern United States. The scale of the development associated with the expansion of the gas industry is noticeable in all aspects of the economies where this activity is occurring. It should be no surprise that many potential unwanted events can threaten a development of this scale and scope. Arguably one of the greatest threats to environmental protection, safety, and production during development in integrity of wellbores, especially in the upper portion of the drill hole where groundwater aquifers are prevalent. As such, the Pennsylvania Department of Environmental Protection (PADEP) requires significant engineering controls for the wellbore in deep Marcellus gas wells. One focus of these controls is on the performance of grouts used to isolate the fluids and gases contained within the wellbore from the surrounding rock formations. Isolation can be a particular challenge due to the geology of some regions, especially in the northern tier gas fields of Tioga and Bradford counties, where these rock formations have been known to release gas into the wellbores (Figure 1). The petroleum industry has long recognized the deleterious impact of gas injection during the placement of the slurry used to isolate the drill holes from the groundwater aquifers. These conditions have been called 'gas migration,' or 'stray gas,' two terms that have the same meaning to some, while to others they represent two different phenomenon.

For some, gas migration is associated with gas flowing into the wellbore during grouting operations, which has the potential to alter the physical properties of the grout. For instance, formation gas has the ability to invade the cement when it is transitioning from a liquid to a gelled state. If this occurs, the integrity of the curing cement can be lessened. It is possible that this phenomenon may have occurred in a small number of Pennsylvania's Marcellus wells, especially in the northeastern part of the state where formations close to the surface have long been known to contain high pressure natural gas. Alternatively, in Pennsylvania gas migration problems in Marcellus shale-gas well bores are often referred to as stray gas. In particular, the term stray gas has recently been used to describe gases migrating from rock formations and entering structures.

The Marcellus stray gas problem seems to be most significant when drilling mud has not been entirely removed from the wellbore prior to grouting. In these instances some of the mud remains along the wellbore skin during the placement of grout. The residue drilling mud can facilitate the development of gas flow paths along the wellbore annulus. Therefore, proper well-bore conditioning prior to cementing operations is critical in reducing the risk of stray gas problems. Otherwise, gas can enter these annular pathways

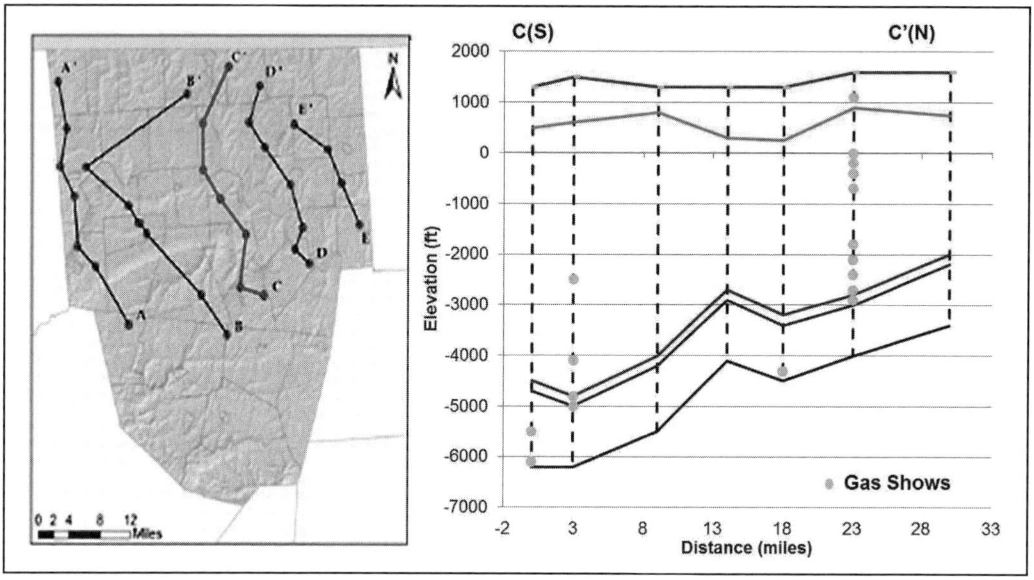

Figure 1. Shallow gas sands in Bradford County; data provided by the PA DEP

and migrate up or down the well to other formations, which is clearly not a desirable outcome.

Both the PA DEP and a number of superior gas-producing companies operating in Pennsylvania have been working diligently to solve the stray gas problem. The PA DEP is developing protocols for collecting significant amounts of data from past and current drilling sites to determine the extent and magnitude of gas migration in wells and to better inform gas well drill operators of proper construction techniques. PA DEP is also interested in determining if modifications to existing regulations are needed to address these issues. It should be noted that the PA DEP has already significantly enhanced its drilling, casing and cementing regulations to, in part, address wellbore integrity and stray gas–related incidents. The current regulations became effective on February 5, 2011. Additionally, the PA DEP continues to re-evaluate and update its regulations, policies and procedures to proactively address and manage the continuing evolution of drilling, casing and cementing practices for all oil and gas wells both drilled and produced in the Commonwealth.

The gas companies and their suppliers have brought in drilling and cementing experts from all over the world to find new processes and techniques to prevent (a) gas migration through the grout, and (b) gas movement along the annulus of the wellbores. In moving forward, some operators have stopped using drilling mud in the shallow parts of their wellbores to eliminate the negative effect of drilling mud on the formation of annulus pathways. In addition, Schlumberger is a leader in workingt to research, formulate and produce cement mixes that lessen the potential for formation gas migrating into the grout as it cures and hardens. The progress being made is very encouraging and a significant factor contributing to advancement is the cooperation between government and industry in addressing these issues.

In general, it is difficult to know exactly what the conditions are down hole. There are some excellent geophysical tools that the industry uses to assess the performance the annulus but these tools have certain limitations. At the University of Pittsburgh, we are currently simulating the complex wellbore conditions with a specially constructed high pressure chamber (Figure 2).

It is our aim to generate information on how wellbore cement annulus is being affected by gas migration/stray gas. This is being accomplished

Figure 2. Simulation chamber used to simulate wellbore conditions

by simulating in-situ conditions within our test chamber. The conditions generated by the chamber cover a range of temperatures and pressures with controlled gas injection situations. After each test, a series of forensic analysis are performed to map the migration paths of the injected gas within the grout. In addition, the permeability and mechanical properties, i.e., strength, stiffness, Poisson's ratio, etc., are being determined for each test condition. Finally, analytical and/or numerical simulations will be validated with respect to the experimental tests and will be utilized to investigate a wider range of conditions. Preliminary results will be presented at the symposium.

The overall goal is to generate information on how the cement annulus is being affected by the gas migration/stray gas. Once a better understanding of these processes is achieved, additional engineering solutions can be developed to insure groundwater aquifers will not be adversely impacted.

Experiences in Reservoir Model Calibration for Coal Bed Methane Production in Deep Coal Seams in Russell County, Virginia

Foteini Vasilikou
Virginia Center for Coal and Energy, Virginia Tech, Blacksburg, Virginia

Cigdem Keles
Virginia Center for Coal and Energy, Virginia Tech, Blacksburg, Virginia

Zach Agioutantis
Virginia Center for Coal and Energy, Virginia Tech, Blacksburg, Virginia

Nino Ripepi
Virginia Center for Coal and Energy, Virginia Tech, Blacksburg, Virginia

Michael Karmis
Virginia Center for Coal and Energy, Virginia Tech, Blacksburg, Virginia

ABSTRACT

Injection and storage of carbon dioxide (CO_2) on the coal seam matrix has two benefits: mitigation of greenhouse gas emissions and enhanced recovery of coalbed methane (ECBM). The theory of ECBM is based on the natural affinity for the porous coal matrix to preferentially adsorb CO_2 over methane. According to laboratory measurements on bituminous Appalachian coals, the coal matrix can adsorb almost twice the amount of CO_2 by volume as methane at typical reservoir pressures. Based on the fact that coalbed methane (CBM) is currently economically produced in many areas in the world, presents the likelihood that CO_2 injected under the appropriate conditions could sequester the CO_2 while displacing and producing CBM.

The Coal Seam Group of the Southeast Regional Carbon Sequestration Partnership (SECARB) has recently completed the injection of 1,000 tons of carbon dioxide into multiple deep unmineable coal seams as part of a field validation test at their Russell County, VA, site in Central Appalachia and is planning the injection of 20,000 tons of CO_2 in a nearby field in Buchanan County, VA. This paper presents the reservoir modeling procedure and the parameters developed in order to obtain an accurate history match for both gas and water production for a group of eight mature CBM wells that were utilized in the Russell County field test. Optimal parameters were determined both for the single well models as well as for the combined eight-well model. This model will form the basis for simulating CO_2 injection and subsequently predicting post-injection behavior for the upcoming injection test at the Buchanan county site.

INTRODUCTION

The Coal Seam Group of the Southeast Regional Carbon Sequestration Partnership (SECARB) has recently completed the injection of 1,000 tons of carbon dioxide into multiple deep unmineable coal seams as part of a field validation test at their Russell County, VA, site in Central Appalachia (Ripepi et al. 2009).

Figure 1 shows an areal view of the general area, while Figure 2 shows an areal view of the injection well (BD114), two monitoring wells

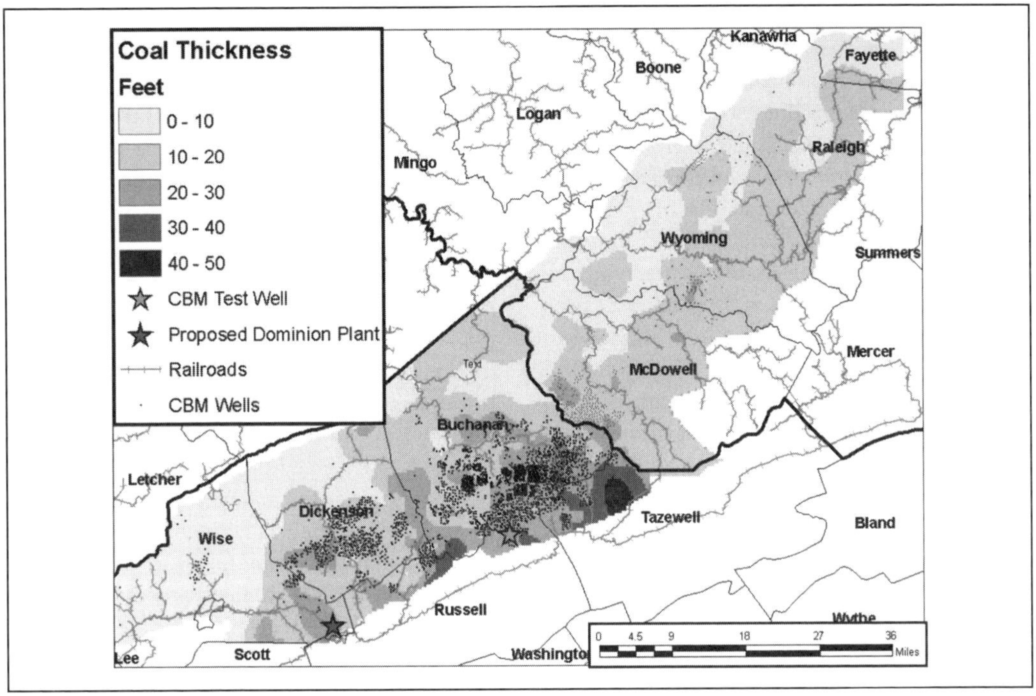

Figure 1. Location of field test site (Ripepi et al., 2009)

(black dots) and the closest offset producing CBM wells.

Prior to conducting the injection test, a reservoir simulator was built that predicted the extent of the CO_2 plume based on history matching of CH_4 and water production data (VCCER 2011). In this paper the original reservoir simulation is extended by using new knowledge regarding seam geometry, coal properties (such as gas content, adsorption isotherms, anisotropy in permeability), butt and face cleat orientation and by considering well interference.

GEOLOGY

In 2008, a carbon dioxide injection site was selected in Russell County, Virginia, where an existing CBM well was donated by CNX Gas Corporation, a subsidiary of CONSOL Energy Inc. The injection well may be found referenced by its company well name, BD114, or by the State of Virginia designation, RU-0084. BD114 is located at an elevation of 2,523 feet in the Honaker District of Russell County, Virginia.

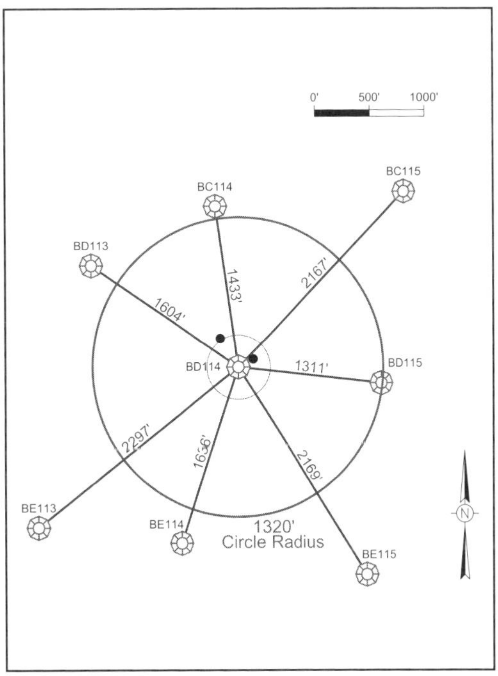

Figure 2. Layout of injection and offset wells (Karmis et al., 2008)

The Central Appalachian Basin is a northeast-to-southwest trending basin encompassing approximately 10,000 square miles in southwestern Virginia, southern West Virginia and eastern Kentucky (Conrad et al. 2006). Production of CBM began in 1988 with the development of the Nora Field in Dickenson County, Virginia, followed by CONSOL Energy developing the Oakwood Field in Buchanan County, Virginia. Since that time, over 5,600 CBM wells have been drilled and brought on-line as producing gas wells in southwest Virginia through 2012 (VaDMME 2013). The coals in the region near the injection site include those of the Pocahontas Formation and Lee Formation, which directly overlie the late Mississippian Bluestone Formation. Coal seams of the Pocahontas and Lee Formation are medium to low-volatile bituminous, high rank and high gas content coals that include the Pocahontas No. 1 through Pocahontas No. 9 seams (Pocahontas Formation) and the Upper Seaboard, Greasy Creek, Middle and Lower Seaboard, Upper and Middle Horsepen, War Creek, and Lower Horsepen coals (Lee Formation).

The Upper Horsepen (UH) 2 and 3 coal seams are the thickest single completion at 3.7 feet of net coal at a depth of 1,374 feet, with a 0.2 feet parting between the seams. The Pocahontas No. 3 coal seam is 2,208 feet deep and 2.4 feet thick at the test site and well outside of current deep mining activities.

The primary confining units include multiple layers of low permeability shale and siltstone beds that range in thickness from five to 55 feet in the vicinity of the injection well. Permeability for the shale and siltstone units is expected to range from 0.001 to 0.1 md, with low porosity. Even the sandstone units are considered to provide confining beds due to the well-cemented nature of these rocks. The Lee and Pocahontas Formation sandstones are known to have low permeability and porosity values, and do not produce natural gas in this area. The tight sandstone units range in thickness from five to 60 feet at the injection site. The sandstone units are expected to have porosity values that range from 1.0 to 3.0 percent and permeability values ranging from 0.1 to 1.0 md based on core testing. All lithologies will encounter some natural fractures; however, these fractures are likely to be cemented with quartz and calcite and are not expected to provide permeability pathways based on core analysis, well logging and field testing (Karmis et al. 2008).

Face and butt cleat planes were measured at two coal seam outcrop locations near the site and corresponded to within seven degrees of the known face and butt cleat directions of N18W and N167E, respectively, for the deep mined Pocahontas No. 3 coal seam in Buchanan County. A Rose Diagram (Figure 3) was developed to graphically display the cleat directions from the injection well (VCCER 2011).

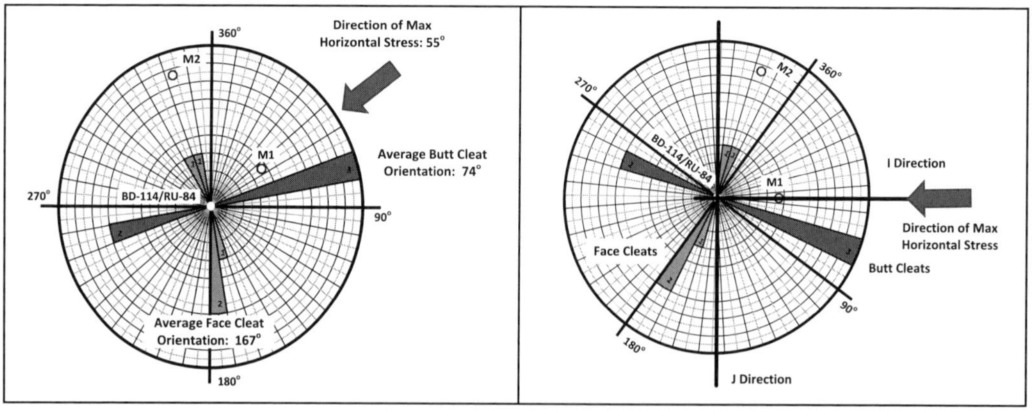

Figure 3. (left) Rose diagram showing face and butt clear orientation; (right) rotation by 35° clockwise

Table 1. BD114 stimulated coal seams (VaDMME, 2002)

Hydraulic Fracture Stage	Coal Seam	Depth to Coal (ft)	Depth to Coal (m)	Coal Thickness (ft)	Coal Thickness (m)	Cumulative Completed Coal Thickness (ft)	Cumulative Completed Coal Thickness (m)	Zone Thickness (ft)	Zone Thickness (m)
Stage 4	Greasy Creek 1	1045	318.4	1.2	0.4	1.2	0.4	9.6	3.0
	Seaboard 2	1191	362.9	0.5	0.2	1.7	0.6		
	Lower Seabord 1&2	1251	381.3	2.0	0.6	3.7	1.2		
	Lower Seaboard 3	1313	400.3	2.2	0.7	5.9	1.9		
	Upper Horsepen 2&3	1374	418.9	3.7	1.1	9.6	3.0		
Stage 3	Middle Horsepen 1	1421	433.0	2.2	0.7	11.8	3.7	9.8	3.0
	Middle Horsepen 2	1497	456.4	0.7	0.2	12.5	3.9		
	Pocahontas 11	1547	471.5	1.9	0.6	14.4	4.5		
	Pocahontas 10	1573	479.5	1.0	0.3	15.4	4.8		
	Lower Horsepen 1	1622	494.4	2.1	0.6	17.5	5.4		
	Lower Horsepen 2	1627	495.9	1.9	0.6	19.4	6.0		
Stage 2	Pocahontas 9	1664	507.0	1.8	0.5	21.2	6.5	9.3	2.8
	Pocahontas 8-1	1710	521.2	2.0	0.6	23.2	7.1		
	Pocahontas 8-2	1725	525.8	1.3	0.4	24.5	7.5		
	Pocahontas 7-1A	1758	535.9	0.8	0.2	25.3	7.7		
	Pocahontas 7-1B	1765	538.0	0.9	0.3	26.2	8.0		
	Pocahontas 7-2	1850	563.9	1.6	0.5	27.8	8.5		
	Pocahontas 7-3	1875	571.5	0.9	0.3	28.7	8.8		
Stage 1	Pocahontas 6	1984	604.8	0.5	0.2	29.2	9.0	7.6	2.3
	Pocahontas 5	2033	619.7	0.7	0.2	29.9	9.2		
	Pocahontas 4-1	2125	647.7	1.9	0.6	31.8	9.8		
	Pocahontas 4-2	2148	654.7	1.1	0.3	32.9	10.1		
	Pocahontas 3-1	2208	673.0	2.4	0.7	35.3	10.8		
	Pocahontas 3-4	2258	688.3	1.0	0.3	36.3	11.1		
Total	24 Coal Seams			36.3	11.1	36.3	11.1	36.3	11.1

WELL STIMULATION

Twenty-four separate coal seams, totaling 36.3 feet, between the depths of 1,044 and 2,259 feet, were perforated through the casing (Table 1) for well stimulation. The coal seams perforated for the four stage well stimulation average 1.5 feet thick.

BD114 was stimulated over perforations ranging in depth from 1,044.5 to 2,258 feet (318.36 to 688.24 m), using a four-stage nitrogen foam hydraulic fracturing treatment (VaDMME 2002). These four stages included 24 separate coal seams with a combined coal thickness of 36.3 feet (11.1 m), which were completed (i.e., perforated and fracked) (Table 1). The first stage of the injection well was not considered completed and was abandoned due to loss of energy and "sanding off." Therefore, a packer was set at 1,400 feet in the casing of BD114 and a string of 2⅞-inch (7.2 cm) tubing was run back to the surface to isolate the injection zone. Thus, the interval for injection comprised only the first three stages of the stimulation program.

PRODUCTION HISTORY

BD114 was brought on-line as a CBM producing well in 2002. Water is produced through a 2⅞-inch string of tubing set below the Pocahontas No. 3 coal seam, and gas is produced between the casing and the tubing (annulus). Up until injection commenced in 2009, BD114 has averaged production of 1.19 thousand cubic meters (Mm^3) (42 thousand cubic feet (Mcf)) of natural gas per day and 2.2 barrels of water per day (Figure 4).

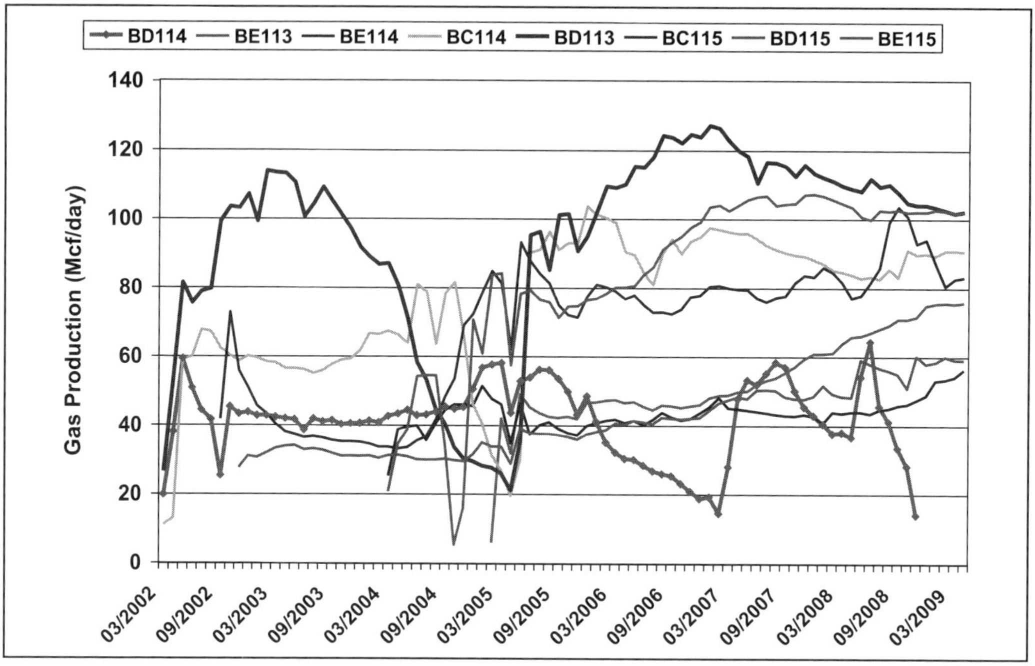

Figure 4. Gas production history at the injection and the offset wells

BD114 produced 2.89 MMm^3 (102 MMcf), and 5,360 barrels of water prior to being taken off-line for conversion to an injection well. Gas production to date is nearly ten percent of the estimated gas in place (VCCER 2011). BD114 is a below average gas producer for this gas field. The average production of the seven offset wells is 1.87 Mm^3 (66 Mcf) per day of natural gas and 2.5 barrels per day of water. Figure 4 presents the gas production history for the injection well and the nearby offset wells (VCCER 2011).

MODELING

Description of Production Model

Preliminary reservoir modeling prior to the actual field test where CO_2 was injected into the coal seams was conducted utilizing Advanced Resources' COMET3 reservoir simulator for CBM. The ultimate goal of this preliminary modeling was to estimate the size of the CO_2 plume as well as the expected CO_2 injection rate that could be achieved during injection operations. As additional laboratory analysis, field data and results from the CO_2 injection field test became available, the need to revise the original reservoir model became apparent. At the same time, Computer Modeling Group Ltd.'s GEM software, a compositional and unconventional reservoir simulator, was utilized since it could simulate the effect of fractures. This paper discusses the setup and parameters for production modeling of the Russell County field test using CMG's reservoir simulator. The ultimate goal of the current study will be to obtain a good history match that will help modeling of Enhanced Coal Bed Methane production after the injection well comes back on line. The latter work is detailed in Vasilikou et al. (2013).

A dual porosity, single permeability model was setup, since coal matrix permeability is considered extremely low. As mentioned before, there were four fracture stages including a total of 24 coal seams. Gas and water rates were available per well and, therefore, only aggregate production was available for all seams. Since coal seams have different characteristics it was decided to model four equivalent coal seams at the highest elevation of each fracture stage. The thickness of these

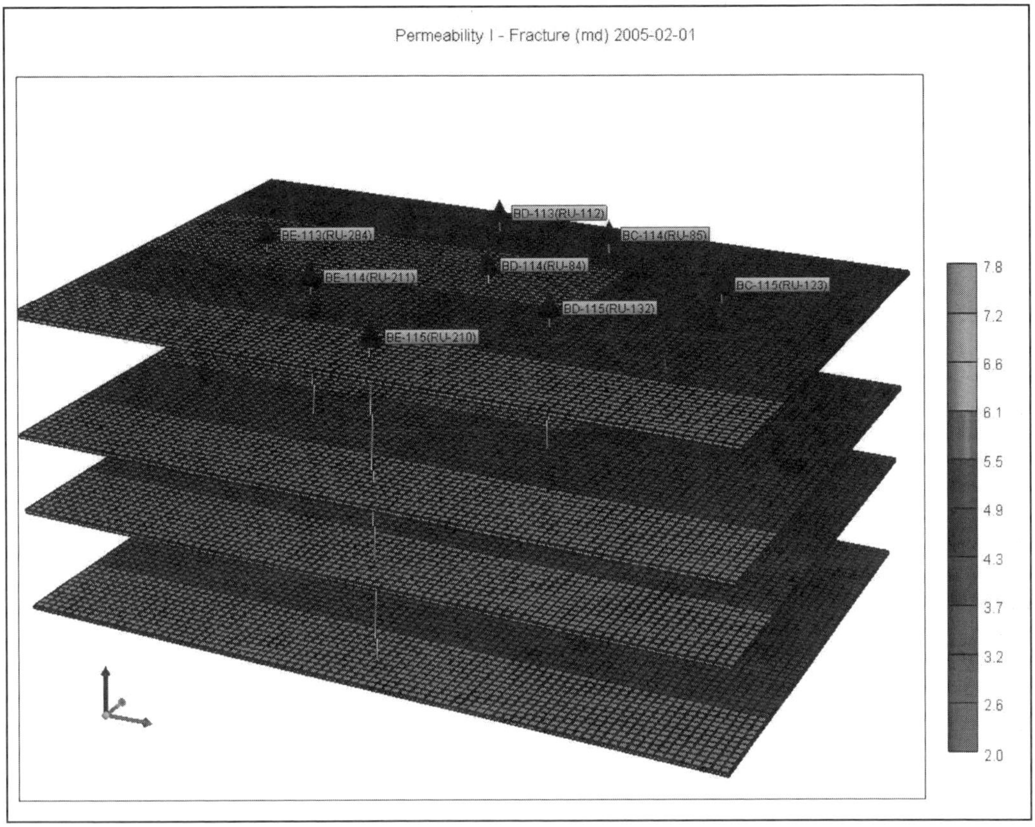

Figure 5. 3-D view of model grid and wells

modeled coal seams correspond to the sum of the thicknesses of each seam within the fracture stage. A grid was generated that covers all wells with nearly the same distance from the boundaries for all wells.

The orientation of the hydraulic fractures was critical in determining grid generation. As the model can only allow for hydraulic fractures along either the I (x) or J (y) direction, and hydraulic fractures were developed along the maximum horizontal stress in the area (Figure 3), the grid was rotated by 35 degrees clockwise (–35°) to orient the maximum horizontal stress in the I direction. In addition, it was assumed that the permeability in the I direction is equal to butt cleat permeability and that in the J direction equal to face cleat permeability.

Furthermore, it was assumed that the modeled group of eight wells can be considered independent of other wells surrounding this grid. The resulting grid for all layers including the eight wells is shown in 3D view in Figure 5.

Although actual fracture modeling was not accomplished within the scope of this study, the foundation was set and results will be presented in the upcoming paper by Vasilikou et al. (2013).

The parameters and material properties shown in Table 2 were assembled utilizing data from (a) collected from monitoring well BD114-M2 (Figure 3), (b) coal sample analysis (Harpalani 2012).

In addition it should be noted that historical production data for gas and water were aggregated on a monthly basis.

Previous work (VCCER 2011) had shown that cleat permeability may be different around each producing well. Under this assumption eight single-well models were setup which comprised

Table 2. Parameters and material properties used in the reservoir model

Number of Grid Blocks per Direction (combined model)	I= 84, J= 56
Number of Grid Blocks per Direction (single-well models)	Variable (typical I=25,J=14)
Grid Block Sizes per Direction (combined model), m	DI=25, DJ=25
Coal Layers and Thicknesses, m	4 Layers;(i) 2.93, (ii) 2.98, (iii) 2.83, (iv) 2.31
Coal Layers Depths- Grid Tops, m	318.4, 433, 507, 604.8
Coal (Matrix) Porosity,%	6.5%, dual porosity model
Cleat (Fracture) Porosity,%	Variable, 1–3%, dual porosity model
Matrix Permeability, mD	Not in use, single permeability model
Cleat Permeability, mD	Variable (domain divided into 6 regions), 2–18.97 mD, face to butt cleats ratio:1–5
Fracture Spacing per Direction, m	I: 0.00508, J: 0.00508, K: 0.00508
Well Radius, m	0.088
Formation Compressibility (Matrix & Fracture), 1/kPa	2.90075E-07 (at reference pressure 4136.8 kPa) (Harpalani, 2012)
Reservoir Temperature, °C	27.2
Water Density, kg/m^3	1000.8 at reference pressure 101.3 kPa
Relative Permeability Curves Gas-Water	Gash et al. 1993, Mavor et al. 1953 (see Figure 6)
Adsorption Isotherms for Methane according to Langmuir. Langmuir Constants VL, gmole/kg; PL, 1/kPa	Layer 1: Pocahontas 11, VL= 0.8201299, PL= 0.0005224 Layer 2: Pocahontas 11, VL= 0.8559385, PL= 0.0005224 Layer 3: Pocahontas 7, VL= 0.8201299, PL= 0.00047818 Layer 4: Pocahontas 3, VL= 0.7700043, PL= 0.00034816 (VCCER, 2011)
Coal Density, kg/m^3	1323.9[VCCER, 2011]
Methane Desorption Time, days	50
Pore Pressure Gradient, kPa/m	7

25 to 50 cells in the X direction, approximately 20 cells in the Y direction and 1 cell in the Z direction for each of the four modeled seams. Single-well models used the cleat permeability regions shown in Figure 7 and Figure 8.

RESULTS AND DISCUSSION

Table 3 shows the optimal parameters determined for the single-well runs. The bottom hole pressure (assumed to be equal to the reservoir pore pressure during production) also follows an expected trend, i.e., increased for higher water production, and with a decreasing trend as gas rate values increase. Note that these models do not include modeling of hydraulic fractures. Models with hydraulic fractures are discussed in Vasilikou et al. (2013).

Figure 9 presents the reservoir simulation history-match results for BD-115 (RU-132) for a single well run. Gas rate predictions match exactly the measured data, while water rate predictions are adequate based on the quality of measured water data. The data for water production by well are estimated by dividing the total water production of a group of wells by the time the water pump on each well operated, unlike gas production which is measured at each well. The bottom hole pressure (assumed to be equal

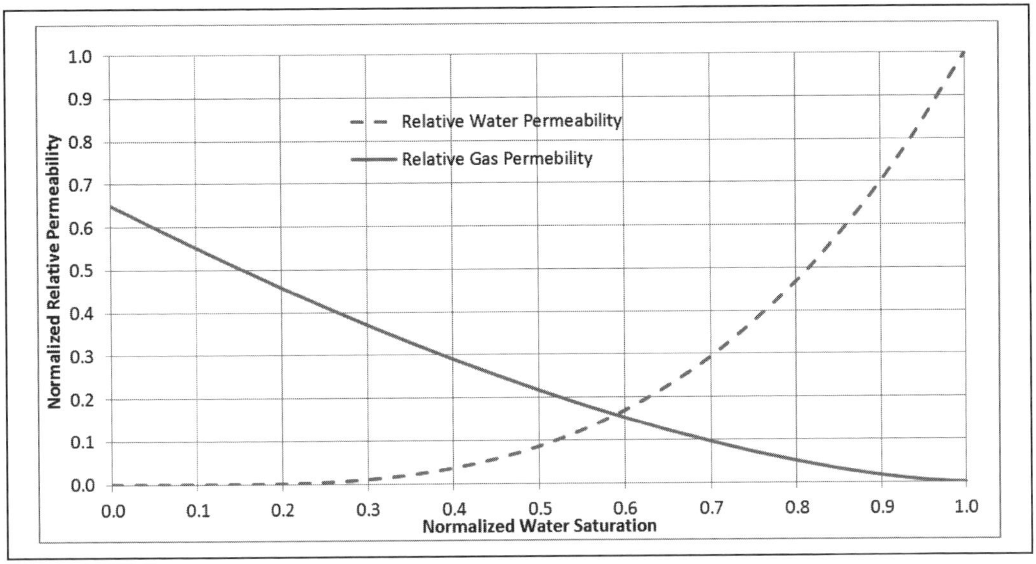

Figure 6. Relative permeability curves using in the single well runs (Gash et al. 1993, Mavor et al. 1953) (SWT: Water relative permeability or water saturation fraction; SLT: Gas relative permeability or gas saturation fraction) (see Table 2)

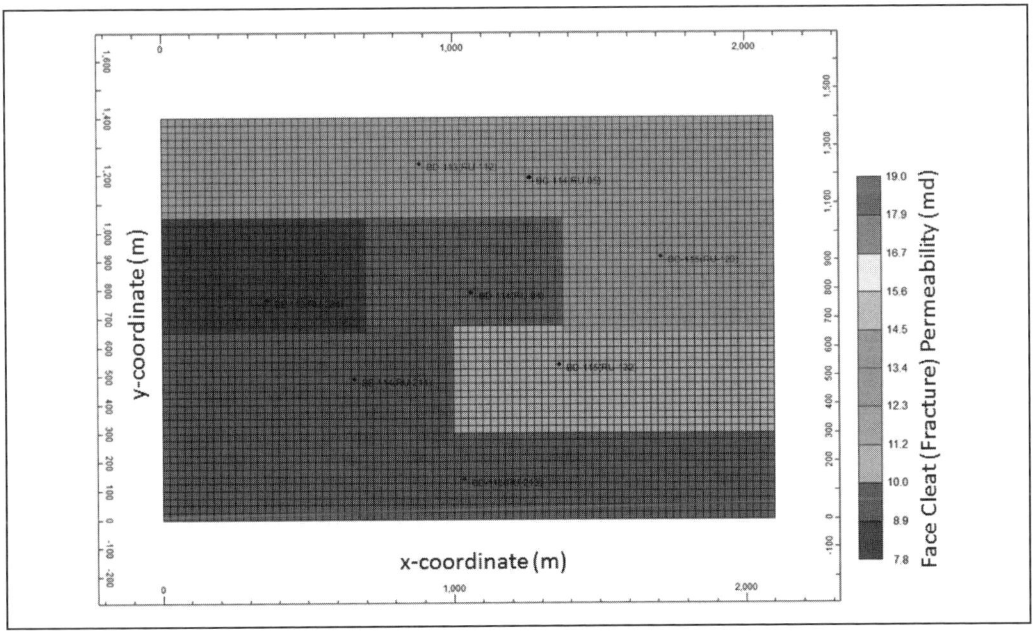

Figure 7. Face cleat permeability regions for single-well models and combined model

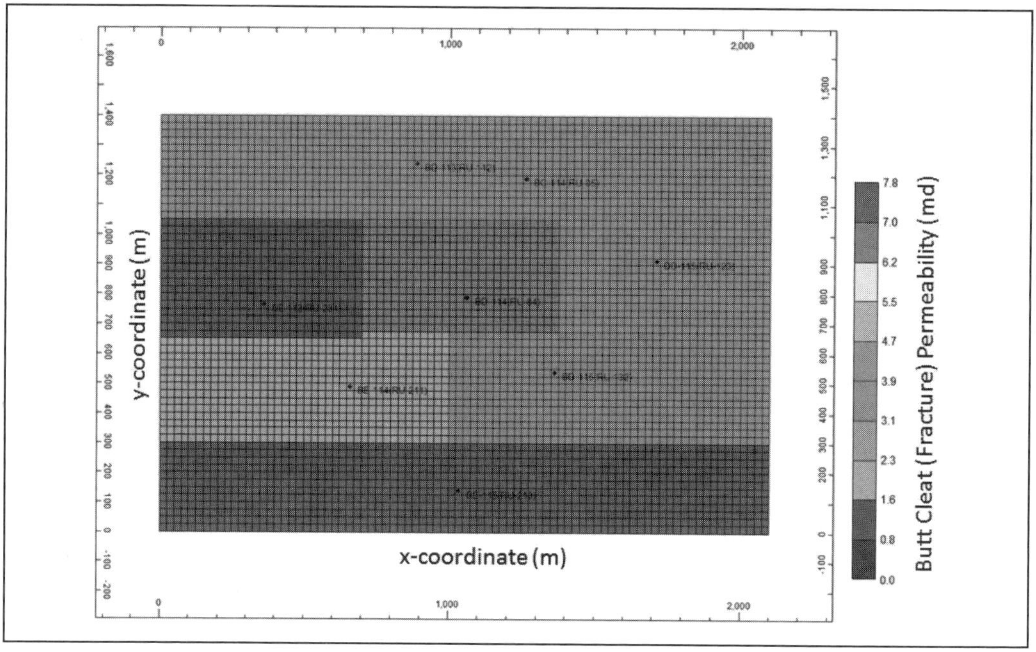

Figure 8. Butt cleat permeability regions for single-well models and combined model

Table 3. Optimal parameters determined for the single-well runs

Well	Fracture Porosity	Face Cleat Permeability (mD)	Butt Cleat Permeability (mD)	Face to Butt Cleat Permeability Ratio	Bottom Hole Pressure (kPa)	Water Saturation in the Cleats
RU-84	0.0013	18.97	6.32	3	200	1
RU-85	0.001	14.42	4.12	3.5	700	1
RU-112	0.0013	14.42	4.12	3.5	1200	0.9
RU-123	0.001	14.42	4.12	3.5	200	0.8
RU-132	0.001	10.42	3.47	3	200	0.9
RU-210	0.001	10	7.8	1.28	1300	1
RU-211	0.001	10	2	5	1400	1
RU-284	0.001	7.8	7.8	1	1250	1

to the reservoir pore pressure during production) also follows an expected trend, i.e., increased for higher water production, and with a decreasing trend as gas rate values increase.

Figure 10 presents the reservoir simulation history-match results for BD-115 (RU-132) for a combined run where the single run optimum parameters were utilized as inputs for each well. In both cases the gas rate match is perfect as the primary simulation constraint comprises matching the gas rate.

The optimal parameters for the combined model were then determined using the optimization algorithm (CMOST) built into the CMG/GEM package. The initial range for each optimized parameter was determined using data ranges that were based on the data in Table 3 for the single-well runs. The actual data ranges entered into the optimizer are shown in Table 4. The optimization algorithm was instructed to find optimum parameter combinations in these ranges. The objective function was set to minimize

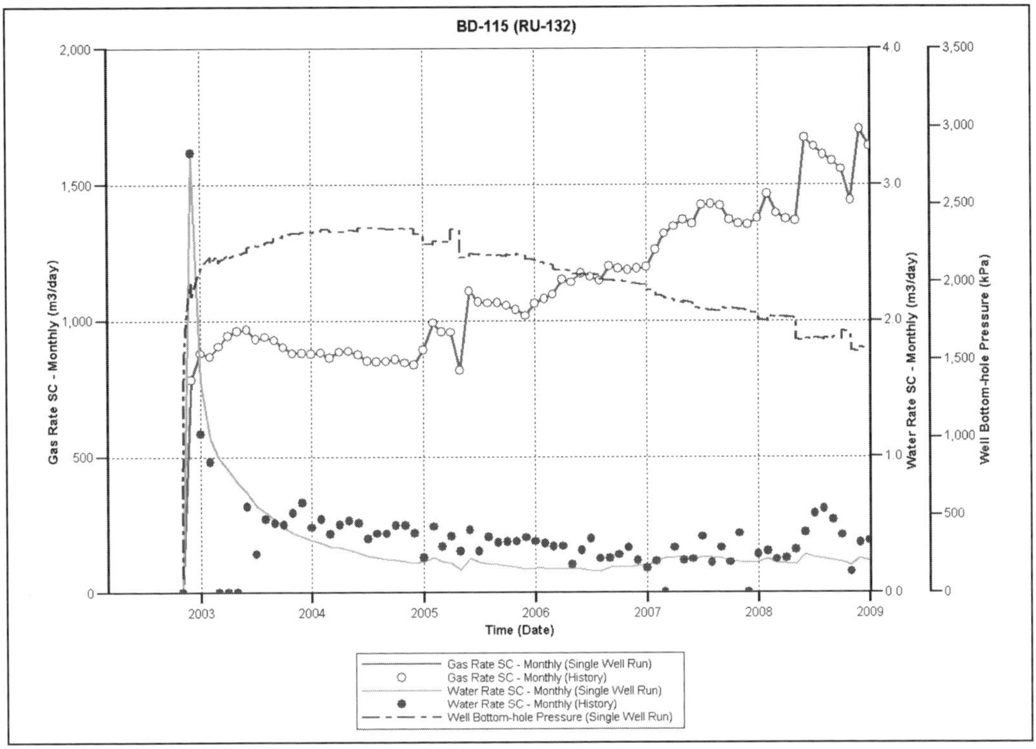

Figure 9. History match on gas and water rates for well BD-115 (RU-132) under a single well run. The bottom hole pressure is also shown.

the error on cumulative water while the operating constraint was to match the gas produced. The face to butt cleat permeability ratio was kept at 3 for all runs because preliminary runs showed that approximate ratio to be valid for all wells. Although the algorithm should examine all possible combinations, some parameter combinations are evidently not tested due to some internal optimization. In total 835 runs were completed using CMOST. Table 5 shows the results obtained for all wells.

In addition the optimum relative permeability curves were determined by the CMOST algorithm. Two of these diagrams are shown in Figure 11. The fact that different optimum curves are calculated for each well indicates that the relative permeability curves have a significant impact on the gas and water rate calculations. Optimum curves differ considerably between wells.

Figure 12 presents the reservoir simulation history-match results for BD-115 (RU-132) for a combined run where the optimal parameters from the combined well optimization were used. It is clear that the calculated water rate production matches the measured data better than in the case of the previous two models (single or combined run). At the same time the maximum pore pressure calculated for this production history match is higher than in previous runs, reaching almost 3,500 kPa at its maximum.

CONCLUSIONS

In this paper, the modeling procedure for coal bed methane production of selected wells in Russell County, VA, is discussed in detail. Through parametric analysis and optimization, the best parameters for this reservoir are identified. Utilizing these parameters, the model can accurately predict the gas production rate for all wells through history matching, while the total water produced from the reservoir matches reasonably well the measured values.

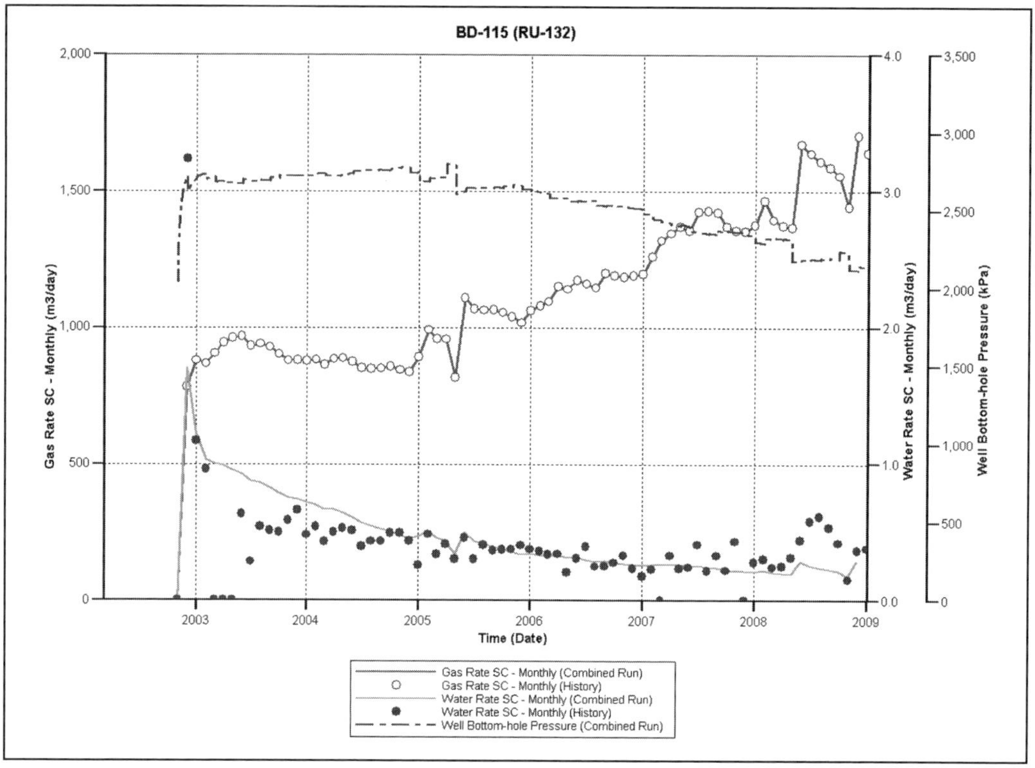

Figure 10. History match on gas and water rates for well BD-115 (RU-132) for the combined run. The bottom hole pressure is also shown.

Table 4. Parameter range used as input in the CMOST model runs

Fracture Porosity	Butt Cleat Permeability (mD)	Bottom Hole Pressure (kPa)	Water Saturation in the Cleats
0.0010	1	200	0.900
0.0015	2	500	0.925
0.0020	3	800	0.950
0.0025	4	1100	0.975
0.0030	5	1400	1.000
	6	1700	
		2000	

In addition, the model identifies the most sensitive parameters in the simulation and how these interact in closely spaced wells. The model developed here along with the optimal parameters that were calculated will be subsequently used for modeling of CO_2 injection and the corresponding enhanced gas recovery potential in the same group of seams.

ACKNOWLEDGMENTS

Financial assistance for this work was provided by the U.S. Department of Energy through the National Energy Technology Laboratory's Program under Contract No. DE-FC26-04NT42590 and DE-FE0006827.

Table 5. Optimal parameters after CMOST runs and manual adjustments for the multi-well runs

Well	Fracture Porosity	Face Cleat Permeability (mD)	Butt Cleat Permeability (mD)	Face to Butt Cleat Permeability Ratio	Bottom Hole Pressure (kPa)	Water Saturation in the Cleats
RU-84	0.0020	12.0	4.0	3	200	0.950
RU-85	0.0011	10.5	3.5	3	200	0.975
RU-112	0.0010	9.0	3.0	3	800	0.950
RU-123	0.0010	18.0	6.0	3	500	0.800
RU-132	0.0026	16.8	5.6	3	1100	0.975
RU-210	0.0015	9.0	3.0	3	1500	0.950
RU-211	0.0025	9.0	3.0	3	1700	0.950
RU-284	0.0010	9.0	3.0	3	800	0.900

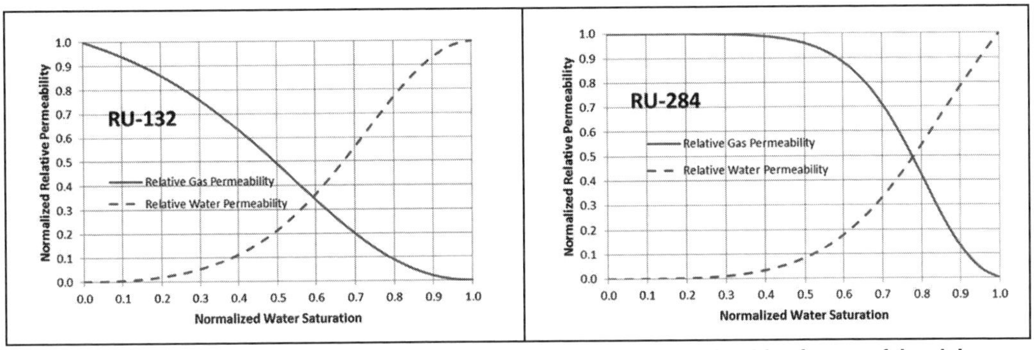

Figure 11. Relative permeability curves as determined by the optimization algorithm for two of the eight production wells (SWT: Water relative permeability or water saturation fraction; SLT: Gas relative permeability or gas saturation fraction)

REFERENCES

Gash, B.W., Volz, R.F., Potter, G., and Corgan, J.M. 1993. The effects of cleat orientation and confining pressure on cleat porosity, permeability and relative permeability in coal. Paper 9321 in Proceedings of the 1993 International CoalBed Methane Symposium. Tuscaloosa: University of Alabama.

Mavor, M.J.and Robinson, J.R., 1993, "Analysis of Coal Gas Reservoir Interference and Cavity Well," Paper SPE, 25860, presented at the Joint Rocky Mountain Regional and Low Permeability Reservoirs Symposium, Denver, CO, April 26–28.

Computer Modelling Group Ltd (CMG), C. (2003). GEM. Calgary, Canada.

Ripepi, N., Karmis, M., Miskovic, I., Shea C., and J.M. Conrad. 2009. Results from the Central Appalachian Basin Field Verification Test in Coal Seams. 26th Annual International Pittsburgh Coal Conference, Pittsburgh, PA, USA, 2009. CD-ROM of Proceedings, Paper # 33-4.

Karmis, M., Ripepi, N., Miskovic, I., Conrad, M., Miller, M., and C. Shea. 2008. CO_2 Sequestration in Unminable Coal Seams: Characterization, Modeling, Assessment and Testing of Sinks in Central Appalachia. Twenty-Fifth Annual International Pittsburgh Coal Conference, Pittsburgh, PA, USA, 2008. CD-ROM of Proceedings, Paper # 29-3.

Conrad, J.M., Miller, M.J., Phillips, J., Ripepi, N. 2006. Characterization of Central Appalachian Basin CBM Development: Potential for Carbon Sequestration and Enhanced CBM Recovery, 2006 International Coalbed Methane Symposium, Preprint 0625, Tuscaloosa, AL.

Virginia Department of Mines, Minerals and Energy (VaDMME). 2013. http://www.dmme.virginia.gov/DGO/Production/2010-County.pdf, accessed: February 2013.

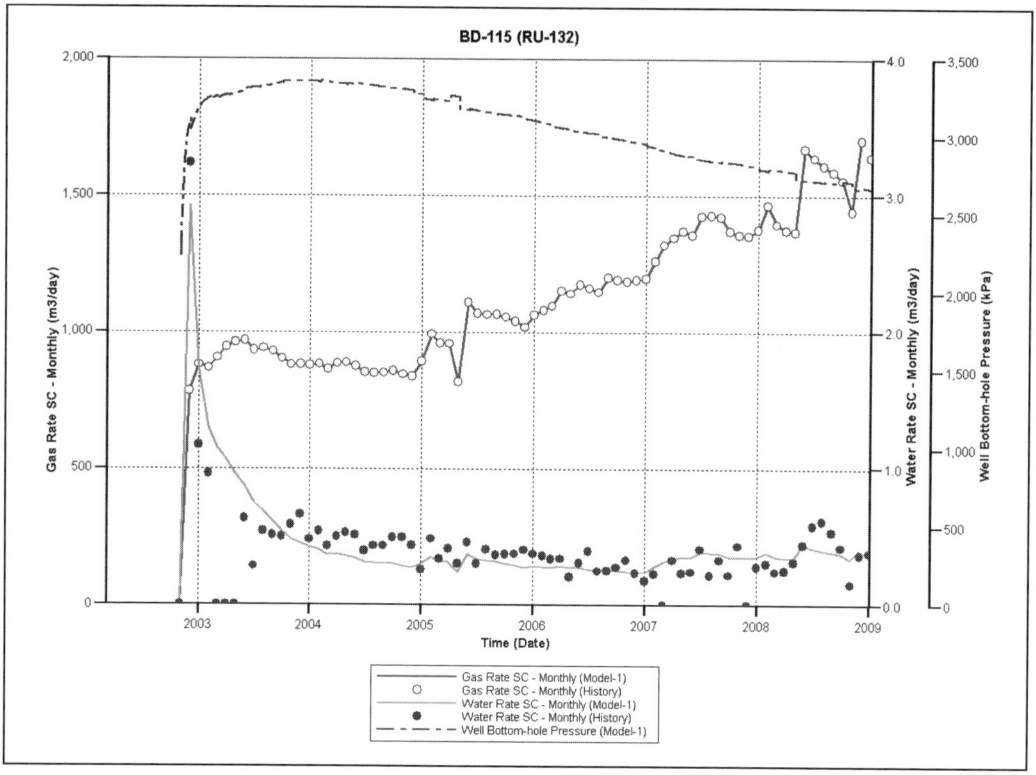

Figure 12. History match on gas and water rates for well BD-115 (RU-132) for the combined run with optimized input parameters. The bottom hole pressure is also shown.

Conrad, J.M. , Miller, M.J., Ripepi N. 2007. Potential for Carbon Sequestration in the Central Appalachian Basin. Air and Waste Management Association Conference, 2007, Paper # 187.

Vasilikou, F.C. Keles, Z. Agioutantis, N. Ripepi and M. Karmis, 2013. Model Verification of Carbon Dioxide Sequestration in Unminable Coal Seams with Enhanced Coal Bed Methane Recovery, to be published in 23rd World Mining Congress, August 11–15, Montreal, Canada.

Virginia Center for Coal and Energy Research (VCCER). 2011. Final Technical Report: Characterization and Field Validation of the Carbon Sequestration Potential of Coal Seams in the Central Appalachian Basin, 2011: Blacksburg, VA.

Virginia Department of Mines, Minerals and Energy (VaDMME). 2002. "CBM-PGP-BD114 Completion Report," Form DGO-GO-15, 105 p.

Harpalani, S, personal communication, December 14, 2012.

Sustainable Degasification Strategies: A Case Study in the Qinshui Basin of China

Steven A. Keim
Cardno MM&A, Blacksburg, Virginia

Kray D. Luxbacher
Department of Mining & Minerals Engineering, Virginia Tech, Blacksburg, Virginia

Michael E. Karmis
Virginia Center for Coal and Energy Research, Virginia Tech, Blacksburg, Virginia

ABSTRACT

China's Qinshui Basin holds coals of significant thickness and gas content. Over the course of the past decade, substantial progress in the development of the Qinshui Basin gas resources has been realized. While several published works have analyzed the geology of the basin, very limited production data has been made public. Both hydraulically stimulated vertical fracture wells and multilaterally drilled horizontal wells have been used in the region, with the latter much less common. This study serves as a framework in analyzing the sustainable production of multiple horizontal and vertical well completion strategies. Sustainability is realized through optimization of environmental, social, and economic factors. This work specifically considers greenhouse gas emissions, mine safety, and economic return. Three-dimensional reservoir modeling software shows that multilaterally drilled horizontal drainage patterns offer the most economic and environmentally friendly results when used in advance of mining. Based on the obtained modeling results, the high capital investment of horizontal wells is quickly offset by high initial gas production rates. Furthermore, high drilling density patterns with closely spaced laterals decrease the variability in seam gas content ahead of mining, providing the potential for increased health and safety of the underground workforce and increased coal production. From a greenhouse gas emission standpoint, a decrease in carbon dioxide equivalent emissions of 15-million metric tons can be achieved over a 20-year mining period for high-capacity longwall operations in the Qinshui Basin through careful degasification planning. Although hydraulically fractured vertical wells have been used extensively, modeling concludes that horizontal completions are more sustainable.

This publication was developed under Assistance Agreement No. XA–833963–01 awarded by the U.S. Environmental Protection Agency. It has not been formally reviewed by EPA. The views expressed in this document are solely those of the authors and EPA does not endorse any products or commercial services mentioned in this publication.

INTRODUCTION

Scope of Study

Multiple studies have been published to describe the coalbed-methane-related geology of the Qinshui Basin (Su et al. 2005; Sang et al. 2009; Wei et al. 2005). Limited information pertaining to actual production data in the Qinshui Basin has been made public, with the majority of published work being generalized and offering ranges of obtained production values. In lieu of production data, detailed three-dimensional reservoir modeling offers the opportunity to predict the production capabilities of multiple degasification strategies. This paper, which examines multiple well completion strategies, serves as modeling work analyzing the benefits of degasification on safety, the environment, and economics.

Modeling inputs for the work have been obtained from a greenfield concession area within the Qinshui Basin. In an effort to maintain confidentiality and protect the mineral lessee, specifics pertaining to the site location are withheld. Outcomes presented in this work can be used on an international basis as a framework for determining global degasification benefits.

Coal Geology of the Study Area

The Southern Shanxi Province (southeastern Qinshui Basin) contains thick, gassy coals targeted both as a coal and coalbed methane resource. The primary coal beds targeted for mining and coalbed methane development in the Qinshui Basin are the No. 3 and No. 15 seams (Zou et al. 2005), both anthracitic in rank and formed from a tectonic thermal event (Su et al. 2005). The seam depths range from 600 to 2,100 feet (Su et al. 2005). An overlying aquifer causes difficulty in dewatering the No. 15 seam. Therefore, the No. 3 seam is used as the basis of modeling for this work.

The No. 3 seam has thicknesses greater than 20 feet in some areas, and has been well explored. Structurally, the region is characterized by many north-trending anticlines and synclines. Faulting is also prominent, with normal and reverse faults of substantial throw and displacement.

Exploration programs have not noted significant water-bearing aquifers proximal to the No. 3 seam. Still, the seam is saturated with water, which must be removed prior to the initiation of gas flow. The No. 3 seam is overlain and underlain by both shales and sandstones, some of which are carboniferous. Overall, the structural setting of the basin is favorable for horizontal well development.

Historical CBM Production in the Qinshui Basin

Surface-drilled completion strategies, which have been effectively implemented in the Qinshui Basin, include traditional hydraulically fractured vertical wells and multilaterally drilled horizontal wells (Luo et al. 2009). Over the course of the past decade, the number of coalbed methane wells in the Qinshui Basin has increased dramatically from less than 250 to over 2,500 (Sang et al. 2009; Luo et al. 2009). The majority of the wells are vertical wells with hydraulically fractured stimulation treatments. As of 2009, less than 30 multilaterally drilled horizontal wells had been developed, attributed to higher capital costs (Luo et al. 2009). Hydraulically fractured vertical wells in the Qinshui Basin produce gas at a rate between 70,000 and 140,000 cubic feet per day (Su et al. 2005). Modeling completed by Luo et al. (2009) estimates that horizontal wells in the basin should produce at rates around 700,000 cubic feet per day.

Overview of Greenhouse Gas Considerations

The production and combustion of fossil fuels as an energy source releases large volumes of greenhouse gases to the atmosphere, particularly carbon dioxide and methane. Carbon dioxide, formed from the combustion of fossil fuels, represents the largest anthropogenic greenhouse gas. Methane, the main constituent of natural gas, possesses radiative trapping capabilities roughly 20 times greater than that of carbon dioxide (Seinfeld and Pandis 2006, USEPA 2011). Gas contents of deep coal seams located in China's Quinsui Basin can exceed 500 cubic feet per ton (Su et al. 2005), resulting in significant volumes of methane being released through the atmosphere during the mining process. The United States Environmental Protection Agency (EPA) indicates that CO_2 equivalent (CO_2E) emissions for the United States and China are estimated to be 72,000 and 130,000 metric tons, respectively (USEPA 2006). Recovering this methane prior to mining provides a viable, clean-burning energy source that can be used in residential and commercial applications.

Additionally, mining health and safety is closely tied to environmental impacts. Effective degasification, coupled with capture of methane at a wellhead and either flaring of the gas or transport to a pipeline, reduces emissions in underground mines, improving health and safety by

Table 1. Modeling inputs and associated ranges in Qinshui Basin

Parameter	Modeled	Measured	Reported Regional Values Min	Reported Regional Values Max	Regional Value Source(s)
Depth (ft)	2000	Yes	650	2,300	(Su et al. 2005)
Coal thickness (ft)	10	Yes	1.6	23	(Wei et al. 2007)
Water saturation (%)	85	No			
Fluid pressure gradient (psi/ft)	0.3	Yes	0.30	0.48	(Zou et al. 2010; Sang et al. 2009)
Reservoir pressure (psi)	600	Calculated			
Permeability (md)	2.0	Yes	0.1	2.0	(Su et al. 2005; Yao et al. 2008)
Porosity (%)	3	No	1.15	7.69	(Zou et al. 2010)
Langmuir volume (ft^3/ton)	1000	Yes	1450	1870	(Zou et al. 2010)
Langmuire pressure (psi)	300	Yes	422	467	(Zou et al. 2010)
Gas content (ft^3/ton)	500	Yes	280	880	(Lu et al. 2010; Yao et al. 2008; Su et al. 2005)

reducing the risk of explosion and also reducing the emission of ventilation air methane (VAM) to the atmosphere.

Mine safety is a substantial issue in China, and the impact of government regulations on the safety of the Chinese coal workforce has appeared regularly in the literature (Shi 2009). While the Chinese safety record has improved recently, grave numbers are still evident: between 2000 and 2007, the fatality rate in China was 3.04 per million tons of mined coal, compared to 0.027 in the US (Wei 2009; Lu and Li 2011). Methane-related explosions are the root cause of the majority of Chinese coal-related fatalities. According to 2007 data, methane was linked to 51.2 percent of large incidents (three to nine fatalities) and 80.3 percent of extremely large accidents (> ten fatalities) (Cheng et al. 2011). Effective degasification has the potential to significantly decrease the staggering coal-related fatality rates in China.

The impacts of degasification as they relate to mining, as well as the path methane takes to the atmosphere, whether combusted as fuel or released as ventilation air methane (VAM), is analyzed in the following sections.

RESERVOIR MODELING

Geological Modeling Inputs

Reservoir modeling utilizes Advanced Resources International's (ARI) COMET3 reservoir simulator. Modeling inputs, as assumed or obtained from the greenfield property, are shown below in Table 1. Other values were derived from the literature.

As evident in Table 1, all obtained geological inputs from the greenfield property are representative of published values for the region with the exception of the Langmuir coefficients. The Langmuir isotherm curve and initial reservoir conditions are plotted in Figure 1. The initial reservoir pressure of 600 psi is calculated by multiplying the fluid pressure gradient (0.3 psi per foot) by the coal seam depth (2,000 feet). While the measured Langmuir coefficients are not within the range of those published by Zou et al. (2010), the degree of saturation is within the general range stated by Su et al. (2005), who find the degree of saturation of Qinshui coals to range between 56 and 100 percent. Based on the obtained Langmuir coefficients, initial gas content, seam depth, and fluid pressure gradient, the critical desorption pressure is calculated to be 300 psi. This requires a decrease in reservoir pressure from 600 to 300 psi for gas to begin desorbing from the coal matrix.

Relative permeability curves obtained by Kissell and Edwards (1975) study were used in modeling. While the relative permeability curves were based on a Pocahontas No. 3 coal sample, the Qinshui reservoir exhibits similar geological qualities as the Pochahontas No. 3 seam, with a

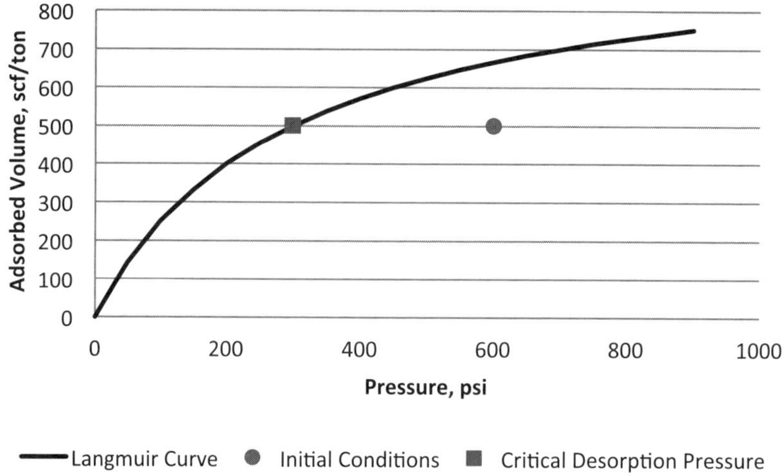

Figure 1. Langmuir isotherm with initial conditions and critical desorption pressure

similar depth, fluid pressure gradient, methane concentration, and fracture network. Although relative permeability will vary by coal rank among other geologic parameters, no data was available pertaining to relative permeability within the Qinshui basin. The variations in coal rank between the Pocahontas No. 3 and Qinhsui coals could introduce error in modeling output. Still, the Kissell and Edwards (1975) Pocahontas No. 3 curves are included and considered representative based on the aforementioned geologic consistencies between the two basins. Additionally, permeability anisotropy is excluded from modeling, with the face cleat and butt cleat directions modeled with the same permeability value of 2.0 md.

Modeled Well Geometry

An infinite number of horizontal wellbore patterns exist. Common drainage patterns include the single lateral, dual lateral, quadrilateral, and Pinnate (Maricic et al. 2008; Palmer 2010). Pitchfork patterns, as analyzed by Keim et al. (2011), offer an alternative to Pinnate-style patterns with the opportunity for higher gas recovery. Success has been noted in the Shanxi Province through herringbone or fishbone patterns, which are geometrically similar to the Pinnate pattern (Zhiming and Zhang 2009). To provide a comparative basis, the wellbore patterns depicted in Figure 2 are modeled. Their corresponding drainage areas and required in-seam drilling length are shown in Table 2.

The wells depicted in Figure 1 and Table 2 provide a comparative basis in terms of drainage area and drilling length. The modeled vertical fracture well can be compared to a single lateral well, as it is designed to drain the same area (79 acres). The single lateral and dual lateral wells both provide drilling densities of 41 feet per acre. The Turkey Foot and Pinnate-style patterns are modeled to drain 250 acres, with the Pinnate pattern requiring substantially more in-seam drilling footage. The presented wells allow for multiple conclusions to be drawn with respect to drainage area, drilling density, and completion strategy when wells are analyzed on mining, environmental, and economic criteria.

Grids for the modeled wells utilize individual blocks, which are 100 feet by 100 feet. As the vertical fracture well production will be highly dependent on the stimulated fracture length, half fracture lengths were modeled as a worst (lowest), expected, and best (highest) case. Half fracture lengths for the vertical wells were set to 500 (worst), 700 (expected), and 900 (best) feet.

The model is governed based on water production limitations and well pressure constraints. All wells are assumed to produce water at a

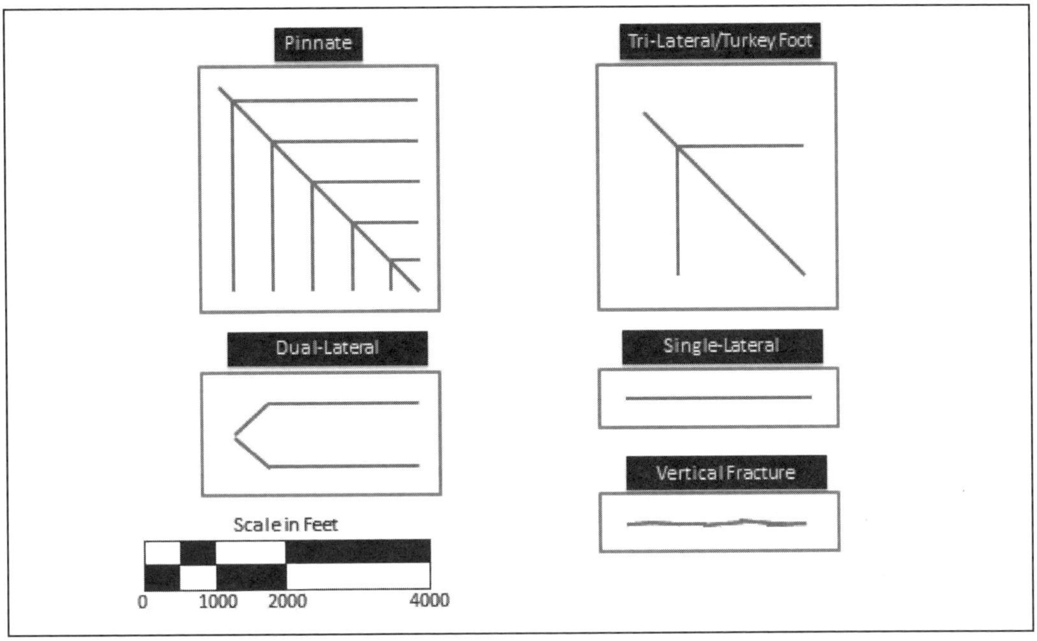

Figure 2. Plan view of modeled well geometry

Table 2. Drainage area and required in-seam drilling for modeled wells

Parameters	Pinnate	Turkey Foot	Dual Lateral	Single Lateral	Vertical Fracture
In-seam drilling length (ft)	20,000	7,100	6,800	3,200	—
Drainage area (acres)	250	250	166	79	79
Drilling density, in seam footage per acre drained (ft/acre)	80	28	41	41	—

maximum rate of 250 barrels per day until bottom hole pressure at each well is reduced from the initial reservoir pressure to atmospheric pressure. Upon reaching the atmospheric conditions, water is continually pumped at a sufficient rate to maintain atmospheric conditions.

Modeling Results

Production rates over a 20-year period for the horizontal wells and the expected 700-ft half fracture length vertical well are displayed in Figure 3. Figure 4 displays the production rates for the worst-, expected-, and best-case vertical fracture wells. Cumulative production charts for the same well layouts are depicted in Figure 5 and Figure 6.

As shown in Figure 3 and Figure 5, the Pinnate well out-produces all other horizontal wells in terms of production rate and cumulative production. The rapid reservoir depressurization from the high in-seam drilling length of the Pinnate pattern recovers methane faster than any other horizontal pattern, particularly for this low permeability application. Compared to the Turkey Foot pattern which drains a similar area, the Pinnate produces nearly double the gas over a 20-year span. Modeling shows that the gas production rate of the Turkey Foot pattern is more sustained over the 20-year span, with a lower total cumulative production figure. As expected, the dual lateral configuration produces roughly double the gas compared to the single lateral completion. Having the same drainage area, the single lateral well produces roughly 80 percent more gas than the expected case for the vertical fracture

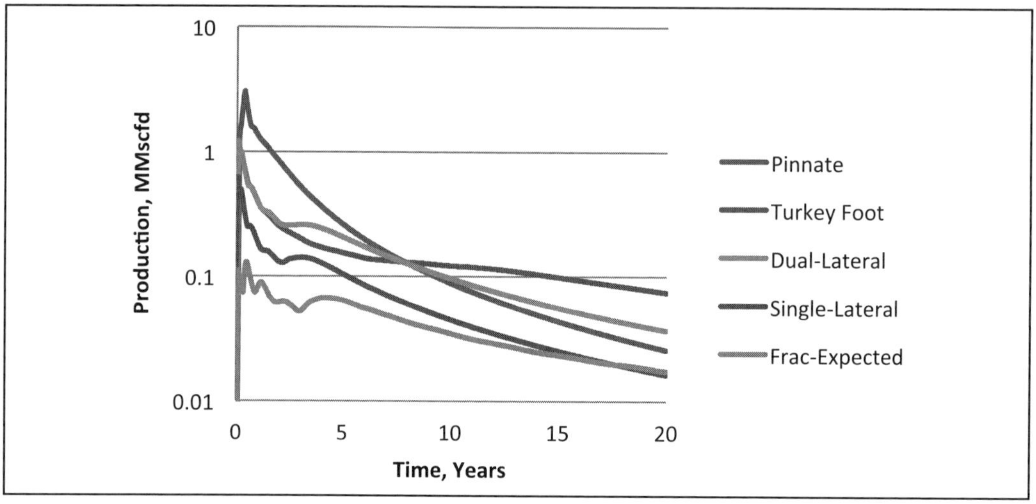

Figure 3. Production rates for horizontal well layouts and expected fracture well

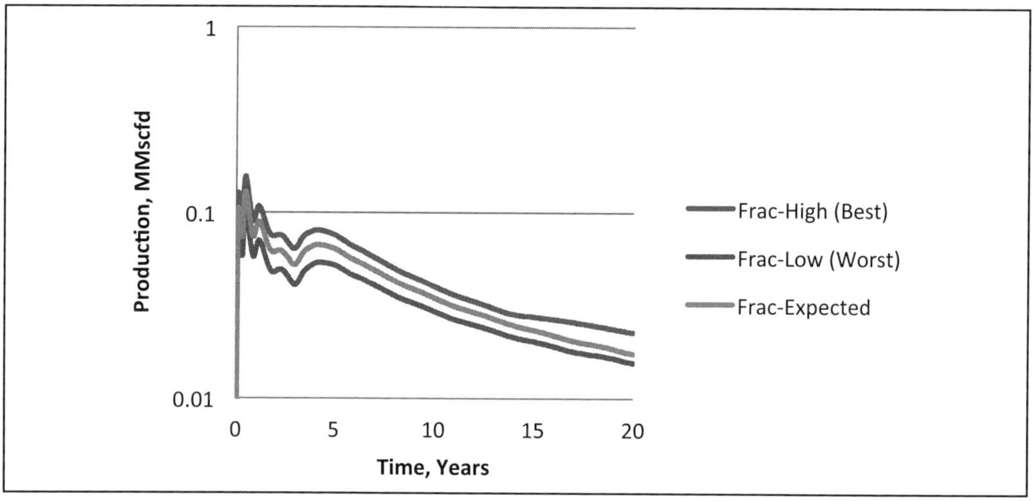

Figure 4. Production rate comparison for best, expected, and worst case vertical fracture completions

completion. Modeling shows that the half fracture length dictates both the production rate and cumulative production figure for hydraulically fractured vertical well treatments. The production rates and cumulative production curves are directly proportional to modeled half fracture length. Over a 20-year period, fracture length can be attributed to a nearly 150 million cubic foot increase in production when comparing the best-case and worst-case models.

While the production figures obtained from modeling provide a comparative basis to analyze well performance, additional aspects of each well layout must be analyzed to optimize performance. The following sections examine the obtained modeling results based on greenhouse gas emissions and economic considerations.

GREENHOUSE GAS CONSIDERATIONS

Analysis of Reservoir Modeling

Regulatory standards require that the gas content of Chinese coal seams be lowered to a threshold value of 280 cubic feet per ton prior

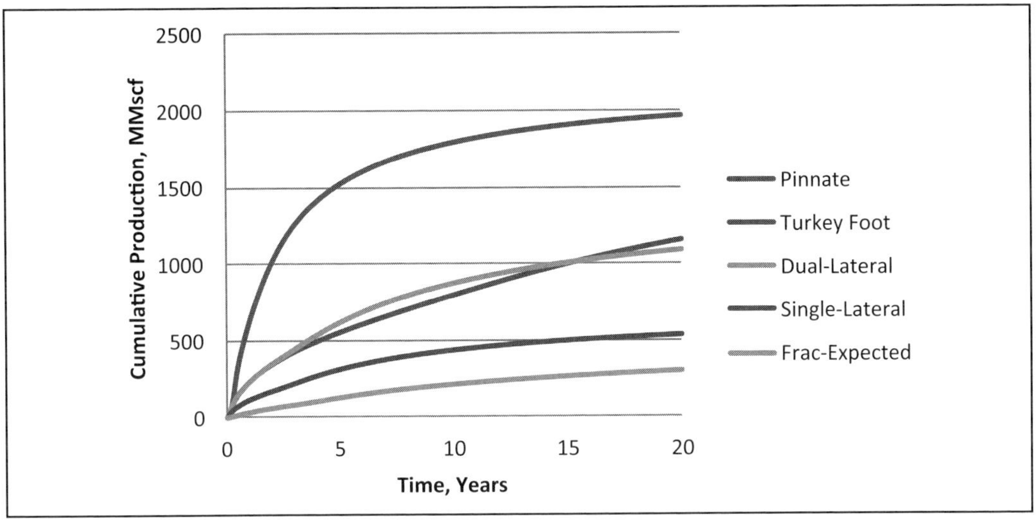

Figure 5. Cumulative production for horizontal well layouts and expected fracture well

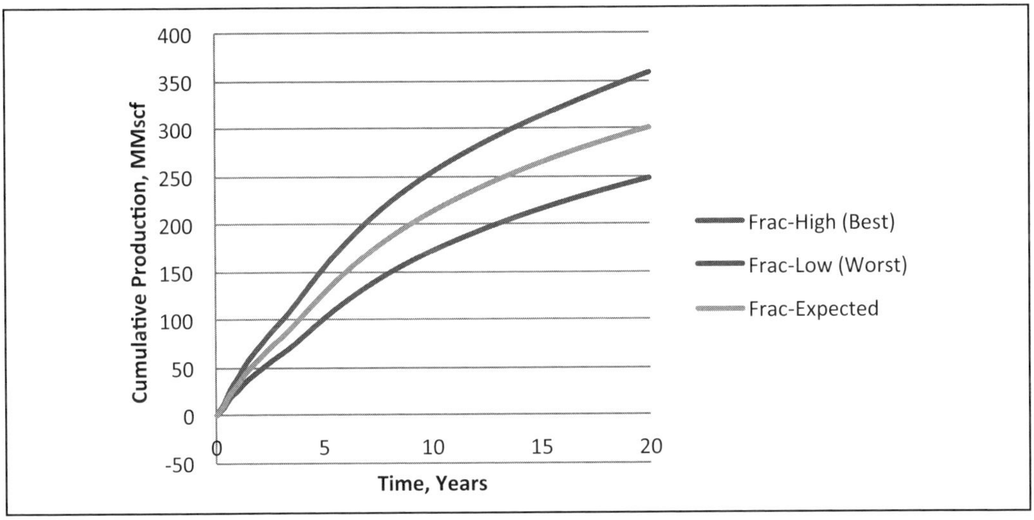

Figure 6. Cumulative production comparison for best, expected and worst case vertical fracture completions

to mining, corresponding to a gas recovery rate of approximately 57 percent for this case study. When analyzing the recovery of coalbed methane, total recovery values for an entire drainage area can be skewed, with large changes in recovery at varying distances from the producing wellbore. An average recovery of 57 percent actually represents a higher recovery proximal to the wellbore and significantly lower recoveries away from the wellbore. This creates a potential safety hazard, because the gas recovery gradient could actually represent large variations in coal seam gas content. Figures 7 and 8 contain schematics comparing the spatial variation in recoveries of Pinnate- and Turkey Foot-style patterns after two years and 20 years of production, respectively. In Figure 7, both wells display large variations in gas content. Figure 8 depicts much less variation in gas content for the Pinnate well, with the entire drainage area exhibiting about

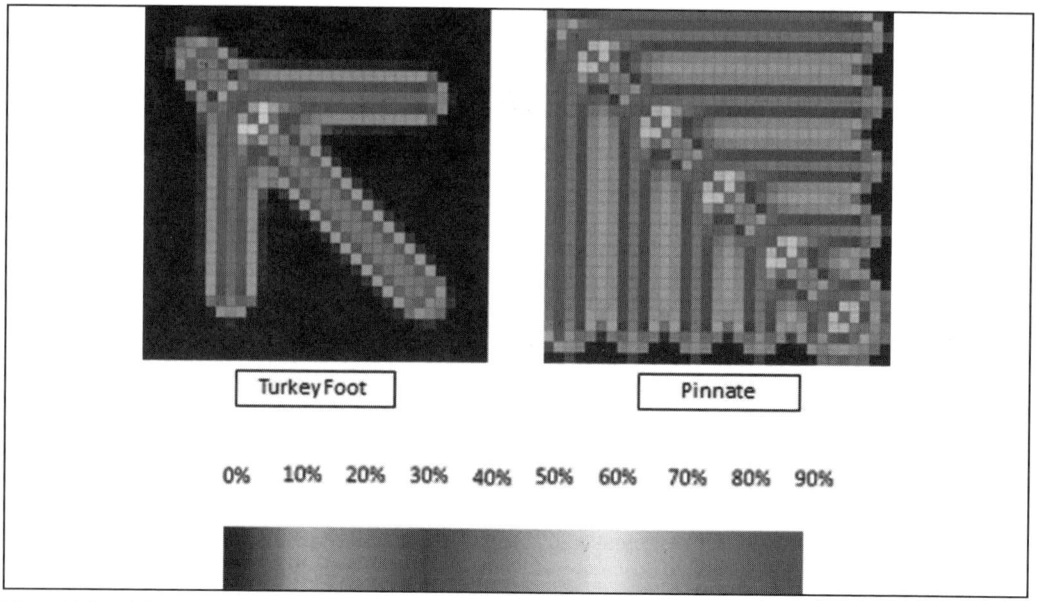

Figure 7. Spatial recovery comparison of Turkey Foot and Pinnate wells after 2 years of production

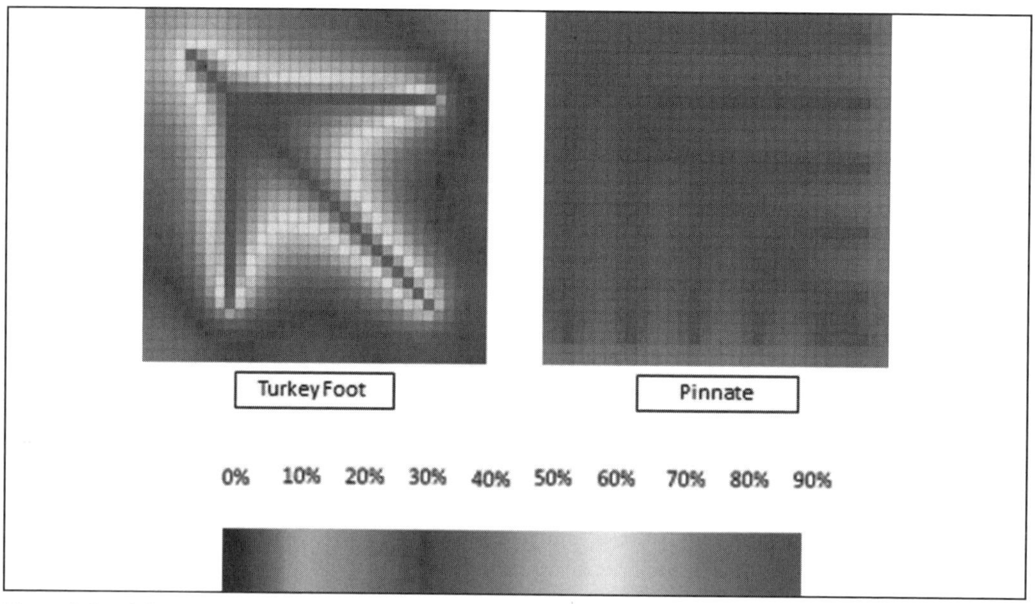

Figure 8. Spatial recovery comparison of Turkey Foot and Pinnate wells after 20 years of production

90-percent recovery. The Turkey Foot-style well shows high recoveries proximal to the wellbore with significantly lower recoveries away from the wellbore. Based on these figures, the Pinnate-style pattern shows much more favorable characteristics for degasification ahead of mining, despite the added expense related to more in-seam drilling.

In an effort to compare the spatial recoveries of all modeled well layouts, a histogram was created which displays the percentage of modeled grid blocks for each well layout with specified gas

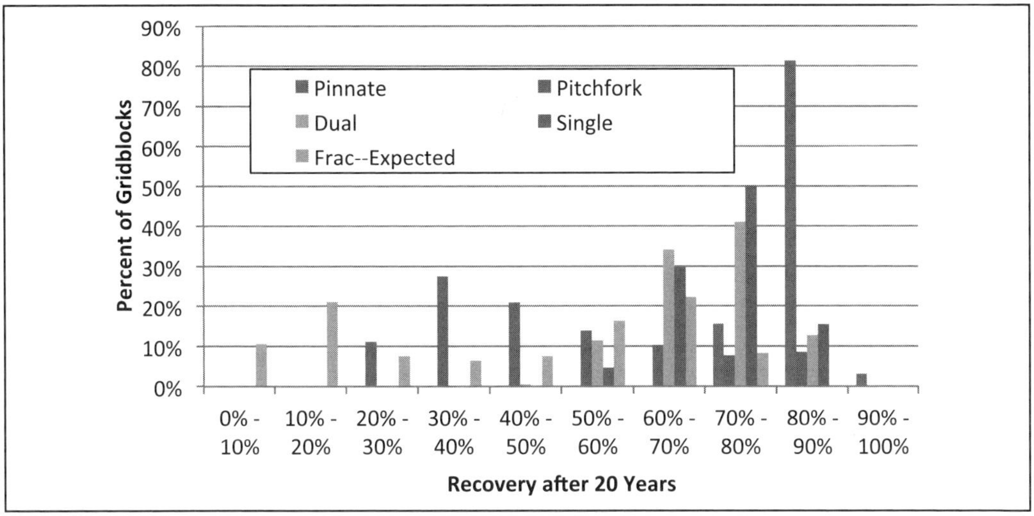

Figure 9. Recovery analysis by grid blocks after 20 years of production

recoveries after 20 years of production. Figure 9 displays the histogram.

As shown in Figure 9, the hydraulically fractured well exhibits a very large variation in gas recovery, similar in distribution to the Turkey Foot pattern. Gas recovery values for the hydraulically fractured well vary from 0 to 80 percent. The Turkey Foot pattern recovery values range from 20 to 90 percent. In both cases, there are large areas that are not effectively drained. By decreasing the drainage areas and subsequently increasing the number of wells in a given area, these distributions will shift to the right with more areas exhibiting higher recoveries. The single lateral and dual lateral wells show similar distributions, with at least 50-percent recovery across the entire drainage area. The close horizontal lateral spacing of the Pinnate well shows the most impressive performance. After a 20-year period, all targeted coal has been drained to at least 70-percent recovery, with the majority in the 80- to 90-percent range. Modeling results for the Qinshui's geological conditions suggest that multilaterally drilled horizontal patterns offer consistent recoveries across a drainage area.

Effective degasification provides the opportunity for improved health and safety of miners and increased productivity. Wells with a high drilling density per unit area, such as the modeled Pinnate well, exhibit the highest recoveries and smallest variations over the drainage area. As applied to the Qinshui Basin, after a 20-year period, the Pinnate well is the only modeled well that drains all coal to the regulatory standard of 280 cubic feet per ton. Modeling shows that the fast recovery time of the Pinnate (or a similar pattern) is unquestionably the best method for degasification ahead of coal mining in the Qinshui Basin.

Calculating and Forecasting Emission Reductions

The following sections outline the potential carbon dioxide equivalent emission reductions associated with degasification ahead of a hypothetical coal operation in the Qinshui Basin. The methodology of determining equivalent emissions is presented as applied to the hypothetical case study. Feasible recovery figures are discussed based on reservoir modeling output.

Provided that either: (a) detailed reservoir modeling has been used to predict well production, or (b) good estimates of production for degasification wells can be made based on regional data where geologic conditions are consistent, emission reduction estimates can be made using a simple methodology. For the purpose of this study, carbon dioxide equivalent

Table 3. Hypothetical Qinshui longwall mine model inputs

Parameter	Value
Study period (years)	20
Coal thickness (ft)	10
Clean coal annual production (million tons/year)	4
Clean coal specic gravity	1.55
Extraction ratio	90%
Plant yield	85%
Affected area (acres)	4964
Total affected coal (million tons)	104.6
Total affected volume of gas (bcf)	52.3

emission reductions are calculated on a coalbed methane recovery basis. Recovered methane will be combusted and released to the atmosphere in the form of carbon dioxide. Methane that is not recovered is vented to the atmosphere, resulting in equivalent carbon dioxide emissions. In an effort to determine the true environmental impact of degasification as applied to a pre-mining situation, a hypothetical longwall mine is utilized.

Mining-based assumptions for the hypothetical Qinshui longwall mine are outlined in Table 3. Over a period of 20 years, clean coal is produced at a constant rate of 4 million tons per year. Coal is assumed to have a thickness of ten feet, consistent with degasification modeling in a previous section. Clean coal specific gravity is estimated to be 1.55, higher than bituminous coals due to its anthracitic rank. Extraction ratio and plant yield are respectively assumed to be 90 percent and 85 percent. Using these inputs, the longwall mine will encompass an area of nearly 5,000 acres over the 20-year study period. Within this 5,000-acre area, 52.3 billion cubic feet of gas will be exposed with the opportunity for recovery prior to mining or ventilation to the atmosphere during mining.

Calculations of equivalent carbon dioxide emissions on a recovery basis require knowledge of four significant factors:

1. Concentration of methane in produced gas. The concentration of methane in produced gas for the Qinshui Basin is relatively high. For the purpose of this study, a methane concentration of 90 percent is utilized. The remaining ten percent is comprised of constituents assumed to have an insignificant atmospheric heat-trapping impact. Of the four discussed significant factors, this is the only factor which will vary on a site-specific basis.
2. Conversion factor to convert volume of methane (ft^3) to mass of methane (metric tons). One cubic foot of methane has a mass of 1.93 × 10–5 metric tons (USEPA 2011).
3. Global warming potential (GWP) of methane to convert mass of methane (metric tons) to mass of carbon dioxide (metric tons). The global warming potential represents the heat trapping capabilities of a compound in relation to the heat trapping capabilities of an equivalent mass of carbon dioxide over a 100-year period. For the purpose of this study, the value of 21 is used (USEPA 2011), although values of up to 23 may also be found in the literature (Seinfield and Pandis 2006).
4. Stoichiometric conversion to convert combusted methane mass (metric tons) to carbon dioxide mass (metric tons). The combustion of methane results in the creation of one molecule of carbon dioxide for each molecule of methane. Based on the molar masses of each molecule, 1 metric ton of combusted methane forms 2.75 metric tons of carbon dioxide.

The equivalent emissions of ventilated methane are outlined in Equation 1.

$$M_E = V_R C(1.93 \times 10^{-5})(21) \quad (1)$$

where M_E represents the metric tons of equivalent carbon dioxide emissions by ventilating a volume of gas, V (cubic feet), with a specified methane concentration, C (fraction). As previously discussed, the remaining constants convert the volume of methane to a mass of methane, then to an equivalent mass of carbon dioxide.

Recovered, combusted methane will emit the mass of carbon dioxide presented in Equation 2.

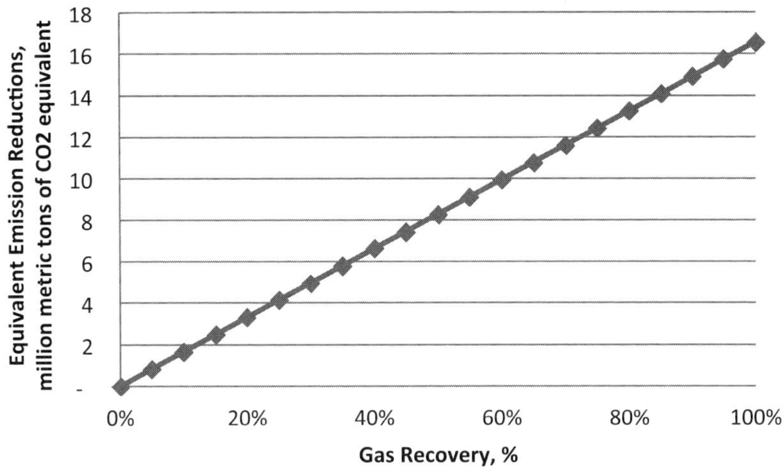

Figure 10. Emission reductions for hypothetical Qinshui longwall operation at various gas recoveries

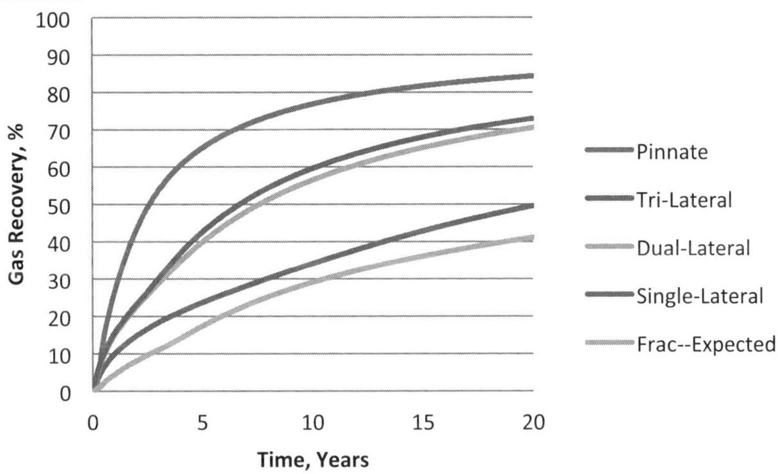

Figure 11. Recovery vs. time for horizontal and expected vertical fracture wells

$$M_C = V_R C(1.93 \times 10^{-5})(2.75) \qquad (2)$$

where M_C represents the metric tons of emitted carbon dioxide by combusting a specified volume of recovered gas, V_R (cubic feet), at a given methane concentration, C (fraction).

By applying Equations 1 and 2 to the volume of exposed gas presented in Table 3, emission reductions based on various gas recoveries can be calculated. Figure 10 displays emission reductions in million metric tons of carbon dioxide equivalent emissions based on gas recoveries from zero to 100 percent for the hypothetical Qinshui longwall mine.

Reservoir modeling was conducted over a 20-year period, during which each well layout reaches a different level of recovery. Figure 11 contains the required drainage time for wells to reach specified recoveries. This offers a planning tool, as cross referencing target emission reductions in Figure 10 with gas recoveries in Figure 11 show the range of potential drainage patterns which can be used.

The Pinnate-style well exhibits the highest recovery over a 20-year period, followed by the single lateral, dual lateral, trilateral, and vertical fracture well. Increasing the recovery of each well can be obtained by decreasing the drainage area, subsequently requiring more wells to be drilled on a per-unit-area basis. The Pinnate-style well becomes asymptotic to a recovery value of nearly 90 percent. This corresponds to a decrease CO_2E of roughly 15 million metric tons when using multiple Pinnate wells ahead of a 20-year longwall mining plan in the Qinshui Basin.

A variety of wellbore patterns and degasification strategies can be implemented, as drainage time ahead of mining will vary on an individual well basis in accordance to the mine plan. For example, 50-percent recovery can be obtained in 20 years with a Turkey Foot-type pattern, 7 to 8 years with dual and single lateral setups, and in less than 3 years with a Pinnate-style well. A detailed financial analysis can be used to optimize this strategy, as briefly described in the following section.

ECONOMIC EVALUATION

A detailed financial analysis represents a key factor in the design and selection of a wellbore pattern. The following sections outline the process in determining the economically optimum drainage pattern based on those patterns discussed in the preceding sections. Annual production values obtained through reservoir modeling in a previous section are used in the discounted cash flow study. A sensitivity analysis is also included to identify those factors which have the most significant impact on the financial assessment.

Financial Model Inputs

Table 4 outlines the inputs as used for the financial modeling assessment. Cash flows are discounted over a 20-year project period at a rate of ten percent. The financial model inputs are assumed to be accurate for the Qinshui Basin based on the authors' past affiliations with producing entities in the region. Gas prices start at a value of $5.60 per million BTU. Processed gas is assumed to have a heating value of 950 BTU per cubic foot. Gas prices escalate at a rate of three percent over the first 15 years of the project, and then become constant for the final five years. A gas shrinkage volume equivalent to seven percent of the processed gas is incorporated to account for the lost produced gas utilized for gas compression and transmission. Gas gathering costs and overhead costs have respective values of $0.33 and $0.45 per thousand cubic feet, with costs escalating at an annual rate of 3 percent throughout the project life. The financial assessment represents a pre-tax and pre-royalty model.

Required capital expenditures for each well configuration are summarized in Table 5. The Pinnate-style well requires 20,000 feet of in-seam drilling and is assumed to have a capital expenditure of $1.85 million. The remaining horizontal wells incorporate the baseline Pinnate cost and an incremental price reduction of $40 per linear foot of in-seam drilling. This reduction accounts for decreases in required capital expenditures for horizontal wells with lower in-seam drilling lengths, as they will require less drilling time. Vertical fracture wells are assumed to cost $400,000. Table 5 also includes operating costs and maintenance costs, which vary for horizontal and vertical wells.

Financial Modeling Results

Incorporating the discussed financial modeling inputs with production values obtained from reservoir modeling results yield the cash flows summarized in Table 6. Each well configuration's internal rate of return (IRR) and net present value

Table 4. Financial modeling inputs

Parameter	Value Assumed
Cash flow discount rate	10%
Initial gas price ($/MMbtu)	5.60
Heating value adjustment (btu/scf)	950
Gas price escalation (years 1–15)	3%
Gas price escalation (years 16–20)	0%
Gas shrinkage (compressor fuel)	7%
Gas gathering cost ($/mcf)	0.33
Overhead cost ($/mcf)	0.45
Cost escalation	3%

Table 5. Capital and operating expenditures by well

	Pinnate	Turkey Foot	Dual Lateral	Single Lateral	Fracture
In-seam drilling length (ft)	20,000	7,100	6,800	3,200	X
Drainage area (acres)	250	250	166	79	79
Base capital (million $)	1.185	1.185	1.185	1.185	0.4
Capital adjustment ($/ft)	40	40	40	40	X
Capital adjustment	-	516,000	528,000	672,00	X
Adjusted capital (million $)	*1.85*	*1.334*	*1.322*	*1.178*	*0.4*
Year 1 fixed operating expenses ($/month)	7,500	7,500	7,500	7,500	4,000
Year 2 fixed operating expenses ($/month)	5,000	5,000	5,000	5,000	2,500
Year 3+ fixed operating expenses ($/month)	2,000	2,000	2,000	2,000	1,500

Table 6. Annual cash flow, project NPV, and IRR by well

Year	Pinnate	Turkey Foot	Dual Lateral	Single Lateral	Frac (Low)	Frac (Expected)	Frac (High)
0	$(1,850,000)	$(1,334,000)	$(1,322,000)	$(1,178,000)	$(400,000)	$(400,000)	$(400,000)
1	$2,500,583	$879,792	$879,509	$386,548	$58,664	$87,264	$115,863
2	$1,598,605	$421,624	$438,019	$179,964	$60,770	$85,321	$109,872
3	$1,039,033	$336,820	$390,418	$194,811	$56,318	$76,252	$96,185
4	$692,827	$279,224	$396,709	$202,992	$61,910	$83,783	$105,650
5	$496,716	$252,236	$358,276	$173,471	$73,351	$95,205	$117,054
6	$375,778	$235,085	$308,142	$142,120	$68,210	$87,325	$106,442
7	$295,583	$223,571	$264,792	$117,364	$60,571	$77,119	$93,670
8	$239,211	$222,459	$230,178	$98,237	$53,399	$67,855	$82,314
9	$197,848	$222,855	$202,274	$83,260	$47,342	$60,093	$72,850
10	$166,462	$221,600	$179,496	$71,360	$43,339	$54,708	$66,088
11	$141,996	$221,987	$160,713	$61,741	$39,101	$49,353	$59,616
12	$122,504	$222,475	$145,043	$53,849	$36,113	$45,447	$54,798
13	$106,680	$220,055	$131,809	$47,267	$33,984	$42,547	$51,124
14	$93,616	$215,330	$120,510	$41,711	$31,467	$39,379	$47,363
15	$82,686	$209,444	$110,786	$36,970	$29,749	$37,108	$45,976
16	$70,084	$195,462	$97,980	$30,921	$27,073	$33,709	$43,994
17	$59,468	$182,196	$86,869	$25,710	$23,935	$29,945	$41,596
18	$50,443	$169,697	$77,169	$21,187	$21,552	$27,022	$39,011
19	$42,701	$157,960	$68,639	$17,233	$19,818	$24,698	$36,283
20	$35,229	$144,727	$59,981	$13,260	$17,126	$21,345	$32,858
NPV	$4,142,627	$1,421,528	$1,469,992	$56,700	$35,733	$179,260	$328,447
IRR	100%	34%	38%	11%	12%	18%	24%

(NPV) are also included as calculated over the 20-year project period.

Based upon the financial assessment of each well, the Pinnate well provides the highest level of economic feasibility. While the Pinnate configuration shows the highest NPV, project IRR is a better indicator of investment quality assuming that compared projects have a similar scale and duration. In addition to having the most favorable IRR, the Pinnate well provides the fastest project payback. The high gas production rates associated with the Pinnate well allow a payback period of less than one year, largely contributing to its high IRR.

While the Pinnate-style configuration shows the best economic investment criteria, it requires a significantly larger investment than the other wells. Still, it should be noted and understood

Table 7. Sensitivity study parameters

Parameter	Percent-Adjustment	Baseline Value	Adjustment
Initial gas price	±40%	$5.60	±$2.24
Gas price escalation (years 1–15)	±50%	3%	±1.5%
Costs (gathering and overhead) ($/mcf)	±50%	$0.78	±$0.39
Cost escalation	±50%	3%	±1.5%
Capital (horizontal and vertical wells)	±20%	varies	varies
Vertical well operating expenses (year 1)	±50%	$4,000	±$2,000
Vertical well operating expenses (year 2)	±50%	$2,500	±$1,250
Vertical well operating expenses (year 3)	±50%	$1,500	±$750
Horizontal well operating expenses (year 1)	±50%	$7,500	±$3,750
Horizontal well operating expenses (year 2)	±50%	$5,000	±$2,500
Horizontal well operating expenses (year 3)	±50%	$2,000	±$1,000

that the Pinnate-style configuration drains an equivalent area of the Turkey Foot, 50-percent more area than the dual-lateral configuration, and double the area of the discussed single lateral and vertical fracture wells. In analyzing the capital costs associated with drilling additional wells of the alternate configurations, the Pinnate well requires the least capital per unit area, making it more financially attractive.

Sensitivity Analyses

The financial modeling inputs which have the most significant influence on project NPV and IRR are determined through a sensitivity analyses. For the purpose of this study, financial parameters with the highest likelihood of variation were discretely changed to analyze their impact on the NPV and IRR for each well scenario. Through this process, the framework of the financial model remained the same as in previous sections. As with the initial financial analysis, gas price escalation is only assumed to occur over the first 15 years of the project. Similarly, well maintenance and operating costs only vary during the initial three years of production. Table 7 outlines those parameters which are altered and their respective likely minimum and maximum values as incorporated in the sensitivity study. For simplicity in displaying the sensitivity results, gathering and operating costs were combined and changed simultaneously. The same methodology was applied to well operating and maintenance expenses for different project years: for example, the expenses for all project years were changed by a factor of 50-percent simultaneously as opposed to an independently varying expense for individual years.

A graphical depiction of the sensitivity analysis results is shown in Figure 12 and Figure 13. The column charts contain the baseline respective NPV and IRR values for each well pattern as obtained in a previous section and displayed in Table 6. Error bars are included to represent the impact of the anticipated maximum differences of the financial modeling inputs as displayed in Table 7.

As expected, initial gas price data plays the most significant role in the financial analyses for both the NPV and IRR calculations. While gas price escalation shows a much lesser impact than initial gas price, the sensitivity analysis only analyzes gas price escalation on a positive basis with gas prices increasing through the life of the project. A downturn in gas prices could pose a more significant impact, especially in the early years of a well's life.

After initial gas price, the relative significance of financial modeling inputs on NPV vary based on well pattern. For all well patterns, cost escalation shows the least significance. Gas price escalation, gathering and overhead costs, capital expenditures and operating expenses demonstrate varying importance depending on the well pattern. Still, the impact of these parameters is minor in comparison to the impact of initial gas price.

Based on the sensitivity analysis results, second to initial gas price, project IRR is most

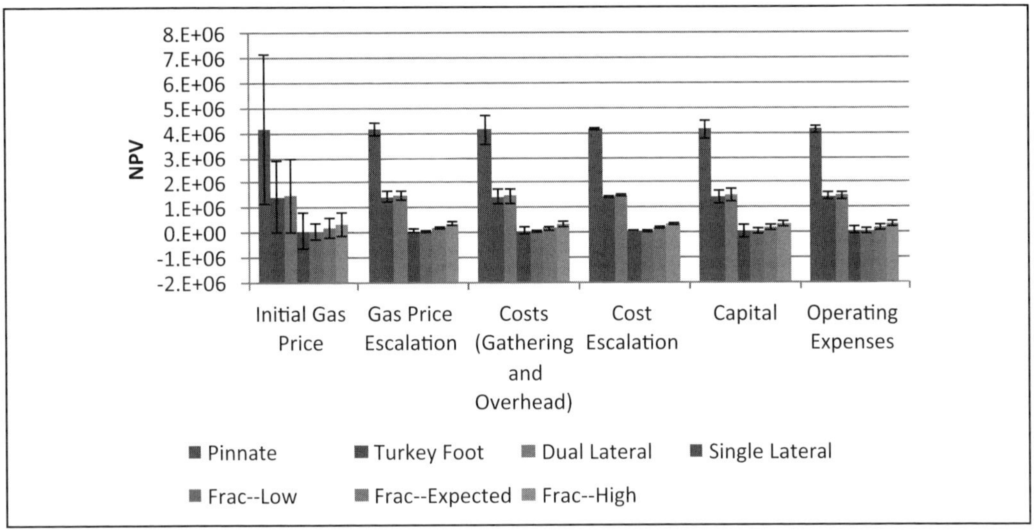

Figure 12. NPV sensitivity analysis results

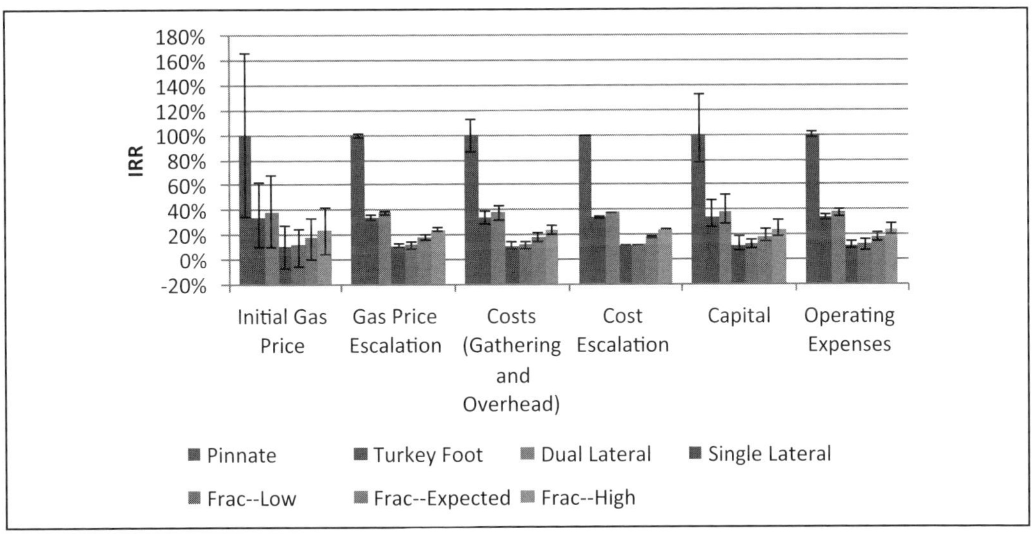

Figure 13. IRR sensitivity analysis results

sensitive to capital expenditures, followed by gathering and overhead costs, operating expenses, gas price escalation and cost escalation. The early years of a project largely dictate the IRR, so that gas price escalation and cost escalation hold the least significance of all input parameters. At a small incremental annual increase, price and cost escalation do not become significant until the later years of the well's producing life. The Pinnate wellbore pattern is the only configuration which maintains a financially viable IRR value at the least anticipated initial gas price of $2.24 per MMBTU, with a corresponding IRR of 35 percent.

CONCLUSIONS

The preceding sections serve as a framework for determining the optimum degasification pattern based on coalbed methane drainage ahead of mining, incorporating health and safety issues, greenhouse gas emissions and financial modeling. Of the analyzed well patterns, Pinnate-style wells provide the best characteristics for economic production while simultaneously decreasing greenhouse gas emissions and offering the potential for increased mine productivity. While the Pinnate pattern is the only studied multilaterally drilled design, variations of this design should provide similar results.

Contrarily, while traditional hydraulically fractured vertical wells require a significantly lower capital investment, they do not effectively drain CBM from coal ahead of mining and result in a lower project rate of return. In the event that the required capital is not available to develop large Pinnate-style patterns, single lateral and dual lateral configurations provide the next best economic return but result in lower recoveries than the Pinnate-style wells. Additionally, to successfully develop Pinnate wells, higher operator skill and state-of-the-art drilling technology is required. The multiple branches and associated geometry can be difficult to drill without technically advanced drilling equipment. However, they do have the advantage of few drill sites and less surface disturbance.

As carbon dioxide equivalent emission reductions are directly proportional to the volume of produced gas by a well or series of wells, Pinnate wells offer the best option of drainage in advance of mining to decrease greenhouse gas emissions. Gas recoveries of upwards of 80 percent are feasible through the use of Pinnate wells when operated over a 20-year period. As applied to a longwall operation in the Qinshui Basin, an 80-percent recovery corresponds to a decrease in carbon dioxide equivalent emissions of approximately 15 million metric tons over a 20-year mine plan. The remaining modeled well layouts cannot reach a recovery of 80-percent over a 20-year period. To boost recovery values for the remaining wells, drainage area can be decreased, subsequently increasing the number of required wells to drain a given area.

Raw recovery values are useful in assessing the effectiveness of a well as applied on an economic and greenhouse gas emission basis. Spatial distributions of recovery across a drainage area offer a better analysis method for determining the impact of a degasification well on mine safety. Minimizing the variation in recovery will result in more consistent conditions during mining and minimizes isolated gassy zones. The closely spaced laterals of Pinnate-style wells help maintain a consistent recovery across the drainage area. The minimal drilling footage per unit area of Turkey Foot-style patterns results in a large variation, similar to the variation exhibited by hydraulically fractured wells. Single lateral and dual lateral configurations provide a fairly consistent distribution of recoveries, albeit lower recoveries than that of the Pinnate well. A case study analysis of a producing longwall in the Qinshui Basin serves as an example of successful drainage through surface-drilled horizontal and in-mine horizontal drainage. Production has increased 10-fold throughout the evolution of the degasification system at this operation.

In conclusion, increasing production of coalbed methane is the key factor to enhancing the profitability of drainage wells and improving reductions in equivalent carbon dioxide emissions. Larger drainage patterns make the best use of capital. The high initial production rates of larger drainage patterns help offset capital investment and increase a project's internal rate of return. Further, closely spaced laterals decrease variability in recovery across a drainage area, enhancing mine safety and potentially boosting productivity. While this work is applied to China's Qinshui Basin, the framework for assessing the effectiveness of coalbed methane drainage wells is applicable on a global scale. The discussed results are applicable for coalbed methane basins exhibiting similar geologic characteristics to those of the Qinshui Basin, including those with low permeability, gassy, thick coal seams. Sustainability as applied to fossil fuels seems paradoxical, but, given existing technology, the production of fossil fuels is vital to sustaining society as we know it. Fossil fuels can and should be produced with careful attention to sustainability, including community and worker health, environmental impacts, and economic considerations. This work provides a framework for considering a few points of each of these pillars of sustainability. There are certainly other considerations for a comprehensive evaluation including the production and disposal of water, community involvement in infrastructure construction, economic development and the implications of additional domestic energy sources, just to name a few.

REFERENCES

Cheng, Y., Wang, L., Zhang., X. 2011. Environmental impacts of coal mine methane emissions and responding strategies in China. International Journal of Greenhouse Gas Control. 5: 157–166.

Karacan, C. 2008. Evaluation of the relative importance of coalbed methane reservoir parameters for prediction of methane inflow rates during mining of longwall development entries. Computers and Geosciences. 34(9): 1039–1114.

Keim, S., Luxbacher, K., Karmis, M. 2011. A numerical study on optimization of multilateral horizontal wellbore patterns for coalbed methane production in Southern Shanxi Province, China. International Journal of Coal Geology. 86(4): 306–317.

Lu Y., Li., X. 2011. A study on a new hazard detecting and controlling method: The case of coal mining companies in China. Safety Science. 49: 279–285.

Lu, Y., Liu, Y., Li, X., and Kang, Y. 2010. A new method of drilling long boreholes in low permeability coal by improving its permeability. International Journal of Coal Geology. 84: 94–102.

Luo, D., Chu, W., Wu, X., Li, W. 2009. Analysis of economic benefits of coalbed methane drilling technologies. Petroleum Exploration and Development. 36(3): 40407.

Maricic, N., Mohaghegh, S.D., and Artun, E. 2008. A Parametric Study on the Benefits of Drilling Horizontal and Multilateral Wells in Coalbed Methane Reservoirs. SPE Reservoir Evaluation and Engineering. 11(6): 976–983.

Palmer, I. 2010. Coalbed methane completions: A world view. International Journal of Coal Geology. 82: 184–195.

Sang, S., Liu, H., Li, Y., Li, M., Li, L. 2009. Geological controls over coal-bed methane well production in southern Qinshui Basin. Procedia Earth and Planetary Sciences. 1: 917–922.

Seinfield, J., Pandis, S. 2006. Atmospheric Chemistry and Physics, 2nd Edition. published by John Wiley & Sons, Inc., Hoboken, NJ.

Shi, X. 2009. Have government regulations improved workplace safety? A test of the asynchronous regulatory effects on China's coal industry, 1995–2006. Journal of Safety Research. 40. 207–213.

Stermole, F., Stermole, M. 2006. Economic Evaluation and Investment Decision Methods, 11th Edition. Published by Investment Evaluations Corporation, Lakewood, CO.

Su, X., Lin, X., Liu, S., Zhao, M., Song, Y. 2005. Geology of coalbed methane reservoirs in the Southeast Qinshui Basin of China. International Journal of Coal Geology. 62. 197–210.

U.S. Environmental Protection Agency. 2006. Global anthropogenic non-CO2 greenhouse gas emissions: 1990–2020.

U.S. Environmental Protection Agency. 2011. Coal Mine Outreach Program. Interactive Units Converter. <http://www.epa.gov/cmop/resources/converter.html>.

Wei, C., Qin, Y., Wang, G., Fu, X., Jiang, B., Zhang, Z. 2007. Simulation study on evolution of coalbed methane reservoir in Qinshui Basin, China. International Journal of Coal Geology. 72. 569.

Yao, Y., Liu, D., Tang, D., Huang, W., Tang, S., and Che, Y. 2008. A Comprehensive Model for Evaluating Coalbed Methane Reservoirs in China. Acta Geol. Sin-Engl. 82 (6). 1251270.

Zhiming, W., Zhang, J., 2009. CBM reservoir thickness affects production from horizontal wells. Oil and Gas Journal. 107(16). 42–46.

Zou, M., Wei, C., Pan, H., Sesay., K.S., Cao, J. 2010. Productivity of coalbed methane wells in southern of Qinshui Basin. Mining Science and Technology. 20(5). 765–769.

3.2
Public Health and Energy Production

Evaluation of Research Addressing Chronic Health in Coal-Dependent Communities in Central Appalachia 172

Comparison of Mortality Rates Between Appalachian Coal Mining Counties with Non-Coal Mining 189

Population and Exposure Characteristics of Coal Mining and Non-Coal Mining Counties in West Virginia 202

Inventory of Bacterial Impairments in Central Appalachia 214

Identification of Toxic Agents and Potential Exposure Routes to Appalachian Coal Mining Communities 229

Evaluation of Research Addressing Chronic Health in Coal-Dependent Communities in Central Appalachia

S. Meacham
Edward Via College of Osteopathic Medicine, Blacksburg, Virginia

S. Sukpraprut
Edward Via College of Osteopathic Medicine, Blacksburg, Virginia

T. Taber
Edward Via College of Osteopathic Medicine, Blacksburg, Virginia

D. Metzger
Edward Via College of Osteopathic Medicine, Blacksburg, Virginia

ABSTRACT

In recent years peer-reviewed articles reporting on the health effects of coal mining in central Appalachian populations have appeared with increasing frequency. Studies reported on various health conditions, ranging from cancer, cardiovascular disease and diabetes, to health related measures of quality of life, to birth defects and tooth decay. The objective of this study was to evaluate original research articles on the basis of relevance and assess the strength of evidence elements. Peer-reviewed research publications on chronic diseases in coal dependent communities in central Appalachia began to appear around the year 2000. We scored identified publications using a rubric developed to objectively and consistently rate study relevance. An integrated method of evaluating study protocols was adapted from commonly used methods of rating health related publications. From more than 60 articles initially identified, 16 were written relating coal mining to human health and ten addressed the concerns of diabetic patients, but without concern for the environmental exposure to coal mining activities. The mean relevance scores for the articles focused on the prevalence of chronic diseases, the management of diabetes, and the third grouping of all other studies were 95%, 77% and 53%, respectively. Almost all studies utilized research protocols that were ecological, descriptive, retrospective, cross-sectional and most often dependent on secondary public data sources. The publications selected were limited by protocol design, thus limiting the strength of the evidence. Future studies are needed that can (1) directly assess environmental exposures in humans, (2) relate exposure directly to chronic disease etiology, and (3) overcome other geographic, temporal and human behavior factors. Epidemiological and, more specifically, ecological studies have the advantage of focusing attention on 'the big picture' and highlighting areas for future research. Experimental protocols are needed to provide science-based evidence to support regulations and policies impacting the balance between our national and global needs for energy and human health.

INTRODUCTION

In a recent publication we reported on original research articles appearing since about 2000 relating coal industry activities in Appalachian communities to increased prevalence rates of chronic diseases (Meacham et al. 2012). The increase in articles on this topic parallels the heightened policy debates surrounding Environmental Protection Agency (2012) permitting for coal mining activities, particularly, mountaintop removal coal mining. Legitimate concerns exist related to human health and environmental

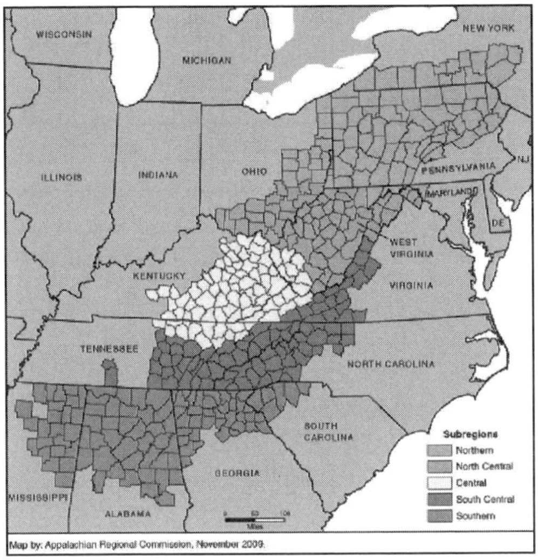

Figure 1. Central Appalachia is defined by the Appalachian Regional Commission (ARC) as the area that includes counties in Kentucky, Tennessee, Virginia, and West Virginia (ARC 2012)

issues with all large scale industries, including the coal industry. It is imperative that science based evidence guide decision making regarding coal production, human health, and environmental impacts. In this communication we report the process used to identify relevant publications and evaluate research methodologies in publications reporting on the effects of coal mining on human health in central Appalachia.

Central Appalachia is defined by the Appalachian Regional Commission (ARC) as the area that includes counties in Kentucky, Tennessee, Virginia, and West Virginia. Much of the central Appalachian area is considered economically distressed and 'at risk' with high poverty rates, high unemployment rates and low high school and college completion rates (Figure 1) (ARC 2012). The primary economic driver in this region has been the coal industry. Central Appalachia has the highest volume of mountain top removal coal mining in the United States. Mountaintop removal mining, more than other types of surface mining and underground mining, has spurred public debates regarding environmental and health concerns (Sayler 2012).

For years various government agencies and non-profit organizations have released information about human health in the area through annual reports, intervention program outcomes, etc. Relatively few original research studies have been published and most rely on protocol designs that can suggest, but not confer, cause-effect relationships about environmental exposures and the development of chronic disease. Research protocol designs have great bearing on the strength of results and, in turn, the strength of the evidence used to support policy development.

Systematic evaluations of original research are customary and serve to inform health policies, i.e., the U.S. recommended dietary allowances (Dietary Reference Intakes 1997; Meacham 2003; Meacham et al. 2008). Nursing professionals also developed a ranking rubric to systematically evaluate protocol designs for various treatment regimens (Melnyk, B.M. and Fineout-Overholt, E. 2005). Systematic literature reviews are generally performed by panels of experts who distill the science based evidence for various applications, including policy development.

Systematic reviews are also specific to the question, often adaptations of previous models. When a review focused on disease etiology in cancer an expert panel noted that hierarchal systems were inadequate. Consequently, an inclusive system was created to utilize data produced by all types of original research data, including descriptive and experimental, retrospective and prospective, and epidemiological, clinical and laboratory based (World Cancer Research Foundation/American Institute for Cancer Research 2007).

For our objective and consistent analysis of data on coal and human health, we adopted an inclusive strategy to review epidemiological literature. The objectives of our study were to identify relevant studies and assess the strength of the evidence based on protocol designs to answer the question, "Do coal mining activities contribute to chronic health disparities in central Appalachian communities?"

METHODOLOGY

An initial literature search of computerized databases, PubMed, EBSCO and CiNAHL was performed using the root search terms 'Appalachian,' 'coal' and 'health'. In a previous publication we reported that over 60 articles were identified with 38 articles written in English addressing health in Appalachian communities. The most stringent inclusion criteria limited the search to peer-reviewed, original research articles and found most articles to have been published since about 2000. A relevance rating was assigned to each study based on geographic region, chronic disease and community health and the inclusion of coal as an environmental exposure. Due to the limited number of publications meeting all the inclusion criteria special notations were made of publications written as commentaries, editorials or introductions to special editions. Likewise, noted were articles that were not peer reviewed, were outside our geographic scope or reported on mental health, occupational health or sociocultural concepts (Meacham et al. 2012).

Publications identified were scored for relevance based on the following factors: original research, geographic location, human study, chronic disease, and coal as an environmental exposure. Each factor was rated on a scale of 1 to 5 points, with 5 points being the highest possible rating. An overall relevance score was assigned as a percentage of the total possible, 25 points.

In addition to relevance, publications were reviewed on the basis of study design: study type, study level, data source, and data level. Study types were generally epidemiological and were further described as ecological studies if the analyses were performed with units of measure at levels other than individual levels, i.e., county levels. Retrospective and prospective were also used to identify study types. Using an integrated rating system studies were aligned according to study level using a 1–7 hierarchal scale (Table 1). Controlled randomized trials had higher rankings than cohort studies or cross-sectional studies. Data from public data bases were described as public and self-reported survey responses were considered sources of individual data. Data collected from individuals for the purpose of the study were referred to as primary data. Data obtained from public databases or originally collected through another protocol were referred to as secondary data.

Table 1. Hierarchal ranking of study types used to evaluate the strength of findings from studies at each level (Modified from Melnyk, B.M., and Fineout-Overholt, E. 2005)

Study Level	Study Type
7	Systematic reviews of randomized, controlled trials
6	Controlled studies, randomized trails
5	Controlled studies, non-randomized trials
4	Case controlled and cohort studies
3	Systematic reviews of descriptive and qualitative or cross-sectional studies
2	Single descriptive, qualitative studies
1	Committee or authority reports

RESULTS

As reported previously more than 60 articles were grouped into three categories, those primarily addressing the prevalence of chronic diseases, a second group of articles addressing the management of diabetes mellitus, hereafter referred to as diabetes, and a third group of other publications related but with a less specific focus (Meacham et al. 2012). Michael Hendryx, Ph.D. at West Virginia University (WVU) appeared as an author on 24 of 61 articles; all 16 articles on chronic diseases (Table 2), one of ten articles on diabetes (Table 3) and seven of 35 other articles (Table 4). Sharon Denham, Ph.D., R.N., at Ohio University, appeared as an author on 14 articles, including nine articles on diabetes (Table 3) and five other articles (Table 4).

The mean relevance score for the first group of articles was 95% and addressed specifically chronic disease in coal dependent communities in central Appalachia (Table 2).

Table 3 lists the publications identified in the systematic literature review that addressed diabetes, most found through the CiNAHL database search. The mean relevance score for the publications on diabetes was 77%. Nine publications

Table 2. Relevance scores for studies published by Dr. Michael Hendryx and others reporting on health in Appalachian communities that included coal as an environmental exposure variable

Reference	Original Research	Geographic Area: National	Geographic Area: Appalachian	Geographic Area: Central Appalachia	Human Study	Community Chronic Disease	Coal	Sum	%
	5 points	3 points	4 points	5 points	5 points	5 points	5 points	25 points	
1. Ahern, M., Hendryx,M., Conley, J., Fedorko, E., Ducatman, A., and Zullig, K.J. 2011. The association between mountaintop mining and birth defects among live births in central Appalachia, 1996–2003. Environ. Res. 111(6):838–846.	5			5	5	3	5	23	92%
2. Epstein P.R., Buonocore, J.J., Eckerl, K Hendryx, Stout, B.M., Heinberg, R., Clapp, R.W., May, B., Reinhart, N.L., Ahern, M.M., Doshi, S.K. and Glustrom, L. 2011. Full cost accounting for the life cycle of coal. Ann. N.Y. Acad. Sci. 1219:73–98.	5	3			5	3	5	21	84%
3. Esch, L. and Hendryx, M.S. 2011. Chronic cardiovascular disease mortality in mountaintop mining areas of central Appalachian states. J. Rural Health. 27(4):350–357.	5			5	5	5	5	25	100%
4. Hendryx, M.S. 2009. Mortality from heart, respiratory, and kidney disease in coal mining areas of Appalachia. Int. Arch. Occup. Environ. Health. 82(2):243–249.	5		4		5	5	5	24	96%
5. Hendryx, M.S. 2011. Poverty and mortality disparities in central Appalachia: mountaintop mining and environmental justice. J. Health Disparities Research and Practice. 4(3):44–53.	5			5	5	5	5	25	100%
6. Hendryx, M.S. and Ahern, M.M. 2008. Relations between health indicators and residential proximity to coal mining in West Virginia. Am. J. Public Health. 98(4):669–671.	5			5	5	5	5	25	100%
7. Hendryx, M.S. and Ahern, M.M. 2009. Mortality in Appalachian coal mining regions: the value of statistical life lost. Public Health Rep. 124(4):541–550.	5		4		5	4	5	23	92%
8. Hendryx, M.S., Ahern, M.M. and Nurkiewicz, T.R. 2007. Hospitalization patterns associated with Appalachian coal mining. J. Toxicol. Environ. Health, Part A. 70(24):2064–2070.	5			5	5	5	5	25	100%
9. Hendryx, M.S., Fedorko, E. and Anesetti-Rothermel, A. 2010. A geographical information system-based analysis of cancer mortality and population exposure to coal mining activities in West Virginia, United States of America. Geospat. Health. 4(2):243–256.	5			5	5	5	5	25	100%
10. Hendryx, M.S., Fedorko, E. and Halverson, J. 2010. Pollution sources and mortality rates across rural-urban areas in the United States. J. Rural Health. 26(4):383–391.	5	3			5	5	5	23	92%
11. Hendryx, M.S, O'Donnell, K. and Horn, K. 2008. Lung cancer mortality is elevated in coal-mining areas of Appalachia. Lung C ancer . 62(1):1–7.	5		4		5	5	5	24	96%
12. Hendryx, M., Wolfe, L., Luo, J. and Webb, B. 2011. Self- reported cancer rates in two rural areas of West Virginia with and without mountaintop coal mining. J. Community Health. DOI 10.1007/s10900–011–9448–5.	5			5	5	5	5	25	100%
13. Hendryx, M.S. and Zullig, K.J. 2009. Higher coronary heart disease and heart attack morbidity in Appalachian coal mining regions. Prev. Med. 49(5):355–359.	5		4		5	5	5	24	96%
14. Hitt, N.P. and Hendryx, M.S. 2010. Ecological integrity of streams related to human cancer mortality rates. Ecohealth . 7(1):91–104.	5			5	5	5	5	25	100%
15. Zullig, K.J. and Hendryx, M.S. 2010. A comparative analysis of health-related quality of life for residents of U.S. counties with and without coal mining. Public Health Rep. 125(4):548–555.	5	3			5	3	5	21	84%
16. Zullig, K.J. and Hendryx, M.S. 2011. Health-related quality of life among central Appalachian residents in mountaintop mining counties. Am. J. Public Health. 101(5):848–853.	5			5	5	3	5	23	92%

Table 3. Relevance scores for studies on diabetes in Appalachian communities

Reference	Original Research	Geographic Area: National	Geographic Area: Appalachian	Geographic Area: Central Appalachia	Human Study	Community Chronic Disease	Coal	Sum	%
	5 points	3 points	4 points	5 points	5 points	5 points	5 points	25 points	
1. Barker, L.C., Gerzoff, R., Denham, S., Shrewsberry, M. and Cornelius–Averhart, D. 2010. Residence in a distressed county in Appalachia as a risk factor for diabetes, Behavioral Risk Factor Surveillance System, 2006–2007. *Preventing Chronic Disease: Public Health Research, Practice, and Policy.* 7(5):1–9.	5		4		5	5	0	19	76%
2. Denham, S.A., Manoogian, M. and Schuster, L. 2007. Managing family support and dietary routines: Type 2 diabetes in rural Appalachian families. *Families, Systems & Health.* 25(1):36–52.	5			5	5	5	0	20	80%
3. Denham, S.A., Remsburg, K. and Wood, L. 2010. Diabetes education in the Appalachian region: providers' views. *Rural Remote Health.* 10(2):1321.	5		4		5	5	0	19	76%
4. Denham, S.A., Ware, L.J., Raffle, H. and Leach, K. 2011. Family inclusion in diabetes education: a nationwide survey of diabetes educators. *Diabetes Educ.* 37(4):528–535.	5	3			5	5	0	18	72%
5. Denham, S.A., Wood, L.E. and Remsberg, K. 2010. Diabetes care: provider disparities in the US Appalachian region. *Rural Remote Health.* 10(2):1320.	5		4		5	5	0	19	76%
6. Manoogian, M.M., Harter, L.M. and Denham, S.A. 2010. The storied nature of health legacies in the familial experience of type 2 diabetes. *J. Family Communication.* 10:40–56.	5		4		5	5	0	19	76%
7. O'Brien, T. and Denham, S.A. 2008. Diabetes care and education in rural regions. *Diabetes Educ.* 34(2):334–347.	5	3			5	5	0	18	72%
8. Pollard, C., Bailey, K.A., Petitte, T., Baus, A., Swim, M. and Hendryx, M. 2009. Electronic patient registries improve diabetes care and clinical outcomes in rural community health centers. *J. Rural Health.* 25(1):77–84.	5			5	5	5	0	20	80%
9. Schwartz, F.L., Denham, S.A., Heh, V., Wapner, A. and Shubrook, J. 2010. Experiences of children and adolescents with Type I diabetes in school: Survey of children, parents, and schools. *Diabetes Spectrum.* 23(1):47–55.	5			5	5	5	0	20	80%
10. Schwartz, F.L., Ruhil, A.V., Denham, S.A., Shubrook, J., Simpson, C. and Boyd, S.L. 2009. High self–reported prevalence of diabetes mellitus, heart disease, and stroke in 11 counties of rural Appalachian Ohio. *J. Rural Health.* 25(2):226–230.	5			5	5	5	0	20	80%

appearing since 2007 addressed the management of diabetes in Appalachian communities and were coauthored by Sharon Denham. Another publication, by Pollard et al. (2009), included Hendryx as a coauthor and reported on the effectiveness of using electronic patient records as an important part of self-management in diabetes care.

In Table 4 there were 35 additional studies identified in the original search that discussed chronic diseases in coal dependent communities in central Appalachia. The mean relevance score was 53%. Borak et al. (2012) had a relevance score of 96% for the publication reassessing data previously reported in three publications with Hendryx as a coauthor; Hendryx (2009), Hendryx and Ahern (2009), and Hendryx et al. (2008). Three studies by Blackley et al. (2011), McGarvey et al. (2011), and Schoenberg et al. (2011) had relevance scores of 76%, 80% and 76%, respectively. These authors addressed chronic health conditions in Appalachian communities but did not consider coal as an environmental variable, nor did they specifically address diabetes management. Likewise, articles by Hendryx and Fedorko (2011), Luo and Hendryx (2011), and Luo et al. (2011) had relevance scores of 72%, 76% and 76%, respectively. The authors reported on the incidence of cancer associated with toxic chemicals released in the environment but did not consider toxins released specifically from coal related activities.

In the list of additional studies (Table 4) there were seven publications where Hendryx appeared as an author. These articles were not included with the "original 16" in Table 2 which listed work conforming to the most stringent inclusion criteria, chronic disease in coal dependent communities in Appalachia. Hendryx's publications listed in Table 4, as mentioned above, addressed toxic release inventory discharges, as well as other aspects of health, i.e., mental health (Hendryx 2008) or access to health care (Hendryx et al. 2002). One article was an introduction to a special section of a scientific meeting (Hendryx 2011) and another, a commentary, published in the policy section of *Science* (Palmer et al. 2010).

Articles by Denham also appeared on the list of additional studies (Table 4), but these works did not address coal or diabetes management. In these publications other sociocultural concepts were the main points of interest, i.e., family health (Denham 2003); health education (Denham et al. 2004); spirituality (Diddle and Denham 2010); and alcohol and tobacco use (Meyer et al. 2008). The additional studies identified (Table 4) reported on related concerns, i.e., food environments (Ahern et al. 2011), occupational health and international mining activities (Bilban 2005; Donoghue 2004; Feng and Qiu 2008; Gold et al. 2008; Huang and Finkelman 2008; Joy 2004; Kang and Kim 2010).

The 'original 16' in Table 2 were evaluated based on features of the protocol design. Study types, study levels, data types and data levels appear in Table 5. Studies were rated using a 1 to 7 hierarchal scale with a score of 7 reflecting the most rigorous research protocol design (Table 1). The original 16 Hendryx group studies were categorized as study level 2; all were single studies rather than systematic reviews, all were descriptive rather than experimental, and all were cross-sectional rather than cohort.

Table 5 also indicates reports that all but one study was considered retrospective. Only one study had a prospective method of collecting data with a door-to-door survey in a community-based participatory study (Hendryx et al. 2011). An earlier survey study utilized individual survey data from a state telephone study conducted years earlier; thus the data was referred to as secondary, having been collected with intentions other than those for which the data were used in the current study (Hendryx and Ahern, 2008). Study types were also generally described as ecological studies if the analyses were performed with units of measure at county levels or another geographic area rather than individual levels, i.e., Hendryx et al. (2007); Hendryx and Ahern (2008); Hendryx (2009); and Esch and Hendryx (2011). Exceptions were seen with studies referred to as exploratory, i.e, the analysis of birth defect data (Ahern et al. 2011) and with studies we described as modeling with unique presentations that monetized the

Table 4. Relevance scores for other studies identified and less directly associated with chronic diseases in Appalachian mining communitiest

Reference	Original Research	Geographic Area: International	Geographic Area: Appalachian	Geographic Area: Central Appalachia	Human Study	Community Chronic Disease	Coal	Sum	
	5 points	2 points	3 points	4 points	5 points	5 points	5 points	5 points	%
1. Ahern, M., Brown, C., and Dukas, S. 2011. A national study of the association between food environments and county-level health outcomes. *J. Rural Health.* 27(4):367–379.	5		3			5			52%
2. Bernhardt, E.S. and Palmer, M.A. 2011. The environmental costs of mountaintop mining valley fill operations for aquatic ecosystems of the Central Appalachians. *Ann. NY Acad. Sci.*1223:39–57.					5			5	40%
3. Bilban, M. 2005. Occupational medicine in the Slovene area. *J. Occup. Health.* 47(3):193–200.		2				5	0	5	48%
4. Blackley, D., Behringer, B. and Zheng, S. 2011. Cancer mortality rates in Appalachia: descriptive epidemiology and an approach to explaining differences in outcomes. *J. Community Health.* DOI:10.1007/s10900–011–9514-z.	5			4		5	5		76%
5. Borak, J., Salipante-Zaidel, C., Slade, M.D. and Fields, C.A. 2012. Mortality disparities in Appalachia: reassessment of major risk factors. *J. Occup. Environ. Med.* 54(2):146–156.	5			4		5	5	5	96%
6. Cohen, R.A., Patel, A. and Green, F.H. 2008. Lung disease caused by exposure to coal mine and silica dust. *Semin. Respir. Crit. Care Med.* 29(6):651–661.			3			5	0	5	52%
7. Denham, S.A. 2003. Relationships between family rituals, family routines and health. *J. Family Nurs.* 9(3):305–330.			3			5			32%
8. Denham, S.A. 1999. Part 3: Family health in an economically disadvantaged population. *J. Family Nurs.* 5(2):184–213.	5			4		5	2		64%
9. Denham, S.A., Meyer, M.G., Toborg, M.A. and Mande, M. J. 2004. Providing health education to Appalachia populations. *Holist. Nurs. Pract.* 18(6):293–301.	5			4		5	2		64%
10. Diddle, G. and Denham, S.A. 2010. Spirituality and its relationships with the health and illness of Appalachian people. *J. Transcult. Nurs.* 2010;21(2):175–182.				4		5			36%
11. Donoghue, A.M. 2004. Occupational health hazards in mining: an overview. *Occup. Med.* 54(5):283–289.		2				5	0	5	48%
12. Feng, X. and Qiu, G. 2008. Mercury pollution in Guizhou, southwestern China–an overview. *Sci. Total Environ.* 400(1–3):227–237.		2						5	28%
13. Gold, L.S., De Roos, A.J., Waters, M., and Stewart, P. 2008. Systematic literature review of uses and levels of occupational exposure to tetrachloroethylene. *J. Occup. Environ. Hyg.* 5(12):807–839.			3				0		12%
14. Gulumian, M., Borm, P.J., Vallyathan, V. Castranova, V., Donaldson, K., Nelson, G. and Murray, J. 2006. Mechanistically identified suitable biomarkers of exposure, effect, and susceptibility for silicosis and coal-worker's pneumoconiosis: a comprehensive review. *J. Toxicol. Environ. Health B. Crit. Rev.*9(5):357–395.		2				5	0	5	48%
15. Hendryx, M.S. 2008. Mental health professional shortage areas in rural Appalachia. *J. Rural Health.* 24(2): p.179–182.	5			4		5	3		68%
16. Hendryx, M.S. 2011. Introduction to special section: Environmental health for rural populations. *J. Rural Health. 27(4): 339–341.*									
17. Hendryx, M.S., Ahern, M.M., Lovrich, N.P. and McCurdy, A.H. 2002. Access to health care and community social capital. *Health Serv. Res.* 2002. 37(1):87–103.	5		3			5	3		64%
18. Hendryx, M.S. and Fedorko, E. 2011. The relationship between toxics release inventory discharges and mortality rates in rural and urban areas of the United States. *J. Rural Health* . 27(4):358–366.	5		3			5	5		72%

(table continues)

Table 4. Relevance scores for other studies identified and less directly associated with chronic diseases in Appalachian mining communities (continued)

Reference	Original Research	Geographic Area: International	Geographic Area: Appalachian	Geographic Area: Central Appalachia	Human Study	Community Chronic Disease	Coal	Sum	%
	5 points	2 points	3 points	4 points	5 points	5 points	5 points	5 points	
19. Huang, X. and Finkelman, R.B. 2008. Understanding the chemical properties of macerals and minerals in coal and its potential application for occupational lung disease prevention. *J. Toxicol. Environ. Health B Crit. Rev.* 11(1):45–67.	5		3			5	0	5	72%
20. Joy, J. 2004. Occupational safety risk management in Australian mining. *Occup Med (Lond).* 54(5):311–315.		2				5	0		28%
21. Kang, S.K. and Kim, E.A. 2010. Occupational diseases in Korea. *J. Korean Med. Sci.* 25(Suppl):S4–12.		2				5	5	5	68%
22. Kavanagh, A.M., Bentley, R., Turrell, G., Broom, D.H.and Subramanian, S.V. 2006. Does gender modify associations between self rated health and the social and economic characteristics of local environments? *J. Epidemiol. Community Health* . 60(6):490–495.	5	2				5			48%
23. Lee, S.Y., Chen, W.L. and Weiner, B.J. 2004. Communities and hospitals: social capital, community accountability, and service provision in U.S. community hospitals. *Health Serv. Res.* 39(5):1487–1508.	5		3			5			52%
24. Li P., Feng, X.B., Qiu, G.L., Shang, L.H. and Li, Z.G. 2009. Mercury pollution in Asia: a review of the contaminated sites. *J. Hazard. Mater* . 168(2–3):591–601.		2				5			28%
25. Luo, J., and Hendryx, M.S. 2011. Environmental carcinogen releases and lung cancer mortality in rural-urban areas of the United States. *J. Rural Health.* 27(4):342–349.	5		3			5	5	1	76%
26. Luo, J., Hendryx, M.S. and Ducatman, A. 2011. Association between six environmental chemicals and lung cancer incidence in the United States. *J. Environ. Public Health.* 2011: Article ID 463701, 9 pages. doi:10.1155/2011/463701463–701.	5		3			5	5	1	76%
27. Ma Y., Prasad, M.N., Rajkumar, M. and Freitas, H. 2011. Plant growth promoting rhizo bacteria and endophytes accelerate phytoremediation of metalliferous soils. *Biotechnol. Adv.* 29(2):248–258.			3			5			32%
28. McGarvey, E.L., Leon-Verdin, M., Killos, L.F., Guterbock T. and Cohn W.F. 2011. Health disparities between Appalachian and non-Appalachian counties in Virginia USA.*J. Community Health.* 36(3):348–356.	5				5	5	5		80%
29. Meyer, M.G., Toborg, M.A., Denham, S.A. and Mande, M.J. 2008. Cultural perspectives concerning adolescent use of tobacco and alcohol in the Appalachian mountain region. *J.Rural Health.* 24(1):67–74.	5			4		5	2		64%
30. Palmer, M.A., Bernhardt, E.S., Schlesinger, W.H., Eshleman, K.N., Foufoula-Georgiou, E., Hendryx, M.S., Lemly, A.D., Likens, G.E., Loucks, O.L., Power, M.E., White, P.S. and Wilcock, P.R. 2010. Science and regulation. Mountaintop mining consequences. *Science* . 327:148–149.					5	5			40%
31. Ross, M.H. and Murray, J. 2004. Occupational respiratory disease in mining. *Occup. Med. (Lond).* 54(5):304–310.		2				5	0	5	48%
32. Schoenberg, N.E., Bardach, S.H., Manchikanti, K.N. and Goodenow, A.C. 2011. Appalachian residents' experiences with and management of multiple morbidity. *Qual. Health Res.* 21(5):601–611.	5			4		5	5		76%
33. Sharma, S. and Rees, S. 2007. Consideration of the determinants of women's mental health in remote Australian mining towns. *Aust. J. Rural Health.* 15(1):1–7.		2				5			28%
34. Vearrier, D. and Greenberg, M.I. 2011. Occupational health of miners at altitude: adverse health effects, toxic exposures, pre-placement screening, acclimatization, and worker surveillance. *Clin. Toxicol. (Phila).*49(7):629–640.			3			5	0	5	52%
35. Zipper, C.E., Burger, J.A., Skousen, J.G., Angel, P.N., Barton, C.D., Davis, V. and Franklin, J.A. 2011. Restoring forests and associated ecosystem services on appalachian coal surface mines. *Environ. Manage,* 47(5):751–765.				4				5	36%

Table 5. Relevance scores, study type, study level, data source and data level for studies published by Dr. Michael Hendryx and colleagues reporting on health in Appalachian communities that included coal as an environmental exposure variable

Study Reference	Relevance Score	Study Type (Study Level)	Health Data Source (Data Level)
Ahern, M., Hendryx, M., Conley, J., Fedorko, E., Ducatman, A., and Zullig, K.J. 2011. The association between mountaintop mining and birth defects among live births in central Appalachia, 1996–2003. *Environ. Res.* 111(6):838–846.	92%	ecological, descriptive, retrospective, cross-section (2)	public, secondary
Epstein P.R., Buonocore, J.J., Eckerl, K. Hendryx, Stout, B.M., Heinberg, R., Clapp, R.W., May, B., Reinhart, N.L., Ahern, M.M., Doshi, S.K. and Glustrom, L. 2011. Full cost accounting for the life cycle of coal. *Ann. N.Y. Acad. Sci.* 1219:73–98.	84%	descriptive, modeling, retrospective, (2)	public, secondary
Esch, L. and Hendryx, M.S. 2011. Chronic cardiovascular disease mortality in mountaintop mining areas of central Appalachian states. *J. Rural Health.* 27(4): 350–357.	100%	ecological, descriptive, retrospective, cross-section (2)	public, secondary
Hendryx, M.S. 2011. Poverty and mortality disparities in central Appalachia: mountaintop mining and environmental justice. *J. Health Disparities Research and Practice.* 4(3):44–53.	100%	ecological, descriptive, retrospective, cross-section (2)	public, secondary
Hendryx, M.S. 2009. Mortality from heart, respiratory, and kidney disease in coal mining areas of Appalachia. *Int. Arch. Occup. Environ. Health.* 82(2):243–249.	96%	ecological, descriptive, retrospective, cross-section (2)	public, secondary
Hendryx, M.S. and Ahern, M.M. 2009. Mortality in Appalachian coal mining regions: the value of statistical life lost. *Public Health Rep.* 124(4):541–550.	92%	descriptive, modeling retrospective, cross-section (2)	public, secondary
Hendryx, M.S. and Ahern, M.M. 2008. Relations between health indicators and residential proximity to coal mining in West Virginia. *Am. J. Public Health.* 98(4): 669–671.	100%	ecological, descriptive, retrospective, cross-section (2)	self-report/ secondary
Hendryx, M.S., Ahern, M.M. and Nurkiewicz, T.R. 2007. Hospitalization patterns associated with Appalachian coal mining. *J. Toxicol. Environ. Health, Part A.* 70(24):2064–2070.	100%	individual/ecological descriptive, retrospective, cross-section (2)	public/ individual, prim/second
Hendryx, M.S., Fedorko, E. and Anesetti-Rothermel, A. 2010. A geographical information system-based analysis of cancer mortality and population exposure to coal mining activities in West Virginia, United States of America. *Geospat. Health.* 4(2):243–256.	100%	ecological, descriptive, retrospective, cross-section (2)	public, secondary
Hendryx, M.S., Fedorko, E. and Halverson, J. 2010. Pollution sources and mortality rates across rural-urban areas in the United States. *J. Rural Health.* 26(4):383–391.	92%	ecological, descriptive, retrospective, cross-section (2)	public, secondary
Hendryx, M.S, O'Donnell, K. and Horn, K. 2008. Lung cancer mortality is elevated in coal-mining areas of Appalachia. *Lung Cancer.* 62(1):1–7.	96%	ecological, descriptive, retrospective, cross-section (2)	public, secondary
Hendryx, M., Wolfe, L., Luo, J. and Webb, B. 2011. Self-reported cancer rates in two rural areas of West Virginia with and without mountaintop coal mining. *J. Community Health.* DOI 10.1007/s10900-011-9448-5.	100%	descriptive, prospective, cross-section (2)	self-report, primary
Hendryx, M.S. and Zullig, K.J. 2009. Higher coronary heart disease and heart attack morbidity in Appalachian coal mining regions. *Prev. Med.* 49(5):355–359.	96%	ecological, descriptive, retrospective, cross-section (2)	public, secondary
Hitt, N.P. and Hendryx, M.S. 2010. Ecological integrity of streams related to human cancer mortality rates. *Ecohealth.* 7(1):91–104.	100%	ecological, descriptive, retrospective, cross-section (2)	public, secondary
Zullig, K.J. and Hendryx, M.S. 2011. Health-related quality of life among central Appalachian residents in mountaintop mining counties. *Am. J. Public Health.* 101(5):848–853.	92%	ecological, descriptive, retrospective, cross-section (2)	public, secondary
Zullig, K.J. and Hendryx, M.S. 2010. A comparative analysis of health-related quality of life for residents of U.S. counties with and without coal mining. *Public Health Rep.* 125(4):548–555.	84%	ecological, descriptive, retrospective, cross-section (2)	public, secondary

public health burdens of environmental hazards (Epstein et al. 2011) or calculated the statistical value of life (Hendryx and Ahern 2009).

DISCUSSION

Legitimate concerns exist related to human health and environmental issues with all large scale industries such as coal. Science based evidence is more important than ever to guide decision making regarding coal production, human health and environmental impacts. Strength of evidence relies on various elements of protocol design; study quality, quantity, consistency impact and generalizability. In a previous article we discussed research publication quantity, quality and consistency regarding chronic diseases in coal dependent communities in central Appalachia. We noted an increase in publishing frequency paralleling the EPA permit controversies and the number of publications with common authors. The use of journal impact factors as an objective, yet indirect, proxy for publication quality was acknowledged and the limitations stated (Meacham et al. 2012). While the researchers have discussed the study limitations within their publications, others, as presented below, expressed additional concerns that hamper the impact of findings produced through study protocols that may not produce the strongest evidence.

Our comprehensive, computerized electronic search identified articles in three commonly used databases, PubMed, EBSCO and CiNAHL. CiNAHL promotes being the largest electronic database collection of nursing and allied health literature available and produced articles on diabetes not detected in reviews of PubMed or EBSCO. Through analysis of research study design, we further examined research quality on the basis of scientific design and validity.

We designed a rubric tool to objectively and consistently evaluate each article on relevance and then we classified the articles based on strength of evidence elements. We began the process of grouping the articles by convenience as described earlier; prevalence of chronic disease, those that addressed the management of diabetes, and other articles related but less specifically focused (Meacham et al. 2012). The publications appearing in Table 2 had the highest mean relevance scores (95%) and all with a common author, Michael Hendryx, Ph.D. As seen in Table 5 our classification of study protocols confirmed a consistency in study levels and study types; almost all of the WVU studies were ecological, descriptive, retrospective, cross-sectional and reliant on secondary data. This consistency does not highlight successive improvements in study designs to elevate the strength of evidence elements or highlight distinctive study characteristics as seen in the study ranking rubric in Table 1.

Conversely, within this study design context, various methodology strategies were demonstrated by the WVU group to quantify the impact of coal on health. The studies conducted by the WVU group portrayed logical and creative stepwise advances in the methods employed with each successive study reported. Zullig and Hendryx (2010, 2011) reported on the impact of coal mining on health-related quality of life indicators in Appalachian communities. A number of publications reported 'monetized' measures of the impacts of coal on the environment and human life. Hendryx and Ahern (2009) examined the mortality rates in Appalachian coal mining areas and estimated a corresponding 'statistical life lost' value. Similarly, Hendryx and coworkers estimated the full cost accounting for the life cycle of coal based on environmental and health factors (Epstein et al. 2011). The theoretical models used to derive this economic measure have been used by government agencies to convey the public health burden of environmental exposures, such as the suggested negative health consequences of coal mining.

Caution should prevail when mathematical models of human disease conditions are presented, as this type of study design can easily lead to faulty conclusions. In a recent presentation, Stahl et al. (2012) warned of the dangers of "proofiness" in the evaluation of the impact of mountaintop removal coal mining. Stahl used the term "proofiness" to refer to the use or misuse of mathematical arguments to arrive at misleading conclusions. They discussed the need for caution when determining the appropriateness of data and methodologies when making controversial environmental assessments.

In general, prospective studies have a number of advantages over retrospective studies. One study in the "suite of studies" from WVU (Table 2) can be considered prospective, having used a community based participatory research project collecting data from 773 adults in a door-to-door survey (Hendryx et al. 2011). The self-reported health data presented by Hendryx and Ahern (2008) had been collected retrospectively with a telephone survey conducted in 2001. In these two survey studies, the data collected were considered primary data while most other data sources were typically secondary. Primary data is generally preferred over secondary data. However, self-reported health data introduce their own inherent limitations. Personal biases, reporting errors, etc., are difficult to overcome.

The first experimental study has recently been published by Hendryx and colleagues, progressing beyond observational and descriptive study designs (Knuckles et al. 2012). They reported on the physiological impact of simulated mountain top mining particulate matter exposure on rats, demonstrating impaired microvascular function. They concluded that a similar physiological effect may contribute to cardiovascular disease in humans exposed to particulate matter as a result of coal mining activities in Appalachia. These findings have not been confirmed by clinical trials in humans and are not yet translatable to realistic exposure levels in natural environments.

Publications in Table 3 had a mean relevance score of 77% and focused on managing diabetes in Appalachian populations. Sharon Denham, Ph.D., R.N., and coworkers produced nine of the ten articles. While diabetes is a chronic disease, the publications on managing diabetes deserved special attention; as such they were grouped separately from the articles addressing other health conditions. Diabetes, in contrast to other chronic conditions, can in many cases be managed and allow patients a high quality of life despite their condition. Thus, the opportunities for successful and effective intervention programs are greater for this condition than for most, i.e., end stage renal failure or chronic obstructive pulmonary disease. The diabetes literature also preferentially reported on self-management education for patients with support from care givers, health care providers and health education programs (Denham et al. 2007; O'Brien and Denham 2008; Manoogian et al. 2010; Denham et al. 2010ab; Denham et al. 2011). These studies were not evaluated further due to variability in study protocol, which made using a rubric less meaningful, and the Denham studies have not been called into question.

All other articles (Table 4) identified in the initial literature search did not reference coal or focus on diabetes directly and typically addressed occupational health in international settings. However, the articles do offer insights into the generalizability and global significance of the problem. While the mean relevance score for articles listed in Table 4 was 53% there were several regionally appropriate studies that had relevance scores similar to those authored by the Hendryx and Denham groups; McGarvey et al. (2011) from Virginia and Blackley et al. (2011) from Tennessee. We previously discussed these studies and acknowledged their discussions of cultural sensitivity as important components of the health care process in Appalachian communities (Meacham et al. 2012).

Our conservative lists of articles grouped into defendable categories provided a logical means of discussing the original research. A more liberal literature search would yield a larger number of articles. Without the explicit reference to coal mining, for example, many articles would be included, such as those authored by Nancy Schoenberg and colleagues from Kentucky, reporting primarily on cultural aspects of health and cancer. A recent publication by this team shared research findings that inform community interventions (Kruger et al. 2012). Other reports confronted managing multiple morbidities (Schoenberg et al. 2011) and 'deconstructing fatalism' in underserved Appalachian residents (Drew and Schoenberg 2011; Paskett et al. 2011). Another recent publication by a research team in Kentucky has also examined geographic patterns of lung cancer incidence (Christian et al. 2011). Again, the limitations of these ecological studies are fundamental to the study designs and, as stated by the authors, require further research with biological and environmental samples to confirm cause-effect relationships.

Thus, to date the majority of studies on health in coal dependent communities in Appalachia have increased awareness on negative health impacts by reporting on increased rates of a broad spectrum of conditions: diabetes, diseases of the cardiovascular, respiratory and digestive systems, and even birth defects and tooth loss. However, the inherent limitations in study designs cannot be overridden by publishing more studies, collecting more data on robust subject numbers or refining measurement techniques, i.e., geospatial analysis. Likewise, as discussed, the studies performed by others with similarly high relevance scores were also limited to these same protocols, i.e., McGarvey et al. 2011; Blackley et al. 2011; Schoenberg et al. 2011) (Table 4).

The authors have acknowledged their study limitations, those inherent to the research protocols as well as the shortcomings in implementation, i.e., small sample sizes, geographically and temporally appropriate measures for diseases that progress over time in residents that move in and out of communities, etc. The results have also been found to differ between, and even within, author groups.

Concurrently, reassessments and critiques have cast doubt, primarily on the results presented by Hendryx and coworkers (Borak et al. 2012; Exponent 2012; Kirby 2013). Borak et al. (2012) performed a statistical reassessment of the publicly available data used in three of the articles published by Hendryx and others (Hendryx 2009, Hendryx and Ahern 2009 and, Hendryx et al. 2008). Borak et al. reassessed the predictive value of coal mining and other risk factors for explaining reported higher mortality rates in Appalachia. Their results confirmed poverty, income, education, location, obesity, and gender as factors independently related to age-adjusted, all-cause mortality. Coal mining, as well as other factors, i.e., unemployment, insurance, and access to care, were not per se independent risk factors for increased mortality in Appalachia. Borak et al. (2012) reported instances in the studies reassessed where incomplete or inconsistent definitions of covariates introduced analytical complications. In another critique, the statistical assessment of the association between mountaintop mining and birth defects in central Appalachia conveyed that Ahern et al. (2011) failed to fully recognize and account for limitations in U.S. natality files, used an internet blog to provide mining data, lacked valid and reliable data on confounding factors, and thus the quality and type of data was insufficient to support the conclusions made by the authors (Exponent, Health Sciences 2011).

More recently Kirby (2013) cautioned readers using the results of the study on birth defects and mountaintop removal coal mining by Ahern et al. 2011. Kirby's critique conveyed that the appreciation for detail was lacking in terms of both birth defect research and environmental epidemiological methods. The use of better data sources and analytical methods was needed as the current procedural flaws that render the conclusion, "the observation that risk for birth defects may be increased for residents of counties with mountaintop mining," suspect at best.

While there is doubt about the statistical and practical validity of the state-of-the-science at this time, the reports appear to have had an impact. For example, EPA news releases report changes in regulations for permitting mining operations in central Appalachia on the basis of reported health claims (EPA, 2010, 2012). Most of the current news releases relate to water quality data. It is imperative that study limitations and, thus, weak evidence, is accurately and fairly portrayed when reporting chronic health data, particularly data that raise serious doubts as just described in reassessments, critiques and editorial retorts.

The generalizability of disputed findings provides additional concerns. Reports provided to international commissions to inform policy makers on the health effects and environmental consequences of coal mining have incorporated the studies currently disputed. The independent report commissioned by Beyond Zero Emissions in response to concerns in the Hunter Region of New South Wales (Colagiuri et al. 2012) is lengthy and addresses four major questions. For the question, "What specific diseases or other health problems are associated with coal mining in local communities?" the evidence table in the report cited 13 articles as evidence; 11 were research publications by the Hendryx group (Kirby 2013). Kirby cautioned that superficially the efforts of Hendryx and others appeared plausible, but the detail in methodology was lacking to produce conclusions supported by sound evidence.

Science based evidence is more important than ever to guide decision making regarding coal production, human health and environmental impacts. National and international debates surrounding energy and coal production are anticipated, and are expected to be politically controversial, tied to economic and environmental issues and vulnerable to subjective and emotional public reactions, not to mention easily inferred, and preferred, interpretations of data. While the authors clearly acknowledged the limitations of their studies, the repeated protocol strategies, the consistency in conclusions and the successive publications with a consistently contributing researcher are not capable of moving science to cause-effect conclusions. Consistent conclusions are generally regarded as a strong point in research, with the publications representing a wide array of independent researchers and varying research protocols. However, the consistent findings in this instance were reported with a common researcher on all publications. In actuality, this weakens rather than strengthens the argument whether the conclusions are free of serious doubt or not.

The inferences made from cause-effect relationships combined with the assumptions drawn by readers too easily, and possibly conveniently, propel authors' carefully stated concluding statements to erroneously accepted conclusions. Public biases toward optimism taken from inferences are hard to redirect and correct. The 'fine print,' or more appropriately, the 'final print,' the copious limitations proclaimed at the end of a publication are often overlooked, the caveats lost and the context of science dwarfed once a media sensation builds, bending to public biases toward expected rather than science supported conclusions. Epidemiological studies have limitations and come to a point where an inference can be made about cause-effect relationships but they, as well as any other single type of study design, cannot determine or 'prove' whether an exposure caused a specific disease. Strong study designs are needed to produce valid support for cause-effect relationships with statistically significant findings. Direct measures of biomarkers indicating environmental exposures affecting chronic disease rates are needed.

CONCLUSION

Epidemiological studies are concerned with the incidence of disease and direct attention to the etiology of disease progression. The majority of studies reviewed serve as an important 'spring board' for the design and implementation of experimental and intervention studies as 'next steps' in the quest to identify health concerns associated with coal mining. Regardless of the specific etiology of the chronic diseases studied, Appalachia remains a region with documented health disparities. There is a need for further research to bend the trends that report increasing prevalence rates of chronic diseases, accelerate the implementation of effective intervention programs and anticipate the successes to be realized through health education, cultural awareness, increased access to primary care providers and utilization of emerging technologies to reach rural areas in Appalachian communities.

ACKOWLEDGMENT

This study was sponsored by the Appalachian Research Initiative for Environmental Science (ARIES). The views, opinions and recommendations expressed herein are solely those of the authors and do not imply any endorsement by ARIES employees, other ARIES-affiliated researchers or industrial members. Information about ARIES can be found at http://www.energy.vt.edu/ARIES.

REFERENCES

Ahern, M.M., Hendryx, M.S., Conley, J., Fedorko, E., Ducatman, A., and Zullig, K.J. 2011. The association between mountaintop mining and birth defects among live births in Central Appalachia, 1996–2003. *Environ Res.* 111(6):838–846.

Ahern, M., Brown, C., and Dukas, S. 2011. A national study of the association between food environments and county-level health outcomes. *J. Rural Health.* 27(4):367–379.

Appalachia Community Cancer Network. *Addressing the Cancer Burden in Appalachian Communities 2010.* Appalachia Community Cancer Network, National Cancer Institute. Available at: http://www.accnweb.com/docs/2010/AddressingtheCancerBurdenAppCommunities2010.pdf. Accessed April 2012.

Appalachian Regional Commission. *The Appalachian Region.* Washington, DC: Appalachian Regional Commission; 2011. Available at: http://www.arc.gov/appalachian_region/TheAppalachianRegion Accessed January 2013.

Barker, L., Crespo, R., Gerzoff, R., Denham, S., Shrewsberry, M., and Cornelius-Averhart, D. 2010. Residence in a distressed county in Appalachia as a risk factor for diabetes, Behavioral Risk Factor Surveillance System, 2006–2007. *Prev. Chronic Dis.* 7(5):A104:1–9.

Bernhardt, E.S., and Palmer, M.A. 2011. The environmental costs of mountaintop mining valley fill operations for aquatic ecosystems of the Central Appalachians. *Ann. NY Acad. Sci.* 1223:39–57.

Bilban, M. 2005. Occupational medicine in the Slovene area. *J. Occup. Health.* 47(3):193–200.

Blackley, D., Behringer, B., and Zheng, S. 2011. Cancer mortality rates in Appalachia: Descriptive epidemiology and an approach to explaining differences in outcomes. *J Community Health.* doi:10.1007/s10900–011–9514-z.

Borak, J., Salipante-Zaidel, C., Slade, M.D., and Fields, C.A. 2012. Mortality disparities in Appalachia: Reassessment of major risk factors. *J. Occup. Environ. Health.* 54(2):146–155.

Colaguiri, R., Cochrane, J., and Girgis, S. 2012. Health and social harms of coal mining in local communities; Spotlight on the Hunter Region. Beyond Zero Emissions, Melbourne,October 2012.

Cohen, R.A., Patel, A., and Green, F.H. 2008. Lung disease caused by exposure to coal mine and silica dust. *Semin. Respir. Crit. Care Med.* 29(6):651–661.

Denham, S.A. 2003. Relationships between family rituals, family routines and health. *J. Family Nurs.* 9(3):305–330.

Denham, S.A. 1999. Part 3: Family health in an economically disadvantaged population. *J. Family Nurs.* 5(2):184–213.

Denham, S.A., Meyer, M.G., Toborg, M.A., and Mande, M.J. 2004a. Providing health education to Appalachia populations. *Holist. Nurs. Pract.* 18(6):293–301.

Denham, S.A., Manoogian, M., and Schuster, L. 2007. Managing family suppport and dietary routines: Type 2 diabetes in rural Appalachian families. *Families, Systems, & Health.* 25(1):36–52.

Denham, S.A., Remsburg, K., and Wood, L. 2010a. Diabetes education in the Appalachian Region: providers' views. *Rural Remote Health.* 10(2):1321.

Denham, S.A., Ware, L.J., Raffle, H., and Leach, K. 2011. Family inclusion in diabetes education: a nationwide survey of diabetes educators. *Diabetes Educ.* 37(4):528–535.

Denham, S.A., Wood, L.E., and Remsberg, K. 2010b. Diabetes care: provider disparities in the US Appalachian Region. *Rural Remote Health.* 10(2):1320.

Diddle, G., and Denham, S.A. 2010. Spirituality and its relationships with the health and illness of Appalachian people. *J. Transcult. Nurs.* 21(2):175–182.

Dietary Reference Intakes for Calcium, Phosphorus, Magnesium, Vitamin D, and Fluoride. 1997. National Academy of Sciences. Institute of Medicine. Food and Nutrition Board.

Donoghue, A.M. 2004. Occupational health hazards in mining: an overview. *Occup. Med.* 54(5):283–289.

Drew, E.M., and Schoenberg, N.E. 2011. Deconstructing fatalism: ethnographic perspectives on women's decision making about cancer prevention and treatment. *Med. Anthropol. Q.* 25(2):164–182.

Environmental Protection Agency. 2012. Surface Coal Mining Activities under Clean Water Act Section 404. http://water.epa.gov/lawsregs/guidance/wetlands/mining.cfm#memo20100401. Accessed January 2013.

Environmental Protection Agency. 2010. http://www.epa.gov/sciencematters/august2010/mountaintop.htm. Accessed October 2012.

Epstein, P.R., Buonocore, J.J., Eckerle, K., Hendryx, M.S., Stout, B.M., Heinberg, R., et al. 2011. Full cost accounting for the life cycle of coal. *Ann. NY Acad. Sci.* 1219:73–98.

Esch. L., and Hendryx, M.S. 2011. Chronic cardiovascular disease mortality in mountaintop mining areas of central Appalachian states. *J. Rural Health.* 27(4):350–357.

Exponent IHSG. 2011. Critical Review: Ahern M.M. et al. The association between mountaintop mining and birth defects among live births in Central Appalachia, 1996–2003. *Env. Res.* June 21. Washington, DC; pages 1–8.

Feng, X., and Qiu, G. 2008. Mercury pollution in Guizhou, southwestern China–an overview. *Sci. Total Environ.* 400(1–3):227–237.

Gold, L.S., De Roos, A.J., Waters, M., and Stewart, P. 2008. Systematic literature review of uses and levels of occupational exposure to tetrachloroethylene. *J. Occup. Environ. Hyg.* 5(12):807–839.

Gulumian, M., Borm, P.J., Vallyathan, V. Castranova, V., Donaldson, K., Nelson, G., and Murray, J. 2006. Mechanistically identified suitable biomarkers of exposure, effect, and susceptibility for silicosis and coal-worker's pneumoconiosis: a comprehensive review. *J. Toxicol. Environ. Health B. Crit. Rev.* 9(5):357–395.

Hendryx, M.S. 2008. Mental health professional shortage areas in rural Appalachia. *J. Rural Health.* 24(2):179–182.

Hendryx, M.S. 2009. Mortality from heart, respiratory, and kidney disease in coal mining areas of Appalachia. *Int. Arch. Occup. Environ. Health.* 82(2):243–249.

Hendryx, M.S. 2011. Introduction to special section: Environmental, health for rural populations. *J. Rural Health.* 27(4):339–341.

Hendryx, M.S. 2011. Poverty and mortality disparities in Central Appalachia: Mountaintop mining and environmental justice. *J. Health Disparities Res. Pract.* 4(3):44–53.

Hendryx, M.S., and Ahern, M.M. 2008. Relations between health indicators and residential proximity to coal mining in West Virginia. *Am. J. Public Health.* 98(4):669–671.

Hendryx, M.S., and Ahern, M.M. 2009. Mortality in Appalachian coal mining regions: the value of statistical life lost. *Public Health Rep.* 124(4):541–550.

Hendryx, M.S., Ahern, M.M., Lovrich, N.P., and McCurdy, A.H. 2002. Access to health care and community social capital. *Health Serv. Res.* 37(1):87–103.

Hendryx, M.S., Ahern, M.M., and Nurkiewicz, T.R. 2007. Hospitalization patterns associated with Appalachian coal mining. *J Toxicol Environ Health A.* 70(24):2064–2070.

Hendryx, M.S., and Fedorko, E. 2011. The relationship between toxics release inventory discharges and mortality rates in rural and urban areas of the United States. *J. Rural Health.* 27(4):358–366.

Hendryx, M.S., Fedorko, E., and Anesetti-Rothermel, A. 2010. A geographical information system-based analysis of cancer mortality and population exposure to coal mining activities in West Virginia, United States of America. *Geospat. Health.* 4(2):243–256.

Hendryx, M.S., Fedorko, E., and Halverson, J. 2010. Pollution sources and mortality rates across rural-urban areas in the United States. *J. Rural Health.* 26(4):383–391.

Hendryx, M.S., O'Donnell, K., and Horn, K. 2008. Lung cancer mortality is elevated in coal-mining areas of Appalachia. *Lung Cancer.* 62(1):1–7.

Hendryx, M.S., Wolfe, L., Luo, J., and Webb, B. 2011. Self-reported cancer rates in two rural areas of West Virginia with and without mountaintop coal mining. *J. Community Health.* 37(2):320–327.

Hendryx, M.S,. and Zullig, K.J. 2009. Higher coronary heart disease and heart attack morbidity in Appalachian coal mining regions. *Prev. Med.* 49(5):355–359.

Hitt, N.P., and Hendryx, M.S. 2010. Ecological integrity of streams related to human cancer mortality rates. *Ecohealth.* 7(1):91–104.

Huang, X., and Finkelman, R.B. 2008. Understanding the chemical properties of macerals and minerals in coal and its potential application for occupational lung disease prevention. *J. Toxicol. Environ. Health B Crit. Rev.* 11(1):45–67.

Joy, J. 2004. Occupational safety risk management in Australian mining. *Occup Med (Lond).* 54(5):311–315.

Kang, S.K., and Kim, E.A. 2010. Occupational diseases in Korea. *J. Korean Med. Sci.* 25(Suppl):S4–12.

Kavanagh, A.M., Bentley, R., Turrell, G., Broom, D.H., and Subramanian, S.V. 2006. Does gender modify associations between self-rated health and the social and economic characteristics of local environments? *J. Epidemiol. Community Health.* 60(6):490–495.

Kirby, R. 2013. Birth defects and environmental exposures: Logical, diabolical or the devil is in the details? *Environ. Res.* 20:140–141.

Knuckles, T.L., Stapleton, P.A., Minarchick, V.C., Esch, L., McCrawley, M., Hendryx, M.S., and Nurkiewicz, T.R. 2012. Air pollution particulate matter collected from an Appalachian mountain top mining site induces microvascular dysfunction. doi:10.1111/micc.12014.

Kruger, T.M., Swanson, M., Davis, R.E., Wright, S., Dollarhide, K., and Schoenberg, N.E., 2012. Formative research conducted in rural Appalachia to inform a community physical activity intervention. *Am J Health Promot.* 26(3):143–151.

Lee, S.Y., Chen, W.L., and Weiner, B.J. 2004. Communities and hospitals: social capital, community accountability, and service provision in U.S. community hospitals. *Health Serv. Res.* 39(5):1487–1508.

Li P., Feng, X.B., Qiu, G.L., Shang, L.H., and Li, Z.G. 2009. Mercury pollution in Asia: a review of the contaminated sites. *J. Hazard. Mater.* 168(2–3):591–601.

Luo J., and Hendryx, M.S. 2011. Environmental carcinogen releases and lung cancer mortality in rural-urban areas of the United States. *J. Rural Health.* 27(4):342–349.

Luo, J., Hendryx, M.S., and Ducatman, A. 2011. Association between six environmental chemicals and lung cancer incidence in the United States. *J. Environ. Public Health.* 463701. Epub 2011 Jul 10.

Ma Y., Prasad, M.N., Rajkumar, M., and Freitas, H. 2011. Plant growth promoting rhizobacteria and endophytes accelerate phytoremediation of metalliferous soils. *Biotechnol. Adv.* 29(2):248–258.

Manoogian, M.M., Harter, L.M., and Denham, S.A. 2010. The storied nature of health legacies in the familial experience of type 2 diabetes. *J. Family Communication.* 10:40–56.

McGarvey, E.L., Leon-Verdin, M., Killos, L.F., Guterbock, T., and Cohn, W.F. 2011. Health disparities between Appalachian and non-Appalachian counties in Virginia USA.*J. Community Health.* 36(3):348–356.

Meacham, S.L. 2003. New dietary reference intakes set for energy, carbohydrates, fiber, fat, fatty acids, cholesterol, proteins, and amino acids. *ACSM's Health and Fitness J.* 7(1):25–27.

Meacham, S.L., Grayscott, D., Chen, J.J., and Bergman, C.J. 2008. Review of the Dietary Reference Intakes for calcium: Where do we go from here? *Crit. Rev. Food Sci. Nutr.* 48(5): 378–384.

Meacham, S.L., Sukpraprut, S., Taber, T., and Metzger, D. 2012. Publications on chronic disease in coal dependent communities in central Appalachia. *J. Health Disparities Res. Pract.* 5(3):62–79.

Melnyk, B.M., and Fineout-Overholt, E. 2005. *Evidence-Based Practice in Nursing and Healthcare: A Guide to Best Practice.* Philadelphia, PA: Lippincott Williams and Wilkins.

Meyer, M.G., Toborg, M.A., Denham, S.A., and Mande, M.J. 2008. Cultural perspectives concerning adolescent use of tobacco and alcohol in the Appalachian mountain region. *J. Rural Health.* 24(1):67–74.

O'Brien, T., and Denham, S.A. 2008. Diabetes care and education in rural regions. *Diabetes Educ.* 34(2):334–347.

Palmer, M.A., Bernhardt, E.S., Schlesinger, W.H., Eshleman, K.N., Foufoula-Georgiou, E., and Hendryx, M.S., et al. 2010. Science and regulation. Mountaintop mining consequences. *Science* (Commentary). 327:148–149.

Paskett, E.D., Fisher, J.L., Lengrerich, E.J., Schoenberg, N.E., Kennedy, S.K., Conn, M.E. et al. 2011. Disparities in underserved white populations: the case of cancer-related disparities in Appalachia. *Oncologist.* 16(8):1072–1081.

Pollard, C., Bailey, K.A., Petitte, T., Baus, A., Swim, M., and Hendryx, M.S. 2009. Electronic patient registries improve diabetes care and clinical outcomes in rural community health centers. *J. Rural Health.* 25(1):77–84.

Ross, M.H., and Murray, J. 2004. Occupational respiratory disease in mining. *Occup. Med. (Lond).* 54(5):304–310.

Salipante-Zaidel, C., and Borak, J. 2010. Hendryx Critique; National Mining Association Press Release. Washington, DC, May 6.

Sayler, K.L. 2012. United States Geological Society. Land cover trends project, Central Appalachians. landcovertrends.usgs.gov/east/eco69Report.html. Accessed January 2013.

Schoenberg, N.E., Bardach, S.H., Manchikanti, K.N., and Goodenow, A.C. 2011. Appalachian residents' experiences with and management of multiple morbidity. *Qual. Health Res.* 21(5):601–611.

Schwartz, F.L., Denham, S.A., Heh, V., Wapner, A., and Shubrook, J. 2010. Experiences of children and adolescents with Type I diabetes in school: Survey of children, parents, and schools. *Diabetes Spectrum.* 23(1):47–55.

Schwartz, F., Ruhil, A.V., Denham, S., Shubrook, J., Simpson, C., and Boyd, S.L. 2009. High self-reported prevalence of diabetes mellitus, heart disease, and stroke in 11 counties of rural Appalachian Ohio. *J. Rural Health.* 25(2):226–230.

Sharma, S., and Rees, S. 2007. Consideration of the determinants of women's mental health in remote Australian mining towns. *Aust. J. Rural Health.* 15(1):1–7.

Stahl, C., Cimorelli, A., Babendreier, J., and Kremer, J. 2012. The dangers of "proofiness" in the evaluation of mountaintop removal coal mining impacts. Abstract #151. Society of Environmental Toxicology and Chemistry, North America 33rd Annual Meeting, Long Beach, CA, November.

Vearrier, D., and Greenberg, M.I. 2011. Occupational health of miners at altitude: adverse health effects, toxic exposures, pre-placement screening, acclimatization, and worker surveillance. *Clin. Toxicol. (Phila).* 49(7):629–640.

Virginia Diabetes Council. *Virginia Diabetes Plan 2008–2017.* Virginia Diabetes Council. Available at: http://www.diabetesappalachia.net/diabetes_appa.html Accessed April 2012.

Virginia Rural Health Association Data Portal. Available at: http://www.vrhrc.org/dataportal/index.htm. Accessed October 2012.

Virginia Department of Health. Available at: http://www.vdh.state.va.us/Vital_Records. Accessed July 2012.

World Cancer Research Foundation/American Institute for Cancer Research. 2007. Second expert report, food, nutrition, physical activity and the prevention of cancer: a global perspective. November.

Zipper, C.E., Burger, J.A., Skousen, J.G., Angel, P.N., Barton, C.D., Davis, V., and Franklin, J.A. 2011. Restoring forests and associated ecosystem services on Appalachian coal surface mines. *Environ. Manage.* 47(5):751–765.

Zullig, K.J., and Hendryx, M.S. 2010. A comparative analysis of health-related quality of life for residents of U.S. counties with and without coal mining. *Public Health Rep.* 125(4):548–555.

Zullig, K.J., and Hendryx, M.S. 2011. Health-related quality of life among central Appalachian residents in mountaintop mining counties. *Am. J. Public Health.* 101(5):848–853.

Comparison of Mortality Rates Between Appalachian Coal Mining Counties with Non-Coal Mining

Jeanine M. Buchanich
Department of Biostatistics, Graduate School of Public Health, University of Pittsburgh, Pittsburgh, Pennsylvania

Evelyn O. Talbott
Department of Epidemiology, Graduate School of Public Health, University of Pittsburgh, Pittsburgh, Pennsylvania

Ada O. Youk
Department of Biostatistics, Graduate School of Public Health, University of Pittsburgh, Pittsburgh, Pennsylvania

Andrew N. Potter
Department of Biostatistics, Graduate School of Public Health, University of Pittsburgh, Pittsburgh, Pennsylvania

Lynne P. Marshall
Department of Epidemiology, Graduate School of Public Health, University of Pittsburgh, Pittsburgh, Pennsylvania

ABSTRACT

Recent studies have indicated that adverse health outcomes occur at higher rates in Appalachia, particularly in West Virginia. However, conflicting evidence has been found regarding whether these disparities are due in part to coal mining in Appalachia or to other social, economic, or health factors. In this study, we computed age-adjusted mortality rates and the associated confidence intervals for multiple causes of death. These were then used to conduct a graphical analysis comparing the mortality rates for the coal mining and non-coal mining counties. For all cause mortality rates, coal mining counties had higher mortality rates than non-coal mining counties. However, there are no consistent trends for across the specific causes of mortality in this study. Additional studies are needed to refine the nature of these findings.

INTRODUCTION

Recent studies have indicated that adverse health outcomes, including mortality, occur at higher rates in the Appalachian region of the US (Hendryx, et al. 2008, Wingo, Tucker et al. 2008, Hendryx 2009, Christian, et al. 2011, Esch and Hendryx 2011, Borak, et al. 2012). However, conflicting evidence has been found regarding whether these disparities are due in part to coal mining in Appalachia (Hendryx, et al. 2008, Hendryx 2009, Hendryx and Ahern 2009, Hendryx, et al. 2010, Esch and Hendryx 2011, Borak,et al. 2012). In an effort to compare mortality in coalmining versus non-coal mining counties, we performed an ecological evaluation of total and select cause-specific mortality rates of West Virginia (WV) coal mining counties compared to mortality in Appalachian non-coal mining counties.

METHODS

We conducted descriptive and statistical analyses of the total and select cause-specific mortality rates in WV coal mining and non-coal mining counties. We attempted to control for the possible confounding of socioeconomic status (SES) by selecting control counties for comparison based on county-to-county frequency matching. We controlled for age, time, race, and sex by utilizing age-time-race-specific mortality rates in our statistical comparisons. This study was exempt from review by the University of

Pittsburgh Institutional Review Board because it only involved aggregate statistical analyses of mortality data.

Identification of WV Mining Counties and Appalachian Comparison Counties

We used ever/never coal mining status in 1994 as a proxy for the whole study period, hypothesizing that areas of recent coal mining also have historical coal mining (Energy Information Agency 1994). We defined Appalachian counties as those indicated by the Appalachian Region Commission (ARC) (Appalachian Regional Commission), a regional economic development agency established by an act of Congress in 1965.

Before creating matched pairs of WV coal mining counties and a non-mining county, we gathered demographic information for all counties in WV from Census 2000 DP-3: Profile of Selected Economic Characteristics and QT-P33: Income in 1999 by Selected Household, Family, and Individual Characteristics for median income (U.S. Census Bureau). We also gathered mining status in 1994 (Energy Information Agency 1994) and the same demographic data (U.S. Census Bureau) for Pennsylvania, Maryland, Virginia, North Carolina, Tennessee, Kentucky, and Ohio; all of the Appalachian non-coal mining counties in these states, plus WV, became our pool of potential matches.

The matching process was conducted iteratively. The first step was to calculate an income range (median income ±5%) for each coal mining and non-coal mining county eligible for inclusion. For each coal mining county, potential non-coal mining matches were identified by median income and geographic proximity, with counties within WV given higher priority. Next, we chose the non-coal mining county with the nearest median income to the coal mining county as a candidate match. We did not match adjacent counties with the exception of Mineral County (Hampshire County) and Marshall County (Jackson County).

Generation of Mortality Rates

We used the Mortality and Population Data System (MPDS) (Marsh, Youk et al. 2005) to generate the mortality rates and standard errors for the causes of interest. The MPDS contains the underlying cause of death code (using International Classification of Diseases (ICD) four-digit codes in effect at the date of death) for all persons who died of cancer causes in the US between 1950 and 2007 and 1960–2007 for non-cancer causes, including total mortality. The specific causes of death evaluated were: total mortality (International Classification of Disease (ICD) 9th revision codes 001–999), mortality from all cancers (ICD 9th revision codes 140–209), respiratory system cancer (ICD 9th revision codes 160–165), kidney cancer (ICD 9th revision code 189), diabetes mellitus (ICD 9th revision code 250), heart disease (ICD 9th revision codes 390–398, 402, 404, 410–429), cerebrovascular disease (ICD 9th revision codes 430–438), non-malignant respiratory disease (ICD 9th revision codes 460–519), and all external causes (ICD 9th revision codes 800–949).

Data Analysis

We performed a graphical analysis of the age-adjusted mortality rates for males and females in the coal mining counties as compared to the non-coal mining counties for 5-year time periods from 1950–2007 (cancer causes) or 1960–2007 (non-cancer causes). Graphs were created for each cause of death of interest and included both the age-adjusted mortality rate and its associated 95% confidence interval (CI) for each time period; also shown for select causes of death are detailed tables with the mortality rate and 95% CI for each time period. For each time period, the rates for the coal mining counties and non-coal mining counties were considered statistically significantly different if the 95% confidence intervals were not overlapping. These comparisons do not account specifically for the matching of non-coal mining county to coal mining county, but instead aggregate all of the 31 coal mining and non-mining counties.

RESULTS

The figures and tables show the age-race-adjusted mortality rates per 100,000 people in the coal mining counties as compared to the non-coal mining counties for each of the five year time periods. As seen in Figure 1, all cause mortality rates are statistically significantly higher in coal mining counties compared to non-coal mining counties across all time periods for males and females. For all cancer mortality (Figure 2), rates for males in both groups reach a peak in the late 1980s/early 1990s and then begin declining; rates for females in both groups appear more stable. Table 1 shows that male rates in the coal mining counties are statistically significantly higher from the 1960s through the 1980s but are not consistently statistically significantly higher from the 1990s through the present. For females, all cancer mortality is statistically significantly higher in the 1980s and 1990s, although appears to be converging in the most recent time period examined.

The patterns for all respiratory system cancer mortality (Figure 3) and all respiratory system cancer mortality are similar to the trends for all cancer, although the rates in women appear to be rising through the 1990s and 2000s.

Figure 4 shows that the patterns for kidney cancer are less consistent, with more variability and elevations by the non-coal mining counties in both males and females. The increased variability is due to the small number of kidney cancer cases in the counties involved, and is reflected in Table 2 which shows no statistically significantly different confidence intervals. Table 2 also reveals occasionally higher mortality rates in non-coal mining counties, especially among males. Rates for males in both groups are rising while rates for women seem to have remained stable.

Mortality rates for heart disease (Figure 5) are decreasing in both non-coal mining and coal mining counties, although coal mining county rates are consistently statistically significantly higher than non-coal mining counties.

Stroke mortality (Figure 6) is declining for both males and females in both coal mining and non-coal mining counties. As shown, non-coal mining rates for males and females are statistically significantly higher compared to the coal mining counties through the 1970s, when non-coal mining remains higher although the difference is no longer statistically significant. Table 3 emphasizes the dramatic reduction in mortality rates for stroke and, in the 2005–2007 period, rates for all four groups are virtually identical.

Diabetes mortality (not shown) showed a great deal of variability, with excesses in both non-coal mining and coal mining counties for males and females, although none of the differences were statistically significant until the 1990s. From that point forward, coal mining counties had statistically significantly higher rates for males and females. The rates for all external causes (not shown) are nearly identical between the coal mining and non-coal mining counties with no statistically significant differences

Rates for non-malignant respiratory disease (Figure 7) have remained relatively stable over time for males and females. As shown in Table 4, rates for males are statistically significantly higher in coal-mining counties throughout all time periods. For females, rates are statistically significantly higher in the coal mining counties in 1975–79 and from 1985–94; from 1960–69 rates for females in the non-coalling mining counties were not statistically significantly higher.

DISCUSSION

This study found limited evidence of elevated total and cause-specific mortality rates when comparing coal mining to non-coal mining counties in Appalachia. For several causes of death, coal mining counties had statistically significantly higher mortality rates compared to non-coal mining counties. Total mortality was statistically significantly higher across all time periods for men and women in coal mining compared to non-coal mining counties. While Hendryx and Ahern (Hendryx and Ahern 2009) also found higher mortality among coal-mining counties in Appalachia, their comparison group was non-mining, non-Appalachia which may not account for factors and culture unique to Appalachia (Lengerich, et al. 2004, Lengerich, et al. 2006). We also found higher rates of heart disease in coal

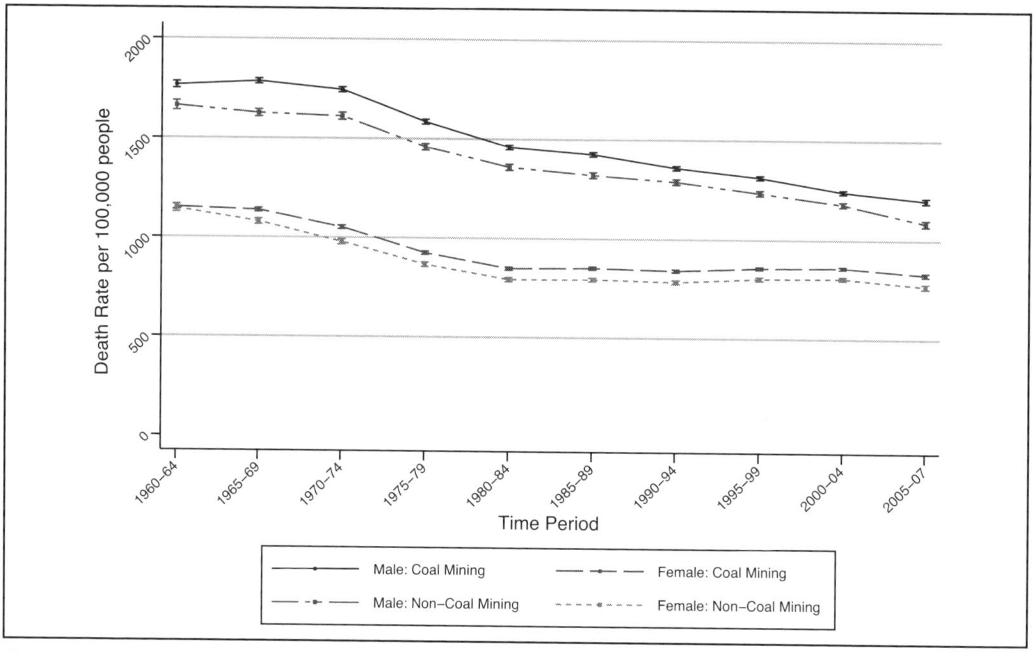

Figure 1. All cause age-race adjusted mortality rates males and females in the West Virginia coal mining counties and the income matched non-coal mining Appalachian counties with the 95% confidence intervals plotted for each time period

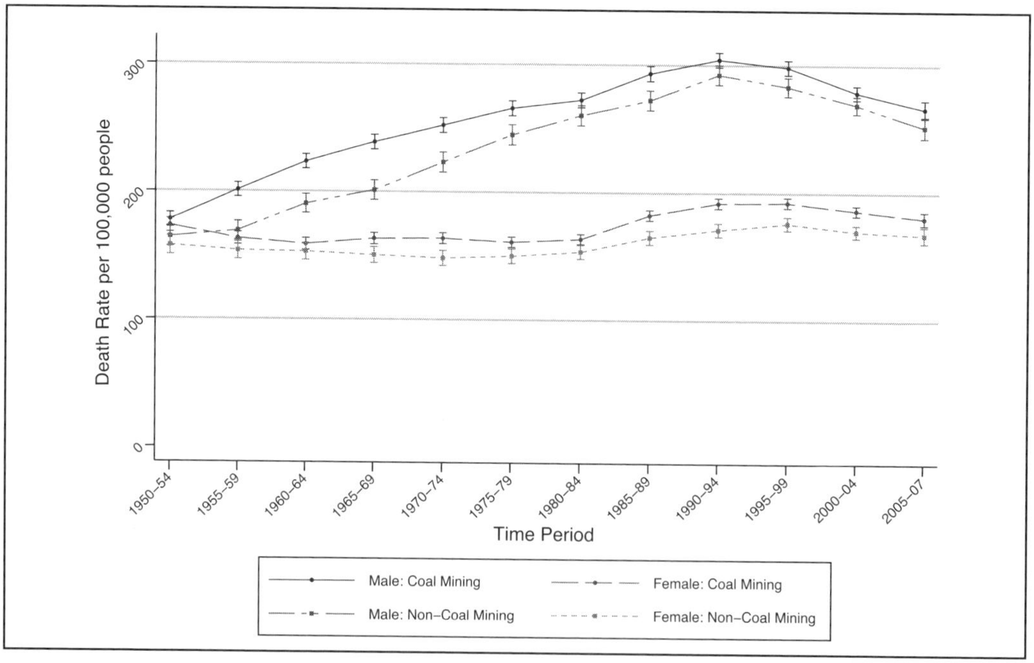

Figure 2. All cancer age-race adjusted mortality rates for males and females in the West Virginia coal mining counties and the income matched non-coal mining Appalachian counties with the 95% confidence intervals plotted for each time period

Table 1. All cancer mortality rates*

	Females					Males				
	Coal Mining		Non-Coal Mining			Coal Mining		Non-Coal Mining		
Time Period	Rate†	CI‡	Rate	CI		Rate	CI	Rate	CI	
1950–54	172.87	167.60,178.15	157.35	150.15,164.55	*	177.64	172.28,183.00	164.06	156.58,171.54	*
1955–59	162.98	158.16,167.80	153.59	146.82,160.37		201.06	195.59,206.53	168.96	161.65,176.28	*
1960–64	158.67	154.13,163.21	152.77	146.23,159.30		223.29	217.71,228.86	190.33	182.86,197.79	*
1965–69	163.08	158.63,167.53	150.29	144.00,156.57	*	238.70	233.01,244.38	201.20	193.60,208.80	*
1970–74	163.54	159.25,167.82	148.11	142.22,154.01	*	252.08	246.29,257.86	223.19	215.32,231.05	*
1975–79	160.78	156.73,164.83	149.88	144.21,155.54	*	265.65	259.84,271.46	244.89	236.99,252.80	*
1980–84	163.31	159.36,167.27	153.38	147.97,158.80	*	272.28	266.51,278.05	260.29	252.47,268.12	
1985–89	182.29	178.18,186.39	164.90	159.47,170.33	*	293.34	287.40,299.28	272.37	264.62,280.11	*
1990–94	191.98	187.85,196.11	171.04	165.74,176.35	*	304.36	298.48,310.23	292.53	284.81,300.26	
1995–99	192.56	188.49,196.63	176.25	171.01,181.49	*	298.42	292.73,304.10	283.31	275.99,290.64	*
2000–04	186.37	182.39,190.35	170.14	165.17,175.11	*	278.84	273.44,284.24	269.28	262.51,276.05	
2005–07	180.63	175.47,185.78	167.65	161.37,173.93	*	266.32	259.51,273.12	252.03	243.69,260.37	

* The table lists age-adjusted mortality rates for West-Virginia coal mining counties and the income matched non-coal mining counties. Statistically significant results are indicated by an * after the non-coal mining county 95% confidence interval.
† Rate is an age-adjusted mortality rate per 100,000 population.
‡ CI is a 95% confidence interval around the mortality rate.

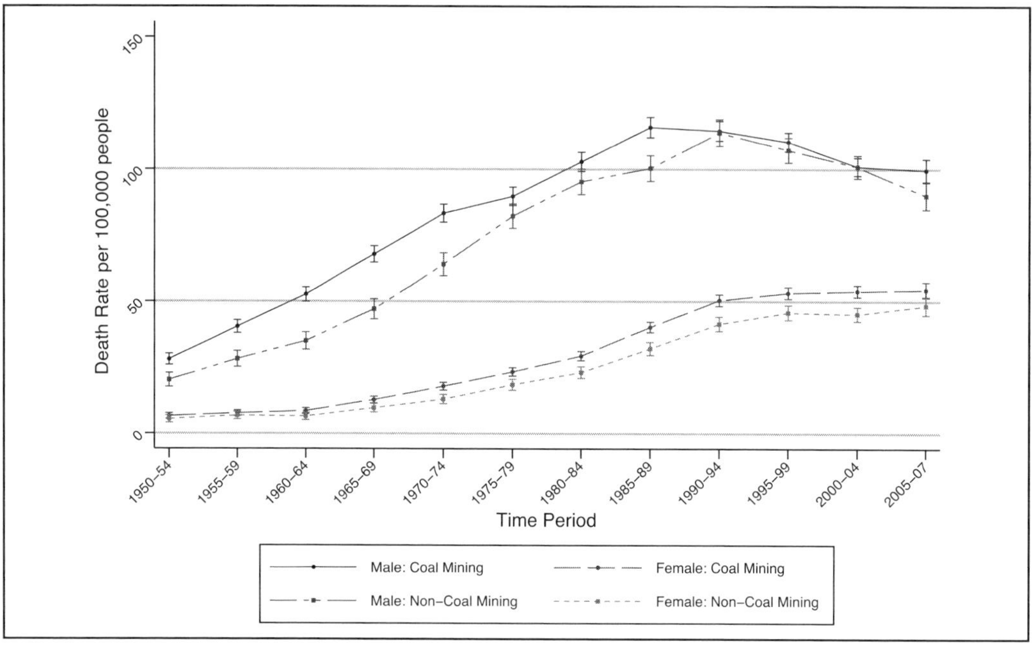

Figure 3. Respiratory cancer age-race adjusted mortality rates for males and females in the West Virginia coal mining counties and the income matched non-coal mining Appalachian counties with the 95% confidence intervals plotted for each time period

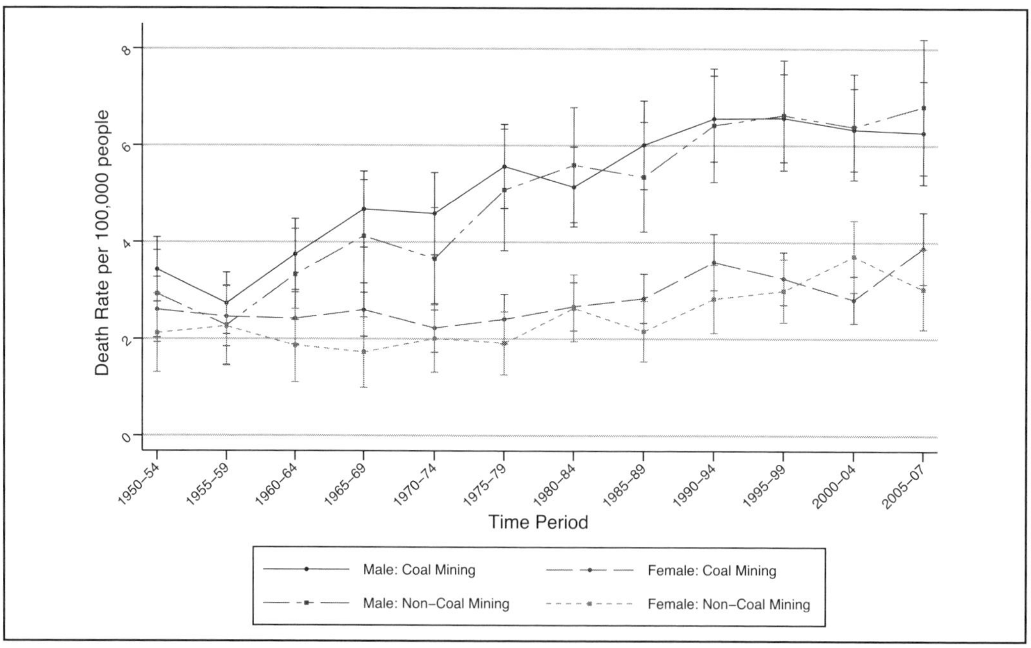

Figure 4. Kidney cancer age-race adjusted mortality rates for males and females in the West Virginia coal mining counties and the income matched non-coal mining Appalachian counties with the 95% confidence intervals plotted for each time period

Table 2. Kidney cancer mortality rates*

	Females				Males			
	Coal Mining		Non-Coal Mining		Coal Mining		Non-Coal Mining	
Time Period	Rate†	CI‡	Rate	CI	Rate	CI	Rate	CI
1950–54	2.60	1.93,3.28	2.12	1.30,2.94	3.43	2.77,4.10	2.93	2.03,3.83
1955–59	2.46	1.84,3.08	2.26	1.44,3.08	2.73	2.10,3.37	2.28	1.46,3.10
1960–64	2.42	1.88,2.96	1.86	1.10,2.62	3.74	3.01,4.48	3.33	2.39,4.27
1965–69	2.60	2.05,3.15	1.72	0.99,2.45	4.68	3.89,5.46	4.12	2.96,5.28
1970–74	2.22	1.72,2.73	2.00	1.31,2.70	4.59	3.73,5.44	3.65	2.60,4.71
1975–79	2.41	1.90,2.92	1.91	1.26,2.56	5.56	4.69,6.43	5.08	3.82,6.34
1980–84	2.67	2.17,3.17	2.64	1.94,3.33	5.14	4.32,5.97	5.59	4.41,6.78
1985–89	2.84	2.33,3.35	2.16	1.53,2.78	6.01	5.10,6.92	5.35	4.22,6.48
1990–94	3.59	3.01,4.17	2.83	2.13,3.53	6.55	5.67,7.44	6.42	5.24,7.59
1995–99	3.25	2.71,3.79	3.00	2.35,3.65	6.57	5.66,7.48	6.62	5.49,7.76
2000–04	2.81	2.33,3.30	3.71	2.97,4.45	6.33	5.48,7.18	6.38	5.29,7.47
2005–07	3.88	3.14,4.62	3.03	2.20,3.86	6.26	5.20,7.33	6.80	5.40,8.20

* The table lists age-adjusted mortality rates for West-Virginia coal mining counties and the income matched non-coal mining counties. Statistically significant results are indicated by an * after the non-coal mining county 95% confidence interval.

† Rate is an age-adjusted mortality rate per 100,000 population.

‡ CI is a 95% confidence interval around the mortality rate.

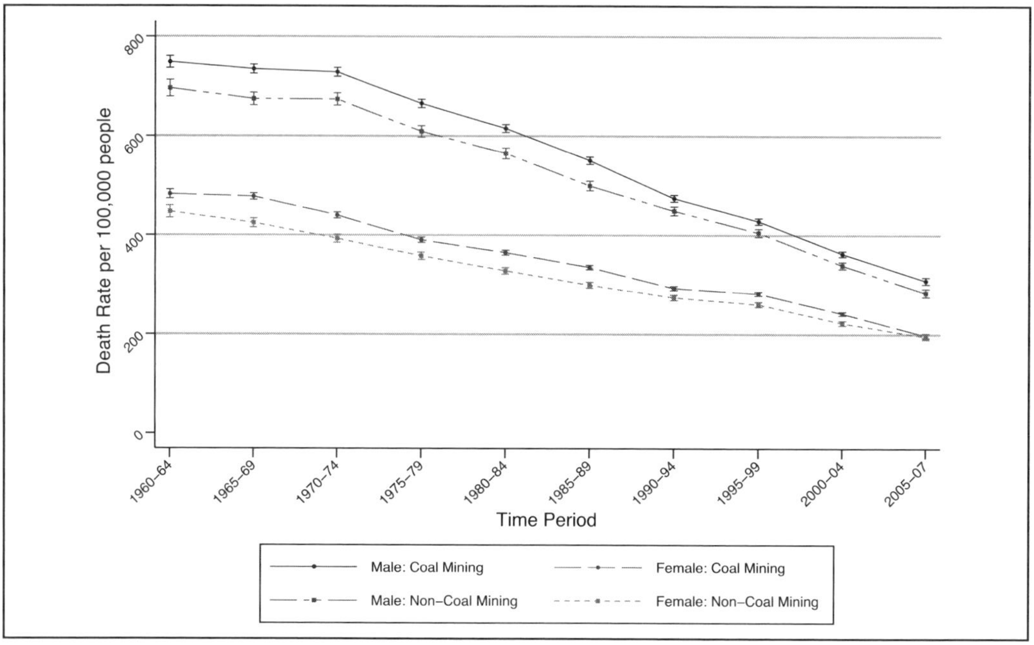

Figure 5. All heart disease age-race adjusted mortality rates for males and females in the West Virginia coal mining counties and the income matched non-coal mining Appalachian counties with the 95% confidence intervals plotted for each time period

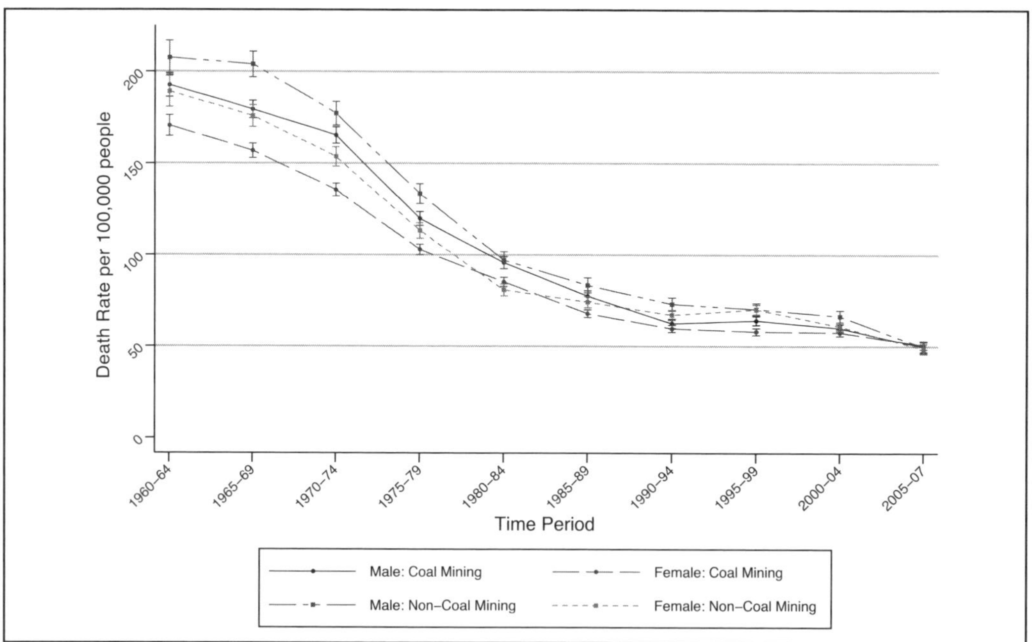

Figure 6. Stroke age-race adjusted mortality rates for males and females in the West Virginia coal mining counties and the income matched non-coal mining Appalachian counties with the 95% confidence intervals plotted for each time period

Table 3. Stroke mortality rates*

	Females					Males				
	Coal Mining		Non-Coal Mining			Coal Mining		Non- Coal Mining		
Time Period	Rate†	CI‡	Rate	CI		Rate	CI	Rate	CI	
1960–64	170.60	164.91,176.29	189.16	180.77,197.55	*	192.64	186.18,199.10	207.58	198.13,217.03	
1965–69	156.84	152.87,160.81	175.79	169.85,181.73	*	179.40	174.60,184.20	203.97	196.90,211.03	*
1970–74	135.51	132.01,139.01	153.62	148.32,158.92	*	165.30	160.83,169.78	177.20	170.75,183.65	*
1975–79	102.88	100.04,105.72	113.20	108.95,117.44	*	119.78	115.96,123.60	133.44	127.98,138.91	*
1980–84	85.30	82.80,87.80	80.92	77.62,84.21		95.75	92.46,99.04	97.13	92.51,101.74	
1985–89	68.02	65.91,70.13	74.26	71.28,77.23	*	77.54	74.57,80.52	83.46	79.37,87.55	
1990–94	59.81	57.86,61.75	67.24	64.51,69.98	*	62.58	60.02,65.14	73.16	69.60,76.72	*
1995–99	58.17	56.29,60.05	70.24	67.60,72.89	*	64.22	61.77,66.67	70.44	67.13,73.75	*
2000–04	57.70	55.88,59.52	61.03	58.59,63.48		60.12	57.80,62.43	66.62	63.43,69.81	*
2005–07	50.76	48.49,53.03	48.67	45.84,51.49		49.76	46.98,52.55	49.70	46.29,53.11	

* The table lists age-adjusted mortality rates for West-Virginia coal mining counties and the income matched non-coal mining counties. Statistically significant results are indicated by an * after the non-coal mining county 95% confidence interval.

† Rate is an age-adjusted mortality rate per 100,000 population.

‡ CI is a 95% confidence interval around the mortality rate.

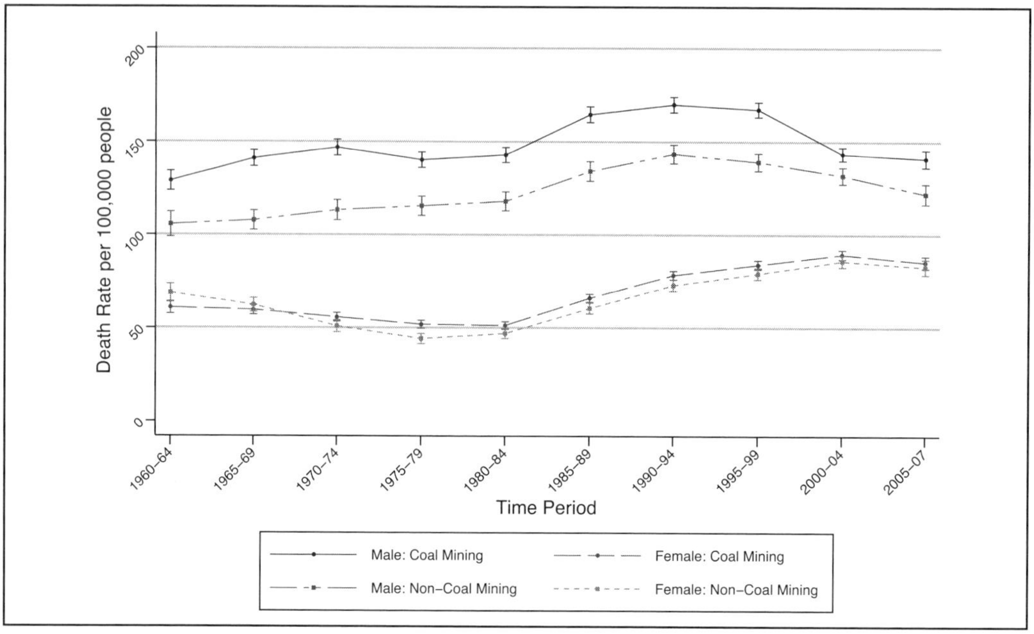

Figure 7. Non-malignant respiratory diseases age-race adjusted mortality rates for males and females in the West Virginia coal mining counties and the income matched non-coal mining Appalachian counties with the 95% confidence intervals plotted for each time period

mining compared to non-coal mining counties, similar to Hendryx and Zullig (Hendryx and Zullig 2009), which identified higher morbidity from heart disease among Appalachian coal mining and non-coal mining counties, compared to non-Appalachian, non-coal mining counties in the US.

However, higher mortality in coal mining counties was not consistently found across all causes examined, including kidney cancer and stroke, where non-coal mining counties had higher mortality. Hendryx et al. (Hendryx, et al. 2008, Hendryx 2009, Hendryx and Ahern 2009, Esch and Hendryx 2011) identified higher mortality risks for heart disease, respiratory disease, and kidney disease. The results found here do not entirely support those findings, although Hendryx did not use an Appalachian non-coal mining comparison for those analyses. While we did find elevated heart disease mortality risks among coal mining counties, we did not find excesses of kidney cancer. Our findings from NMRD are not at all consistent with the Hendryx (2009) findings.

The categories in which we found excesses are arguably heavily influenced by personal behaviors and risk factors, including heart disease (Schnohr, et al. 2002, Yong, et al. 2012), diabetes (Garber 2012) and lung cancer (Peto, et al. 2000). We were not able to control for personal risk factors in these analyses, including no control for confounding by smoking for causes of death highly affected by smoking such as heart disease and respiratory system cancer. Non-malignant respiratory disease had statistically significantly higher rates for males compared to females in all time periods and, among males, for coal mining counties compared to non-coal mining counties. Rather than being reflective of a pervasive effect of coal mines on the community, this pattern seems to reflect the effect of occupational lung diseases, such as pneumoconiosis, among males working in the coal mines (Suarthana et al. 2011).

Our findings also reflect the general trends in national mortality rates (Howlader et al., 2012). These include declines in total, heart disease, and stroke mortality. All cancer, respiratory, and

Table 4. Non-malignant respiratory disease mortality rates*

	Females					Males				
	Coal Mining		Non-Coal Mining			Coal Mining		Non- Coal Mining		
Time Period	Rate†	CI‡	Rate	CI		Rate	CI	Rate	CI	
1960–64	60.78	57.53,64.03	68.67	63.77,73.57		128.89	123.64,134.14	105.53	98.82,112.24	*
1965–69	59.59	57.08,62.10	62.22	58.50,65.94		140.91	136.62,145.20	107.69	102.48,112.91	*
1970–74	55.68	53.35,58.00	50.87	47.68,54.05		146.77	142.41,151.12	113.16	107.74,118.58	*
1975–79	51.84	49.72,53.96	44.18	41.45,46.90	*	140.06	135.91,144.22	115.38	110.18,120.57	*
1980–84	51.29	49.29,53.30	46.99	44.32,49.67		142.78	138.73,146.84	117.91	112.84,122.97	*
1985–89	66.00	63.81,68.19	60.51	57.60,63.42	*	164.56	160.33,168.78	134.04	128.76,139.32	*
1990–94	78.37	76.08,80.66	72.96	69.96,75.97	*	169.87	165.72,174.02	143.34	138.28,148.40	*
1995–99	84.00	81.62,86.38	79.20	76.14,82.26		167.23	163.23,171.23	139.08	134.31,143.85	*
2000–04	89.52	87.04,92.00	86.00	82.81,89.19		143.42	139.83,147.01	131.78	127.33,136.22	*
2005–07	85.35	82.12,88.59	82.81	78.73,86.89		140.88	136.21,145.54	121.90	116.48,127.33	*

* The table lists age-adjusted mortality rates for West-Virginia coal mining counties and the income matched non-coal mining counties. Statistically significant results are indicated by an * after the non-coal mining county 95% confidence interval.
† Rate is an age-adjusted mortality rate per 100,000 population.
‡ CI is a 95% confidence interval around the mortality rate.

Table 5. Demographic data for coal mining and non-coal mining counties

Demographic Quantity	Coal Mining Counties		Non-Coal Mining Counties	
	average	range	average	range
Median income ($)	32,853	20,496–43,628	33,020	22,261–44,302
Non-white residents (%)	4.5	1.0–15.9	3.7	1.2–11.1
Individuals over 65yrs (%)	16.8	10.2–20.9	17.4	11.1–22.5

Source: US Census 2010.

kidney cancer mortality rates have not seen the same declines and have increased or remained stable over time (Howlader et al., 2012). The patterns found here in our comparison of WV coal mining counties with Appalachian non-coal mining counties are consistent with changes in rates in the US over time.

Thirty-one WV coal mining and their non-coal mining county median family income matches were used in these analyses, although the results were aggregated by mining status not by matched pair. We chose to use income when determining a match leading to some coal-mining counties being matched to non-coal mining counties several counties away, including in neighboring states. While not used in matching, other demographic factors were similar. As shown in our companion paper (Talbott, et al., submitted) and highlighted in Table 5, the average proportion of non-white residents in coal mining and non-coal mining counties is similar (4.5% (range 1.0–15.9%)and 3.7% (range 1.2–11.1%), respectively). Overall the proportion of individuals over 65 is also very similar in the coal and non-coal mining counties (16.8% (range 10.2–20.9%) and 17.4% (range 11.7–22.5%), respectively). A total of 16.2% of West Virginians are aged 65 and older which is considerably higher than the national average level of 13.3% (US Census, 2010). However, differences in potential mortality confounding factors, such as education, personal risk factors, access to health care, and health beliefs, could be affecting mortality patterns (Lengerich, et al. 2004, Lengerich, et al. 2006).

This study had some limitations. Because we used one-to-one matching for the counties, slight differences in demographic factors could disproportionately affect the mortality rates, although the counties were aggregated here, lessening that potential. We defined coal mining as being present in 1994 based on EIA data, similar to what was done in the Hendryx analysis (Hendryx 2009); this may not reflect historical mining patterns or subsequent mine closings in the counties assessed. Further analyses using more complete historical data, including the amount of coal mined and the type of mining (surface or underground) should be conducted. For some cause of death categories, including kidney cancer, the small numbers of death led to imprecise estimates with larger confidence intervals, although mortality patterns could still be detected. In such cases, we placed less emphasis on the estimate and its confidence interval and more on the patterns found in the rates.

The results of these analyses indicate that, without control for some potential confounding factors, total and cause-specific mortality is only elevated in WV coal-mining counties relative to Appalachian non-coal mining counties for certain causes of death. Additional studies of Appalachian mortality are required to understand the complex interactions of factors in that area. These studies should include full consideration to the many confounding factors associated with mortality in these areas, and have a full understanding of the environmental conditions in Appalachia, which include not only the coal mining industry but also the chemical industry.

ACKOWLEDGMENT

This study was sponsored by the Appalachian Research Initiative for Environmental Science (ARIES). The views, opinions and recommendations expressed herein are solely those of the authors and do not imply any endorsement by ARIES employees, other ARIES-affiliated researchers or industrial members. Information about ARIES can be found at http://www.energy.vt.edu/ARIES.

REFERENCES

Borak, J., et al. 2012. Mortality disparities in appalachia: reassessment of major risk factors. *J Occup Environ Med.* 54(2): 146–156.

Christian, W.J., et al. 2011. Exploring geographic variation in lung cancer incidence in Kentucky using a spatial scan statistic: elevated risk in the Appalachian coal-mining region. *Public Health Rep.* 126(6): 789–796.

Energy Information Agency 1994. *Coal Industry Annual* 1994. U.S. Department of Energy.

Esch, L. and M. Hendryx 2011. Chronic cardiovascular disease mortality in mountaintop mining areas of central Appalachian states. *J. Rural Health.* 27(4): 350–357.

Garber, A.J. 2012. Obesity and type 2 diabetes: which patients are at risk? *Diabetes, Obesity and Metabolism.* 14(5): 399–408.

Hendryx, M. 2009. Mortality from heart, respiratory, and kidney disease in coal mining areas of Appalachia. *Int. Arch. Occup. Environ. Health.* 82(2): 243–249.

Hendryx, M. and M.M. Ahern 2009. Mortality in Appalachian coal mining regions: the value of statistical life lost. *Public Health Rep.* 124(4): 541–550.

Hendryx, M., et al. 2010. A geographical information system-based analysis of cancer mortality and population exposure to coal mining activities in West Virginia, United States of America. *Geospat Health.* 4(2): 243–256.

Hendryx, M., et al. 2008. Lung cancer mortality is elevated in coal-mining areas of Appalachia. *Lung Cancer.* 62(1): 1–7.

Hendryx, M. and K.J. Zullig 2009. Higher coronary heart disease and heart attack morbidity in Appalachian coal mining regions. *Preventive Medicine.* 49(5): 355–359.

Howlader N, et al. (eds) 2012. *SEER Cancer Statistics Review,* 1975–2009 (Vintage 2009 Populations). http://seer.cancer.gov/csr/1975_2009_pops09/, based on November 2011 SEER data submission, posted to the SEER web site, 2012. Bethesda, MD: National Cancer Institute.

Lengerich, E.J., et al. 2006. Images of Appalachia. *Prev Chronic Dis.* 3(4): A112.

Lengerich, E.J., et al. 2004. The Appalachia Cancer Network: cancer control research among a rural, medically underserved population. *J. Rural Health.* 20(2): 181–187.

Marsh, G., et al. 2005. Mortality and population data system (MPDS): University of Pittsburgh, Department of Biostatistics technical support.

Peto, R., et al. 2000. Smoking, smoking cessation, and lung cancer in the UK since 1950: combination of national statistics with two case-control studies. *BMJ.* 321: 323–329.

Schnohr, P., et al. 2002. Coronary heart disease risk factors ranked by importance for the individual and community. A 21 year follow-up of 12000 men and women from The Copenhagen City Heart Study. *European Heart Journal.* 23(8): 620–626.

Suarthana, E., et al. 2011. Coal workers' pneumoconiosis in the United States: regional differences 40 years after implementation of the 1969 Federal Coal Mine Health and Safety Act. *Occup Environ Med.* 68(12): 908–913.

Talbott, E.O., et al., 2013. Population and Exposure Characteristics of Coal Mining and Non-Coal Mining Counties in West Virginia. *SME Transactions.* submitted.

U.S. Census Bureau. 2000 Census Summary File 3; Tables DP-3 and QT-P33.; generated by Andrew Potter; using American FactFinder; http://factfinder2.census.gov; 23 January 2012.

Wingo, P.A., et al. 2008. Cancer in Appalachia, 2001–2003. *Cancer.* 112(1): 181–192.

Yong Long, et al. 2012. Analysis of the Prevalence of Cardiovascular Disease and Associated Risk Factors for European-American and African-American Populations in the State of Pennsylvania 2005–2009. *The American Journal of Cardiology.* In press.

Population and Exposure Characteristics of Coal Mining and Non-Coal Mining Counties in West Virginia

Evelyn O. Talbott
Department of Epidemiology, University of Pittsburgh, Graduate School of Public Health, Pittsburgh, Pennsylvania

LuAnn L. Brink
Department of Epidemiology, University of Pittsburgh, Graduate School of Public Health, Pittsburgh, Pennsylvania

Lynne P. Marshall
Department of Epidemiology, University of Pittsburgh, Graduate School of Public Health, Pittsburgh, Pennsylvania

Jeanine M. Buchanich
Department of Epidemiology, University of Pittsburgh, Graduate School of Public Health, Pittsburgh, Pennsylvania

ABSTRACT

The objective of this study was to provide a description of the demographic information, amount of coal production, the distribution of cardiovascular and respiratory hospitalizations, and environmental exposures by county in West Virginia. In order to make this assessment, the state was divided into two groups, coal mining counties and non-coal mining counties, based on the presence of coal production in 2005 identified from the West Virginia Office of Miners' Safety and Training report. There were marked differences in rates of hospitalizations between coal and non-coal counties, but as this initial effort is ecological and descriptive in nature, statistical tests for differences between coal mining and non-coal mining counties were not conducted.

INTRODUCTION

There is a very large and significant literature related to occupational exposures in the coal mining industry. Exposures such as occupational noise exposure, accidental injuries, bronchial and respiratory exposure due to coal dust, and increased risk of lung and stomach cancer are well documented in early studies (Coggon 1998; Donoghue 2004; Fernández-Navarro 2012). Relatively few investigations have considered the relationship between residential proximity to coal mining sites (underground and surface) and health outcomes (Howel 2001). In such studies to date, the metric most often employed to define exposure has been annual volume of coal production (Hendryx 2007). No studies have reported on a comparison of information on exposures to ambient air pollutants within proximity to the community itself or have considered exposures of other industries within these communities.

Using individual pollutants as environmental exposures is the next step in the progression of investigating possible associations to the health and well-being of coal mining communities in West Virginia (WV). This study is an exploration of population features, coal production, health outcomes, and environmental exposures of the state as a whole and by county in order to provide an overview of exposure and health characteristics of WV coal mining and non-coal mining counties. These descriptions will inform future health studies.

WEST VIRGINIA DEMOGRAPHICS AND COAL PRODUCTION

Table 1 contains for the state of WV and each county, information on population, the total number of residents who work in the coal

Table 1. West Virginia population characteristics by county

County	Coal Mining 2005 (Y/N)*	Population (per 2011 Census Estimate)†	# Employed in Coal Mining Industry (%)*	% Non-white†	% 65+†
Statewide		1,855,364	17,992 (1.0)	5.9	16.2
Barbour	Y	16,520	176 (1.1)	3.2	16.8
Boone	Y	24,444	3,614 (14.8)	1.4	14.3
Braxton	Y	14,485	118 (0.8)	1.9	17.7
Brooke	Y	23,844	34 (0.1)	3.0	19.0
Clay	Y	9,357	146 (1.6)	1.4	16.1
Fayette	Y	45,699	507 (1.1)	6.4	16.8
Grant	Y	11,891	127 (1.1)	2.0	19.0
Greenbrier	Y	35,800	77 (0.2)	5.1	19.4
Harrison	Y	69,436	650 (0.9)	4.0	16.7
Kanawha	Y	192,315	1,512 (0.8)	10.8	16.8
Lincoln	Y	21,550	88 (0.4)	1.0	15.3
Logan	Y	36,457	1,347 (3.7)	3.5	15.5
Marion	Y	56,586	568 (1.0)	5.6	17.0
Marshall	Y	32,800	1,189 (3.6)	2.0	17.5
McDowell	Y	21,729	947 (4.4)	10.6	16.6
Mercer	Y	62,465	7 (0.01)	8.4	18.0
Mineral	Y	28,192	12 (0.04)	4.8	17.7
Mingo	Y	26,563	1,551 (5.8)	3.2	13.8
Monongalia	Y	98,528	1,064 (1.1)	8.9	10.2
Nicholas	Y	26,268	459 (1.7)	1.8	17.5
Preston	Y	33,723	236 (0.7)	2.4	16.2
Raleigh	Y	79,127	1,136 (1.4)	11.1	16.1
Randolph	Y	29,465	61 (0.2)	2.7	18.3
Tucker	Y	7,021	30 (0.4)	1.4	20.9
Upshur	Y	24,322	327 (1.3)	2.4	16.9
Wayne	Y	42,126	422 (1.0)	1.5	17.0
Webster	Y	9,143	347 (3.8)	1.5	17.8
Wyoming	Y	23,419	1,240 (5.3)	1.9	15.3
Berkeley	N	105,750		11.1	11.7
Cabell	N	96,653		8.4	15.9
Calhoun	N	7,652		1.6	18.5
Doddridge	N	8,171		3.1	15.8
Gilmer	N	8,705		15.9	14.0
Hampshire	N	23,812		2.7	16.9
Hancock	N	30,571		4.1	18.9
Hardy	N	13,912		4.9	17.2
Jackson	N	29,241		1.8	17.7
Jefferson	N	54,225		10.9	12.2
Lewis	N	16,416		2.2	18.3
Mason	N	27,298		2.4	17.3
Monroe	N	13,534		2.6	20.1
Morgan	N	17,535		3.0	19.0
Ohio	N	44,246		6.7	18.4

(table continues)

Table 1. West Virginia population characteristics by county (continued)

County	Coal Mining 2005 (Y/N)*	Population (per 2011 Census Estimate)†	# Employed in Coal Mining Industry (%)*	% Non-white†	% 65+†
Pendleton	N	7,673		3.8	22.5
Pleasants	N	7,611		2.8	16.4
Pocahontas	N	8,786		2.3	19.6
Putnam	N	56,008		3.2	14.7
Ritchie	N	10,295		1.4	17.9
Roane	N	14,858		1.7	17.8
Summers	N	13,867		7.0	19.6
Taylor	N	16,916		2.4	16.1
Tyler	N	9,121		1.3	18.7
Wetzel	N	16,351		1.2	19.8
Wirt	N	5,762		1.7	15.9
Wood	N	87,120		3.4	17.1

* WV Office of Miners' Health Safety and Training, 2005 Coal Production by County.
† US Census Bureau.

mining industry, the percent of residents who are non-white, and the percent of residents who are 65 years of age and older. 'Coal Mining' is defined as any active mining in 2005 per the WV Office of Miners' Safety and Training (http://www.wvminesafety.org/cnty2005.htm) report. Population and demographic information was identified using the 2010 US Census (http://quickfacts.census.gov/qfd/states/54000.html). The number of residents who work in the coal mining industry was provided by the 2005 WV Office of Miners' Safety and Training (http://www.wvminesafety.org/cnty2005.htm) report and is only available for counties that mine coal, although it is reasonable to assume that there are employees of coal mines who reside in non-coal mining counties. The total population of WV is 1,855,364, with 1,103,275 (59.5%) residing in coal mining communities and 752,089 residing in non-coal mining areas.

Kanawha County is the most populous in WV with 10.4% of the total population. Boone County has the highest percentage of residents who are employed by the coal mining industry (14.8%). The average proportion of non-white residents in coal mining counties is 4.1% and ranges from 1.0–11.1%, compared to 4.2% with a range from 1.2–15.9% in non-coal mining counties. The average proportion of residents 65 years of age or older in coal mining counties is 16.8% (range 10.2–20.9%) in non-coal mining counties the average proportion is 17.3% (range 11.7–22.5%). A total of 16.2% of West Virginians are aged 65 and older which is considerably higher than the national average level of 13.3% (US Census, 2010). Overall the proportion of individuals over 65 is very similar in the coal and non-coal mining counties.

Table 2 provides coal production information for the state of WV as a whole and by county. The production amounts are given as a breakdown, in tons, of surface mining, underground mining, and total production for 2005 (http://www.wvminesafety.org/cnty2005.htm). Total coal mining by county can also be found in Figure 1. WV is a major coal producing state both for surface and underground coal production. Boone County had the highest coal mining production of any WV county in 2005. With almost 20% of the entire state's production, Boone County had more than twice the production of any other county. Kanawha and Marshall Counties are the second and third highest coal producing counties in WV respectively. Statewide, WV produced more than 1.5 times the amount of coal from underground mining than

Table 2. 2005 West Virginia coal production by type and county

County	Surface Mining–Tons* (% of Total)	Underground Mining–Tons* (% of Total)	Total Mining–Tons*
Statewide	60,891,662 (38.2)	98,606,407 (61.8)	159,498,069
Barbour	146,579 (16.3)	752,190 (83.7)	898,769
Boone	15,589,659 (50.9)	15,058,730 (49.1)	30,648,389
Braxton	0	1,153,785 (100)	1,153,785
Brooke	220,638 (100)	0	220,638
Clay	3,942,353 (100)	0	3,942,353
Fayette	2,240,474 (61.0)	1,432,766 (39.0)	3,673,240
Grant	0	269,207 (100)	269,207
Greenbrier	0	401,573 (100)	401,573
Harrison	109,852 (1.6)	6,716,569 (98.4)	6,826,421
Kanawha	5,632,859 (39.3)	8,692,381 (60.7)	14,325,240
Lincoln	22,174 (2.6)	839,743 (97.4)	861,917
Logan	9,508,489 (73.5)	3,431,698 (26.5)	12,940,187
Marion	94,367 (1.5)	6,359,281 (98.5)	6,453,648
Marshall	0	13,924,916 (100)	13,924,916
McDowell	2,240,730 (45.9)	2,636,742 (54.1)	4,877,472
Mercer	17,738 (26.8)	48,534 (73.2)	66,272
Mineral	93,316 (100)	0	93,316
Mingo	7,276,897 (54.1)	6,165,813 (45.9)	13,442,710
Monongalia	432,756 (4.1)	10,229,601 (95.9)	10,662,357
Nicholas	3,420,643 (84.2)	641,339 (15.8)	4,061,982
Preston	20,144 (1.3)	1,509,216 (98.7)	1,529,360
Raleigh	3,374,763 (34.1)	6,507,617 (65.9)	9,882,380
Randolph	0	473,571 (100)	473,571
Tucker	0	114,582 (100)	114,582
Upshur	7,101 (0.6)	1,191,323 (99.4)	1,198,424
Wayne	813,279 (16.3)	4,191,105 (83.7)	5,004,384
Webster	3,403,298 (69.9)	1,462,088 (30.1)	4,865,386
Wyoming	2,283,553 (34.2)	4,402,037 (65.8)	6,685,590

* WV Office of Miners' Health Safety and Training, 2005 Coal Production by County.

from surface mining in 2005. However, this ratio varies greatly from county to county, with most counties having both surface and underground mines. Brooke, Clay, and Mineral counties have their entire coal production from surface mines. Braxton, Grant, Greenbrier, Marshall, Randolph and Tucker counties have their entire coal production from underground mines.

WEST VIRGINIA HOSPITALIZATIONS

The WV Health Care Authority provided hospitalization data for the entire state of West Virginia from 2005–2009. Table 3 contains counts of circulatory hospitalizations (ICD-9 codes 390–459.9) and respiratory hospitalizations (ICD-9 codes 460–519.9) by county for 2007. Using the population data found in Table 1, Table 3 also includes crude rates, which do not take the age distribution of the population into account, by county for these outcomes. Statewide, there were 23.6/1000 hospitalizations for circulatory diseases. These ranged from 7.0 (Mineral County) to 40.7 (Logan County) in coal mining counties and from 2.4 (Hampshire County) to 45.2 (Lewis County) in non-coal mining counties. For respiratory diseases, the statewide hospitalization rate was 15.9/1000. For coal mining counties, the range was from 7.0 (Monongalia County) to 35.5

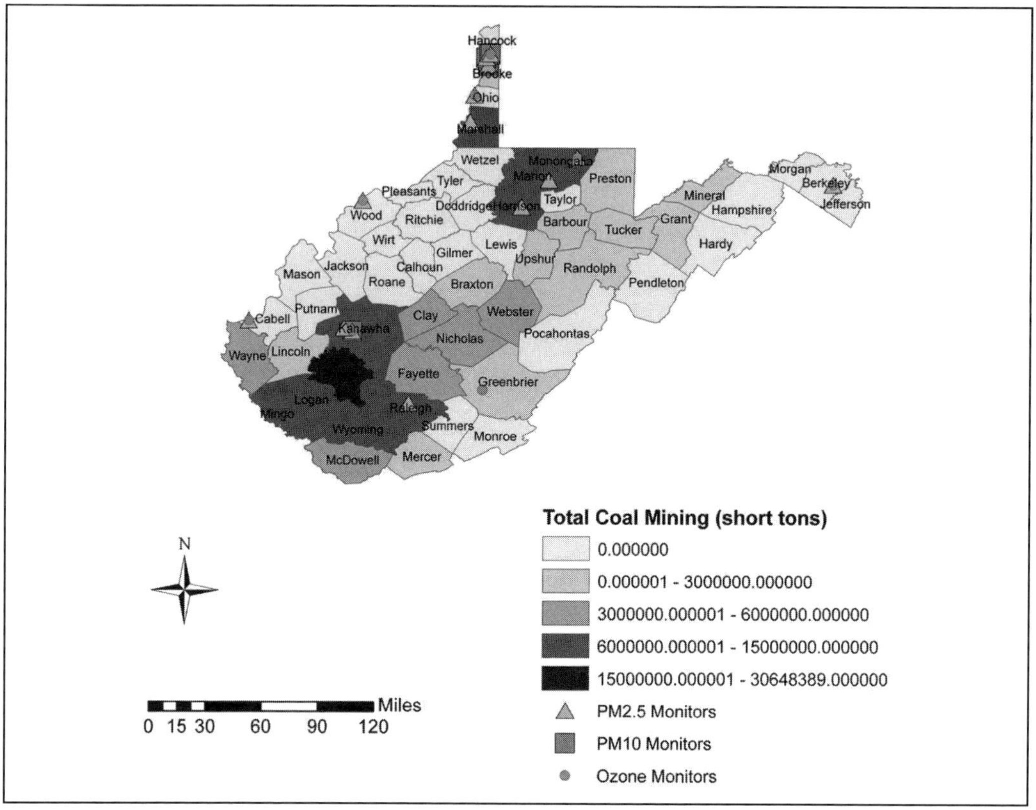

Figure 1. Air monitors and coal production by county in West Virginia, 2005

(Logan County) and for non-coal mining counties the range was from 3.4 (Hampshire County) to 29.9 (Lewis County).

EXPOSURES

The location of Criteria Air Pollutant (CAP) monitors in WV was provided by the US EPA (http://www.epa.gov/reg3artd/airquality/r3monitors-tabular.htm) and is shown in Figure 1. There are currently 5 PM_{10} monitors, 14 $PM_{2.5}$ monitors, and 8 ozone monitors throughout the state. These are typically placed near high traffic areas and significant point sources of ambient pollution.

PM, at both the less than 10 μm in diameter (PM_{10}) and less than 2.5 μg in diameter ($PM_{2.5}$) sizes, is measured and regulated by the EPA. Major sources of PM are industrial and vehicular. The current standard for PM_{10} is 150 μg/m^3, not to be exceeded in a 24-hour span more than once per year, on average, over 3 years. The current standard for $PM_{2.5}$ is 35 μg/m^3 for the 98th percentile averaged over 3 years, or a 15 μg/m^3 annual mean averaged over 3 years (http://www.epa.gov/air/criteria.html). The average annual level of $PM_{2.5}$ by county for West Virginia in 2005 can be seen in Figure 2. These data were provided by the CDC's Environmental Tracking Network (http://ephtracking.cdc.gov/QueryPanel/EPHTNQuery/EPHTQuery.html). Clearly, high $PM_{2.5}$ levels coincide with urban areas of WV. The coal mining counties (in bold) have heterogeneous concentrations of $PM_{2.5}$. Although Boone County is the highest coal producing county in the state, it had mid-range $PM_{2.5}$ concentration. In 2005, Brooke and Kanawha Counties exceeded the 15 μg/m^3 limit. There are multiple sources of $PM_{2.5}$ including both vehicular traffic and industrial emissions based on fossil fuel combustion.

Table 3. 2007 West Virginia circulatory and respiratory hospitalizations by county

County	Coal Mining 2005 (Y/N)*	# Circulatory Hospitalizations†,‡	Crude Circulatory Hospitalization Rate (per 1,000)	# Respiratory Hospitalizations†,§	Crude Respiratory Hospitalization Rate (per 1,000)
Statewide		43,725	23.6	29,420	15.9
Barbour	Y	406	24.6	346	20.9
Boone	Y	741	30.3	499	20.4
Braxton	Y	271	18.7	179	12.4
Brooke	Y	428	18.0	343	14.4
Clay	Y	316	33.8	196	20.9
Fayette	Y	1,452	31.8	946	20.7
Grant	Y	154	13.0	114	9.6
Greenbrier	Y	767	21.4	563	15.7
Harrison	Y	1,993	28.7	1,542	22.2
Kanawha	Y	5,698	29.6	2,987	15.5
Lincoln	Y	555	25.8	370	17.2
Logan	Y	1,483	40.7	1,296	35.5
Marion	Y	1,781	31.5	1,024	18.1
Marshall	Y	936	28.5	529	16.1
McDowell	Y	522	24.0	486	22.4
Mercer	Y	1,483	23.7	1,497	24.0
Mineral	Y	198	7.0	205	7.3
Mingo	Y	753	28.3	837	31.5
Monongalia	Y	1,530	15.5	690	7.0
Nicholas	Y	629	23.9	458	17.4
Preston	Y	795	23.6	391	11.6
Raleigh	Y	3,034	38.3	1,765	22.3
Randolph	Y	872	29.6	717	24.3
Tucker	Y	143	20.4	85	12.1
Upshur	Y	579	23.8	327	13.4
Wayne	Y	819	19.4	605	14.4
Webster	Y	224	24.5	178	19.5
Wyoming	Y	708	30.2	503	21.5
Berkeley	N	793	7.5	566	5.4
Cabell	N	1,856	19.2	1,290	13.3
Calhoun	N	158	20.6	104	13.6
Doddridge	N	137	16.8	73	8.9
Gilmer	N	272	31.2	163	18.7
Hampshire	N	56	2.4	81	3.4
Hancock	N	792	25.9	609	19.9
Hardy	N	102	7.3	74	5.3
Jackson	N	850	29.1	627	21.4
Jefferson	N	323	6.0	312	5.8
Lewis	N	742	45.2	491	29.9
Mason	N	912	33.4	521	19.1
Monroe	N	280	20.7	240	17.7
Morgan	N	113	6.4	134	7.6
Ohio	N	1,174	26.5	800	18.1
Pendleton	N	72	9.4	53	6.9

(table continues)

Table 3. 2007 West Virginia circulatory and respiratory hospitalizations by county (continued)

County	Coal Mining 2005 (Y/N)*	# Circulatory Hospitalizations†,‡	Crude Circulatory Hospitalization Rate (per 1,000)	# Respiratory Hospitalizations†,§	Crude Respiratory Hospitalization Rate (per 1,000)
Pleasants	N	137	18.0	100	13.1
Pocahontas	N	168	19.1	174	19.8
Putnam	N	1,148	20.5	580	10.4
Ritchie	N	244	23.7	180	17.5
Roane	N	342	23.0	177	11.9
Summers	N	329	23.7	205	14.8
Taylor	N	324	19.2	197	11.6
Tyler	N	208	22.8	117	12.8
Wetzel	N	595	36.4	293	17.9
Wirt	N	141	24.5	94	16.3
Wood	N	2,187	25.1	1,487	17.1

* WV Office of Miners' Health Safety and Training, 2005 Coal Production by County
† West Virginia Health Care Authority
‡ ICD-9 codes 390–459.9
§ ICD-9 codes 460–519.9

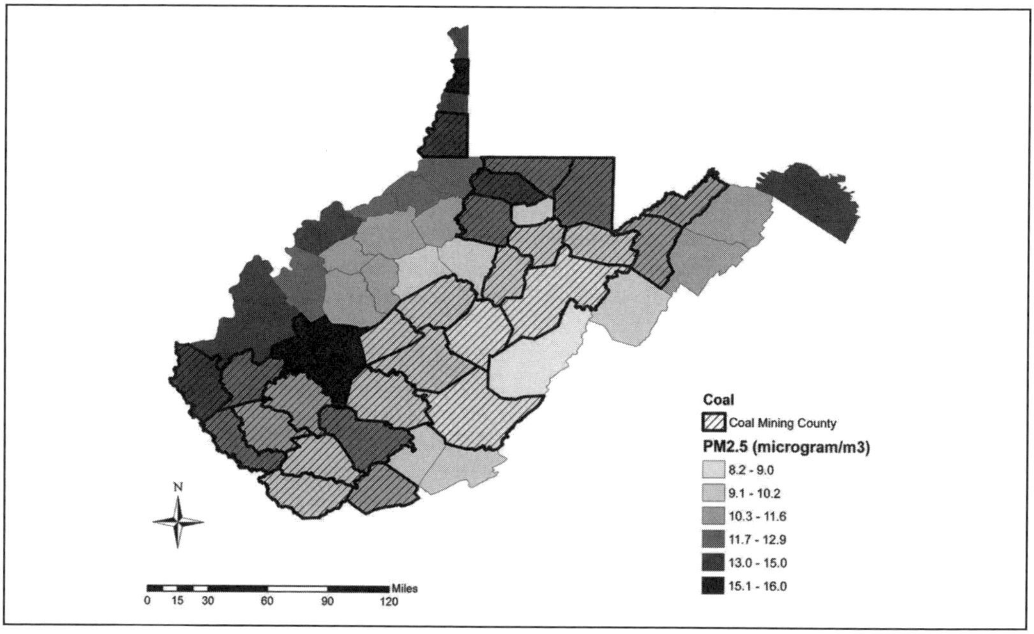

Figure 2. Monitored and modeled average annual $PM_{2.5}$ by county in West Virginia, 2005

There are several sources of information on exposures available by county, census tract and by location that can be used to quantify potential exposures in WV. These represent all sources both vehicular and industrial including coal mining related exposures. Table 4 presents the results of a literature search for coal mining related exposures. For purposes of this current study, four example exposures were investigated: lead, arsenic, cadmium, and particulate matter (PM).

Table 4. Possible exposures of interest

Pollutant	Ref1*	Ref2	Ref3	Ref4	Ref5	Ref6	Ref7	Ref8	Ref9	Data Available
PM	✓							✓	✓	✓†
Ozone										✓†
Lead		✓			✓	✓				✓‡
Arsenic		✓	✓							✓‡
Cadmium					✓	✓				✓‡
PAH							✓		✓	✓‡
Chromium				✓	✓					✓‡
Nickel		✓			✓					✓‡
NO_x										
Benzene										✓‡
Selenium		✓	✓							✓‡
Diesel PM									✓	✓‡
Aluminum	✓	✓								
Calcium	✓	✓			✓					
Iron	✓	✓		✓	✓	✓				
Potassium	✓	✓								
Magnesium	✓	✓			✓					
Sodium	✓	✓			✓					
Titanium	✓	✓								
Zinc	✓	✓		✓	✓	✓				
Sulfate/sulfur	✓	✓								
Silicates/silicon		✓			✓					
Chlorine		✓								✓‡
Phosphorus		✓			✓					✓‡
Manganese		✓			✓	✓				✓‡
Copper		✓		✓	✓	✓				
Bromine		✓								
Strontium		✓			✓					
Gallium		✓			✓					
Zirconium		✓			✓					

* Ref = reference.

† EPA air monitors.

‡ 2005 NATA.

1. Knuckles TL, Stapleton PA, Minarchick VC, Esch L, McCawley M, Hendryx M, et al. 2012. Air pollution particulate matter collected from an Appalachian mountaintop mining site induces microvascular dysfunction. *Microcirculation.* Doi: 10.1111/micc.12014.
2. Lee CY, Lee SL, Sheehan CE and Wang Y. 1996. *Composition of coal dusts and their cytotoxicity on alveolar macrophages.* US. Army Armament Research, Development and Engineering Center. Technical Report ARCCB-TR-96026.
3. Ebdon L and Wilkinson JR. 1987. The determination of arsenic and selenium in coal by continuous flow hydride-generation atomic absorption spectrometry and atomic fluorescence spectrometry. *Analytica Chimica Acta.* 194: 177–187.
4. Dalal NS, Newman J, Pack D, Leonard S and Vallyathan Val. 1995. Hydroxyl radical generation by coal mine dust: possible implication to coal workers' pneumoconiosis (CWP). *Free Radical Biology & Medicine.* 18(1):11–20.
5. Freedman RW and Sharkey AG. 1972. Recent advances in the analysis of respirable coal dust for free silica, trace elements, and organic constituents. *Annals of the New York Academy of Sciences.* 200:7–16.
6. Spencer S. 2001. Effects of coal dust on species composition of mosses and lichens in an arid environment. *Journal of Arid Environments.* 49:843–853.
7. Yunker MB, Macdonald RW, Vingarzan R, Id R, Mitchell H, Goyette D, et al. 2002. PAHs in the Fraser River basin: a critical appraisal of PAH ratios as indicators of PAH source and composition. *Organic Geochemistry.* 33:489–515.
8. Pless-Mulloli T, King A, Howel D, Stone I, Merefield J. 2000. PM_{10} levels in communities close to and away from opencast coal mining sites in Northeast England. *Atmospheric Environment.* 34:3091–3101.
9. Litton CD. 2002. Studies of the measurement of respirable coal dusts and diesel particulate matter. *Measurement Science and Technology.* 13:365–374.

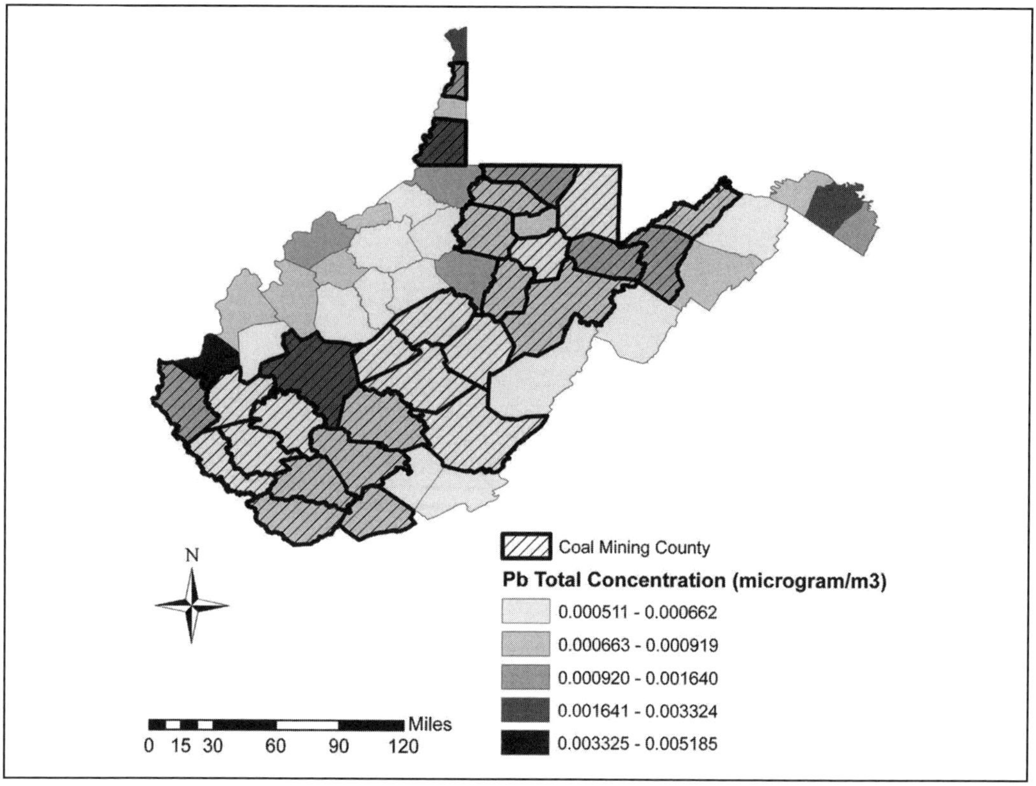

Figure 3. Lead concentrations by county in West Virginia (per 2005 NATA)

Lead, arsenic, and cadmium have all been associated with coal mining (Table 4) as they are natural constituents of coal and of the earth's crust. Levels of these pollutants have been modeled by the EPA for the National-scale Air Toxics Assessment (NATA). NATA uses information from emission sources (both stationary and mobile) to estimate ambient exposures of 187 hazardous air pollutants. These reports have been published for the years 1996, 1999, 2002, and 2005 (http://www.epa.gov/ttn/atw/natamain/). Maps of ambient air levels of lead, arsenic, and cadmium (estimated by the 2005 NATA) by county in WV can be found in Figures 3, 4, and 5, respectively. These three figures illustrate that coal mining counties run through the center of the state of WV, and have varying levels of lead, arsenic, and cadmium. Counties with coal mining have exposure levels of these pollutants which range from the lowest concentrations to the highest concentrations detected in WV. Boone County, the highest coal producing county in WV, is in the lowest concentration range of lead, arsenic, and cadmium. The second and third highest coal producing counties (Kanawha and Marshall) are in the mid to high ranges for these pollutants.

However, NATA estimates are known to underestimate exposures and are not meant to be compared to regulatory standards (Talbott 2011). Of these three air pollutants, only lead is EPA monitored and regulated. By this standard, the average level of lead over a three month period is not to exceed 0.15 $\mu g/m^3$ (http://www.epa.gov/air/criteria.html). For reference, the 2005 annual mean of air lead measured from the one CAP monitor (located in Ohio County) in WV that measures this pollutant was 0.0100 ± 0.0066 $\mu g/m^3$. This one monitor represents a "ground level" reading yearly average for this one area in Ohio County.

Although the EPA does not have a standard for arsenic or cadmium, the California

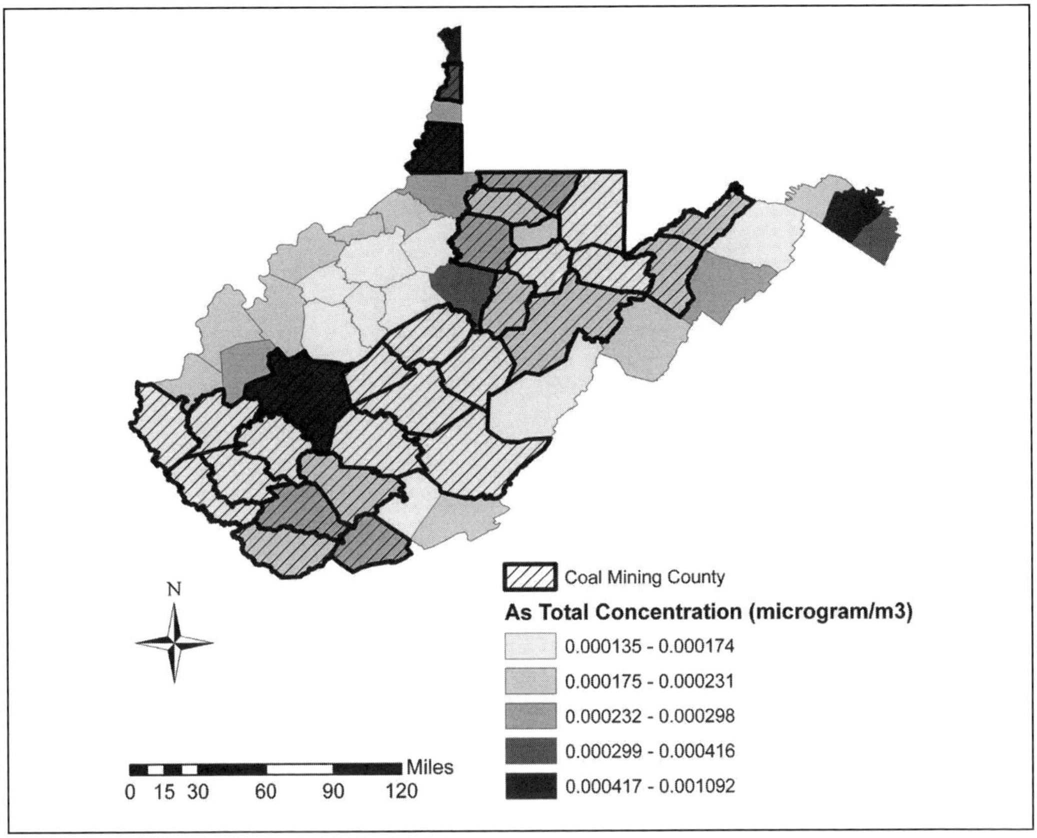

Figure 4. Arsenic concentrations by county in West Virginia (per 2005 NATA)

Environmental Protection Agency (CalEPA) has established a chronic inhalation reference level of 0.03 μg/m^3 for arsenic based on developmental effects in mice (http://www.epa.gov/ttn/atw/hlthef/arsenic.html) and 0.01 μg/m^3 for cadmium based on kidney and respiratory effects in humans (http://www.epa.gov/ttn/atw/hlthef/cadmium.html). Levels of these metals are modeled by NATA on a three year basis by both census tract and county, but it is important to note these levels are not coal mining specific; they represent numerous sources of both industrial as well as consumer-produced levels. The same can be said for all of the ambient air pollutants. As previously mentioned, all other sources of such emissions must be considered when evaluating coal and non-coal mining communities.

DISCUSSION

This study is an initial effort at descriptive demographics, coal production, and possible exposures for coal producing and non-coal producing counties in WV. However, it is limited by its descriptive and ecological nature. Associations between exposures and health outcomes have not been investigated. The information gathered and reported in this study should be used to inform further research into the health of coal mining communities in WV.

Subsequent research should include rates for specific health outcomes including Acute Myocardial Infarction (AMI), stroke, Chronic Obstructive Pulmonary Disease (COPD), and cancer. Additionally, the environmental exposures reported must be investigated for association with these health outcomes.

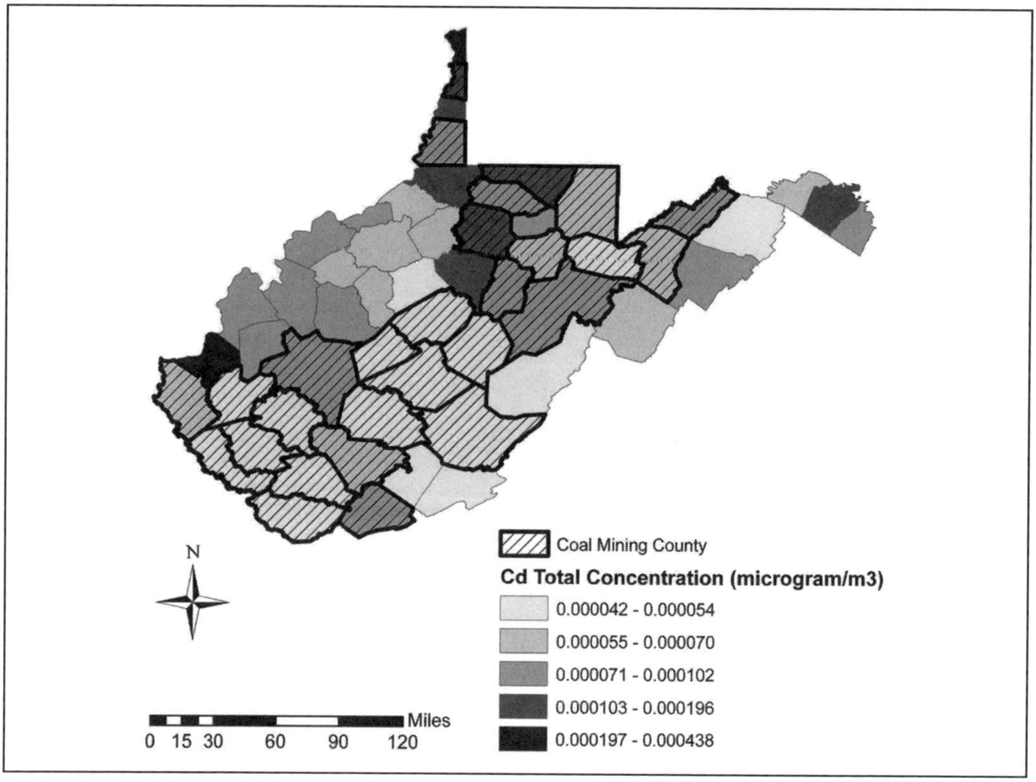

Figure 5. Cadmium concentrations by county in West Virginia (per 2005 NATA)

Crude rates of hospitalizations for circulatory and respiratory disease for one year (2007) have been presented here. Crude rates do not take into account the different age and sex distribution of different areas. Age is a primary determinant of health status; different age compositions between populations will influence morbidity rates (Lilienfeld 1976). Many health outcomes also differ by gender (Annandale 1990; Macintyre 1996). In order to compare rates of hospitalization in the counties of WV, age and sex adjusted rates must be calculated.

County level annual exposure estimates are broad characterizations and not specific to a personal residence. Environmental exposures actually vary by season, time of day, topography, and specific location within a county (Hoek 2008; Schwartz 1999). While it may not be possible at this time to achieve personal exposure estimates, smaller geographic areas such as zip code or census tract could be investigated to consider community level exposures.

ACKOWLEDGMENT

This study was sponsored by the Appalachian Research Initiative for Environmental Science (ARIES). The views, opinions and recommendations expressed herein are solely those of the authors and do not imply any endorsement by ARIES employees, other ARIES-affiliated researchers or industrial members. Information about ARIES can be found at http://www.energy.vt.edu/ARIES.

REFERENCES

Annandale, E., Hunt, K. 1990. Masculinity, femininity and sex: an exploration of their relative contribution to explaining gender differences in health. *Sociology of Health & Illness.* 12(1):24–46.

Beckett, W.S. 1991. Ozone, Air Pollution, and Respiratory Health. *The Yale Journal of Biology and Medicine.* 64:167–175.

Coggon, D., Taylor, A.N. 1998. Coal mining and chronic obstructive pulmonary disease: a review of the evidence. *Thorax.* 53:398–407.

Donoghue, A.M. 2004. Occupational health hazards in mining: and overview. *Occupational Medicine.* 54:283–289.

Environmental Protection Agency. National Ambient Air Quality Standards (NAAQS). http://www.epa.gov/air/criteria.html. (Accessed September 2012).

Environmental Protection Agency. Ozone (O_3) Standards–Table of Historical Ozone NAAQS http://www.epa.gov/ttn/naaqs/standards/ozone/s_o3_history.html. (Accessed September 2012).

Environmental Protection Agency. Fused Discrete Air Quality Surfaces. http://www.epa.gov/esd/land-sci/lcb/lcb_fdaqs_archive.html. (Accessed September 2012).

Environmental Protection Agency. National Air Toxics Assessment. http://www.epa.gov/ttn/atw/natamain/. (Accessed September 2012).

Environmental Protection Agency. Text Version: Mid-Atlantic Air Quality Monitors. http://www.epa.gov/reg3artd/airquality/r3monitors-tabular.htm. (Accessed September 2012).

Environmental Protection Agency. Technology Transfer Network Air Toxics Web Site. Arsenic Compounds. http://www.epa.gov/ttn/atw/hlthef/arsenic.html. (Accessed September 2012).

Environmental Protection Agency. Technology Transfer Network Air Toxics Web Site. Cadmium Compounds. http://www.epa.gov/ttn/atw/hlthef/cadmium.html. (Accessed September 2012).

Fernández-Navarro, P., García-Pérez, J., Ramis, R., Boldo, E., López-Abente, G. 2012. Proximity to mining industry and cancer mortality. *Science of the Total Environment.* 435–436:66–73.

Hendryx, M., Ahern, M.M., Nurkiewicz, T.R. 2007. Hospitalization Patterns Associated with Appalachian Coal Mining. *Journal of Toxicology and Environmental Health, Part A.* 70:2064–2070.

Hoek, G., Beelen, R., de Hoogh K., Vienneau, D., Gulliver, J., Fischer, P., Briggs, D. 2008. A review of land-use regression models to assess spatial variation of outdoor air pollution. *Atmospheric Environment.* 42:7561–7578.

Howel, D.H., Pless-Mulloli, T., Damell, R. 2001. Consultations of Children Living Near Open-Cast Coal Mines. *Environmental Health Perspectives.* 109(6):567–571.

Lilienfeld, A.M. 1976. Mortality Statistics. In *Foundations of Epidemiology.* Edited by Oxford University Press, Inc.

Macintyre, S., Hunt, K., Sweeting, H. 1996. Gender Differences in Health: Are Things Really as Simple as They Seem? *Soc. Sci. Med.* 42(4):617–624.

National Environmental Public Health Tracking Network Reporting Tool. http://ephtracking.cdc.gov/QueryPanel/EPHTNQuery/EPHTQuery.html. (Accessed September 2012).

Schwartz, J. 1999. Air Pollution and Hospital Admissions for Heart Disease in Eight U.S. Counties. *Epidemiology.* 10(1):14–22.

Talbott, E.O., Brink, L.L., Sharma, R.K., Marsh, G.M., Rager, J.R. *A Draft Report of the Comparison of NATA Air Lead Estimates with EPA Air Lead Monitoring Sites.* Annual Report to CDC in fulfillment of Contract 200-2010-374440. University of Pittsburgh Graduate School of Public Health, Pittsburgh, Pa 15261. Submitted September 29, 2011.

US Census Bureau. State & County Quick Facts: West Virginia. http://quickfacts.census.gov/qfd/states/54000.html. (Accessed September 2012).

West Virginia Office of Miners' Health Safety and Training. 2005 Coal Production by County http://www.wvminesafety.org/cnty2005.htm. (Accessed September 2012).

Inventory of Bacterial Impairments in Central Appalachia

N.A. Cook
Department of Biological Sciences, Virginia Tech, Blacksburg, Virginia

L.A. Krometis
Department of Biological Sciences, Virginia Tech, Blacksburg, Virginia

E.A. Sarver
Department of Mining and Minerals Engineering, Virginia Tech, Blacksburg, Virginia

ABSTRACT

Although it has been over four decades since water-related regulations such as the CWA, SDWA, and NEPA were passed, significant water quality issues in the United States remain unresolved. In the eastern U.S. in particular, unacceptably high concentrations of pathogen indicator organisms (e.g., *Escherichia coli*, fecal coliforms) are the primary contaminant responsible for state-identified surface water impairments requiring Total Maximum Daily Load (TMDL) development. Anecdotal reports suggest that inadequate or non-existent wastewater treatment may represent a significant contributor to surface water contamination in remote or underserved regions of Appalachia. In some areas where residences are relatively close to creeks or streams, it is not uncommon for household sewage to be discharged directly to receiving waters, a practice known as "straight-piping."

Despite the widespread incidence of fecal indicator bacteria (FIB) impairments, a broad assessment of the extent of identified impairments and their relationship with land use, co-contaminants, and proximity to high-density populations has not been completed for Appalachia. As a preliminary review of these potential relationships, publicly available GIS data were compiled and analyzed for an area of interest that includes six geographically central states in the region, which also share coal production as a current and historical common economic driver. These data were analyzed for the purpose of extracting potential correlations between water quality designations and topographic or demographic factors, and for initiating dialogue on potential solutions that involve a variety of regional stakeholders. Given the relative significance of coal production in the area of interest, specific discussion is offered on the possible requirements and benefits that might be associated with strategies that involve coal industry members as key stakeholders.

INTRODUCTION

Although water-related regulations such as the Clean Water Act (CWA), the Safe Drinking Water Act (SDWA) and the National Environmental Policy Act (NEPA) have been in place for over four decades, significant water quality issues in the United States remain unresolved. In accordance with the federal Clean Water Act (CWA) of 1972, states are required to monitor all surface waters within their borders and report any failures to adhere to water quality standards appropriate to their given uses (e.g., fishing and swimming). Section 305(b) of the law requires that waters be classified by their state-designated use(s). This classification includes a set of accompanying numerical limits for various pollutants (e.g., bacteria) of potential concern published by the US Environmental Protection Agency (USEPA). States must monitor their waters and report those that fail to meet standards associated with specified designated uses pursuant to Section 303(d) of the CWA; hence the list of these impaired waters nationally is frequently referred to as the "303(d) list." States are required to address waters on this list through the development of Total Maximum Daily Load (TMDL) plans, which identify minimum source-specific pollutant loading reductions required to achieve the designated use for each 303(d)-listed water body. States maintain the right to determine their own schedules of monitoring, methods of data collection, decisions on water body classification, and alternative standards (which may or may not be more stringent than federal standards) based on local land uses, stream designations, or regional hydrology characteristics (e.g., use attainability analyses). If any of these efforts are deemed unsatisfactory, states can be subject to penalty under federal law. This non-centralized decision-making paradigm means that a river that might be listed as impaired in one state may not be so listed in another. It can also lead to discrepancies when comparing data from multiple states. In an effort to eradicate data inconsistencies, the USEPA often issues guidance on best practices for TMDL development for specific pollutants; however, evolving priorities and technology can render goals a shifting target (NRC 2001, Keller and Cavallaro 2008).

In the eastern United States, unacceptably high concentrations of pathogen indicator organisms (e.g., *Escherichia coli*, fecal coliforms) are the primary contaminant responsible for state-identified surface water impairments requiring TMDL development (NRC, 2001; Gaffield et al., 2003; USEPA, 2011). Although generally non-pathogenic themselves, the presence of indicator organisms is associated with contamination by mammalian fecal material and public health risk (Paruch and Mæhlum, 2012). Exposure to waters containing high concentrations of coliform or enterococci bacteria have been linked with an elevated risk of gastroenteritis, respiratory infections, and skin infections in exposed human populations (Prüss, 1998). Common sources of pathogen loadings to receiving waters include inadequately treated wastewater discharges, septic leakage, wildlife, domestic animals, and runoff from manure-amended agricultural soils (Gerba and Smith, 2005; Ishii and Sadowsky, 2008). The degree of contamination by any given source is often related to overall watershed land use and topography (Mallin et al., 2000; Field and Samadpour, 2007).

Development of TMDLs generally follows a four step process: impairment identification; determination of the maximum pollutant load the water body can assimilate while maintaining minimum standards for its designated uses; establishment of links between pollutant sources and receiving water concentrations; and source load allocation to inform implementation plan development. A 2008 survey of TMDLs across the United States determined the leading challenges of TMDL development to be a lack of funding and lack of monitoring data to inform watershed characterization and predictive modeling. Successful implementation of existing TMDLs on private lands generally hinged on the engagement of local stakeholders and the availability of funds (Benham, 2008).

The Appalachian region comprises a wide variety of land uses associated with various potential sources of fecal loadings to surface waters, including: agricultural (livestock sources), forest (wildlife sources), and human communities (wastewater sources).As loadings of human waste are generally considered most likely to contain human-infectious pathogens and therefore of greatest health risk (Ishii and Sadowsky, 2008; Soller et al., 2010), reports of inadequate sanitation and discharges of untreated sewage in this region are of particular concern (Gasteyer and Vaswani, 2004; Wescoat et al., 2007). Many communities in the geographically central part of this region have evolved from coal camps in narrow mountain hollows along streams. Soils along these hollows are notoriously thin, and often inappropriate for septic systems, and housing density often makes appropriate siting of drain fields difficult. Even properly installed septic systems can contribute human waste due to the karst (cracked rock) geology of the region. Groundwater flows quickly through this geological strata, preventing the intended treatment by the system's drain field (Katz et al., 2010; Johnson et al., 2011). Some areas in the region simply discharge the grey water from their septic systems into trenches that run to the streams. In many cases, no septic system exists at all, and waste from household toilets, showers, and other plumbed fixtures is piped directly to the creek or into similar trenches ("straight-piping").

While historical precedent (i.e., the justification that "it's always been this way") or a failure to recognize potential consequences (i.e., lack of education) may contribute to the prevalence of straight-piping, inadequate water and sanitation issues in the United States are often strongly related to economic and/or political challenges (Gasteyer and Vaswani, 2004; Wescoat et al., 2007). According to the latest economic report published by the Appalachian Regional Commission (ARC; www.arc.gov), this region significantly lags the national average in employment, personal income per capita, investment income, and proprietor average income; and much of the central, coal-producing region lags even further behind (ARC, 2011). (The ARC is a federally-established partnership between state and local governments created by Congress in 1956 for the purpose of stimulating economic development in the Appalachian region.

Moreover, many of the rural counties in this area are comprised of isolated communities without a sufficiently large central municipality to organize widespread infrastructure efforts. Under such conditions, it is not surprising that centralized or advanced wastewater treatment systems are often absent or implausible. Collaboration between public and private regional stakeholders is likely necessary to improve sanitation and prevent related health risks.

The concept of a collaborative effort requires that parties affected by such decisions be identified. In Appalachia these might include municipal and county governments, taxpayers and citizens, private industries, and groups within the general citizenry that advocate for abstract entities, such as the environment or a specific political party. There are also a large number of NGOs (non-government organizations) that seek to influence the region through advocacy for their own specific interests as well. The relationships within and between these groups are often complex, making solutions to any problem in the region non-trivial. Often, ideological conflicts between opposing groups can overshadow concrete problems that can be more easily addressed with existing technology, such as sewerage issues in rural communities.

Much of the current environmental focus with respect to water quality in the central Appalachian region is on the abundance and diversity of stream macro-invertebrates, which are often used as a proxy for the general degradation of stream ecology (Pond et al., 2008; Echols et al., 2009). Relatively little attention, however, has been paid to pathogen indicator concentrations in surface waters and related human health risks. The present study describes an initial review of publically available GIS data related to surface water impairments, population density, and common land uses, including coal mining. The goal of this analysis is to examine the relationships between FIB impairments and other factors at the inventory scale in order to prioritize future more finely targeted analyses. To complete this initial inventory-level analysis, data were collected from the various state and federal agencies and projected onto a common GIS coordinate system for comparative analysis using the varied geoprocessing abilities of ArcGIS software. (ArcGIS software is distributed by ESRI Products, Redlands, California; www.esri.com.)

METHODS

Appalachia broadly refers to a cultural region in the eastern U.S., which the ARC has geographically defined to include a total of 420 counties in 13 states spanning from New York in the north to Alabama and Mississippi in the south. For the purposes of this study, the area of interest (AOI) was limited to the ARC counties within six geographically central states: Kentucky, Maryland, Ohio, Pennsylvania, Virginia, and West Virginia (Figure 1). This AOI is termed

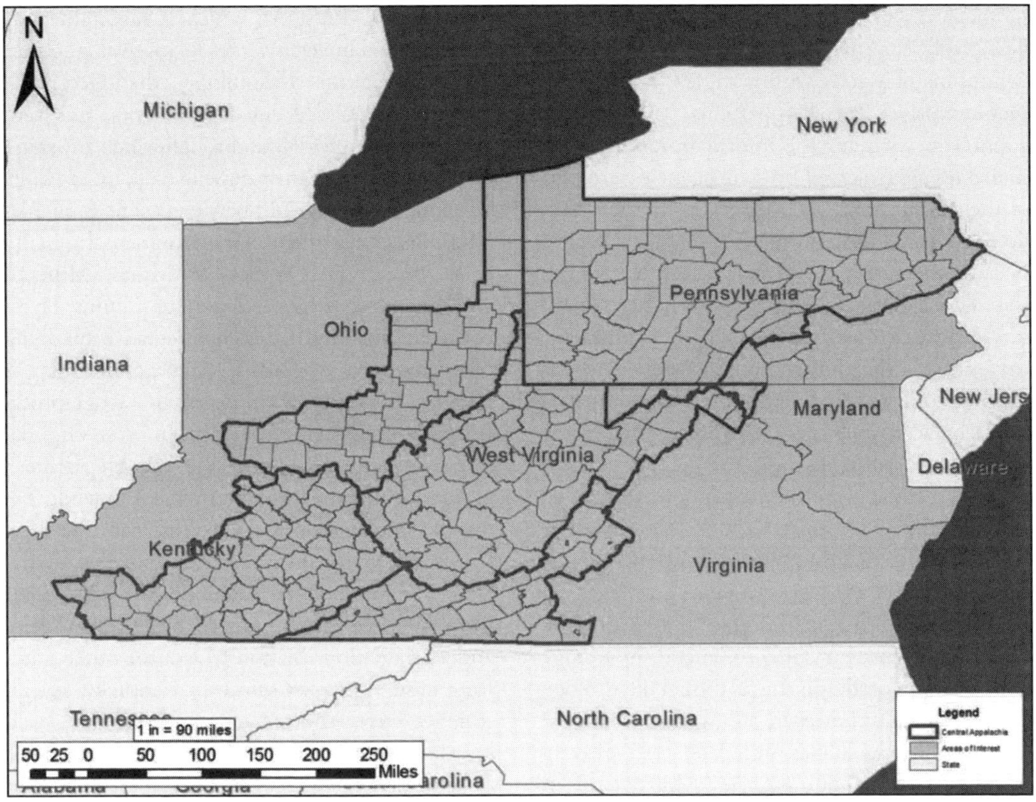

Figure 1. Central Appalachian region area of interest

"Central Appalachia" here. (The ARC has defined five sub-regions in Appalachia; the AOI in the current study includes all of Central Appalachia, as well as parts of Southern, Northern, and North Central Appalachia as defined by the ARC.)

This region accounts for the majority of the energy resource production (i.e., coal and natural gas) in Appalachia; Alabama and Mississippi do produce significant coal and gas, respectively, however, they are non-contiguous with the AOI and do not share many of the same regional geographic and cultural characteristics.

The following datasets were sought for the six states in the Central Appalachian study region as defined above:

- Census tract information
- Government unit boundaries (state and county boundaries)
- Hydrology data (stream and watershed shapefiles)
- Land use characteristics
- Surface coal mining activity (AMLs, active mining, permitted areas)
- Stream impairment information (303(b) and/or 303(d) shapefiles)

The two main goals governing data selection were to examine any potential relationships between stream impairments and other activities at the watershed level, as well as to examine the proximity of coal mining operations to such impairments as a methodology for quantifying the potential significance of targeted mitigation efforts by mine operators. Stream impairments in the form of 303(d) shapefiles represent the best publicly available dataset that describe the conditions of waters within the AOI, the states that comprise Central Appalachia, as well as the country as a whole. The information within the shapefiles are collected by state agencies as a part of their 305(b) and 303(d) responsibilities mandated by the CWA and reviewed by the USEPA. Because these data are formally accepted by both state and federal governments as a descriptor of surface water quality for regulatory purposes, they provide the most practical avenue towards making comparative assessments between watershed characteristics and impairments of interest (in this case, pathogen indicators).

The census tract information was acquired through the US Census TIGERlines database (United States Census Bureau, 2010), which provides borders of government entities such as counties, municipalities, and census districts (including census results). General hydrology data were secured via the National Hydrology Dataset (NHD); however, 303(d) data that were used for most analyses were collected directly from the relevant state agencies. This data often includes more updated hydrography than general NHD data (USGS, 2011). Land use characteristics were acquired as a raster (set of 30m × 30m pixels) from the NLCD (National Land Cover Database), which was initially completed in 2006 and is based on imagery collected in the early 2000s (Frye, 2011). Much of the remaining data collection required personal contact with relevant state agencies (e.g., Departments of Environmental Quality) to secure the most up-to-date information. While impairment information is submitted to the USEPA database regularly in keeping with the requirements of the CWA, states often have more recent assessments available locally.

For many states, boundaries of current or permitted coal mining areas were not readily available via generic internet searches. Mining information was alternatively obtained from either the state environmental agency or the state land management agency. For Pennsylvania, all data could be obtained via the Pennsylvania Spatial Data Access (PASDA), which is maintained by Penn State University, the Pennsylvania Office for Information Technology, the Geospatial Technologies Office, and collaborators. This website contains an FTP link through which any state-released GIS data can be downloaded. In states like Virginia, data were split between agencies such as Virginia's Department of Environmental Quality (DEQ) and the Division of Mines, Minerals and Energy (DMME). Data for mining layers could be obtained through interactive mapping programs such as DMME's Geoexplorer map or DEQ's Virginia Environmental Geographical Information Systems (VEGIS). In West Virginia, some data were acquired via the Department of Environmental Protection (DEP), while the mining information was acquired from the West Virginia GIS Technical Center (uploaded by the West Virginia Natural Resource Analysis Center from digitized data produced by DEP's Mining and Reclamation division). Data for other states were more difficult to find; in Kentucky, several searches were required to narrow down a web location for GIS data. Eventually the search led to the Kentucky Geological Survey (KGS) at the University of Kentucky, where links to mining GIS data and hydrology data (303d layers) could be found. Some states, such as Maryland and Ohio, presented greater challenges. Access to Ohio's data required direct electronic communication with specific state agencies, and it was necessary to interpret the data further upon receipt, which meant further person-to-person communication. Ohio data were separated into A, B, C, and D-Law categories based on their status as AMLs or as active mining sites. The data were found through the Ohio's Geographic Information Management Systems (GIMS), which did not allow data download, but provided electronic mail contacts for data requests. Maryland mining data were never recovered, despite multiple attempts to communicate with the relevant state agency. These missing data likely have a minimal impact on the GIS analysis as only three counties in the state of Maryland fall within the Central Appalachian region, and the state is responsible for little coal production.

With respect to surface water impairment data, most states had data readily available via online access at their respective environmental agencies. As mentioned previously, Kentucky, Virginia, Pennsylvania and West Virginia had readily available 303(d) GIS datasets for download. Two states (Maryland and Ohio) required direct contact to receive information, and the impairment information was only available at watershed resolution (as opposed to stream-segment level resolution). Maryland has scheduled the release of stream-segment level data for late 2012 (after the time of writing), but Ohio has no currently set timetable for improved data. In Pennsylvania, data were downloaded via PASDA, but the actual data were missing fields that were described in the metadata. This omission required contact with the appropriate contacts to resolve, whereupon the corrected data were downloaded via an FTP link.

Once all of the datasets were secured, the challenges associated with combining data from individual states with their own impairment terminology and categorization schemes became increasingly apparent. In the case of Kentucky, numerical rather than textual codes are reported for specific impairment classifications, so further communication with state employees was required to decipher the codes. Ohio and Kentucky have over seventy categories of impairments, while both Maryland and West Virginia have less than fifteen. Many states listed multiple impairments for a single stream (e.g., pathogen indicators and turbidity, etc.) while some states only listed one. Even the category of "FIB impairments" varied greatly between the six states, and the application of different indicator standards (fecal coliform standard versus *E. coli* standard) may result in differences in classification. Table 1 shows the different terms by state. The difference in resolution of the data as well as the difference in attribute naming required some intensive parsing to compare accurately. Data for which impairment entries were left blank or labeled "Insufficient Data" were regarded as non-assessed for the purposes of this study.

Table 1. Terms for FIB impaired waters by state

State	Terms Indicating FIB Impairment
Kentucky	Fecal Coliform, *Escherichia Coli*
Maryland	Bacteria
Ohio	*E. coli*
Pennsylvania	Pathogens
Virginia	*Escherichia Coli*, Fecal Coliform
West Virginia	Fecal/Bacteria

GIS Methodology

Following data collection and inventory, the following analyses were completed:

- Comparison of total impairments for the Central Appalachian and non-Central Appalachian sections of each state;
- Quantification of FIB impairments in Central Appalachia vs. non-Central Appalachia (in terms of number of watersheds or stream miles, as appropriate);
- Categorization of FIB-impaired watersheds by dominant land use;
- Comparison of population density in FIB-impaired watersheds; and
- Identification of mining operation proximity to FIB-impaired stream segments or watersheds.

Before performing any analysis on this collection of GIS information, a common projection was required; most of the AOI (Central Appalachia) falls within the NAD 1983 UTM Zone 17, therefore all data was projected to match this zone (most data was initially projected in local state plane formats). Additionally, for all processes requiring clipping, selecting or intersecting commands, geometry was re-calculated before summaries were completed. For most of these analyses, a simple command in ArcGIS software extracts the necessary information. To find the total impairments, the "select by location" command allowed the program to select stream impairments by their geographical location. While this process is not sufficient for high-accuracy raster processes required at the stream level, it sufficed for a large-scale approach where regions of interest were 12-digit watersheds or where a Central Appalachian versus non-Central Appalachian classification was desired. These impairments were tallied by stream mileage (where possible) for comparison. Applied to a subset of the impairment dataset (only FIB impairments), this same command gave a summary of the streams impaired for fecal coliform or *E. coli*. This step allowed the comparison of total FIB impairments in Central Appalachia versus those in non-Central Appalachia.

To complete the land use comparison, the 12-digit NHD watershed boundaries were used as zones for tabulating the pixels in the NLCD dataset. This information was joined with impairment information to connect both the impairments and the land use data to one attribute table. These data were exported to Microsoft Excel, where formulas were used to determine what the major land use per watershed was for all watersheds that contained FIB impairments. The original NLCD classifications present in the Central Appalachian region include a list of twenty different categories, which were condensed to eight for the purposes of this project. The categories of land use within the GIS analysis were barren, agricultural, developed, forest, herbaceous, shrub, open water and wetland.

It should be noted that some limitations are inherent in studying NLCD for such a large area, such as the inability to account for the effects of land usage in upstream watersheds on downstream basins. For this initial inventory of bacterial impairments over a relatively large study region, such effects were assumed to be negligible; but a more in-depth statistical study would be necessary to draw definitive conclusions about land usage at a finer scale. Another point of potential concern for this summary method is the scale resolution itself. For small stream branches, pixels at a 30 meter resolution may not be appropriate for fine-scale analysis. However, here comparisons were made using 303(d)-listed stream data, which generally only involve the main stem of 12-digit watersheds—thus 30 meter scale NLCD data are adequate for the purposes of this preliminary study.

In order to complete the population density comparison, census districts were intersected with watershed boundaries to determine the fractional population density of each watershed. A weighted sum was then calculated for each watershed using the census district fragments to find the average population density for each watershed. From this calculation, watersheds could be selected based on the presence of a FIB impairment using the "select by location" command once again. The population density factor was then binned by 2,590 people/km^2 (1,000 people/mi^2). While this method is limited by its assumptions of even population distribution within districts, an efficient inventory-level study necessitated a rather general approach for reviewing the more than 6,500 12-digit watersheds that comprise the six-state region of interest. In order to more accurately tie population densities to specific impaired streams (i.e., rather than uniformly spread population over an entire census district), a much more in-depth method for geo-locating population would be necessary. This necessity is based on the fact that census districts span several 12-digit watersheds in some cases, while several census districts may fall within a single 12-digit watershed in more developed areas. For a more accurate analysis, individual pixels could be assigned weighted population values according to the level of development within that pixel—and thus population density could be estimated at higher-resolution along impaired streams specifically, as opposed to generalizing about the 12-digit watershed.

A two-step method was used to quantify mining operation proximity to FIB impairments. First, a comparison was made based on the presence of both within the same watershed. Next, FIB impairments within a 25 km (~15 mile) search radius of mining permit boundaries were used to select watersheds that could be mitigation targets. The purpose of this section of the analysis was to quantify the potential effect targeted mitigation efforts by mining operations could potentially have on FIB impairments.

RESULTS AND DISCUSSION

For review of the following relationships, it should be noted that these examinations are for inventory purposes and preliminary analysis. To more closely examine any of these issues, formalized hypothesis testing and geostatistical analysis must be performed—which is the aim of ongoing work by the authors. Data collection and organization across six states in and of itself presented several challenges worthy of discussion. Regardless of specific project objectives, locating accurate and readily comparable environmental data are often a nontrivial challenge prior to the actual analyses, particularly when working across state boundaries. Between the six states reviewed in this study, data comparison required the translation of over 160 different impairment terms into a common terminology. States like Maryland simplified their categories into ten main terms, while states like Kentucky and Ohio listed over 50. Extensive person-to-person discussion with contacts from the individual state agencies was required to ensure the meaning of terms and the compatibility of terms from different states. A central database from which regional GIS data can be acquired would be of great benefit in facilitating geospatial studies on a regional scale. In addition, standardization of reporting formats would facilitate more efficient and accurate analyses.

Beyond basic impairment classification, analysis of available state 303(d) lists reveal that the six states in this region vary widely in their relative levels of completion of surface water assessment (Table 2) and in the level of detail associated with those assessments. Some coverage statistics are difficult to compare across states; Ohio and Maryland assess at the watershed level rather than stream segment level, resulting in an overestimation of completed stream assessment. In contrast, Pennsylvania's assessment coverage extends beyond the borders of their state, perhaps in an effort track specific pollution issues originating within their state borders. It should be noted that the impairments included in datasets summarized in Table 2 likely represent only a small fraction of the water quality impairments that actually exist in some states, like Kentucky and Virginia, where the assessment of only a small portion (<20%) of surface waters has been officially completed.

An examination of impairment causes indicates that elevated concentrations of *E. coli* or fecal coliforms account for a large proportion of identified impairments in assessed Central Appalachian surface waters (Table 3) but reveals no immediate contiguous trend throughout the region as a whole. Some states have more identified FIB impairments in their Central Appalachian regions versus their non-Central Appalachian regions, while in others the inverse is true. It is important to note that region-specific factors can have a large effect on these results. Assessment coverage (Table 2) shows that many streams have not been assessed, so making direct conclusions from this information is difficult. In West Virginia, only one impairment is listed per GIS shapefile feature, but a single stream may have spatially duplicate features that contain additional impairments for the same stream segment. While sweeping statements about FIB impairments cannot be made, it is clear that elevated concentrations of fecal indicator bacteria

Table 2. Impaired stream miles in states comprising central appalachia

State	Assessment Coverage	Central Appalachia Impaired Streams, Any Pollutant (km)	Non-Central Appalachia Impaired Streams, Any Pollutant (km)
Kentucky	11%	1,832.10	2,070.85
Maryland	61%	25 Watersheds*	113 Watersheds*
Ohio	80%	407 Watersheds*	697 Watersheds*
Pennsylvania	110%	8,176.74	3,480.32
Virginia	16%	1,905.27	5,621.00
West Virginia	41%	18,856.07	N/A

*Data only available at the watershed level.

Table 3. Fib impairments in states comprising Central Appalachia

State	Pathogen Impaired Streams (km)	% Regional Impairments Attributed to Fecal Indicator Bacteria	
		Central Appalachia	Non-Central Appalachia
Kentucky	1,813	47	46
Maryland	5,058*	39	31
Ohio	4,123*	22	21
Pennsylvania	3,266	5	12
Virginia	3,364	77	64
West Virginia	787	26	N/A
Total	18,413	32	25

*Numbers calculated by summing NHD stream layers within impaired watersheds.

Table 4. Statewide FIB impairments (and overall land area) vs. primary land use

State	% of Watersheds Containing FIB Impairments Where Land Use Category Represents Majority of Pixels in Raster (Overall State %)							
	Barren	Agricultural	Developed	Forest	Herb.	Shrub	Water	Wetland
Kentucky	0	37.99	4.17	57.60	0	0	0.25	0
	(.37)	(32.89)	(7.25)	(52.14)	(4.15)	(0.32)	(1.83)	(1.05)
Maryland	0	0	33.62	59.48	0	0	0	6.90
	(.40)	(32.01)	(17.84)	(33.63)	(1.76)	(1.76)	(1.18)	(12.89)
Ohio	0	68.66	10.45	20.90	0	0	0	0
	(.19)	(50.15)	(14.51)	(31.18)	(1.62)	(0.38)	(1.03)	(0.94)
Pennsylvania	0	0	23.14	48.76	0	0	0	28.10
	(1.30)	(2.22)	(11.37)	(51.87)	(6.56)	(3.42)	(0.00)	(23.26)
Virginia	0	11.73	4.86	81.64	0	0	0	1.78
	(.27)	(21.40)	(9.42)	(58.23)	(1.93)	(2.57)	(1.03)	(5.14)
West Virginia	0	3.17	1.46	95.37	0	0	0	0
	(.12)	(9.41)	(7.66)	(79.86)	(0.09)	(1.73)	(0.00)	(1.12)
Total	0	15.92	7.02	74.11	0	0	0.05	2.90
	(.44)	(24.68)	(11.34)	(51.15)	(2.44)	(1.70)	(0.85)	(7.40)

are not uncommon, and in some states, a major surface water quality issue.

Table 4 suggests that, within the six states that comprise the Central Appalachian region as defined in the AOI, FIB impairments typically are related to a few specific land uses. Watersheds with forest areas listed as the primary land use contain the overwhelming majority of listed FIB impairments—with forested areas accounting for nearly three quarters of all such impairments. It is important to note that while a watershed may be primarily forested, a small rural community with inadequate sewage could still be the primary cause for FIB impairment. It may be that the relationship between a largely forested watershed and FIB impairments parallels the increased incidence of non-centralized sewage disposal practices (septic tanks, straight-piping) in less developed watersheds, though substantial further study would be required to confirm this hypothesis. In Table 4, values are given for the percentage of watersheds in each state that contain FIB impairments, classified by land use category. For a watershed to be assigned to a specific land use category, the majority of the area (based on zonal tabulation of NLCD pixels) in that watershed was classified within the given land use category (a condensed version of NLCD classification). In parenthesis, the total percentage of land area in each state is given for comparison.

Table 5. Statewide FIB impairments vs. population density in states comprising Central Appalachia

	% of FIB impairments in Watersheds vs. Population Density ppl/mi^2 [ppl/km^2]						
State	>1000 [>2560]	1000–2000 [2560–5120]	2000–3000 [5120–7680]	3000–4000 [7680–10420]	4000–5000 [10420–12800]	5000–6000 [12800–15360]	6000–7000 [15360–17920]
Kentucky	93.60	3.20	2.20	.99	0.00	0.00	0.00
Maryland	80.49	4.88	4.88	3.66	1.22	2.44	2.44
Ohio	91.04	5.97	0.00	2.99	0.00	0.00	0.00
Pennsylvania	57.27	36.00	4.69	2.31	0.00	0.00	0.00
Virginia	93.47	3.98	1.74	.50	.31	.50	.31
West Virginia	97.76	2.23	0.00	0.00	0.00	0.00	0.00
Total	85.61	9.38	2.25	1.74	.26	.49	.46

Based on this data, it does not appear that the trends in FIB-land use category can be simply explained by the overall percentages of watersheds in each category. For example, while approximately 58% of land area in Virginia is classified as forested, nearly 82% of total FIB impairments occur within watersheds that are classified as mostly forested. Indeed, the data gathered thus far make a strong case for examining this relationship further as part of any real efforts to reduce bacterial impairments within this region. Both Maryland and Pennsylvania showed a higher proportion of FIB-impaired waters related to development than the other states. This relationship mirrors the results of studies that point to urban stormwater as another source of pathogens in surface waters (Gaffield et al., 2003). On the other hand, Ohio has a far greater instance of FIB impairments near agricultural land uses, which may point to a variation in agricultural practices or methods of stream assessment between Ohio and other states.

Because the primary land use of a given watershed may not be directly responsible for observed concentrations of pathogen indicators, it is also critical to determine relationships between water quality and population density, as anthropogenic influence is often associated with surface water impairments. By mapping census tracts to 12-digit watersheds, a relationship between the population density of a watershed and the prevalence of FIB impairments could be investigated. The results presented in Table 5 along with Table 4 suggest that FIB impairments are common in rural areas. Over 85% of FIB impairments in the six states that comprise Central Appalachia were located within watersheds that had fewer than 2,560 people/km^2 (1,000 people/mi^2). It is important to note that despite this classification, a watershed with a large proportion of forestland could be impaired by pathogen loading from a small, high-density developed area. However, these data again support the idea that a concerted effort to solve rural issues in water quality would most likely target a large proportion of these impairments.

The widespread nature of FIB impairments in rural Appalachia, in conjunction with anecdotal reports of inadequate or nonexistent wastewater treatment in rural communities suggests that an intervention is necessary to improve surface water quality. Given the stakeholders in this region and their associated resources, it is clear that expedient and realistic solutions will require novel strategies. Considering the significant presence of coal mining activities in Central Appalachia, and the increasing emphasis on environmental protection and corporate social responsibility by the mining industry, it appears that one viable alternative may include engaging mine operators as stakeholders with a key role to play (Craynon et al., 2012). Such an approach would, however, necessitate identification of and agreement on incentives for all stakeholders involved.

Given the spatial and temporal scales required for surface mining, and the complex topography and hydrology of Central Appalachia, interactions between mining activities and water

Table 6. FIB Impairment incidence with mining permits in AOI (Central Appalachia)

	Comparison of Mining Presence and FIB Impairments				
State	Total Watersheds	Impaired Watersheds	Watersheds w/ Mining	Watersheds w/FIB Impairment	Watersheds w/Mining and FIB Impairments
Kentucky	641	406	145	81	26
Maryland*	N/A	N/A	N/A	N/A	N/A
Ohio	663	407	142	143	15
Pennsylvania	1,177	708	163	69	14
Virginia	377	293	50	259	39
West Virginia	742	633	352	413	206
Total	3,600	2,447	852	965	300

*Comparison not made due to lack of mining GIS data.

resources are inevitable. At present, the coal industry faces serious challenges in gaining permits under the CWA—particularly since a new focus has been placed on heretofore unconsidered dissolved solids loadings from mine spoil discharges, and their potential effects on aquatic macro-invertebrates (Cox and Sarver, 2012). The CWA allows for a variety of compensatory mitigation schemes under some circumstances, including water quality impacts of discharged fill material into wetlands or streams (USEPA, 2008), although "in-kind" mitigation has been preferred.

Under an alternative strategy for improving overall regional environmental and human health, the underlying question becomes one of allowing and encouraging stakeholder participation in addressing FIB impairments. Should flexibility be integrated into the CWA permitting process to allow mitigation trade-offs in order to promote holistic water quality management? Such a strategy would require careful prioritization of water quality goals, and careful implementation in terms of trade-off credit valuation and meaningful stakeholder participation, but may well provide net positive impacts for ecological and public health. In relevant cases where scientific data and stakeholder input support holistic water quality management, an opportunity may exist for mine operators and NPDES and other CWA permittees to contribute to efforts to reduce FIB impairments as part of their overall commitment to environmental and social responsibility.

GIS analysis of the relationship between mining activity and FIB impairments is a cursory method of supporting the argument for flexible, targeted mitigation to support overall water quality improvement for the region. Generally, mitigation allowances are limited to the permitted area of mining operation or the stream reach on which the mining operation is located (i.e., only the drainage area potentially impacted by the mine). This narrow focus does not provide flexibility to mine operators(or other CWA permitees) to contribute to sustainable development in the larger region by addressing community issues. A "watershed approach," as increasingly emphasized in contemporary environmental management practices, would certainly broaden potential impacts.

Although FIB impairments are not directly associated with mining, Table 6 reveals a significant number of watersheds with mining activities do coincide with FIB impairments. Under an even more holistic "community approach" significant improvements might be realized with long-term benefits to human (and ecological) health and well-being. This type of approach would be aimed at social responsibility, wherein corporate citizens act to benefit the *communities* in which they operate; and it must be recognized that communities have economic, cultural, social and environmental ties to local industries, whether or not those industries operate within the specific drainage area or watershed occupied by community members. As the "community approach" is far outside the current regulatory framework

Table 7. Potential FIB impairment targets for mining operations in AOI (Central Appalachia)

State	Kentucky	Maryland	Ohio	Pennsylvania	Virginia	West Virginia	Total
% FIB Addressed (In-Watershed)	41.0%	N/A	10.5%	19.0%	14.5%	49.2%	38.6%
% FIB Addressed (25 km radius)	63.5%	N/A	37.8%	69.4%	32.9%	94.4%	76.8%

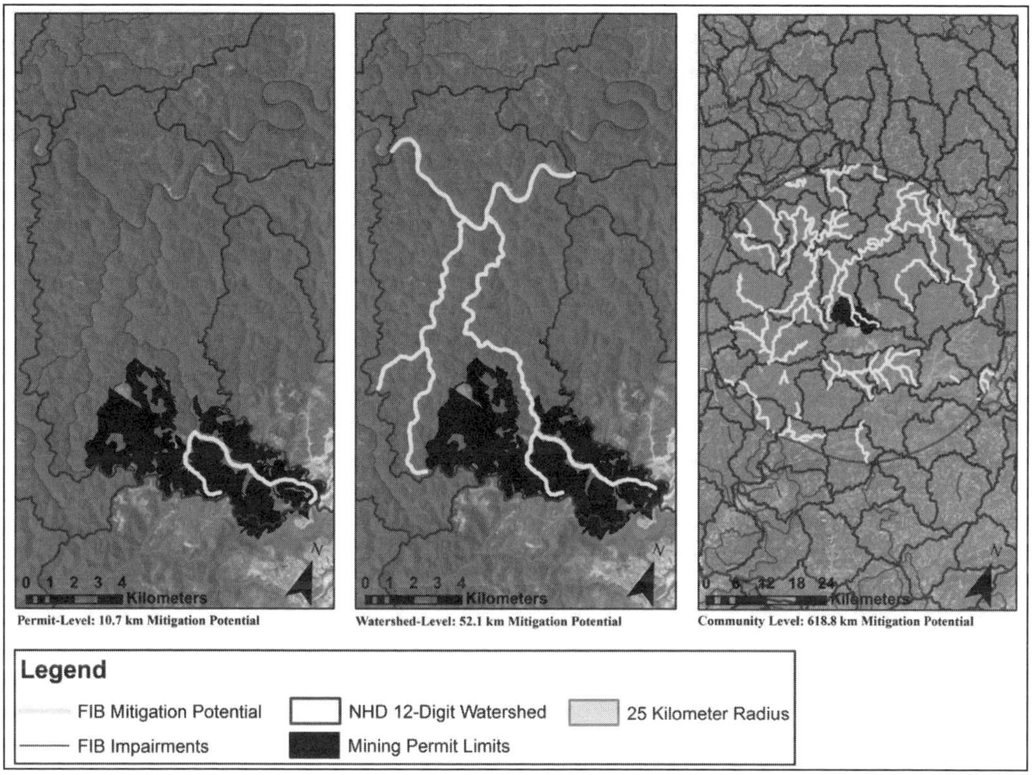

Figure 2. Drainage level, watershed level, and community level mitigation strategies at a West Virginia example location

and typical operating procedures, it is duly noted that adoption would require substantial changes in environmental policy and corporate culture—both of which should be underpinned by significant further scientific analysis.

Table 7 and Figure 2 illustrate the above arguments for FIB mitigation approaches based on corporate responsibility over those based on narrower criteria in the study region. While a "watershed approach" could potentially impact about one third of the total FIB impairments in the AOI, a "community approach" could significantly increase this impact. For example, by applying a generic 25 km (15 mile) radius to define the "community" around a given mine, it is expected that overt three quarters of the total FIB-impairments could be potentially addressed in the AOI. Although this definition of community may be too narrow or too wide on an individual-mine basis (i.e., community-mine associations are in reality very site- and factor-specific), the point is to demonstrate that a more flexible and holistic approach offers greater possibilities for results. In many cases, the economic, social and environmental impacts of mining operations (both positive and negative) cross watershed boundaries; therefore, it seems plausible that serving a community with mitigation efforts for FIB impairments should also extend beyond the watershed of physical operation. Indeed, the case for any flexibility is especially compelling in West Virginia, where even a watershed approach could potentially address half of that states' FIB impairments.

Although sewerage issues may be a primary environmental health concern in some rural watersheds in the Appalachian region, it is worth noting that EPA currently states that the majority of FIB impairments nationwide are associated with agricultural practices (e.g., livestock, manure amendment of soils). Therefore in a broader context, compensatory mitigation, in either a "watershed" or "community" approach, need not be limited to human sewage, but might be utilized to address a wide variety of water pollution sources that are detrimental to ecological stream health. Stream ecology could equally benefit from treatment structures removing pathogens, organic material and nutrients contributed by agricultural or livestock land uses. Overall, both human health and stream ecology might be improved through the permit-related mitigation process required of mine operators or other permittees. Resources available for mitigation could support the engineering work required to solve relevant water quality issues by implementing solutions (such as installing localized package plants), which could address problems such as *E. coli* presence in streams and poor health scores in benthic macroinvertebrate communities. While the current emphasis on "in-kind" mitigation may be appropriate in terms of hydrology, it overlooks some opportunities for improvement in both human and stream ecology health that a more diverse array of environmental remediation efforts would likely provide.

CONCLUSIONS

This initial inventory-level review of bacterial surface water impairments in the Central Appalachian region demonstrated significant challenges associated with the compilation of freely available GIS data at the multi-state level. In both the mining resources and the water quality arenas, differences in data availability (e.g., accessibility, resolution or non-existence) and formatting proved to be nontrivial pre-analysis hurdles. A collaborative database for natural resource and environmental data would greatly benefit all those who need to analyze data across state lines. Current collection of data in a patchwork fashion ultimately renders geoprocessing difficult and forces significant assumptions during the comparison of heterogeneous datasets. Data manipulation to facilitate comparisons across state lines inevitably introduces some error and uncertainty, as subjective evaluations are required. National formatting and labeling standards would greatly benefit regional studies.

Even if uniform and complete data were available for the targeted study region, the methods and relatively coarse data resolution (i.e., 12-digit watersheds) used here result in certain limitations. To more carefully study many of the relationships of interest would require a statistics-driven approach that includes hypothesis testing and completely raster-based geoprocessing techniques to increase analysis resolution from a general watershed inventory to a localized cause-and-effect study at the stream level. Despite these limitations, some useful initial conclusions can still be drawn from this inventory-level review of the data.

It is suggested by the literature that many FIB impairments are the result of a wide variety of sources, both urban and rural. In the Central Appalachian region, it appears that agricultural practices, wildlife populations, and the use of sub-standard sewage practices (failing septics and straight-piping) are all potential contributors to elevated in-stream pathogen concentrations. By analyzing the data contained in state-available GIS datasets, this variety of stressors seems to be supported by the relationship between land use types (mainly forest and agricultural, some developed) and FIB impairment presence. The GIS data also suggest a link between low population density and the presence of FIB stream impairments. It only takes a single small community lacking sewerage to impair a significant portion of a watershed for pathogens. This phenomenon is both a result of physical geography and economics in the region, and is reflected by the GIS data.

As discussed previously, these potential relationships between water quality and land use suggest possible strategies to address these impairments. Without a municipal tax base, solutions to sewerage issues in these rural areas will have to come from a collaborative effort supported by funding from a joint effort between public and private stakeholders in the region. The proximity analysis highlighted in Tables 6 and 7 (as well as Figure 2) suggests that coal mine operators in Central Appalachia could make a significant positive impact on environments and communities if encouraged to do so by the relevant regulatory agencies. Moreover, engaging local corporate citizens (i.e., mine operators in the case of Central Appalachia) to participate in local problem solving is critical to sustainable development: near-term economic and societal benefits of industry must be balanced with long-term health and well-being of environments and communities.

ACKOWLEDGMENT

This study was sponsored by the Appalachian Research Initiative for Environmental Science (ARIES). The views, opinions and recommendations expressed herein are solely those of the authors and do not imply any endorsement by ARIES employees, other ARIES-affiliated researchers or industrial members. Information about ARIES can be found at http://www.energy.vt.edu/ARIES.

The authors thank C. Davis for their contributions to this paper.

REFERENCES

ARC, 2011. Economic Overview of Appalachia–2011.

Benham, B., 2008. Lessons Learned from TMDL Implementation Case Studies. Water Practice 2.

Cox, A., Sarver, E.S., 2012. Salt: an Emerging Concern for the Global Mining and Minerals Industries. Annual Meeting of the Society of Mining Professors, Wroclaw, Poland.

Craynon, J.R., Sarver, E.A., Robertson, D.P., 2012. Could a public ecology approach help resolve the mountaintop mining controversy? Resources Policy.

Echols, B.S., Currie, R.J., Cherry, D.S., 2009. Preliminary results of laboratory toxicity tests with the mayfly, *Isonychia bicolor* (Ephemeroptera: Isonychiidae) for development as a standard test organism for evaluating streams in the Appalachian coalfields of Virginia and West Virginia. Environmental Monitoring and Assessment 169, 487–500.

Field, K.G., Samadpour, M., 2007. Fecal source tracking, the indicator paradigm, and managing water quality. Water Research 41, 3517–3538.

Frye, J.e.a., 2011. Completion of the 2006 National Land Cover Database for the Conterminous United States. PE&RS.

Gaffield, S.J., Goo, R.L., Richards, L.A., Jackson, R.J., 2003. Public Health Effects of Inadequately Managed Stormwater Runoff. American Journal of Public Health 93, 1527–1533.

Gasteyer, S., Vaswani, R., 2004. Still Living Withouthe Basics in the 21st Century: Analyzing the Availability of Water and Sanitation Services in the United States. Rural Community Assistance Partnership, Washington, DC.

Gerba, C.P., Smith, J.E., 2005. Sources of Pathogenic Microorganisms and Their Fate during Land Application of Wastes. J. Environ. Qual. 34, 42–48.

Ishii, S., Sadowsky, M.J., 2008. *Escherichia coli* in the Environment: Implications for Water Quality and Human Health. Microbes and Environments 23, 101–108.

Johnson, T.B., McKay, L.D., Layton, A.C., Jones, S.W., Johnson, G.C., Cashdollar, J.L., Dahling, D.R., Villegas, L.F., Fout, G.S., Williams, D.E., Sayler, G., 2011. Viruses and Bacteria in Karst and Fractured Rock Aquifers in East Tennessee, USA. Ground Water 49, 98–110.

Katz, B.G., Griffin, D.W., McMahon, P.B., Harden, H.S., Wade, E., Hicks, R.W., Chanton, J.P., 2010. Fate of Effluent-Borne Contaminants beneath Septic Tank Drainfields Overlying a Karst Aquifer. Journal of Environment Quality 39, 1181.

Keller, A.A., Cavallaro, L., 2008. Assessing the US Clean Water Act 303(d) listing process for determining impairment of a waterbody. Journal of Environmental Management 86, 699–711.

Mallin, M.A., Williams, K.E., Esham, E.C., Lowe, R.P., 2000. Effect of Human Development on Bacteriological Water Quality in Coastal Watersheds. Ecological Applications 10, 1047–1056.

NRC, 2001. Assessing the TMDL approach to water quality management. Committee to Assess the Scientific Basis of the TMDL Approach to Water Pollution Reduction, The National Academies Press, Washington, DC.

Paruch, A.M., Mæhlum, T., 2012. Specific features of *Escherichia coli* that distinguish it from coliform and thermotolerant coliform bacteria and define it as the most accurate indicator of faecal contamination in the environment. Ecological Indicators 23, 140–142.

Pond, G.J., Passmore, M.E., Borsuk, F.A., Reynolds, L., Rose, C.J., 2008. Downstream effects of mountaintop coal mining: comparing biological conditions using family- and genus-level macroinvertebrate bioassessment tools. Journal of the North American Benthological Society 27, 717–737.

Prüss, A., 1998. Review of epidemiological studies on health effects from exposure to recreational water. International Journal of Epidemiology 27, 1–9.

Soller, J.A., Schoen, M.E., Bartrand, T., Ravenscroft, J.E., Ashbolt, N.J., 2010. Estimated human health risks from exposure to recreational waters impacted by human and non-human sources of faecal contamination. Water Research 44, 4674–4691.

United States Census Bureau, 2010. 2010 Census TIGER/Line® Shapefiles (http://www.census.gov/geo/www/tiger/tgrshp2010/tgrshp2010.html).

USEPA, 2008. Wetlands Compensatory Mitigation.

USEPA, 2011. National Section 303(d) List Fact Sheet (http://oaspub.epa.gov/waters/national_rept.control).

USGS, 2011. Nationaly Hydrography Dataset (NHD): http://nhd.usgs.gov/.

Wescoat, J.L., Headington, L., Theobald, R., 2007. Water and poverty in the United States. Geoforum 38, 801–814.

Identification of Toxic Agents and Potential Exposure Routes to Appalachian Coal Mining Communities

S.D. Whitacre
School of Environment and Natural Resources, The Ohio State University, Columbus, Ohio

N.T. Basta
School of Environment and Natural Resources, The Ohio State University, Columbus, Ohio

C.J. Everett
Department of Crop and Soil Environmental Sciences, Virginia Tech, Blacksburg, Virginia

Kristen Minca
School of Environment and Natural Resources, The Ohio State University, Columbus, Ohio

W.L. Daniels
Department of Crop and Soil Environmental Sciences, Virginia Tech, Blacksburg, Virginia

ABSTRACT

Recently there has been increased concern about the health of Appalachian communities in the vicinity of coal mining operations. An important consideration in community health is the physical health of its residents. Several studies have reported that coal mining communities have a higher incidence of disease, including; asthma, birth defects, black lung, cardiovascular disease, cardiopulmonary disease, hypertension, kidney disease, lung disease, and cancer as well as mortality, than non-coal mining communities.

The objectives of this study were to conduct an extensive literature search in order to (1) identify environmental agents linked to asthma, birth defects, cardiovascular disease, hypertension, kidney disease, lung disease, and cancer, (2) identify toxic agents and potential exposure pathways associated with coal mining and (3) assemble the literature into an associations matrix that allows for the presence and absence of overlaps between health effect and toxic agents linked with coal mining to be identified.

The health effects literature search was performed using the Medline: ProQuest database. Search terms included "environmental health," "air pollution," and "tap water," in conjunction with the specific health effect/disease being targeted in the search. The search results were limited to review articles in English. All of the articles found were published in 2001 or later, with the majority published between 2008 and 2011. The potential sources literature search was performed using Science Citation Index, Web of Science, and PubMed databases. Search terms included "coal mining contaminants," "coal mining toxic agents," "coal mining dust," and "coal mining and ground water." The search results were limited to articles that identified toxic agents that coal mining communities may have had elevated exposure to because of an increased concentration of toxic agents at the coal mining site, an increased transmission of agents due to coal mining activities, or a combination of the two.

The health effects literature search resulted in 108 papers which identified over 200 toxic agents associated with asthma, birth defects, cardiovascular disease, hypertension, kidney disease, lung disease, and cancer. The number of toxic agents was reduced to 23 when screened by literature references linking the toxic agents to coal mining. The resulting associations matrix allows for potential exposure routes of plausible toxic agents to be identified. Given the toxic agents identified, the primary routes of exposure are ingestion and inhalation for soil/dust and ingestion for water. Fugitive dust in Appalachian coal mining communities is the major exposure pathway of concern for surface coal mining operations. Further research is needed to quantify the amount of dust leaving coal mining operations and determine the total content and bioavailable fractions (via ingestion and inhalation) of dust constituents, in order to determine if exposure to coal mining communities is sufficiently elevated to cause adverse health effects.

INTRODUCTION

Recently there has been increased concern about the health of Appalachian communities in the vicinity of coal mining operations. An important consideration in community health is the physical health of its residents. Several studies have reported that coal mining communities have a higher incidence of disease, including; asthma, birth defects, black lung, cardiovascular disease, cardiopulmonary disease, hypertension, kidney disease, lung disease, and cancer as well as mortality, than non-coal mining communities (Ahern et al. 2011; Hendryx 2009; Hendryx and Ahern 2008; Hendryx et al. 2007; Hendryx et al. 2010; Hendryx et al. 2008; Hendryx et al. 2011; Hendryx and Zullig 2009). In addition to environmental exposures, several factors have been identified as associated with chronic disease including lack of physical activity, poor nutrition, obesity, excessive alcohol consumption, smoking, awareness of health status (i.e., blood pressure, cholesterol, etc.), and access to health care (Centers for Disease Control and Prevention 2012). However, previous studies conclude that the increased incidence of disease is due to coal mining activity with few if any toxic agents identified.

When the source of a toxic agent is not in the immediate vicinity of a person or population, pathways of exposure must be established. Environmental pathways of exposure can be complex, with the extent of exposure determined by exposure factors such as amount, frequency, and duration. The modes in which toxic agents enter the body are limited to inhalation, ingestion, and dermal absorption (US EPA 2011). Depending on its chemical properties, the toxic agents that enter the body will have specific sites where the agent will inflict damage on the body tissues.

The toxic effects of many environmental agents have been extensively studied. As a result, literature linking environmental chemicals to specific diseases can be used to identify agents associated with diseases reported to be elevated in coal mining communities. The objectives of this study were to conduct an extensive literature search in order to (1) identify environmental agents linked to asthma, birth defects, cardiovascular disease, hypertension, kidney disease, lung disease, and cancer, (2) identify toxic agents and potential exposure pathways associated with coal mining and (3) assemble the literature into an associations matrix that allows for the presence and absence of overlaps between health effect and toxic agents linked with coal mining to be identified.

METHODS

Literature reviews targeting (1) health effects and (2) potential sources of toxic agents carried in soil/dust and water as a result of coal mining activities were performed. The health effect literature search was performed to identify environmental factors associated with specific diseases reported in the literature of Appalachian coal mining (Ahern et al. 2011; Hendryx 2009; Hendryx and Ahern 2008; Hendryx et al. 2007; Hendryx et al. 2010; Hendryx et al. 2008; Hendryx et al. 2011; Hendryx and Zullig 2009). The health effects literature search was performed using the Medline: ProQuest database. Search terms included "environmental health," "air pollution," "tap water," in conjunction with the specific the health effect/disease being targeted in the search. The search results were limited to review articles in English. All of the articles found were published in 2001 or later, with the majority published between 2008 and 2011. The potential sources literature search was performed using Science Citation Index, Web of Science, and PubMed databases.

Search terms included "coal mining contaminants," "coal mining toxic agents," "coal mining dust," and "coal mining and ground water." The search results were limited to articles that identified toxic agents that coal mining communities may have had elevated exposure to because of an increased concentration of toxic agents at the coal mining site, an increased transmission of agents due to coal mining activities, or a combination of the two.

HEALH EFFECTS LITERATURE SEARCH RESULTS

Asthma

Search results for asthma yielded nine papers, with results summarized in Table 1 (Breysse et al. 2010; Cecchi et al. 2010; De Luca et al. 2010; Dozor 2010; Holgate et al. 2010; Sly 2011; Stapleton et al. 2011; Tzivian 2011; Weinmayr et al. 2010).

Birth Defects and Low Birth Weight

Search results for birth defects and low birth weight yielded five papers with results summarized in Table 2 (Wang and Pinkerton 2007; Wigle et al. 2008; Wigle et al. 2007; Wu 2010; Vrijheid et al. 2011).

Cancer

Search results for cancer yielded 35 papers with results summarized in Table 3 (Brody 2010; Carpenter 2010; Chameides 2010; Field 2010; Fucic et al. 2010; Gamble 2010; Gibb et al. 2011; Guha et al. 2010; Hajdu 2011; Hansen and Lassen 2011; Hendryx et al. 2010; Hendryx et al. 2008; Hendryx et al. 2011; Hosgood et al. 2011; IARC 2011; Kligerman and White 2011; Kunzli

Table 1. Literature search results for toxic agents associated with asthma

Active Smoking	Diesel exhaust
Environmental tobacco smoke	Indoor air pollution
Ambient air toxicants	Traffic exhaust
• Particulate matter undefined	Insecticides
• Particulate matter—$PM_{2.5}$	• DDT/DDE
• Particulate matter—$PM_{2.5-10}$	• Organophosphate insecticides
• Particulate matter—PM_{10}	• Other or unspecified insecticides
• Particulate matter—Ultrafine particles	Maternal use of paracetamol (an analgesic)
• Ozone	Personal care products
• Nitrogen dioxide	Exposure to biologically active allergens
• Sulfur dioxide	Airborne mouse allergen
Air pollution episodes	Climate change × pollen
Chemical domestic products for cleaning	Respiratory viruses

Table 2. Literature search results for toxic agents associated with birth defects and low birth weight

Active Smoking	PCBs
Environmental tobacco smoke	Phthalates
Ambient air toxicants	TCDD
• Particulate matter undefined	Pesticides
• Carbon monoxide	Solvents
• Ozone	• Chlorinated solvents
• Nitrogen dioxide	• Glycol ethers
• Sulfur dioxide	• Unspecified solvents

(table continues)

Table 2. Literature search results for toxic agents associated with birth defects and low birth weight (continued)

Arsenic	Drinking water nitrate, nitrite (NO_2 + NO_3)
Cadmium	Aeroallergens
Lead	Power-frequency magnetic fields
Mercury	Proximity to hazardous waste disposal sites
Methyl mercury	Radiofrequency radiation
Mixed metals	Solid waste
Chlorination disinfection by-products	Sunlight (UV radiation)
Dioxin	

Table 3. Literature search results for toxic agents associated with cancer

Active Smoking	Acrylonitrile
Environmental tobacco smoke	Benzene
Ambient air toxicants	Benzidine
• Particulate matter—$PM_{2.5}$	Bisphenol A
• Particulate matter—PM_{10}	Brominated flame retardants
• Particulate matter—Ultrafine particles	Carbon tetrachloride
Air pollution episodes	Dioxin
Coke oven emissions	Ethylene oxide
Cooking fumes from high temperature oils	Formaldehyde
Diesel exhaust	Hydrazine
Household coal combustion	Napthalene
Household biomass burning	PCBs
Metal dust	Perchloroethylene
Nanoparticles	Persistent organic pollutants
Night shift work	Polycyclic aromatic hydrocarbons
Non-ferrous metal dust and fumes	Vinyl chloride
Occupation as a painter	Pesticides
Traffic emissions	Solvents
Socioeconomic risk factors	Nutraceuticals
Education	Water disinfection by-products
Occupational class	Human papillomavirus
Housing characteristics	Human immunodeficiency virus
Deprivation index	Chlamydia pneumonia
Arsenic	Prior lung disease
Beryllium	Electricity (extremely low frequency)
Cadmium	Electromagnetic fields
Chromium	Ionizing radiation
Cobalt	Microwave radiation
Nickel	Non-ionizing radiation
Silica	Radionuclides
Asbestos	Radon
Bitumen	Sunlight (UV radiation)
1,3 butadiene	Wireless devices (radiofrequency)

and Tager 2005; Kutting and Drexler 2010; Linet and Inskip 2010; Lubin 2010; Mates et al. 2010; McErlean and Ginsberg 2011; Merletti et al. 2011; Mossman et al. 2011; Nielsen and Wolkoff 2010; Samet 2010; Samet 2011; Samet et al. 2009; Soto and Sonnenschein 2010; Stefanovic and Polenakovic 2009; Suk 2010; Sung et al. 2011; Vena 2010; Weichenthal et al. 2010; Wigle et al. 2008; Wigle et al. 2007).

Cardiovascular Disease

Search results for cardiovascular disease yielded 30 papers with results summarized in Table 4 (Bernal-Pacheco and Roman 2007; Bhatnagar 2006; Brook 2007; Brook et al. 2010; Cole and Freeman 2009; Delfino et al. 2005; Fang et al. 2010; Faught et al. 2009; Galimanis et al. 2009; Gill et al. 2011; Hassing et al. 2009; Hendryx and Ahern 2008; Hendryx and Zullig 2009;

Table 4. Literature search results for toxic agents associated with cardiovascular disease

Active Smoking	Chromium
• Carbon monoxide	Copper
• Vapor phase components	Iron
• Particulate phase 'tar'	Lead
• Nicotine	Mercury
Environmental tobacco smoke	Nickel
Ambient air toxicants	Selenium
• Particulate matter undefined	Sulfur
• Particulate matter—$PM_{2.5}$	Vanadium
• Particulate matter—$PM_{2.5-10}$	Zinc
• Particulate matter—PM_{10}	Silica
• Particulate matter—Ultrafine particles	Quartz
• Carbon monoxide	Inorganic dust
• Ozone	Sand
• Ozone × allergen level	Cotton
• Nitrogen dioxide	Aldehydes
• Sulfur dioxide	Dioxin
• Sulfates	Endotoxin
Asphalt fumes	PCBs
Black Carbon	Polyaromatic hydrocarbons
Diesel exhaust	Semiquinones
Distance to major road	Styrene
In vehicle $PM_{2.5}$	Volatile organic substances
Metal	Pesticides
Metal fumes	• Chlorinated pesticides
Occupational and Indoor pollutants	• Unspecified pesticides
Organic carbon species	Solvents
Respirable PM	• Chlorinated solvents
Traffic emissions	• Glycol ethers
Welding and soldering fumes	• Unspecified solvents
Welding fume $PM_{2.5}$	Maternal drugs
Heavy metals and elements	Maternal toxins
Arsenic	Maternal infection
Cadmium	Radiation

Table 5. Literature search results for toxic agents associated with hypertension

Active Smoking	Particulate matter undefined
• Carbon monoxide	Arsenic
• Vapor phase components	Cadmium
• Particulate phase 'tar'	Lead
• Nicotine	Mercury
Ambient air toxicants	PCBs

Table 6. Literature search results for toxic agents associated with kidney disease

Aluminum	Chlorinates
Arsenic	Chloroform
Beryllium	Dinitrotoluene
Bismuth	Ethylene glycol
Cadmium	Halogenated alkenes
Chromium	Hydrocarbon-induced nephrotoxicity
Germanium	Methylene chloride
Gold	Napthenes
Lead	Organic chemicals and solvents
Mercury	Toluene
Silicon	Trichloroethylene
Uranium	Volatile hydrocarbons
Aliphatic aromatics	Polycyclic aromatic hydrocarbons in water
Aristolochic acid	Aromatic amines in water
Arsine gas	Viruses
Bipyridyl herbicides	Ochratoxin A
Carbon tetrachloride	Mycotoxins

Hoffmann et al. 2009; Houston 2007; Houston 2011; Humblet et al. 2008; Mastin 2005; Navas-Acien et al. 2007; O'Toole et al. 2008; Peters 2005; Sauvant and Pepin 2002; Soto and Sonnenschein 2010; Su et al. 2011; Sun et al. 2010; Vardavas and Panagiotakos 2009; Vaziri 2008; Vermylen et al. 2005; Virdis et al. 2010; Zanobetti et al. 2011).

Hypertension

Search results for hypertension yielded nine papers with results summarized in Table 5 (Virdis et al. 2010; Sun et al. 2010; Brook 2007; Delfino et al. 2005; Chen et al. 2007; Houston 2007; Gallagher and Meliker 2010; Vaziri 2008; Houston 2011; Everett et al. 2011).

Kidney Disease

Search results for cancer yielded 15 papers with results summarized in Table 6 (IPCS 1991; Cannata-Andia and Fernandez-Martin 2002; Soderland 2010; Voice et al. 2006; Gallagher and Meliker 2010; Bandara et al. 2010; Edwards and Prozialeck 2009; Prozialeck and Edwards 2010; Ekong et al. 2006; Evans and Elinder 2011; Vicente-Vicente et al. 2010; Brautbar 2004; Stefanovic and Polenakovic 2009; Rankin 2004; Bruning 2002).

Lung Disease

Search results for lung disease yielded four papers with results summarized in Table 7 (Kunzli and Tager 2005; Soto-Martinez and Sly 2010; Peden and Bush 2011; Po et al. 2011).

COAL MINING AS A POTENTIAL SOURCE OF TOXIC AGENTS

Soil and Dust

The potential source literature search yielded twelve papers that indicate potential increased exposure to toxic agents carried in soil and or dust brought to the surface by coal mining activities and are summarized in Table 8. Studies indicated an increase in ambient air toxicants in the form of undefined particulate matter (Ghose 2007; Trivedi et al. 2010; Ghose and Majee 2007; Chaulya 2004), particulate matter less than 10 μm in diameter (PM_{10}) (Trivedi et al. 2010; Ghose and Majee 2007; Chaulya 2004; Dubey et al. 2011; Mukherjee et al. 2005), carbon monoxide (Ghose 2007), nitrogen dioxide (Ghose 2007; Ghose and Majee 2007; Dubey et al. 2011), and sulfur dioxide (Ghose 2007; Ghose and Majee 2007; Dubey et al. 2011). Elevated levels of elemental toxicants that have been reported to be associated with coal mining include: arsenic (As) (Black and Craw 2001; Šebestová et al. 1996; De and Mitra 2004), Beryllium (Be) (Šebestová et al. 1996), cadmium (Cd) (Dubey et al. 2011; Das and Chakrapani 2011), chloride (Cl) (Ghose and Majee 2007), chromium (Cr) (Dubey et al. 2011; De and Mitra 2004), copper (Cu) (Dubey et al. 2011; Šebestová et al. 1996; De and Mitra 2004; Das and Chakrapani 2011), iron (Fe) (Dubey et al. 2011; Brook et al. 2010), lead (Pb) (Dubey et al. 2011; De and Mitra 2004), manganese (Mn) (Brook et al. 2010; Dubey et al. 2011), nickel (Ni) (Dubey et al. 2011; Das and Chakrapani 2011; De and Mitra 2004), zinc (Zn) (Dubey et al. 2011; Šebestová et al. 1996; De and Mitra 2004; Das and Chakrapani 2011), and silica (Dubey et al. 2011; Onder and Yigit 2009). In addition, organic toxicants in the form of aldehydes (Ghose and Majee 2007; Ghose 2007), benzene (Ghose and Majee 2007), and polyaromatic hydrocarbons (PAH) (Ghose and Majee 2007) have also been identified.

Table 7. Literature search results for toxic agents associated with lung disease

Ambient air toxicants
- Particulate matter undefined
- Carbon monoxide
- Ozone
- Ozone × allergen level
- Nitrogen dioxide
- Sulfur dioxide
- Biomass smoke

Ground Water

Surface water was not considered due to the infrequency in which surface water is consumed directly and therefore not a significant exposure pathway. However, ground water is a potential drinking water source for many Appalachian residents and therefore the ingestion of water from wells is a major exposure pathway of concern. The potential source literature search yielded four papers that indicate potential increased exposure to toxic agents released in ground water as a result coal mining activities and are summarized in Table 9. The studies indicate increased groundwater concentrations of organic carbon (McAuley and Kozar 2006), aluminum (Al) (McAuley and

Table 8. Literature search results for toxic agents potentially associated with soil/dust due to coal mining

Ambient air toxicants	Copper
• Particulate matter undefined	Iron
• Particulate matter—PM_{10}	Lead
• Carbon monoxide	Manganese
• Nitrogen dioxide	Nickel
• Sulfur dioxide	Zinc
Arsenic	Silica
Beryllium	Aldehydes
Cadmium	Benzene
Chloride	Polyaromatic hydrocarbons
Chromium	

Table 9. Literature search results for toxic agents potentially associated with ground water due to coal mining

Organic carbon species	Iron
Aluminum	Lead
Ammonia	Manganese
Arsenic	Nickel
Calcium	Potassium
Chromium	Sulfur
Cobalt	Zinc
Copper	Quartz

Kozar 2006), ammonia (McAuley and Kozar 2006), As (Wigginton et al. 2008; Gupta 1999), calcium (Ca) (McAuley and Kozar 2006), cobalt (Co) (De and Mitra 2004), chromium (Cr) (McAuley and Kozar 2006), copper (Cu) (Gupta 1999), iron (Fe) (Wigginton et al. 2008; McAuley and Kozar 2006), potassium (K) (McAuley and Kozar 2006), manganese (Mn) (McAuley and Kozar 2006; Wigginton et al. 2008; Gupta 1999), nickel (Ni) (McAuley and Kozar 2006), lead (Pb) (Wigginton et al. 2008), quartz (McAuley and Kozar 2006), sulfur (S) (McAuley and Kozar 2006), and zinc (Zn) (Wigginton et al. 2008; De and Mitra 2004).

ASSOCIATIONS MATRIX RESULTS

The results of the two literature searches were combined into a single associations matrix, which allowed for overlaps in toxic agent induced health effects and potential sources of toxic agents from coal to be observed. The full associations matrix that lists each overlap with citations separately is too large for publication in this manuscript. A summary of only the overlaps between single associations and toxic agents are presented in Table 10. Of particular importance are the agents associated with cancer. The International Agency for Research on Cancer (IARC 2011) has categorized agents into five groups: (1) carcinogenic to humans, (2A) probably carcinogenic to humans, (2B) possibly carcinogenic to humans, (3) not classifiable as to its carcinogenicity to humans, and (4) probably not carcinogenic to humans. Of the agents identified with potential links to coal mining, arsenic, beryllium, cadmium, chromium (VI), silica, and benzene are categorized as group 1, and cobalt and nickel as group 2B.

DISCUSSION

The associations matrix summarizes important information that ties only plausible toxic agents to previously cited health effects of concern. Increases in particulate matter and ambient air toxicants are due primarily to open cast surface coal mining (Trivedi et al. 2010; Ghose and Majee 2007). Activities that generate fugitive dust (dust that moves off site) include dust created from wind erosion of disturbed land and stockpiles, drilling, and blasting. Some studies indicate that up to 80% of total dust was attributed to transportation, including loading/unloading of coal and overburden and dust generated from unpaved haul roads (Ghose 2007). However, the exact source or mining practice that may contribute to these increases can vary from mining site to mining site (Ghose 2007).

Elevations in specific toxicants arise from multiple sources and practices. The sources of carbon monoxide, nitrogen dioxide, sulfur dioxide, aldehydes, benzene, and other PAHs in coal mining areas are due to such things as mine fire, burning of coal, blasting, and movement of vehicles (Ghose 2007; Ghose and Majee 2007). Elevation of As in soil/dust is a primarily a result of As co-occurrence with pyrite. Pyrite (FeS_2) is an iron sulfide mineral common in coal deposits and may be a source of As enrichment that is released into the environment in coal removal processes. Concentration of As in pyrite has been documented to be up to 25,525 mg/kg in eastern Kentucky coal fields and up to 3,900 mg/kg in West Virginia coal fields (Diehl et al. 2012).

While enrichment of Cd, Cr, Cu, Fe, Pb, Ni, Zn, and silica in coal mining compared to non-coal mining soil/dust have been reported (De and Mitra 2004; Šebestová et al. 1996; Das and Chakrapani 2011; Onder and Yigit 2009), most of the soil/dust concentrations fall well below the 95th percentile of U.S soils (Smith et al. 2005). Rather than drastic enrichment of soil/dust concentrations of Cd, Cr, Cu, Fe, Pb, Ni, Zn, and silica due to coal mining, evidence suggests

Table 10. Associations matrix of health effects and toxic agents in ground water (W) and soil/dust (S) potentially due to coal mining

Associations/Causes	Asthma	Birth Defects	Cardiovascular Disease	Hypertension	Kidney Disease	Lung Disease	Cancer
Ambient air toxicants							
• Particulate matter undefined	S	S	S	S		S	
• Particulate matter—PM_{10}	S		S				S
• Carbon monoxide		S	S			S	
• Nitrogen dioxide	S	S	S			S	
• Sulfur dioxide	S	S	S			S	
Organic carbon species			W				
Aluminum					W		
Arsenic		S,W	S,W	S,W	S,W		S,W
Beryllium					S		S
Cadmium		S	S	S	S		S
Chromium			S,W		S,W		S,W
Cobalt							W
Copper			S,W				
Iron			S,W				
Lead		S,W	S,W	S,W	S,W		
Nickel			S,W				S,W
Sulfur			W				
Zinc			S,W				
Silica			S				S
Quartz			W				
Aldehydes			S				
Benzene							S
Polyaromatic hydrocarbons			S				

primarily crustal origins, and elevated exposure is due in large part to increased amounts of dust generated by coal mining as discussed previously (Dubey et al. 2011).

Potential enrichment of ground water with Al, As, Cr, Co, Cu, Fe, Pb, Ni, Zn, silica, and quartz may be attributed to acid mine drainage that leaches these constituents from soil. Coal mining exposes pyrite to atmospheric conditions and allows it to react with oxygen, resulting in hydrolysis of water molecules and the release of acidic protons (Black and Craw 2001). In northern and central Appalachian coal mining regions, mine-drainage constituents may become elevated compared to levels in un-mined areas because of exposure of acidic mine drainage to naturally occurring calcareous materials. The result is neutral pH ground water with elevated levels of Al, Ca, Fe, Zn, and silica (McAuley and Kozar 2006); which have been identified as agents of concern in the associations matrix (Table 10). Elevated As levels, above the US EPA drinking water standard (10 µg/L) have been detected in ground water samples collected from the Appalachian region. Wigginton et al. (2008) reported elevated levels of As, Fe, Mn, and Pb in some wells in Big Sandy Region of Kentucky and West Virginia. However, the authors concluded that linking the elavated levels to coal mining was not possible due to naturally higher concentrations occuring in Big Sandy Region rock.

While the associations matrix indicates an indirect link between health effects and coal mining source using literature not limited to Appalachia and health effects, there are two studies which support these results with direct biological measures. A study by Johnson et al. (2011), using toenails collected from colorectal cancer patients as a biomarker, documented potentially elevated exposure to As, Cr, and Ni in Appalachian Kentucky coal mining counties over a non-mining reference county. The study also found that lung cancer incidence, mortality rates, and colorectal mortality rates are higher in Appalachian coal mining counties than non-coal mining counties in Kentucky. However, further research is needed in order to determine the source of As, Cr, and Ni and establish an exposure pathway that links the source to individuals with evidence of increased exposure. In a study by Knuckles et al. (2012), rats' lungs were exposed to dust collected one mile from an active mountain top mining site. The exposure resulted in microvascular dysfunction, which may contribute to cardiovascular disease. However, there was no evidence provided that can confirm that the dust collected for the exposure study can be sourced to the coal mine. Characterization of potential source material from the coal mining site sufficient to confirm that coal mining is the source is necessary in order to link coal mining to the study conclusions.

CONCLUSION

This study provides a broad look at possible toxic agents that could be implicated in health effects reported to be elevated in Appalachian coal mining communities. The list of toxic agents was significantly narrowed when screened by literature references linking the toxic agents to coal mining. The resulting associations matrix allows for potential exposure routes of plausible toxic agents to be identified. Given the toxic agents identified, the primary routes of exposure are ingestion and inhalation for soil/dust and ingestion for water.

Fugitive dust in Appalachian coal mining communities is the major exposure pathway of concern for surface coal mining operations. In addition to the potential for adverse health effects associated with particulate matter, constituents in the dust may also need to be considered. In particular, As, Cd, and Pb are associated with many of the health effects reported to be elevated in coal mining communities. Ingestion of potentially toxic constituents in ground water may also be a concern, but the variability in regional hydrology and natural geology make it difficult to establish a link between elevated levels of potentially toxic agents and coal mining in Appalachia.

While this study provides potential associations between adverse health effects and toxic agents, it is impossible to establish a causal linkage without site specific exposure characterization in the Appalachian region. Further research is needed to quantify the amount of dust leaving coal mining operations and determine the total content and bioavailable fractions (via ingestion and inhalation) of dust constituents, in order to determine if exposure to coal mining communities is sufficiently elevated to cause adverse health effects.

ACKOWLEDGMENT

This study was sponsored by the Appalachian Research Initiative for Environmental Science (ARIES). The views, opinions and recommendations expressed herein are solely those of the authors and do not imply any endorsement by ARIES employees, other ARIES-affiliated researchers or industrial members. Information about ARIES can be found at http://www.energy.vt.edu/ARIES.

REFERENCES

Ahern, M.M., M. Hendryx, J. Conley, E. Fedorko, A. Ducatman, and K.J. Zullig. 2011. The association between mountaintop mining and birth defects among live births in central Appalachia, 1996–2003. *Environ Res.* 111 (6):838–846.

Bandara, J.M., H.V. Wijewardena, J. Liyanege, M.A. Upul, and J.M. Bandara. 2010. Chronic renal failure in Sri Lanka caused by elevated dietary cadmium: Trojan horse of the green revolution. *Toxicol Lett.* 198 (1):33–39.

Bernal-Pacheco, O., and G.C. Roman. 2007. Environmental vascular risk factors: new perspectives for stroke prevention. *J Neurol Sci.* 262 (1-2):60–70.

Bhatnagar, A. 2006. Environmental cardiology: studying mechanistic links between pollution and heart disease. *Circ Res.* 99 (7):692–705.

Black, A., and D. Craw. 2001. Arsenic, copper and zinc occurrence at the Wangaloa coal mine, southeast

Otago, New Zealand. *International Journal of Coal Geology.* 45:181–193.

Brautbar, N. 2004. Industrial solvents and kidney disease. *International Journal of occupational and environmental health.* 10 (1):5.

Breysse, P.N., G.B. Diette, E.C. Matsui, A.M. Butz, N.N. Hansel, and M.C. McCormack. 2010. Indoor air pollution and asthma in children. *Proc Am Thorac Soc* 7 (2):102–106.

Brody, J.G. 2010. Everday exposures and breast cancer. *Reviews on environmental health.* 25 (1):1–7.

Brook, R.D. 2007. Why physicians who treat hypertension should know more about air pollution. *Journal of Clinical Hypertension.* 9 (8):629–635.

Brook, R.D., S. Rajagopalan, C.A. Pope, 3rd, J.R. Brook, A. Bhatnagar, A.V. Diez-Roux, F. Holguin, Y. Hong, R.V. Luepker, M.A. Mittleman, A. Peters, D. Siscovick, S.C. Smith, Jr., L. Whitsel, J.D. Kaufman, E. American Heart Association Council on, C.o.t.K.i.C.D. Prevention, P.A. Council on Nutrition, and Metabolism. 2010. Particulate matter air pollution and cardiovascular disease: An update to the scientific statement from the American Heart Association. *Circulation* 121 (21):2331–2378.

Bruning, T. 2002. Nephrotoxicity and nephrocarcinogenicity of dinitrotoluene; new aspects to be considered. *Reviews on environmental health.* 17 (3):10.

Cannata-Andia, J., and J. Fernandez-Martin. 2002. The clinical impact of aluminum overload in renal failure. *Nephrology Dialysis Transplantation.* 17 (2):4.

Carpenter, D.O. 2010. Electromagnetic fields and cancer; the cost of doing nothing. *Reviews on environmental health.* 25 (1):75–80.

Cecchi, L., G. D'Amato, J.G. Ayres, C. Galan, F. Forastiere, B. Forsberg, J. Gerritsen, C. Nunes, H. Behrendt, C. Akdis, R. Dahl, and I. Annesi-Maesano. 2010. Projections of the effects of climate change on allergic asthma: the contribution of aerobiology. *Allergy.* 65 (9):1073–1081.

Chameides, W.L. 2010. Environmental factors in cancer; focus on air pollution. *Reviews on environmental health.* 25 (1):17–22.

Chaulya, S.K. 2004. Spatial and temporal variations of SPM, RPM, SO2 and NOx concentrations in an opencast coal mining area. *J Environ Monit.* 6 (2):134–142.

Chen, C.J., S.L. Wang, J.M. Chiou, C.H. Tseng, H.Y. Chiou, Y.M. Hsueh, S.Y. Chen, M.M. Wu, and M.S. Lai. 2007. Arsenic and diabetes and hypertension in human populations: a review. *Toxicol Appl Pharmacol.* 222 (3):298–304.

Cole, M.P., and B.A. Freeman. 2009. Promotion of cardiovascular disease by exposure to the air pollutant ozone. *Am J Physiol Lung Cell Mol Physiol.* 297 (2):L205–208.

Das, S.K., and G.J. Chakrapani. 2011. Assessment of trace metal toxicity in soils of Raniganj Coalfield, India. *Environ Monit Assess.* 177 (1-4):63–71.

De Luca, G., F. Olivieri, G. Melotti, G. Aiello, L. Lubrano, and A.L. Boner. 2010. Fetal and early postnatal life roots of asthma. *J Matern Fetal Neonatal Med.* 23 Suppl 3:80–83.

De, S., and A.K. Mitra. 2004. Mobilization of heavy metals from mine spoils in a part of Raniganj coalfield, India: Causes and effects. *Environmental Geosciences.* 11 (2):65–76.

Delfino, R.J., C. Sioutas, and S. Malik. 2005. Potential Role of Ultrafine Particles in Associations between Airborne Particle Mass and Cardiovascular Health. *Environ Health Perspect.* 113 (8):934–946.

Diehl, S.F., M.B. Goldhaber, A.E. Koenig, H.A. Lowers, and L.F. Ruppert. 2012. Distribution of arsenic, selenium, and other trace elements in high pyrite Appalachian coals: Evidence for multiple episodes of pyrite formation. *International Journal of Coal Geology.* 94 (0):238–249.

Dozor, A.J. 2010. The role of oxidative stress in the pathogenesis and treatment of asthma. *Ann N Y Acad Sci.* 1203:133–137.

Dubey, B., A.K. Pal, and G. Singh. 2011. Assessment of ambient particulate matter in coal mining area, Eastern India. *International Journal of Applied Sciences.* 3 (1):1–11.

Edwards, J.R., and W.C. Prozialeck. 2009. Cadmium, diabetes and chronic kidney disease. *Toxicol Appl Pharmacol.* 238 (3):289–293.

Ekong, E.B., B.G. Jaar, and V.M. Weaver. 2006. Lead-related nephrotoxicity: a review of the epidemiologic evidence. *Kidney Int.* 70 (12):2074–2084.

Evans, M., and C.-G. Elinder. 2011. Chronic renal failure from lead: Myth or evidence based fact? *Kidney International.* 119 (8):272.

Everett, C.J., I. Frithsen, and M. Player. 2011. Relationship of polychlorinated biphenyls with type 2 diabetes and hypertension. *J Environ Monit.* 13 (2):241–251.

Fang, S.C., A. Cassidy, and D.C. Christiani. 2010. A systematic review of occupational exposure to particulate matter and cardiovascular disease. *Int J Environ Res Public Health.* 7 (4):1773–1806.

Faught, B.E., A.D. Flouris, and J. Cairney. 2009. Epidemiological Evidence Associating Secondhand Smoke Exposure with Cardiovascular Disease. *Inflammation and Allergy–Drug Targets.* 8:321–327.

Field, R.W. 2010. Environmental factors in cancer; radon. *Reviews on environmental health.* 25 (1):23–31.

Fucic, A., M. Gamulin, Z. Ferencic, D.S. Rokotov, J. Katic, A. Bartonova, I.B. Lovasic, and D.F. Merlo. 2010. Lung cancer and environmental chemical exposure: a review of our current state of knowledge with reference to the role of hormones and hormone receptors as an increased risk factor for developing lung cancer in man. *Toxicol Pathol.* 38 (6):849–855.

Galimanis, A., M.L. Mono, M. Arnold, K. Nedeltchev, and H.P. Mattle. 2009. Lifestyle and stroke risk: a review. *Curr Opin Neurol.* 22 (1):60–68.

Gallagher, C.M., and J.R. Meliker. 2010. Blood and urine cadmium, blood pressure, and hypertension: a systematic review and meta-analysis. *Environ Health Perspect.* 118 (12):1676–1684.

Gamble, J. 2010. Lung cancer and diesel exhaust: a critical review of the occupational epidemiology literature. *Crit Rev Toxicol.* 40 (3):189–244.

Ghose, M.K. 2007. Generation and quantification of hazardous dusts from coal mining in the Indian context. *Environ Monit Assess.* 130 (1-3):35–45.

Ghose, M.K., and S.R. Majee. 2007. Characteristics of hazardous airborne dust around an Indian surface coal mining area. *Environ Monit Assess.* 130 (1-3):17–25.

Gibb, H., C. Haver, D. Gaylor, S. Ramasamy, J.S. Lee, D. Lobdell, T. Wade, C. Chen, P. White, and R. Sams. 2011. Utility of recent studies to assess the National Research Council 2001 estimates of cancer risk from ingested arsenic. *Environ Health Perspect.* 119 (3):284–290.

Gill, E.A., C.L. Curl, S.D. Adar, R.W. Allen, A.H. Auchincloss, M.S. O'Neill, S.K. Park, V.C. Van Hee, A.V. Diez Roux, and J.D. Kaufman. 2011. Air pollution and cardiovascular disease in the Multi-Ethnic Study of Atherosclerosis. *Prog Cardiovasc Dis.* 53 (5):353–360.

Guha, N., F. Merletti, N.K. Steenland, A. Altieri, V. Cogliano, and K. Straif. 2010. Lung Cancer Risk in Painter: A Meta-Analysis. *Environmental Health Perspectives.* 118 (3):303–312.

Gupta, D.C. 1999. Environmental aspects of selected trace elements associated with coal and natural waters of Pench Valley coalfield of India and their impact on human health. *International Journal of Coal Geology.* 40:133–149.

Hajdu, S.I. 2011. Much overlooked causes of lung cancer. *Annals of clinical and laboratory science.* 41 (1):97–101.

Hansen, J., and C.F. Lassen. 2011. Occupation and cancer risk by use of Danish registers. *Scandinavian journal of public health* 39 (7):136–140.

Hassing, H.C., T.B. Twickler, J.J.P. Kastelein, M.J.M. Cramer, and F.R. Cassee. 2009. Air pollution as noxious environmental factor in the development of cardiovascular disease. *The Netherlands Journal of Medicine.* 67 (4):116–121.

Hendryx, M. 2009. Mortality from heart, respiratory, and kidney disease in coal mining areas of Appalachia. *Int Arch Occup Environ Health.* 82 (2):243–249.

Hendryx, M., and M.M. Ahern. 2008. Relations between health indicators and residential proximity to coal mining in West Virginia. *Am J Public Health.* 98 (4):669–671.

Hendryx, M., M.M. Ahern, and T.R. Nurkiewicz. 2007. Hospitalization patterns associated with Appalachian coal mining. *J Toxicol Environ Health A.* 70 (24):2064–2070.

Hendryx, M., E. Fedorko, and J. Halverson. 2010. Pollution sources and mortality rates across rural-urban areas in the United States. *J Rural Health.* 26 (4):383–391.

Hendryx, M., K. O'Donnell, and K. Horn. 2008. Lung cancer mortality is elevated in coal-mining areas of Appalachia. *Lung Cancer.* 62 (1):1–7.

Hendryx, M., L. Wolfe, J. Luo, and B. Webb. 2011. Self-Reported Cancer Rates in Two Rural Areas of West Virginia with and Without Mountaintop Coal Mining. *J Community Health.*

Hendryx, M., and K.J. Zullig. 2009. Higher coronary heart disease and heart attack morbidity in Appalachian coal mining regions. *Prev Med.* 49 (5):355–359.

Hoffmann, B., S. Moebus, N. Dragano, S. Mohlenkamp, M. Memmesheimer, R. Erbel, K.H. Jockel, and G. Heinz Nixdorf Recall Investigative. 2009. Residential traffic exposure and coronary heart disease: results from the Heinz Nixdorf Recall Study. *Biomarkers.* 14 Suppl 1:74–78.

Holgate, S.T., H.S. Arshad, G.C. Roberts, P.H. Howarth, P. Thurner, and D.E. Davies. 2010. A new look at pathogenesis of asthma. *Clinical Science.* 118 (7):439–450.

Hosgood, H.D., 3rd, H. Wei, A. Sapkota, I. Choudhury, N. Bruce, K.R. Smith, N. Rothman, and Q. Lan. 2011. Household coal use and lung cancer: systematic review and meta-analysis of case-control studies, with an emphasis on geographic variation. *Int J Epidemiol.* 40 (3):719–728.

Houston, M.C. 2007. The role of mercury and cadmium heavy metals in cardiovascular disease, hypertension, coronary heart disease, and myocardial infarction. *Alternative Therapies in Health and Medicine.* 13 (2):S128-S133.

Houston, M.C. 2011. Role of mercury toxicity in hypertension, cardiovascular disease, and stroke. *J. Clin. Hypertens. Journal of Clinical Hypertension.* 13 (8):621–627.

Humblet, O., L. Birnbaum, E. Rimm, M.A. Mittleman, and R. Hauser. 2008. Dioxins and cardiovascular disease mortality. *Environ Health Perspect.* 116 (11):1443–1448.

IARC. 2011. Agents classified by the IARC Monographs, Volumes 1–102, edited by I. A. f. R. o. Cancer.

IPCS. 1991. Principles and methods for the assessment of nephrotoxicity associated with exposure to chemicals, edited by I. P. o. C. Safety.

Johnson, N., B.J. Shelton, C. Hopenhayn, T.T. Tucker, J.M. Unrine, B. Huang, W. Christian, Z. Zhang, X. Shi, and L. Li. 2011. Concentrations of arsenic, chromium, and nickel in toenail samples from Appalachian Kentucky residents. *J Environ Pathol Toxicol Oncol.* 30 (3):213–223.

Kligerman, S., and C. White. 2011. Epidemiology of lung cancer in women; risk factors, survival, and screening. *American Journal of Roentgenology.* 196 (2):287–295.

Knuckles, T.L., P.A. Stapleton, V.C. Minarchick, L. Esch, M. McCawley, M. Hendryx, and T.R. Nurkiewicz. 2012. Air pollution particulate matter collected from an appalachian mountaintop mining site induces microvascular dysfunction. *Microcirculation.*

Kunzli, N., and I.B. Tager. 2005. Air pollution ; from lung to heart. *Swiss Medical Weekly.* 135:697–702.

Kutting, B., and H. Drexler. 2010. UV-induced skin cancer at workplace and evidence-based prevention. *Int Arch Occup Environ Health.* 83 (8):843–854.

Linet, M.S., and P.D. Inskip. 2010. Cellular (Mobile) Telephone Use and Cancer Risk. *Reviews on environmental health.* 25 (1):51–55.

Lubin, J.H. 2010. Environmental factors in cancer; radon. *Reviews on environmental health* 25 (1):33–38.

Mastin, J.P. 2005. Environmental Cardiovascular Disease. *Cardiovascular Toxicology* 5:91–94.

Mates, J.M., J.A. Segura, F.J. Alonso, and J. Marquez. 2010. Roles of dioxins and heavy metals in cancer and neurological diseases using ROS-mediated mechanisms. *Free Radic Biol Med.* 49 (9):1328–1341.

McAuley, and Kozar. 2006. Ground water quality in unmined areas and near reclaimed surface coal mines in the Northern and Central Appalachian coal regions, Pennsylvania and West Virginia, edited by USGS.

McErlean, A., and M.S. Ginsberg. 2011. Epidemiology of lung cancer. *Seminars in roentgenology.* 46 (3):173–177.

Merletti, F., C. Galassi, and T. Spadea. 2011. The socioeconomic determinants of cancer. *Environmental Health.* 10 (1):S7.

Mossman, B.T., M. Lippmann, T.W. Hesterberg, K.T. Kelsey, A. Barchowsky, and J.C. Bonner. 2011. Pulmonary endpoints (lung carcinomas and asbestosis) following inhalation exposure to asbestos. *J Toxicol Environ Health B Crit Rev.* 14 (1-4):76–121.

Mukherjee, A.K., S.K. Bhattacharya, and H.N. Saiyed. 2005. Assessment of respirable dust and its free silica contents in different Indian coalmines. *Industrial Health.* 43:277–284.

Navas-Acien, A., E. Guallar, E.K. Silbergeld, and S.J. Rothenberg. 2007. Lead exposure and cardiovascular disease—a systematic review. *Environ Health Perspect.* 115 (3):472–482.

Nielsen, G.D., and P. Wolkoff. 2010. Cancer effects of formaldehyde: a proposal for an indoor air guideline value. *Arch Toxicol.* 84 (6):423–446.

O'Toole, T.E., D.J. Conklin, and A. Bhatnagar. 2008. Environmental risk factors for heart disease. *Reviews on environmental health.* 23 (3):167–202.

Onder, M., and E. Yigit. 2009. Assessment of respirable dust exposures in an opencast coal mine. *Environ Monit Assess.* 152 (1-4):393–401.

Peden, D.B., and R.K. Bush. 2011. Advances in environmental and occupational respiratory disease in 2010. *J Allergy Clin Immunol.* 127 (3):696–700.

Peters, A. 2005. Particulate matter and heart disease: evidence from epidemiological studies. *Toxicol Appl Pharmacol* 207 (2 Suppl):477–482.

Po, J.Y.T., J.M. FitzGerald, and C. Carsten. 2011. Respiratory disease associated with solid biomass fuel exposure in rural women and children: systematic review and meta-analysis. *Thorax.* 66 (3):232–239.

Prozialeck, W.C., and J.R. Edwards. 2010. Early biomarkers of cadmium exposure and nephrotoxicity. *Biometals.* 23 (5):793–809.

Rankin, G.O. 2004. Nephrotoxicity induced by C- and N-arylsuccinimides. *J Toxicol Environ Health B Crit Rev.* 7 (5):399–416.

Samet, J.M. 2010. Radiation and cancer risk; more epidemiological research is needed. *Reviews on environmental health.* 25 (1):47–50.

Samet, J.M. 2011. Radiation and cancer risk: a continuing challenge for epidemiologists. *Environmental Health.* 10 (1):S4.

Samet, J.M., E. Avila-Tang, P. Boffetta, L.M. Hannan, S. Olivo-Marston, M.J. Thun, and C.M. Rudin. 2009. Lung cancer in never smokers: clinical epidemiology and environmental risk factors. *Clin Cancer Res.* 15 (18):5626–5645.

Sauvant, M.-P., and D. Pepin. 2002. Drinking water and cardiovascular disease. *Food and Chemical Toxicology.* 40:1311–1325.

Šebestová, E., V. Machovič, H. Pavlíková, J. Lelák, and L. Minarík. 1996. Environmental impact of brown coal mining in Sokolovo basin with especially trace metal mobility. *Journal of Environmental Science and Health. Part A: Environmental Science and Engineering and Toxicology.* 31 (10):2453–2463.

Sly, P.D. 2011. The early origins of asthma: who is really at risk? *Curr Opin Allergy Clin Immunol.* 11 (1):24–28.

Soderland, P. 2010. Chronic kidney disease associated with environmental toxins and exposures. *Advances in chronic kidney disease.* 17 (3):11.

Soto-Martinez, M., and P.D. Sly. 2010. Relationship between environmental exposures in children and adult lung disease: The case for outdoor exposures. *Chronic respiratory disease.* 7 (3):173–186.

Soto, A.M., and C. Sonnenschein. 2010. Environmental causes of cancer: endocrine disruptors as carcinogens. *Nature Reviews Endocrinology.* 6:363–370.

Stapleton, M., A. Howard-Thompson, C. George, R.M. Hoover, and T.H. Self. 2011. Smoking and asthma. *J Am Board Fam Med.* 24 (3):313–322.

Stefanovic, V., and M. Polenakovic. 2009. Fifty years of research in Balkan endemic nephropathy: where are we now? *Nephron Clin Pract.* 112 (2):c51–56.

Su, T.C., S.Y. Chen, and C.C. Chan. 2011. Progress of ambient air pollution and cardiovascular disease research in Asia. *Prog Cardiovasc Dis.* 53 (5):369–378.

Suk, W.A. 2010. Environmental factors in cancer; radiation. *Reviews on environmental health.* 25 (1):57–62.

Sun, Q., X. Hong, and L.E. Wold. 2010. Cardiovascular effects of ambient particulate air pollution exposure. *Circulation.* 121 (25):2755–2765.

Sung, B., S. Prasad, V.R. Yadav, A. Lavasanifar, and B.B. Aggarwal. 2011. Cancer and diet: How are they related? *Free Radic Res.* 45 (8):864–879.

Trivedi, R., A. Mondal, M.K. Chakraborty, and B.K. Tewary. 2010. A statistical analysis of ambient air quality around opencast coal projects in Wardha Valley Coalfields, Western Coal Fields Limited, India. *Indian Journal of Environmental Protection.* 30 (12):969–977.

Tzivian, L. 2011. Outdoor Air Pollution and Asthma in Children. *Journal of Asthma.* 48 (5):470–481.

Vardavas, C.I., and D.B. Panagiotakos. 2009. The causal relationship between passive smoking and inflammation on the development of cardiovascular disease: a review of the evidence. *Inflammation and Allergy–Drug Targets.* 8:328–333.

Vaziri, N.D. 2008. Mechanisms of lead-induced hypertension and cardiovascular disease. *Heart and Circulatory Physiology.* 295:H454-H465.

Vena, J.E. 2010. Lung, breast, bladder and rectal cancer. *Reviews on environmental health.* 25 (1):39–45.

Vermylen, J., A. Nemmar, B. Nemery, and M.F. Hoylaerts. 2005. Ambient air pollution and acute myocardial infarction. *Journal of Thrombosis and Haemostasis.* 3:1955–1961.

Vicente-Vicente, L., Y. Quiros, F. Perez-Barriocanal, J.M. Lopez-Novoa, F.J. Lopez-Hernandez, and A.I. Morales. 2010. Nephrotoxicity of uranium: pathophysiological, diagnostic and therapeutic perspectives. *Toxicol Sci.* 118 (2):324–347.

Virdis, A., C. Giannarelli, M. Fritsch Neves, S. Taddei, and L. Ghiadoni. 2010. Cigarette smoking and hypertension. *Current Pharmaceutical Design.* 16:2518–2525.

Voice, T., D. Long, Z. Radovanovic, J. Atkins, S. McElmurry, N. Niagolova, P. Dimitrov, E. Petropoulos, and V. Ganev. 2006. Critical evaluation of environmental exposure agents suspected in the etiology of balkan endemic nephropathy. *International Journal of occupational and environmental health.* 12:8.

Vrijheid, M., D. Martinez, S. Manzanares, P. Dadvand, A. Schembari, J. Rankin, and M. Nieuwenhuijsen. 2011. Ambient air pollution and risk of congenital anomalies: a systematic review and meta-analysis. *Environ Health Perspect.* 119 (5):598–606.

Wang, L., and K.E. Pinkerton. 2007. Air pollutant effects on fetal and early postnatal development. *Birth Defects Res C Embryo Today.* 81 (3):144–154.

Weichenthal, S., C. Moase, and P. Chan. 2010. A review of pesticide exposure and cancer incidence in the Agricultural Health Study cohort. *Environ Health Perspect.* 118 (8):1117–1125.

Weinmayr, G., E. Romeo, M. De Sario, S.K. Weiland, and F. Forastiere. 2010. Short-term effects of PM10 and NO2 on respiratory health among children with asthma or asthma-like symptoms: a systematic review and meta-analysis. *Environ Health Perspect.* 118 (4):449–457.

Wigginton, A., J. Mitchell, G. Evansc, and S. McSpirit. 2008. Assessing the Impacts of Coal Waste on Residential Wells in the Appalachian Region of the Big Sandy Watershed, Kentucky and West Virginia: An Exploratory Investigation. *Journal of the Kentucky Academy of Science.* 69 (2):152–163.

Wigle, D.T., T.E. Arbuckle, M.C. Turner, A. Berube, Q. Yang, S. Liu, and D. Krewski. 2008. Epidemiologic evidence of relationships between reproductive and child health outcomes and environmental chemical contaminants. *J Toxicol Environ Health B Crit Rev.* 11 (5-6):373–517.

Wigle, D.T., T.E. Arbuckle, M. Walker, M.G. Wade, S. Liu, and D. Krewski. 2007. Environmental hazards: evidence for effects on child health. *J Toxicol Environ Health B Crit Rev.* 10 (1-2):3–39.

Wu, C. 2010. Overview of developmental and reproductive toxicity research in China: history, funding mechanisms, and frontiers of the research. *Birth Defects Res B Dev Reprod Toxicol.* 89 (1):9–17.

Zanobetti, A., A. Baccarelli, and J. Schwartz. 2011. Gene-air pollution interaction and cardiovascular disease: a review. *Prog Cardiovasc Dis.* 53 (5):344–352.

4.1 Total Dissolved Solids and Environmental Impacts 1

Comparison of Groundwater Seepage Modeling in Conventional and Geomorphic Valley Fill Design 246

Predicting Total Dissolved Solids Release from Overburden in Appalachian Coal Fields 255

Natural Attenuation of TDS in Mountaintop Mines 265

Predicting TDS Release from Appalachian Mine Spoils 275

Long-Term Trends of Electrical Conductivity in Streams Flowing Out of Headwater Valley Fills in Virginia 286

Comparison of Groundwater Seepage Modeling in Conventional and Geomorphic Valley Fill Design

John D. Quaranta
Civil and Environmental Engineering, West Virginia University, Morgantown, West Virginia

Leslie Hopkinson
Civil and Environmental Engineering, West Virginia University, Morgantown, West Virginia

Paul Ziemkiewicz
West Virginia Water Research Institute, West Virginia University, Morgantown, West Virginia

ABSTRACT

Excess spoil generated during surface mining in southern West Virginia is generally placed in headwater valleys. Known as valley fills, these structures are designed to move water rapidly through constructed drains to maximize geotechnical stability. Seepage from valley fills tends to be elevated with respect to calcium, magnesium, alkalinity and sulfates and there is evidence that, in high concentrations, these ions can contribute to stream degradation. New fluvial geomorphic principals are being researched to aid in reclamation alternatives to valley fill designs. Geomorphic designs have proven successful in semi-arid regions; however, there has been little research performed into the application of this approach for eastern surface mining reclamation.

This research investigated the differences in seepage quality and quantity between the conventional valley fill method and geomorphic designs on a permitted valley fill in southern West Virginia. The computational modeling involved geomorphic design for a proposed valley fill in southern West Virginia using commercial software. A comprehensive seepage analysis was then developed using a finite element method numerical model for assessing the groundwater flow characteristics of the fill rock for a 10 year time period. This analysis was performed for both the conventional and geomorphic fill designs.

Differences in seepage for the conventional and geomorphic fill were presented and discussed as a comparison of the two designs. Analysis criteria were chosen as a way to compare the results of the two fills in order to investigate if an advantage for one fill design was apparent. If an advantage of one fill was apparent, the magnitude of the advantage was quantified using a percent change in results. The results projected that higher water velocities (decreased residence times) and less water storage occur through mine spoil reclaimed according to the geomorphic fill design. Shorter hydraulic transit times through the spoil are expected to result in lower ionic concentrations in discharge water.

INTRODUCTION AND BACKGROUND

Surface coal mining is a dominant land use in the central Appalachian region of West Virginia. In surface mining, the overburden from mountains is removed to expose buried coal, which is then extracted. The bulk of mined rock (spoil) is placed on the mined surface while excess spoil is placed in external dumps known as valley fills. Mountaintop mining reclamation is often accomplished through the design of valley fills. The West Virginia Department of Environmental Protection (WVDEP) stated that valley fill design must meet the following requirements: long-term static factor of safety of 1.5 or higher, 2:1 slopes with 20-ft wide benches every 50 vertical feet, a rock core, and properly designed drainage for 100-yr, 24-hr rain event (WVDEP 1999). Two areas of concern are associated with the design of valley fills: slope stability and the behavior of surface and groundwater (seepage and runoff).

Geomorphic design seeks to develop a landscape that can be managed to achieve improvements in water quality. Specifically, geomorphic design aims to produce valley fills that are more aesthetically pleasing and have a more natural look than conventional valley fills by incorporating natural landform shapes in design. Also, geomorphic design may provide greater stability through more natural slope profiles. In natural landforms, erosive forces balance with resistive forces over time. Geomorphic design creates a more natural landform that could have long-term equilibrium and low erosion rates. Another goal of geomorphic design is to improve runoff and groundwater seepage. By increasing runoff and reducing infiltration, groundwater seepage could be improved, which could aid in slope stability and contaminant transport.

New fluvial geomorphic principals are being developed to aid in reclamation alternatives to conventional valley fill designs. So far, this method has only been field tested in the United States in semi-arid regions. Its use in the central Appalachian region that dominates West Virginia has not been analyzed. Designing natural landforms in the steep terrain of this region can prove to be very complex and require higher initial construction costs. Regulations for geomorphic design are lacking and differ from current regulations for valley fill design (Michael et al. 2009). If successful, geomorphic design could counteract higher initial construction costs with lower long-term costs, but an experimental project using geomorphic design in central Appalachia is needed to provide a better comparison to conventional valley fill design.

MATERIALS AND METHODS

Before modeling could be completed, the material and site characteristics had to be defined. This consisted of obtaining the hydraulic conductivity of the fill material and defining the hydrology of the site.

Hydraulic Conductivity

The material property of particular interest for numerical modeling was hydraulic conductivity. Hydraulic conductivity was chosen from a range of values found from previous work done with waste rock tailings. These ranged from an unsaturated hydraulic conductivity of 1×10^{-5} m/s (Abdelghani et al. 2009) to a saturated hydraulic conductivity of 1×10^{-7} m/s (Aubertin et al. 1996). These previously published values were chosen because they represented actual data used in practice.

Site Hydrology

Rainfall intensities for specific storm types were taken from the IDF (intensity-duration-frequency) table from National Oceanic and Atmosphere Association's (NOAA) Precipitation Frequency Data Server (NOAA 2012). This data was taken from the Logan, WV station because it was the closest NOAA station to the valley fill site of interest. For the transient condition, infiltration was determined as a function of time. The first step in developing this function was to find the infiltration function for different storm events. Frequencies of 2, 10, 25, 50, 100, and 500 years were analyzed for both 1 hr and 24 hr durations. These storms were modeled using the Horton infiltration capacity equation. For the 1 hr storms, the initial infiltration capacity was equal to the unsaturated hydraulic conductivity (1×10^{-5} m/s) for all storms except the 2 year storm. This is because the intensity of every storm other than the 2 year storm was greater than the unsaturated hydraulic conductivity. As the soil became saturated, the rate at which the water was able to infiltrate decreased until all storms approached the same final infiltration capacity at the end of 1 hour. This final infiltration capacity was equal to the saturated hydraulic conductivity (1×10^{-7} m/s). For the 24-hr storms, the initial infiltration capacity was equal to the rainfall intensity was all frequency storms. This was because each storm's rainfall intensity was less than the unsaturated hydraulic conductivity. Each storm reached the final infiltration capacity (saturated hydraulic conductivity of the soil) within 1 hr.

NUMERICAL MODELING

A numerical model was used to analyze the fills using the finite element method. This required three steps. First, the geometry of the model was identified and discretized into elements. After that was done, material properties were defined. Thirdly, boundary conditions were applied. Modeling was performed for both a conventional valley fill and a geomorphic fill, which are described in detail in the section. Each fill was analyzed with regards to seepage.

Groundwater seepage was modeled in a two-dimensional state. This consisted of taking a profile slice along the centerline of the fill. Seepage for the valley fill was analyzed in a transient time step condition over the span of 10 years. 10 time steps were modeled to produce results for each of the 10 years. The geometry of the conventional fill was taken from a permit fill of an actual valley fill design. The conventional fill profile had a length of 1,370 meters and height of 215 meters. The geometry of the geomorphic fill was taken from a developed geomorphic design. The geomorphic fill profile had a length of 1,370 meters and a height of 245 meters.

Once materials and site hydrology were defined, boundary conditions were determined to fully define the model. To do this, an infiltration function spanning 10 years was developed from the NOAA precipitation data. It is important to note that the analysis did not take into account any stream infiltration, only infiltration from precipitation. For modeling a valley fill, two different infiltration functions had to be developed: a function for water infiltrating on a flat surface and a function for water infiltrating on a sloped surface. These functions corresponded to varying land use for the conventional and geomorphic fills. For each of the functions, a different percentage of precipitation translated into infiltration. The function for a flat surface considered 85% of precipitation infiltrating (Sharma et. al 1983) and was applied to the top of the conventional fill. The function for a sloped surface considered 55% of precipitation infiltrating (Ziemkiewicz et al. 2011; Meek and O'dell 2012) and was applied to the face of the conventional fill and to the entire geomorphic fill. The geomorphic fill's lower percentage infiltration function meant that a lower volume of water infiltrated into the fill than into the conventional fill.

RESULTS

In order to analyze the conventional and geomorphic fill on a year-to-year basis for each of the 10 years, parameters had to be chosen by which to analyze and compare. These parameters included water velocity (m/s) at the toe and storage volume (m^3) in the fill. Water velocity at the toe described movement of water at the toe of the fill. Storage volume within the fill defined the volume of water that was being stored in the fill at any given time. Each fill was modeling as a profile slice of the total volume. This consisted of a two-dimensional profile with a 1 meter unit thickness. The conventional fill's profile had a fill volume of 72,632 m^3, whereas the geomorphic fill's profile had a fill volume of 55,403 m^3. Since the behavior of groundwater seepage was dependent on the amount of fill material, each criterion was normalized by dividing the result by the fill's respective fill volume. This provided results that could be compared directly between the two fills.

Figures 1 and 2 show the plan view of the conventional fill and the visual results for year 4 for the corresponding profile slice. Figures 3 and 4 show the plan view of the geomorphic fill and the visual results for year 4 for the corresponding cross-section slice. All distances are in meters.

The shaded contours of each result showed levels of pore-water pressure. Higher pore-water pressure was depicted by a darker shade. Contours for pore-water pressure ranged from minimum to maximum pore-water pressure (kPa) and used an interval of 200 kPa for each contour. The dotted line within the fill showed the transition from negative pore-pressure to positive pore-pressure. Within the dotted line were zones with positive pore-pressure. These were the saturated zones

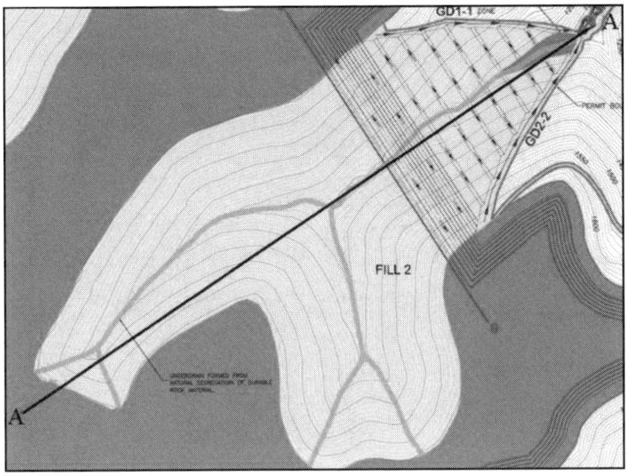

Figure 1. Conventional fill plan view with profile slice location (length of A-A = 1,370 m)

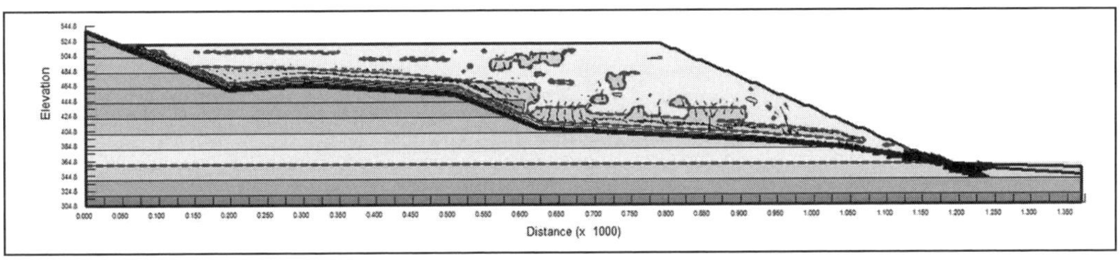

Figure 2. Conventional fill year 4 profile results (cross-section A-A)

within the fill. Water flow was depicted by black vectors. Larger vectors indicated faster water flow and respective direction.

For both the conventional and geomorphic fill, the majority of flow was both within the core drain and in the saturated zone above the core drain. Water within the drain caused a negative pore-pressure in the area above the drain and below the saturated zone. As the zone between the drain and the saturated zone dissipated water, negative pore pressure kept the saturated zone in tension and at a level above the drain. For the conventional fill, the negative pore pressure above the drain ranged from approximately –150 kPa to 740 kPa. For the geomorphic fill, the negative pore pressure above the drain ranged from approximately –170 kPa to –1,000 kPa. Over time, the level of saturation lowered towards the level of the drain.

Once results for the conventional fill and geomorphic fill were collected, they were compared on a yearly basis with regards to chosen analysis criteria. For each year of the model, the result of each criterion for the conventional fill was compared directly to the same criterion for the geomorphic fill using a "percent change." This percent change calculated the difference between the geomorphic value and conventional value as a percentage of the original conventional fill value. The percent change was calculated for each criterion at each time step by Equation 1.

Positive values indicated a point where the geomorphic fill had a property with higher magnitude than the conventional fill, while negative values indicated a point where the conventional fill had a property of higher magnitude than the geomorphic fill. This type of difference was

$$\text{Percent Change}\left(\%\right) = \frac{(\text{Geomorphic value} - \text{AOC value})}{\text{AOC value}} * 100 \qquad (1)$$

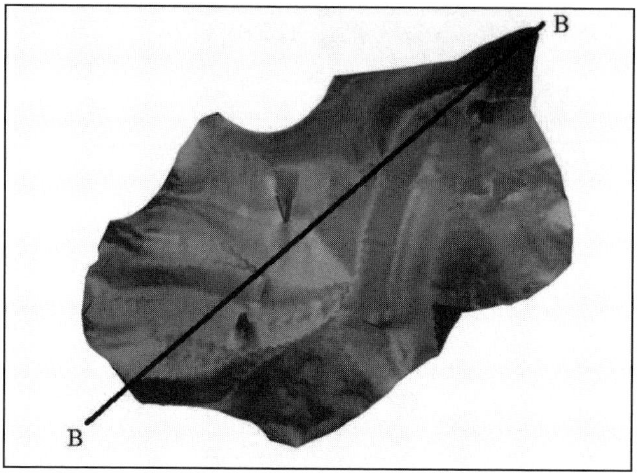

Figure 3. Geomorphic fill plan view with profile slice location (length of B-B = 1,370 m)

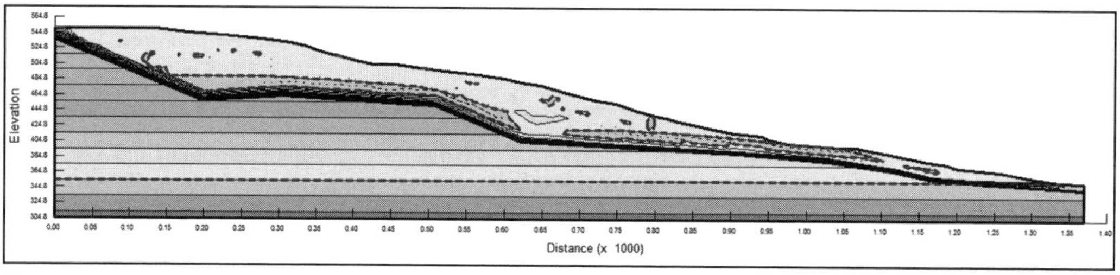

Figure 4. Geomorphic fill year 4 profile results (cross-section B-B)

chosen because a positive or negative difference could show an advantage of either the conventional or geomorphic fill, depending on what criterion was being compared. The advantage was in faster drainage through the fill as a result of higher seepage velocities and less water storage through lower storage volumes. Since the initial value was from the conventional model and the value of interest came from the geomorphic fill, calculating the change as a percentage of the conventional value best presented the magnitude of advantage for either fill. Also, a percent change approach provided a common scale by which to compare the magnitude of the advantage of different analysis criteria. A higher percent change corresponded to a higher advantage to one fill.

Higher values with respect to water velocity at the toe meant that the water was seeping through the fill at a faster rate. Lower values with respect to storage volume meant that less water was being stored within the fill. Faster water movement through the fill was advantageous because less water in the fill resulted in less chance of slope instability due to buildup of water. Lower storage volume was advantageous because it meant less water contact with internal fill materials. This is important when analyzing contaminant transport, because more contact with the material results in more contaminant desorption (Ziemkiewicz et al. 2011). Selenium is a naturally present trace metal that at high concentrations can be toxic to humans and ecosystems and can be released during mining. It is present in coal overburden. Selenium enters ground and surface water by being desorbed from coal overburden by water (Ziemkiewicz et al. 2011). An advantage with regards to seepage could translate into an advantage in contaminant transport. Therefore, a positive percent change with respect to water velocity was a time at which the geomorphic fill had an advantage over the conventional fill, and a negative difference related to an advantage for

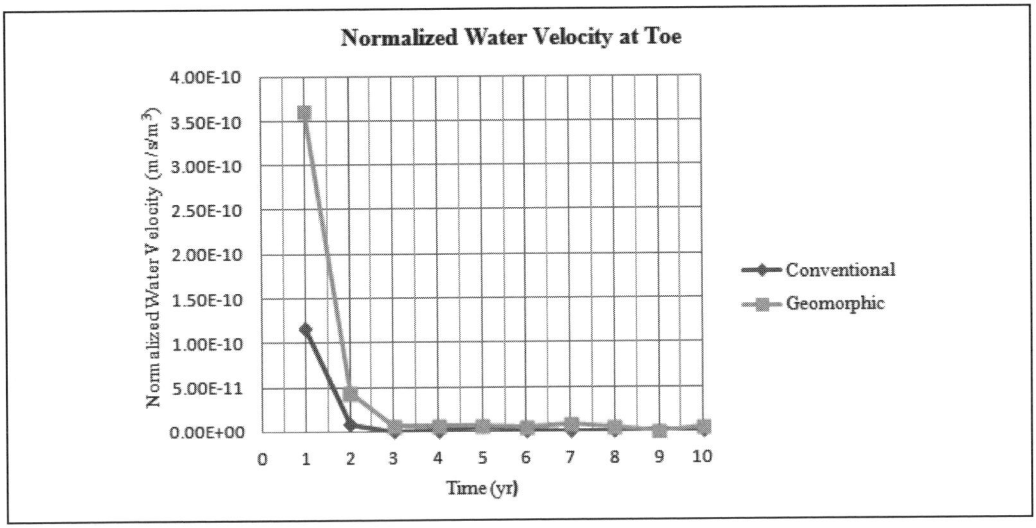

Figure 5. Normalized water velocity at toe over time

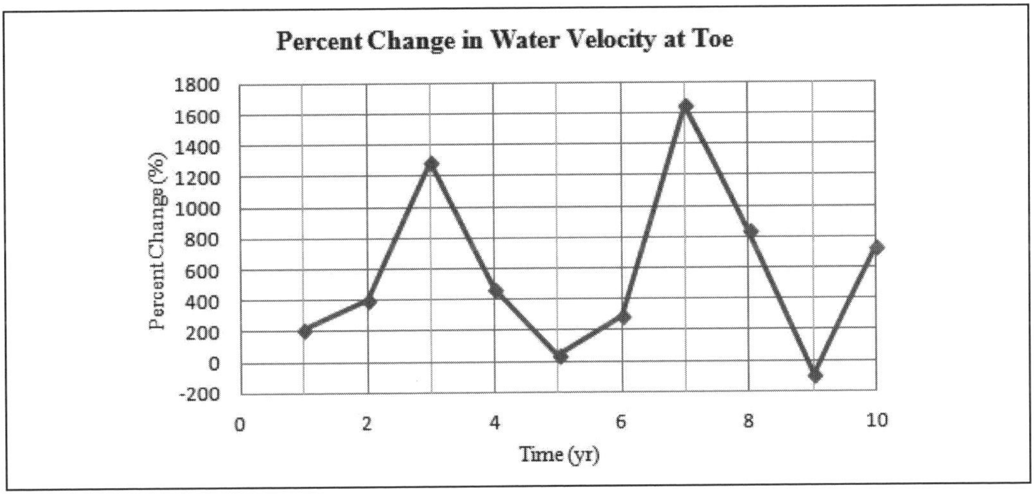

Figure 6. Percent change in water velocity at toe between fills

the conventional fill. For storage volume, a negative percent change corresponded to an advantage for the geomorphic fill, whereas a positive percent change corresponded to an advantage for the conventional fill.

Water Velocity at Toe

The variation of normalized water velocity at the toe is plotted in Figures 5 and 6.

The water velocity of the toe was lower for the conventional fill than for geomorphic fill for every year of the 10 year model except for year 9. Water velocity at the toe showed an advantage for the geomorphic fill for the entire duration other than a short time in year 9, where the conventional fill had a 98.2% advantage. The geomorphic fill had peaks in advantage of 1,290% at year 3 and 1,650% at year 7. The magnitude of the geomorphic fill's advantage was much higher than the advantage for the conventional fill.

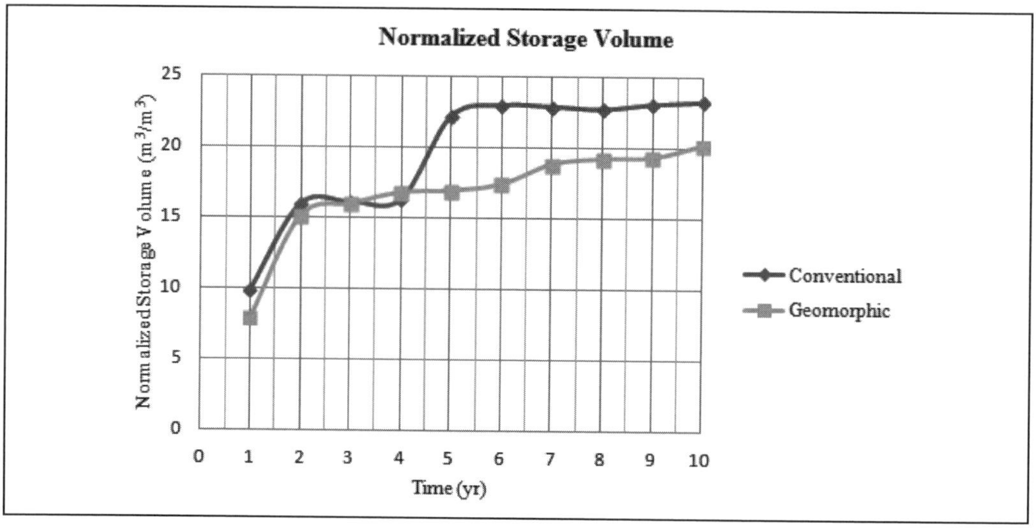

Figure 7. Normalized storage volume over time

Storage Volume

The variation of normalized storage volume is plotted in Figures 7 and 8.

The volume of water stored by the conventional fill was higher than the geomorphic fill for the entire duration except between years 3 and 4, where it showed a slight drop in storage percentage. The geomorphic fill showed a much steadier and smoother increase in storage percentage over time. The geomorphic fill had an advantage for the duration of the 10 years except between years 3 and 4, where the conventional fill showed an advantage of 3.20%. The peak in advantage for the geomorphic fill was 24.3% in year 6. The magnitude of the advantage for the conventional fill between years 3 and 4 was much lower than the advantage for the geomorphic fill for all the other years.

CONCLUSIONS

The following conclusions are the result of analyzing each fill individually, collecting results, and comparing those results:

- In the computer modeling study, for both the conventional and geomorphic fill, the majority of flow was both within the core drain and in the saturated zone above the core drain. Water within the drain caused a negative pore-pressure in the area above the drain and below the saturated zone. As the zone between the drain and the saturated zone dissipated water, negative pore pressure within the fill volume maintained the saturated zone in tension and at a level above the drain. For the conventional fill, the negative pore pressure above the drain ranged from approximately −150 kPa to 740 kPa. For the geomorphic fill, the negative pore pressure above the drain ranged from approximately −170 kPa to −1,000 kPa. Over time, the level of saturation lowered towards the level of the drain.
- Water velocity at the toe showed an advantage through higher discharge velocity for the geomorphic fill for the entire duration other than a short time in year 9, where the conventional fill had a 98.2% advantage. The geomorphic fill had peaks in advantage of 1,290% at year 3 and 1,650% at year 7.
- The geomorphic fill had an advantage with respect to storage volume for the duration of the 10 years except between

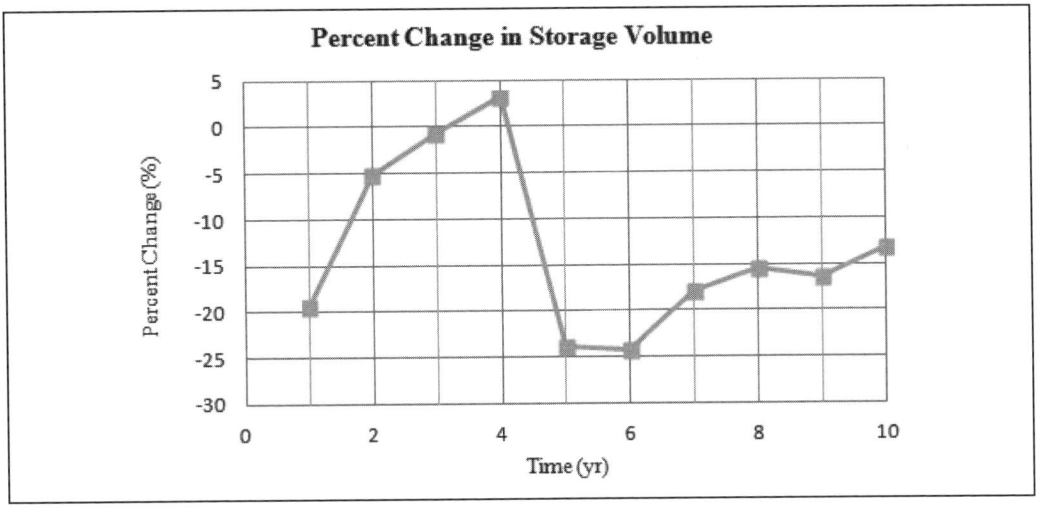

Figure 8. Percent change in storage volume between fills

years 3 and 4, where the conventional fill showed an advantage of 3.20%. The peak in advantage for the geomorphic fill was 24.3% in year 6.

- For both water velocity at the toe and storage volume, the magnitude of the advantage for the geomorphic fill was much higher than the magnitude of the advantage for the conventional fill. The advantage was a result of the geomorphic fill allowing less water to infiltrate, moving water through the fill at a faster rate, having less storage volume, and lower pore pressure buildup.

All of these conclusions pointed to supporting the idea of the geomorphic design of valley fills having advantages over conventional valley fill design. The geomorphic fill showed a distinct advantage in the behavior of groundwater seepage by moving water through the fill at a faster rate and storing less water. Faster water movement through the fill and less water storage within the fill were advantageous because less water in the fill resulted in less chance of slope instability due to buildup of water and less contact with the internal fill materials, which could translate into an advantage in contaminant transport. With the limited amount of work that has been done with geomorphic fills in the region of central Appalachia, this research has provided a sound initial analysis to compare with previously used design techniques. Further research must be done in order to make a fully informed decision as to whether or not geomorphic design would be feasible to implement in the reclamation of surface mines in central Appalachia.

ACKNOWLEDGMENTS

The Wheeling Jesuit University–National Technology Transfer Center is the project sponsor for all work in the project and is under contract through the US Center for Disease Control (CDC). The authors wish to thank Mr. J. Davitt McAteer, Program Director, for support in completing this work.

The authors thank Nathan DePriest for his contribution to this paper.

REFERENCES

Abdelghani, F.B., Simon, R., Aubertin, M., Molson, J. & Therrien, R. 2009. Use of the hydrogeosphere code to simulate water flow and contaminants transport through mining wastes disposed in a symmetric open pit within fractured rock.

Proc. Tailings and Mine Waste. Taylor and Francis Group. Vail, CO. 601–611.

Aubertin, M., Bussierre, B. & Chapuis, R. 1996. Hydraulic conductivity of homogenized tailings from hard rock mines. *Can. Geotech. J.*, 33: 470–482.

Meek, A. & O'dell, K. 2012. Selenium Treatment Arch-Eastern, Birch Mine. *Proc., West Virginia Mine Drainage Task Force Symposium*. Morgantown, WV.

Michael, P., Superfesky, M. and Uranoswki, L. 2009. Challenges of applying geomorphic and stream reclamation methodologies to mountaintop mining and excess spoil fill construction in steep slope topography (e.g., central Appalachia). *Proc., Geomorphic Reclamation Forum,* OSM, Bristol, VA.

National Oceanic and Atmospheric Administration. 2012. Precipitation Frequency Data Server (PFDS). *Hydrometeorological Design Studies Center*, http://hdsc.nws.noaa.gov (Feb. 10, 2012).

Sharma, K.D., Singh, H.P. & Pareek, P. 1983. Rainwater infiltration into a bare loamy sand. *Hydrological Sci. J.*, 28(3), 417–424.

West Virginia Department of Environmental Protection, Division of Mining and Reclamation. 1999. Permit Handbook: AOC and excess spoil disposal.

West Virginia Department of Environmental Protection. Permit ID 5500809. Pine Creek No. 2 Surface Mine (July 12, 2007).

Ziemkiewicz, P., O'Neal, M., and Lovett, R. 2011. Selenium leaching kinetics and in-situ control. *Mine Water & the Environment*, 30(2), 141–150.

Predicting Total Dissolved Solids Release from Overburden in Appalachian Coal Fields

Jessica Odenheimer
Plant and Soil Sciences, West Virginia University, Morgantown, West Virginia

Jeff Skousen
Plant and Soil Sciences, West Virginia University, Morgantown, West Virginia

Louis McDonald
Plant and Soil Sciences, West Virginia University, Morgantown, West Virginia

ABSTRACT

The Appalachian coal industry has been successful in developing technologies to identify, handle, treat and isolate potentially acid-forming overburden materials during coal mining to reduce or eliminate water quality problems. However, the techniques used to predict acid mine drainage potential have not been tested for total dissolved solids (TDS) prediction; therefore, new techniques may need to be devised. With a goal to develop methods to predict TDS release from overburden materials, our objective for this paper was to determine the effect of a dilute nitric acid (HNO_3) solution on weathering of overburden materials. Fifteen overburden samples (five strata from three locations) were collected from surface mines in West Virginia and three of these samples were used in this study. Ground samples were mixed with a dilute HNO_3 solution and shaken for two weeks to simulate weathering. Supernatants were extracted and analyzed every other day for pH, EC, and other selected ions. Acid-Base Accounting (ABA) parameters, which measure sulfur and neutralization potential (NP), were compared to results from our weathering tests. Results showed that predictions of TDS release by ABA were similar to the actual release of constituents in our weathering study. Sample WV 2 released the highest concentrations of elements, including high Fe, Al, Mn, and Ca, which are primary elements of concern in acid mine drainage. These results were consistent with this sample's moderate pyrite and NP contents. WV 3 and WV 4 produced lower levels of TDS release, which was consistent with lower pyrite and NP contents in these samples. An additional test using EDTA as a chelator will be performed.

INTRODUCTION

Coal is the primary source of fuel for electricity production in the United States, contributing more than 50% of the electricity generated nationwide. Due to land disturbance activity and environmental impacts that can occur during surface mining, the coal industry has been at the forefront of controversy surrounding human and environmental health and safety. Deforestation, leveling of mountains, acid mine drainage, erosion, subsidence, acid rain, and air pollution are just a few environmental impacts caused by the extraction and burning of coal.

Proper reclamation can lessen and in some cases eliminate the effect of many of these impacts through proper planning and identification of materials being disturbed. Overburden material, which covers the coal seam, is blasted apart and moved in order to reach the coal. It is this mining process (overburden blasting, removal and placement) that allows exposure of the rock materials to physical and chemical weathering, which can release soluble constituents into surrounding water sources. All of the blasted overburden cannot be put exactly back into place because the volume of rock increases due to blasting (Merricks et al. 2007). This excess material is commonly placed in valley fills or other constructed fills (Robins 1979; White and Barata 1995; Peng 2000; Messinger and Paybins 2003; Pond et al. 2008). Depending on the chemical composition of the rocks, weathering can promote acid mine drainage, alkaline mine drainage (Berhhardt et al. 2012), as well as release of total dissolved solids and heavy metals in runoff.

The application of innovative procedures during mining and reclamation has helped alleviate negative impacts of coal mining activities. The Acid-Base Accounting (ABA) procedure is an analytical method that determines the total amount of acidity and alkalinity that may be produced from the weathering of overburden materials (Skousen et al. 1997). ABA is the most common method for predicting post-mining water quality (Perry 1985). Unfortunately, no such method currently exists for the prediction of TDS release from overburden material. ABA does, however, provide information on how reactive the overburden is. Pyritic materials and carbonates are the two most common reactive components of overburden material, and if present in high quantities a large quantity of total dissolved solids is a probable outcome. TDS are defined as the total sum of cations and anions in solution, in units of mg/L. Common ions are SO_4^{2-}, HCO_3^-, Cl^-, Ca^{2+}, K^+, Na^+, and Mg^{2+}, but any dissolved ion that is present in solution will contribute to TDS. Since ion concentrations control electrical conductivity (EC) in water, EC can be used to estimate TDS with a conversion factor of 0.64. This conversion factor is generally used in waters dominated by chloride; however, this conversion factor underestimates TDS in sulfate-rich solutions (Evangelou 1998).

TDS and Influences on Human and Aquatic Health

Congress passed the Safe Drinking Water Act in 1974, which requires the United States Environmental Protection Agency (USEPA) to determine safe levels of chemicals in drinking water. TDS is considered a secondary pollutant with a maximum criteria limit of 500 ppm. TDS is not expected to harm human health at this level, but may be toxic to aquatic life and can damage water treatment equipment (USEPA 2012). Appalachian streams not influenced by coal mining commonly have TDS concentrations less than 75 mg/L and TDS is primarily comprised of bicarbonate, calcium, and chloride. In streams that are impacted by coal mining, TDS levels can range from 400 to 2,000 mg/L and TDS is primarily made up of sulfate, bicarbonate, calcium and magnesium (Timpano et al. 2010; Pond et al. 2008).

Aquatic organisms are commonly used in lab toxicity tests to evaluate the levels at which TDS becomes harmful. Common indicator species are benthic macroinvertebrates. Pond et al. (2008) found that mining activity impacted benthic macroinvertebrate communities with a strong correlation to a gradient of ionic strength. Pond et al. (2008) showed that EC in streams in unmined areas of the Appalachian region is generally less than 75 μS/cm, compared to up to 30 times unmined levels downstream of mountain top mining sites. Aquatic insects make up a large portion of benthic macroinvertebrates. Ephemeroptera (mayflies), Plecoptera (stoneflies), and Trichoptera (caddisflies) are orders of aquatic insects that are collectively termed EPT. These three orders in particular are commonly analyzed because of their sensitivity to pollution. The EPT Index is not a standard method of water quality assessment used within the Natural Resourses Conservation Service, but is useful in monitoring the species richness of a particular stream (USDA EPT Index). These organisms are the foundation for many food chains and are critical for the processing and cycling of nutrients (WVDEP 2012).

TDS is not the only water quality parameter that creates stress on organisms in aquatic ecosystems and therefore is not independent of other stressors in central Appalachian streams, which makes it difficult to monitor its direct effects (Timpano et al. 2010). A quick laboratory or field technique for estimating constituent release potentials from coal overburden material would be helpful for mine planning, placement and abatement procedures for coal operators, which also could help to minimize impacts to mined streams.

Our objective was to create a simple laboratory weathering technique to accurately predict the amount of TDS released from overburden materials. To do this, we compared a weathering/shaking procedure using dilute HNO_3 to predictions from ABA parameters.

MATERIALS AND METHODS

Field Overburden Collection

Fifteen overburden samples from each of three West Virginia surface coal mines were collected. These 15 samples represented a specific type of rock with an expected high, medium, or low TDS potential based on our experience with ABA or other methods of inference (e.g., local water quality data). From these 15 samples, three were chosen for this weathering study. In selecting our three overburdens, a preference was placed upon sampling rock units with variation in sulfur and alkalinity contents, and that would most likely be disposed of in valley fills or backfills instead of those strata that will be isolated as toxic strata and presumably hydrologically isolated.

Sample Processing

For each sample, all material was crushed and passed through a 1.25 cm (0.5-in.) sieve. All samples were further ground using a BICO pulverizing mill, until roughly half of the sample could pass through a 250 µm (0.0098-in.) sieve.

Laboratory Analysis of Mine Spoils

Three grams (3.00 g) of each of the three overburden samples were mixed with 2,000 mL of a dilute nitric acid (HNO_3) (0.0159 M) solution (prepared as a 1/1000 dilution of 15.9 M trace metal grade HNO_3). The dilute HNO_3 had a blank EC of 5.93 mS/cm and a pH of 2.0. The overburden samples were collected from the top of a 250 µm (0.0098-in.) sieve to control for particle size. Containers with 3-g samples in HNO_3 solution were replicated three times and placed on a reciprocating shaker for 14 days. After 24 hours, a 15-mL aliquot of solution was extracted and analyzed. Additional aliquots were taken and analyzed approximately every other day for two weeks. Electrical conductivity and pH were determined using standard methods. Ca, Mg, Na, K, P, Fe, Al, Mn, B, Sr, Pb, Cu, and Ni were determined by ICP-OES (Optima DV-2100, Perkin-Elmer, Norwalk, CT), but only data for Fe, Al, Mn, and Ca are displayed here.

RESULTS AND DISCUSSION

Under current thinking, rocks with high TDS potential may be shales or sandstones with high sulfur contents or high carbonate (neutralization potential, NP) contents or those containing both. Rock units with medium potential may be shales or other rocks with low to moderate sulfur and/or carbonate content. Rock units with low potential may be hard sandstones or oxidized/weathered sandstones or soil-like materials. In standard ABA, the relative amount of acid-producing rock is compared to the amount of acid-neutralizing rock to predict the water quality of a site (Skousen et al. 2002). We predicted, using the mining operation's overburden ABA analysis, that sample WV 2 would have the highest TDS release because of the moderate sulfur concentration (0.15%) and high NP (30 g/kg) (Table 1). WV 3 would be expected to release intermediate TDS because of the low sulfur concentration (0.001%), but high NP (60 g/kg). Sample WV 4 had moderate sulfur (0.10%) but very low NP (3 g/kg), which we expected to release less TDS than the other two samples.

Predicted TDS release using ABA parameters was proportional to final cumulative concentrations of elements released after two weeks of shaking in dilute HNO_3. Sample WV 2 had the highest concentrations released during weathering, WV 3 was intermediate and WV 4 produced the lowest concentration of constituents released (Figure 1). However, ABA only provides relative TDS release predictions (high, medium or low) without describing or identifying actual levels or concentrations, nor does it provide a rate of release. Therefore, additional analyses were needed through the use of weathering studies.

Table 1. Acid-base account parameters for three overburden samples from West Virginia surface coal mining sites

Sample	Color	Rock Type	pH	S %	MPA g/kg	NP g/kg	NET NP g/kg	Predicted TDS Release
WV 2	2.5Y 6/1	Black shale	6.0	0.15	30.0	30.0	0	High
WV 3	5Y 7/1	Gray shale	8.1	0.001	0.25	60.0	60.0	Medium
WV 4	2.5Y 3/1	Sandstone	5.0	0.10	4.0	3.0	−1.0	Low

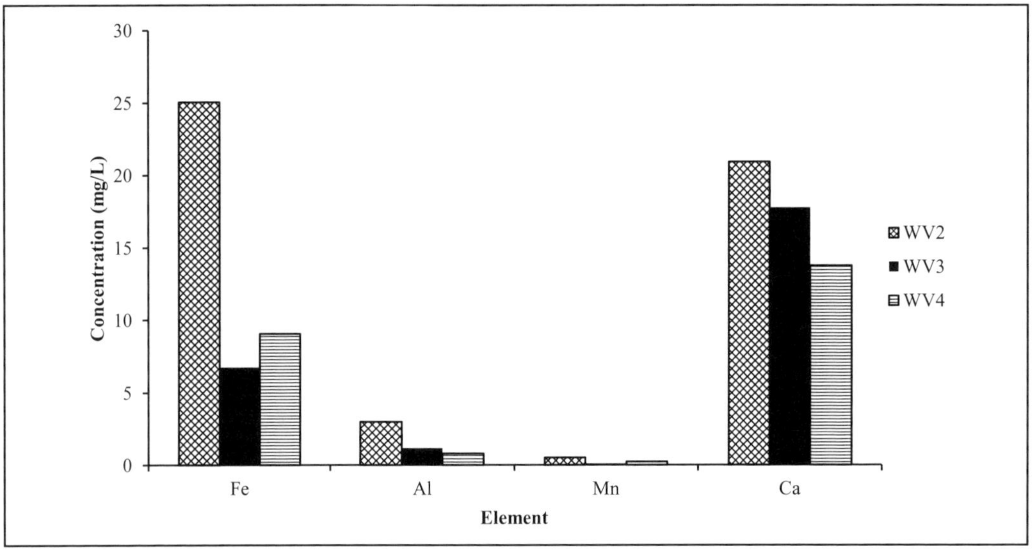

Figure 1. Concentrations of Fe, Al, Mn, and Ca from three West Virginia surface coal mine overburden samples at the end of the two-week shaking period with dilute HNO_3 extracting solution

Ions

The pH of the extracted solution stayed strongly acidic over the two weeks (Figure 2), which provided us with the ability to assume that once the ions went into solution, they would remain in solution.

The oxidation of pyrite (FeS_2) is a significant factor in the formation of acid mine drainage (AMD). AMD is produced when sulfide minerals associated with coal seams react with oxygen and water to form low pH, sulfate rich, and high iron solutions (Mack et al. 2010). As expected, the largest Fe concentrations in solutions from our weathering studies were observed after 24 hours (Figure 3) and followed the expected pattern of rapid pyrite oxidation with samples containing sulfur (Table 1). WV 2, with the highest sulfur concentrations, had much higher Fe concentrations during the first shaking period compared to the other two samples.

The largest Al concentrations were observed after shaking for 192 hours for all three samples. Al concentrations dropped rapidly after 192 hours and increased again after 240 hours (Figure 4). Low pH conditions accelerate the release of aluminum and other toxic elements from minerals (Bigham et al. 1996; Kittrick et al. 1982) and the pH of our solutions stayed at about 2.0 during the course of the two-week period. These time periods corresponded to the length of time required for the dilute HNO_3 solution to promote the breakdown of aluminum- containing minerals in the samples and no alkalinity was present to neutralize the acidity (Seidel et al. 1998). Additional time was required for more Al to be dissolved during the 240- to 288-hr time period.

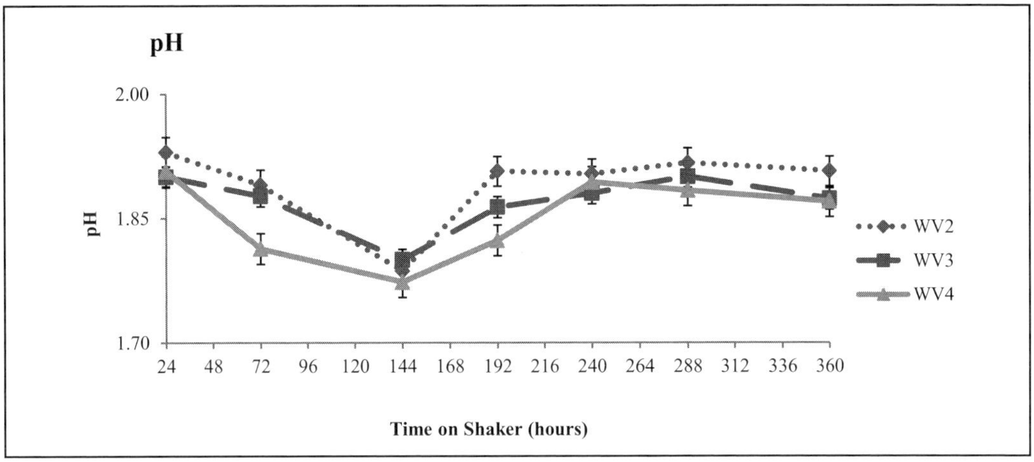

Figure 2. pH of solution over two weeks for the three overburden samples

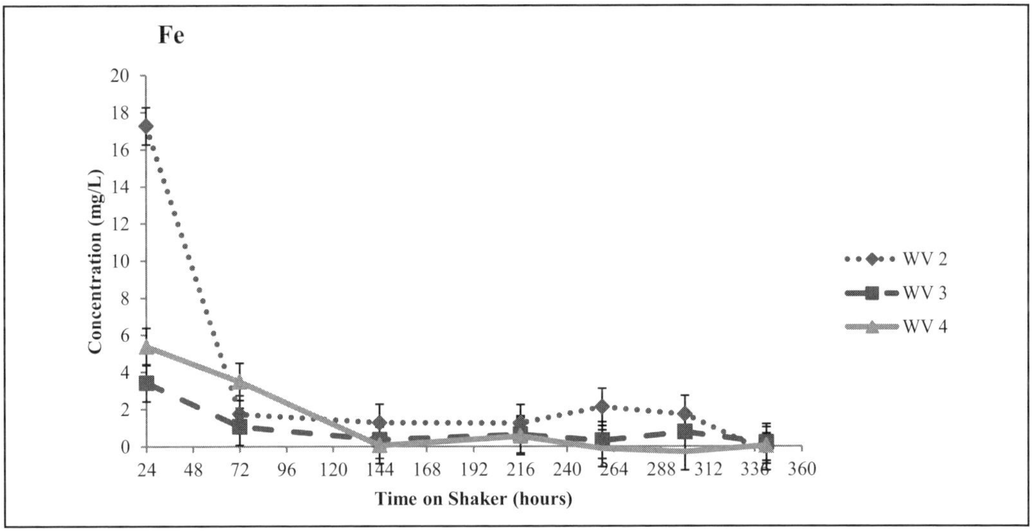

Figure 3. Iron concentrations released over time from three overburden samples with a dilute HNO_3 extracting solution from a West Virginia surface coal mine

Very little Mn was released from these samples with dilute HNO_3 (Figure 1), but sample WV 2 showed higher amounts of Mn release than the other two samples, even though the total amount was small (Figure 5). Manganese has been found to be directly related to carbonate carbon and NP (Morrison et al. 1990). The trends in Mn release are consistent with the ABA results from these samples (Table 1).

Most alkaline material is in the form of calcareous minerals containing primarily calcium carbonate ($CaCO_3$) (Johnson 2005). Limestone is composed of $CaCO_3$ and is commonly found in the Appalachian region. From the NP listed in Table 1, we predicted that WV 2 and WV 3 could potentially release more Ca compared to WV 4, which had little NP. Both WV 2 and WV 3 released high levels of Ca, with higher amounts of Ca from WV 2, which we presumed to be because of rapid weathering of $CaCO_3$ due to the acidity generated by pyrite oxidation and the acidity from the dilute HNO_3 solution.

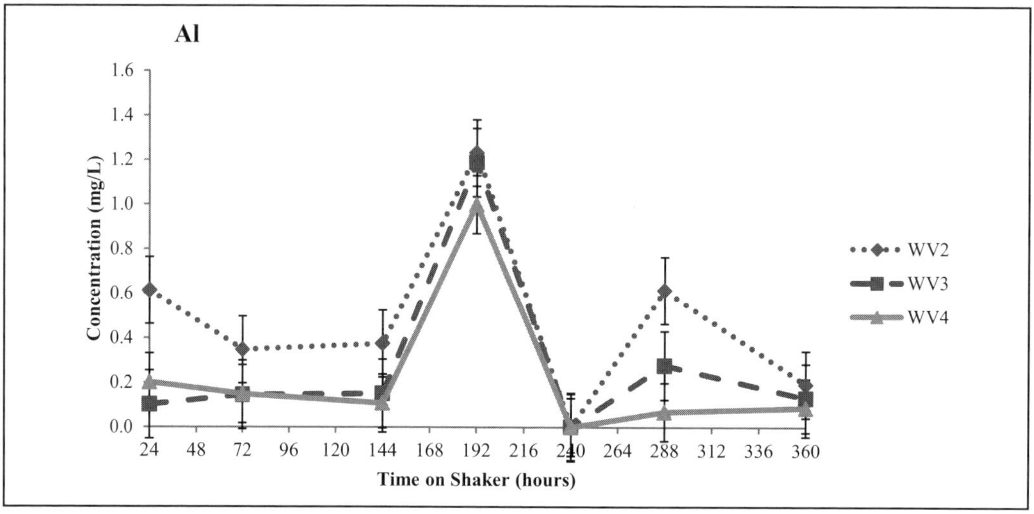

Figure 4. Aluminum concentrations released over time from three overburden samples with a dilute HNO_3 extracting solution from a West Virginia surface coal mine

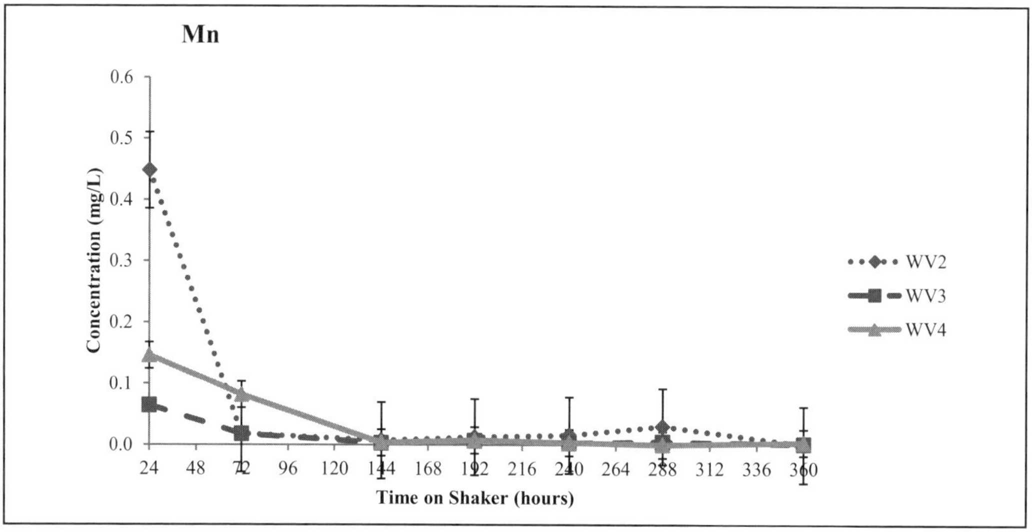

Figure 5. Manganese concentrations released over time from three overburden samples with a dilute HNO_3 extracting solution from a West Virginia surface coal mine

Although there was no pyrite in sample WV 3, high levels of Ca were released due to the low pH of the dilute HNO_3 solution. Most of the Ca was released after shaking for 144 hours (Figure 6). Ca must be mostly dissolved prior to the onset of any significant leaching of Al (Seidel et al. 1998). Al began dissolving after 144 hours, which is the same time in which Ca dissolution was complete (Figure 4).

EC and TDS

Soluble salt content as measured by EC from overburden material is important for the planning and placement of overburden material because of its relationship to TDS. Accurate determination of TDS is a slow and tedious procedure, but TDS can be estimated by a common function provided by the U.S. Salinity Laboratory Staff (1954) using

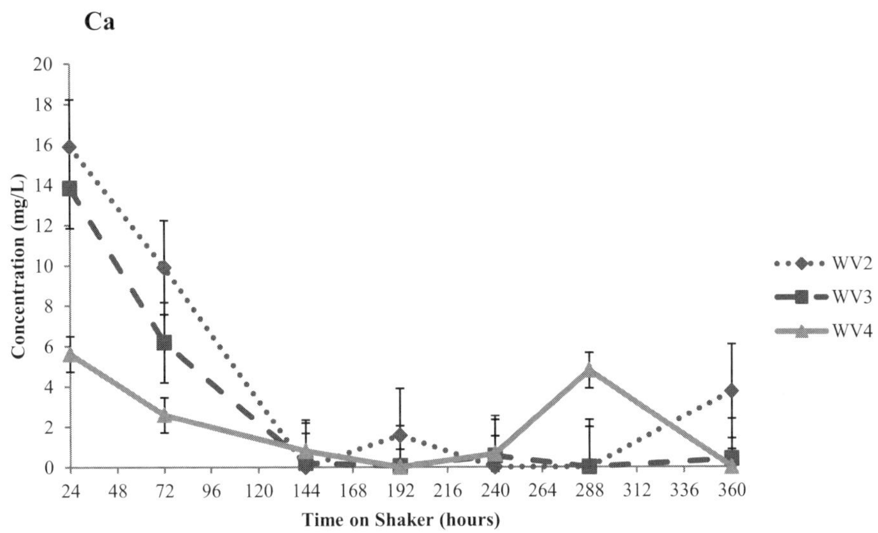

Figure 6. Calcium concentrations released over time from three overburden samples with a dilute HNO_3 extracting solution from a West Virginia surface coal mine

a conversion factor between TDS and EC. This conversion factor is 0.64 if the EC is less than 1 mmho/cm (Evangelou 1998). Equation 2 is utilized when EC is greater than 1 mmho/cm (Evangelou 1998). These conversion factors are generally used in waters dominated by chloride; however this may underestimate TDS in sulfate-rich solutions (Evangelou 1998). The highest EC we encountered was about 0.85 mmho/cm (850 μS/cm), implying that the conversion factor in Equation 1 is suitable for these samples.

$$TDS = 0.64 * EC\ (\mu S/cm) \quad (1)$$

$$TDS = 0.64\ [EC\ (\mu S/cm)]^{1.087} \quad (2)$$

Figure 7 displays the trend in net EC (the measured EC of the solutions minus the initial blank EC) from our three overburden samples over our two- week weathering procedure. All three overburden samples showed a rapid release of soluble salts and high EC (800 to 875 μS/cm) after 24 hours, then a decrease to less than 100 μS/cm after shaking for 144 hours. Small increases were found between 240 and 288 hours. Since all three overburden samples showed similar trends, the average net EC treadline was plotted in Figure 8 against the treadline for predicted TDS using Equation 1. The average net EC and estimated TDS fell below the TDS maximum criteria limit (MCL) of 500 mg/L within the first 24 hours.

CONCLUSIONS

Depending on the physical and chemical composition of overburden from surface coal mining sites, weathering can promote the release of TDS into the environment. Some studies have indicated that high TDS has detrimental effects on aquatic life. Therefore, TDS release from disturbed overburden must be better understood, and techniques to predict TDS levels after disturbance need to be developed.

Three samples from strata at West Virginia coal mining sites were collected and compared to ABA parameters. Based on the amount of sulfur and neutralization potential from the ABA, we predicted that sample WV 2 should generate the highest TDS release of the three, while WV 4 should generate the least TDS. The predictions of TDS release from ABA were proportional to the total amounts of elements released during a two-week weathering study with a dilute HNO_3 extracting solution. WV 2, having moderate

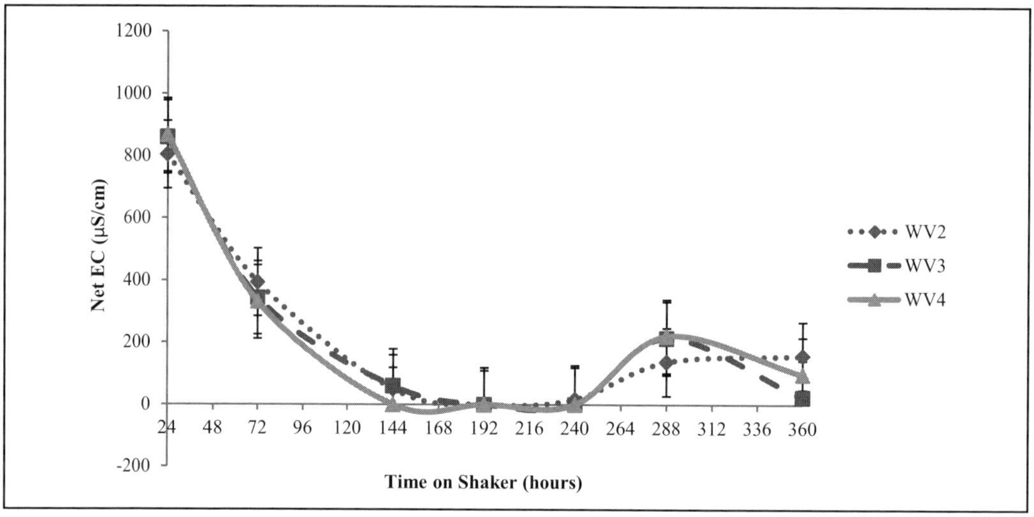

Figure 7. Net electrical conductivity of three overburden samples after shaking and weathering in a dilute HNO_3 extracting solution from a West Virginia surface coal mine over time

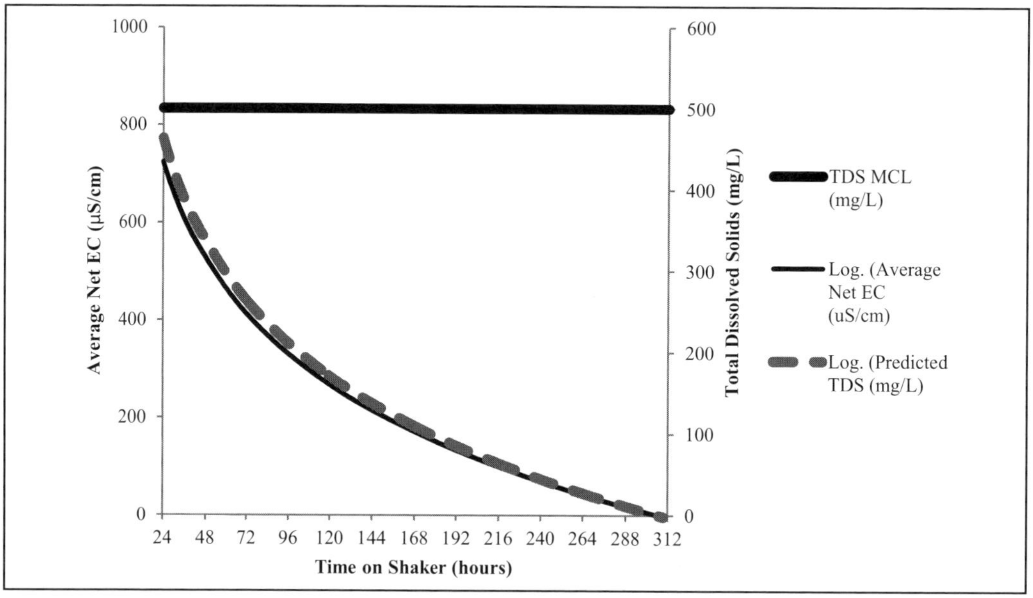

Figure 8. A comparison between the average net EC and estimated TDS from three overburden samples from a West Virginia surface coal mine over time

levels of sulfur and NP, displayed the highest concentrations of elements released during shaking and extraction, including high Fe, Al, and Mn, and Ca. WV 3 and WV 4 released lower levels of TDS due to lower amounts of sulfur and NP. EC was converted to a predicted TDS concentration using equations in the literature. We found that within the first 24 hours, TDS release from these three rock samples decreased to levels below the TDS MCL of 500 mg/L. Release rates from our weathering study show that TDS concentrations decline rapidly during weathering.

The next step of our research will be to repeat this shaking experiment using EDTA as the extracting solution on a greater number and variety of overburden samples. Although the solid: solution used in this experiment is very different from field leaching, we are hoping to develop an index that can be correlated to column leaching and field leaching experiments for future prediction of TDS release from overburden material in Appalachian coal fields.

ACKOWLEDGMENT

This study was sponsored by the Appalachian Research Initiative for Environmental Science (ARIES). The views, opinions and recommendations expressed herein are solely those of the authors and do not imply any endorsement by ARIES employees, other ARIES-affiliated researchers or industrial members. Information about ARIES can be found at http://www.energy.vt.edu/ARIES.

REFERENCES

Bernhardt, E., D. Lutz, R. King, J. Fay, C. Carter, A. Helton, D. Campagna, and J. Amos. 2012. How many mountains can we mine? Assessing the regional degradation of central Appalachian rivers by surface coal mining. Environ. Sci. Technol. 46(15):8815–8122.

Bigham, J.M., U. Shwertmann, S.J. Traina, R.L. Winland, and M. Wolf. 1996. Schwertmannite and the chemical modeling of iron in acid sulfate waters. Geochim. Cosmochim. Acta 12:2111–2121.

Evangelou, V.P. 1998. Environmental Soil and Water Chemistry. New York: John Wiley & Sons, Inc., 1998.

Johnson, D., and K. Hallberg. 2005. Acid mine drainage remediation options: a review. Science of the Total Environment 338:3–14.

Kittrick, J., D. Fanning, and L. Hossner. 1982. Acid Sulfate Weathering. SSSA Spec. Publ. 10. SSSA, Madison, WI.

Mack, B., L.M. McDonald, and J. Skousen. 2010. Acidity decay of above-drainage underground mines in West Virginia. *J. Environ. Qual.* 39: 1043–1050.

Merricks T.C., D.S. Cherry, C.E. Zipper, R.J. Currie, and T.W. Valenti. 2007. Coal-mine hollow fill and settling pond influences on headwater streams in southern West Virginia, USA. *Environ. Monit. Assess.* 129:359–378.

Messinger T., and K.S. Paybins. 2003. Relations between precipitation and daily and monthly mean flows in gaged, unmined and valley-filled watersheds, Ballard Fork, West Virginia, 1999–2001. Water Resources Investigation Report 03-4113. Charleston, WV: U.S. Geological Survey; 2003. Available from: http://pubs.usgs.gov/wri/wri034113/pdfs/wrir034113.pdf.

Morrison, J.L., S.D. Atkinson, and B.E. Scheetz. 1990. Delineation of potential manganese sources in the coal overburdens of western Pennsylvania. In: Proceedings, American Society for Mining and Reclamation Conference and Exhibition, Charleston, WV. April 23–26, 1990.

Peng, S.S. 2000. Mountaintop removal controversy slows West Virginia coal mining. Min. Engineering 52:53–58.

Perry, E. 1985. Overburden analysis: An evaluation of methods. p. 369–375. In *Proc. Symposium on Surface Mining, Hydrology, Sedimentology, and Reclamation*. 9–13 Dec. 1985. Univ. of Kentucky, Lexington.

Pond, G., M. Passmore, F. Borsuk, L. Reynolds, and C. Rose. 2008. Downstream effects of mountaintop coal mining: comparing biological conditions using family- and genus level macroinvertebrate bioassessment tools. *J. North Am. Benthol. Soc.* 27:717–737.

Robins J.D. 1979. Constructing spoil storage fills in the steep terrain Appalachian coalfields. p. 463–70. In: Brawner, C.O. and I. Dorling, eds. Stability in coal mining: proceedings of the first international symposium on stability in coal mining, Vancouver, British Columbia, Canada. San Francisco, CA: Miller Freeman Publications.

Seidel, A., A. Sluszny, G. Shelef, and Y. Zimmels. 1998. Self inhibition of aluminum leaching from coal fly ash by sulfuric acid. Chemical Eng. J. 72: 195–207.

Skousen, J., Renton, J., Brown, H., Evans, P., Leavitt, B., Brady, K., Cohen, L., Ziemkiewicz. 1997. Neutralization potential of overburden samples containing siderite. J. Environ. Qual. 26: 673–681.

Skousen, J., J. Simmons, and P. Ziemkiewicz. 2002. Acid-Base Accounting to Predict Post- Mining Drainage Quality on Surface Mines. J. Environ. Qual. 31:2034–2044.

Timpano, A., S. Schoenholtz, C. Zipper, and D. Soucek. 2010. Isolating effects of total dissolved solids on aquatic life in central Appalachian coalfield streams. p. 1284–1302. In: Proceedings, American Society for Mining and Reclamation, Pittsburgh, PA, June 5–11, 2010.

Tissue, B. 2009. Basics of Analytical Chemistry and Chemical Equilibria. Fall 2009. Blacksburg, VA: The Tissue Free Press.

White, M.A., and F.A. Barata. 1995. State-of-the-art in mountaintop removal and contour mining. Mining Engineering 47:1111–1114.

WVDEP, 2012. West Virginia Department of Environmental Protection. Benthic Macroinvertebrate and Fish Information. 2012 State of West Virginia. Available from: http://www.dep.wv.gov/WWE/watershed/bio_fish/Pages/Bio_Fish.aspx.

USEPA. 2012. Water Monitoring and Assessment: Total Solids. Available from: http://water.epa.gov/type/rsl/monitoring/vms58.cfm>.

USDA EPT Index. NRCS Watershed Science Institute, Watershed Condition Series. Technical Note 3. The EPT Index. Available from: ftp://ftp.wcc.nrcs.usda.gov/wntsc/strmRest/wshedCondition/EPTIndex.pdf.

U.S. Salinity Laboratory Staff. 1954. Diagnosis and improvement of saline and alkali soils. U.S. Dep. Agric. Handb. 60:160.

Natural Attenuation of TDS in Mountaintop Mines

Ben Mack
West Virginia Water Research Institute, Morgantown, West Virginia

Jason Fillhart
West Virginia Water Research Institute, Morgantown, West Virginia

Brady Gutta
West Virginia Water Research Institute, Morgantown, West Virginia

Paul Ziemkiewicz
West Virginia Water Research Institute, Morgantown, West Virginia

ABSTRACT

A large amount of research has been performed on Mountain Top Mining (MTM) and its effects. However, specific ion concentrations of valley fill discharges and how they are related to overburden geology and valley fill age has not been well-studied. This study researched the effects of valley fill age and flow regime.

Water samples, flow, and field data were collected at two southern West Virginia mine complexes during a low flow period and three mine complexes during a high flow event. Samples were collected from the toes of 20 valley fills during the low flow sampling event and 38 fills during the high flow sampling event. Data were sorted by valley fill age and sampling event. Regression analysis was used to determine correlation between valley fill age and parameter concentrations.

Sulfate, bicarbonate, calcium, and magnesium made up the largest percentage of TDS. Sulfate, bicarbonate, and magnesium increased slightly over time, while calcium changed very little over time. Selenium concentrations showed a decrease over time, although the R^2 value was very low. High concentrations between 0–5 years and at 15 years after fill construction greatly influenced the trend in selenium concentrations over time.

Future sampling likely could improve the correlation between valley fill age and parameter concentrations by increasing the number of sampled fills, gathering additional data from existing datasets, or widening the geographical area to be more inclusive of regional geology. Increasing the size of the study area also may show that the data trends found from these fills are vastly different than other fills.

INTRODUCTION

Mountain Top Mining (MTM) began in West Virginia in the late 1960s as a way to economically extract multiple thin seams of coal from difficult terrain. Associated with this practice are constructed valley fills where excess spoil materials are deposited. These fills are typically placed in the heads of hollows where headwater streams are formed. Although a great deal of research has been performed on various aspects of MTM and its effects, there is little documented research to indicate the specific ion compositions of valley fill toe discharges, how they relate to the mine's geology, or how those ionic compositions change over time. Ions found in the largest concentrations (Ca, Na, Mg, SO_4, HCO_3 and Cl) constitute the overwhelming majority of Total Dissolved Solids (TDS) and Electrical Conductivity (EC) in this region of Appalachia. Compliance with selenium

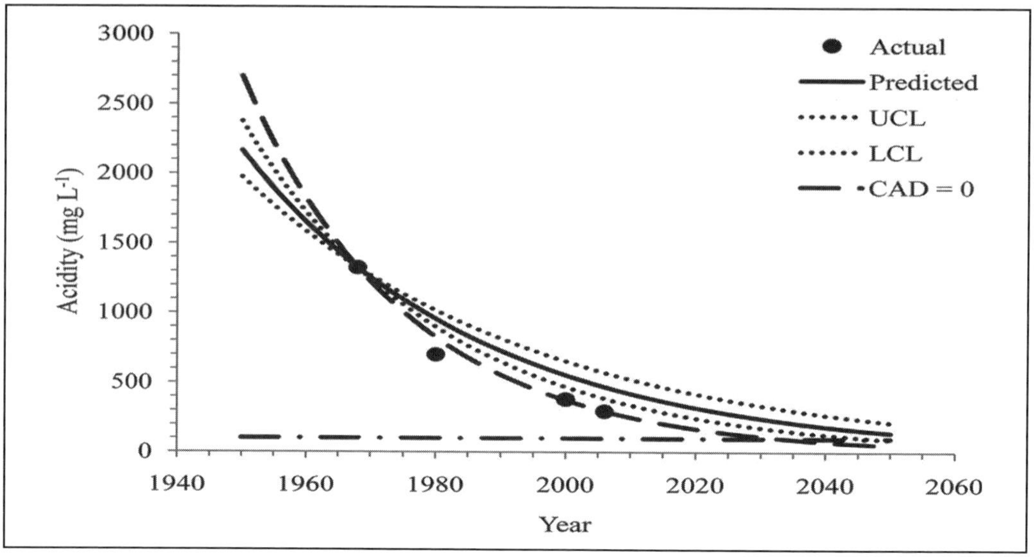

Figure 1. Acidity (mg/L) for Upper Freeport sites extrapolated to the year 2050. Horizontal dashed line represents the cutoff acidity to begin passive treatment (100 mg/L) CAD, cumulative average difference; LCL, lower confidence limit; UCL, upper confidence limit.

criteria has also become a potential regulatory issue in the southern coalfields. Characterization of these chemical profiles is particularly important in light of West Virginia's new narrative water quality standard for TDS. It is our hypothesis that ion composition and concentration are functions of overburden geology and valley fill age. We expect that ion leaching will be relatively rapid, given the high permeability of these valley fills.

LITERATURE REVIEW

Previous research has estimated the longevity of specific mine pollutants. Investigations of surface mine discharges have shown that acidic water may be released at a consistent level for 10 to 20 years (Meek 1996), while Wood et al. (1999) determined that surface mines attenuated much more quickly than underground mines. However, underground mine discharges are more difficult to predict. Younger (1997) estimated that acidic drainage may continue for 10 to 100 years, dependent upon hydrologic factors, pyrite reaction rate, and amount of pyrite. Other researchers have found that the worst pollution occurs within the first 40 years after mine closure (Wood et al. 1999) and that several decades must elapse before water quality will improve significantly (Jones et al. 1994).

Research by West Virginia University has studied the longevity of mine water pollutants from underground mines in the Pittsburgh and Upper Freeport coal seams. Mack et al. (2010) found that acidity decreased, on average, 80% over 40 years (Figure 1). Demchak (2005) studied 20 underground mines and found that acidity, metals, and sulfate decreased 50–80% over 32 years. A partially flooded Pittsburgh seam mine was also studied by Demchak (2005). She found that sulfate concentrations showed a very rapid rate of exponential decline (Figure 2).

Additional research by the West Virginia Water Research Institute (WVWRI) has estimated the rate of decline for other mine drainage parameters, such as selenium. Figure 3 shows changes in selenium concentrations as a function of time after mining began. Selenium concentrations began to decline approximately eight years after mining began and reached a concentration of less than 1 ug/L after 25 years.

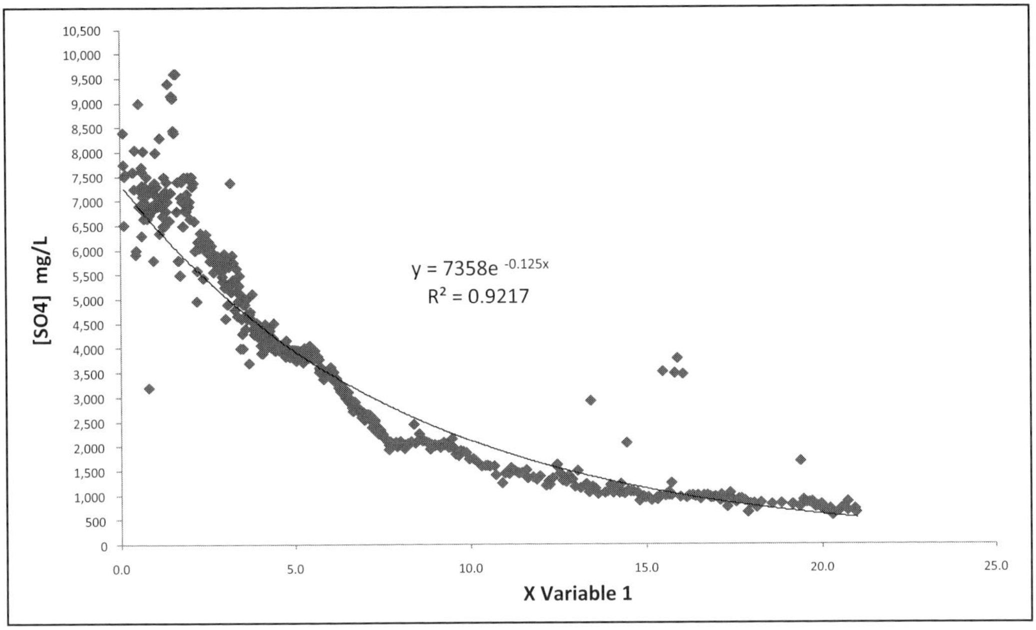

Figure 2. Sulfate decline during the 21 years after flooding of a mine in the Pittsburgh coal seam

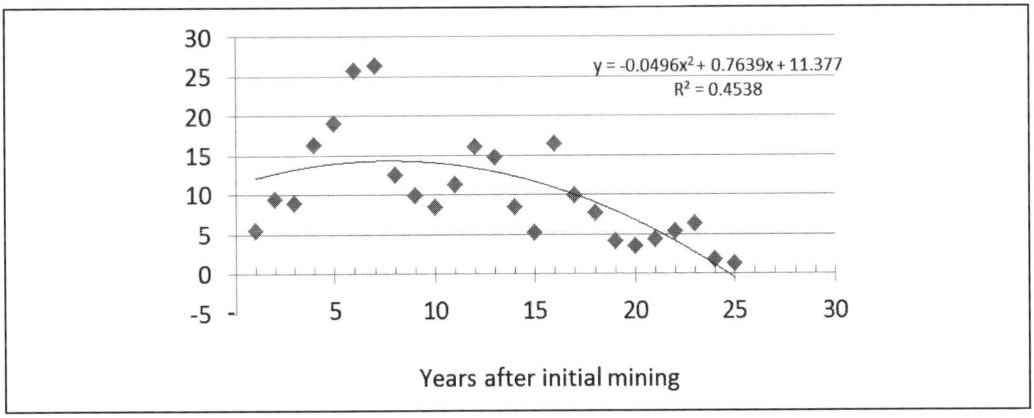

Figure 3. Average Se concentrations (µg/L) as a function of years since initial mining

METHODS

The WVWRI conducted field sampling events at two southern West Virginia mine complexes during a low flow period and three mine complexes during a high flow event. These large complexes have been in existence for a long time and have numerous valley fills of different sizes and ages. Effluent samples were collected from the toes of 20 valley fills during the low flow sampling event and 38 fills during the high flow sampling event (Figure 4).

Individual fills were selected so that there were at least five within each age class:

- Age class 0–5 years
- Age class 6–10 years
- Age class 11–15 years
- Age class 16–25 years

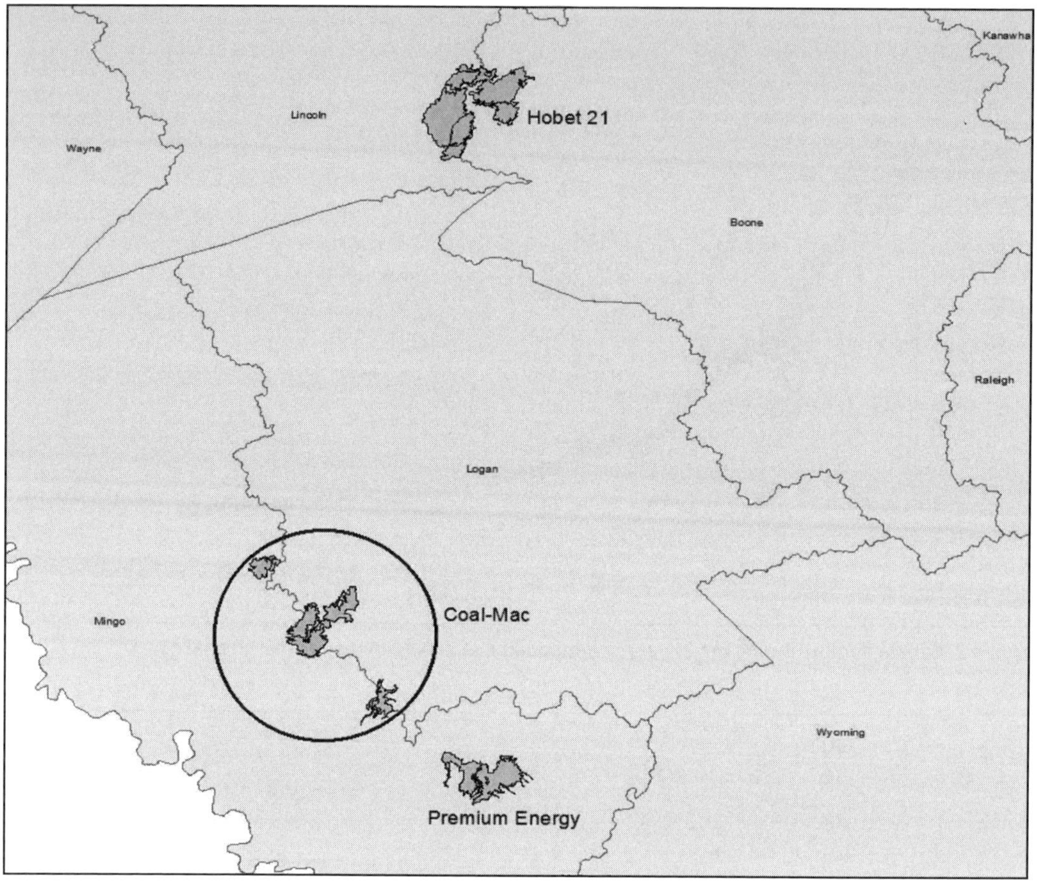

Figure 4. Mine sites that were sampled during two sampling events. Samples were collected from Hobet 21, Coal-Mac, and Premium Energy. The three circled polygons are all part of the Coal-Mac mine complex.

Chemical sampling parameters and field parameters included:

- Flow, pH, Electrical Conductivity (EC), Total Dissolved Solids (TDS), Temperature, Dissolved Oxygen (DO)
- Aluminum (Al), Iron (Fe), Manganese (Mn)
- Calcium (Ca), Magnesium (Mg), Sodium (Na), Bicarbonate Alkalinity (HCO_3), Sulfate (SO_4), Chloride (Cl)
- Selenium (Se)

Field and lab data were collected and analyzed using protocols approved by the US EPA. Flow values were determined using a Marsh-McBirney Flo-Mate 2000 (Marsh-McBirney, Loveland, CO) and the cross-sectional area of the discharge. All other field parameters were determined by using an YSI 556 multiprobe. All water samples were sent to REIC Laboratory in Beaver, West Virginia for analysis. Al, Fe, Mn, Ca, Mg, and Na were determined using EPA method 200.7. Cl and SO_4 were found using method number E300.0 and HCO_3 was determined using SM 2320 B. Both total and dissolved Se were ascertained using method SM 3114 B. Data were sorted by valley fill age and sampling event, and data trends were determined using regression analysis.

RESULTS

Data were categorized into field and lab data for final analysis. The highest R^2 values were used to determine the amount of correlation between parameter concentrations and valley fill age for field parameters (Table 1) and lab parameters (Table 2). The highest R^2 values were used in all graphical analyses to ensure that the greatest correlation between time and parameter concentration was determined.

Field Data

All valley fills were sampled twice in an attempt to characterize the current effluent conditions from these fills. Data for EC, TDS, and pH were separated by sample events (Figures 5, 6, and 7). The data were then analyzed using regression to determine if trends were present.

TDS and EC data followed the same general trend. Regardless of sampling event, concentrations of both parameters increased slightly as the valley fills aged. TDS and EC increased the fastest between 0–5 yrs, possibly due to initial saturation of the fill after construction. However, the trend line begins to level off 5+ yrs after construction, especially for Sample 1 (Figures 5 and 6).

TDS and EC concentrations increased more quickly and had a higher R^2 value during sampling event #1 (Figures 5 and 6). Parameter concentrations were not impacted by sample time when the valley fills reached roughly ten years old. Overall, EC and TDS concentrations changed very little between sampling events.

pH values over time were also not affected by changes in sampling events. Some pH variability was observed between sampling events (Figure 7). Sampling event had no effect on pH when the fill was approximately two years old. In general, pH values changed very little as a function of valley fill age.

Laboratory Data

Nine ions (HCO_3, SO_4, Cl, Fe, Mn, Ca, Mg, Na, and Se) were used as indicators of changes in TDS over time. Each ion was graphed as a

Table 1. R^2 values and trend type for field parameters

Parameter	Sample 1 vs. VF Age	Sample 2 vs. VF Age
EC	Power 0.230	Exponential 0.092
TDS	Polynomial 0.242	Exponential 0.102
pH	Polynomial 0.011	Polynomial 0.020

Table 2. R^2 values and trend type for lab parameters

Parameter	All Fills vs. VF Age
SO_4	Exponential 0.201
HCO_3	Exponential 0.070
Ca	Polynomial 0.005
Mg	Power 0.301
Se	Exponential 0.136

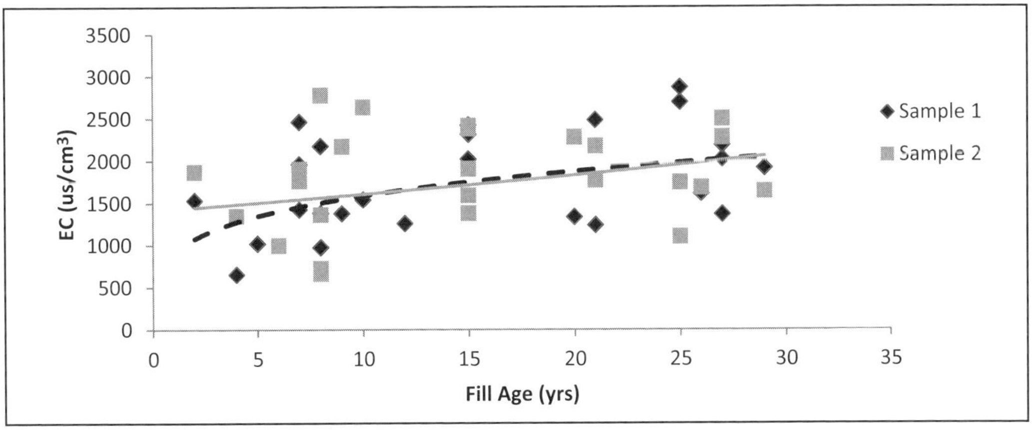

Figure 5. Changes in EC as a function of valley fill age. The darker line is the trendline for Sample 1.

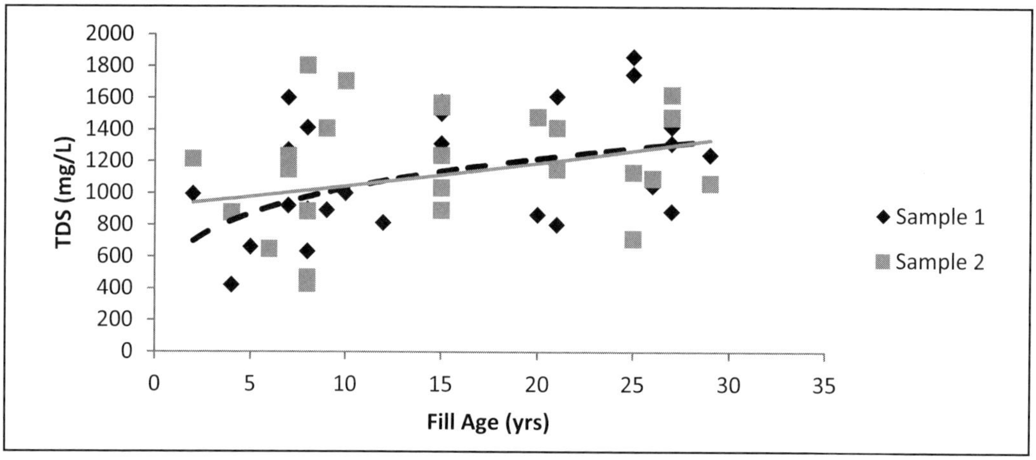

Figure 6. Changes in TDS as a function of valley fill age. The darker line is the trendline for Sample 1.

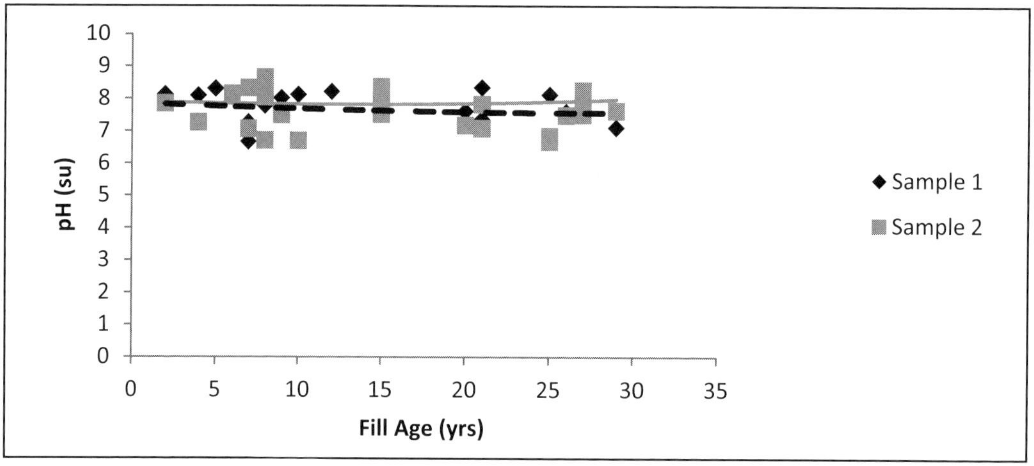

Figure 7. Changes in pH as a function of valley fill age. The darker line is the trendline for Sample 1.

function of valley fill age. The data were analyzed using regression analysis. Figure 8 details percentage of contribution of each ion to TDS.

Sulfate concentrations showed an increase over time (Figure 9). Sulfate ions constitute 54% of the TDS chemistry of these fills (Figure 8), which means that sulfate concentrations will typically be higher in the first several years after fill completion. A lack of data for fills age 0–5 years could explain the increase in sulfate concentrations. Sulfate concentrations could be greater in the first five years after valley fill completion, which could change the shape of the curve from an increasing trend to a decreasing trend. Sulfate is highly soluble and the majority of it should be flushed from the valley fill within the first five years after construction is complete.

Similar to sulfate, HCO_3 (alkalinity) concentrations increased over time (Figure 10). However, HCO_3 increased more slowly than sulfate. The low R^2 value indicated a weak increasing trend (Table 2). Production of alkalinity increased past the age of the oldest fill. However, a 29 year-old valley fill with an HCO_3 concentration of 1,060 mg/L significantly affected the rate of HCO_3 increase. We expect a decline in HCO_3 production as alkaline material is leached out of the fills.

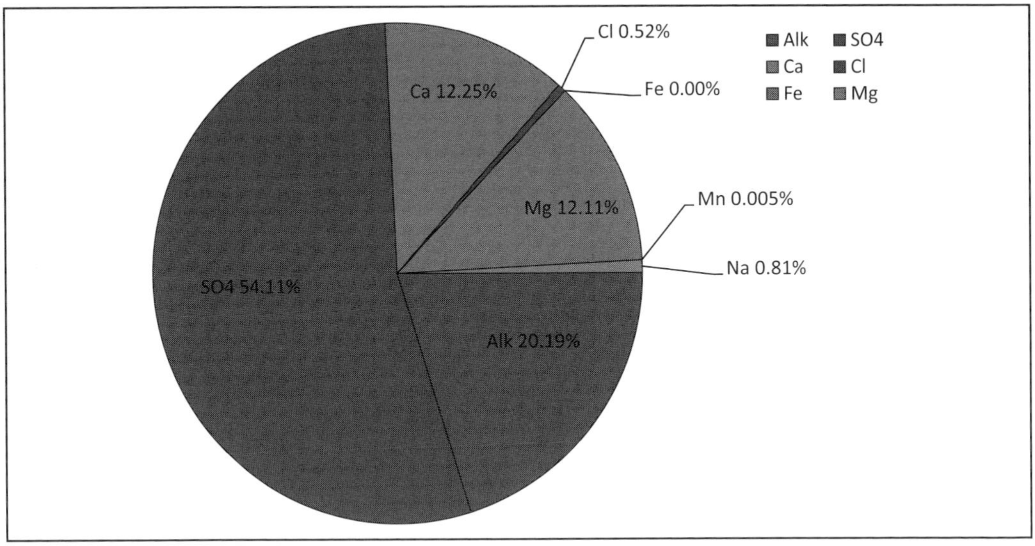

Figure 8. Major ions contributing to TDS

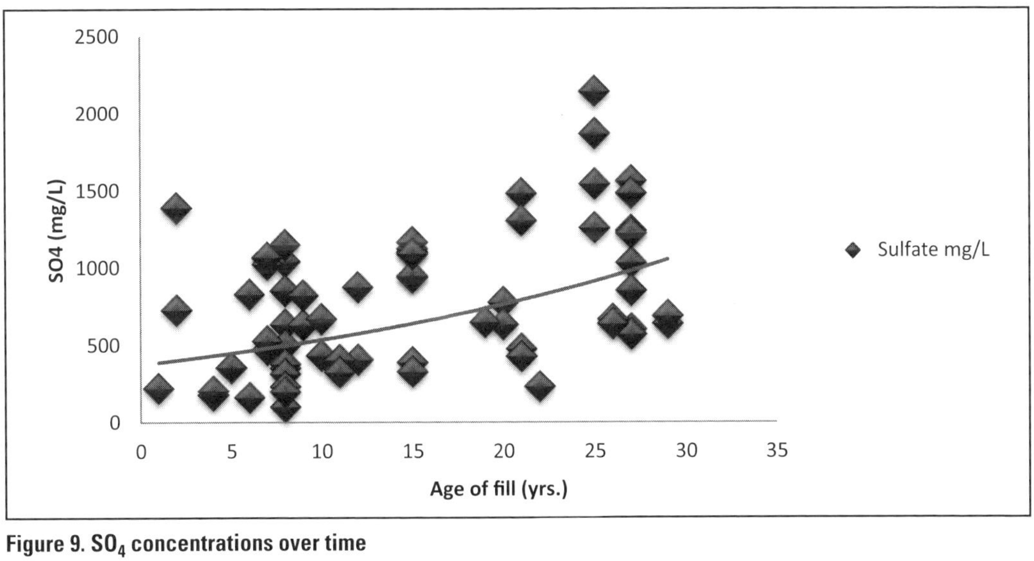

Figure 9. SO_4 concentrations over time

Calcium and magnesium each contributed roughly 12% of the total makeup of TDS (Figure 8). Calcium concentrations ranged from 75 to 425 mg/L. Calcium concentrations changed very little due to valley fill age as evidenced by the low R^2 value (Table 2 and Figure 11). Calcium's low solubility could explain why the there was no observed change in calcium concentrations.

Magnesium concentrations ranged from approximately 40 to 450 mg/L (Figure 12). Concentrations increased past the age of the oldest fill. This trend indicates that flushing of magnesium minerals continues to occur past 30 years.

Although selenium is not a large contributor to TDS in this region, it is a regulated parameter in mine discharges. Selenium concentrations showed a decreasing trend over time, although the R^2 value is only 0.136 (Figure 13 and Table 2). The data is dominated by higher Se concentrations between 5–10 years and at 15 years. More

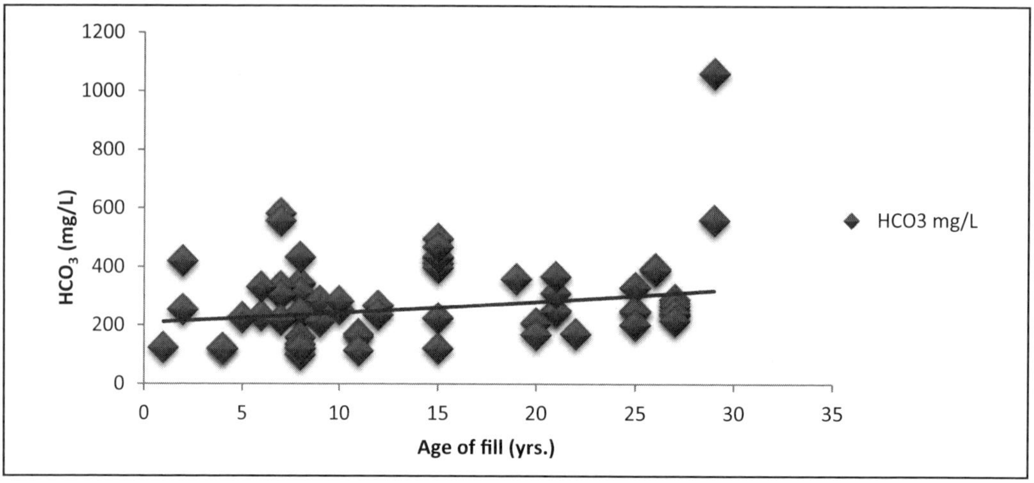

Figure 10. HCO_3 concentrations over time

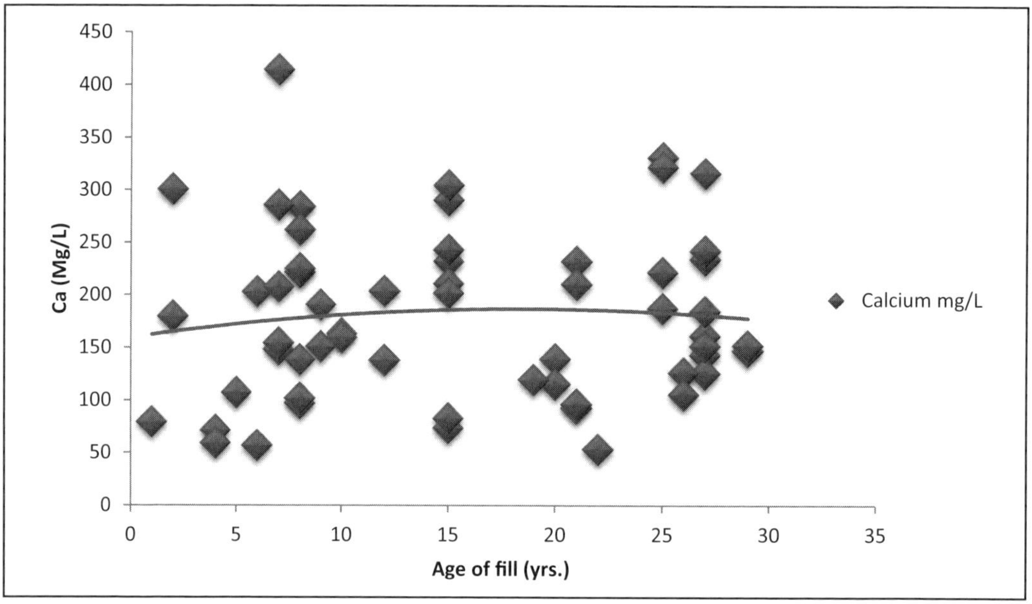

Figure 11. Calcium concentrations over time

data is required to determine if the current trend would continue.

CONCLUSIONS

Field data and flow values were taken from 38 valley fills to determine TDS makeup in southern WV. EC and TDS concentrations were found to react similarly between sampling events. Both parameters increased slightly between 0–5 years after construction and leveled off after five years. R^2 values showed very weak correlations between EC and TDS concentrations over time during both sampling events.

pH values decreased slightly over time (R^2= 0.011 for Sample Event 1 and 0.020 for Sample Event 2), although there was less of an effect between sampling events than with EC or TDS.

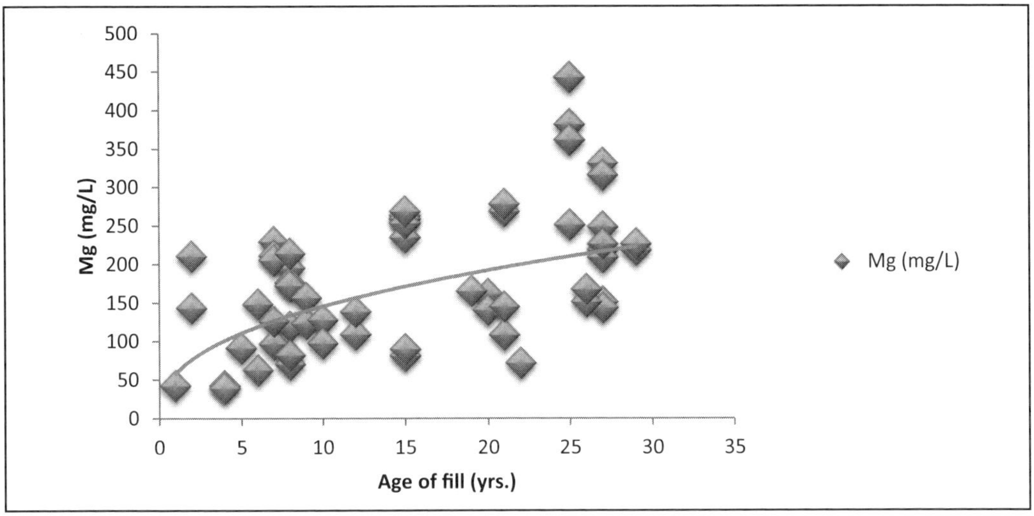

Figure 12. Magnesium concentrations over time

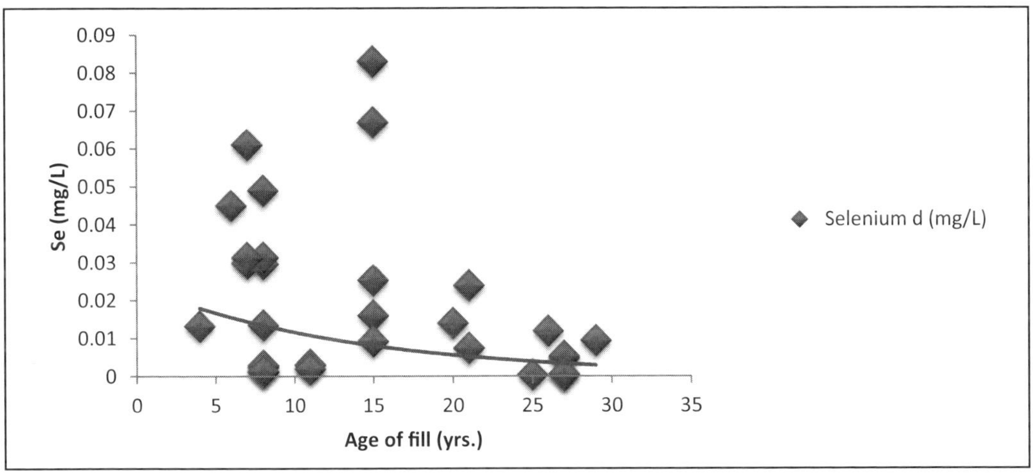

Figure 13. Selenium concentrations over time

pH showed no change between sampling events when the fill was roughly two years old.

Water samples were also collected at the same time as field parameters and flow. Sulfate, bicarbonate, calcium, and magnesium made up the largest percentage of TDS. Sulfate, bicarbonate, and magnesium increased slightly over time. The large concentrations of these ions available within a fill indicate that a longer flushing period may be required before concentrations begin to decrease. A lack of more data for fills 0–5 years old could explain concentration increases over time. Calcium changed very little over time. Calcium's low solubility rate is likely impacting its leaching rate.

Selenium was also included in this study because it is regulated in mine discharges. Selenium concentrations showed a decrease over time, although the R^2 value is low. High concentrations between 0–5 years and at 15 years after fill construction greatly influenced the trend in selenium concentrations over time.

This is an ongoing study and these results are only for the first year. One limitation in the current data set is a lack of sampling data involving valley fills between the ages of 0–5 years. The low R^2 values found in the current data do not indicate that a noteworthy correlation between valley fill age and parameter concentration exists. The target R^2 value that determines a good correlation between fill age and parameter concentration is difficult to establish. However, a R^2 value of at least 0.5 is preferred (Buda and Jarynowski 2010).

Better fits likely could be established by increasing the number of sampled fills, gathering additional data from existing datasets, or widening the geographical area to be more inclusive of regional geology. Increasing the size of the study area also may show that the data trends found from these fills are vastly different than other fills.

These curves are the best model fit for the current sampling. However, the trends may change with continued sampling. We expect many of the studied parameters to decrease over time. The rate of the expected decrease will be dependent upon multiple factors, including solubility and initial concentration of the parameter. Further research is needed to determine the effect of valley fill age on parameter concentrations.

REFERENCES

Buda, A. and A. Jarynowski, 2010. *A Lifetime of Correlations and Their Applications: Vol. 1.* Poland. Wydawnictwo Niezalezne.

Demchak, J. 2005. Water Quality Changes of Underground Mines in Northern West Virginia. PhD dissertation. West Virginia University, 2005.

Jones, P.M., Mulvay, S.M., and Fish, D. 1994. *The Role of Sulfate and Ionic Strength on the Shift from Acid to Alkaline Mine Drainage in Southwest Pennsylvania.* pp. 289–295 *In* Proceedings of the International Land Reclamation and Mine Drainage Conference and the Third International Conference of the Abatement of Acid Mine Drainage. Vol. 2: Mine Drainage. US Bureau of Mines Special Publication SP 06B-94.

Mack, B., L.M. McDonald and J. Skousen. 2010. *Acidity Decay of Above-Drainage Underground Mines in West Virginia.* J. Environ. Qual. 39:1–8 (2010).

Meek, F.A. 1996. *Evaluation of Acid Prevention Techniques Used in Surface Mining. Chapt. 11. In* Acid Mine Drainage Control and Treatment. National Mine Land Reclamation Center, Morgantown, WV.

Wood, S.C., P.L. Younger, and N.S. Robins. 1999. *Long-Term Changes in the Quality of Polluted Minewater Discharges from Abandoned Underground Coal Workings in Scotland.* Quarterly J. Engineering Geology 32: 69–79.

Younger, P.L. 1997. *The Longevity of Minewater Pollution: A Basis for Decision-Making.* The Science of the Total Environment. 194–195: 457–466.

Predicting TDS Release from Appalachian Mine Spoils

W. Lee Daniels
Department of Crop and Soil Environmental Science, Virginia Tech, Blacksburg, Virginia

Zenah Orndorff
Department of Crop and Soil Environmental Science, Virginia Tech, Blacksburg, Virginia

Matt Eick
Department of Crop and Soil Environmental Science, Virginia Tech, Blacksburg, Virginia

Carl Zipper
Department of Crop and Soil Environmental Science, Virginia Tech, Blacksburg, Virginia

ABSTRACT

The Appalachian coal industry faces increasing regulatory pressures to limit discharge of total dissolved solids (TDS) from surface runoff and fill drainage into headwater streams. Unweathered overburden materials in the region contain varying amounts of reactive carbonates, feldspars and traces of pyrite which can oxidize, react and leach to produce significant TDS loads under both low and circumneutral pH conditions. Thus, there is a clear need for the development of new efficient laboratory techniques to predict the TDS elution of mine spoils. The objective of this study was to evaluate a simple column leaching approach for this purpose and to study acid-base interactions and overall TDS elution relationships from a variety of spoil materials. Ten spoils typical of those used as topsoil substitutes or placed into valley fills were characterized and subjected to leaching in columns for 22 weeks under saturated and unsaturated conditions. Leachates were analyzed for pH, electrical conductance (EC), Ca, sulfate and other major ions. The pH of the bulk samples (prior to leaching) was typically near-neutral to alkaline. Bulk leachate EC produced by fresh spoils was initially moderate (≤1,000 μS/cm), although one value ranged up to ~3,500. Relatively low total S (<0.15%) in conjunction with calcium carbonate equivalent (CCE) levels of 1.3–6% resulted in low (or no) potential acidity levels. Leachate pH increased noticeably for most materials over the first few leachings, but after approximately one month, the majority of samples showed no major changes. Leachate EC decreased over the first several leachings and most samples achieved a relatively steady state with relatively low levels (<500 to 1,000 μS/cm) within several pore volumes. Leachate pH, EC, Ca, and sulfate elution were directly related to internal acid-base reactions and to a lesser extent, water saturation regime. While the column technique evaluated here may provide a viable option for TDS prediction, it is both resource- and time-intensive and needs to be tested and calibrated against actual field scale data from the same or similar strata. Current research conducted by the Appalachian Research Initiative for Environmental Science (ARIES) is focused on (1) conducting column leaching trials on a much wider array of regional spoils, (2) developing new quick static tests to predict both peak and long-term TDS production and (3) determining appropriate scaling factors to relate both column and static test data to predict field conditions.

INTRODUCTION AND BACKGROUND

The Surface Mine Control and Reclamation Act (SMCRA) of 1977 contained a number of contentious provisions including the allowance for use of pre-selected overburden materials as topsoil substitutes when the native A+E horizon materials are less than 15 cm (6") thick, and the physical and chemical properties of the proposed substitute spoil materials are deemed suitable for such use. Since native topsoil layers throughout the Appalachian coalfields are usually less than 15 cm thick, and removing them from steep slopes is difficult and expensive, the vast majority of coal mined lands in the region have employed topsoil substitutes. Thus, the final graded and revegetated surfaces usually consist of overburden derived topsoil substitutes that are commonly much higher in pH and weatherable minerals, including carbonates, than native soil profiles and rock saprolites. Similarly, SMCRA requires the isolation of acid-forming materials (≥5 tons CCE/1,000 tons net acid-forming) away from contact with surface runoff and isolation away from internal drainage in mining fills. However, where acid-forming strata are thin relative to adjacent net alkaline strata or where they can be blended with sufficient volumes of other net alkaline materials in the blasting-hauling-placement sequence, the industry has relied upon appropriate application of acid-base-accounting procedures (Skousen et al. 2002) to ensure that the drainage from highwall backfills and valley fills is non-acidic and below discharge limits for Fe, Mn and SO_4^{2-}. As described below, both of these practices (topsoil substitution and acid/alkaline spoil blending) have more than likely led to enhanced levels of total dissolved solids (TDS) in Appalachian coal mine discharge waters vs. undisturbed pre-mine conditions.

The dominant constituent ions of TDS released by weathering of central Appalachian mine spoil materials are typically Ca, K, Mg, Na, SO_4^{-2} and HCO_3^-, in circumneutral waters and higher amounts of Al, Fe and Mn in strongly acidified systems (Daniels et al. 2009; Skousen et al. 2000). All of these components are released by various chemical weathering reactions, particularly when deeper unweathered and unoxidized strata are unearthed and exposed to rapid oxidation, hydrolysis, hydration and dissolution reactions in the surfaces and fills of active coal mines. The upper 5 to 15 m of overburden is typically weathered and oxidized to some extent by long-term geochemical weathering (Haering et al. 2004) and leaching and is therefore much less likely to generate significant TDS after blasting and placement. However, deeper strata commonly contain significant amounts of reactive sulfides, carbonates, feldspars and other mineral phases that rapidly produce varying loads of soluble ions to percolating water and runoff, particularly as they interact with their initial pore volumes of bathing waters.

Essentially, surface coal mining in the Appalachian region creates a set of "weathering conditions" whereby generation of significant loads of TDS to leaching waters and runoff is significant and largely unavoidable. That being said, the fact that current regulations and mining practice mandate that all potentially acid forming materials (potential acidity >5 tons/1,000 tons CCE net acid) must be isolated away from the surface and from significant ground water interaction certainly eliminates the largest potential source of net TDS release. However, the fact that many strata contain significant amounts of reactive sulfides that are essentially balanced by carbonates in terms of their acid-base account (ABA) means that these materials will contribute substantially to TDS loadings (particularly Ca, Mg, and SO_4^{-2}) as they weather while maintaining moderate to circumneutral pH conditions in that drainage (Daniels et al. 2009). In addition to these acid-base reactions, fresh spoils can also contribute significantly to TDS due to simple dissolution reactions of carbonate cements, hydrolysis weathering of feldspars and other primary mineral grains and traces of entrained Cl salts.

Over the past decade, the concept of topsoil substitution has been criticized from a number of perspectives including potential effects on water quality due to inclusion of higher pH and more chemically reactive overburden in contact with surface water runoff. Similar concerns about elevated long-term emission of TDS from backfills and valley fills have also emerged. In a series of actions between 2009 and 2011, the US EPA attempted to use its Clean Water Act Section 404 authority that to establish 500 µS/cm electrical conductivity (equivalent to about 350 mg/L TDS) as a potential or *de facto* standard, at least within heavily mined watersheds. For example, citing Pond et al. (2008), an EPA official testified to the US Congress that "These dissolved ions are not readily sequestered by the surrounding geology and may ultimately emanate from the fills for decades. … This impairs the use of the streams and ultimately leads to listing of these streams as 'impaired water bodies' in EPA's water quality reports …" (Pomponio, 2009). Several recent studies (Chapman et al. 2000; Goodfellow et al. 2000; Pond et al. 2008; Timpano 2010) have addressed the issue of TDS as a major stressor upon receiving streams in mined watersheds. More recent EPA guidance (issued in 2011) was based on the belief that mine discharges with EC levels <300 µS/cm generally will not cause significant degradation of the aquatic ecosystem, while instream conductivity levels >500 µS/cm are likely to be associated with significant adverse impacts on the aquatic ecosystem. The proposed guidelines suggested that projects expected to increase conductivity to levels >300 µS/cm should require adaptive remedial action to prevent conductivity from rising to levels that may contribute to water quality degradation and sites with discharges >500 µS/cm should not be allowed to continue to operate. While the ability of US EPA to directly implement this guidance was recently rejected in U.S. Federal Court, the decision (Walton 2012) left it open for OSM and the states to develop and establish TDS regulatory limits based upon the "best available science." Thus, we feel it likely that coal mining discharges will be directly regulated for TDS (via measurement of EC) at some point in time.

Historically, pre-mining overburden analysis to minimize water quality impacts has focused upon conventional ABA (Sobek et al. 2000) since discharge pH, SO_4^{-2}, Fe and Mn levels have been the principal compliance parameters. Little, if any, work has been reported to date on the utility of either ABA or other techniques to predict bulk TDS. We believe that other relatively simple laboratory techniques such as variations upon saturated paste electrical conductance (EC; Rhoades et al. 1982) or abrasion/hydrolysis pH measurements may also be helpful in predicting short-term bulk TDS elution potential for a given material. We also feel it likely that a combination of typical ABA measures such as pyritic-S and CCE coupled with an appropriate EC measurement on a ground sample may give an accurate of both total long-term mass (from the ABA) and short-term initial pore volume (from EC) TDS elution potential.

The primary focus of this paper is to present a summary of our recent preliminary findings on the characteristics and potential leaching behaviors of selected central Appalachian mine spoils in terms of pH, EC, and ions of interest (Ca and SO_4^{2-}) in relation to the TDS discharge issues discussed above. This paper also describes the column leaching technique that we have employed in detail and our current estimation of its predictive capabilities and limitations. Finally, we use this paper to present the current framework for a much larger study on the topic that is currently being conducted by the Appalachian Research Initiative for Environmental Science (ARIES).

MATERIALS AND METHODS

Sample Collection and Geology

Fifteen mine spoils associated with major surface mining activity and valley fills in southwest Virginia and eastern Kentucky were collected from six different coal mine sites. The Virginia strata were derived from the Wise, Norton and Lee Formations and the Kentucky samples from the Breathitt Formation; all middle to upper Pennsylvanian. The original sampled materials were all approved for use as either topsoil substitutes or for placement into durable rock hollow fills. Thus, none of these materials were expected to be net acid-forming. A subset of ten spoil samples was selected for use in a more detailed column leaching study. These subsequent column study samples were selected based on results from static testing and representation of diverse spoil types including sandstones, siltstones and mudstones, as well as different degrees of weathering for paired spoil samples.

Spoil and Refuse Characterization Methods

All analyses were conducted in triplicate unless specifically noted otherwise. Electrical conductance (EC) (Rhoades, 1982) and pH were determined on saturated paste extracts on ground (≤2 mm) subsamples from each spoil. Each spoil or refuse sample was mixed with deionized (DI) water until it formed a glistening paste, equilibrated for ≥2 hours, then filtered and analyzed. As opposed to using a defined quantity of water, this method allows comparison of materials under similar levels of saturation (i.e., coarser materials will require less water than fine materials to achieve saturation). For comparison, pH was also determined on a 2:1 water to sample solution. Peroxide potential acidity (PPA) was determined by a modified hydrogen peroxide oxidation technique as described by Orndorff et al. (2008). The method oxidizes all sulfides with 30% H_2O_2 and assumes net internal reaction of acids formed with carbonates and other neutralizers. Total net potential acidity is determined by whole sample titration following full reaction. Total-S was determined via combustion on an Elementar CNS analyzer. The calcium carbonate equivalent (CCE) of each spoil was determined via AOAC method 955.01 (AOAC International, 2002).

Experimental Procedures for Column Leaching Trial

All samples were air-dried prior to processing. To eliminate preferential flow within the columns, coarse and large fragments of mine spoil and coal refuse were crushed to pass a 1 cm sieve and then carefully back-blended. The leaching columns were built from 7.6 cm PVC pipe with an inside diameter of 7.4 cm, length of 40 cm, and a concave PVC endcap (see Figure 1). The column was drained through a 1 cm PVC pipe nipple fitted and glued with two-component adhesive into the center of the endcap. An attached Tygon

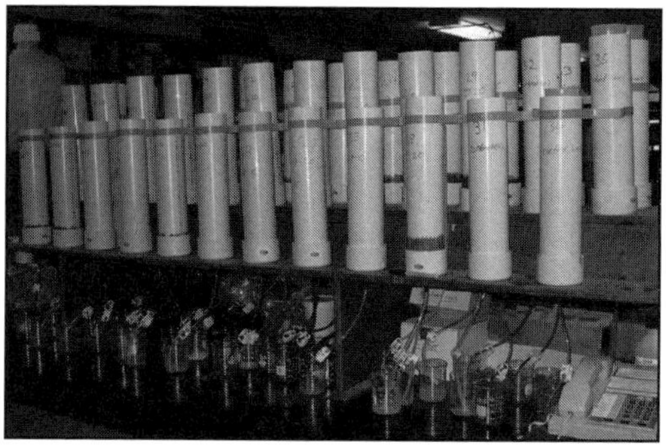

Figure 1. Photo of leaching column array in laboratory. All sample treatments were run in triplicate and dosed with 2.5 cm (1 inch) of simulated precipitation twice per week.

tube drained the leachate and provided the flexibility to create anaerobic conditions in the saturated columns by preventing drainage and gas exchange via a saturated loop at the bottom of the column. To prevent leaching losses of particulate matter and blockage of the drainage hole by particulates, a 0.1 mm nylon mesh circle was glued into the endcap and covered with a 7.6 cm diameter Whatman #1 filter paper. The filter paper was covered with a 2.5 cm layer of acid-washed coarse sand to promote uniform drainage. The sand layer also served as a leachate reservoir to ensure that the bottom section of the sample material remained unsaturated in those treatment columns. The column was then filled with a sample volume of 1,200 cm^3, and capped with a 2.5 cm sand layer to facilitate uniform distribution of the applied leaching solution. Each sample was run in triplicate under saturated and unsaturated conditions (six columns per sample).

Simulated rainfall (pH 4.6; prepared according to Halvorson and Gentry 1990) was applied twice per week to provide 2.5 cm of rainfall per event (100 ml per leaching event). The water was dispensed into a perforated cup placed on top of the sand to uniformly apply the solution to the sample material. During the first week of the experiment the dosing cycle was accelerated to satisfy the hydration demand to a point where leachates were generated in essentially a "piston flow" response over 48 to 72 hours in response to

Table 1. Selected chemical properties of the ten column leaching study samples

Lab-ID	Material	2:1 pH	Saturated Paste pH	Saturated Paste EC dS/m	PPA* Tons CCE/ 1,000 Tons	CCE† %	Total S %
1	Unweathered SS‡	7.98	6.88	1.27	0	2.7	0.06
2	Unweathered SiS	7.14	7.04	3.48	0	4.6	0.23
3	Partly weathered Sh-SS	5.24	6.93	0.94	3.58	1.3	0.07
4	Weathered SiS	5.22	6.46	0.29	0.22	1.6	0.03
6	Unweathered SiS-Sh	7.02	7.26	1.40	0	2.1	0.14
7	Weathered SiS	7.03	7.66	0.20	0.12	4.7	0.03
10	Unweathered SiS	8.14	7.85	0.66	0	6.0	0.11
11	Weathered SS	5.24	6.28	0.56	0.28	3.7	0.03
12	Unweathered SS	8.64	7.84	0.40	0	5.3	0.12
14	Weathered SS	5.74	7.49	0.36	0.12	3.4	0.03

* Potential Peroxide Acidity (PPA) via H_2O_2 oxidation. Values shown represent net acidity/lime demand.
† Calcium Carbonate Equivalent (CCE).
‡ SS = Sandstone; SiS = Siltstone/Mudrocks; Sh = Shale.

the surface dosing regime. Our intent was to keep the bulk of the unsaturated columns at approximately "field capacity" so that after a precipitation event, an equal volume (~100 ml) of leachate would drain freely into the collection bottles. The saturated columns were kept saturated to the surface. After the addition of simulated precipitation, the Tygon tubing at the bottom of the column was opened long enough to collect 100 ml of leachate. As needed, small amounts of simulated rainfall water were added to the saturated columns to compensate for evaporative losses and to assure that the top of the columns remained saturated.

The leachate samples were analyzed and/or preserved after an 18 hour leaching/collection period. This precipitation/sampling procedure was repeated every three or four days; simulated acid precipitation was applied Mondays and Thursdays, samples were collected and analyzed on Tuesdays and Fridays. The samples were analyzed immediately for pH and EC, and subsamples were analyzed by ICPES for several elements including Ca and S. The leachates were also analyzed for all other major ions as described by Daniels et al. (2009), but the full data set is not presented here.

RESULTS AND DISCUSSION

Characterization of Chemical Properties

Selected chemical properties of the sample set are presented in Table 1. The pH of the samples was generally in the neutral to alkaline range which is typical of fresh, relatively unweathered materials from this region due to hydrolysis reactions involving broken primary mineral grains and carbonate dissolution. Only a few of the mine spoils were slightly acidic with no sample pH below 6.2. The soluble salt content, as indicated by saturated paste EC, produced by fresh spoils was generally low (<1.0 dS/m or 1,000 μS/cm), although one sample ranged up to 3,500 μS/cm. Further discussion of the relationship between soluble salt content and EC is provided with the leaching column results. Samples with elevated EC values appeared well-related to either their potential acidity (as estimated by PPA) and/or CCE. Only four out of the ten samples had a total sulfur content of >0.10%, which is not unusual for the region sampled. A comparison of the pH values determined by saturated paste (pH_{sp}) versus in a 2:1 water to sample mix (Table 1) indicated lower pH values by the 2:1 method for near-neutral to acidic samples. This may be due to the difference in equilibration time (much longer for the

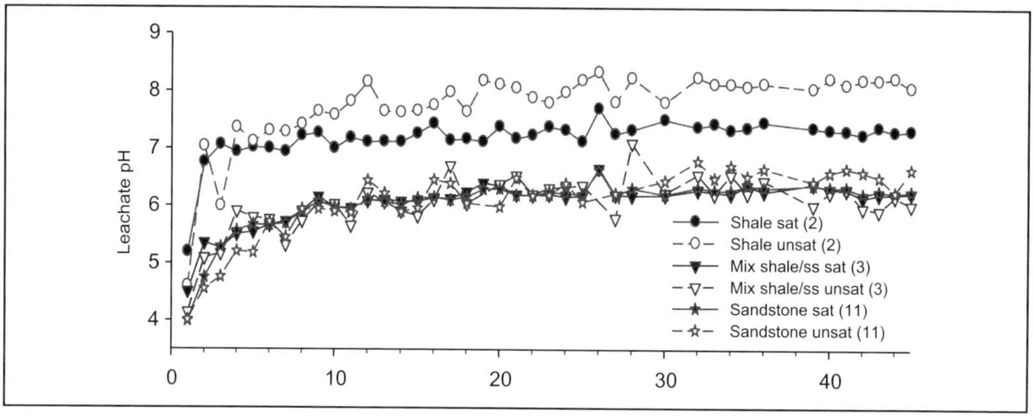

Figure 2. Leachate pH from selected mine spoils under saturated and unsaturated conditions. The 45 leachate events (x-axis) occurred over 22 weeks. One pore volume was approximately equivalent to 7 to 9 leaching doses.

saturated paste method) allowing some of the alkaline mineral surfaces to react with acidic components. Direct comparison of properties from weathered mine spoils versus unweathered mine spoils showed that weathered mine spoils tended towards lower pH (Table 1) as expected. Only one weathered sample (7) had a neutral 2:1 pH, 7.03; the other weathered samples all had 2:1 pH values between 5.22 and 5.74. The pH among the partially weathered/unweathered mine spoils followed the trend that those with higher percentages (via visual observation of rock oxidation/color) of weathered material generated lower pH.

Leaching Column Trial

The data presented in Figures 2–6 report results from 45 leaching events over a 22 week period. The data are presented here are selected from a larger sample set (n=15; Daniels et al. 2009) and demonstrate typical spoil type responses. We highlight pH, EC, Ca and sulfate because they are particularly relevant to acid-base reactions. All data presented in Figures 2–6 represent the mean observations from three replicate columns. Details on column replicability (which was very good) can be found in Daniels et al. (2009). For example, relative treatment differences discussed here for pH and EC by treatment × date were separated by at least two standard errors.

pH

Unweathered overburden in southwestern Virginia can contain significant amounts (0.5 to >3.0%) of carbonates (Howard 1979) which occur primarily as secondary cementing agents and to a more limited extent, reactive pyrite (Sobek et al. 2000). Depending on the distribution and quantity of these minerals, the pH of unoxidized overburden is between 6.5 and 8.0 (Roberts et al. 1988) while that of weathered/partially weathered overburden is generally between 4.5 and 6.0. For most materials used in this study, leachate pH (Figure 2) from the first leaching events was substantially lower than the initial saturated paste pH data on the fresh bulk materials (pH). We presume this was simply due to the initially low pH of the simulated precipitation used and lack of chemical equilibration time in the columns. With few exceptions, leachate pH increased over the first few leaching events, and achieved a relatively stable equilibrium within 10 to 20 leaching events. Under saturated conditions, only three samples (1, 2 and 11) equilibrated to pH values near their saturated paste extract pH, while the remaining seven samples (saturated) maintained pH values that were approximately 0.3 to 1.0 pH unit less than saturated paste EC. The higher pH levels from unsaturated conditions could be due to the effects of increased CO_2 partial pressures

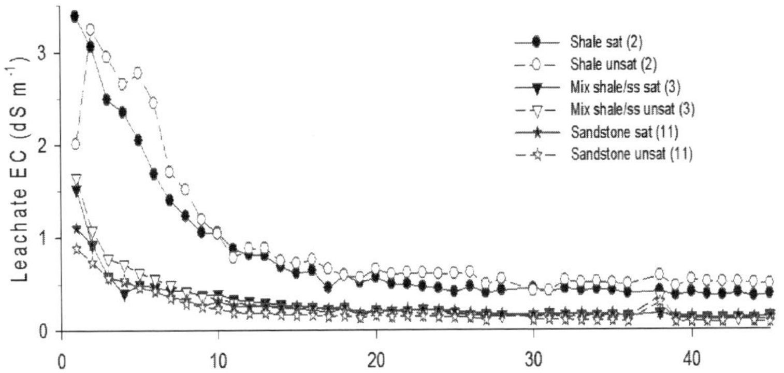

Figure 3. Mine spoil leachate electrical conductance (EC) from selected spoil materials of different rock types. The 45 leachate events (x-axis) occurred over 22 weeks. 1.0 dS/m = 1,000 μS/cm.

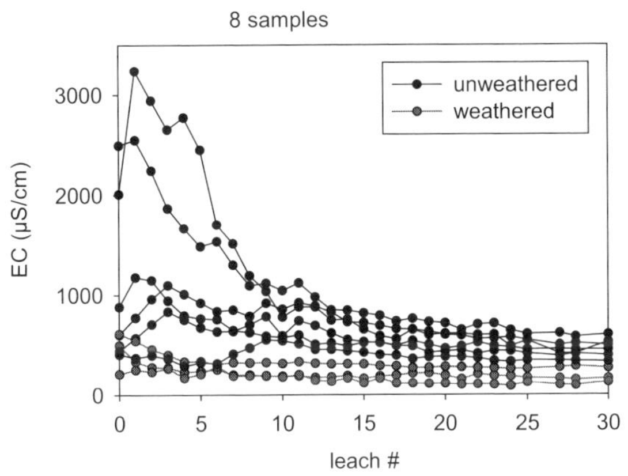

Figure 4. Mine spoil leachate electrical conductance (EC) for finer textured mudstone/ siltstone samples from 30 leaching cycles for unsaturated leaching conditions. Data reported here include several samples in addition to those described in Table 1 from parallel studies sponsored by the Powell River Project.

on carbonate dissolution in the unsaturated columns or perhaps siderite ($FeCO_3$) formation in the saturated columns.

Electrical Conductance (EC)

Parallel studies completed in conjunction with the results presented here (Daniels et al. 2009) indicated a high correlation coefficient (r = 0.98) between EC and TDS, as expected, indicating that EC can be used as an effective proxy for TDS. However, actual TDS recovery values did not always agree well with those predicted by saturated paste EC and/or the sum of ion mass leached. Leachate EC from unweathered mine spoil varied by rock type with finer textured rocks generally producing higher EC levels (Figure 3). There were also dramatic differences between partially oxidized and weathered samples of similar geology. The difference in results between weathered and unweathered samples was most pronounced for siltstone/mudstone (Figure 4) samples. Interestingly, saturated vs. unsaturated conditions had minor effects, except for sample 2, which was clearly more reactive as discussed later. In all samples, initial EC values dropped quickly, and typically achieved a steady state within 10 to 20 leaching events, maintaining relatively low levels for the remainder of the leaching trial. Most

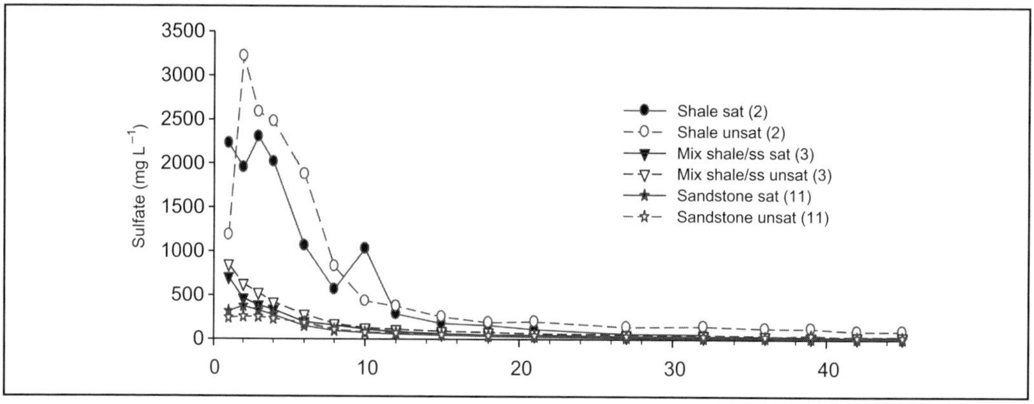

Figure 5. Mine spoil leachate sulfate from selected materials of different types. The 45 leachate events (x-axis) occurred over 22 weeks.

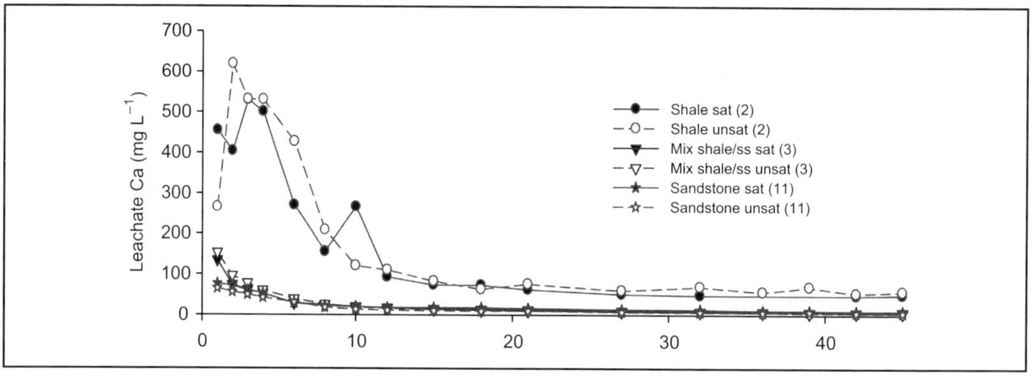

Figure 6. Mine spoil leachate Ca from selected materials of different types. The 45 leachate events (x-axis) occurred over 22 weeks.

notable among the spoil materials were samples 2 (shown) and 6 (not shown), both unweathered mudstones, which maintained values greater >1,000 µS/m through the first 10 leachings.

The high EC/TDS values generated by samples 2 and 6 probably represent the presence of highly reactive sulfides that, even though present in relatively low amounts (0.23% and 0.14% S), reacted quickly with substrate carbonates to produce prolonged sulfate release over the extent of this experiment. We presume that pyrite in these samples may have been finely divided and framboidal. Collectively, this acid-base reaction control on leachate chemistry was also reflected in leachate sulfate and Ca concentrations as discussed below.

Sulfate

Sulfate release patterns (Figure 5) reflect the acid-base reactions due to fundamental differences in the geology/mineralogy among the samples and primarily reflect the total-S content of the materials (Table 1). In SW Virginia, where most of these samples originated, the vast majority of strata within the Pennsylvanian system are fluvial-deltaic facies which are generally low in pyritic-S. Many of the massive sandstones that dominate the Lee, Norton, and Wise formations contain secondary carbonate cementing agents (Howard et al., 1988) which offset the relatively minor amounts of sulfidic minerals found in most geologic sections (like #11). Significant accumulations of sulfides do occur in coal seams and underclays throughout the region; however, these seams are relatively thin (<3 m). For most spoil samples used in this study, the majority of sulfate release occurred within the first pore volume (7 to 9 leaching events). The greatest sulfate release occurred from sample 2, a mudstone from the Breathitt Formation in eastern Kentucky. This formation has been noted and studied for its acid-forming potential (Barnhisel and Massey, 1969). For this material (#2), saturated conditions did appear to somewhat retard the rate of pyrite oxidation and S release as also reflected in the EC results shown in Figure 3.

Calcium and Other Ions

Calcium leaching patterns (Figure 6) for the mine spoils followed the pattern of sulfate very closely. This further amplifies our conclusion that the majority of net TDS release in these materials is controlled by acid-base neutralization reactions. As expected, unweathered spoil materials released greater amounts of Ca than did the weathered materials (data not shown). Sulfate and Ca generally accounted for over 70% of the mass of ions leached, although significant amounts of Mg and K were eluted from most treatments and bicarbonate release was also significant in the saturated unweathered mine spoil treatments (Daniels et al. 2009).

SUMMARY AND CONCLUSIONS

The mine spoils sampled in this study were typical of surface mined overburden employed in the central and southern Appalachians for topsoil substitutes and placed into valley fills in that they were non-acid forming and moderate (>5.5) in initial pH and prolonged leachate pH. All of the mine spoils maintained a moderate pH (5.7 to 8.0) in leachates over the full study period (22 weeks). As expected, mine spoils that were significantly pre-weathered were lower in pH than unweathered materials of similar geology. Most unweathered spoil samples eluted moderate initial EC levels (≥1.0 dS/m or 1,000 μS/cm) over their initial leaching cycles. Although EC values dropped rapidly for most samples, this behavior would produce TDS loadings to discharge waters on active operations that continue to disturb and place fresh spoils in direct contact with surface runoff and ground water over extended periods of time. Differences in TDS release appear to be clearly related to sulfide oxidation reactions with subsequent generation of sulfate and other reaction and dissolution products. The relative concentrations of sulfate, Ca and other major ions in the leachates over time were controlled primarily by acid-base reactions, pH solubility controls, and water saturation regime to a more limited extent. Calcium and sulfate release patterns were very similar to each other; this was most apparent in sample 2 due to its noticeable fluctuations.

Overall, results from this study and others cited herein indicate that identification and isolation of high TDS producing strata could potentially be utilized as a part of pre-mine planning and permitting procedures to minimize TDS release to receiving streams. We suggest that active mine operations should be modified to place high TDS producing materials in ways that reduce contact with percolating drainage waters. It is also clear from our data and from related studies that near-surface pre-oxidized and weathered strata will be much lower in their TDS producing potentials and their use as topsoil substitutes should be emphasized and encouraged. We believe that this practice would minimize TDS of surface runoff waters.

Of course, all of these suggested practices need to be tested and validated in full scale field settings. The effects of field hydrogeologic conditions such as seasonal water flux vs. the constant leaching conditions imposed by our column technique also need to be assessed. Regardless of these concerns, we believe that the utilization of simple column leaching procedures such as those employed in this study can provide viable and useful data for the purposes of TDS prediction.

Unresolved Questions and Further Research Needs Addressed by ARIES

While the column technique evaluated here provides a viable option for TDS prediction, it serves primarily as a useful research tool and may not be practical for routine application in pre-mine overburden analysis. First of all, the technique requires at least three to six months to complete, with considerable laboratory labor and analytical costs. Even if converted over to commercial scale and run without replication and with limited effluent analyses (e.g., EC and pH), we doubt that the cost per sample could be lowered below $500 (U.S.) per sample. Secondly, in order for the columns to run without a dominance of macropore flow (to ensure uniform replication), the spoil must be sized down to ≤1 cm, which means the reactive surface area per unit mass is much higher than would be found in run-of-mine spoils. Thus, we assume that the technique over-predicts the short term TDS peak concentrations and may also over-predict the longer term equilibrium TDS release potentials for a given spoil. Finally as noted earlier, our conclusions regarding the predictive nature of our column technique need to be validated against larger scale field leaching data with the same or similar spoils.

Therefore, a major new research program currently being conducted by the Appalachian Research Initiative for Environmental Science (ARIES; http://www.energy.vt.edu/index.html) is focused on (1) conducting column leaching trials on a much wider array of regional spoils, (2) developing new quick lab and field static tests to predict both peak and long-term TDS production and (3) determining appropriate scaling factors to better relate and scale both column and static test data to actual field discharge conditions. The current ARIES work effort is cooperative with researchers at the University of Kentucky and West Virginia University.

ACKNOWLEDGMENTS

This study was sponsored by the Appalachian Research Initiative for Environmental Science (ARIES). The views, opinions and recommendations expressed herein are solely those of the authors and do not imply any endorsement by ARIES employees, other ARIES-affiliated researchers or industrial members. Information about ARIES can be found at http://www.energy.vt.edu/ARIES.

Portions of this column research work reported here were also supported by the OSM Applied Science Program with oversight by Nancy Pointon and Lois Uranowski of the Appalachian Regional Office in Pittsburgh. Components of this work were also supported by the Powell River Project. Current work in this program is supported by the Appalachian Research Initiative for Environmental Science (ARIES). We appreciate the assistance of Julie Burger, Pat Donovan, and Athena Tilley in our labs at Virginia Tech. Finally, we want to thank our numerous industry cooperators in the Virginia and Kentucky coalfields who assisted us with sample collection.

The authors thank Mike Beck for his contribution to this paper.

REFERENCES

AOAC International. 2002. Official methods of analysis of AOAC International, 17th Ed. AOAC International, Gaithersburg, MD, USA, Official Method 955.01.

Barnhisel, R.I., and Massey, H.G. 1969. Chemical, mineralogical and physical properties of eastern Kentucky acid-forming coal spoil materials. *Soil Science*, 108:367–372.

Chapman, P.M., Bailey, H.B., and Canaria, E. 2000. Toxicity of total dissolved solids associated with two mine effluents to chironomid larvae and early life stages of rainbow trout. *Environmental Toxicology and Chemistry*, Vol. 19(1), pp. 210–214.

Daniels, W.L., Beck, M., Eick, M.J., and Orndorff, Z.W. 2009. *Predicting Contaminant Leaching Potentials for Central Appalachian Overburden and Coal Refuse Materials*. Final Report to OSM Applied Science Research Program, December 2009. http://www.techtransfer.osmre.gov/NTTMainSite/appliedscience/2007/Projects/VTLeachingPotentialNPointon07FR.pdf.

Goodfellow, W.L., Ausley, L.W., Burton, D.T., Denton, D.L., Dorn, P.B., Grothe, D.R., Heber, M.A., Norberg-King, T.J. and Rodgers Jr., J.H. 2000. Major ion toxicity in effluents: a review with permitting recommendations. *Environ. Toxicol. Chem.* 19: 175–182.

Haering, K.C., Daniels, W.L., and Galbraith, J.M. 2004. Appalachian mine soil morphology and properties: Effects of weathering and mining method. *Soil Sci. Soc. Am J.* 68:1315–1325.

Halvorson, H.G., and Gentry, C.E. 1990. Long-term leaching of mine spoil with simulated precipitation. In Skousen, J., J. Sencindiver, and D. Samuel ed., *Proc. of the 1990 Mining and Reclamation Conf. and Exhib., Vol. 1*; pp. 27–32. 1990 National Meeting of the Amer. Soc. of Mining and Rec., Published by ASMR, 3134 Montavesta Rd., Lexington, KY 40502. http://www.asmr.us.

Howard, J.L. 1979. *Physical, Chemical, and Mineralogical Properties of Mine Spoil Derived from the Wise Formation, Buchanan County*, Virginia. M.S. thesis. Virginia Polytech. Inst. and State Univ., Blacksburg, VA.

Howard J.L., Amos D.F., and Daniels W.L. 1988. Phosphorus and potassium relationships in southwestern Virginia mine spoils. *J. Environ. Qual.* 17(4):695–700.

Pond, G.J., Passmore, M.E., Borsuk, F.A., Reynolds, L., and Rose, C.J. 2008. Downstream effects of mountaintop coal mining: comparing biological conditions using family- and genus-level macroinvertebrate bioassessment tools. *J. N. Am. Benthol. Soc.* 27:717–737.

Pomponio, J. 2009. *Testimony to US Senate Committee on Environment and Public Works*, Subcommittee on Water and Wildlife, 25 June 2009. http://epw.senate.gov/public/index.cfm?FuseAction=Files.View&FileStore_id=53b87b86-805f-4a7f-a7e3-79ff7e5b3eb8.

Orndorff, Z.W., Daniels, W.L., and Fanning, D.S. 2008. Reclamation of acid sulfate soils using lime-stabilized biosolids. *J. Env. Qual.* 37:1447–1455.

Roberts, J.A., Daniels, W.L., Bell, J.C., and Burger, J.A. 1988. Early stages of mine soil genesis in a Southwest Virginia spoil lithosequence. *Soil Sci. Soc. Am. J.* 52:716–723.

Rhoades, J.D. 1982. Soluble salts. *In Methods of Soil Analysis, Part 2: Chemical and Microbiological Properties*–Agronomy Mono. No. 9. ASA-SSSA, Madison, WI, USA.

Skousen, J.G., Sexstone, A., and Ziemkiewicz, P.F. 2000. Acid mine drainage control and treatment. P. 131–168 *In* R.I. Barnhisel et al. (ed.) *Reclamation of Drastically Disturbed Lands*. Agron. Mono. 41. ASA, CSSA, and SSSA, Madison, WI.

Skousen, J., Simmons, J., McDonald, L., and Ziemkiewicz, P. 2002. Acid-base accounting to predict post-mining drainage quality on surface mines. *J. Env. Qual.* 31:2034–2044.

Sobek, A.A., Skousen, J.G., and Fisher Jr., S.E. 2000. Chemical and physical properties of overburdens and minesoils. p. 77–104. In R.I. Barnhisel et al. (ed.) *Reclamation of Drastically Disturbed Lands*. Agron. Mono. 41. ASA, CSSA, and SSSA, Madison, WI.

Timpano, A., Schoenholtz, S., Zipper, C., and Soucek, D. 2010. Isolating effects of total dissolved solids on aquatic life in central Appalachian coalfield streams. p. 1284–1302 in R.I. Barnhisel (Ed.), *Proc., 2010 Nat. Meeting of the Amer. Society of Mining and Rec.*, Pittsburgh, PA, June 5–11, 2010. Published by ASMR, 3134 Montavesta Rd., Lexington, KY 40502. http://www.asmr.us.

Walton, R.B. (2012). *Final decision, National Mining Association vs. USEPA and Sierra Club, Case* 1:10-cv-01220-RBW Document 167 Filed 07/31/12, United States District Court, Washington, DC.

Long-Term Trends of Electrical Conductivity in Streams Flowing Out of Headwater Valley Fills in Virginia

D.M. Evans
Crop and Soil Environmental Sciences Department, Virginia Tech, Blacksburg, Virginia

P.F. Donovan
Crop and Soil Environmental Sciences Department, Virginia Tech, Blacksburg, Virginia

C.E. Zipper
Crop and Soil Environmental Sciences Department, Virginia Tech, Blacksburg, Virginia

W.L. Daniels
Crop and Soil Environmental Sciences Department, Virginia Tech, Blacksburg, Virginia

ABSTRACT

Mining in the central Appalachian coal fields impacts water quality and quantity and can lead to degraded stream structures and functions. Environmental regulations designed to limit impacts to streams have recently focused on valley fill operations and their tendency to raise stream water electrical conductivity (EC). However, the relative impact of the valley fills, compared to surface and underground mining operations in the watershed above the valley fill, is poorly understood. The duration of the impacts of valley fills on stream water EC is also uncertain. We analyzed water quality monitoring records from 212 valley fills in southwest Virginia (USA) for temporal trends in electrical conductivity. Water quality and associated data were obtained from the Virginia Department of Mines, Minerals and Energy. Records for each valley fill ranged from one to 21 years of data with 30 to 1,164 samples and mean standard conductivity (at 25°C) ranging from 227 to 2,866 μS/cm. Data were analyzed during valley fill construction, after valley fill construction, and after vegetation had reclaimed the site. We analyze data describing site factors in an effort to determine which are associated with high peak EC and sites with longer duration of high EC. We also identify site factors and valley fill attributes that are associated with streams that have relatively low EC.

ACKNOWLEDGMENTS

This study was sponsored by the Appalachian Research Initiative for Environmental Science (ARIES). The views, opinions and recommendations expressed herein are solely those of the authors and do not imply any endorsement by ARIES employees, other ARIES-affiliated researchers or industrial members. Information about ARIES can be found at http://www.energy.vt.edu/ARIES.

4.2
Environmental Considerations in Underground Mining

Event Tree Failure Mode Development for Mining Near Bodies of Water 288

Applications of Subsurface Subsidence Model to Study Longwall Subsidence Influences on Overburden Hydrological System 297

Addressing the Effect of Sloping Terrain on Ground Movements Due to Underground Mining 308

Application of Subsidence Prediction Methodologies for Sizing Barrier Pillars for Stream Protection in Appalachia 319

Preliminary Assessment of Factors Affecting Springs Undermined by Longwall Mines in Pennsylvania 336

Event Tree Failure Mode Development for Mining Near Bodies of Water

Thomas Wachtel
Civil and Environmental Engineering, West Virginia University, Morgantown, West Virginia

John D. Quaranta
Civil and Environmental Engineering, West Virginia University, Morgantown, West Virginia

Hema Siriwardane
Civil and Environmental Engineering, West Virginia University, Morgantown, West Virginia

ABSTRACT

In the 1970s there was an increase in demand for coal for energy production. During this time, the Bureau of Mines contracted two engineering firms, Wardell (1976) and Skelly and Loy (1976), to develop safety guidelines for mining near surface bodies of water. These guidelines became IC 8741 by Babcock and Hooker (1977), which determined the proximity that mining can occur to a water body without compromising the safety of the miners. This report exists today as the standard for such instances. While the report protects the mine itself, it does not consider the effects of mining on the reservoirs or dams.

Subsurface coal mining creates changes in the overlying strata through subsidence and changes in hydraulic conductivity. The horizontal distance that these occur is based on the depth of the mine and the type of overburden in the area. In Appalachia, subsidence generally occurs at an angle from vertical between 15–30 degrees (up to over 40 degrees) called the "angle of draw." This is the focus of IC 8741 for protection of the mine against inundation. However, changes in the soil structure extend past this range. Head loss and permeability changes have been recorded to extend up to 60 degrees (Reed and Rauch, 2001). From a mining perspective, the changes in groundwater flow generally occur near the surface and do not affect the integrity of the mine due to the presence of an aquiclude zone. The changes in groundwater levels occur when pools form in the subsidence basin and the effects are only temporary (Luo, 2010). From a civil engineering perspective, the hydraulic conductivity of the soil changes permanently and can lead to internal erosion affecting reservoirs and the dam structure.

Current IC8741 offset distances guidelines are developed for the safety of the miners during operation. It does not factor in the potential damage to the reservoir or the dam. For adequate offset parameters to be used for dam safety, the guidelines need to cover both the safety of the mine and surface water bodies. This research presents the development of event tree failure modes detailing the factors that contribute to changes in strain leading to increased permeability in the overlying strata for room and pillar, and longwall mining methods.

INTRODUCTION

The former US Bureau of Mines performed studies in the mid-1970s to develop guidelines for mining near and under surface bodies of water (Babcock and Hooker, 1977). These guidelines were limited by focusing on the safety of the workings and the miners from water inundation during the lifespan of the mine.

These studies did not detail the affect mining operations would have on changes in groundwater flow and potential for internal erosion caused by subsidence and increased permeability. The likelihood of altered flow paths and flow rates could induce internal soil erosion leading to development of failure modes reducing the integrity of a reservoir or dam. Internal erosion is defined as a movement of fine particles within the soil matrix voids.

RESEARCH AIM AND SCOPE

The purpose of this research was to identify conditions that can initiate erosion and risk for surface bodies of water due to changes from room and pillar, and longwall type extraction coal mining in Appalachia. Determination of potential risk factors can lead to a better understanding of the potential effect mining has on water bodies and groundwater flow. This paper details a method for developing event trees to determine the possible risk to a dam or reservoir from room and pillar, and full extraction longwall mining. The approach presented considers potential subsidence effects from the mining, and changes in groundwater flow that can lead to subsurface erosion.

LITERATURE REVIEW

Current guidelines for mining near surface bodies of water were developed by the Bureau of Mines in the mid-1970s known as Information Curricular 8741 (IC8741) based on reports by Wardell (1976), and Skelly and Loy (1976). This is the most widely referenced document when determining offset distances for mining near surface bodies of water. However, the guidelines are designed to limit water inundation into the mine during operation.

The proposed offset zones are based on subsidence occurring due to mining operations. However, the mine subsidence produces two related zones of influence; these are described as the angle of draw due to subsidence, and the induced angle of groundwater influence. Figure 1 illustrates these zones with correlation to the offset zones recommended by IC8741.

The IC8741criteria recommended for mining near surface bodies of water includes the following criteria: (i) minimum 200 ft horizontal offset from the high water mark for depths up to 350 ft below ground surface, which develops approximately a 29.74 degree angle of draw at that depth; (ii) below 350 ft the offset zone extends outward 25 degrees from the vertical or 65 degrees from horizontal; and (iii) directly below the surface water reservoir, mining can occur at a depth of 60 times the thickness of the seam for a longwall type mine.

With respect to dam structures, the criteria include: (i) mining under the dam structure is not recommended; (ii) mining near a dam requires a minimum offset of 200 ft for depths up to 350 ft; and (iii) to offset zone extends outwards 25 degrees from the vertical or 65 degrees from horizontal beyond 350 ft. The IC8741 criteria identify a permitted strain of 8.75 mm/m (0.00875 mm/mm or 0.875%) below a surface body of water.

Data for the Appalachian coal fields indicate that the subsidence angle of draw can extend more than 40 degrees. The Mining Engineering Handbook listed typical values of angle of draw ranging between 10 to 38 degrees for the eastern United States (Singh 1992). Booth (2006) reports the range for Appalachia and Illinois subsidence generally falls between 20 to 40 degrees. Peng (1995) discussed the range of angle of draw has been reported to be between 4 to 45 degrees. However, in the development of a database for subsidence from longwall mining, Peng (1995) found the maximum angle of draw for 110 cases was 24 degrees.

The angle of groundwater influence correspondingly ranges from 16 to 60 degrees (Booth, 2006). Generally groundwater influence is within 40 degrees unless in areas of steep terrain. However, in one case study, the range of groundwater or "dewatering" influence reached 70 (60±20) degrees (Reed and Rauch, 2001). There is an interrelationship of these two effects; specifically, the subsidence strain causes surface deformation and changes in strata permeability altering the seepage characteristics of the strata and mine–reservoir site. Figure 1 illustrates that at an approximately 200 ft offset the angle of groundwater influence will extend into the body of the surface water reservoir at the high water mark elevation for the full range of influence.

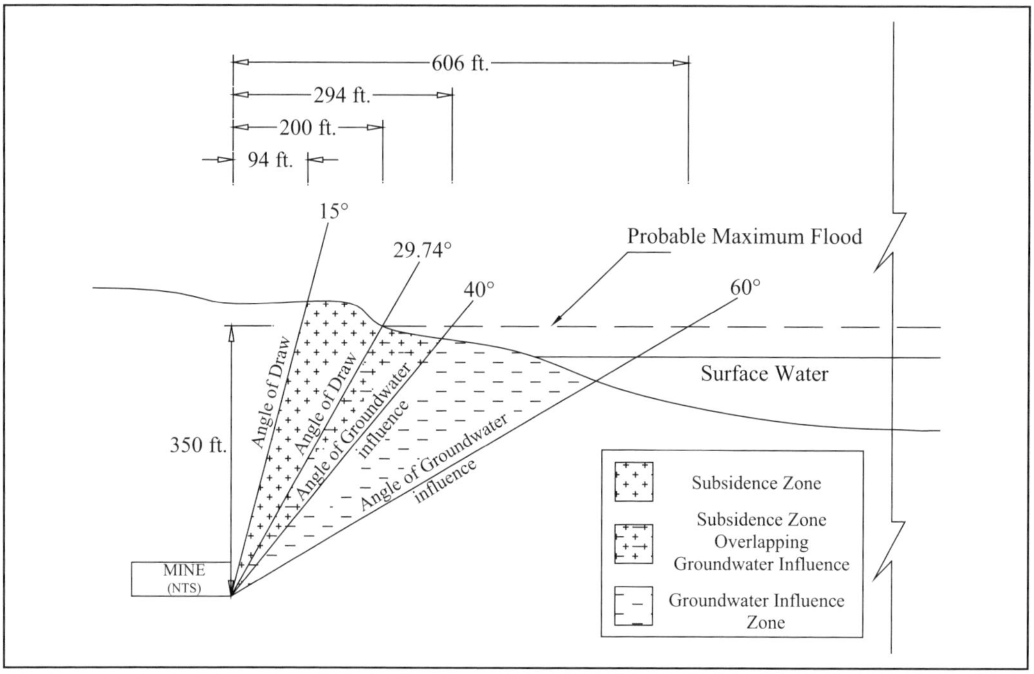

Figure 1. Subsidence angle of draw and angle of groundwater influence

Kendorski (1993) identified that the subsurface strata permeability increases when strains due to mining are greater than 0.001 in/in, or 0.1% for sandstones. Ground surface cracking develops parallel to mine edges directly above the mine with extensional strains ranging 0.006 to 0.009 in/in (0.6 to 0.9%). Field measurements have indicated maximum tensile strains of 0.0134 and 0.021 (1.34 to 2.1%) (Kelleher, 1991). These strains were sufficient to increase the hydraulic conductivity of shale by two to three orders of magnitude and one order of magnitude for sandstone (Kelleher, 1991).

From the hydrogeology perspective presented, research on the effects of mine subsidence related to issues including the advance of the subsidence wave. This induces tension and compression strains leading to zones of subsidence extension, compression and fractured zones (Booth et al. 2000). The overburden strata strain effects culminate in permeability changes occurring in the near surface soils. The changes occurring during mining are the highest. Booth (2000) reports that reductions in the subsidence compression have been shown at approximately –10×. While the permeability increases in the extension (dilation) zones range from +10× to +1,000×. These changes have permanent effects on the groundwater system.

In summary, the contrast between IC8741 and research from the areas of subsidence and hydrogeology indicate the following:

- The angle of subsidence draw in Appalachia ranges upwards of 40 degrees while the angle of draw recommended by IC8741 using a 350 (107.69 m) ft deep mine offset at 200 ft (61.54 m) is a 29.74 degree angle of draw. Therefore IC8741 can underestimate the subsidence effect on strain.
- IC8741 recommends a 8.7 mm/mm maximum tensile strain in overburden strata. Literature indicates that a tensile strain of 0.001 in/in can cause permanent increase in permeability of sandstone (Kendorski, 1993).
- IC8741 does not address development of the groundwater angle of influence. Literature indicates that in Appalachia this

angle can extend up to 79 degrees from the vertical edge of a mine affecting surface water head and flow. This occurred in a stream with bare bedrock and lack of fine sediment which could seal subsidence cracks (Reed and Rauch, 2001).
- The IC8741 guidelines are intended only for the life of the mine. No information is provided for the following conditions when pillar deterioration occurs, when pillars punch through roof or floor layers, or for full extraction retreat mining.

RISK-BASED APPROACH FOR DAM SAFETY

The important problem is that underground mining for energy extraction near or under bodies of water at USACE and Reclamation projects may raise the risk to that project. The USACE methodology involves fully describing and evaluating site specific Potential Failure Modes (PFM) then risk analysis using event trees or other approaches incorporate the identification of a dam's site-specific potential failure modes. The method requires a Failure Mode Identification of seepage effects due to partial extraction room and pillar and full extraction longwall mining event initiated near a reservoir rim. The remainder of this paper is dedicated to development of an alternative risk based analysis approach at USACE reservoirs due to underground coal mining.

METHOD

The probability of subsurface soil erosion to occur due to mining near surface bodies of water is controlled by the angles of draw (subsidence) and angle of groundwater influence (hydrogeology). This is linked to increase in soil permeability based on the type of overburden found in the shallow overburden aquifers typical in Appalachia. There are initial factors that must be considered that are site specific when developing a Failure Mode Analysis using the event tree approach.

EVENT TREES

Event tree analysis for this paper was designed based on Quantitative Risk Assessment (QRA) to estimate the probability of failure (Fell et al. 2000). Quantitative Risk Assessment is used for determining failure mode identification, analysis for probability of failure and calculation of losses that would occur in case of failure. For this paper the event tree analysis is only concerned with the likelihood of failure. Potential damages and loss of life are not considered given they are on a case by case study. For example, a reservoir located in a highly populated zone would have higher risk for loss of life while the risk to life loss in a sparsely populated zone would be minimal. The potential for failure for either reservoir may be similar. This analysis is designed to estimate the potential for such failure.

The event tree is designed based on initial conditions governing the mine. For this paper there are two mine conditions to be evaluated: (1) room and pillar mining with less than 70 percent mineral extraction, and (2) longwall. The initiating event is considered to be what begins the potential for erosion to occur. This is shown in columns called levels. An example is shown in Figure 1. Each level has a predetermined probability of occurrence based on that event. The probabilities for each level are multiplied for each possible path. Failure is tallied at the end for only the paths that affect the last level. Such that an event that causes an occurrence of one level would not necessarily cause the failure in the last level.

The following narrative presents a potential failure mode analysis for longwall and room and pillar mining near the rim of a USACE reservoir. It develops the structure of an event tree for a potential failure mode analysis of a reservoir due to subsurface seepage

Initial Conditions for Room and Pillar Mining

These are the initial conditions of the mining operation including the strata and overall type of terrain where mining takes place.

Pool Level Elevation: The pool level of a reservoir is important in determining the risk of subsurface erosion occurrence. There are three typical pool levels considered: summer, winter and maximum pool level.

Time Intervals: Room and pillar subsidence can occur over long periods of time. There is minimal initial subsidence during the mining operation and as long as the pillars remain intact there will not be significant surface subsidence. However, surface subsidence for room and pillar mining can be delayed for decades. The potential for continuation of subsidence to occur remains until the mined voids are filled by caved strata (Singh, 1992).

Terrain Type: The surface terrain directly above a room and pillar mine is important to determine the total stress on an individual pillar. In pillar design, stress is assumed to be evenly distributed throughout the mine plan (Farmer, 1992). Pillar failure occurs when the stress exceeds the compressive strength of the pillar.

Depth of Coal Seam: The depth of the extracted coal seam has a direct effect on the angles of subsidence and accordingly the angle of groundwater draw. Hence: the deeper the mine the greater the potentially affected area by introducing horizontal strains affecting the groundwater draw. The coal seam depth is a risk factor to consider in order to recognize the potential to develop horizontal strains in the subsurface.

Retreat Mining: Retreat mining is a common method used to extract maximum amount of coal in a room and pillar mine. Once a mine has been excavated through traditional room and pillar methods, the pillars are then harvested, either partially or fully. The percentage of total coal mined dictates how subsidence will occur.

Angle of Dip: The angle of dip is the angle in which a coal seam rises or falls with respect to horizontal. Room and pillar mining is not recommended above angles greater than 30 degrees. In the Appalachian mountain range, the angles of dip generally fall less than 10 degrees.

Pillar Size: The width-to-height ratio of pillars affects the likeliness of how the pillars fail.

Geologic Makeup of Mine Roof or Floor: The type of rock found directly above or below a mine seam will impact the ability for support structures to function properly. Pillar design requires the pillar stress does not exceed the maximum stress capacity of the rock it is in contact with.

Figures 2 and 3 outline a generalized event tree developed for this failure mode due to room and pillar mining while Figure 4 shows the progressive event tree for longwall mining.

Initial Conditions for Longwall Mining

Characteristics of subsidence for longwall mining initiate immediately after critical width is reached and references cite full subsidence effects realized within a few weeks to 10 years post mining (Singh, 1992).The initial conditions for potential reservoir failure, where failure is considered as uncontrolled reservoir pool level, are predetermined based on the following items.

Type of Geology: The type of soils and strata located around the reservoir determines the depth location of any aquifers in the region. It also has a direct affect on the increase of permeability and type or erosion expected.

Pool level Elevation: The horizontal and vertical distances of the edge of the longwall mine to the rim of the reservoir at a given time.

The Time Frame of Operations: Determination of whether mining is currently taking place or the length of time that has passed since mining operations have occurred. Affects the probability of the change in order of magnitude in permeability (order of magnitude 1×–1000×).

Level 1 Angle of Draw: The angle of draw determines the horizontal surface distance from the mine that subsidence occurs. The typical values of angle of draw are between 10 and 38 degrees (Singh, 1992). In this range there is an expectancy of surface fracturing and strains greater than 0.001 in/in. The probability of continuation for failure is based on the proximity of this zone to the pool rim. If the reservoir edge lies within the angle of draw the probability of seepage will lie near 100%. If it lies outside of the angle of draw, the probability will become based on the expectancy of groundwater draw.

Level 2 Angle of Groundwater Draw: The angle of groundwater draw determines the horizontal distance from the mine that change in groundwater flow is likely to occur. This angle is correlated to the angle of draw but not dependant

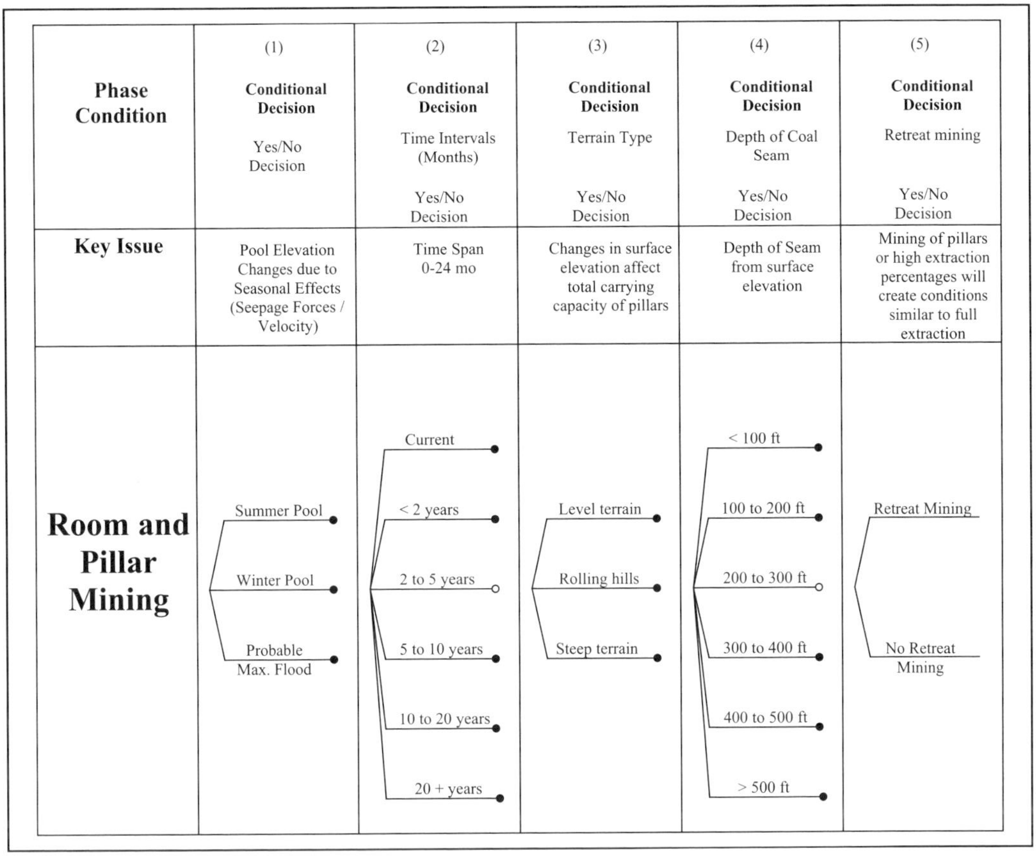

Figure 2. Failure mode event tree for room and pillar mining

on it. However, in all conditions it is expected to extend past the angle of subsidence. In the Appalachian mountain range this angle generally falls between 30 and 40 degrees with maximum angles up to 60 degrees. The angle of groundwater draw may reduce water levels outside of this range due to proximity of aquifers and horizontal groundwater flow. New paths for groundwater flow may occur creating seepage changes exceeding the limits of groundwater draw.

Level 3 Permeability Increase: The probability of permeability increase at the reservoir rim is based on the distance between the extents of subsidence and groundwater draw to the pool. The closer either of these extents are, the greater the likelihood of continuation for failure. The characteristics of the strata will have an effect on the amount of change in permeability. During mining operations a condition called a subsidence wave occurs that creates zones of tension and compression. Typically compression occurs within the mine trough while tension occurs at the edges extending outwards. However changes from compression to tension or vice versa will occur at mining passes or concludes. Typical changes in permeability range from 10× to 1000× normal conditions in tensional areas and –10× in compressional areas (Booth, 2006). Once subsidence has concluded, the permeability tends to self-heal and reduces in magnitude to pre-subsidence conditions but generally never fully recovers. Therefore the probability of permeability increase is also a function of time.

The probability of increased permeability is dependent on the type of overburden so it is site specific. Sandstones tend to increase in

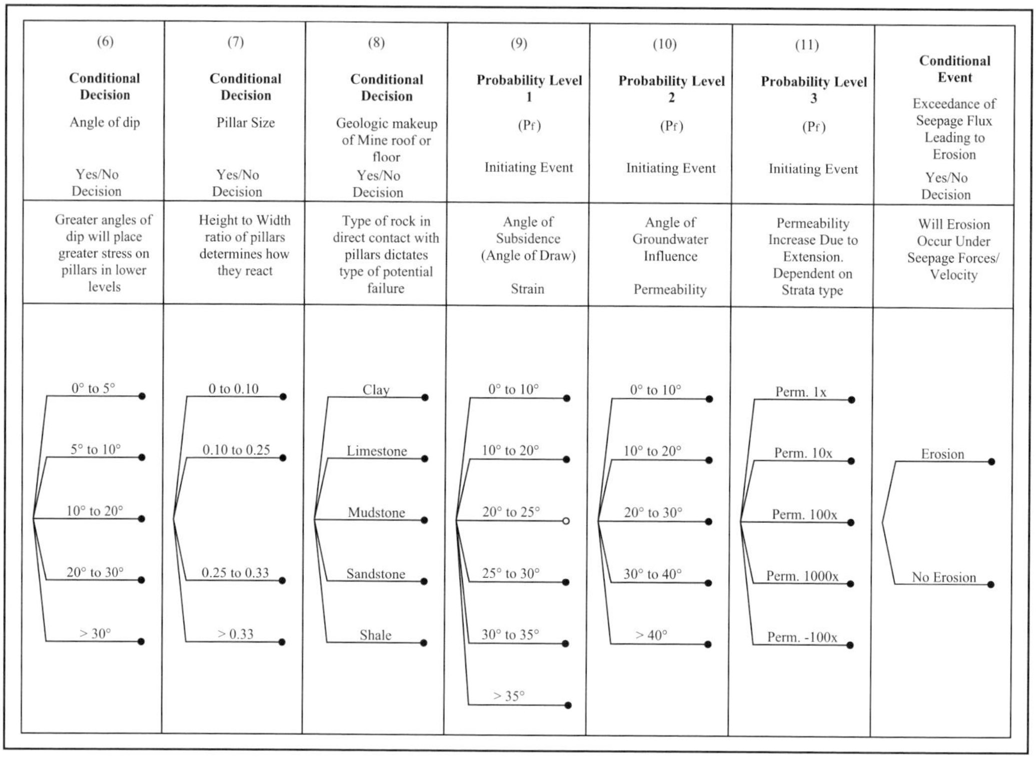

Figure 3. Continuation of failure mode event tree for room and pillar mining

permeability in the range of one order of magnitude while shale increases by a magnitude of 2 to 3 times (Van Roosendaal et al., 1995). These are the maximum changes in the permeability of the rock and will decrease over time. However, permanent change for hydraulic conductivity is normally an increase in 1 order of magnitude.

Each of these initial conditions determines the probability of occurrence for each level in the event tree. These probability levels occur for both longwall (full extraction) and room and pillar (partial extraction) mining. Once mining has occurred, pillars will begin to deform due transmission of weight (Singh, 1992). As time increases, probabilities of these levels for partial extraction will become similar to longwall mining as pillars fail and maximum subsidence occurs. A failure mode analysis based on a longwall mine initiating event was developed by Wachtel et. al 2012. In this parametric study, event probabilities were included to develop a relationship with mining offset distances.

CONCLUSIONS

The event trees presented here provide a step by step method for determining the progression of subsidence to the possibility of sub-surface erosion to occur. Current guidelines based off of IC8741are designed to shield mines from water intrusion during the lifespan of the mine. They do not reflect the potential for changes in groundwater flow in the overlying strata or the extent of subsidence once a mine is closed.

Once pillars begin to fail, full subsidence from partial extraction will imitate that from longwall mining. Maximum subsidence will dictate the extents a mine will have on groundwater draw.

To determine appropriate offset distances, both angles of subsidence and groundwater influence need to be considered as shown in Figure 5.

Literature has shown that strains induced by subsidence can cause changes in groundwater flow. The increased permeability within the

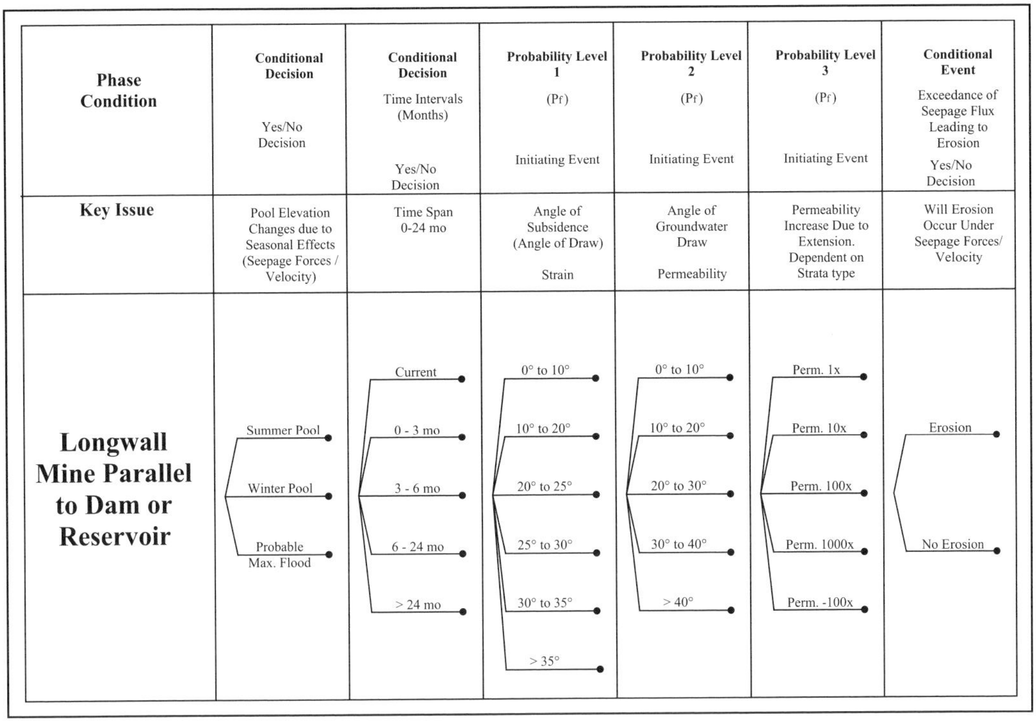

Figure 4. Failure mode event tree for longwall mining

overburden may lead to erosion and potential risk to the integrity of reservoirs and dams. This risk can occur at distances far beyond the subsidence zone which current guidelines implement for off-set distances. Further research need to be done to determine the effect subsidence from mining has on changes in permeability to accurately determine safe offset distances for mining near surface bodies of water.

ACKNOWLEDGMENT

The authors thank Dirk van Zyl for his contribution to this paper.

REFERENCES

Army Corps of Engineers (USACE) (2008), Establishment of Risk Management Center (RMC) as an Expertise Center (CX) and Directory of Expertise (DX) at the U.S. Army Institute for Water Resources (CEIWR). Department of the Army, U.S. Army Corps of Engineers, Washington, D.C., November 20, 2008.

Babcock, C.O., Hooker, V.E. (1977). Results of Research to Develop Guidelines for Mining Near Surface and Underground Bodies of Water, IC 8741. [Washington, D.C.] US Department of Interior, Bureau of Mines Information Circular.

Bai, M., Kendorski, F., Van Roosendaal, D. (1995). Chinese and North American High-Extraction Underground Coal Mining Strata Behavior and Water Protection Experience and Guidelines, Proc. 14th Conf. on Ground Control in Mining, West Virginia University, Morgantown, pp. 209–217.

Booth, C.J. (2006). Groundwater as an environmental constraint of longwall coal mining. *Environmental Geology* 49(6): 796–803.

Farmer, Ian (1992). Room and Pillar Mining. *SME Mining Engineering Handbook*. H.L. Hartman. Littleton, Colorado, Society for Mining, Metallurgy, and Exploration, Inc. Volume 2: pp. 1681–1701.

Fell, R., Bowles, D.S., Anderson, L.R., Bell, G.(2000). The status of methods for estimation of the probability of failure of dams for use in quantitative risk assessment. Proc. International Commission on Large Dams 20th Congress, Beijing, China.

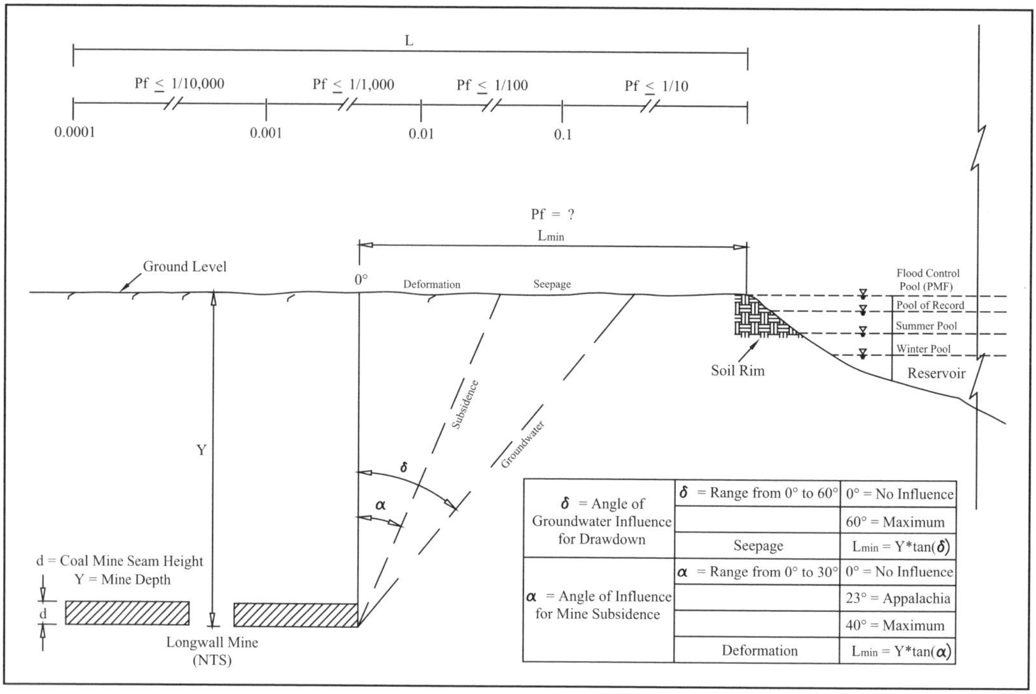

Figure 5. Range of tolerable risk offsets for subsurface mining

Kelleher, J.T., Ven Roosendaal, D.J., Mehnert, B.B., Brutcher, D.F., Bauer, R.A. (1991), Overburden Deformation and Hydrologic Changes Due to Longwall Coal Mine Subsidence on the Illinois Basin. Land subsidence. Proc 4th Int. Symp. Land subsidence, Houston, TX 12–17 May. IAHS Publication No. 200, pp. 1985–204.

Kendorski, F.S.(1993), Effect of High-Extraction Coal Mining on Surface and Ground Waters, Proc. 12th Conf. on Ground Control in Mining, West Virginia University, Morgantown, p. 412–425.

Luo, Yi, Peng, S. (2011), Subsurface subsidence prediction model and its potential application to the study of longwall subsidence effects on the hydrologic system, Transactions of the Society of Mining, Metallurgy, and Engineering, Vol. 328, pp. 458–465.

Reed, C., Rauch, H.(2001), Mining Induced Streamflow Loss for Spruce Laurel Fork in Boone County, West Virginia, Proc. 2001 National Association of Abandoned Mine Land Programs Annual Conference, Athens, Ohio, 24p.

Singh, M.M. (1992). Mine Subsidence. *SME Mining Engineering Handbook*. H.L. Hartman. Littleton, CO, Society for Mining, Metallurgy, and Exploration, Inc. Volume 1: p. 939–971.

Skelly and Loy (1976). “Guidelines for Mining Near Water Bodies, Phase III—Final Report.” U.S. Department of Interior, Bureau of Mines, Project No. HO252083, Denver, CO.

Van Roosendaal, D.J., Kendorski,F.S., Padgett, J.T. (1995). Application of Mechanical and Groundwater-Flow Models to Predict the Hydrogeologic Effects of Longwall Subsidence- A Case Study. 14th Conference on Ground Control in Mining. Morgantown, WV: 252–260.

Wardell, K., and Partners (1976). Guidelines For Mining Under Surface Water. Phase III and Final Report, U.S. Bureau of Mines—Contract No. HO.25201, K. Wardell & Partners, Newcastle Staffs [England].

Wachtel, T., Quaranta, J., Van Zyl, D., Siriwardane, H. (2012), Seepage Risk Sensitivity Study for a Potential Failure Mode Analysis due to Mining near Bodies of Water. Proceedings of Tailings and Mine Waste 2012, Keystone, CO.

Applications of Subsurface Subsidence Model to Study Longwall Subsidence Influences on Overburden Hydrological System

Biao Qiu
Department of Mining Engineering, West Virginia University, Morgantown, West Virginia

Yi Luo
Department of Mining Engineering, West Virginia University, Morgantown, West Virginia

ABSTRACT

In order to protect water resources and to avoid significant mining influences on environment, it is very important to have a tool to quantify the impacts of longwall subsidence on surface and subsurface hydrological systems. The quantitative study depends heavily on subsurface deformations, geological and hydrological system of the overburden strata. The subsurface subsidence prediction model developed by the authors can be applied in the studies of the effects of longwall mining on the hydrological system. This model is able to consider the effects of the massive hard strata on the distributions of subsidence and total strain in the overburden strata. The voids created by differential subsurface subsidence in the upper aquifer strata can serve as additional storage locations for water from surface water streams and subsurface aquifers. The differential subsurface vertical and horizontal movements can create contiguous zones with high void intensity in the overburden strata near the edges of the longwall panel as well as a short distance behind the moving longwall face. The subsidence-induced total strain would change the permeability in the overburden strata. This paper attempts to link the predicted subsurface total strain to the permeability. A computer program incorporating the distribution field of intact and disturbed permeability is developed for studying the longwall subsidence effects of the hydrological system. The model is examined and compared with a case study.

INTRODUCTION

Longwall coal mining causes movements and deformations from the coal seam level up to the surface level. Subsurface subsidence affects the permeability and porosity of the overburden strata. The potential longwall mining subsurface subsidence impacts on the overburden hydrological system include water pooling on the surface area, temporary redistribution of the hydrological system, and dewatering of surface streams and subsurface aquifers. For protecting the water resources and reducing the environmental effects during coal mining, it is very important to model and quantify the impacts of the longwall mining on the potential disruption of water resources.

A newly developed subsurface subsidence prediction model by the authors offers a useful tool to realistically simulate rock mass response to longwall mining operations. Compared to previously developed subsurface subsidence prediction models, this model makes it possible to consider the effects of the thick hard rock (i.e., limestone and sandstone) layers on the magnitudes and distributions of subsurface movements and deformations. This paper describes recent application cases of this model to investigate the effects of longwall mining on surface and subsurface hydrological system in coal mines in the eastern United States.

Coupling analyses of the rock stress and fluid is conducted to examine the permeability of the fractured rock mass. A hydro-mechanical model

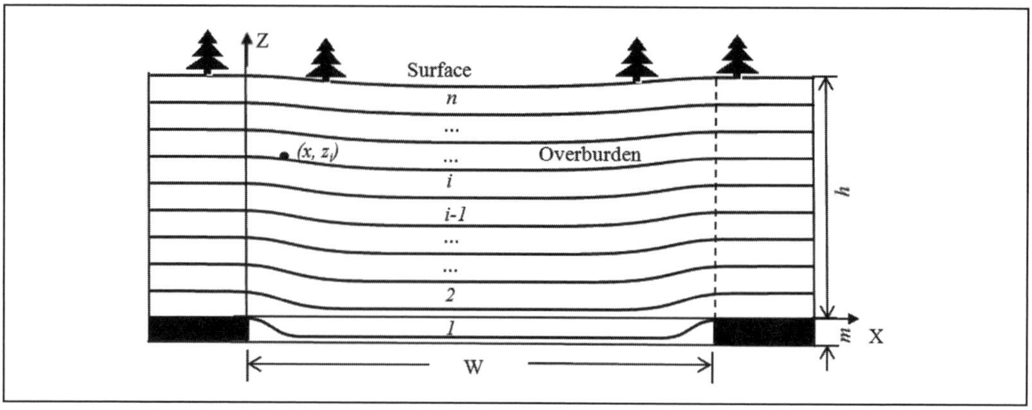

Figure 1. Subsurface subsidence prediction model (*h*—Overburden depth; *m*—Mining height; *W*—Panel width; n—Number of overburden layers)

is developed to determine the strain-permeability relationships. A two stage approach is followed to develop the models of longwall mining impacts on the overburden hydrological system. The first stage is to use the subsurface subsidence prediction model to predict the geo-mechanical response of the rock mass to longwall mining. The second stage is to calculate the subsurface permeability distribution based on the strain-permeability relationships, and to assess longwall mining impacts on the hydrological system. A computer program is developed to facilitate this process, and a case study is presented here to demonstrate the program.

SUBSURFACE MOVEMENTS AND DEFORMATIONS PREDICTION

In this subsurface subsidence prediction method, the overburden strata of the longwall gob are divided into a finite number (n) of layers of equal thickness. The layers are numbered from the immediate roof stratum to the surface as 1, 2, …, n as shown in Figure. 1. The subsidence on the top surface of a given layer can be determined by the following procedure: (1) transferring the overburden load above it into an equivalent uniformly distributed load on the layer; (2) defining the subsidence influence function at a prediction point using the equivalent load, layer thickness, percent of hard rock in the layer and vertical movement at the layer bottom directly under the prediction point, which is the mining height for first layer and the previously predicted subsidence for the subsequently layers, and (3) integrating the defined influence function within proper horizontal interval to obtain the final subsidence on top of the current layer. This procedure is repeated, layer by layer upwards, until the ground surface is reached.

The detailed subsurface subsidence prediction model is described elsewhere (Luo and Qiu 2012). The main mathematical equations of the prediction model are presented in this section. The most important step is to define the influence functions that describe the induced movements in the upper layer by the subsidence in its underlying layer. Based on the Knothe theory and focal point theorem, the derived influence functions for subsidence and horizontal displacement along any major cross-section are shown in Equations 1 and 2, respectively.

$$f_s(x', z_i) = \frac{S(x + x', z_{i-1}) \cdot a_i}{R_i} e^{-\pi\left(\frac{x'}{R_i}\right)^2} \qquad i = 1, 2, \ldots, n \tag{1}$$

$$f_u(x', z_i) = -2\pi \frac{S(x + x', z_{i-1}) \cdot a_i \cdot n}{R_i \cdot h} \cdot x' e^{-\pi\left(\frac{x'}{R_i}\right)^2} \qquad i = 1, 2, \ldots, n \tag{2}$$

$$S(x, z_i) = \frac{a_i}{R_i} \int_{d_{i1}-x}^{W_i - d_{i2} - x} S(x + x', z_{i-1}) \cdot e^{-\pi \left(\frac{x'}{R_i}\right)^2} dx' \qquad i = 1, 2, \ldots, n \tag{3}$$

$$U_i(x, z_i) = -2\pi \frac{a_i \cdot n}{h \cdot R_i} \int_{d_{i1}-x}^{W_i - d_{i2} - x} S(x + x', z_{i-1}) \cdot x' \cdot e^{-\pi \left(\frac{x'}{R_i}\right)^2} dx \qquad i = 1, 2, \ldots, n \tag{4}$$

The final subsidence and horizontal displacement at a prediction point in the overburden are determined by integrating the respective influence functions within the computing zone as shown in Equations 3 and 4. In these equations, a_i, R_i, d_{i1} and d_{i2} are the final subsidence parameters for the *i*th layer that are very important for subsidence prediction accuracy. The methods to determine these final subsidence parameters are discussed in a previous publication (Luo and Qiu 2012).

The differential strata movements in both horizontal and vertical directions induce deformations in the subsurface strata. In surface subsidence studies, surface deformations are traditionally described by slope, strain and curvature. However, for applications dealing with subsurface subsidence, the distributions of horizontal, vertical and total strains in the overburden strata are very useful for assessing the subsidence influences on subsurface structures, hydrological system and gob well degasification operations. Horizontal strain (ε_x) is defined as the first derivative of horizontal displacement with respect to *x* (Equation 5). Sufficient horizontal strain could cause vertical fractures or even cracks in the strata. The vertical strain (ε_z) is defined as the first derivative of subsurface subsidence with respect to *z* (Equation 6). Sufficient vertical strain could cause bed separations along the strata bedding planes or even step cracks. The total strain (ε_t), defined in Equation 7, is an indicator of the severity of expansion or shrinkage of a volume of rock strata under the influence of the subsidence process. It should be noted that the expansive type of total strain (in positive value), reflecting the intensity of voids, is an indicator of the increase in porosity and permeability for seepage flows of gases or water in the subsurface strata. For simplicity, the expansive total strain is also called void intensity.

$$\varepsilon_x(x, z) = \frac{dU(x, z)}{dx} \tag{5}$$

$$\varepsilon_z(x, z) = \frac{dS(x, z)}{dz} \tag{6}$$

$$\varepsilon_t(x, z) = \varepsilon_x + \varepsilon_z + \varepsilon_x \cdot \varepsilon_z \tag{7}$$

MINING INDUCED SUBSURFACE PERMEABILITY CHANGE

Permeability is a property of soil or rock, which describes the ease with which water can move through the pore spaces or fractures. In Darcy's law, it relates the water flow rate through a unit cross-sectional area under a unit gradient of hydraulic head. The permeability for groundwater flow in a particular rock is heavily dependent both on the type and micro-structure of the rock, as well as the stress conditions applied to the rock (Rutqvist and Stephansson 2003). It is apparent that application of a sufficient tensile stress could increase the porosity of the rock and consequently its permeability (Singhal and Gupta 2010). Therefore, stress effects on permeability should be considered in the groundwater flow simulations.

To examine the effects of subsurface subsidence on overburden hydrological systems, the first effort is to establish the stress-permeability relationship (Bai and Elsworth 1994). Longwall mining induces stress and strain changes in the overburden strata. Tension and compression zones will be formed around the edges of the panel from the mining horizon to the ground surface. Consequently, additional pores will be created in the tension zones while the original pores will be shrunk in the compression zones. These changes can have a profound effect on the rock mass permeability. Field studies have shown both increase and decrease of approximately one

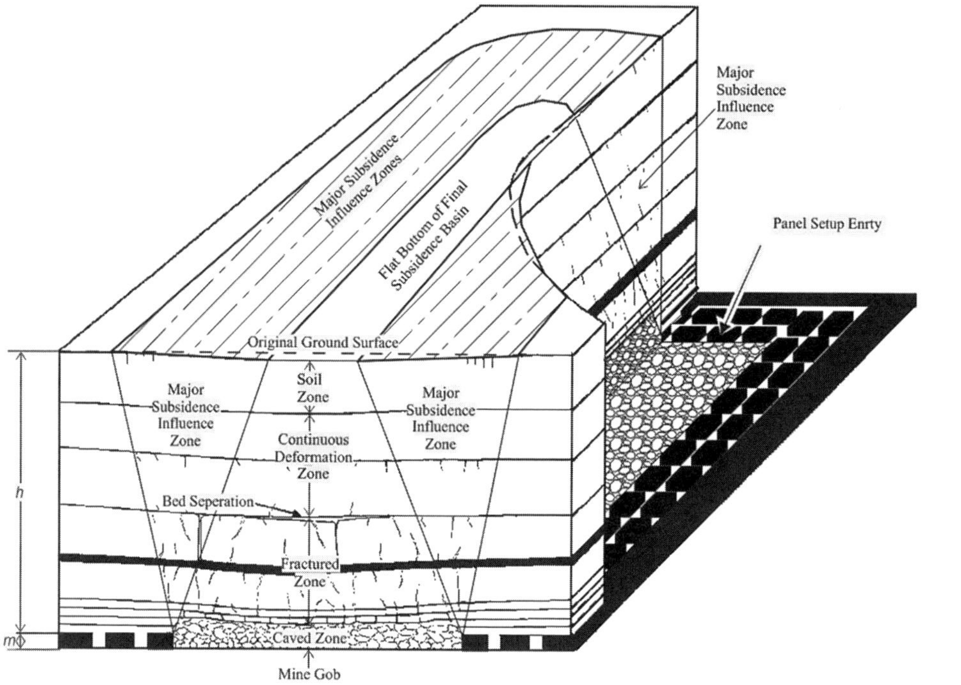

Figure 2. Overburden deformation zones caused by longwall subsidence

order of magnitude in the permeability of the rock mass above a longwall panel (Hasenfus et al. 1988; Esterhuizen and Karacan 2005).

Longwall subsidence can cause varying degrees of disturbance to the overburden strata, as shown in Figure 2. The caved zone ranges in thickness from two to ten times the mining height (Peng 2006). It is characterized by irregular rock fragments as well as high void ratios and permeability. Laboratory tests have shown that the void ratio in the caved zone can be on the order of 30–45% (Esterhuizen and Karacan 2005).

The fractured zone is located above and around the caved zone, which ranges in thickness from 28 to 52 times the mining height. It is characterized by vertical and/or sub-vertical fractures and horizontal cracks due to bed separation. In this zone, water can drain directly into the caved zone and then to the mine gob. Measurements of permeability in the fractured rock have shown up to forty fold increases in permeability (Hasenfus et al. 1988).

Between the fractured zone and the ground surface is the continuous deformation zone. In this zone, the strata deform gently without causing any major cracks that extend far enough to cut through the thickness of the strata, as in the fractured zone. The mine subsidence has insignificant effects on permeability and water storage in this zone.

On the surface, there is a soil zone of varying depth, depending on the location. In this zone, cracks open and close as the longwall face comes and goes. In general, cracks on and near the panel edges tend to remain open permanently, whereas those in and around the center of the panel will close back up when the longwall face has passed for a sufficient distance. The surface fracture zone potentially has vertically transmissive surface cracks and disruptions, which can cause the dewatering of surface water bodies (Kendorski 2006).

Under increased compressive stress, the voids in and the permeability of an elastic rock will be reduced. However, when the stress on the rock exceeds the elastic limit, the rock will behave plastically, in which state the stress changes insignificantly while the strain increases quickly. Therefore, most stress-permeability models can not accurately predict the permeability change for large deformation conditions such as those in the caved and fractured zones. For this reason, the strain-permeability model is a better approach for estimating the permeability change in problems dealing with large rock deformation.

Reiss (1980) links a rock's permeability to its porosity by Equation 8. In the equation, K is the permeability under mining influence, K_0 is the initial permeability, φ is the porosity under mining influence, φ_0 is the initial porosity.

$$\frac{K}{K_o} = \left(\frac{\varphi}{\varphi_0}\right)^3 \tag{8}$$

The porosity under mining influence can be determined by the initial porosity and the total strain (e_t) using the following equation.

$$\varphi = \frac{\varphi_0 + \varepsilon_t}{1 + \varepsilon_t} \tag{9}$$

By combining Equations 8 and 9, the resulting mathematical formula to determine the permeability change at a given subsurface point is shown as Equation 10. In this equation, the total strain (e_t) at the specified point can be predicted using the subsurface subsidence prediction model.

$$K = K_o\left(\frac{1 + \frac{\varepsilon_t}{\varphi_0}}{1 + \varepsilon_t}\right)^3 \tag{10}$$

Figure 3 shows the relationship between the subsidence-induced total strain and the permeability under different initial porosity conditions (0.05, 0.10, 0.15, and 0.20). The permeability increases with the total strain. The permeability of the rock with lower porosity will be more significantly affected by total strain.

APPLICATION CASE

The proposed methodology is applied to assess the subsidence influence to a large water reservoir that provides water needs to a nearby small city. Mining operation in one longwall panel is planned to be conducted directly under the tail half of the reservoir while its earth dam is located beyond the panel over the solid coal as shown in Figure 4. The longwall panel width is 1,425 ft

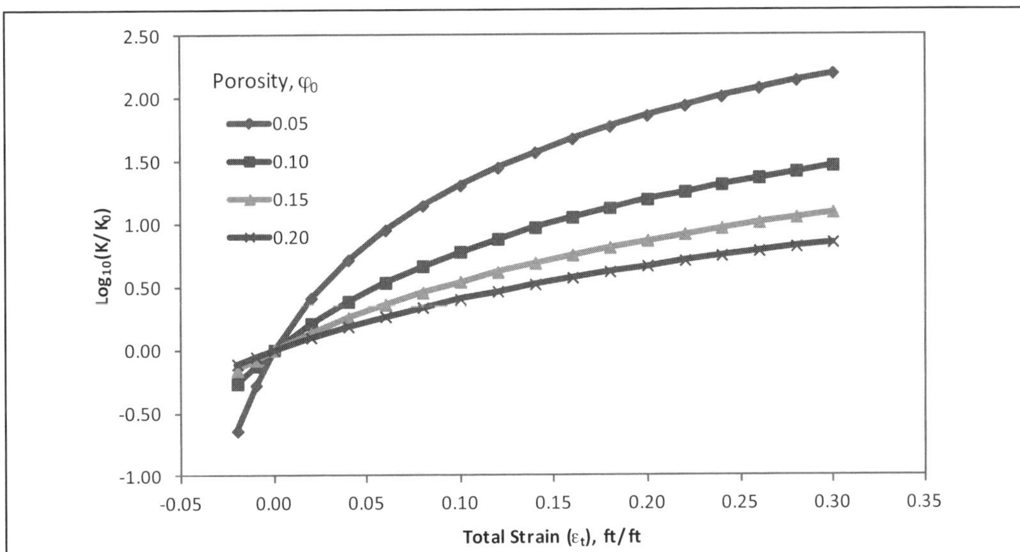

Figure 3. Relationship of total strain and permeability

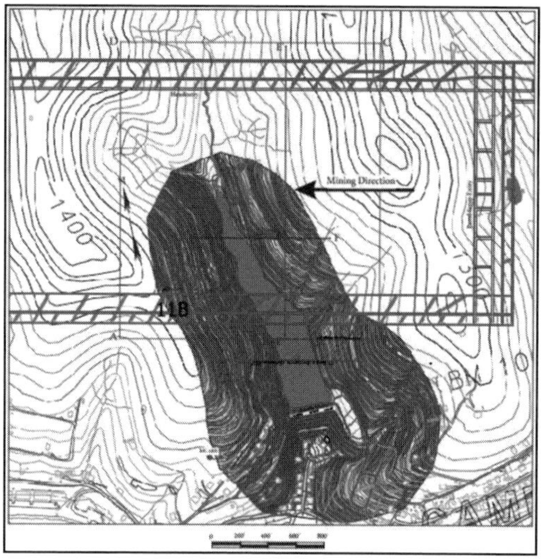

Figure 4. The reservoir and the longwall panels under it

(434 m). The smallest overburden depth under the reservoir area is about 720 ft (219 m).

Two main important concerns for this project are: (1) whether the longwall mining subsidence can induce leakage problem and change the water-holding capacity of the reservoir, and (2) whether water leaked from the reservoir would affect the underground longwall operation. In order to address these two concerns, a comprehensive subsidence study is conducted. In the study, it is important to find whether contiguous highly-fractured zones will be formed in the overburden strata to connect the surface reservoir and the underground longwall gobs.

After the construction of the reservoir, a layer of silt with varying thickness, up to 15 ft (4.6 m), has been deposited on the bottom of the reservoir after its construction, decreasing its water-holding capacity. The provided core log information from a nearby geological exploration hole is shown in Figure 5. It shows that about 39 ft (12 m) of claystone and shale layer is located right below the reservoir bottom. After a 7.9-ft (2.4 m) sandstone layer, another 36 ft (11 m) of claystone layer follows. The next 150 ft (46 m) overburden strata contains claystone layers of 14, 10, 9 and 28 ft (4.3, 3.0, 2.7 and 8.5 m) thick. Therefore, the impermeable claystone layers account for about

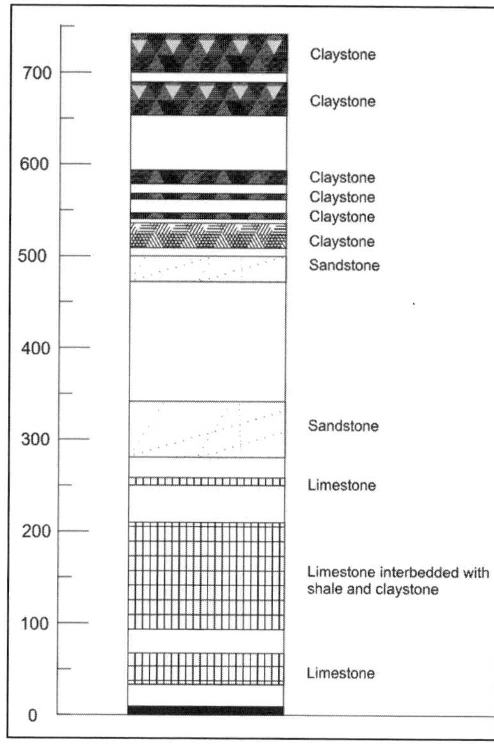

Figure 5. Geological column at mine site

58% of the top 230 ft (70 m) overburden strata. Two layers of sandstone (28 and 60 ft or 8.5 and 18.3 m thick) are located about 240 and 400 ft (73 and 122 m) below the bottom of the reservoir, respectively. It is also important to note the presence of substantial limestone beds in the roof of the coalbed. From the bottom up, the maximum height of the coal seam to be mined is 9.1 ft (2.8 m). Three limestone layers, 30, 120 and 8 ft (9.1, 36.6 and 2.4 m) thick, are located about 40, 90 and 250 ft (12, 27, 76 m) above the coal seam. The limestone is stiffer and stronger than the shale, claystone and siltstone strata.

In the study, surface final and dynamic predictions using CISPM-W, a comprehensive surface subsidence program package, are performed. The predicted maximum final and dynamic surface tensile strains at the bottom of the reservoir are just capable of initiating hairline cracks on the ground surface. However, due to the thick silt deposit on the reservoir bottom, such hairline

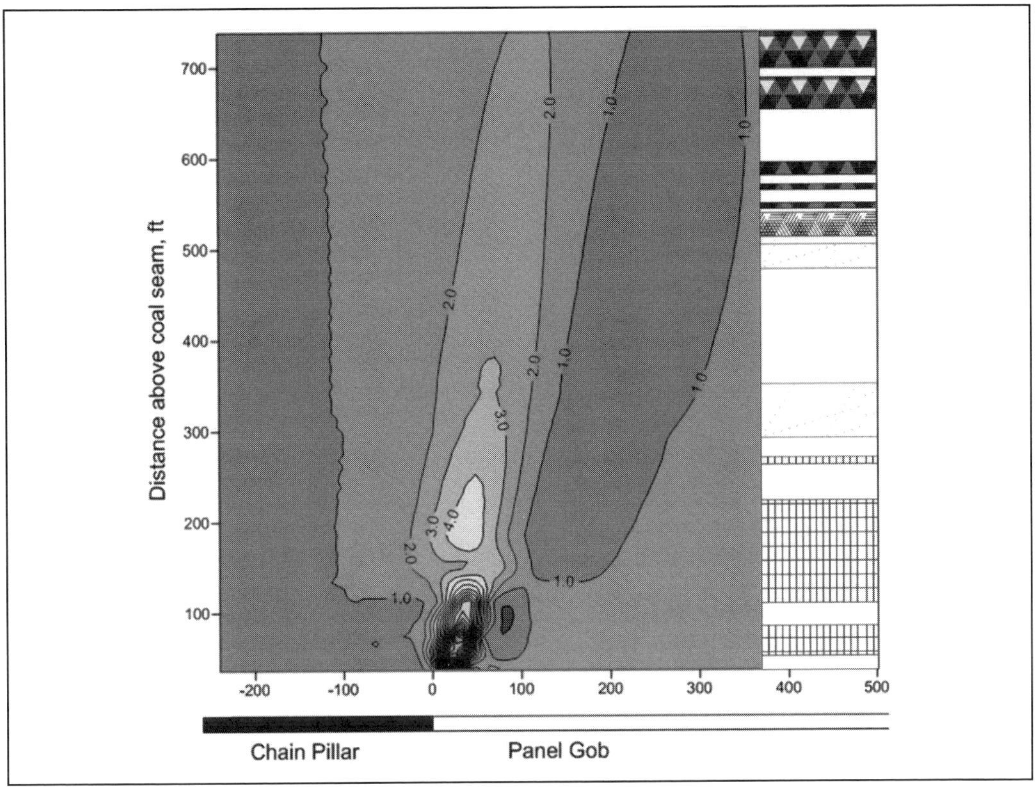

Figure 6. Permeability change (k/k_0) caused by the mining of the longwall panel

cracks are likely to be filled with the silt once they are created.

Using the new subsurface subsidence prediction model, the final movements and deformations in the subsurface strata are predicted. After substituting the predicted final void intensity into Equation 10, the subsidence influence to the permeability of overburden strata under the reservoir can be determined and the results are plotted in Figure 6. The bridging effect of the 120-ft (36.6-m) thick limestone layer located about 90 ft (27 m) above the coal seam is clearly shown in the figure. This thick competent rock layer considerably changes the distribution patterns of strata deformations and the permeability below and above it. The prediction shows that the maximum subsidence influence on the strata permeability at the bottom of reservoir is that the initial permeability will be doubled. A zone of high permeability increase occurred in the area a short distance inside the longwall panel edge.

In order to assess the possibility for the reservoir water to leak into the longwall gob in large quantity, a numerical simulation study has been conducted. As shown in Figure 6, the most possible water leakage path is along the high permeability increase zone from the surface to the mine level near the panel edge. The numerical study focused on assessing the possibility of the water leakage through this path. One half width of the longwall panel was selected for the numerical simulation. The 2-D numerical model consisted of about 2,300 finite elements and simulated the overburden above the longwall panel of 748 ft (228 m) in width and 738 ft (225 m) in depth, as shown in Figure 7. Element sizes varied, but were selected so that the element size was about 3 ft (0.91 m) in the zone of interest, near the panel edge. Element sizes increase with increasing distance from the area of interest.

In building the numerical model, the ground water elevation at mine gob was set to be zero,

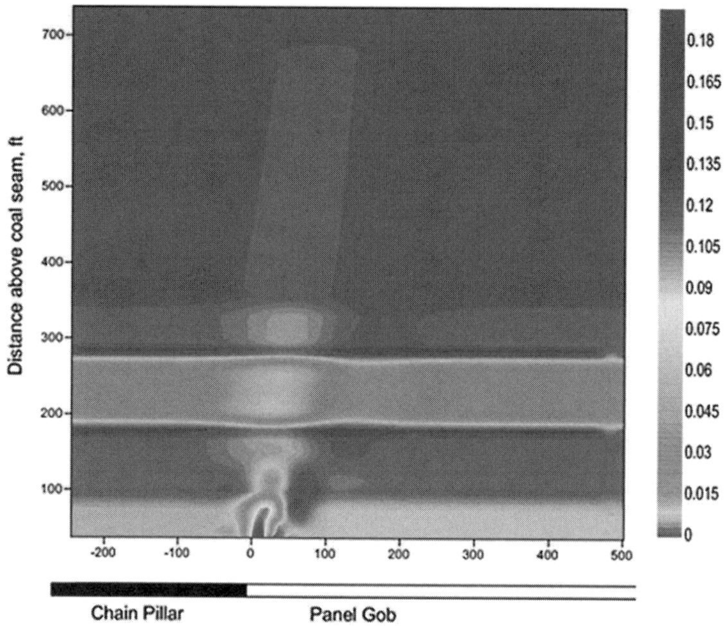

Figure 7. Post-mining vertical hydraulic permeability distribution over the longwall panel

Table 1. Initial hydraulic property of coal measure rocks

Rock Types	Permeability, ft/day (mm/day)		Porosity, %
	Horizontal	Vertical	
Soil	2.74E-04 (0.08)	2.74E-04 (0.08)	0.15
Claystone	5.48E-04 (0.17)	5.48E-04 (0.17)	0.10
Shale	2.74E-03 (0.83)	1.37E-03 (0.42)	0.05
Coal	2.74E-03 (0.83)	2.74E-04 (0.83)	0.05
Limestone	5.47E-03 (1.67)	5.47E-03 (1.67)	0.10
Sandstone	2.74E-02 (8.35)	2.74E-02 (8.35)	0.15

and the ground water elevation at the surface equaled the reservoir water elevation minus the coal seam elevation (738 ft or 225 m). The vertical boundaries of the model were set to be impermeable. The actual overburden sequence has been simplified, combining lithological layers to represent rock characteristics of primary importance to obtain an average response. Two numerical models were built to represent the pre-mining and post-mining conditions, respectively.

Table 1 presents the initial hydraulic property of the coal measure rocks used in the numerical simulation, which is determined from published values (Esterhuizen, and Karacan 2005). The hydraulic permeability of the overburden strata after mining is calculated based on the predicted final total strain distribution; the results are plotted in Figure 7. It shows that a zone of high permeability is located in the sandstone strata between 200 and 300 ft (61 and 91 m) above the coal seam. The longwall subsidence induces high total strain and high permeability in the area located near but inside the longwall panel edge. In this area, the zones with high permeability are limited within the 350 ft (106 m) from the coal seam. Above this level, the subsidence influence to the permeability is insignificant due to much lower total strain and the low permeability of the

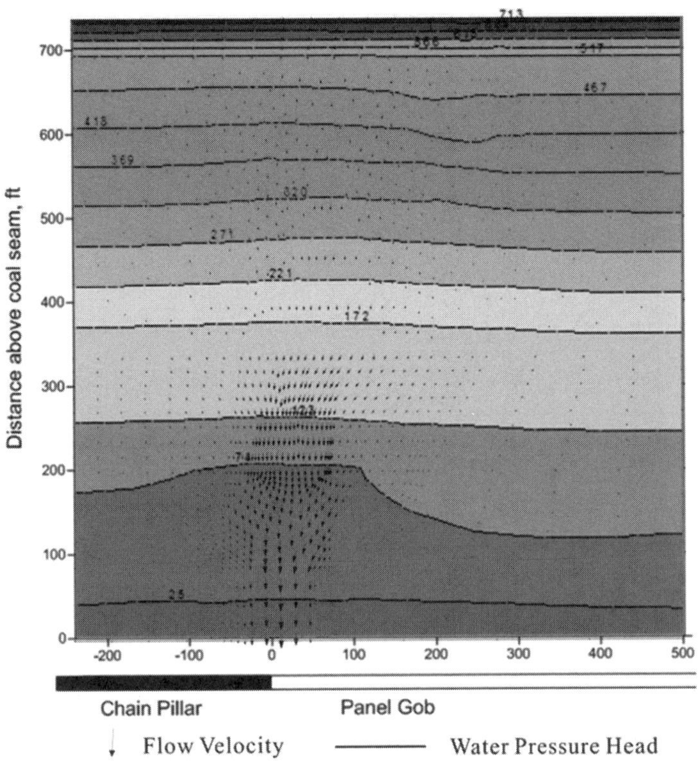

Figure 8. Post-mining ground water pressure head and flow velocity distribution over the longwall panel

claystone strata and thick silt at the bottom of the reservoir. The maximum vertical permeability after mining reaches 0.18 ft/day (54.86 mm/day) in the area directly above the longwall panel edge. However, the vertical permeability in most of the other areas is less than 0.09 ft/day (27.43 mm/day).

The distribution of the simulated post-mining ground water pressure head in reference to the mine level is plotted in Figure 8. In the zone where increased permeability and water flow are induced by mine subsidence, the water head contour lines bent upward. It shows that the head at a point has decreased from its original level. The vectors of water flow velocity in the overburden strata after subsidence are also plotted in Figure 8. The zone with higher flow velocities is located a short distance inside the panel edge. Based on the simulation, the water leakage rate from the surface reservoir to the mine gob is about 1.42 ft^3/day per foot (0.132 m^3/day per meter) of distance along the panel longitudinal direction. From Figure 4, the equivalent average width of the reservoir water surface along the panel longitudinal direction is determined to be 83.8 ft (25.5m). The daily water leakage from the reservoir to underground longwall gob is estimated to be 119 ft^3/day (3.4 m^3/day). For the underground mining operation, the increased in-flow of water (0.62 gallons/min or 0.0024 m^3/min) to the mine from the surface reservoir is very insignificant to the mine pumping system. Therefore, the water leaked from the reservoir to the underground mine through the subsidence-disturbed overburden strata should not create a safety concern.

As shown in the previous simulation, the longwall subsidence is very unlikely to draw any significant water from the reservoir in comparison to its water supplies. The only remaining

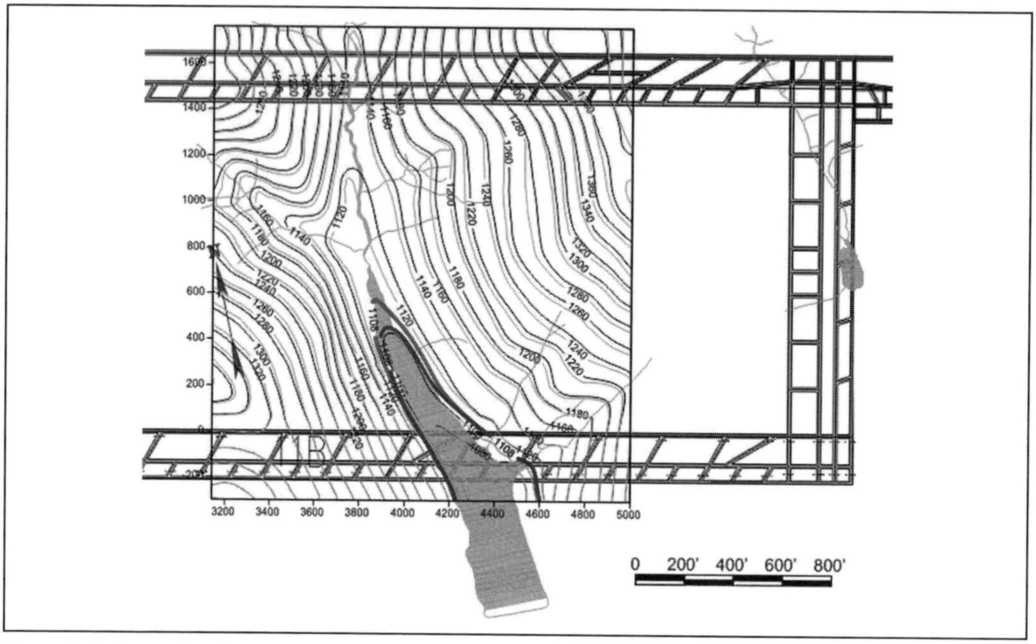

Figure 9. Surface elevation contours before (light blue) and after (black) mine subsidence

question is whether the subsidence would reduce the water-holding capacity of the reservoir. For this assessment, the final subsidence prediction is performed in a rectangular area around the reservoir. The surface coordinates and elevations at a large number of selected points in the specified area before mine subsidence are read from the mine map. After the predicted final subsidence at each of the selected points is subtracted from its original elevation, it results in the new elevation at that point after subsidence. The surface topography contour lines before (light blue) and after (black) subsidence are plotted in Figure 9. The subsidence event moves the surface contour lines in the area above the longwall panel away from the reservoir.

The water surface boundaries at the current water level of 1,108 ft (338 m) before (dark blue) and after (red) mine subsidence are also plotted in Figure 9. It shows that the elevation changes would expand the tail water side of the reservoir edge for a distance of about 150 ft (46 m) to the upper stream. The volume calculation performed shows that the longwall subsidence event will increase the water-holding capacity of the reservoir by 291,863 ft^3 (8,265 m^3) if the water surface level is maintained at current level. The increased capacity should be actually an added benefit to the nearby city.

CONCLUSIONS

The enhanced subsurface subsidence prediction model provides a good tool for analyzing the longwall mining influence on surface and subsurface hydrological systems. The total strain calculated in the subsurface subsidence prediction represents the shrinkage or expansion of the rock volume in the overburden strata. Such change would affect the permeability of the overburden strata. The relationship between the total strain and the permeability change is established. A case study is presented to demonstrate the capability of the subsurface subsidence prediction model and its application in studying the subsidence effects on surface and subsurface hydrological systems. Due to relatively large overburden depth and overburden composition in the case, the longwall mining operation will have very insignificant influence on the surface reservoir and the underground mining operation.

REFERENCES

Bai, M., D. Elsworth, 1994, "Modeling of Subsidence and Stress-dependent Hydraulic Conductivity of Intact and Fractured Porous Media," Rock Mechanics and Rock Engineering, Vol. 27, No. 4, pp. 209–234.

Esterhuizen, G.S., and C.O. Karacan, 2005, "Development of Numerical Models to Investigate Permeability Changes and Gas Emission around Longwall Mining Panel," *The 38th U.S. Symposium on Rock Mechanics (USRMS)*, Anchorage, Alaska, pp. 1–13.

Hasenfus, G.J., K.L. Johnson, and D.W. Su, 1988, "Hydrogeomechanics Study of Overburden Aquifer Response to Longwall Mining," *Proceedings 7th International Conference on Ground Control in Mining*, Ed. S.S. Peng, West Virginia University, Morgantown, WV, pp. 149–162.

Kendorski, F.S., 2006, "Effect of Full-Extraction Underground Mining on Ground and Surface Waters: A 25-Year Retrospective," *Proc. 25th Int'l Conference on Ground Control in Mining*, Ed. S.S. Peng et al., West Virginia University, WV, pp. 425–430.

Luo, Y., S.S. Peng, C.T. Holland, and Y.Q. Zhang, 2001, "Simulation of water seepage through, and stability of, coal mine barrier pillars, " *SME Annual Meeting Pre-print 01-131*, pp. 1–7.

Luo, Y., and B. Qiu, 2012, "Enhanced Subsurface Subsidence Prediction Model Considering Overburden Stratifications," *Mining Engineering*, Vol. 64, No. 10, pp. 78–84.

Peng, S.S., 2006, "Coal Mine Ground Control, Third Edition," West Virginia University, Morgantown, WV, 764 p.

Rutqvist, J., O. Stephansson, 2003, "The Role of Hydro-mechanical Coupling in Fractured Rock Engineering," Hydrogeology Journal, Vol. 11, No. 1, pp. 7–40.

Reiss, L, 1980, "The Reservoir Engineering Aspects of Fractured Formations," Houston, TX: Gulf Publishing Co., 108 p.

Singhal, B.B.S., R.P. Gupta, 2010, "Applied hydrogeology of fractured rocks, second edition," Springer Dordrecht Heidelberg London New York, 408 p.

Addressing the Effect of Sloping Terrain on Ground Movements Due to Underground Mining

Z. Agioutantis
Virginia Center for Coal and Energy, Virginia Tech, Blacksburg, Virginia

M. Karmis
Virginia Center for Coal and Energy, Virginia Tech, Blacksburg, Virginia

ABSTRACT

The prediction of ground movements due to underground mining is now a mature methodology that can be applied to perform accurate and realistic predictions of potential impacts to surface structures and facilities. Through the application of the influence function method, a number of deformation indices can be calculated using average or conservative parameters, including subsidence, slope, horizontal strain and surface curvature, at any point on the surface, or at any other elevation above the extracted seam.

In the case of horizontal displacement calculations, research has identified the effect of sloping terrain on the displacement vectors. Surface points, once disturbed by the subsidence process, tend to displace downwards, due to sloping surface terrain, as well as towards the subsidence trough.

In this paper, a new tool will be presented that provides a methodology to allow for site specific adjustments of horizontal displacements due to sloping terrain. This method has been incorporated into the Surface Deformation Prediction System (SDPS) package developed at Virginia Tech. A case study is presented where the application of the terrain slope adjustment in ground movement predictions is validated by field measurements.

INTRODUCTION

It is well documented in the literature (Peng and Luo 1989; Luo et al. 1996) that horizontal displacements induced by underground mining under sloping terrain may be different from horizontal displacements measured on a flat surface. Furthermore, Luo et al. (1996) conclude that the main factors that affect the magnitude and orientation of horizontal displacement of surface points are the total movement at each point and the angle of the natural slope at each point. According to their analysis, total movement includes both vertical and horizontal incremental displacements, prior to applying any corrections due to the sloping terrain. At the same time, work by Khair et al. (1987) indicates that topography has no effect on the amount of vertical subsidence, but "the direction of horizontal displacements exclusively depends on topography when the slope is significantly large."

It is evident that at a particular point, the displacement mechanism relates to soil properties and conditions (Karmis et al., 1990; 1991). At a larger scale, mathematical and/or numerical models as described in slope stability evaluation procedures may be applied in order to obtain an overview of movements. This usually involves precise knowledge of soil and overburden mass properties and elastic/plastic response constants, as well as parameters that describe potential sliding failure such as cohesion and angle of internal friction. In this work, a simple mathematical algorithm relying on a single parameter is derived and implemented that allows for more accurate calculations of horizontal displacement over undermined terrain.

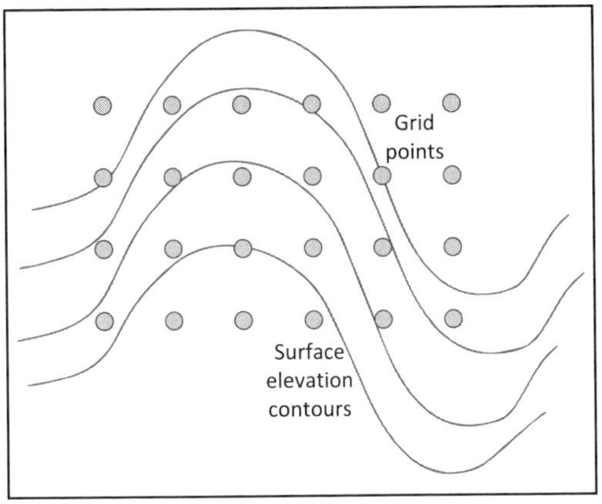

Figure 1. Surface elevation contours and grid points

REPRESENTING SURFACES WITH DIGITAL MODELS

'Digital elevation model' (DEM), 'digital terrain model' (DTM) and 'digital surface model' (DSM) are all terms used to represent a surface terrain (with or without vegetation, structures, etc.) digitally. These models are created from terrain elevation data. Commonly, a DEM can be represented as a vector-based triangular irregular network (TIN) or as a raster (i.e., a set of rectangular grids). In this paper, the term 'digital elevation model' will be used to refer to the digital representation of the topographic surface used in the ground deformation calculations above an undermined area.

In general, more detailed local representations of surfaces are mathematically represented by polynomial equations. Depending on the complexity of such equations, elevation data at several points may be needed to obtain a complete and accurate mathematical model. Elevations at grid points (points on the surface that are evenly spaced as shown in Figure 1. Surface elevation contours and grid points) are commonly used to provide data for such equations.

Although any order of polynomial may be used to model local surface patches, a quadratic function can be regarded as a successful compromise between the number of data points required to uniquely model the surface and the accuracy with which the model fits the true surface. By the nature of such models, it is evident that the equation will be most accurate in the vicinity of these data points. Planar surfaces would only require three points to define the polynomial equation; therefore, they could be modeled using four grid points defining a local window. However, a planar model is rarely sufficient to model actual surfaces in a large or local scale. Polynomials of order three (3) or higher can model more convoluted surfaces including breaks of slope (points of inflection), but require ten points to define the surface. Since this would require a local window of more than 3 by 3 grid points (2 by 2 cells), it is usually impractical for common applications. Thus, the quadratic function, shown in Equation 1, appears the most widely and appropriately used model for general geomorphometric approximation (Wood, 1996). To determine the coefficients of this equation, six known elevation values are needed.

$$z = ax^2 + by^2 + cxy + dx + ey + f \tag{1}$$

Alternatively, Equation 1 can be written in the form of the general conic:

$$ax^2 + by^2 + 2hxy + 2jx + 2ky + m = 0 \tag{2}$$

where $h=c/2$, $j=d/2$, $k=e/2$ 2, and $m=f\text{-}z$.

Instances of the general conic fall into one of three types (hyperbolic, parabolic or elliptic), depending on the values of the coefficients a, b and h. Terrain parameters such as slope angle and slope aspect can be defined by considering the partial differential equations of the general quadratic Equation 1 or the general conic Equation 2, as detailed in the following section.

Slope Angle and Slope Aspect

The rate of change of elevation in both the x and y directions can be used to identify the direction and magnitude of the steepest gradient. These two parameters can be found by taking the partial first order derivatives of (1) above, with respect to x and y:

$$\frac{\partial z}{\partial x} = 2ax + cy + d \tag{3}$$

$$\frac{\partial z}{\partial y} = 2by + cx + e \tag{4}$$

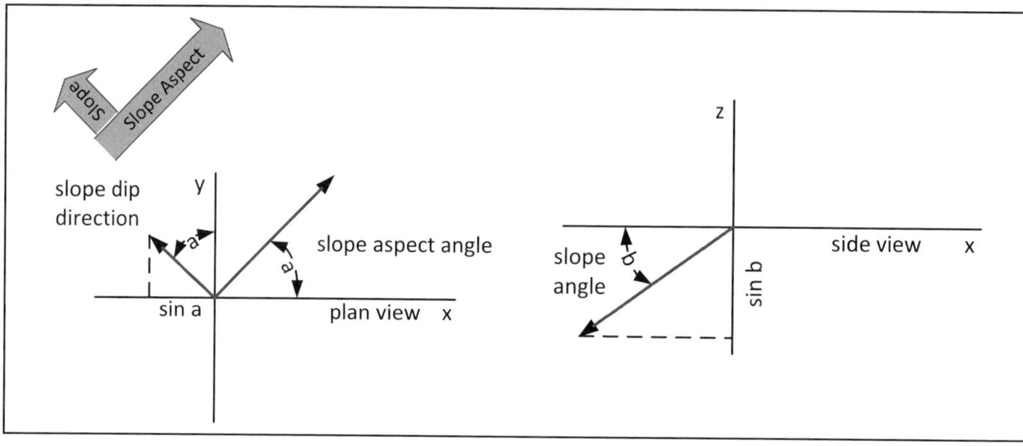

Figure 2. (a) Slope aspect angle and slope dip direction in the x-y axes; (b) slope angle in the x-z axes

The tangent of the maximum slope angle at a point can be found by combining Equations 3 and 4:

$$\text{Slope Tangent} = \sqrt{\left[\frac{\partial z}{\partial x}\right]^2 + \left[\frac{\partial z}{\partial y}\right]^2} \tag{5}$$

By adopting a local coordinate system with the origin located at the point of interest on a specific slope (i.e., x=0, y=0), Equation 5 becomes:

$$\text{Slope Tangent} = \sqrt{d^2 + e^2} \tag{6}$$

where slope is expressed as the tangent of the (maximum) slope angle. Equation 7 shows the conversion of the slope tangent to the slope angle in degrees or radians. The slope angle represents the maximum (natural) slope at the surface location, which is represented by the quadratic Equation 1:

$$\text{Slope Angle} = \arctan[\sqrt{d^2 + e^2}] \tag{7}$$

In a similar fashion, slope aspect (also known as slope direction or slope bearing) is a line defined by its polar angle, which is described in Equation 9 by the two orthogonal partial derivatives (Srinivasan and Engel, 1991).

$$\text{Slope Aspect} = \arctan\left[\frac{e}{d}\right] \tag{8}$$

The slope dip direction is the actual direction of the maximum slope, and it is always orthogonal to the slope aspect angle. Observing the right-hand coordinate system, the slope dip direction can be calculated by Equation 9:

$$\text{Slope Dip Direction} = \text{Slope Aspect} + 90 \tag{9}$$

Note that Equations 6–9 can also be given in terms of the coefficients of the general conic Equation 2. Figure 2a shows the slope aspect and slope dip direction, while Figure 2b shows the slope angle.

Figure 3 illustrates maximum slope angle and slope direction (aspect) for four cases of planar surfaces digitized by square grids. The numbers in the circles represent elevations in units of length. The dashed lines represent the respective elevation contours, while the arrows show the slope aspect or slope direction and the direction of the maximum slope angle for each case, respectively.

By applying Equations 7 and 8 to the first planar surface, it can be easily calculated that the slope aspect is –45°, the direction of the maximum slope angle is +45°, and the maximum slope angle is about 35°.

CALCULATION OF HORIZONTAL DISPLACEMENTS DUE TO UNDERGROUND MINING USING THE INFLUENCE FUNCTION METHOD

Influence function methods have been widely applied for the calculation of a number of surface deformation indices including subsidence and horizontal displacement. Utilizing the

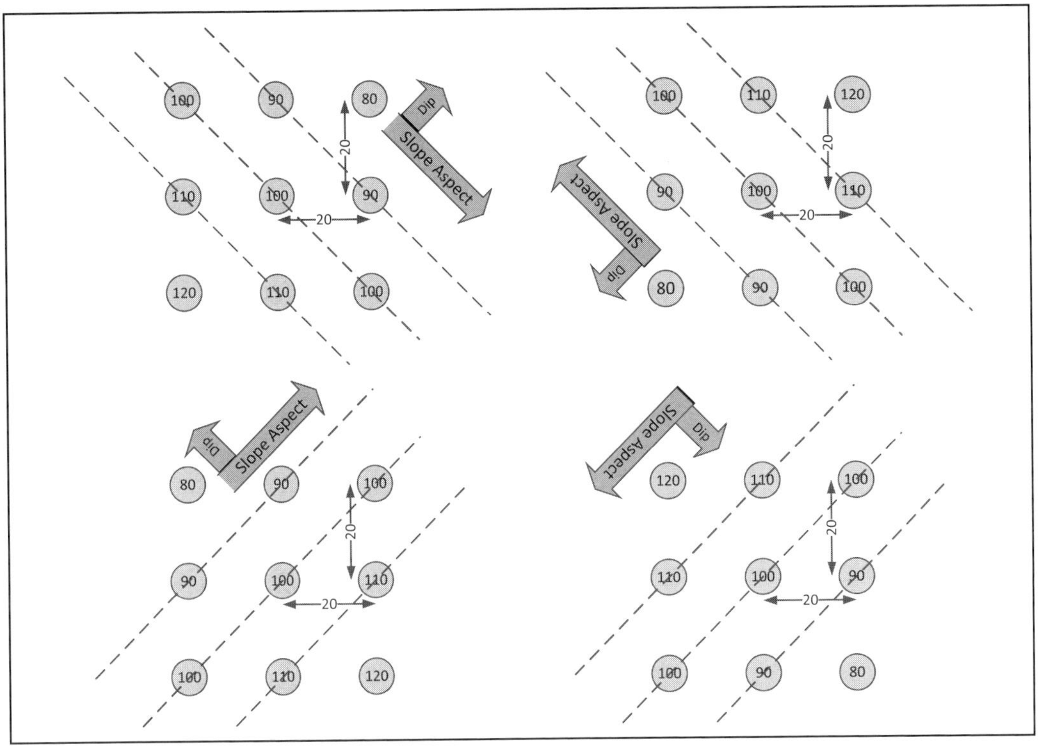

Figure 3. Slope dip direction and slope aspect for four sloping planar surfaces

bell-shaped Gaussian function, subsidence $g(x, s)$ at any point can be given by:

$$g(x,s) = \frac{So(x)}{r} e^{\left(-\pi \frac{(x-s)^2}{r^2}\right)} \qquad (10)$$

where

r = the radius of principal influence = $h/\tan(\beta)$;
h = the overburden depth;
β = the angle of principal influence;
s = the coordinate of the point, $P(s)$, where subsidence is considered;
x = the coordinate of the infinitesimal excavated element; and
$So(x)$ = the convergence of the roof of an infinitesimal excavated element

Therefore, subsidence at any point $P(s)$ can be expressed by the following equation:

$$S(x,s) = \frac{1}{r}\int_{-\infty}^{+\infty} So(x)\, e^{\left(-\pi \frac{(x-s)^2}{r^2}\right)} dx$$

and for finite extraction limits (x_1, x_2)

$$S(x,s) = \frac{S\max}{r}\int_{x_1}^{x_2} e^{\left(-\pi \frac{(x-s)^2}{r^2}\right)} dx \qquad (11)$$

In the case of the half-infinite horizontal plane excavation, extraction is assumed to occur from x to $+\infty$. By identifying the coordinate s with the coordinate x and by selecting the origin of the coordinate system to be at the edge of the extracted area, Equation 11 can be written as:

$$S(x) = \frac{S\max}{r}\int_{x}^{+\infty} e^{\left(-\pi \frac{\lambda^2}{r^2}\right)} d\lambda \qquad (12)$$

Furthermore, Equations 14 and 15 relate the horizontal displacements $U(x)$ and $U(y)$ to the first derivative of subsidence, i.e., tilt for the case a finite size rectangular parcel defined by $(x_1, y_1 - x_2, y_2)$ (see also Karmis et al., 1990):

$$
\begin{aligned}
U(x) &= -Bs\frac{\partial S(x,y)}{\partial x} \\
&= Bs\frac{S\max}{r^2}\left[e^{\left(-\pi\frac{x_2^2}{r^2}\right)} - e^{\left(-\pi\frac{x_1^2}{r^2}\right)}\right] \\
&\quad \times \int_{y_1}^{y_2} e^{\left(-\pi\frac{y^2}{r^2}\right)} dy
\end{aligned} \tag{13}
$$

$$
\begin{aligned}
U(y) &= -Bs\frac{\partial S(x,y)}{\partial y} \\
&= Bs\frac{S\max}{r^2}\left[e^{\left(-\pi\frac{y_2^2}{r^2}\right)} - e^{\left(-\pi\frac{y_1^2}{r^2}\right)}\right] \\
&\quad \times \int_{x_1}^{x_2} e^{\left(-\pi\frac{x^2}{r^2}\right)} dx
\end{aligned} \tag{14}
$$

here *Bs* is the strain coefficient in length units.

Using Equations 14 and 15 the horizontal displacement at any surface point *(x,y)* can be calculated. This calculation depends on the actual location of the surface point, but the point itself is assumed to lie on a horizontal plane at that location.

ADJUSTMENT OF HORIZONTAL DISPLACEMENTS DUE TO SLOPING TERRAIN

As already discussed, the main factors that affect the magnitude and orientation of horizontal displacements at surface points are the total movement at each point (i.e., subsidence and/or horizontal displacement) as well as the direction and angle of the natural slope at each point.

Thus, based on Equations 14 and 15 the adjusted horizontal displacements may be calculated as follows:

$$
\begin{aligned}
U'(x) = U(x) &+ SC * ABS(S(x,y)) \\
&* \cos(90 + \text{slope aspect}) \\
&* \sin(\text{slope angle})
\end{aligned} \tag{15}
$$

$$
\begin{aligned}
U'(y) = U(y) &+ SC * ABS(S(x,y)) \\
&* \sin(90 + \text{slope aspect}) \\
&* \sin(\text{slope angle})
\end{aligned} \tag{16}
$$

where *S(x,y)* is the subsidence at point *(x,y)*, *SC* is a unitless slope coefficient related to the type of top soil and the local conditions.

For the implementation of the above procedure, a detailed digital representation of surface elevations is necessary. If a DEM or a TIN is available, then a detailed grid can be generated and used for calculating the slope aspect and the slope angle.

This methodology has been incorporated into the influence function formulation of the SDPS package to facilitate calculations and visual representation of the results.

DEMONSTRATION EXAMPLE

For this demonstration example, a fully extracted panel of supercritical geometry is defined with the long dimension to the north. A transverse line of monitoring points is defined at mid panel on a planar surface sloping at 10 degrees to the

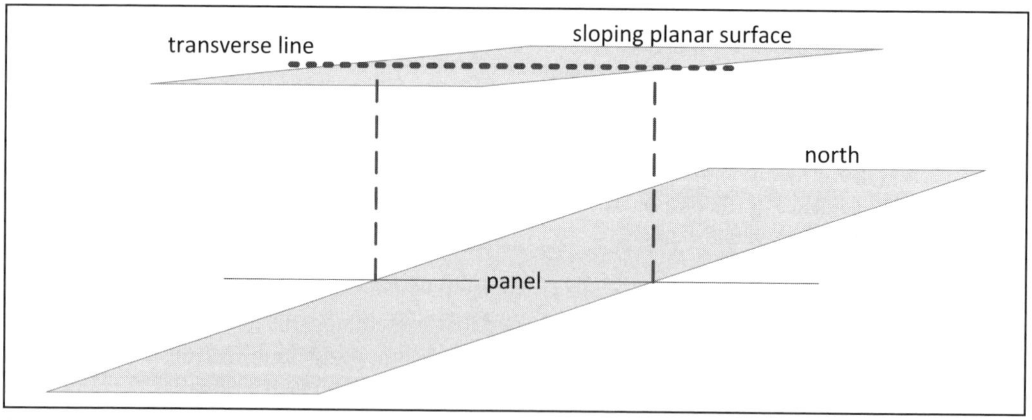

Figure 4. Conceptual diagram of fully extracted panel under a sloping surface

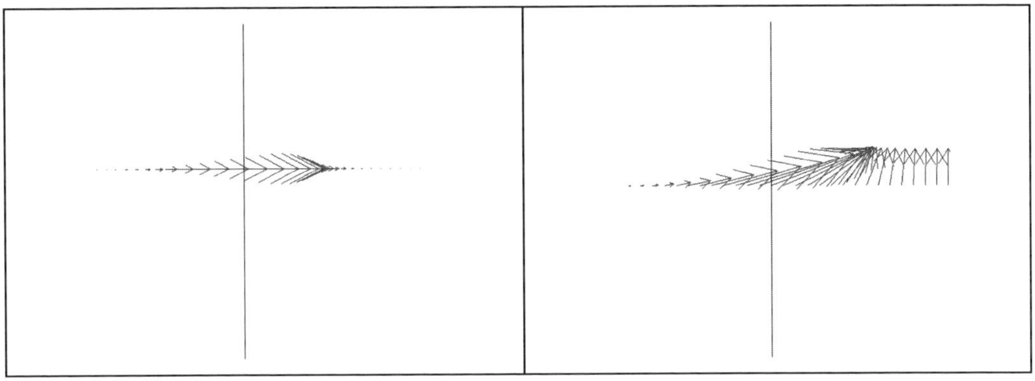

Figure 5. (a) Calculated horizontal displacements without adjustment; (b) calculated horizontal displacements with adjustment; note that only the half panel is shown

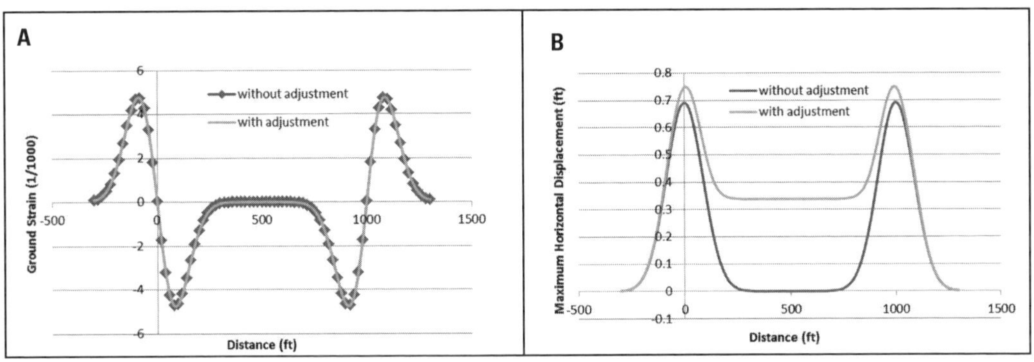

Figure 6. (a) Comparison of ground strains with and without adjustment (the two curves are identical); (b) comparison of horizontal displacement magnitude with and without adjustment

north. A conceptual diagram is given in Figure 4. In order to use the SDPS software, the extraction thickness is set to 5 ft, the percent hardrock to 50%, the tangent of the influence angle (tanb) to 2.31, the strain coefficient (Bs) to 0.35, and the slope coefficient to 0.5. (Note that a slope coefficient of zero is equivalent to no adjustment.)

Figure 5 presents the calculated horizontal displacements before and after the adjustment along the monitoring line. As expected, the displacement vectors point downward, towards the foot of the sloping surface. Figure 6a presents the ground strains calculated before and after the adjustment. As expected the change in the strain is negligible. Figure 6b shows the magnitude of the horizontal displacements. It is interesting to note that although the horizontal displacements were zero at the middle of the panel before the adjustment, they assume a positive value after the adjustment. This is to be expected since the sliding mechanism translates these points more than the less-disturbed surface points located at the sides of the trough.

CASE STUDY

The adjustment of horizontal displacements due to sloping terrain using the above methodology is demonstrated for a case of a longwall panel under hilly terrain. Figure 7 shows the layout of the panel, a monitoring line and the surface contours. Overburden depth under the transverse monitoring line to the west of the panel varies from about 350 ft to about 650 ft and seam thickness is about 6.5 ft. Monitoring point spacing is about 100 ft. An edge effect of 150 ft was applied on both sides of the panel. Figure 8 shows the calculated and

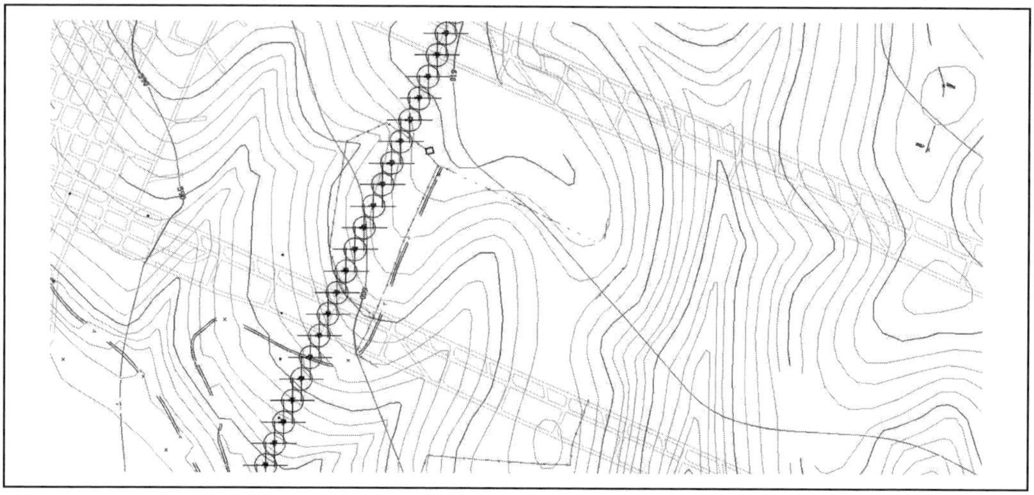

Figure 7. Layout of longwall panel, transverse monitoring line, and surface contours

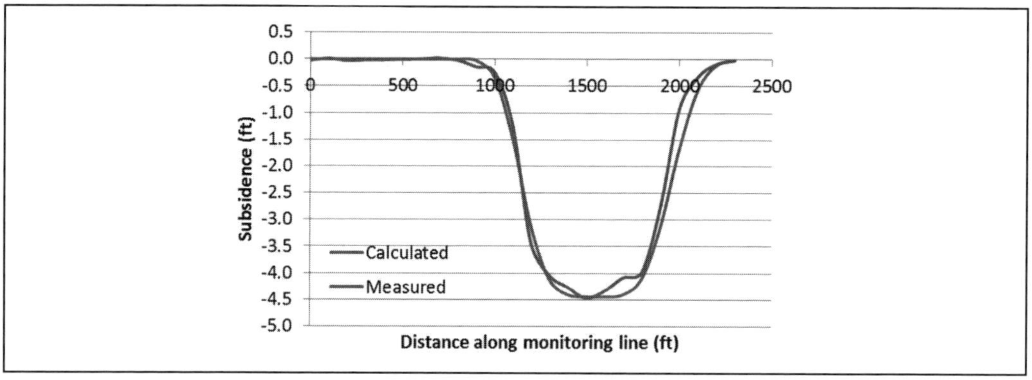

Figure 8. Calculated and measured subsidence along monitoring line

measured subsidence along the monitoring line starting from south going north, assuming full extraction of the panel.

Figure 14 compares the calculated maximum displacement values before and after the adjustment. The horizontal axis represents distance along the monitoring line calculated from the point furthest away from the south west edge of the mine panel. It can be easily observed that the calculated values match very well the measured data for the vertical displacement.

In order to apply the slope adjustment methodology described earlier, a 20×20 ft grid pattern was created based on the surface contour elevations. The spacing of the grid was selected based on the spacing of the available contours, which were set at 20 ft intervals (see Figure 9). The slope aspect and maximum slope were then calculated at the closest grid point to each point on the monitoring line, based on the surface grid shown in Figure 9. The results are shown in Figure 10. Note that the aspect angles are given in a counter-clockwise direction from the positive x axis.

Figure 11a presents a plan view of the calculated maximum horizontal displacement vectors over the panel before the adjustment; Figure 11b shows an overlay of the calculated vectors and the measured values. Figure 12 shows a plan view of the adjusted horizontal displacements using Equations 16 and 17 and a slope coefficient equal

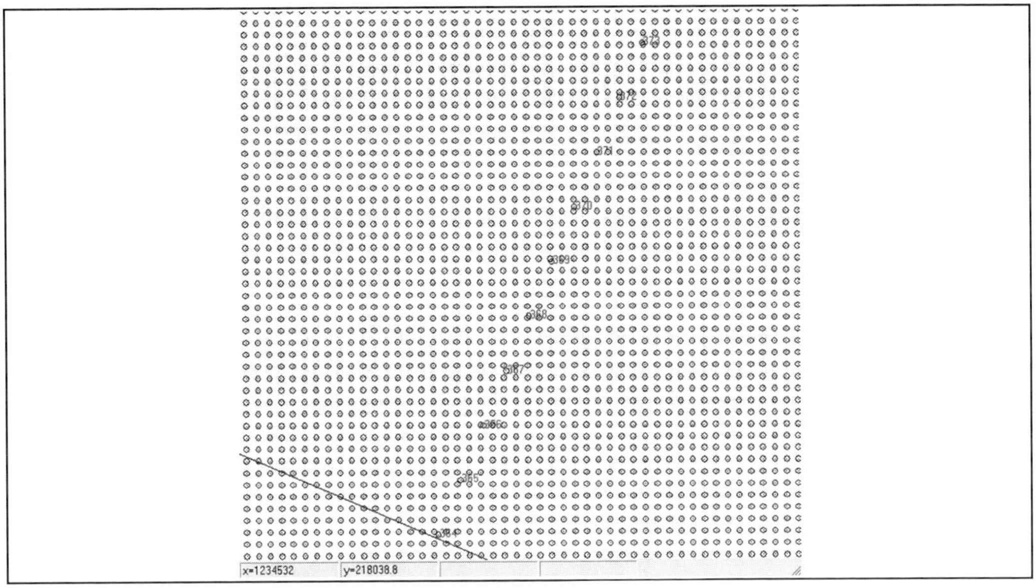

Figure 9. Surface grid points and monitoring points

Point No	Maximum Slope	Slope Aspect
354	13.23	142.44
355	12.69	150.89
356	13.95	152.75
357	17.64	148.34
358	15.17	148.68
359	15.79	145.35
360	14.28	139.19
361	13.05	127.70
362	13.31	122.01
363	14.97	117.81
364	15.50	114.65
365	13.66	110.77
366	12.35	102.03
367	11.23	93.79
368	10.81	87.70
369	12.97	76.77
370	15.20	81.59
371	16.78	96.24
372	11.34	116.94
373	10.58	128.01
374	5.71	259.86
375	4.42	180.98
376	3.55	210.27
377	4.74	342.82

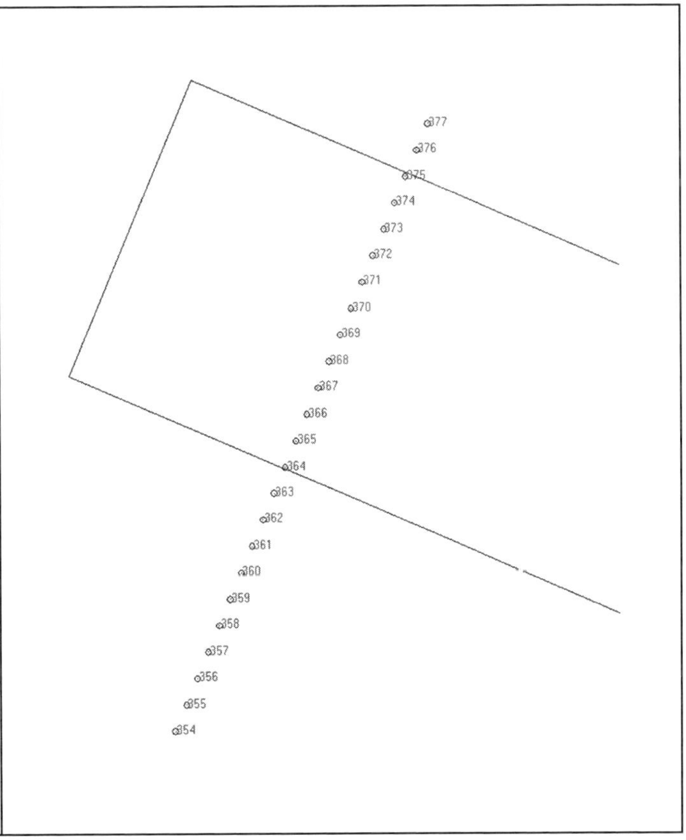

Figure 10. Slope aspect and maximum slope angle calculated for each point on the monitoring line

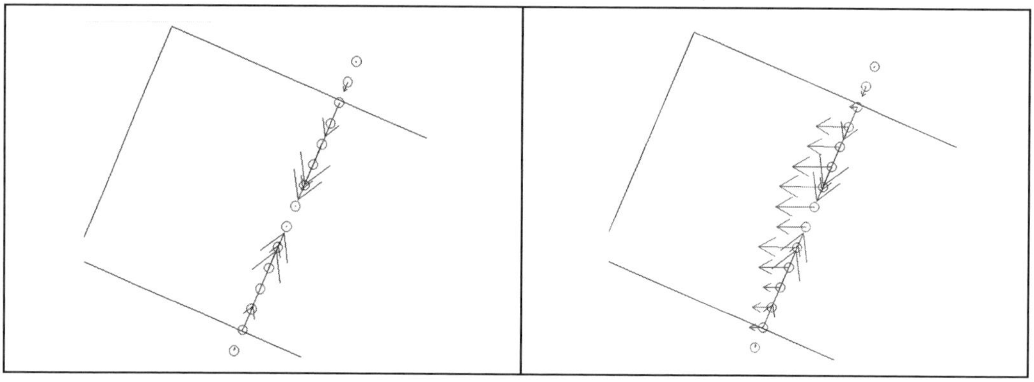

Figure 11. (a) Plot of maximum horizontal displacement vectors without slope adjustment; (b) comparison of horizontal displacement vectors to measured values

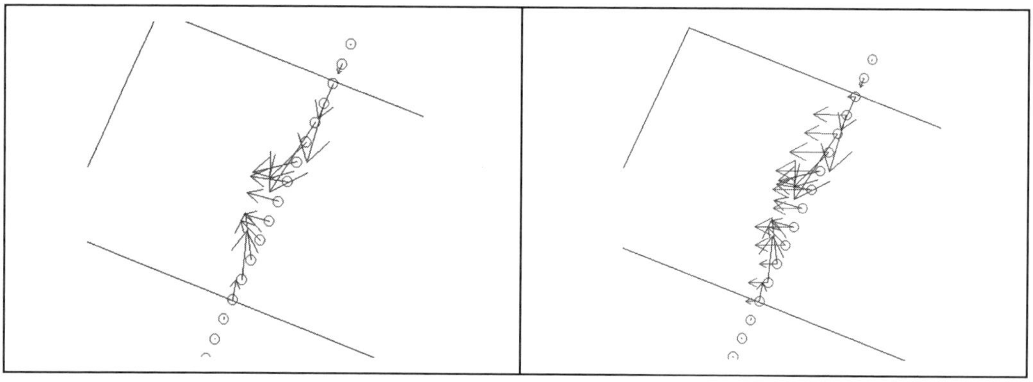

Figure 12. (a) Plot of maximum horizontal displacement vectors with slope adjustment; (b) comparison of horizontal displacement vectors to measured values

to 0.2. Figure 12b presents an overlay of the measured and calculated displacements. It is clear that after applying the adjustment the calculated values follow the natural slope downhill.

Franks and Geddes (1986) suggest that with sloping surfaces the horizontal strain may not be the most useful measure; instead, they propose using the in-plane ground strain as the basic parameter for comparison. Figure 13 compares the calculated ground strain profiles before and after the adjustment. Ground strain is calculated based on directional strain along the monitoring profile adjusted for terrain variation along the same profile. Ground strain is more representative of the strain developed on the ground surface, taking into account the natural terrain and displacements due to the subsidence process. It is evident that the strain profiles before and after the adjustment do not vary significantly as was also demonstrated in the theoretical case.

It can be observed, however, that not all monitoring points follow that trend. This can be attributed to a number of reasons, such as:

- The grid representation of the surface is not as accurate either due to the gridding procedure or due to the original elevation contours. Clearly for this type of calculation a more accurate grid is needed.
- Although the direction of the maximum slope seems to be the major factor, it is possible that other factors contribute to the lateral adjustment of the horizontal displacements calculated at each surface point.

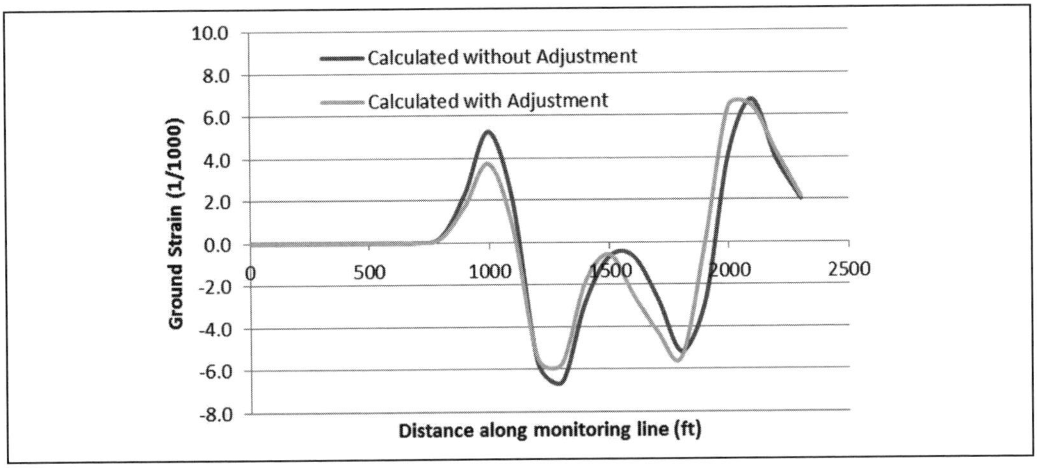

Figure 13. Comparison of calculated ground strain profiles along the monitoring line for displacement vectors with and without adjustment

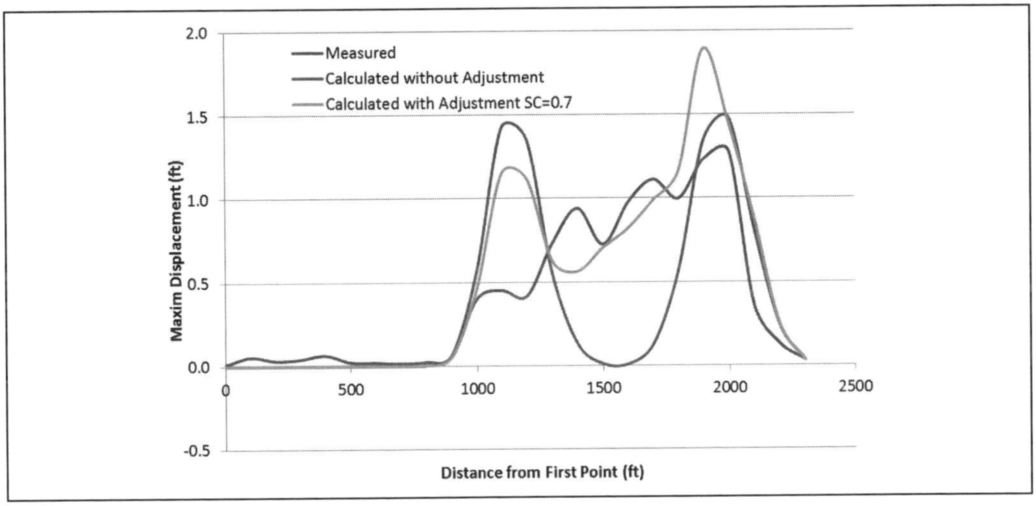

Figure 14. Comparison of maximum horizontal displacement values for measured data, calculated without adjustment and calculated using a slope coefficient SC=0.2

Figure 14 compares the maximum displacement value before and after the adjustment for three cases; measured, calculated without adjustment of the horizontal displacement due to the sloping terrain and calculated after adjusting the horizontal displacements with a slope coefficient equal to 0.7. The horizontal axis represents distance along the monitoring line calculated from the point furthest away from the south west edge of the mine panel.

The high displacement values are calculated in both cases at the high tensile value offsets (Figure 13). However, the adjusted curve matches very well the measured maximum horizontal displacement between the edges of the panel, although the match is not as accurate around the edges of the panel. Results are consistent with the calculations presented for the demonstration example.

CONCLUSIONS

The main factors that affect the magnitude and orientation of horizontal displacements at surface points are the total movement at each point (i.e., subsidence and/or horizontal displacement) as well as the direction and angle of the natural slope at each point. The natural slope conditions at each surface point of interest can be calculated from digital surface models. The simplest such model can be represented by a surface grid generated from surface contours. A quadratic equation is then used for the representation of the natural slope at the cell level.

This paper presents a procedure to adjust the horizontal displacements based on natural slope conditions using a single coefficient. The same correction algorithm is applied to the calculated horizontal displacements of all surface points. The algorithm has been implemented in the Influence Function formation of the SDPS package to facilitate calculations. The algorithm is verified using an actual example of a monitoring line, above a longwall panel, in hilly terrain.

Results of comparing adjusted horizontal displacements to measured data both as vectors on the horizontal plane and as maximum horizontal displacement magnitudes are presented. In both instances the comparison between calculated and adjusted values show that adjusted horizontal displacements using the procedure presented in this paper better match measured values, both in direction and magnitude.

ACKOWLEDGMENT

This study was sponsored by the Appalachian Research Initiative for Environmental Science (ARIES). The views, opinions and recommendations expressed herein are solely those of the authors and do not imply any endorsement by ARIES employees, other ARIES-affiliated researchers or industrial members. Information about ARIES can be found at http://www.energy.vt.edu/ARIES.

REFERENCES

Franks, C.S.M., and Geddes, J.D., 1986, Subsidence on Steep Slopes Due to Longwall Mining, *International Journal of Mining and Geological Engineering*, Vol. 4, No. 4, pp. 291–301.

Karmis, M., Agioutantis Z., and Jarosz, A., 1990, Subsidence Prediction in Steep Slope Terrain. Proceedings, Annual Symposium, Powell River Project, September, pp. 43–55.

Karmis, M., Agioutantis Z., and Jarosz, A., 1991, Subsidence Prediction in Steep Slope Terrain–Model Formulation and Application, Proceedings, Annual Symposium, Powell River Project, September, pp. 46–54.

Karmis, M., Agioutantis Z., and Jarosz, A., 1990, Recent Developments in the Application of the Influence Function Method for Ground Movement Predictions in the U.S., Mining Science and Geotechnology, Vol. 10, pp. 233–245.

Khair, A.W., Quinn, M.K., and Chaffins, R.D., 1987, Effects of Topography on Ground Movement Due to Longwall Mining, Paper presented at SME-AIME annual meeting, Denver, CO, Preprint No. 87–142, 10 p.

Luo, Y., Peng, S.S., and Chen H.J., 1996, Identification of Factors Affecting Horizontal Displacement In Subsidence Process, In *Proceedings 15th International Conference on Ground Control in Mining*, Ed. L. Ozdemir et al., Colorado School of Mines, Golden, CO, pp. 155–165.

Peng, S.S., and Luo, Y., 1989, Slope stability under the influence of ground subsidence due to longwall mining, *Mining Science and Technology*, 8, pp. 89–95.

Srinivasan, R., and Engel, B.A., 1991, Effect of Slope Prediction Methods on Slope and Erosion Estimates. *Applied Engineering in Agriculture*, 7, pp. 779–783.

Wood, J.D., 1996, The geomorphological characterisation of digital elevation models, PhD Thesis, University of Leicester, UK, http://www.soi.city.ac.uk/~jwo/phd.

Application of Subsidence Prediction Methodologies for Sizing Barrier Pillars for Stream Protection in Appalachia

Z. Agioutantis
Virginia Center for Coal and Energy Research, Virginia Tech, Blacksburg, Virginia

M. Karmis
Virginia Center for Coal and Energy Research, Virginia Tech, Blacksburg, Virginia

L. Kirby
Virginia Center for Coal and Energy Research, Virginia Tech, Blacksburg, Virginia

ABSTRACT

The prediction of ground movements due to underground mining is a mature methodology in subsidence engineering used to assess potential impacts to surface structures, facilities and water resources and to implement appropriate methods of control. Predictions of anticipated surface ground movements can be correlated to impact threshold levels for buildings, bodies of water, pipelines, railway and power lines, tailings dams, etc., to estimate the potential impact zone and the magnitude of damage, and to develop risk assessment plans. One methodology widely used for such predictions is the application of influence functions to calculate a number of deformation indices, including subsidence, slope, horizontal strain and surface curvature, at any point on the surface or at any elevation above the extracted seam. This methodology has been successfully implemented in the Surface Deformation Prediction System (SDPS) package.

One specific application of subsidence control is stream protection, particularly in shallow depths. While the stream must be protected, a very conservative approach may leave unnecessarily large blocks of unmined coal (i.e., barrier pillars or barrier pillar systems), without any particular benefit to the protection of surface waters. In this paper, typical barrier pillar sizing methodologies are compared and an approach is recommended for optimum pillar design that can provide stream protection from permanent dewatering. This approach is based on subsidence principles and sound empirical knowledge of ground movement characteristics, which have been developed for different mining, geological and topographic conditions in Appalachia.

INTRODUCTION

Mining Near-Surface Water Structures

A number of studies have been reported in the published literature discussing the factors influencing the extent and magnitude of subsidence impacts on surface water resources due to underground mining. The discussion below focuses mostly on research results related to the Appalachian coalfields in the eastern USA, where this particular issue is receiving considerable attention.

Several studies have addressed dewatering of stream resources in relation to the type of underground mining and overburden geology and thickness. The deformation of the overburden strata above the mine results in changes in hydraulic properties and may impact both the overlying underground and surface water bodies in addition to causing surface subsidence. Typically, impacts to water resources are greater above, and in close proximity to, the subsided area (Walker 1988; Rauch 1985; Hill and Price 1983). Considering the close association of the effects of underground mining on both the hydrologic regime and surface deformations, researchers traditionally have employed similar conceptual frameworks for describing and assessing their extent and magnitude. Such criteria may be defined on the basis of an angle or zone of influence (Tieman and Rauch, 1987). From a study of the effects of deep (500–600 ft, or 152–183 m) longwall mining on three nearby streams in northern West Virginia (Dixon and Rauch, 1988), the most significant stream flow change occurred before the longwall panel edge reaches the stream. However, it was reported that the streams recovered in a period of time ranging from eight months to five years. This study also showed that the greater the overburden thickness, the less the extent of dewatering and the greater the potential for complete recovery of the water resource.

In addition to overburden thickness, studies have also addressed the angle of hydrologic influence and the distance between the mine edge and the surface water body and how this parameter can influence form and function (i.e., discharge capacity, sediment transport, etc.), of the surface water resource. Angle of hydrologic influence is defined in a similar manner to the angle of draw, as the angle between the vertical and the line joining the edge of the mine panel to the limits delimiting the extent of hydrologic response. Several authors have reported values for angles of hydrologic influence, ranging from 24 to 45 degrees (Cifelli and Rauch, 1986; Dixon, 1988; Tieman and Rauch 1987; Walker, 1988; Dixon and Rauch, 1990; Tieman and Rauch, 1992). The extent of hydrologic effects of mining may also be defined in terms of the zone of influence as the buffer area above the mine panel that has been hydrologically affected. Typically these affects extend from about one to five hundred meters (Rauch, 1985; Booth, 1986; Walker et al., 1986; Tieman and Rauch, 1987; Dixon, 1988; Walker, 1988; Matetic and Trevits, 1990; Booth 2003). With regard to the horizontal distance of a water supply from the mine, Booth (1986) denotes that as the face of the panel approaches, the effects of dynamic strata movement increase, followed by a decrease of the hydraulic conductivities, due to the "traveling wave" extension and compression. This produces a drop and partial recovery of water levels with the highest impact above the centerline of the mine panel. In addition, a study that focused on the effects of deep mines on perennial streams in Greene County, Pennsylvania, concluded that, locally, the water levels of shallow aquifers dropped near the mine edge, but had no effect beyond a distance of 581–1,270 ft (~177–387 m) (Moebs and Barton, 1985).

In summary, studies show that longwall panels, or room-and-pillars sections with secondary pillar extraction, have the potential to affect surface water resources in the Appalachian coal basin. Overburden geologic structure and thickness, distance from the mine edge, mine conditions and extent of surface movements all are key drivers in determining the extent of impacts on streams and the recovery cycle of the water resource.

Guidelines for Mining Near-Surface Water Bodies

When considering the impacts of ground movements on surface bodies of water, the most comprehensive analyses of damage criteria and threshold values are derived from case studies from Britain, Australia and the USA. Different criteria have been proposed and can be divided into three broad categories: (a) mining geometry parameters, (b) surface deformation threshold values and (c) combinations of (a) and (b).

Table 1 summarizes different guidelines for mining under or near bodies of water. In many cases, mining under bodies of water depends on the depth between the mine and the surface water feature and/or the tensile strains that will develop on the surface, under the mass of the water or the water retaining structure. Where appropriate, the mining system is also given.

The guidelines presented in Table 1, can be summarized as follows:

- The British National Coal Board, as early as 1968, has recommended a maximum tensile strain value of 10 mm/m when mining under the ocean floor. Whittaker

Table 1. Summary of different guidelines for mining under or near bodies of water (adapted from a report prepared for MSHA by D'Apollonia Engineering, 2009)

Criteria	Description	Underground Mining	Reference
Criteria based on mine geometry			
10t or 5s	Minimum solid overburden	Room and Pillar first mining	Babcock and Hooker (1977)
<10t or 5s provided competent bed is > 1.75s	Minimum solid overburden with competent bed	Room and Pillar first mining	Babcock and Hooker (1977)
max (3w or 270 ft)	Minimum solid overburden	Room and Pillar	Babcock and Hooker (1977)
w ≤ h/3	Maximum width of extracted panel	Room and Pillar	Babcock and Hooker (1977)
max (100 t or > 700 ft)	Minimum overburden	Total extraction	Skelly and Loy (1977)
60t	Minimum solid overburden	Total extraction	Babcock and Hooker (1977)
60t to 117t (worst case) 37t to 105t (limited potential)	Minimum overburden	Total extraction	Kendorski et al. (1979)
Criteria based on surface ground deformation			
< 0.010	Surface tensile strain	Total extraction	Skelly and Loy (1977)
≤ 0.010 (worst case) ≤ 0.015 (limited potential)	Surface tensile strain	Total extraction	Kendorski et al. (1979)
0.010	Surface tensile strain	Mining under the sea	NCB (1968)
0.010	Surface tensile strain	Mining under the sea	Whittaker and Reddish (1989)
Criteria based on mine geometry and surface ground deformation			
≤0.005; overburden ≥60t	Surface tensile strain and minimum overburden		Singh and Bhattacharya (1984)
≤0.005; suggested overburden ≥ 60t	Surface tensile strain and minimum overburden		Singh (1992)

where, h = overburden depth, w = maximum panel width, t = extraction thickness, s = entry width (all units in ft).

and Reddish (1989) have provided further discussion in support of this limit.

- Asimilar threshold value (8.75 mm/m) was also recommended by Babcock and Hooker (1977) in the Bureau of Mines Information Circular 8741, a document generally accepted as the "best practice" guide in the USA. The information in this circular was compiled based on the results of two research projects completed by Skelly and Loy and by Wardell and partners under contract to the USBM.
- Tensile strains on the surface should generally be less that 10 mm/m (0.010).
- A conservative maximum strain value between 5 mm/m and 7 mm/m (0.005–0.007) can be assumed for design purposes depending on the specific conditions.
- The minimum overburden for total extraction under a body of water is approximately 60 times the extraction thickness (60t).

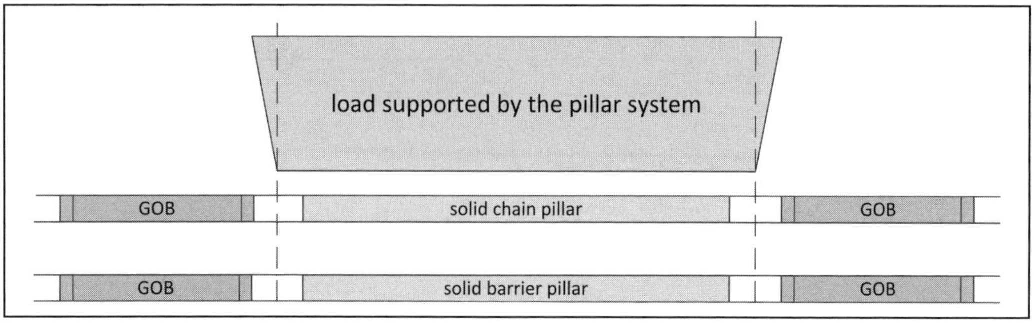

Figure 1. Conceptual diagram of supported load over solid barrier pillars using the ALPS methodology

DESIGN OF STABLE BARRIER PILLARS FOR STREAM PROTECTION

Barrier pillars for stream protection (or stream protection pillars) are designed based on the strength of the material compared to the expected loads from the overburden and the abutment pressures generated due to mining on one or both sides of the pillar. Such pillars are usually designed so that mine operations in adjacent panels are safe for the duration of mining. It should be noted, however, that stable pillars underground will not necessarily prevent ground deformations on the surface.

Analytical Methods

The width of barrier pillars can be calculated using analytical methods that account for the abutment angles due to the gob areas adjacent to the pillars. For the case of full extraction panels adjacent to solid barriers, the analytical methodology found in the Analysis of Longwall Pillar Stability (ALPS) can be utilized. The ALPS methodology was developed by NIOSH (Mark, 1992) and allows calculation of stability factors on chain pillars designed to support headgate and tailgate configurations during longwall extraction. In the case of fully extracted room-and-pillar section, the ARMPS methodology may be used (Mark and Chase, 1977; Mark et al. 2011).

For this application, ALPS was used to simulate the loads on a barrier pillar as follows: A very long chain pillar between two entries was created for a longwall panel width of 1,000 ft and an extraction thickness of 5 ft (Figure 1). The stability factor for isolated loading was calculated. The chain pillar was then assumed to be the barrier pillar between two full extraction panels.

Figure 2 shows a nomogram that was generated using the ALPS Classic formulation where the curves correspond to stability factors. Given the pillar width and the overburden depth, the corresponding stability factor for the solid pillar can be easily estimated for an extraction thickness of 5 ft. The ALPS Classic formulation calculates lower pillar stability factors than the ALPS Revised formulation and it is thus considered more conservative. The stability factors calculated in the following sections are based on the Classic formulation of ALPS.

Pillars between two full extraction panels may be partially mined ensuring, however, that the pillar system still remains stable. ALPS can be utilized to estimate stability factors for these pillar systems as well. Figure 4 shows a nomogram that was generated using the ALPS Classic formulation (using the isolated loading calculations) where the curves correspond to stability factors for pillar systems with a 50% extraction ratio and an extraction thickness of 5 ft. Pillar Stability Factor is defined under ALPS as the ratio of the pillar bearing capacity to the actual pillar load. When this ratio is greater than one then the pillar is considered stable (Figure 3).

For a given overburden depth, the stability factor of the pillar system remains constant as the pillar system width increases, as long as the extraction ratio remains the same.

Based on the ARMPS 2010 analysis (Mark et al. 2011), NIOSH recommends minimum stability factors for solid barrier pillars ranging from

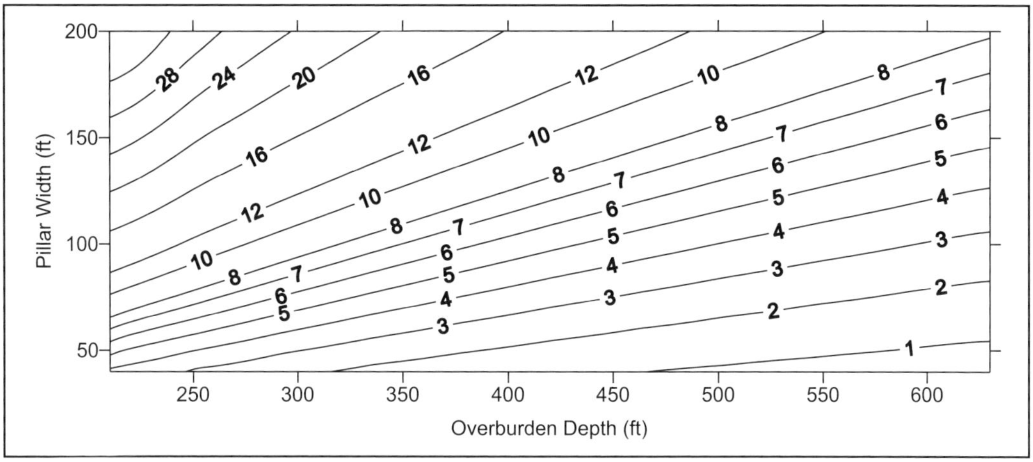

Figure 2. Nomogram for the calculation of solid barrier pillar stability factors based on pillar width and overburden depth using the ALPS Classic formulation (extraction thickness = 5 ft)

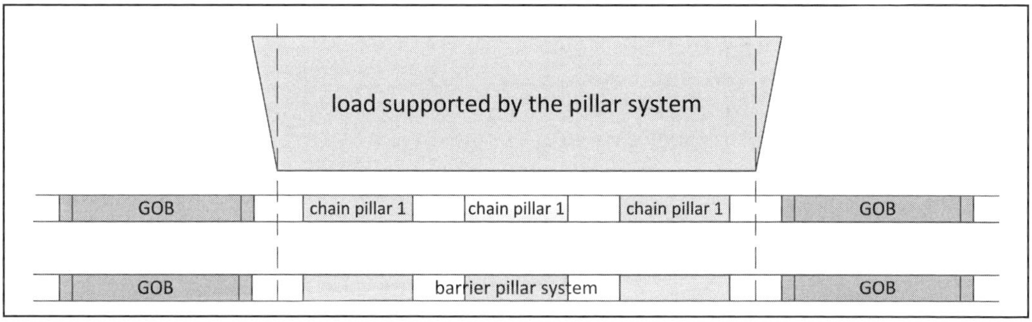

Figure 3. Conceptual diagram of supported load over barrier pillar systems using the ALPS methodology

1.5 to 2.0, depending on mine layout conditions and parameters. Table 2 lists the pillar widths calculated using the ALPS methodology for two different depths for solid pillars and pillar systems with 50% extraction.

Empirical Methods

A number of empirical design methods used to estimate the width of a solid barrier pillar are detailed in Kohler and Tadolini (1995). Kendorski and Bunnell (2007) published a study where these formulations are utilized for the design of an underground water barrier pillar. The methods referenced in these studies do not consider strains on the surface and are based on the mechanical loading of the pillar.

These formulations include:

- The Pennsylvania Mine Inspector's formula, where $W = 20 + 4t + 0.1h$
- The Pressure Arch method, where $W = 2.625^*(h/20 + 20)$
- The British Coal Rule of Thumb, where $W = (h/10) + 45$
- The North American Method, where $W = (h^*P)/(7{,}000 - h)$ and P is the adjacent panel width

Table 3 summarizes the estimated pillar widths for the above mentioned four methods for an overburden depth of 300 ft and 350 ft and an extraction thickness of 5 ft. In the case of the North American Method, an adjacent panel width of 2,000 ft was applied. Pillar widths range from 70 ft to 92 ft for a depth of 300 ft and from 75 ft to 105 ft for a depth of 350 ft. Also the

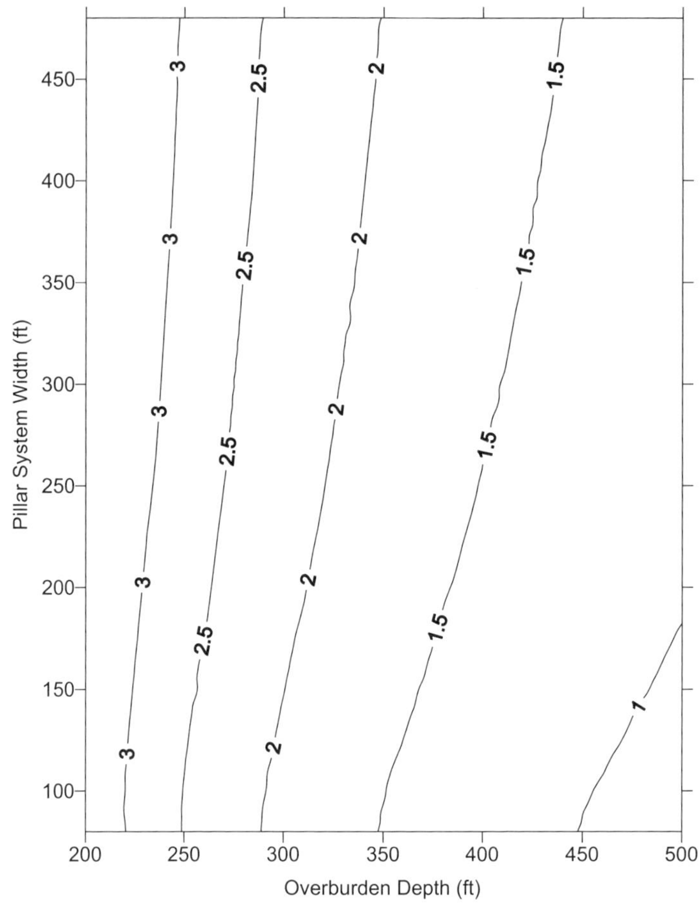

Figure 4. Nomogram for the calculation of barrier system stability factors based on pillar system width and overburden depth using the ALPS Classic formulation (extraction thickness = 5 ft, extraction ratio = 50%)

Table 2. Summary of stability factors for barrier pillars and barrier pillar systems using ALPS and extraction thickness = 5 ft

Depth (ft)	Barrier Pillar System Width (ft)	ALPS SF for Solid Barriers	ALPS SF for Barrier Systems at 50% Extraction
300	50	3.0	—
300	80	6.3	1.88
300	100	8.9	1.91
300	120	11.6	1.94
350	50	2.4	—
350	80	5.0	1.48
350	100	7.0	1.51
350	120	9.3	1.55

ALPS stability factor (Classic formulation) is given as calculated in the previous section.

DESIGN OF BARRIER SYSTEMS FOR STREAM PROTECTION

Mine planning engineers and regulators have often applied empirical rules for the protection of surface structures. For example, the Pennsylvania Bituminous Mine Subsidence Act of 1966, provided protection for certain surface structures by using an offset from the structure and a protection angle assumed to be 15°. Although this specific formulation is not used for stream protection, it is of interest as in a similar manner, a protection angle and offset distance are also used to develop an empirical stream protection barrier method.

Other "rules of thumb" are applied to determine the size of barrier pillars or barrier systems for stream protection and the distance required between the edge of the mine panel or mined area and the surface stream above, to ensure that potential impacts on the surface water structure are minimized. Such formulae are usually empirical, conservative, inflexible and limiting, as they ignore a number of parameters affecting the propagation of movement from the mine level to the surface water structures. A commonly utilized empirical rule that has been accepted in the Appalachian region (Mullins 2012) is depicted in Figure 5.

Table 3. Solid barrier pillar widths estimated using pillar design formulations based on mechanical loading for an extraction thickness of 5 ft and two different overburden depths

Formulation	Solid Pillar Width (ft) for h = 300 ft	ALPS SF
Pennsylvania Mine Inspector's formula	70	5.2
Pressure Arch method	92	7.8
British Coal Rule of Thumb	75	5.7
The North American Method	90	7.6
	Solid Pillar Width (ft) for h = 350 ft	
Pennsylvania Mine Inspector's formula	75	4.5
Pressure Arch method	98	6.8
British Coal Rule of Thumb	80	5.0
The North American Method	105	7.6

This rule assumes both a horizontal and a vertical buffer zone, and uses mining and subsidence empirical parameters, i.e., overburden thickness, coal seam height, and an upper limit value for the angle of draw. The boundaries of the horizontal buffer area are defined from the horizontal distance ($W/2$) between the panel rib and the centerline of the stream (as in Figure 5). This horizontal distance is a function of the angle of draw (γ), the overburden depth (h) and a constant offset ($B/2$) (from Figure 5). The empirical rule assumes a minimum offset distance ($B/2$) of 50 ft (~17 m) from the stream centerline, an acceptable limit for most Appalachian headwater streams. However, for very wide streams this width can increase accordingly, to better protect the stream bed, bank and adjacent alluvial floodplain. For the Appalachian region, the angle γ is taken as the upper limit of the angle of draw for the region, which assumes a value of γ =28 degrees (Karmis et al., 1983). Thus, the rule of thumb equation is given below:

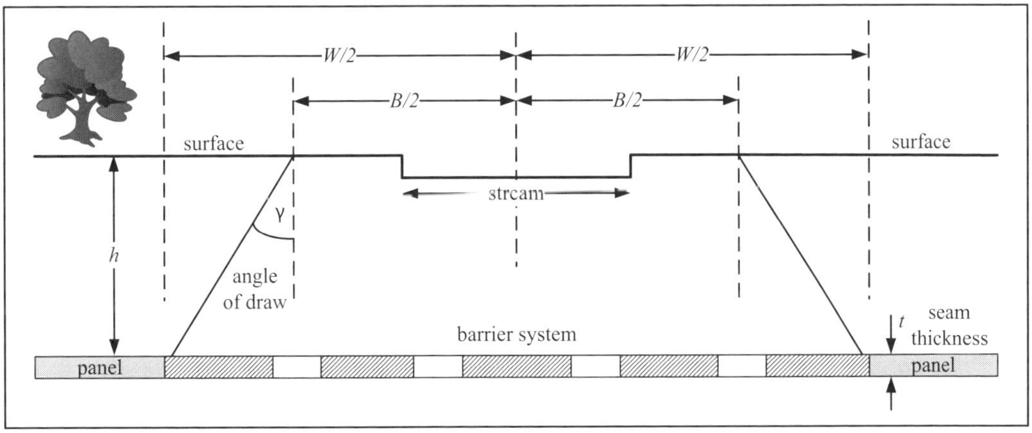

Figure 5. Buffer zone for full extraction panels in the proximity of streams

$$\frac{W}{2} = \frac{B}{2} + h * \tan(\gamma) \leq> W = B + 2h * \tan(\gamma) \qquad (1)$$

The vertical buffer zone is used to ensure that the stream is at a safe distance from the fracture zone (FZ) that develops around the caved area of the mine panel (Peng 1992). The vertically applied buffer zone determines whether complete, or partial, extraction of the coal seam below a water structure is allowed. Figure 6 shows a conceptual diagram of the different zones that develop over a fully extracted area. The caving zone that develops immediately above the panel has a height ranging from $5t$ to $10t$, where t is the extraction thickness. This is followed by the FZ, which may extend from $30t$ to $50t$ (Peng and Chiang 1984). From the top of the FZ to the surface is the continuous deformation zone (CDZ) which often includes (particularly on valley bottoms) cracks, fractures, etc., because of tectonic stresses and structural features. It is assumed that water can be drained in this zone to a distance of about 100 ft (~31 m).

If the overburden depth of the seam under a stream is less than $50t$ + 100 ft (where t is in ft) then a protection pillar should be left in place as defined in Figure 1. First mining can occur in this pillar up to 50% extraction, or to an extraction ratio that will avoid surface movements. For depths greater than $50t$ + 100 ft, the stream can be fully undermined, since an intermediate zone (Figure 6) is present and the drainage zone does not communicate with the FZ to allow dewatering towards the mine level. The above logic is given mathematically below.

$$H_c = 50t + 100\,ft \begin{cases} \text{if } h > H_c \text{ then full undermining is possible} \\ \text{if } h \leq H_c \text{ then only first mining is allowed} \end{cases} \qquad (2)$$

According to the rule of thumb for a 5-foot extraction, the CDZ is in contact with the FZ when $H_c + 50t + 100 \Rightarrow H_c\ 50*5 + 100 = 350$ ft, i.e., the maximum depth for which a protection pillar may be required under the stream. In this particular layout, for mining depths over 350 ft, full extraction under the stream is allowed, while for less depth, only partial extraction is allowed. As a result, the overburden depths for the two case scenarios are set to 300 ft (~91 m) and 350 ft (~107 m), respectively, as in Figure 3. The two depth values selected also satisfy the $60t$ rule, i.e., $60t = 60 \times 5 = 300$ ft, as given in Table 1.

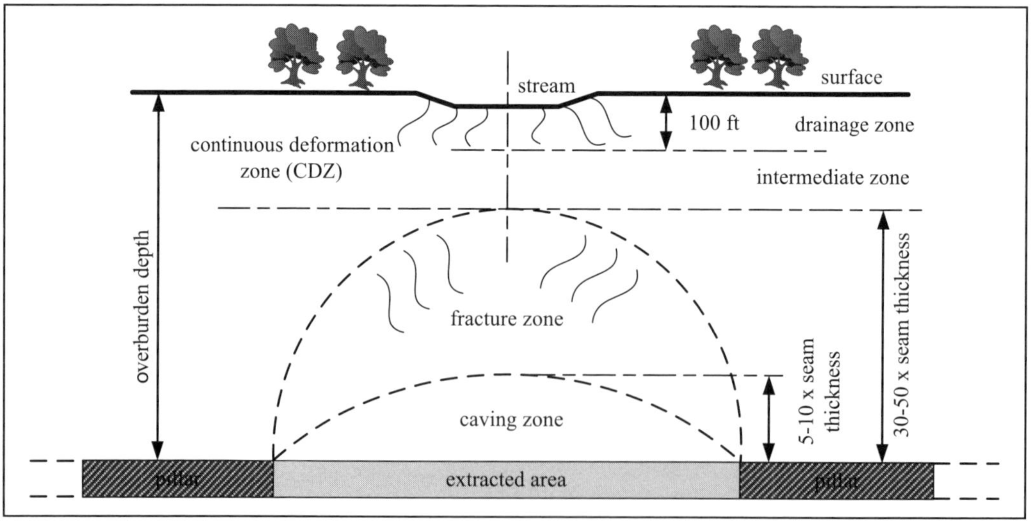

Figure 6. Caving zone, fracture zone and continuous deformation zone (adapted from Pengand Chiang 1984)

Table 4. Barrier pillar widths estimated using pillar design formulations based on rule of thumb formulations for water body protection for an extraction thickness of 5 ft and different overburden depths

Depth (ft)	Pillar System Width	ALPS SF for 50% Extraction Along the Whole Pillar System
300	420	2.35
350	472	1.98

The minimum (W_{min}) values for the barrier pillar half-widths using the empirical protection rule are determined using Equation 1 as follows:

$$\frac{W_{\min}}{2} = \frac{B}{2} + h * \tan(28°)$$

$$=> \frac{W_{\min}}{2} = \begin{cases} \text{for } h = 300 \text{ ft} \rightarrow \frac{W_{\min}}{2} \\ \quad = 210 \text{ ft} => W_{\min} = 420 \text{ ft} \\ \\ \text{for } h = 350 \text{ ft} \rightarrow \frac{W_{\min}}{2} \\ \quad = 236 \text{ ft} => W_{\min} = 472 \text{ ft} \end{cases} \quad (3)$$

The above results are summarized in Table 4. Also the ALPS stability factor (Classic formulation) is presented to allow comparison between the different designs.

DESIGN OF PILLARS AND BARRIERS FOR STREAM PROTECTION USING A SURFACE MOVEMENT APPROACH

Prediction of Ground Deformations Due to Underground Mining

The development of rigorous and well-accepted ground deformation prediction methodologies for assessing mining impacts on surface structures, facilities and water resources is important for subsidence control. This task can be extremely complex because of the number and nature of the parameters affecting ground deformation induced by underground mining, including surface morphology, mine plan, coal structure elevation and characteristics, rate of mining, overburden lithology, and the type of surface facility or water resource to be protected.

A number of subsidence prediction methodologies have evolved and have reached considerable maturity and acceptance in recent years. This can be attributed to both a need for more accurate estimation of potential impacts and damages, as well as the ability to execute more sophisticated calculations. In addition, subsidence impacts and protection of surface and underground water resources have received considerable attention by the mining industry and the regulatory and permitting agencies.

The Surface Deformation Prediction Software (SDPS) is a suite of software modules that can address both surface deformations due to underground mining and mine stability issues. This package was originally developed by Virginia Tech in the 1980s, but since then has been constantly updated and expanded. Prediction of ground deformations is mainly accomplished using the influence function method (Karmiset al. 1990). Several static or dynamic (i.e., dependent on the rate of mining) surface deformation indices can be calculated at any location, given a digitized mine plan, digitized surface topography and knowledge of appropriate subsidence parameters. Calculations are based on several empirical relationships, developed through the statistical analysis of data from a number of case studies (VPI&SU1987; Karmiset al. 1992). Calculated deformation indices include horizontal and ground strain (i.e., strain on a horizontal plane at a given surface point, or strain that accounts for the aspect and slope of the surface in the vicinity of the point), principal strains and maximum strains, either compressive or tensile.

Ground deformation prediction can also be applied in the case of surface water resources. Maximum allowable tensile strains on the surface is considered one of the best indicators for controlling the development of new cracks or extending existing cracks on the surface. Using SDPS, surface deformations including tensile strains as well as final surface profiles can be calculated at any surface point, e.g., the bottom of a stream. It should be noted that strain on the surface

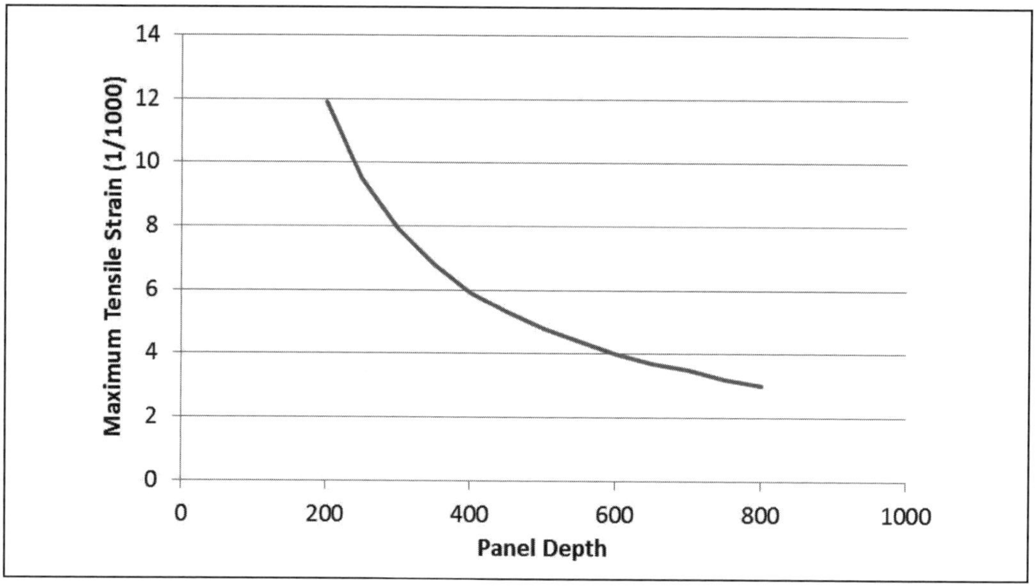

Figure 7. Maximum tensile strain with depth for extraction thickness = 5 ft

decreases with the increasing depth of cover to the extracted seam. Figure 7 presents the decrease of the maximum tensile strain on the surface for different overburden depths in the case of a supercritical fully extracted area and an extraction thickness of 5 ft. The calculations were performed using default subsidence parameters for the Appalachian region (Agioutantis and Karmis 2012).

In the next section tensile strains are calculated for a range of stream protection barrier pillar sizes and an approach is recommended for optimum pillar design given Appalachian underground mining conditions. All calculations given below refer to final subsidence effects after all mining has ceased. Movements while mining is in progress (dynamic movements) are expected to be much less than final movements (Karmis et al. 2008).

Pillar System Design Based on Surface Deformations

Pillar design methodologies based on loading and stability do not account for surface deformations. As already mentioned, pillars or pillar systems may be stable, but deformations may occur on the surface above the pillar. In this section, ground movements on the surface due to underground mining are related to barrier pillar geometries.

An example is presented below to illustrate this concept using the following assumptions:

- Calculations are completed using one mine geometry for two different overburden depths. In each case the maximum ground strain is calculated for different barrier pillar widths under the protected surface body.
- The overlying surface is horizontal. A set of surface points in a dense grid is set to correspond to a stream bed. The location of the edges of two high extraction panels (i.e., longwall panels) located on either side of the stream are varied with respect to the stream centerline.
- The rectangular longwall mine panels are assumed to have the same geometrical characteristics, i.e., width = 1,000 ft (~300 m), length = 1,800 ft (~550 m) and extraction thickness of t = 5 ft (~1.5 m). The dimensions of the panels are selected such that they will be supercritical, i.e., width-to-depth > 1.2.

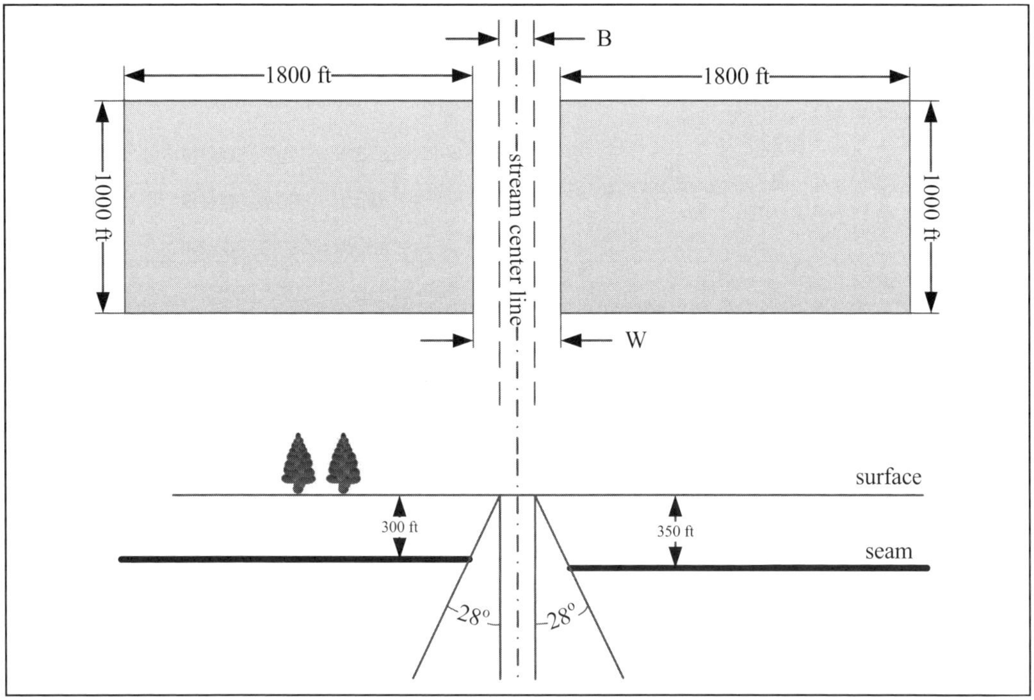

Figure 8. Plan view and section of two mine panels in the proximity of a stream (distances in ft). Protection/buffer zones are calculated using the "rule of thumb" for γ = 28° and for depths of 300 ft (~100 m) and 350 ft (~117 m).

- The overburden geology is assumed to be represented by 50% hard rock. The supercritical subsidence factor for these conditions is about 40%.
- The influence angle used in this analysis is the default value for the eastern Appalachian coalfields (i.e., tanβ=2.31 or β≈67°).
- The strain coefficient used in this analysis is the default value for the eastern Appalachian coalfields (i.e., Bs=0.35 ft).
- It is assumed that the stream section of interest has a width of 20 ft (~6 m), an initial depth of 0.5 ft (~0.15 m) and vertical banks.

The parametric analysis was run by varying the width of the barrier pillar (*W*), as in Figure 8. The analysis was completed using the influence function method available in the latest version of the SDPS software (Agioutantis and Karmis 2012), which can easily be applied to any mining geometry, in order to calculate the ground strain at the stream centerline for different *stable* barrier pillar system widths (*W*).

Figure 9 presents the calculated values of maximum strain plotted for the different depths of 300 ft and 350 ft (~100 m and ~117 m). Under those conditions, the recommended pillar width assuming a tensile strain threshold value of 5 mm/m is approximately 80 ft. It should be noted that the parametric analysis was conducted assuming that the barrier pillar underneath the stream is stable (i.e., not yielding). When stable pillar systems are developed next to full extraction areas, then an edge effect develops for that extraction area, which shifts the inflection point of the resulting subsidence profile towards the gob. This edge effect accounts for the cantilevering of the overburden strata above the gob. The edge effect was automatically estimated using the SDPS package as a function of the overburden (Agioutantis and Karmis 2012). It is immediately

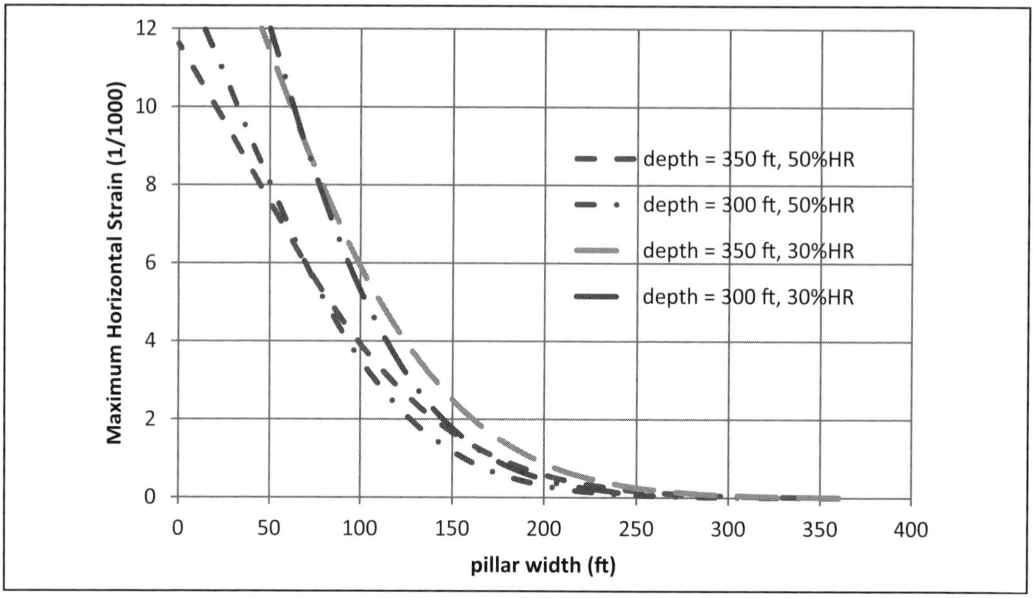

Figure 9. Maximum strains at the stream centerline as a function of barrier pillar width for a 5 ft extraction thickness under 300 ft and 350 ft of overburden for 50% and 30% hardrock

evident that pillar system widths greater than 300 ft (~100 m) result in zero surface strains in the area under examination.

Using the SDPS capabilities, alternative assumptions and scenarios can be calculated and compared. For example, Figure 9 also shows the calculated tensile strain on the surface for the same selected depths, assuming a different geology, i.e., 30% hardrock. In this case, due to the softer overburden, higher strains will be encountered and, therefore, assuming a horizontal strain threshold value of 5 mm/m, the recommended barrier pillar width is in the order of 110 ft.

Figure 10 presents a nomogram where barrier pillar width (*W*) can be calculated as a function of depth (*h*) for different maximum values for tensile strains on the surface. Figure 11 presents a similar nomogram relating pillar width and overburden depth for different subsidence values on the surface for an extraction thickness of 5 ft. The subsidence values are given here in inches.

DISCUSSION

Barrier pillar widths for the protection of surface water resources were calculated using an analytical approach as given by the ALPS formulation. Nomograms were generated for solid barriers as well for barrier pillar systems with 50% extraction. In addition solid barrier pillar dimensions using several empirical formulas were also calculated. Pillar system widths were estimated using a rule of thumb for water body protection, applied in the Appalachian region. Crossing such pillar systems via mains, or other low extraction works, will not compromise the overall stability of the barrier system. All of the above calculations were mainly performed for typical mine plan geometries, i.e., extraction thickness of 5 ft and depths ranging from 300 ft to 350 ft.

Due to the rationale behind these formulations, none of these approaches can estimate the potential surface tensile strains that may lead to drainage of a surface water resource. The application of a well-accept ground deformation prediction methodology allows the calculation of barrier system pillar widths, for given maximum allowable surface tensile strains. To facilitate this task, nomographs were generated relating barrier system pillar width and overburden depth to surface tensile strains, as well as surface subsidence. Thus, pillars can be designed by minimizing the potential surface effects. In all cases these pillars

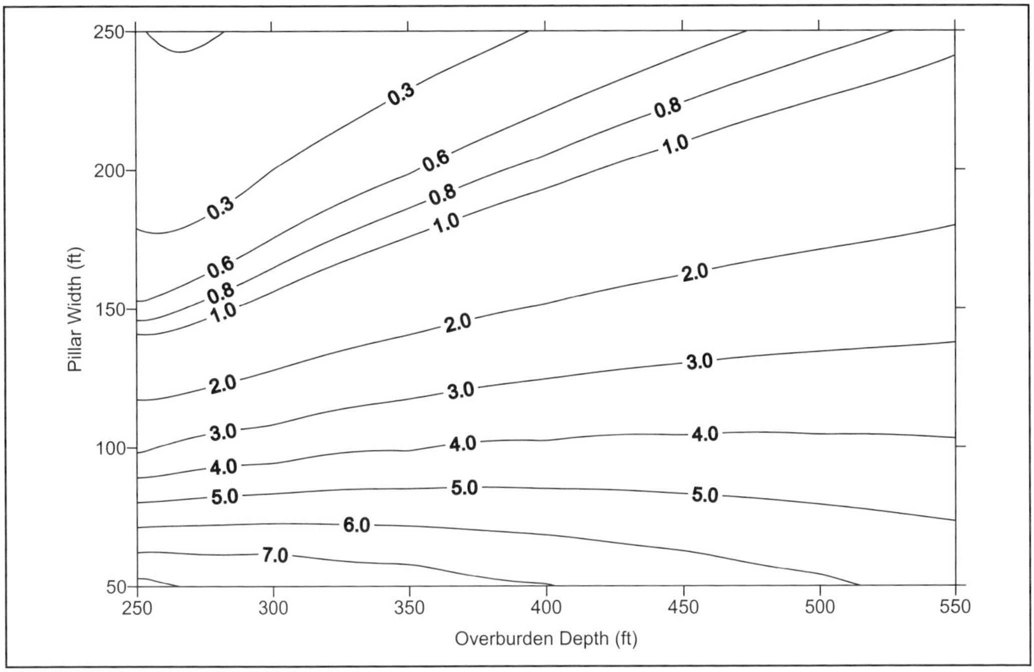

Figure 10. Nomogram relating pillar width and overburden depth for different maximum strains on the surface (1/1000) for an extraction thickness of 5 ft

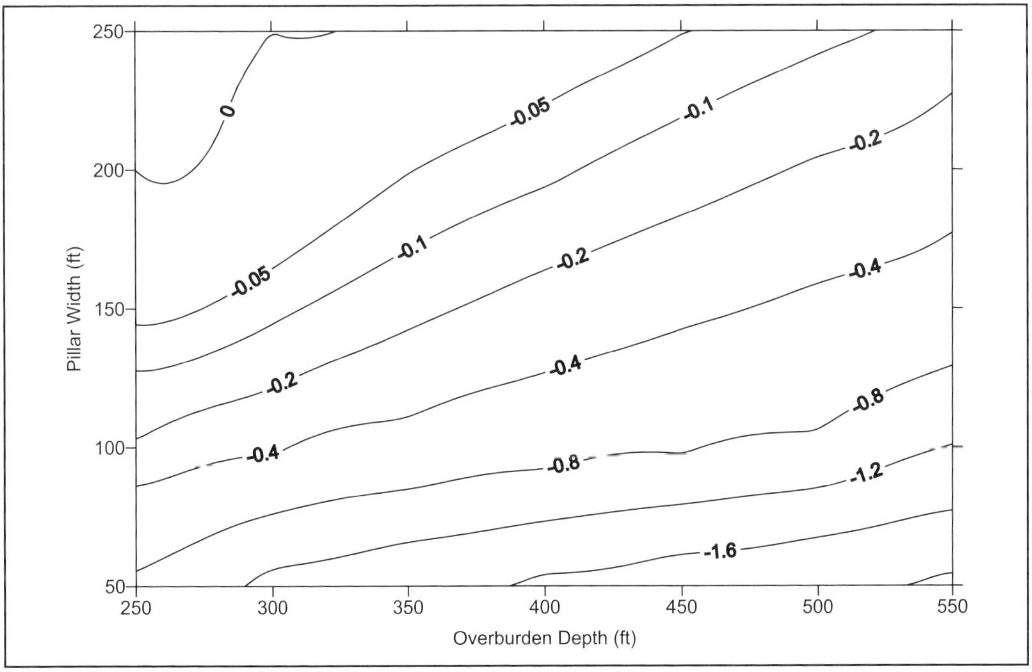

Figure 11. Nomogram relating pillar width and overburden depth for different subsidence values on the surface (in inches) for an extraction thickness of 5 ft

Table 5. Summary of barrier pillar design formulations that can be considered for water body protection. Data refer to extraction thickness = 5 ft, 50% hardrock and overburden depth ranging between 300–350 ft.

Pillar Design Approach	Depth (ft)	Solid Barrier Pillar Width (ft)	Barrier System Width (ft)	Strain on the Surface	Subsidence (in)	Comments
Empirical methods for water body protection	300		420	—		An overall extraction of 50%, ALPS SF = 2.35
	350		472	—		An overall extraction of 50%, ALPS SF = 1.98
Empirical methods based on mechanical loading	300	70–92		—		ALPS SF = 5.2–7.8
	350	75–105		—		ALPS SF = 4.5 – 7.6
Analytical (ALPS)	300	50–120	80–120	—		ALPS SF (solid)= 3.0 – 11.6 ALPS SF (50%)= 1.88–1.94
	350	50–120	80–120	—		ALPS SF (solid)= 2.4 – 9.3 ALPS SF (50%)= 1.48–1.55
Ground deformation calculations	300	~60		0.007	1.1 in	ALPS SF = 4.1
		~80		0.005	0.4 in	ALPS SF = 6.3
	350	~60		0.007	1.3 in	ALPS SF = 3.2
		~80		0.005	0.6 in	ALPS SF = 5.0

should also demonstrate an acceptable stability factor.

Table 5 summarizes the findings for all approaches. Results indicate that the width of barrier pillar systems calculated using rules of thumb significantly overestimate the protection zone and thus may sterilize coal reserves. Using ground strain as the driving criterion, coupled with analytical calculations of stability factors for barrier pillar systems, may result in an optimum design of such barriers pillars left underground for the protection of surface water bodies.

Most empirical approaches for stream protection operate on the assumption that a stream will be protected from dewatering if the fractures extending down from the continuous deformation zone will not meet with the fractures propagating up due to the caving/disturbed gob zone. This approach, although simplistic, provides the basis for all empirical formulations for the protection of streams or other surface water bodies. This methodology has allowed estimates of protection pillars to be made and does not in any way replace a full hydrologic investigation for the area of interest. In this paper, the same rationale was used, but in this case, the barrier pillar for stream protection was calculated using surface tensile strains.

Finally, the SDPS model has the ability to provide tensile strain calculations that can be used for stream protection for a range of depths, topographic conditions and lithologic environments.

CONCLUSIONS

Potential impacts on surface water bodies may be avoided when tensile strains on the surface are kept to low values. It is well documented in the literature that surface cracks and water impacts can be contained if the surface tensile strains, due to underground mining, are kept below 10 mm/m (0.010 in/in). In this study, even more conservative values were utilized which ranged from 5 to 7 mm/m.

The methodology presented was developed for the design of barrier pillars and barrier pillar systems for stream protection based on a different approach than the traditional barrier pillar stability. In this case the maximum expected tensile strain on the surface (under static or dynamic conditions) is used to estimate the width of the barrier system as a function of depth. The nomogram generated in this paper corresponds to a specific extraction thickness, but similar nomograms can be generated for different extraction thicknesses using the SDPS package. Thus, mine planners and regulators can use these nomograms to design barrier systems based on a specific maximum tensile strain criterion. Subsidence and stability factor nomograms can also be utilized for a comprehensive analysis of each case. It is also possible to cross such pillar systems via mains or other low extraction configurations which are not expected to compromise the overall stability of the barrier system.

Results indicate that the width of barrier pillar or barrier pillar systems calculated using rules of thumb overestimate the protection zone and thus may sterilize significant coal reserves without offering additional protection benefits. Using ground strain as the design criterion, coupled with analytical calculations of barrier pillar stability factors, may result to a barrier pillar system that optimizes extraction and also provides protection to surface water bodies.

This approach presents a novel method for estimating barrier pillars for stream protection. The examination and analysis of additional case studies will be important to further establish, validate and utilize this methodology in practice.

ACKOWLEDGMENT

This study was sponsored by the Appalachian Research Initiative for Environmental Science (ARIES). The views, opinions and recommendations expressed herein are solely those of the authors and do not imply any endorsement by ARIES employees, other ARIES-affiliated researchers or industrial members. Information about ARIES can be found at http://www.energy.vt.edu/ARIES.

REFERENCES

Agioutantis, Z. and M. Karmis, 2012. Surface Deformation Prediction System for Windows version 6.1, Quick Reference Guide and Working Examples, Virginia Polytechnic Institute and State University, August 17, 2012, 273 p.

Babcock, C.O. and Hooker, V.E. 1977. Results of Research to Develop Guidelines for Mining Near Surface and Underground Bodies of Water. IC 8741, Bureau of Mines Information Circular, 17p.

Booth, C.J. 1986. Strata-Movement Concepts and the Hydrogeological Impact of Underground Coal Mining. *Ground Water,* Vol. 24, No. 4, July-August.

Booth, C.J. 2003. Groundwater as an Environmental Constraint of Longwall Coal Mining. *Materials and Geoenvironment,* vol. 50(1), pp. 49–52.

Cifelli, R.C. and Rauch, H.W. 1986. Dewatering Effects from Selected Underground Coal Mines in North-Central West Virginia. In Proceedings, 2nd Workshop on surface Subsidence Due to Underground Mining, West Virginia University, Morgantown, W.Va., pp. 249–263.

D'Apollonia Engineering, 2009. Engineering and Design Manual, Coal Refuse Disposal Facilities, Second Edition, U.S. Department of Labor, Mine Safety and Health Administration.

Dixon, D.Y. 1988. A Study of the Dewatering Effects at Three Longwall Mines in the Northern Appalachian Coal Field. Unpublished Thesis, Morgantown, West Virginia 250 p.

Dixon, D.Y. and Rauch, H.W. 1988. Study of Quantitative Impacts to Ground Water Associated with Longwall Coal Mining at Three Mine Sites in the Northern West Virginia Area. In S.S. Peng, ed. *Proceedings of the 7th Conference on Ground Control in Mining,* August 3–5, 1988. pp. 321–335.

Dixon, D.Y. and Rauch, H.W. 1990. The Impact of Three Longwall Coal Mines on Streamflow in the Appalachian Coal Field. In S.S. Peng, ed. *Proceedings of the 9th International Conference on Ground Control in Mining,* Morgantown, West Virginia: West Virginia University, pp. 169–182.

Hill, J.G., and Price, D.R. 1983. The Impact of Deep Mining on an Overlying Aquifer in Western Pennsylvania. *Ground Water Monitoring Review,* Vol. 3, No.1, pp. 138–143.

Karmis, M., Agioutantis, Z. and Jarosz, A. 1990. Recent Developments in the Application of the Influence Function Method for Ground Movement Predictions in the U.S. *Mining Science and Geotechnology*, Vol. 10, pp. 233–245.

Karmis, M., Agioutantis, Z., Andrews, K., 2008. Enhancing mine subsidence prediction and control methodologies, In. S.S. Peng ed., Proceedings of the 27th International Conference on Ground Control in Mining, Morgantown, West Virginia University, pp. 131–136.

Karmis, M., Haycocks, C. and Agioutantis, Z. 1992. The Prediction of Ground Movements caused by Mining. In S.S. Peng, ed. Proceedings, 3rd Workshop on Surface Subsidence due to Underground Mining, Morgantown, WV, pp. 1–9.

Karmis, M., Triplett, T., Haycocks, C. and Goodman, G. 1983. Mining Subsidence and its Prediction in the Appalachian Coalfield, In Proceedings, 24th U.S. Symposium on Rock Mechanics (USRMS) College Station, TX., pp. 665–675.

Kendorski, F.S., Bunnell, M.D., 2007. Design And Performance of a Longwall Coal Mine Water-Barrier Pillar, 26th International Conference on Ground Control in Mining, July 31-August 2.

Kendorski, F.S., Khosla I., Singh, M.M., 1979. Criteria for Determining When a Body of Surface Water Constitutes a Hazard to Mining, Final Report, Contract J0285011, U.S. Department of the Interior, Bureau of Mines, Pittsburgh, PA.

Koehler, J.R., Tadolini, S.C. 1995. Practical Design Methods for Barrier Pillars, U.S. Bureau of Mines, Information Circular 9427, 1995, 19p.

Mark C., 1992. Analysis of Longwall Pillar Stability (ALPS)—An Update. Proceedings, Workshop on Coal Pillar Mechanics and Design, USBM IC 9315, pp. 238–249.

Mark C. and Chase F.E. 1997. Analysis of Retreat Mining Pillar Stability. Paper in New Technology for Ground Control in Retreat Mining: Proceedings of the NIOSH Technology Transfer Seminar. NIOSH IC 9446, pp. 17–34.

Mark, C., Gauna, M., Cybulski, J., Karabin, G. 2011. Applications of ARMPS (Version 6) to Practical Pillar Design Problems, 30th International Conference on Ground Control in Mining, July 26–28, 2011, Morgantown, WV.

Matetic, R.J., Trevits, M.A. 1990. Longwall Mining Impact on Near-Surface Water. In: Proceedings, Association of Engineering Geologists, 33rd Annual Meeting, Pittsburgh, PA. Oct. 1–5.

Moebs, N.N. and Barton, T.M. 1985. Short-term Effects of Longwall Mining on Shallow Water Sources, Mine Subsidence Control. US Bureau of Mines Information Circular, IC 9042, pp. 13–24.

Mullins, P. 2012. Private communication.

National Coal Board. 1968. Working under the sea. Production department instruction PI/1968/8, 4p.

Peng, S.S. 1992. *Surface Subsidence Engineering.* Society for Mining Metallurgy and Exploration, Inc., Littleton, CO., 161p.

Peng, S.S. and Chiang, H.S. 1984. *Longwall Mining.* Wiley, New York, 708p.

Rauch, H.W. 1985. A Summary of the Effects of Underground Coal Mines on Quantity of Ground Water and Streamflow in the North-Central Appalachians. Eastern Mineral Law Foundation, Sheraton Hotel at Station Square, Pittsburgh, PA.

Singh, M.M and Bhattacharya S., 1984. Proposed Criteria for Assessing Subsidence Damage To Renewable Resource Lands, SME-AIME Fall Meeting, Denver, Colorado, October 24–26, Preprint 84-391.

Singh, MM., 1992. Mine Subsidence, Chapter 10.6, In SME Mining Engineering Handbook, 2nd Edition, Volume 1, H.L. Hartman, Senior Editor, SME, Society for Mining, Metallurgy, and Exploration, Inc., Littleton, Colorado.

Skelly and Loy, 1977. Guidelines for Mining near Surface Waters, Open File Report 29-77, U.S. Department of the Interior, Bureau of Mines, Washington, DC.

Tieman, G.E. 1986. Study of Dewatering Effects of Two Underground Longwall Mine Sites in the Pittsburgh Coal Seam of the Northern Appalachian Coal Field. Unpublished M.S. thesis, Department of Geology and Geography, West Virginia University, Morgantown, West Virginia, 147p.

Tieman, G.E. and Rauch, H.W. 1987. Study of Dewatering Effects at an Underground Longwall Mine Site in the Pittsburgh Coal Seam of the Northern Appalachian Coal Field. In Proceedings, U.S. Bureau of Mines Technology Transfer Seminar, Information Circular 9137, U.S. Bureau of Mines, Pittsburgh, Pennsylvania, pp. 72–89.

Tieman, G.E., Rauch, H.W. and Carver, L.S. 1992. Study of Dewatering Effects at a Longwall Mine in Northern West Virginia, In S.S. Peng, ed. Proceedings of the 3rd Workshop on Surface Subsidence Due to Underground Mining. Morgantown, WV: West Virginia University, pp. 214–221.

VPI&SU, Department of Mining and Minerals Engineering. 1987. Prediction of Ground Movements due to Underground Mining in the Eastern United States Coalfields. Final report, Contract No. J5140137, Office of Surface Mining, Reclamation and Enforcement, U.S. Department of Interior, Vol I, 205p., and Vol II, 112p.

Walker, J.S. 1988. Case study of the effects of longwall mining induced subsidence on shallow ground water sources in the northern Appalachian Coalfield. U.S. Dept. of the Interior, Bureau of Mines, RI 9198, 20p.

Walker, J.S., Green, J.B. and Trevits, M.A. 1986. A case study of water level fluctuations over a series of longwall panels in the Northern Appalachian coal region. In Proceedings, 2nd Workshop on surface Subsidence due to Underground Mining, West Virginia University, Morgantown, WV, pp. 264–269.

Whittaker, B.N. and Reddish, D.J. 1989. *Subsidence: Occurrence, Prediction and Control, Elsevier.* Amsterdam, pp. 429–431.

Preliminary Assessment of Factors Affecting Springs Undermined by Longwall Mines in Pennsylvania

Anthony Iannacchione
University of Pittsburgh, Pittsburgh, Pennsylvania

Daniel Bain
University of Pittsburgh, Pittsburgh, Pennsylvania

Michael Keener
University of Pittsburgh, Pittsburgh, Pennsylvania

Christopher Nealen
University of Pittsburgh, Pittsburgh, Pennsylvania

ABSTRACT

In Western Pennsylvania, springs provide a direct measure of the characteristics of perched groundwater aquifers. This paper examines how springs, classified as water sources by the Pennsylvania Department of Environmental Protection (PA DEP), are *affected* by coal mine subsidence. Our research aim is to determine important factors controlling spring performance, i.e., *affected* versus unaffected by longwall subsidence in Washington and Greene Counties, PA.

INTRODUCTION

Data analyzed in this report was collected by the PA DEP from the coal companies during the 3rd ACT 54 Assessment conducted by the University of Pittsburgh (Iannacchione, et al., 2011) in Pennsylvania. These data were contained within the permit files from seven Pennsylvania longwall coal mines operating from August 2003 to August 2008. The location of the areas mined by these seven longwall mines during the study period are shown in Figure 1a.

Of the more than 3,000 water sources, i.e., wells, springs, ponds, etc., identified in the 3rd ACT 54 Assessment, 1,400 are correlated with one of the seven longwall operations (Figure 1b). The database was refined to focus on springs undermined by longwall mining. Of 716 springs identified, 148 (21%) were *affected* by longwall mining. Figure 2 shows the distribution of *affected* springs (white dots) and unaffected springs (black dots). The term *affected* reflects a designation by the PA DEP and establishes company liability, i.e., diminution or contamination of the water source as a result of undermining, producing an action that

- repairs the water source,
- provides an equivalent water source, or
- compensates the land owner.

In Pennsylvania, a water supply can supply water for domestic drinking, farmland irrigation, livestock consumption or any other demonstrated use.

POTENTIAL FACTORS AFFECTING SPRINGS SUBJECTED TO LONGWALL SUBSIDENCE

This analysis focused on the relationships between the development of the subsidence basin and an *affect* to the springs designated as water sources by the PA DEP. The following factors are thought to impact the relationship between the formation of the subsidence basin and springs.

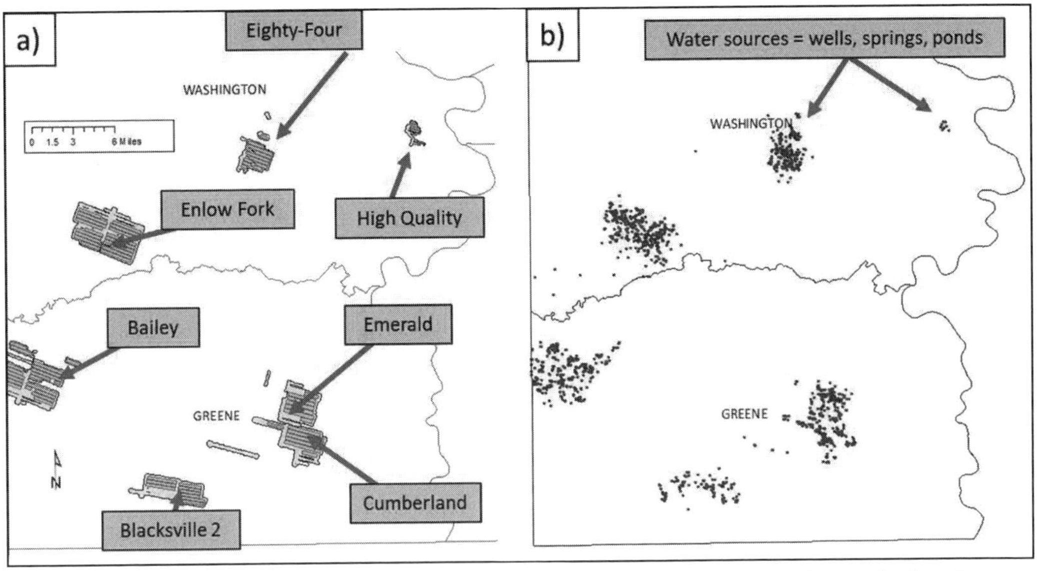

Figure 1. (a) Extent of mining that occurred from August 2003 to August 2008 at seven longwall mines in Washington and Greene Counties, Pennsylvania. The longwall panels are blue in color. Green represents the extent of all mining. (b) Location of the more than 1,400 water supplies (ponds, wells, and springs) undermined by longwall mining operations.

Factor 1, Geographic Location Within the Mining Environment

Surface deformations from longwall mining are capable of altering the character of fracture systems that feed springs in Greene and Washington Counties, PA. In some cases vertical deformations applied to bedding plane separations alter the behavior of fracture aquifer systems. The same can occur when horizontal deformations are applied to vertical joint systems.

To illustrate this, the Surface Deformation Prediction System or SPDS (Agioutantis and Karmis, 2012) was used to model the subsidence basin under a 1,500-ft wide panel at 800-ft of overburden. These are longwall panel dimension often used in current mining operations. Figure 3 demonstrates the vertical and horizontal deformation across the center of the modeled panel with a horizontal surface profile. A seven foot thick Pittsburgh Coalbed mine with 20-pct hardrock in the overlying strata will produce almost 3-ft of vertical subsidence in the center of the panel, reducing to zero over the adjacent gate entries.

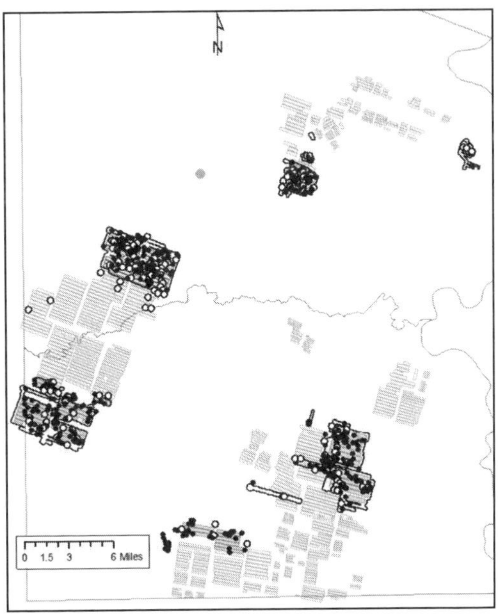

Figure 2. 716 springs located above or close to seven longwall mines operating between August 2003 and August 2008. Note the occurrence of *affected* springs (white dots) outside the mine's RPZ (thick black lines).

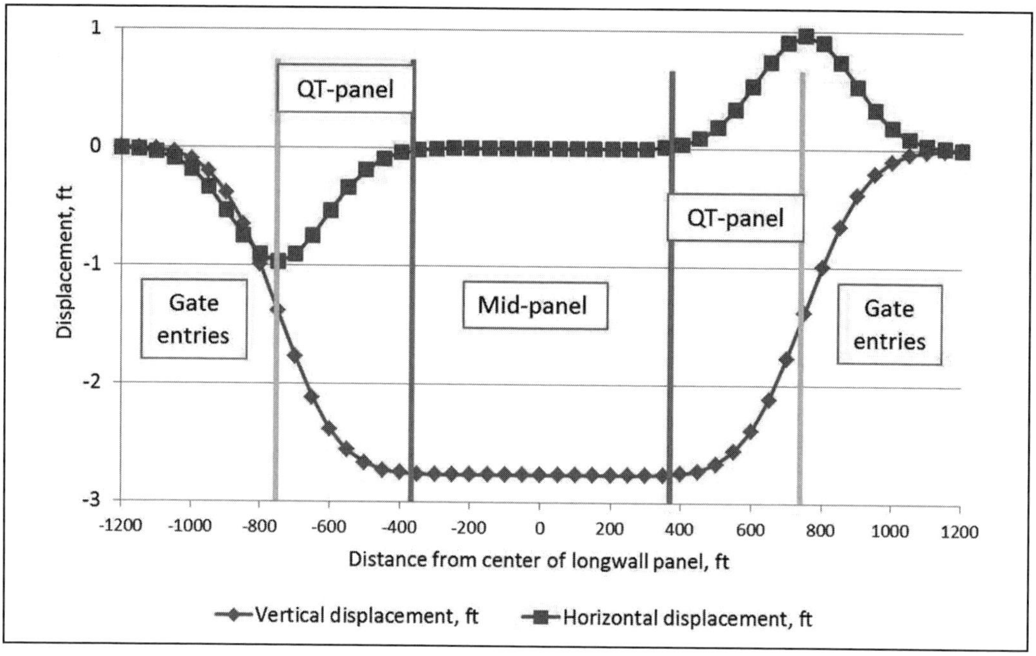

Figure 3. Relationship between vertical and horizontal deformations and the location within the subsidence basin

Table 1. Springs undermined and springs *affected* by longwall mining from August 2003 until August 2008 categorized by underlying mining environment

Mining Environment	Affected	Total	Percent of Total
Mid panel	43	188	23%
Quarter panel	32	187	17%
Gate road	29	127	23%
Outside of the mine	44	214	21%
Total	148	716	

Significant horizontal deformations are principally above the quarter-panel and gate entries typically moving into the center of the panel. The relationship between these deformation patterns and the occurrences of affected springs represents an area of future research.

Each spring was categorized by the underlying mining environment (Table 1). Four mining environments (Figure 4) were established: (1) *Mid Panel* or middle (inner) half of the longwall panel, (2) *Quarter Panel* or edge (outside) half of the longwall panel, (3) *Gate Road* including the longwall gate road entries as well as the adjacent room-and-pillar developments for the main and bleeder entries, and (4) *Outside the Mine* or within the region between the outside of the mine and the Rebuttable Presumption Zone (RPZ), Figure 4. The RPZ is a liability buffer created by projecting a 35-deg line from vertical, from the edge of mining to the surface (Anon, 1999). If the water supply falls within this 35-deg angle, the mining company is assumed liable for the impact, and it is their responsibility to dispute this assumption. All springs within the RPZ were analyzed. In addition, springs determined to be *affected* by mine subsidence outside the RPZ, albeit infrequent in occurrence, were also tracked and analyzed (Figure 5).

When analyzing the data from Table 1, we must realize that the category of Outside the Mine contained a wide variety of conditions, such as springs located over a longwall panel mined prior to August 2003. But what is most interesting about this data is that springs affected by longwall mining are not significantly influenced by the characteristics of the underlying mining environment, i.e., Quarter Panel, Mid

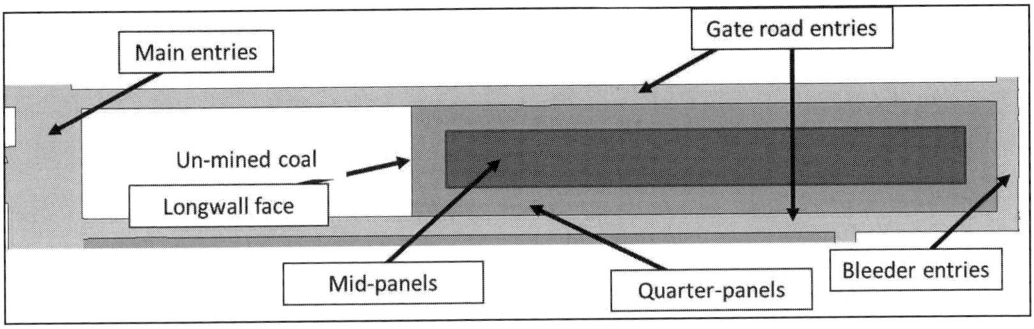

Figure 4. Mine layouts associated with mining environment categories used in Table 1

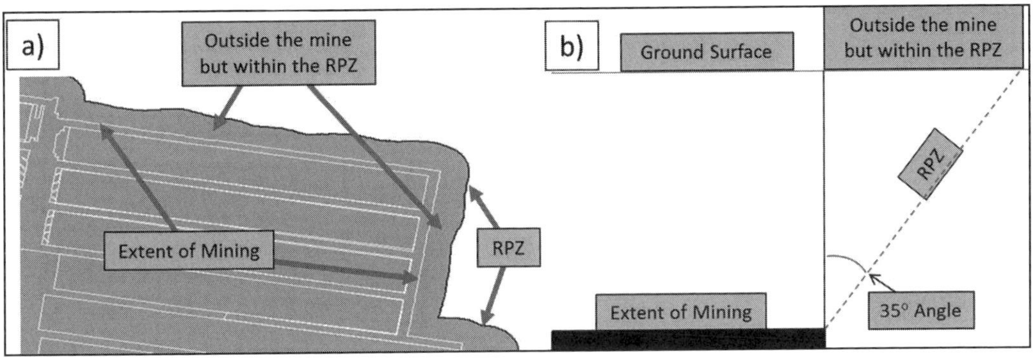

Figure 5. The category of "Outside the Mine" refers to the space between the outer limits of the mine and RPZ, (a) plan view, (b) cross-sectional view

Panel, Gate Road, or Outside the Mine. This is not what we would have suspected since springs located above the Quarter Panel and Gate Road are influenced by higher vertical and horizontal deformations then the Mid Panel (Figure 3), yet they are statistically equivalent to other underlying mining environments.

Factor 2, Topographic Position on the Surface

Each spring was categorized by the topographic position it occupied on the surface (Table 2). Three topographic positions were established: (1) Hilltop, (2) Hillside, and (3) Valley Bottom.

The topographic positions that are associated with the most *affected* springs are Hilltops and Valley Bottoms (Table 2). Hilltops are thought to contain the aquifers of the smallest areal extent while the opposite should be true for Valley Bottoms. Clearly more detailed analysis is needed.

Table 2. Topographic position of springs analyzed in this report

Topographic Position	Affected	Total	Percent of Total
Hilltop	22	73	30%
Hillside	76	456	17%
Valley bottom	50	187	27%
Total	148	716	

Factor 3, Geologic Characteristics of the Groundwater Aquifer

The dip of the formation and the characteristics of nature joints are important in determining where springs are located (Figure 6). In the vast majority of the data analyzed, springs occurred where the dip of the formation, within the hillside, causes the strata to rise in elevation away from the spring. The area comprised by these rising strata represent the maximum areal extent of

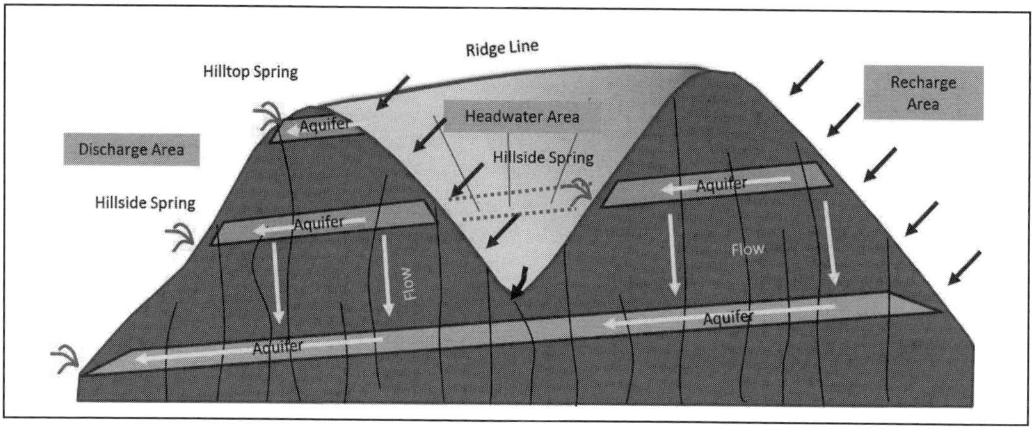

Figure 6. Generalized examples of how springs are influenced by formation dip and fracture orientations

the aquifer. Valley Bottoms should have the largest areal extent to their aquifers, Hilltops the least.

The 716 analyzed springs occurred within four geologic formations (Newport, 1973):

- *Quaternary Alluvium,* Quaternary System (0 to 63-ft thick)—Unconsolidated deposits overlying the bedrock in the major stream valleys. The material includes clay, silt, sand, gravel and some boulders and is generally permeable, and, where saturated, will yield moderate to large supplies of water.
- *Greene Formation,* Permian System (61 to 260-ft thick)—Comprised of the Fish Creek Sandstone and Donley Limestone Members. Sandstones grade irregularly into thin-bedded shaley sandstone and enclose several discontinuous thin shaley beds of limestone. A sandstone 10 to 40-ft thick is the most important water-yielding unit in this formation.
- *Washington and Waynesburg Formations,* Permian-Pennsylvania System (36 to 304-ft thick)—Comprised of alternating beds of shale, sandstone and coal. In general, poor water bearing formations.

Factor 4, Depth of Mining

The seven longwall mines varied in depth of mining from a minimum of 83-ft at the High Quality Mine to a maximum of 1,189-ft at the Blacksville No. 2 Mine (Table 3). The average longwall overburden was 687-ft with a standard deviation of 162-ft. Surface springs above shallow mining environments (High Quality) should produce more affects than deep mining environments (Blacksville No. 2). The authors believe the shallow longwall panels are within the fractured zone as defined by Peng (1992) while the deeper longwall panels are within the continuous deformation zone.

Table 3. Depth of mining characteristics for longwall mines

Mine	Avg.	SD*	Min.	Max.
Bailey	648	130	311	1061
Blacksville No. 2	887	106	649	1189
Cumberland	739	88	559	1029
Eighty-Four	627	109	362	885
Emerald	725	109	356	999
Enlow Fork	750	102	505	1036
High quality	338	71	83	544
Shoemaker	784	61	661	936
Total	687	162	83	1189

* SD = Standard Deviation

DETAILED WATERSHED EVALUATIONS

Short Creek Watershed

The Short Creek Watershed is located over the Enlow Fork Mine in Washington County, PA. Thirty-five springs were identified within the Enlow Fork Mine RPZ (Figure 7).

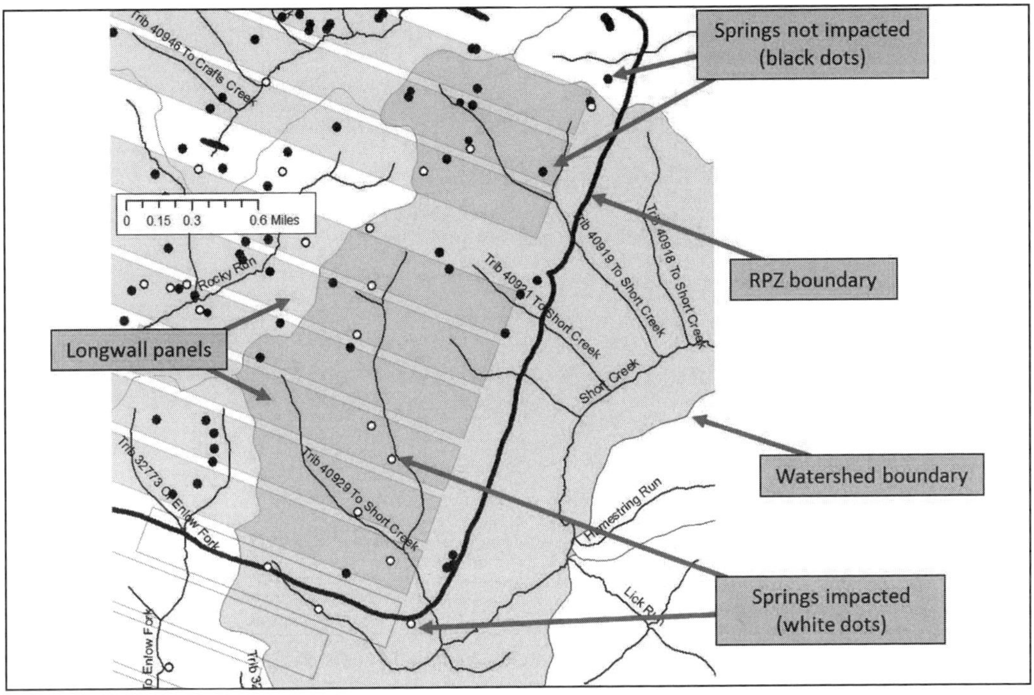

Figure 7. Short Creek Watershed showing the longwall panels, RPZ, and springs

Table 4. Distance of springs from the edge of the longwall panel or its position sorted mining environment within the Short Creek Watershed

Mining Environment	Affected	Total	Percent of Total
Mid panel	5	6	83%
Quarter panel	4	15	27%
Gate road	3	8	38%
Outside of the mine	1	6	17%
Total	13	35	

Geographic Location Within the Mining Environment

The location within the mining environments was determined for 35 springs. Thirteen springs were *affected* by longwall mining subsidence. The majority of the *affected* springs occurred over the Mid Panel and Gate Road mining environments (Table 4).

Topographic Position on the Surface

The topographic position of the springs undermined was analyzed. The majority of the springs

Table 5. Topographic position of springs within the Short Creek Watershed

Topographic Position	Affected	Total	Percent of Total
Hilltop	1	6	17%
Hillside	11	26	42%
Valley Bottom	1	3	33%
Total	13	35	

were located within the hillside topographic position (74%) where 42% of these springs were *affected* (Table 5).

Geologic Characteristics of the Groundwater Aquifer

The majority of the springs are on the hillsides facing the down-dip structural trends of the strata. All springs within the Short Creek Watershed originated within the Greene Formation.

Depth of Mining

The average depth of mining for the 35 springs is 807-ft with a standard deviation of 69-ft.

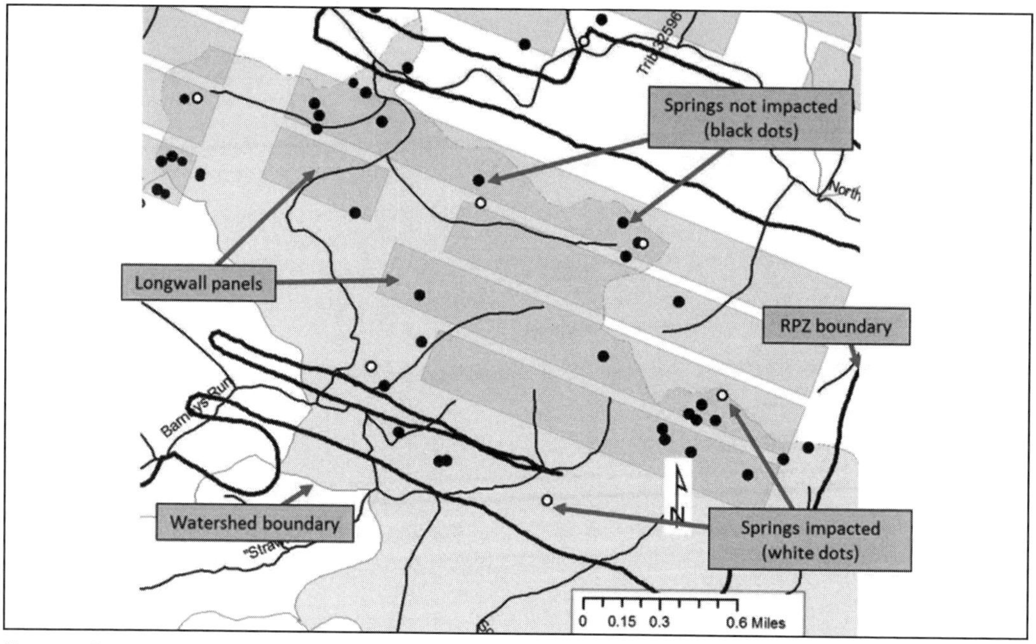

Figure 8. South Fork Dunkard Fork Watershed showing the longwall panels, RPZ, and springs

South Fork Dunkard Fork Watershed

The South Fork Dunkard Fork Watershed is located over the Bailey Mine in Greene County, PA. Forty springs were identified within the Bailey Mine RPZ (Figure 8).

Geographic Location Within the Mining Environment

The locations within the mining environment were determined for 40 springs. Thirteen springs were *affected* by longwall mining subsidence. The majority of the *affected* springs occurred over the Mid Panel and Gate Road mining environments (Table 6).

Topographic Position on the Surface

The topographic position of the springs undermined was analyzed. The percentage of springs *affected* are evenly distributed between the three topographic positions (Table 7).

Geologic Characteristics of the Groundwater Aquifer

While the majority of the springs are on the hillsides facing the down-dip structural trends of the strata several springs are located on the hillsides facing the up-dip structure trends. The water source for these springs are most likely natural joints and not perched aquifers.

Table 6. Distance of springs from the edge of the longwall panel or its position sorted mining environment within the South Fork Dunkard Fork Watershed

Mining Environment	Affected	Total	Percent of Total
Mid panel	2	14	14%
Quarter panel	1	11	9%
Gate road	3	11	27%
Outside of the mine	0	4	0%
Total	6	40	

Table 7. Topographic position of springs within the South Fork Dunkard Fork Watershed

Topographic Position	Affected	Total	Percent of Total
Hilltop	1	6	17%
Hillside	4	29	14%
Valley bottom	1	5	20%
Total	6	40	

Springs within the South Fork Dunkard Fork Watershed occur within the Greene, Washington, and Waynesburg Formations. Most of the springs occurred within the Greene Formation (27); however, only 4% of these were *affected* (Table 8).

Depth of Mining

The average depth of mining for the 40 springs is 691-ft with a standard deviation of 167-ft.

Maple Creek Watershed

The Maple Creek Watershed is located over the High Quality Mine in Washington County, PA. Six springs were identified within the High Quality Mine RPZ (Figure 9a).

Geographic Location Within the Mining Environment

The locations within the mining environment were determined for 6 springs. All six springs were *affected* by longwall mining subsidence. All mining environments within the study area have at least one *affected* spring (Table 9).

Topographic Position on the Surface

The topographic position of the springs undermined was analyzed. Eighty percent of the springs *affected* are within the Hilltop area (Table 10).

Geologic Characteristics of the Groundwater Aquifer

Springs within the Maple Creek Watershed occur in the Waynesburg Formation. Closer examination of local geology, while somewhat limited by available data, suggests that lithology of the water bearing zones may influence spring impacts during or following subsidence (Figure 9b). The test hole data collected for generating this section are

Table 8. Geologic formation associated with springs within the South Fork Dunkard Fork Watershed

Formation	Affected	Total	Percent of Total
Greene	4	27	4%
Washington	1	6	17%
Waynesburg	1	7	14%
Total	6	40	

Table 9. Distance of springs from the edge of the longwall panel or its position sorted mining environment within the Maple Creek Watershed

Mining Environment	Affected	Total	Percent of Total
Mid panel	4	4	100%
Quarter panel	1	1	100%
Gate road	1	1	100%
Outside of the mine	0	0	0%
Total	6	6	

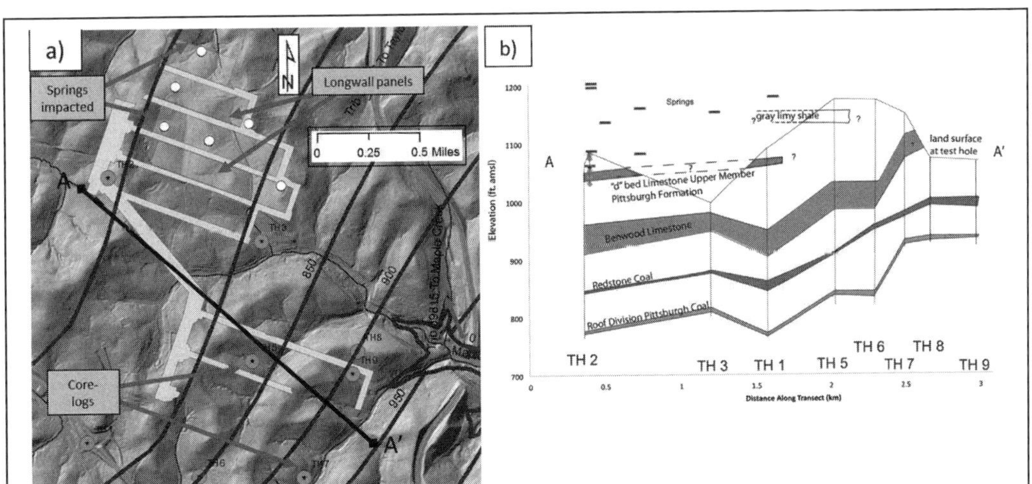

Figure 9. (a) longwall panels, springs affected, core-logs, and structure contour on the Pittsburgh Coalbed at the High Quality mine; (b) stratigraphy in various test holes at location along the transect A-A'

located predominantly in southern, topographically lower portions of the study area. However, historically impacted streams are located mainly in northern, topographically higher portions. Despite this spatial divide, we expect the stratigraphy to be reasonably consistent across both groups of data. Available data suggests the water bearing zone in a water well located near the transect sits just above the 'd' bed of the Limestone Upper Member of the Pittsburgh Formation (grey diamonds in Figure 9b) . In addition, the available data from the higher elevations indicates a relatively thick ~20-ft layer of "gray limy shale" at elevations similar to several of the affected springs. All of the springs seemingly arising from this gray limy shale were affected following long wall mining, suggesting this rock material may be particularly prone to deformation causing substantial changes in hydrologic flowpaths.

Table 10. Topographic position of springs within the Maple Creek Watershed

Topographic Position	Affected	Total	Percent of Total
Hilltop	5	5	100%
Hillside	1	1	100%
Valley bottom	0	0	0%
Total	6	6	

Depth of Mining

The average depth of mining for the 6 *affected* springs is 385-ft with a standard deviation of 49-ft.

Dyers Fork Watershed

The Dyers Fork Watershed is located over the Cumberland Mine in Greene County, PA. Forty-two springs are identified within the Cumberland Mine RPZ (Figure 10).

Geographic Location Within the Mining Environment

The locations within the mining environment were determined for 42 springs. Eight springs were *affected* by longwall mining subsidence. The mining environment locations are evenly distributed (Table 11) with the exception of Gate Roads.

Topographic Position on the Surface

The topographic position of the springs undermined was analyzed. The largest number of springs are located on hillsides (Table 12).

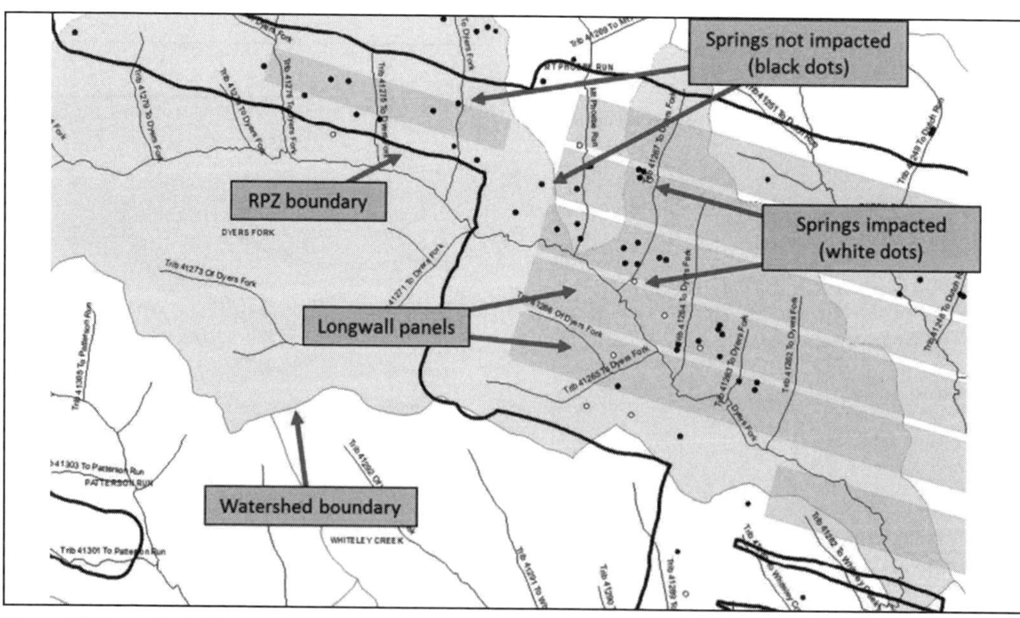

Figure 10. Dyers Fork Watershed showing the longwall panels, RPZ, and springs

Table 11. Distance of springs from the edge of the longwall panel or its position sorted mining environment within the Dyers Fork Watershed

Mining Environment	Affected	Total	Percent of Total
Mid panel	2	14	14%
Quarter panel	2	12	17%
Gate road	3	10	30%
Outside of the mine	1	6	17%
Total	8	42	

Table 12. Topographic position of springs within the Dyers Fork Watershed

Topographic Position	Affected	Total	Percent of Total
Hilltop	1	3	33%
Hillside	3	25	12%
Valley bottom	4	14	29%
Total	8	42	

Table 13. Geologic formation associated with springs within the Dyers Fork Watershed

Formation	Affected	Total	Percent of Total
Greene	2	11	18%
Washington	6	31	14%
Total	8	42	

Geologic Characteristics of the Groundwater Aquifer

While the majority of the springs are on the hillsides facing the down-dip structural trends of the strata several springs are located on the hillsides facing the up-dip structure trends. The water source for these springs are most likely natural joints and not perched aquifers. Springs within the Dyers Fork Watershed occur within the Greene and Washington Formations. Most of the springs occurred within the Greene Formation (27); however, only 4% of these were *affected* (Table 13).

Depth of Mining

The average depth of mining for the 42 springs is 701-ft with a standard deviation of 74-ft.

SUMMARY

Over 1,400 water sources from 7 longwall mining operations were analyzed. Of this total, 716 are springs. 148, or 21%, of these springs were designated as *affected* by longwall subsidence.

Four factors were examined: (1) geographic location within the mining environment, (2) topographic position on the surface, (3) geologic characteristics of the groundwater aquifer, and (4) depth of mining. In addition, four mining environment, i.e., Mid Panel, Quarter Panel, Gate Roads, and Outside the Mine, and three topographic, i.e., Hilltop, Hillside, and Valley Bottoms, categories were established and analyzed.

- When examining the entire database, all mining environments *affected* springs.
- Most springs are located along hillsides (64%).
- The dip of the formation was found to be consistently important in determining the occurrences of springs. Some variations were observed but could not be explained by this analysis.
- The depth of mining was measured for each spring. On average, mining occurred approximately 690-ft below the springs.
- No discernible trends were found when analyzing formation types versus spring frequency or *affects*.

Important Facts

- Springs *affected* by longwall mining are not significantly influenced by the characteristics of the underlying mining environment, i.e., Quarter Panel, Mid Panel, Gate Road, or Outside the Mine.
- The topographic position most associated with *affected* springs are hilltops and valley bottoms (49% of *affected* springs but only 36% of total springs).
- More detailed analysis is needed to understand how a small number of springs are forming up-dip of the strata. It is recommended that additional analysis be completed that would use joint surveys and

Landsat imagery to investigate the role of fracture permeability.

- Details of the geologic members associated with springs are not available and prohibits an appropriate evaluation of this factor.
- The Short Creek Watershed (35 springs) has a relatively moderate number of *affected* springs (37%), a high percentage of Mid Panel *affected* springs (5 of 6), and a spring to mining depth that average 807-ft. All springs are located within the Greene Formation.
- The South Fork Dunkard Fork Watershed (40 springs) has a relatively low number of *affected* springs (15%), a low percentage of Quarter Panel *affected* springs (1 of 11), and a spring to mining depth that averages 691-ft. Springs occurred in all three formations with the most in the Greene Formation.
- The Maple Creek Watershed (6 springs) had 100% of its springs *affected*. The most likely reason for this is the low spring to mining depth (average 385-ft). All of the springs occurred in the Waynesburg Formation. However, this preliminary analysis indicates a more complete understanding of geologic context, when combined with subsidence modeling, may provide important insight into the hydrologic impacts of subsidence resulting from long wall mining.
- Dyers Fork Watershed (42 springs) has a relatively low number of *affected* springs (19%), a moderate percentage of Gate Road *affected* springs (3 of 10), and a spring to mining depth that averages 701-ft. All of the springs occurred in the Greene Formation.

Of the four factors examined, depth of mining below the springs had the most obvious affect. Mining environment and topographic position had only moderate *affects* on the watershed scale and no obvious *affects* on a regional scale. More detailed information on local fracture patterns and geologic conditions associated with springs are needed to fully analyze the characteristics of the groundwater aquifer.

ACKOWLEDGMENT

This study was sponsored by the Appalachian Research Initiative for Environmental Science (ARIES). The views, opinions and recommendations expressed herein are solely those of the authors and do not imply any endorsement by ARIES employees, other ARIES-affiliated researchers or industrial members. Information about ARIES can be found at http://www.energy.vt.edu/ARIES.

REFERENCES

Agioutantis, Z. and Karmis, M. 2012. "Surface Deformation Prediction System for Windows version 6.1, Quick Reference Guide and Working Examples," Virginia Polytechnic Institute and State University, August 17, 2012, 273 p.

Anon. 1999. "Water Supply Replacement and Compliance," PA DEP, Bureau of Mining and Reclamation, Document Number 562-4000-101, 43 p.

Iannacchione, A., Tonsor, S.J., Witkowski, M., Benner, J., Hale, A. and Shendge, M. 2011. "The Effects of Subsidence Resulting from Underground Bituminous Coal Mining on Surface Structures and Features and on Water Resources, 2003 to 2008," PA DEP Website http://www.dep.state.pa.us/dep/deputate/minres/bmr/act54_2008_report/cover.htm, published January 10, 2011, 499 pages and 1 plate.

Newport, T.G. 1973. "Ground-Water Resources of Washington County, Pennsylvania," Water Resource Report 38, Pennsylvania Geologic Survey, 31 p.

Peng, S.S. 1992. *Surface Subsidence Engineering*, Society of Mining, Exploration, and Metallurgy, 161 p.

5.1
Stream Biota Impacts

Methods for Evaluating the Effects of a Simulated Mine Effluent with Elevated Ionic Concentration to Field Collected Benthic Macroinvertebrates 348

Exploring Benthic Impairment and Total Dissolved Solids in the Dumps Creek Watershed 362

Benthic Macroinvertebrate Response to Total Dissolved Solids in Coal Mine Streams 371

How Well Does Electrical Conductivity Predict Stream Impairment? 372

Evaluation of Current Measures of Aquatic Biological Integrity in the Central Appalachian Coalfields: Efficacy and Implications 381

Methods for Evaluating the Effects of a Simulated Mine Effluent with Elevated Ionic Concentration to Field Collected Benthic Macroinvertebrates

Mindy Yeager-Armstead
Integrated Science and Technology Department, Marshall University, Huntington, West Virginia

Mandee Wilson
Integrated Science and Technology Department, Marshall University, Huntington, West Virginia

Lorin Keller
Integrated Science and Technology Department, Marshall University, Huntington, West Virginia

Jeremy Kinney
Integrated Science and Technology Department, Marshall University, Huntington, West Virginia

Kyle McGill
Integrated Science and Technology Department, Marshall University, Huntington, West Virginia

Erin Snyder
Integrated Science and Technology Department, Marshall University, Huntington, West Virginia

ABSTRACT

Increases in dissolved solids and specific conductance result from disturbance in a watershed, which may lead to changes in the biological community of aquatic systems. While previous studies have used specific conductance as a marker for disturbance, only recently has ionic concentration been considered as the causative factor in changes in the biological community. The ionic concentration that limits aquatic communities will be specific to the ionic constituents and the community. Recent literature has attributed mayfly population declines in mining regions to elevated total dissolved solids. However, the causal link between mayfly losses and the water chemistry changes has not been established. The methods described herein were developed to bridge this gap between the suggested impairment threshold from field surveys and the lack of toxicity demonstrated in laboratory testing. These tests will specifically evaluate the benthic macroinvertebrate community response to elevated dissolved solids in a controlled setting. These methods utilize field collected benthic macroinvertebrate communities, which are exposed to simulated mine effluent under laboratory conditions. Specific objectives of the testing described herein were to develop simulated receiving stream water with a composition representative of Appalachian streams receiving mining discharges and to develop laboratory testing methods to evaluate benthic macroinvertebrate community and individual taxa responses to dissolved solids. Measured ionic concentrations of mining influenced streams were used to develop a simulated mine effluent with elevated ionic components similar to that found in West Virginia streams. Benthic macroinvertebrate substrate baskets were colonized in Ash Fork, a relatively undisturbed tributary of Twentymile Creek, in Nicholas County, WV, for 16 days. Additionally, upon retrieval of the substrate baskets, leaf material was collected and placed in mesh bags. The substrate baskets and leaf packs were placed in artificial streams and acclimated to the laboratory for 24 hours prior to testing. The ionic concentration of the streams was then adjusted to target test concentrations (100 µm/cm to 2400 µm/cm) over a 24–48 hour period. Test endpoints were statistical comparisons of standard benthic community evaluation metrics for the substrate baskets and leaf litter packs. The simulated stream exposures of the field collected organisms were successful, with no significant differences between biological communities at test initiation and those exposed to test conditions in dilute or full strength reconstituted moderately hard water. In water with elevated conductance values ranging from 600 µm/cm to 2400 µm/cm, only one of the endpoints was shown to be statistically significant. However, trends were evident that showed benthic macroinvertebrate taxa responsive to elevated to dissolved solids, represented by conductance, in the range of values tested. Additional evaluations are ongoing utilizing the methods described herein.

INTRODUCTION

Recent literature attributing mayfly declines in mining regions to elevated total dissolved solids (TDS) have failed to establish the causal link between mayfly losses and the water chemistry changes. In fact, laboratory testing of mining effluents with elevated dissolved solids have not generated toxicity to the sensitive United States Environmental Protection Agency (USEPA) test organism, *Ceriodaphnia dubia* (Bitzer 2012; Kennedy et al. 2005) at TDS concentrations well above those reportedly limiting biological communities (USEPA 2011). The conductivity threshold developed by the EPA for the protection of aquatic life, 300 µS/cm, is within the range of standard laboratory test waters (USEPA 2002). Ionic constituent concentrations in a solution with this conductivity would be well below reported literature values for toxicity of common ions such as chloride and sulfate (Soucek 2007; Soucek et al. 2011; Elphick et al. 2010a,b). The use of surrogate test organisms is widely accepted in predicting receiving stream effects of industrial discharges and is the cornerstone of the National Pollutant Discharge Elimination System (NPDES) whole effluent toxicity (WET) testing program. Studies have demonstrated the relationship between laboratory testing endpoints and the response of aquatic communities (Eagleson 1990; USEPA 1991) leading to widespread and continued reliance on the WET tests for aquatic resource protection from specific point sources via the NPDES permitting requirements and, in general, via water quality criteria developed using WET tests. However, there are limitations in the ability to predict real-world effects from laboratory testing due to environmental and organism-related variability (Chapman 2000). One comparison of field and laboratory data used to develop chloride benchmarks protective of aquatic life found good agreement between the alternatives (Elphick 2010a). This close agreement demonstrated for a common ion such as chloride makes the discrepancy between the USEPA's conductivity benchmark and the available ceriodaphnid data more puzzling. To bridge this gap between the suggested impairment threshold from field surveys and the lack of toxicity demonstrated in laboratory testing, methods are being developed which more specifically evaluate the benthic macroinvertebrate community response to elevated TDS. These methods utilize field collected benthic macroinvertebrate communities that are exposed to simulated mine effluent under laboratory conditions. Specific objectives are: to develop a simulated receiving stream water with a composition representative of Appalachian streams receiving mining discharges; and to develop laboratory testing methods to evaluate benthic macroinvertebrate community and individual taxa responses to elevated TDS.

METHODS

Developing Simulated High TDS Stream Water

Available water quality information from the mining influenced region in West Virginia was used to develop simulated stream water with elevated TDS representative of streams receiving mining discharges. The use of simulated stream water in laboratory testing is preferable to collecting actual high conductivity stream water to minimize variability in the tests and to allow for comparisons between tests. Also, by developing the water as a reconstituted test solution, interference by other potential stressors in the water was minimized. In order to develop a simulated receiving stream composition representative of Appalachian streams receiving mining discharges, actual in-stream data were utilized. Ten sampling sites receiving mining discharges in excess of 1,500 µS/cm were sampled semi-annually by a coal company discharging into the streams. The 20 samples were analyzed for common ions, which comprise the dissolved solids component of the stream (Table 1). Potassium was not included in this analysis so the concentration of potassium contained in each sample was estimated. This estimate was established for each sample based on the relationship of potassium concentration to specific conductivity in the Coal River Watershed (r^2 = 0.937) using the regression equation, $y = 0.0071x - 0.4845$ (Figure 1). Data from the Coal River was obtained from the West Virginia Department of Environmental Protection Agency's Ambient Water Quality Monitoring database available at the agency website. The Coal River was selected from available watersheds because it had the highest correlation between conductivity and potassium and because the watershed had conductivity values similar to the target range for the testing.

The percentage of the common ions comprising the dissolved solids concentration of the 20 stream samples was calculated by determining the ratio of each ion to the total ionic composition of each solution, summing these values, and taking an average for each ion. The measured contributions of the common ionic constituents to the total dissolved solids are shown in Table 2. The simulated high conductivity stream water was prepared using salts which are routinely used in preparing USEPA's synthetic waters (USEPA 2002), which included reagent grade sodium bicarbonate, potassium chloride, calcium sulfate, magnesium sulfate and calcium carbonate. Sodium chloride was also used.

For preparation of the stream water, a target conductivity of 2,400 µS/cm was selected as the highest test concentration. The TDS concentration providing this conductivity was calculated using an online calculator that estimates electrical conductivity from ionic concentrations. A spreadsheet was constructed that allows for entering the weight of the salt to find the weight of each ion, the ratio in the mixture, and the proximity to the target ion concentration. The mixture best representing field conditions was selected. The percentage contribution of each ion to the weight of the total dissolved solids, which was achieved in the final simulated stream water mixture, is shown with the target concentration in Table 2.

Simulated stream water was prepared in a similar manner to the synthetic water used in toxicity testing. Calcium sulfate was difficult to dissolve so, through trial and error, it was established that when preparing the solution, calcium sulfate will be added to 45% of the volume while the remaining salts are added to the second 45% of the volume. Calcium sulfate was mixed with stirring and aeration or continuous mixing with a submersible pump for at least twenty-four hours prior to combining the solutions. The combined mixture was then aerated vigorously or mixed with the submersible pump for 24 hours prior to use in testing. Because the batches are mixed in 20 gallon carboys, measuring from the carboy wall was not accurate. For this reason, after 24 hours the two volumes described above were combined and brought to a conductivity of 2,400 µS/cm with laboratory water. The laboratory water was prepared by deionization and reverse osmosis and was generally maintained with a TDS concentration less than 2 mg/L. Conductivity was used to confirm accuracy of mixing for the 100% test solutions as well as the dilutions used in testing.

Simulated Stream Environments

Streams are constructed from livestock feeding troughs that are five feet in length and 27 inches in width. A center piece constructed from polyvinyl chloride pipe (PVC) measuring three feet in length are placed longitudinally in the streams to provide opportunity for circular flow in the streams. Where needed, aquarium-grade silicone was used in stream construction. Laminar flow is

Table 1. Concentrations of ionic constituents in mining discharge dominated streams sampled for chronic toxicity testing under high flow (May and June) and low flow (October and November) conditions in 2011. Potassium concentrations were not measured but are estimated based on the relationship of potassium concentration to specific conductivity in the Coal River Watershed ($r^3 = 0.937$) using the regression equation $y= 0.0071x–0.4845$.

	Unit	Site 1	Site 2	Site 3	Site 4	Site 5	Site 6	Site 7	Site 8	Site 9	Site 10
Collection date		5/25/11	5/25/11	5/25/11	5/25/11	5/25/11	6/3/11	6/3/11	6/3/11	6/3/11	6/3/11
Total calcium	mg/L	124.40	50.48	147.20	181.80	9.69	61.53	90.21	84.94	137.50	77.18
Potassium	mg/l	9.16	3.22	8.63	11.80	0.42	6.51	5.96	7.50	8.19	5.94
Magnesium	mg/L	96.17	42.06	103.30	147.40	7.77	36.45	77.58	84.06	107.30	42.49
Total sodium	mg/L	46.70	5.56	6.02	15.50	2.64	137.30	9.40	66.27	11.02	83.15
Chlorides	mg/L	19.42	1.18	11.12	35.68	1.25	15.06	4.85	28.95	2.12	7.26
Total alkalinity	mg/L	54.86	20.17	24.59	413.75	11.89	307.31	93.71	218.07	192.86	139.67
Total sulfates	mg/L	685.04	270.06	697.64	1173.70	44.25	229.73	337.93	471.37	548.34	157.08
Specific conductance	µS/cm	1358	552	1284	1730	127	985	907	1124	1222	905
	Unit	Site 1	Site 2	Site 3	Site 4	Site 5	Site 6	Site 7	Site 8	Site 9	Site 10
Collection date		10/26/11	10/26/11	10/26/11	10/31/11	10/26/11	11/2/11	10/31/11	11/2/11	10/26/11	10/31/11
Total calcium	mg/L	134.47	52.69	168.21	198.65	10.25	66.13	95.64	126.33	79.54	78.52
Potassium	mg/l	11.02	3.80	9.60	9.31	0.67	4.79	4.86	7.60	8.49	2.22
Magnesium	mg/L	88.63	48.67	99.74	155.42	8.15	38.29	79.42	113.21	89.81	48.71
Total sodium	mg/L	49.34	5.41	5.64	17.84	2.15	145.21	9.52	15.54	75.21	91.52
Chlorides	mg/L	18.76	1.25	9.87	21.81	1.14	12.35	5.23	1.87	26.55	8.13
Total alkalinity	mg/L	58.72	22.51	28.51	415.62	12.34	298.45	85.42	211.21	225.63	155.22
Total sulfates	mg/L	712.35	335.27	725.92	789.39	55.21	245.10	375.25	552.16	488.93	226.39
Specific conductance	µS/cm	1621	604	1421	1495	163	743	753	1138	1264	381

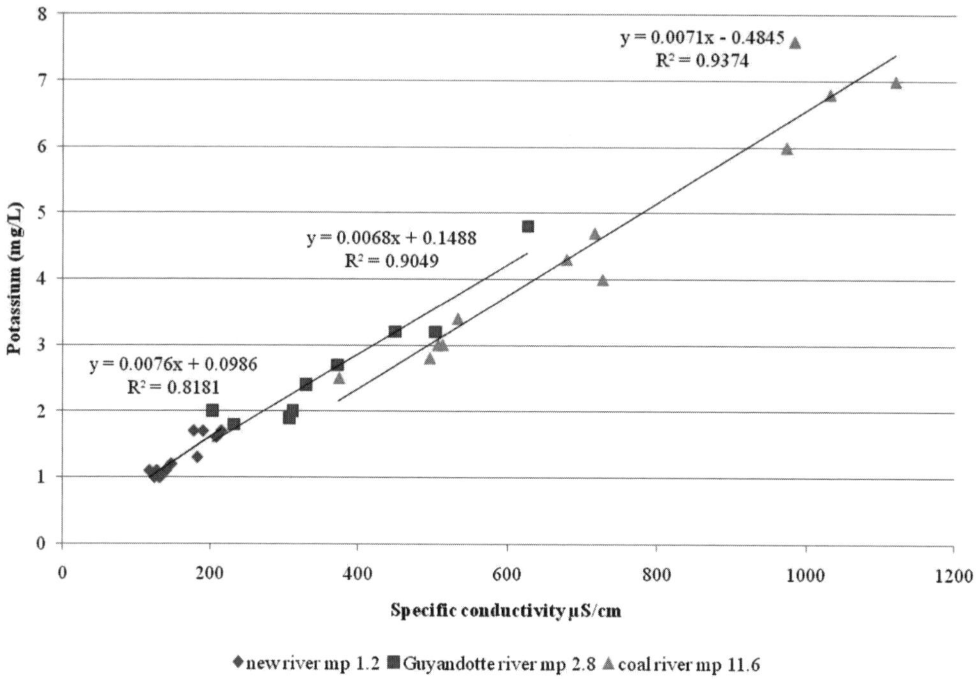

Figure 1. Relationship between potassium and conductance from three WVDEP monitoring stations in mining influenced watersheds. Coal River watershed was selected to estimate potassium due to having the largest dataset and highest correlation coefficient and having conductivity in the range of concentrations being simulated.

Table 2. Estimation of percentage of ionic constituents and the total percentage achieved in the simulated stream water

	% Ionic Constituent Measured (Target)	% Achieved in the Simulated Stream Water
Total Calcium	12.12	11.64
Potassium	0.78	0.61
Magnesium	9.29	7.98
Sodium	5.39	5.55
Chloride	1.32	1.26
Bicarbonate	17.28	13.51
Total Sulfate	54.60	59.45

achieved by using a constructed paddle system, which consists of stainless steel rods attached to a rotisserie motor for consistent flow of approximately 0.01 feet per second. There are paddles that were constructed using five inch diameter plastic lawnmower wheels and four speckling spatulas attached to the rod. The depth of the streams, 7.5 inches, was achieved by placing a draining standpipe in the stream. A gardening shade cloth was suspended over the streams to reduce the light intensity of the ambient laboratory lighting to approximately 45 lux. USEPA's moderately hard water was the dilution water used in testing.

Benthic Macroinvertebrate Community Collection

Benthic macroinvertebrate communities were sampled in streams known to have West Virginia Stream Condition Index (WVSCI) scores in the unimpaired range. In the testing described herein, the community tested was collected in Ash Fork of Twentymile Creek in Nicholas County, West Virginia. Upon stream selection, benthic collection baskets which were 7" × 9" × 3" were deployed in the stream thalwag. The collection baskets were made from bird feeders. The metal casings were covered in rubber to prevent metal exposure to the organisms. The baskets had mesh size of approximately one inch.

At the time of deployment, rocks from the reference site were placed into the baskets and benthic macroinvertebrates were allowed to colonize for 15 days. After the 15 days the baskets were retrieved by placing each individual basket in a two gallon Ziploc bag underwater to ensure there was no drifting of macroinvertebrates out of the baskets. Rock baskets and stream water were transferred to coolers and transferred to the laboratory. Leaf packs were obtained from the stream by randomly selecting available leaf material and placing it into pecan bags. Each bag was then placed in a Ziploc bag and into a cooler. Water samples were collected in laboratory provided bottles with appropriate preservative and taken to BioChem Testing for analysis. After collection, benthic macroinvertebrate baskets were transferred to the laboratory where they were randomly placed gently into the streams. USEPA's moderately hard water was placed in the streams to a depth to cover the baskets. Four benthic macroinvertebrate baskets were transferred into each stream with the stream water and allowed to acclimate for 24 hours prior to initiation of testing. Leaf packs were also transferred into the streams and tied to the stream so they did not move around during testing or become entangled in the paddles. Four replicates of rock baskets and leaf packs were placed into each stream and streams were randomly assigned a conductivity treatment.

After the 24 hour acclimation period, test solution was introduced into the streams using gravity flow and overflow stream water was allowed to flow out the standpipe. Achieving desired conductivity in each stream occurred over a two day period with the target concentration taking longer to achieve at the higher concentrations. The exposure period was 10 days after the target concentration was obtained. The test was a static renewal test with 10 liters of reconstituted stream water changed over twice daily for a total volume of 20 liters per day changed over. In addition to the leaf packs contributing particulate organic material to the steams during testing, each stream received 5 mL of yeast, cereal leaves and trout chow mix (YCT) and 5 mL of *Selenastrum capricornutum* daily after the morning change over. Water chemistry was conducted daily on test solution prior to change over to ensure water was correctly mixed and aerated prior to use. Water draining from streams was also evaluated to ensure water quality parameters were being maintained in an appropriate range. Daily water chemistry included dissolved oxygen, pH, temperature and conductivity.

After the 10-day exposure period, the paddles were stopped and the baskets were gently placed in a bucket for removal from the simulated stream. In the bucket, the basket was opened and the rocks were rinsed. Contents of the bucket were poured through a sieve, which was rinsed into a tray for macroinvertebrate removal. Leaf packs were similarly removed from the streams and rinsed into a sieve. Benthic macroinvertebrates removed were preserved in 70% ethanol and identified to the lowest practical taxa, usually genus. After the baskets and bags were removed, the stream contents were emptied through a sieve to collect any macroinvertebrates that had abandoned their respective baskets during testing. These benthic macroinvertebrates were sorted to evaluate the number of organisms lost from the baskets during testing.

Common descriptive metrics were calculated for rock basket and leaf pack communities. These included total abundance, taxa richness, abundance of mayflies and

Table 3. Water chemistry demonstrating test conditions for each treatment during the 10-day exposure period of the field collected benthic macroinvertebrate community to simulated stream water with elevated conductivity

	100 µS/cm	300 µS/cm	600 µS/cm	1215 µS/cm	1800 µS/cm	2430 µS/cm
Dissolved oxygen (average) mg/L	9.06	8.95	8.78	8.81	8.75	8.96
Dissolved oxygen (range) mg/L	8.39–9.65	8.59–9.56	1.36–9.48	6.70–9.51	7.07–9.31	8.42–9.49
pH (range) S.U.	7.35–7.70	7.38–8.01	7.40–7.89	7.21–8.29	7.41–8.28	7.44–8.36
Temperature (average) °C	17.71	17.73	17.94	17.74	17.87	17.79
Temperature (range) °C	15.43–20.90	15.39–20.90	15.90–20.90	15.50–20.90	15.88–20.90	15.66–20.90

caddisflies (in the leafpacks), richness of the sensitive Ephemeroptera, Plecoptera and Trichoptera organisms, and percentage of sensitive Ephemeroptera, Plecoptera and Trichoptera (EPT) taxa. Data were evaluated for significance using a one way analysis of variance procedure where the assumption of homogeneity and normality were met. A Kruskal-Wallis one-way analysis of variance procedure was use where the assumptions were not met. Statistical comparisons were made for the evaluation of Day 0 community (rock baskets initially collected from the streams) to the end of test communities from the dilute moderately hard USEPA reconstituted water and the standard reconstituted waters with conductivities of 100 µS/cm and 300 µS/cm, respectively. Statistical comparisons were also made for the community metrics from the four exposure concentrations compared to the two reference water communities. Multiple rock baskets and leaf bags were exposed in each treatment stream such that the experimental design, as with many mesocosm experiments, may be considered pseudoreplicated. Differences between the treatments may result from factors other than the TDS exposure concentrations and should be viewed with this limitation.

RESULTS

Water chemistry evaluations showed that dissolved oxygen, pH and temperature were suitable to support the benthic macroinvertebrate communities exposed during testing (Table 3). Target conductance concentrations were achieved within 24 hours of test initiation in the 100 µS/cm, 300 µS/cm, 600 µS/cm and 1200 µS/cm concentrations. Higher conductance values were harder to achieve and were not obtained until days 4 and 6 for the 1,800 µS/cm and 2,400 µS/cm concentrations, respectively (Figure 2).

Overall, abundance indicated a good representation of in-stream conditions with the average Day 0 abundance of 22 organisms (Table 4) corresponding to an in-stream community with approximately 541 individuals per square meter. This is a reasonable approximation of riffle abundance in a stream representing undisturbed conditions. Upon retrieval from Ash Fork, abundance in the baskets ranged from 14 to 30 organisms representing seven to eight taxa. The community had four to five mayfly taxa represented which represented 93 to 100% of the population (11 to 24 organisms). At the end of the exposure period, abundance in the 100 µS/cm and 300 µS/cm conductance treatments averaged 12.5 and 21.25 in the two waters, respectively (Figure 3a). Taxa richness was lowest in the 100 µS/cm concentration (three to seven taxa), but was not significantly different from the Day 0 collection (Table 4; Figure 3b). A similar trend was seen in mayfly abundance and richness, but there were no significant differences (Figure 4a and b).

While no significant differences were found between the community metrics for organisms exposed to the high conductance test concentrations (Table 5), some trends were evident. Abundance was similar at the four test concentration and two reference waters (Figure 5a) as was taxa richness (Figure 5b). A rock basket with 10 Chironomidae and 15 Ephemerellidae in the 300 µS/cm stream treatment created substantial variability in that concentration. Comparison of the sensitive mayfly taxa abundance and

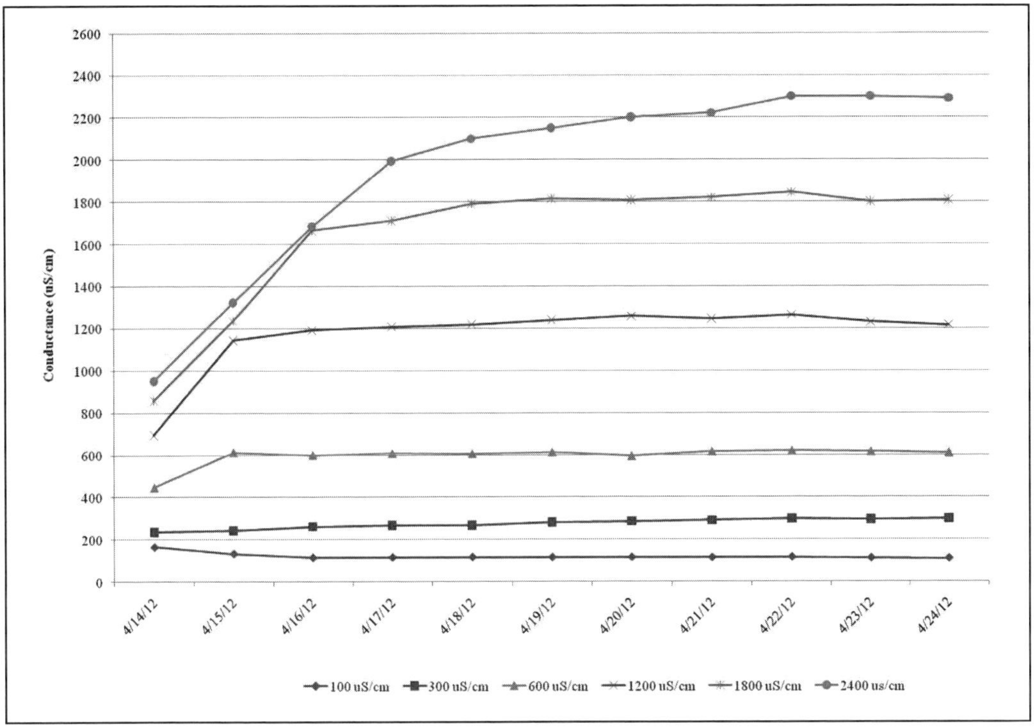

Figure 2. Specific conductance of test solution draining from the simulated stream environments at test initiation (Day 0) and over the 10-day exposure period. Legend value indicates target value. Higher concentrations took longer to achieve due to volumes of change over water necessary to reach the target concentration.

Table 4. Average and range of benthic macroinvertebrate community metrics for the four replicate baskets representing the day of collection compared with reference water communities after the exposure period

	Day 0	100 µS/cm	300 µS/cm	F-Ratio	Prob Level	Chi^2	Prob Level
Abundance	22.5	12.5	21.25	1.61	0.2520		
	14–30	11–17	11–39				
Total # taxa	7.25	5	7.25	2.83	0.1116		
	7–8	3–7	6–7				
# Mayfly taxa	4.25	2.5	3.25	2.64	0.1250		
	4–5	1–4	2–5				
# Stonefly taxa	1.5	1	1.25	0.60	0.5694		
	1–2	1	0–2				
# Caddisfly taxa	0.75	1.25	1.5	1.17	0.3544		
	0–1	0–2	1–2				
Abundance of mayflies	18.25	8	13	3.32	0.0831		
	11–24	4–14	8–21				
% EPT individuals	93–100	82–100	72–100	1.34	0.3106	2.28	0.3200

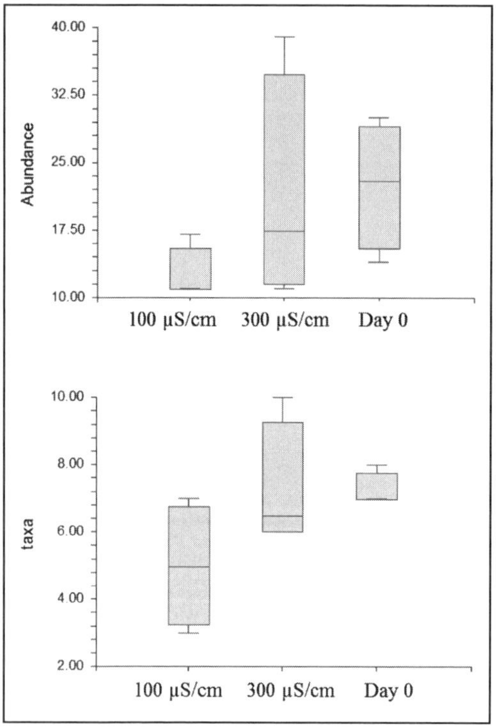

Figure 3. Abundance and taxa richness from the reference streams where field collected benthic macroinvertebrates were exposed to full strength and diluted USEPA reconsituted moderately hard water

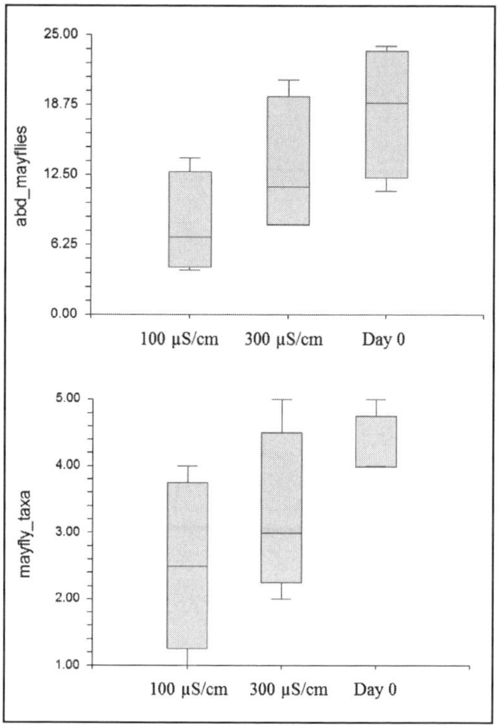

Figure 4. Mayfly abundance and taxa richness from the reference streams where field collected benthic macroinvertebrates were exposed to full strength and diluted USEPA reconstituted moderately hard water

Table 5. Average and range of benthic macroinvertebrate community metrics calculated for benthic macroinvertebrate samples retrieved from rock baskets from each treatment concentration following 10-day exposures to elevated dissolved solids in simulated stream waters

	100 µS/cm	300 µS/cm	600 µS/cm	1215 µS/cm	1800 µS/cm	2430 µS/cm	F-Ratio	Prob Level	Chi^2	Prob Level
Abundance	12	21.25	17.75	13	12	13	1.09	0.4017		
	11–17	11–39	8–24	9–17	6–20	8–19				
Total # taxa	5	7.25	7	7.5	6	8.25	1.04	0.4223		
	3–7	6–7	5–10	6–10	4–10	5–12				
# Mayfly taxa	2.5	3.25	3.5	3.5	2.25	2.25	1.74	0.1763		
	1–4	2–5	3–4	3–5	2–3	2–3				
# Stonefly taxa	1	1.25	0.5	1.5	1.5	1.75	0.64	0.6722		
	1	0–2	0–2	0–4	0–3	1–3				
# Caddisfly taxa	1.25	1.5	1.25	0.5	1	1.75	1.30	0.3066		
	0–2	1–2	1–2	0–1	0–2	1–3				
Abundance of mayflies	8	13	11.5	6.5	6.75	3.5	2.53	0.0668		
	4–14	8–21	5–20	4–10	4–9	2–6				
% EPT Individuals	82–100	72–100	67–92	45–89	83–90	44–89			8.07	0.1526

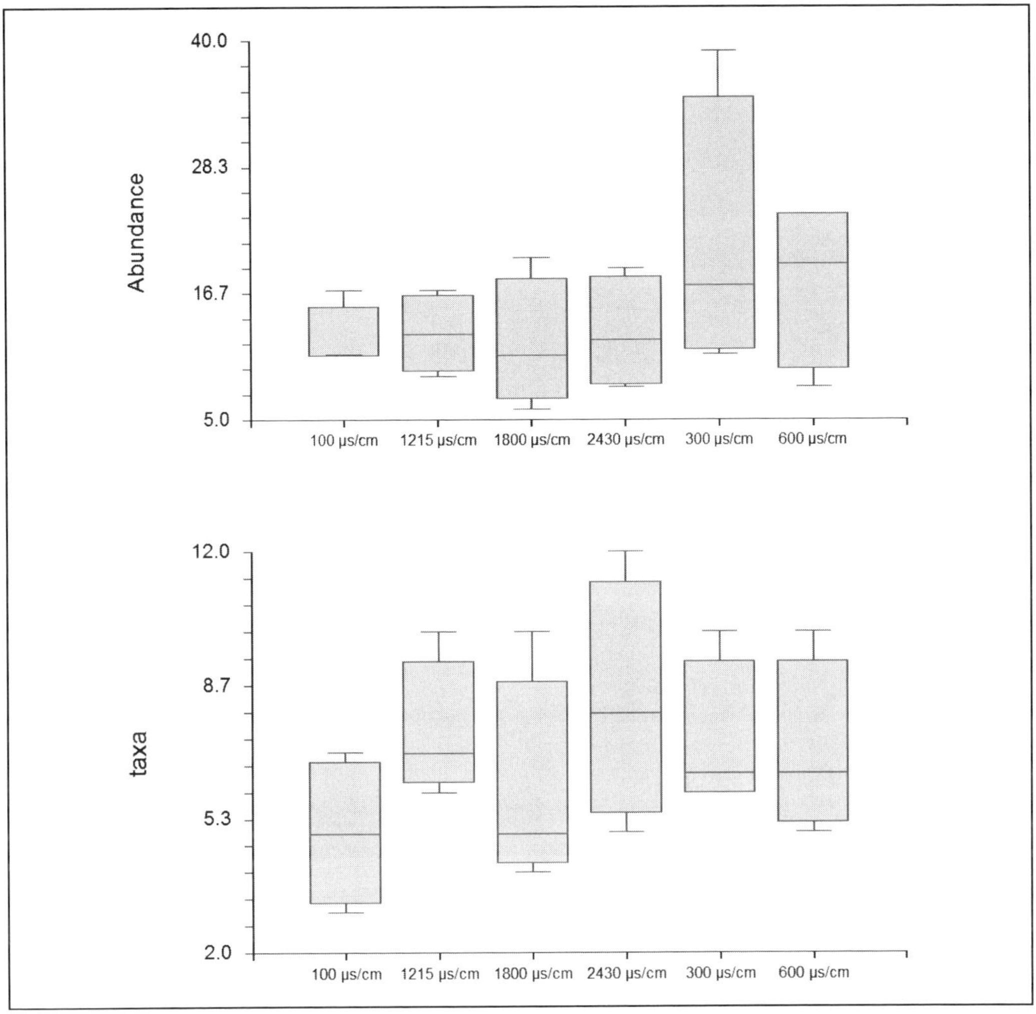

Figure 5. Abundance and taxa richness from the artificial stream exposures of field collected benthic macroinvertebrates to simulated stream waters with elevated conductance

richness values also showed no statistical differences between the treatments concentrations. However, mayfly abundance was lowest at the highest conductance concentration and taxa richness was lowest at the 1,800 µS/cm and 2,400 µS/cm concentrations (Figure 6a and b).

As would be expected, leaf packs were found to have a community substantially different from the rock baskets. The leaf pack community was dominated by Trichoptera, specifically the Limnephiled caddisfly *Pycnopsyche,* and by Oligochaetes. Abundance in the leaf packs ranged from 12 to 100 individuals and was not significantly different between the reference waters (100 µS/cm and 300 µS/cm) and the four treatment concentrations (Table 6). Similarly, there were no differences between the treatments and reference waters with respect to taxa richness. However, mayfly taxa richness was significantly reduced in the leaf packs ($X^2 = 11.04$, $p = 0.05$) although no differences were found in the between treatment comparisons (Figure 7a). Caddisfly abundance was also reduced, although not significantly so, in the 2,400 µS/cm concentration (Figure 7b).

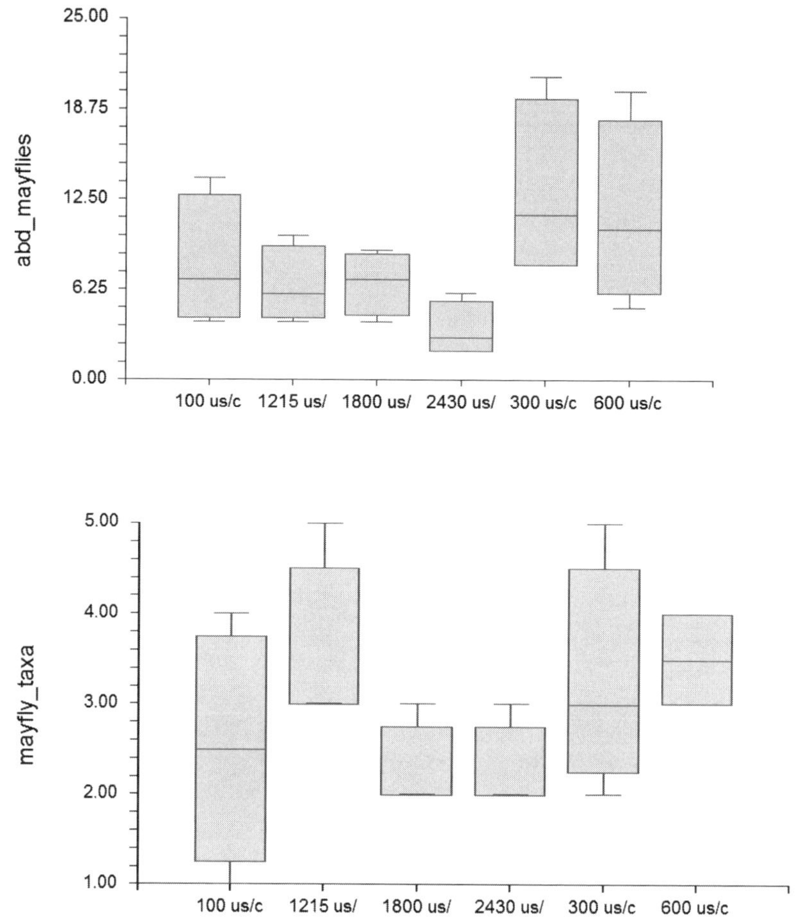

Figure 6. Mayfly abundance and taxa richness from the artificial stream exposures of field collected benthic macroinvertebrates to simulated stream waters with elevated conductance

DISCUSSION

The testing described herein was developed to bridge the gap between the suggested impairment threshold from field surveys summarized in the USEPA benchmark document (USEPA 2011) and the lack of toxicity demonstrated in laboratory testing. These methods are being developed that more specifically evaluate the benthic macroinvertebrate community response to elevated TDS. These methods utilize field collected benthic macroinvertebrate communities that are exposed to simulated mine effluent under laboratory conditions. Specific objectives were to develop a simulated receiving stream composition representative of Appalachian streams receiving mining discharges and to develop laboratory testing methods to evaluate benthic macroinvertebrate community and individual taxa responses to TDS.

Laboratory prepared reconstituted water was successfully developed that simulated natural stream waters receiving mining discharges with elevated conductance. The target concentrations were achieved quickly at lower concentrations but were harder to obtain at higher conductivities.

Table 6. Average and range of benthic macroinvertebrate community metrics calculated for benthic macroinvertebrate samples retrieved from leaf packs from each treatment concentration following 10-day exposures to elevated dissolved solids in simulated stream water

	100 µS/cm	300 µS/cm	600 µS/cm	1215 µS/cm	1800 µS/cm	2430 µS/cm	F-Ratio	Prob Level	Chi^2	Prob Level
Abundance per gram	35.25	79.75	52.25	64.5	50.75	67.25	0.97	0.46		
	12–70	46–99	32–72	49–100	43–58	41–85				
Total # taxa	7	9.5	9	10.25	9	8.75	0.82	0.55		
	4–12	8–11	7–10	9–12	7–13	5–12				
# Mayfly taxa	1.75	2.25	2.25	2	1.25	1	1.52	0.23		
	1–3	2–3	1–3	1–3	0–2	0–2				
# Stonefly taxa	2.5	2.25	2	2	1.75	2	0.16	0.97		
	1–4	2–3	0–4	1–3	1–3	0–4				
# Caddisfly taxa	0.75	1.5	2	1.75	2	2.25	2.22	0.10		
	0–2	1–2	1–3	1–2	1–3	2–3				
Abundance of mayflies	8	14	8.25	4.5	1.75	2	1.26	0.32	11.04	0.05
	1–15	3–25	3–16	2–6	0–4	0–4				
Abundance of caddisflies	18	23	28.5	25.5	38.5	3.25	1.12	0.39		
	2–30	6–37	13–58	2–50	5–75	0–8				
% EPT individuals	54–92	26–92	45–94	42–76	26–74	10–72			2.77	0.74

In future testing, less water will be used in the streams during the 24-hour acclimation period so that less dilution will occur when the streams are being brought to target concentrations.

The simulated streams functioned well with minimal loss of organisms in the reference streams during the course of the 10-day exposure period. Dissolved oxygen was maintained in sufficient range during testing. The feeding regime appears sufficient to maintain the organisms based on maintenance of the communities during the test period. The change over rate of 10 liters twice per day was sufficient to maintain water quality and did not flush the food sources for the organisms as had been noticed in preliminary testing. Benthic macroinvertebrate baskets and leaf packs collected and maintained sufficient abundances; however, some material was lost to the streams and accumulated in the bottom. In future testing, the technique to remove baskets from the stream will be altered to reduce the material lost upon test breakdown. Although not statistically significant, organisms in the 100 µS/cm concentration did not appear to perform as well as the 300 µS/cm concentration in several metrics. This trend will be evaluated in future testing to determine if the invertebrates, particularly the mayflies, have a preference for the water with higher dissolved solids.

In the elevated conductance treatments, with values ranging from 600 µm/cm to 2,400 µm/cm, only one of the endpoints was shown to be statistically significant. However, trends were evident which show benthic macroinvertebrate taxa responsive to elevated to dissolved solids. The mayfly abundance was consistently reduced in the highest conductance treatments, although only significantly so in the leaf packs at the 2,400 µm/cm. Caddisfly abundance was also reduced at this conductance level. These findings indicate the organisms are responsive to the simulated stream water and the elevated TDS in the range of exposures tested. Additional evaluations are ongoing utilizing the methods described herein and expanding on these methods. Specific findings of this preliminary research being more fully evaluated include the poor performance of the organisms in the lowest ionic strenght treatment; the

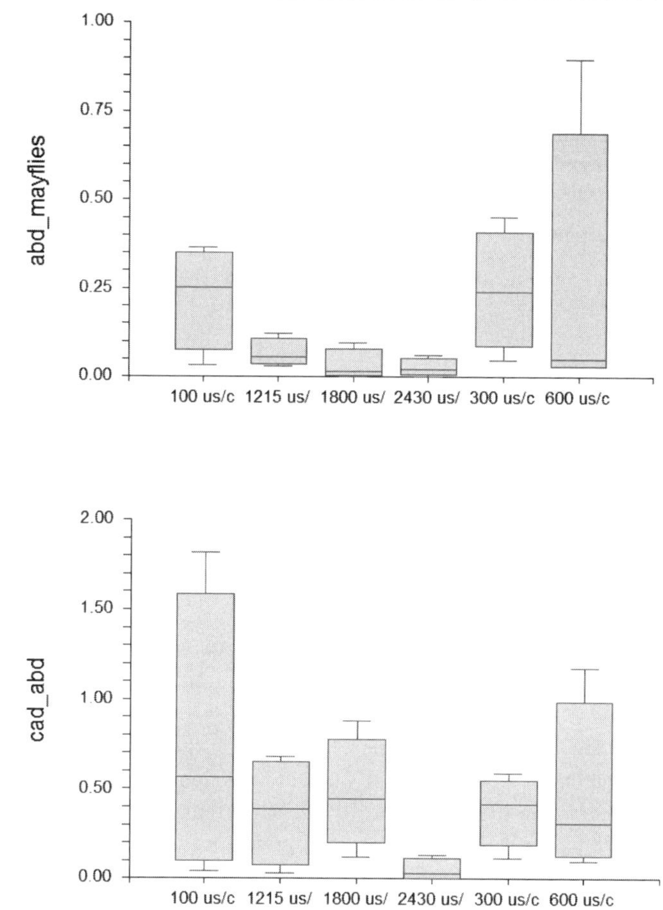

Figure 7. Mayfly abundance and taxa richness from the artificial stream exposures of field collected benthic macroinvertebrates to simulated stream waters with elevated conductance. Abundance values are presented per kilogram of leaf material.

definition of the conductance level where impairment to the biological communities are apparent; consideration of the effects of dissolved solids on individual taxa; and caddisfly sensitivity, which appears in a similar range as mayfly impairment in this limited dataset. Based on this initial assessment, the exposure and testing methods may be modified to increase the number of organisms utilized in the testing and to improve recovery from the simulated stream environments.

ACKOWLEDGMENTS

This study was sponsored by the Appalachian Research Initiative for Environmental Science (ARIES). The views, opinions and recommendations expressed herein are solely those of the authors and do not imply any endorsement by ARIES employees, other ARIES-affiliated researchers or industrial members. Information about ARIES can be found at http://www.energy.vt.edu/ARIES.

The authors thank Alpha Natural Resources and Patriot Coal Corporation for providing the data. Additionally, we are grateful to Adrian Tennant for her invaluable assistance in the lab.

REFERENCES

Bitzer, Leah J., 2012. Chronic Toxicity Testing in Mining Influenced Streams of West Virginia. Thesis, Dissertations and Capstones. Marshall University, Huntington, West Virginia.

Chapman, P.M. 2000. Whole effluent toxicity testing–usefulness, level of protection, and risk assessment. Environ. Toxicol. Chem. 19(1):3–13.

Eagleson, Kenneth W., Lenat, David L., Ausley, Larry W., and Winborne, Ferne B. 1990. Comparison of Measured Instream Biological Responses with Responses Predicted using the *Ceriodaphnia Dubia* Chronic Toxicity Test. Environ. Toxicol. Chem. 9:1019–1028.

Elphick, J.R.F., K.D. Bergh, H.C. Bailey. 2010a. Chronic toxicity of chloride to freshwater species: effects of hardness and implications for water quality guidelines. Environ. Toxicol. Chem. 30(1):239–246.

Elphick, J.R., M. Davies, G. Gilron, E.C. Canaria, B. Lo, H.C. Bailey. 2010b. An aquatic toxicological evaluation of sulfate: the case for considering hardness as a modifying factor in setting water quality guidelines. Environ. Toxicol. Chem. 30(1):247–253.

Kennedy, A.J., Cherry, D.S., Zipper, C.E. 2005. Evaluation of ionic contribution to the toxicity of a coal mine effluent using *Ceriodaphnia dubia.* Arch. Environ. Contam. Toxicol. 49(2):155–162.

Soucek, David John. 2007. Comparison of hardness- and chloride-regulated acute effects of sodium sulfate on two freshwater crustaceans. Environ. Toxicol. Chem. 26(4):773–779.

Soucek, D.J., K.T. Linton, C.D. Tarr, A. Dickinson, N. Wickramanayake, C.G. Delos, L.A. Cruz. 2011. Influence of water hardness and sulfate on the acute toxicity of chloride to sensitive freshwater invertebrates. Environ. Toxicol. Chem 30(4):930–938.

United States Environmental Protection Agency (USEPA). 1991. Technical support document for water quality-based toxics control. EPA/505/2-90-001. Technical Report. Washington, DC.

United States Environmental Protection Agency (USEPA). 2002. Short-term methods for estimating the chronic toxicity of effluents and receiving waters to freshwater organisms. 4th ed. Washington DC; USEPA, Office of Water, EPA-821-R-02-013.

United States Environmental Protection Agency (USEPA). 2011. A Field-Based Aquatic Life Benchmark for Conductivity in Central Appalachian Streams (Final Report). U.S. Environmental Protection Agency, Washington, DC, EPA/600/R-10/023F.

Exploring Benthic Impairment and Total Dissolved Solids in the Dumps Creek Watershed

Laura Kirby
Department of Mining and Minerals Engineering, Virginia Tech, Blacksburg, Virginia

Sara Sweeten
Department of Fisheries and Wildlife Conservation, Virginia Tech, Blacksburg, Virginia

John Craynon
Virginia Center for Coal and Energy Research, Virginia Tech, Blacksburg, Virginia

ABSTRACT

This study looks at the Dumps Creek watershed located in Russell County, Virginia, which has a long history of mining and industrialization. Dumps Creek was placed on the first Commonwealth of Virginia 303(d) List of Impaired Waters in 1994 due to benthic impairment and has not been removed from this list to-date. The U.S. Environmental Protection Agency (EPA) and other regulators have used total dissolved solids (TDS) and/or conductivity as a metric to characterize overall stream health. Historically, other benthic indicators have also been used as a stream characterization metric. This study was conducted in Dumps Creek in Russell County, Virginia, used the Virginia Stream Condition Index (VSCI) as a rating for benthic community health and compared this rating to conductivity measurements in order to assess whether a correlation exists between these two measurements. Both historical data collected by the Commonwealth of Virginia and by mining companies and data collected for this study were used for these comparisons. This work suggests that the current practice of the universal use of conductivity as an indicator of stream health may be premature, without further understanding of how TDS/conductivity correlates to benthic health metrics.

INTRODUCTION

The southwest Virginia watershed of Dumps Creek is currently used for surface and underground mining and has been industrialized since the turn of the century. The watershed was placed on Virginia's 303(d) List of Impaired Waters when in the list was first established in 1994. The watershed was regarded as impaired due to poor benthic health. Due to the 303(d) listing, the state of Virginia also assigns total maximum daily load (TMDL) values for the watershed, in order to facilitate improvement of the impairment. Total dissolved solids (TDS) is one of the metrics that is listed on the TMDL and one that this watershed regularly exceeds due to the need to pump water out of an inactive underground mine. This water is then discharged near the headwaters of Dumps Creek. This study explores the historic and current relationship between TDS/conductivity and benthic health in this watershed.

PREVIOUS STUDIES

Using TDS/conductivity as a metric is a relatively new approach to understanding water chemistry in order to enforce regulation. In particular, based on the work of Pond and Passmore (2008), EPA and state water quality agencies have recently used conductivity in regulating the Appalachian coal mining industry. In contrast, the Illinois

Environmental Protection Agency (2006) and Iowa Department of Natural Resources (2009) concluded from their TDS studies that TDS (and conductivity) is an inappropriate indicator of toxicity to benthic organisms because toxicity is ion-specific. The Illinois study indicates that 2,000 milligrams per Liter (mg/L) TDS with chloride dominance is toxic, but the same level TDS with sulfate dominance is not. Many authors have suggested a direct correlation between conductivity and benthic health, mostly in relation to surface mining; currently the body of literature appears split on whether conductivity is a good stand-alone metric for regulatory purposes.

The EPA contracted various research consultants to study the environmental effects of mountaintop coal mining (MTM) in Appalachia. Conclusions of this research, which focused on a broad array of ecological impacts, were published in both the draft Programmatic Environmental Impact Statement (PEIS) (EPA 2003) and the final PEIS (U.S. EPA 2005) on mountaintop mining and valley fills. Much of this work indicated that surface coal mining resulted in impacts on stream biota, particularly benthic macroinvertebrates.

In a later work, Griffith et al. (2012) focused on studies included in both PEIS publications with respect to stream conductivity, individual ion concentrations and chemistry, in general. Their work concludes that a correlation was found between benthic health and conductivity levels below MTM valley fills as compared to nonmining areas. However, they concluded that, "no studies have quantified the change in conductivity, individual ion concentrations, pH or precipitates as the water progresses downstream" (Griffith et al. 2012). Although correlations are shown through much of the research published in the EPA produced draft and final PEIS, few specifics to understanding this correlation are explored in sufficient detail to explain the mechanisms for the correlations that were found.

Echols, et al. (2009) conducted controlled laboratory tests on a sensitive Ephemeroptera order species, concluding that in exposure to coal processing impoundment effluent, the threshold of impairment-inducing TDS/conductivity concentrations are higher than the benchmark value of 300–500 μS/cm selected by EPA.

Armstead, et al. (2004) collected quantitative and qualitative samples over four seasons. Statistical comparisons were made between the categories of "unmined" (reference), "valley filled," and "valley filled/residential" sampling sites. Results showed that water chemistry did change from a reference condition; however, those changes were not substantial in "valley filled" category streams, and biological impairment was most severe in "valley filled/residential" streams.

The Virginia Stream Condition Index (VSCI) is a multi-metric scoring protocol for benthic macroinvertebrates. This protocol was developed by the Virginia Department of Environmental Quality in order to identify biological impairment of non-coastal streams. The VDEQ used benthic macroinvertebrate data from both impacted and reference streams throughout Virginia in order to determine which family-level macroinvertebrate metrics best evaluated stream condition in all five Virginia ecoregions. The metrics chosen are shown in Table 1.

In order for the VSCI protocol to encompass all streams in Virginia, some resolution is often lost on local scales. For example, the VSCI is used on streams ranging from coastal Piedmont streams (Region 45) to central Appalachian streams (Region 69). Scores in central Appalachian streams, even on reference streams, tend to be lower than other regions of Virginia. This may be due to natural regional differences that the VSCI does not account for or it may be due to the low number of reference sites used in central Appalachia. Of the 62 reference sites used in the development of the VSCI, only five of these sites were in central Appalachia (Burton and Gerritsen 2003).

A study by Northington et al. (2011), assessed the physical habitat, water quality field parameters (including conductivity), and dissolved major ions and metals to evaluate relationships between TDS concentrations and the VSCI. VSCI scores, which are based on the

Table 1. Macroinvertebrate metrics use in 5 Virginia ecoregions (adapted from Burton and Gerritsen (2003))

Metrics, grouped by Category	Definition
Taxonomic Richness: Counts of different taxa within selected taxonomic groups	
Total Taxa	Number of distinct taxa in the entire sample; measures the overall variety of the macroinvertebrate assemblage
EPT Taxa	Sum of distinct taxa in the generally pollution-sensitive insect orders of Ephemeroptera (mayflies), Plecoptera (stoneflies), and Trichoptera (caddisflies)
EPT Taxa less Hydropsychidae	Sum of taxa in the insect orders Ephemeroptera, Plecoptera, and Trichoptera, not including the generally pollution-tolerant caddisfly family Hydropsychidae.
Ephemeroptera taxa	Number of Ephemeroptera taxa (mayfly nymphs)
Plecoptera taxa	Number of Plecoptera taxa (stonefly naiads)
Trichoptera taxa	Number of Trichoptera taxa (caddisfly larvae)
Trichoptera taxa less Hydropsychidae	Number of Trichoptera taxa not including the pollution tolerant caddisfly family Hydropsychidae
Diptera taxa	Number of Diptera taxa ("true" fly larvae and pupae)
Chironomidae taxa	Number of taxa in the family Chironomidae (midge larvae)
Composition: Percent abundance (of individuals in the sample) of…	
%EPT	... Ephemeroptera (mayfly nymphs), Plecoptera (stonefly naiads), and Trichoptera (caddisfly larvae)
%EPT less Hydropsychidae	... Ephemeroptera, Plecoptera, and Trichoptera not including pollution tolerant caddisflies in the family Hydropsychidae
% Ephemeroptera	... mayfly nymphs
% Plecoptera	... stonefly naiads
% Trichoptera	... caddisfly larvae
% Trichoptera less Hydropsychidae	... caddisfly larvae not including those in the pollution tolerant family Hydropsychidae
% Plecoptera plus Trichoptera less Hydropsychidae	... stonefly naiads plus caddisfly larvae not including those in the pollution tolerant family Hydropsychidae
% Diptera	... "true" fly larvae and pupae
% Chironomidae	... Chironomidae (midge) larvae and pupae
% Oligochaeta	... aquatic worms
Trophic Groups: Percent abundance of individuals in the sample, or number of taxa in the sample, whose primary functional mechanism for obtaining food (functional feeding group, FFG) is to…	
% Collectors	... collect/gather depositional organic matter
% Filterers	... filter and collect suspended organic matter
% Predators	... attack prey and ingest whole organisms or their parts
% Scrapers	... graze on substrate- or periphyton-attached algae and associated material
% Shredders	... shred and chew leaf litter and detritus
Scraper taxa	(number of taxa classified primarily as scrapers)
Diversity: Percent abundance in the sample of individuals belonging to…	
% Dominant	... the single most abundant taxon
% 2 Dominant taxa	... the two most abundant taxa
% 5 Dominant taxa	... the five most abundant taxa

(table continues)

Table 1. Macroinvertebrate metrics use in 5 Virginia ecoregions (adapted from Burton and Gerritsen (2003)) (continued)

Tolerance: Counts, proportions, or weighted scores of taxa based on ability to survive exposure to stressors	
Intolerant taxa	Number of taxa with Tolerance Values 3
% Tolerant	Percent abundance of organisms with a Tolerance Value 7
HBI	Abundance-weighted average tolerance of assemblage of organisms (Family taxonomic level)
Habitat: Organisms having the specified dominant behavior for moving and maintaining physical position in their habitat	
% Clingers	Percent abundance of insects having fixed retreats or adaptations for attachment to surfaces in flowing water
Clinger taxa	Number of taxa having fixed retreats or adaptations for attachment to surfaces in flowing water

Commonwealth of Virginia's implementation of stream condition assessment required under the Clean Water Act, assess stream quality based on the population of various organisms in the stream. Findings have shown that certain streams with elevated conductivity values maintain a better VSCI score than some reference streams with very low conductivity.

Pond and Passmore have collaborated on several scientific papers focused on this issue of conductivity. However, even their research appears to yield some different conclusions. Their study in the Straight Creek, Virginia, watershed (Passmore and Pond 2009) reached different results than an earlier Pond and Passmore et al. (2008) study. These differences include the toxicity impacts of conductivity, the comparative impacts of stressors, the ionic makeup of mine-related drainage, and the threshold value of conductivity-associated benthic impairment. The results of this recent investigation, and its contrasting findings to previous studies by the same investigators, shows that the science on these issues is unsettled.

The Passmore and Pond (2009) investigation found that residential land use with improper control of sewage is a major stressor of sensitive benthic life throughout the region. This study also concluded that stream sites near mines were often in better condition than at points farther downstream where residential and other land use impacts occur. The authors offer the opinion that conductivity may be useful as an indicator of general conditions and of trends. They also recommend the construction of sewage systems as an obvious first step to improve the streams studied. Further, the data in this report also confirms that some sites studied in that investigation had unimpaired conditions even with conductivities higher than the levels that had been considered in the previous study to definitely be associated with impairment. Reasons for that difference are discussed, but not fully explained, and may relate to the differences in ionic composition between areas.

Mount et al. (1997) evaluated specific ion effects on aquatic organism toxicity. The report evaluation indicates that sulfate, bicarbonate, calcium, magnesium, sodium, and potassium ions comprise 99% of the total ions in the MTM discharges studied (the ions above are listed from highest concentration to lowest concentration). Sulfate leads the ion composition by approximately six times the concentration of bicarbonate, the next leading ion in the test waters. That being said, the study's toxicity testing indicates that sulfate is the least toxic of the test ions. This study emphasizes the importance of analyzing what individual constituents are being collectively measured by TDS rather than looking at TDS as a singular value.

HISTORIC DATA

Dumps Creek is located in Russell County, Virginia, just northwest of the town of Cleveland. Dumps Creek was placed on the Commonwealth of Virginia's 1994 303(d) List of Impaired Waters

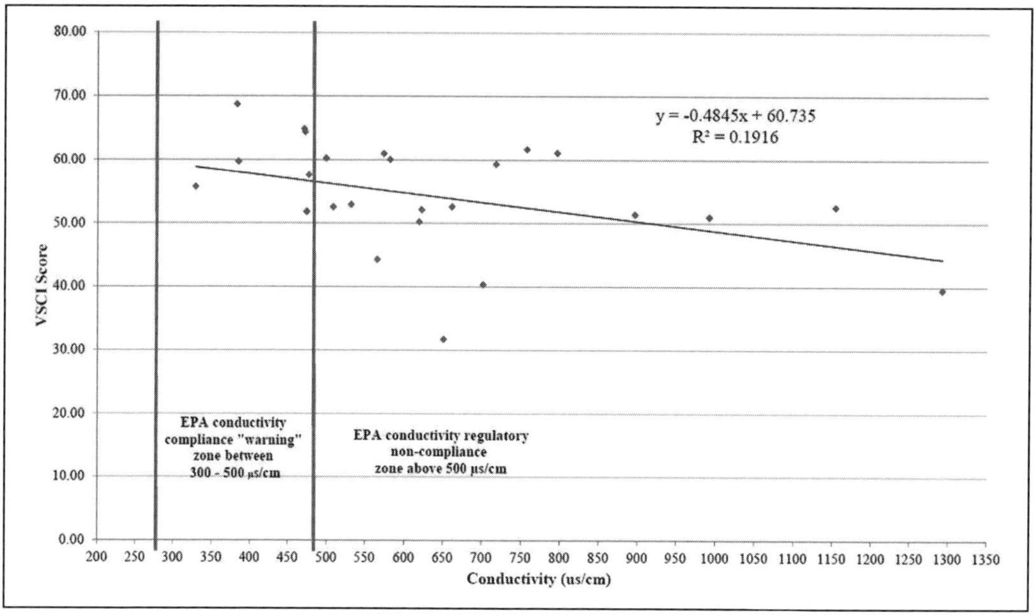

Figure 1. Historic VSCI scores and conductivity in Dumps Creek

because of repeated violations of the general standard based on benthic life criteria and is still on this list today. The impaired stream segment has a length of 3.40 miles, and extends from the Hurricane Fork Confluence to the mouth where Dumps Creek flows into the Clinch River. This stream confluence is near the Appalachian Power Plant in Russell County that discharges to the Clinch River. The land area of the Dumps Creek Watershed is approximately 20,300 acres, with forest and mining as the primary land uses. Dumps Creek has been heavily industrialized since the turn of the century, as a munitions manufacturing facility, and then was heavily surface and underground mined, with a processing facility also located along Dumps Creek. Most mining in Dump's Creek took place before the establishment of environmental protection legislation such as the Clean Water Act (CWA) of 1972 and the Surface Mining and Control Reclamation Act (SMCRA) of 1978.

Currently, there are two pumps staged in the headwaters region of Dumps Creek that help to maintain water levels in an inactive underground mine. The pumping is necessary to control methane levels that rising water could force into an active underground mine that lies stratigraphically above the inactive mine. Water is pumped on an as-needed basis and discharges directly into Dumps Creek.

Conductivity and VSCI data have been collected periodically, although not annually, from 2001 to 2012. We analyzed the historical data available from the Commonwealth of Virginia and other sources. A linear regression analysis of these data show an R^2 of 0.192, which indicates little direct correlation between conductivity and stream impairment as determined by the VSCI. Based on this low level of correlation, no additional regression analyses were conducted (Figure 1).

CURRENT SAMPLING DATA VERIFICATION

To verify the historic data collection and trends, Appalachian Technical Services (ATS) was contracted, as a third party, to collect water chemistry and benthic samples for five locations along Dumps Creek. ATS sent these samples to

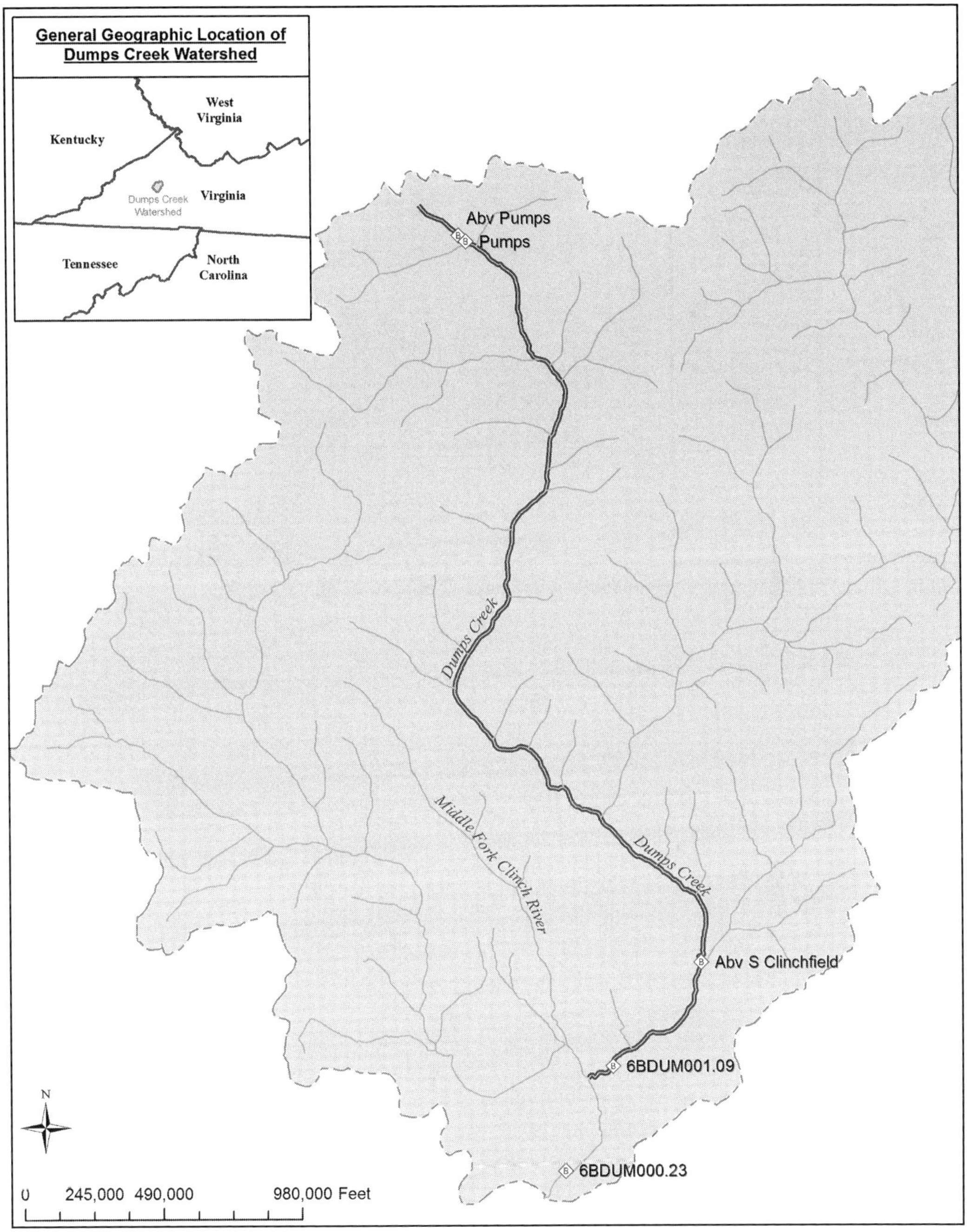

Figure 2. General map of Dumps Creek watershed and current sample locations

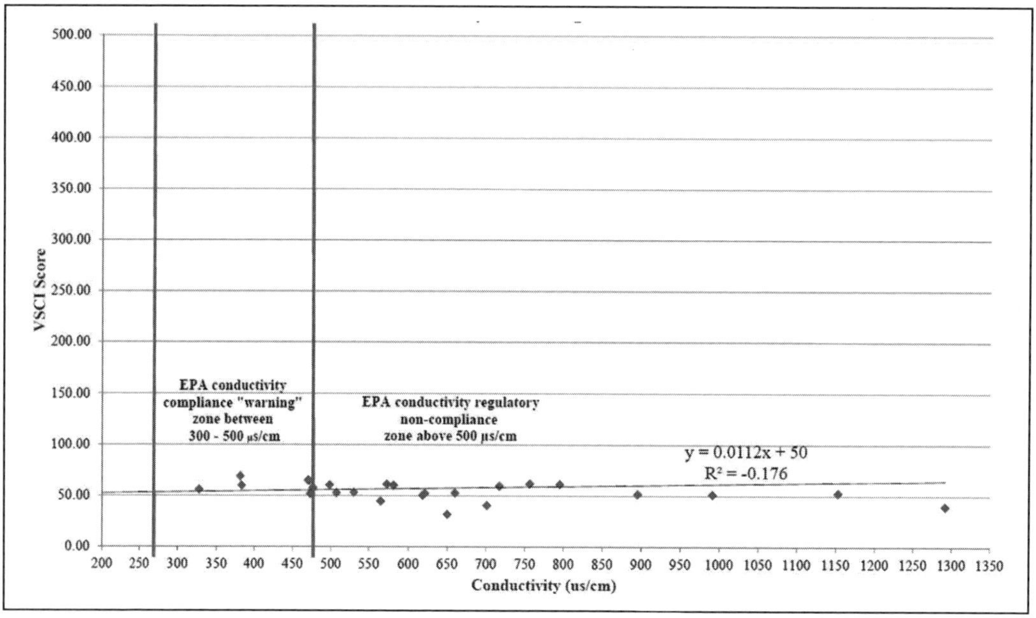

Figure 3. Graph of VSCI scores and conductivity in Dumps Creek including current data

Table 2. Dumps Creek water chemistry, Fall 2012

Sample ID	Date	Iron (mg/L)	Manganese (mg/L)	Aluminum (mg/L)	Chloride (mg/L)	Sulfate (mg/L)	Bicarbonate (mg/L)
6BDUM000.23	10/3/2012	<0.100	<0.100	<0.100	5.23	NA	128
6BDUM001.09	10/3/2012	0.103	<0.100	<0.100	4.54	71.0	132
ABVSClinchfield	10/3/2012	<0.100	<0.100	<0.100	4.39	60.8	132
ABVPUMPS	10/3/2012	0.241	<0.100	<0.100	2.07	89.2	85.6
PUMPS	10/3/2012	0.834	0.118	<0.100	18.1	25.0	540

Note:
mg/L—Milligrams per Liter.
NA—Not Analyzed.
<0.100—This shows that the constituent was less than the Practical Quantitation Limit.

Research Environmental Industrial Consultants, Inc. (REIC), in Beaver, West Virginia, for analysis. REIC is an accredited Virginia laboratory. Current sampling locations are depicted on Figure 2. The southernmost two locations (6BDUM000.23 and 6BDUM001.09) are in close proximity to the locations that are sampled by the Commonwealth of Virginia for the 303(d) listing; the next sample north in the watershed (Abv S Clinchfield) is located above a discharge that drains a community called South Clinchfield, and the two northernmost samples located near the headwaters of Dumps Creek focused on influence associated with pumped water from a nearby underground mine in the watershed. One sample was specifically of the pumped water (Pumps) and one sample was of the surface water in Dumps Creek, prior to reaching the pumps (Abv Pumps).

Water chemistry was taken of the discharged pump water and both water chemistry and benthic macroinvertebrates were sampled above the pump discharged water influence. When this data was added to 2001 to 2012 data, the trend also shows a low correlation between the conductivity and VSCI scores with an R^2 of –0.176 (Figure 3).

Again, based on this low level of correlation, no additional regression analyses were conducted.

WATER CHEMISTRY

The current five samples shown previously on Figure 2 were analyzed for metals, specific conductivity, TDS, and major ions and anions. Table 2 shows the results obtained for the major ions and anions. The data show that this watershed is characterized by bicarbonate anions (rather than sulfate or chloride) and has very low loading in iron, manganese and aluminum. While previous work has concluded that bicarbonate-dominated systems may demonstrate a correlation between conductivity and benthic community health (e.g., Timpano et al. 2011), Dumps Creek appears to show no such correlation.

CONCLUSIONS

The Dumps Creek watershed has a long history of activities that have had potential to cause impairment to water chemistry and the health of stream biota. Both mining operators and regulators have struggled with the impairment of the stream for many years. It has been difficult to adequately quantify and specify what is causing the identified impairment to benthic macroinvertebrate communities in the stream, as evidenced by low VSCI scores over many years.

Recently, the use of conductivity and/or TDS has been suggested as a regulatory metric to quantify stream health. However, historic and current conductivity data and VSCI scores show that, in this specific watershed, these metrics do not appear to have a correlation. This conclusion, while in contrast to the conclusion of Pond and Passmore (2008), is nonetheless consistent with work done by other researchers, including Passmore and Pond (2009). A lack of correlation between conductivity and benthic scores also contrasts with the work of Timpano et al. (2011) in a watershed with similar bicarbonate anion dominance.

Dumps Creek historic benthic and water chemistry data and current, third-party verified, benthic and water chemistry data did not show a correlation between conductivity and benthic health in a bicarbonate anion dominated watershed. This study analyzed data that was available, given the history of this watershed, and assessed whether improving the conductivity and therefore the load to this watershed would improve benthic health in order to remove the watershed from the Virginia 303(d) List of Impaired Waters. These conclusions suggest that benthic health, in this watershed, may be more greatly affected by other influences in the watershed that have yet to be identified, rather than high conductivity.

Finally, the conclusions of this work, again, underscore the inconsistencies between results of studies with this similar topic. There is a need for continued research related to the relationship of water chemistry, as measured by conductivity and specific ion concentrations, and the relative health of benthic macroinvertebrate communities. This work suggests that the current practice of the universal use of conductivity as an indicator of stream health may be premature, without further understanding of how TDS/conductivity correlates to benthic health metrics.

ACKOWLEDGMENT

This study was sponsored by the Appalachian Research Initiative for Environmental Science (ARIES). The views, opinions and recommendations expressed herein are solely those of the authors and do not imply any endorsement by ARIES employees, other ARIES-affiliated researchers or industrial members. Information about ARIES can be found at http://www.energy.vt.edu/ARIES.

REFERENCES

Armstead, M.Y., Yeager-Seagle, J.L., and Emerson, L. 2004. *Benthic Macroinvertebrate Studies Conducted in Mountaintop Mining/Valley Fill Influenced Streams in Conjunction with the USEPA Environmental Impact Study*. Final report at the National meeting of the American Society of Mining and Reclamation and the 25th West Virginia Surface Mine Drainage Task Force. Published by ASMR. Lexington, Kentucky.

Burton, J. and Gerritsen, J. 2003. *A Stream Condition Index for Virginia Non-Coastal Streams*. Prepared for: USEPA Office of Science and Technology,

Office of Water, Washington, DC; USEPA Region 3 Environmental Services Division, Wheeling, WV; and Virginia Department of Environmental Quality, Richmond, VA. Tetra Tech, Inc.

Echols, B.S., Currie, R.J., and Cherry, D.S. 2010. *Preliminary Results of Laboratory Toxicity Tests with the Mayfly, Isonychia bicolor (Ephemeroptera: Isonychiidae) for Development as a Standard Test Organism for Evaluating Streams in the Appalachian Coalfields of Virginia and West Virginia.* Environmental Monitoring and Assessment. 169 (1-4) (October): 487–500. doi:10.1007/s10661-009-1191-3. http://www.ncbi.nlm.nih.gov/pubmed/19888664. Blacksburg, VA: Virginia Tech.

Griffith, M.B., Norton, S.B., Alexander, L.C., Pollard, A.I., and LeDuc, S.D. 2012. *The Effects of Mountaintop Mines and Valley Fills on the Physiochemical Quality of Stream Ecosystems in the Central Appalachians: A review.* The Science of the Total Environment. 417-418:1–12.

Illinois Environmental Protection Agency. 2006. *Preliminary Justification for Changing Water Quality Standards for Sulfates, Total Dissolved Solids and Mixing Zones.*

Iowa Department of Natural Resources. 2009. *Water Quality Standards Review: Chloride, Sulfate and Total Dissolved Solids.* Consultation package.

Mount, D.R.; Gulley, D.D.; Hockett, J.R., Garrison, T.D., and Evans, J.M. 1997. *Statistical Models to Predict the Toxicity of Major Ions to Ceriodaphnia dubia, Daphnia magna, and Pimephales promelas (fathead minnows).* Environmental Toxicology and Chemistry 16(10):2009–2019.

Northington, R.M., Benfield, E.F., Schoenholtz, S.H., Timpano, A.J., Webster, J.R., and Zipper C. 2011. *An Assessment of Structural Attributes and Ecosystem Function in Restored Virginia Coalfield Streams.* Hydrobiologia. 671 (1) (April 16): 51–63. doi:10.1007/s10750-011-0703-7. Available from: http://www.springerlink.com/index/10.1007/s10750-011-0703-7.

Passmore, M., and Pond, G. 2009. *Evaluating Appropriate Existing and Designated Uses of Straight Creek (Lee County, VA) Using Current Macroinvertebrate, Habitat and Water Quality Data.* USEPA Region III, EAID, OMA, Freshwater Biology Team.

Pond, G.J., Passmore, M.E., Borsuk, F.A., Reynolds, L., and Rose, C.J. 2008. *Downstream Effects of Mountaintop Coal Mining: Comparing Biological Conditions Using Family- and Genus-Level Macroinvertebrate Bioassessment Tools.* Journal of the North American Benthological Society 27 (3) (September): 717–737. doi:10.1899/08-015.1. Available from: http://www.bioone.org/doi/abs/10.1899/08-015.1.

Timpano, A., Schoenholtz, S., Zipper, C., and Soucek, D. 2011. *Levels of Dissolved Solids Associated with Aquatic Life Effects in Headwater Streams of Virginia's Central Appalachian Coalfield Region.* Virginia Tech. Prepared for: Virginia Department of Environmental Quality, Virginia Department of Mines, Minerals, and Energy and Powell River Project.

U.S. EPA. 2003. *Draft Programmatic Environmental Impact Statement on Mountaintop Mining/Valley Fills in Appalachia.* Philadelphia, PA: U.S. Environmental Protection Agency, Region 3. Available from: http://www.epa.gov/Region3/mtntop/eis2003.htm.

U.S. EPA. 2005. *Mountaintop Mining/Valley Fills in Appalachia. Final Programmatic Environmental Impact Statement.* Philadelphia, PA: U.S. Environmental Protection Agency, Region 3. Available from: http://www.epa.gov/region3/mtntop/pdf/mtmvf_fpeis_full-document.pdf.

Benthic Macroinvertebrate Response to Total Dissolved Solids in Coal Mine Streams

Stephen Schoenholtz
Virginia Tech, Blacksburg, Virginia

Carl Zipper
Virginia Tech, Blacksburg, Virginia

ABSTRACT

Elevated salinity is considered a potential stressor to the sensitive benthic macroinvertebrate communities in Central Appalachian headwaters streams influenced by coal mining. Virginia Tech has been engaged since 2008 in research intended to describe responses of these organisms to elevated salinity. Completed research used multimetric indices to determine state regulatory impairment thresholds for total dissolved solids (TDS). That study found significant negative correlations between increasing TDS concentration and benthic macroinvertebrate diversity. Ongoing research includes characterizing temporal variability of salinity and biotic condition. Conductivity (a TDS surrogate) is being measured continuously at research sites, which will be compared to seasonal grab sampling as a predictor of benthic macroinvertebrate community structure. Simultaneously, a third study is focused on characterizing the natural seasonal variability of benthic community composition across a gradient of TDS levels. Finally, the most recent effort just begun at Virginia Tech is intended to understand the causal links between salinity and macroinvertebrate responses. This presentation will provide an overview of our recent research results and ongoing efforts to improve understanding of relationships between TDS and benthic macroinvertebrate communities in the Central Appalachians.

ACKNOWLEDGMENT

The authors thank Tony Timpano, Beth Boehme, Damion Drover, and David Soucek for their contributions to this paper.

How Well Does Electrical Conductivity Predict Stream Impairment?

Paul F. Ziemkiewicz
West Virginia Water Research Institute, West Virginia University, Morgantown, West Virginia

Steven Northway
Faculty of Forestry, University of British Columbia, Vancouver, British Columbia

ABSTRACT

Section 303 of the Federal Clean Water Act (CWA) mandates that states develop a process for defining stream impairment. Criteria can be 'numeric,' i.e., water quality parameters, or 'narrative,' which generally involves more qualitative and biological indices such as the West Virginia Stream Condition Index (WVSCI). Recent guidance from USEPA has encouraged the use of an electrical conductivity (EC) based criteria (300 μS/cm) to estimate WVSCI impairment. Despite the enormous regulatory and economic repercussions of this approach, little work has been done to determine the quality of the resulting estimates. This paper reports our efforts to critically evaluate the quality of EC as a predictor of WVSCI scores. We present a case study using West Virginia stream segments to identify error rates as a function of various EC criteria. We found that an EC based regression model, while highly statistically significant, only accounted for 26 percent of the variation of WVSCI scores. This highlights the difference between a statistically significant model and a model that can actually predict a specific outcome with confidence. No evidence was found that an EC value less than 300 μS/cm is a reliable predictor of WVSCI values that meet the West Virginia narrative criteria for stream impairment. This implies that factors other than EC play a significant role in determining WVSCI ranking and that efforts to lower a stream's EC alone will not reliably improve its condition.

INTRODUCTION

Under section 303 of the Federal Clean Water Act, states are obliged to develop and periodically update a list of impaired steam segments. States may employ water quality based numerical standards or biologically based narrative standards to determine impairment. Impaired stream segments may be subject to an improvement program which generally consists of a determination of the causal pollutant(s), a determination of pollutant loadings and sources and, finally, an implementation plan to improve the stream to the extent that it will no longer be impaired. Implementation may consist of restricting loads of the causal pollutants from permitted discharges or denying permits altogether. This may cause significant economic impact on discharge permit holders as their existing treatment systems may require reconfiguration costing tens of millions of dollars. The expectation is that removal of the putative pollutant will correct the impairment. The purpose of this study is to estimate the probability that a change in EC will result in an improvement in the WVSCI score for a stream segment.

West Virginia's impairment level is ideally considered to be a WVSCI score of 68. However, accounting for sampling error, the State's working criterion for impairment is a WVSCI score of 60.6: "For purposes of Federal Clean Water Act compliance under section 303d a West Virginia Stream Condition Index (WVSCI) score of 68 is considered the threshold for impairment. In recognition of the sampling error (7.4) when using the recommended procedure, the State classifies streams impaired at a WVSCI score of 60.6" (WVDEP 2012). Both criteria are employed in this study as appropriate.

The WVSCI based impairment definition leaves the challenge of developing individual discharge standards under section 402 of the CWA such that the additional discharge will not cause the stream to become impaired. There have been attempts to solve this problem by using water quality parameters such as electrical conductivity (EC), dissolved solids or specific ions such as sulfate to predict stream condition index (SCI) scores. Most attempts to correlate water quality with biological assessments, however, rely on EC since it is easy for field samplers to collect and involves no laboratory analytical cost. As a result, large databases containing both WVSCI scores and EC values are available for many stream segments. Recent guidance from USEPA has encouraged the use of an EC criterion (300 µS/cm) to estimate stream impairment (USEPA 2010). In this sense, EC becomes surrogate for WVSCI.

Despite the enormous regulatory and economic repercussions of this approach, little work has been done to determine the quality of the resulting estimates. Quality has two components in this context: the statistical significance of the prediction model and the practical implications of the error range around the predicted values.

Much of the discussion surrounding the effects of surface mining on stream condition revolves around the relationship between EC and WVSCI. It is understood that the same EC value can result from a wide variety of chemistries and that not all combinations yield the same stream benthic effects. Elphick et al. (2011) found a strong negative interaction between hardness and sulfate concentration, reporting that the interaction of sulfate toxicity and water hardness resulted in separate values for soft (10–40 mg/L), moderately hard (80–100 mg/L) and hard water (160–250 mg/L). The resulting values were 129, 644 and 725 mg/L sulfate, respectively, following the SSD approach, and 75, 625 and 675 mg/L sulfate, following the safety factor approach. Dissolved salt concentrations and EC in Appalachian mining district streams are invariably highest during low stream flow periods (Petty et al. 2010), since mines provide significant base flow.

Bernhardt et al. (2012) concluded that EC was a good predictor of WVSCI, suggesting an EC impairment threshold of 308 µS/cm. Their reported upper and lower confidence intervals (CI) were asymmetrical about the mean at 365 and 245 µS/cm at 95 percent probability, or about 22 percent of the mean value. They made no mention of the classification error rate that would have resulted from the application of their model.

METHODOLOGY

This paper reports on our efforts to critically evaluate the quality of EC as a predictor of WVSCI scores. We present a case study using West Virginia stream segments to identify error rates as a function of various EC criteria. The first objective of this study was to identify whether a correlation exists between the WVSCI and EC. Secondly, we identified a best-fit model for predicting WVSCI as a function EC. Third, we identified the precision with which EC predicts WVSCI.

We used a data set consisting of 222 samples containing both EC and WVSCI data for West Virginia stream segments for this study. It was collected by the staff of West Virginia Department of Environmental Protection (WVDEP) as part of the State's Watershed Assessment Program. It will be referred to as the WVDEP data set. This data set was also the basis of a study by Bernhardt et al. (2012). The data were a subset of a larger database collected by WVDEP and winnowed by the agency in an attempt to remove sampling stations with confounding factors. The sampling stations included pristine streams through unmined, but developed, watersheds to watersheds with increasing mining intensities. The data set was collected by WVDEP staff between 1996 and 2008 during the dry period of the West Virginia hydrologic year, April through October.

We used two key parameters for our evaluation: (1) WVSCI, which classifies stream insect families into tolerant and intolerant categories

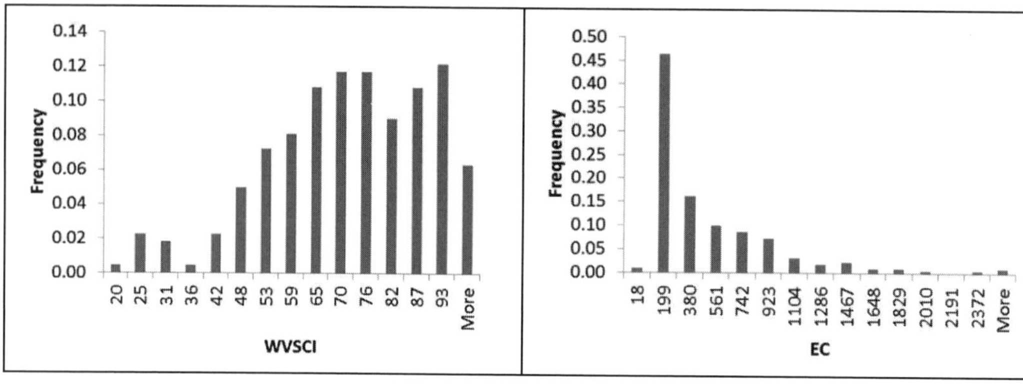

Figure 1. Sample distribution as a function of WVSCI score and EC values. Sample distribution within the WVDEP data set was skewed toward high WVSCI scores and low EC values.

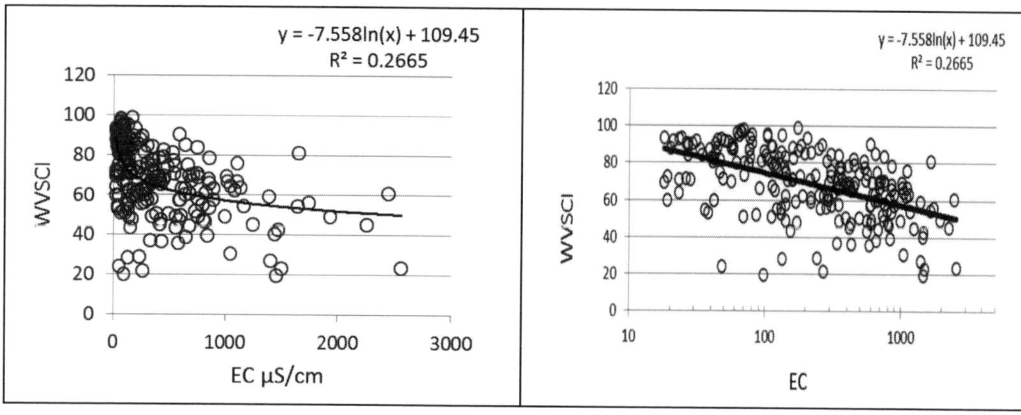

Figure 2. The WVDEP data set was graphed on an non-logarithmic (left) and logarithmic (right) x axis. The black line is the prediction curve given by the model in the upper right corner.

to infer adverse changes in the population due to pollution (Gerritsen et al. 2000); and (2) EC, which is a field based parameter identifying the ability of water to carry an electrical charge. EC increases as ionic concentrations increase. An attempt to identify the relationship between a chemical parameter like EC and WVSCI necessarily infers the behavior of watershed scale processes from limited number of samples. This study attempts to develop a model for predicting a WVSCI outcome based on stream EC. It then estimates the precision with which that prediction can be made, given the inherent variability within the sample and general population.

RESULTS AND DISCUSSION

The observations were not randomly distributed across the range of EC values. Almost two-thirds of samples in the WVDEP data set were in the EC range of 0–399 µS/cm, 21 percent were between 400 and 799 and the remaining 15 percent were distributed between 800 and 2,600 µS/cm. High WVSCI scores and low EC values constituted roughly half of the samples. This results in estimates of higher precision for the low EC ranges and lower precision in the upper EC ranges (Figure 1).

Figure 2 illustrates the relationship between WVSCI and EC in the data set. The two graphs

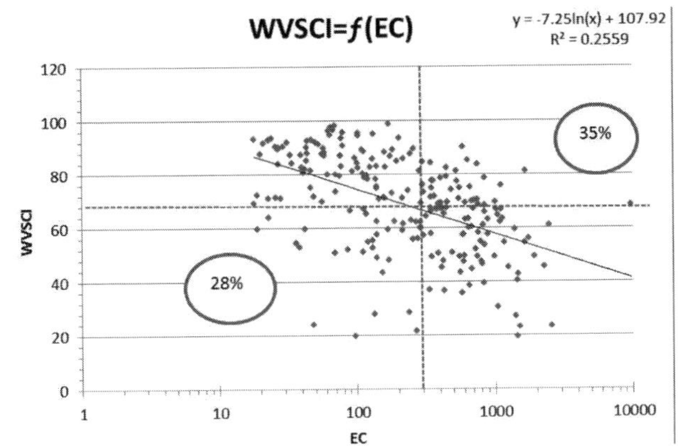

Figure 3. Graph of WVSCI versus EC including mooted thresholds. The vertical red line is an EC value of 300 μS/cm and the horizontal red line is a WVSCI score of 68. The numbers in the red circles represent the proportion of samples that would be misclassified using the thresholds.

in the figure differ only in that the graph on the right utilizes a logarithmic scale for EC. The logarithmic transformation helps to linearize the relationship between WVSCI and EC. The logarithmic transformation also has the advantage of clarifying what is happening at lower EC values. The model yields a weak correlation coefficient of only 0.266.

Model-Free Analysis

Figure 3 overlays the data set with mooted thresholds of 68 for WVSCI and 300 for EC. The upper left quadrant contains samples that would be classified as unimpaired by both (WVSCI and EC) classifications and the lower right quadrant contains samples that would be classified as impaired by both classifications. The upper right and lower left quadrants contain samples where the two measures disagree about impairment classification.

The error rate for a given EC criteria can be calculated by simply plotting the observed values and counting the number of observations which fall outside the predicted responses. For example, if an EC threshold of 300 μS/cm was a perfect predictor of WVSCI impairment, then all WVSCI values would be in either the upper left or lower right quadrants. WVSCI values in the lower left and upper right quadrants represent misclassification. The error rate is a summation of all observations that fall outside the predicted levels at a given EC criterion. Figure 3 indicates that at EC=300 μS/cm (approximately, per Bernhardt et al. 2012) 28 percent of observations with EC<300 also had WVSCI<68 while 35 percent of observations at EC>300 μS/cm had WVSCI scores greater than 68. This constitutes an empirically derived error rate for the WVDEP data set of about 30 percent.

Discretized Analysis

In order to minimize assumptions, we calculated means and confidence intervals for WVSCI scores separately, for separate EC ranges. Table 1 presents the results for this discretized analysis of the relationship between WVSCI and EC. The table presents WVSCI mean values and 95 percent confidence intervals calculated to determine significant differences among means and from WVSCI criteria. Means that are not significantly different from each other are followed by the same letter. Means that are significantly different from the WVSCI criteria, such as 60.6 and 68, are followed by 'yes'; the plus or minus sign indicates whether the WVSCI score is significantly greater or less than the WVSCI criteria. Singular values, with zero degrees of freedom, are indicated by N/A. Only two class mean WVSCI values could be distinguished from the remaining eighteen EC ranges: 0–99 and 1400–1499. One

Table 1. Model-free analysis of WVSCI vs. EC including a test* to discriminate mean values among EC ranges

	WVSCI				Significant	Different from	
EC Range	LCL	Mean	UCL	n	Difference	60.6?	68?
0–99	76.46	80.44	84.43	64	a	Yes+	Yes+
100–199	68.15	73.25	78.35	41	ab	Yes+	No
200–299	56.66	65.41	74.16	19	b	No	No
300–399	61.47	67.26	73.06	19	b	Yes+	No
400–499	56.62	64.36	72.10	14	b	No	No
500–599	51.10	62.37	73.65	11	b	No	No
600–699	54.32	62.26	70.19	12	b	No	No
700–799	52.83	61.69	70.55	9	b	No	No
800–899	50.80	59.31	67.83	9	b	No	Yes–
900–999	N/A	49.33	N/A	1	N/A	N/A	N/A
1000–1099	41.06	58.81	76.55	5	ab	No	No
1100–1199	52.95	65.17	77.38	4	ab	No	No
1200–1299	N/A	45.22	N/A	1	N/A	N/A	N/A
1300–1399	–101.93	43.27	188.46	2	ab	No	No
1400–1499	15.12	31.46	47.80	4	c	Yes–	Yes–
1500–1699	–52.47	67.92	188.30	2	ab	No	No
1700–1999	22.25	52.77	83.28	2	ab	No	No
2000–2599	4.93	43.30	81.68	3	ab	No	No
Total				222			

* Mean tests were carried out using Student's t-test (Ostle 1963).

EC range (1400–1499) fell below the WVSCI criteria of 60.6 and that two EC ranges, (800–899 and 1400–1499) fell below the 68 WVSCI criteria. The latter EC range was followed by ranges that were not significantly different from either WVSCI criterion.

This analysis confirms the general tendency of higher EC to be associated with lower WVSCI. However, the sample size is not sufficient for this approach to demonstrate that the means differ from each other. Most importantly, the inherent lack of precision within the data does not allow a determination of the critical WVSCI values except at extreme levels of EC.

Table 2. WVSCI vs. EC regression models* ranked according to correlation coefficient

n = 222	y = WVSCI, x = EC	
Curve Type	Model	R^2
Logarithmic	y=-7.558ln(x)+109.45	0.2665
Polynomial(2°)	y=8E-06x2+0.0334x+79.454	0.2622
Linear	y=0.0192x+76.784	0.2372
Exponential	y=75.265e-3E-04x	0.2147
Power	y=125.5x-0.118	0.2110

*All models reported in this study were derived using the Microsoft Excel Regression package.

Regression Model of WVSCI vs. EC

An attempt was made to develop regression models to predict WVSCI from EC. None of the models were particularly successful. Table 2 summarizes the performance of five curve types ranked in order of decreasing correlation coefficient (R^2). The highest R^2 was 0.266, indicating that the model only explains 27 percent of the population variation. The poor R^2 results from the large amount of residual variation in WVSCI not explained by EC (Petty et al. 2010) combined with sampling error. We consider this a weak model.

Figure 4 presents the logarithmic regression model between EC and WVSCI. The central solid (red) line is the model. The closest solid lines to the model line (light blue and light green) are the 95 percent confidence interval for the model. The outer solid lines (dark blue and dark green) are the confidence intervals for the population

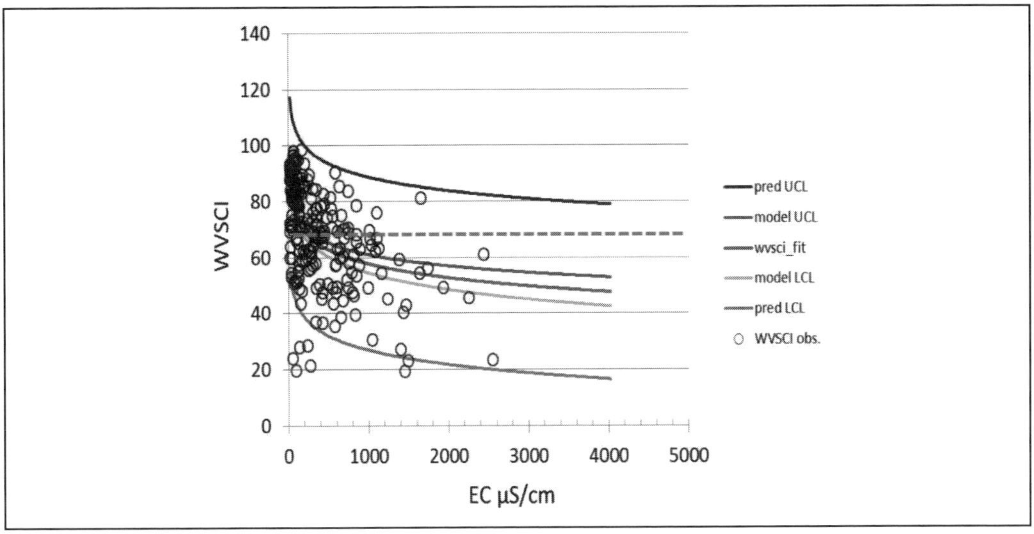

Figure 4. Log-linear model between EC and WVSCI with confidence intervals

(Draper and Smith 1981). The dashed line is the WVSCI criterion of 68.

The model curve and the confidence interval curves for the model in Figure 4 are very similar to the results reported by Bernhardt et al. (2012). Consistent with Bernhardt et al. (2012), a EC value of approximately 300 µS/cm is associated with a model predicted mean WVSCI score of 68. Critically though, Bernhardt et al. (2012) did not report confidence intervals for the sample population. Figure 4 clearly illustrates that observations are readily possible on either side of the WVSC=68 criterion at all levels of EC. This illustrates that no level of EC is suited to predict the divide between WVSCI impairment.

A further distinction is important to make between model and population confidence intervals. A larger sample size will result in smaller confidence intervals around the model, but it will not affect the size of the confidence interval around the population. The latter determines the ability of the model to reliably predict a WVSCI score based on the stream's EC.

Regression Model of WVSCI Impairment vs. EC

WVSCI scores were classified within EC ranges that were arbitrarily set at 100 µS/cm. This allowed a mean and CI to be generated for each EC range. The resulting distribution was determined to be normal and was used in further regression analyses. Using the central points of the EC ranges as the independent variable, regression models were developed to estimate the probability that, at a given EC value, WVSCI scores would be above or below two stream condition criteria: 60.6 (unimpaired gray zone) and 68 (unimpaired, good). The coefficient of correlation (R^2) was used to select the model that best fit the data.

Strong models were developed with the power models providing the highest R^2 values for both WVSCI criteria (Table 3). The results indicate that the probability that a WVSCI score would fall below the 60.6 criteria 50 percent of the time would occur at an EC of 969, while the probability of falling below 68 50 percent of the time would occur at an EC of 389 µS/cm.

Figure 5 illustrates the resulting curves for the two criteria models. The colored horizontal lines represent probabilities of 0.25, 0.50 and 0.75. The lower curve (blue) represents the prediction curve for the 60.6 WVSCI criteria while the upper curve (red) represents the prediction curve for the WVSCI criteria of 68. The polygons are the observed probabilities.

This analysis is again consistent with previous analyses in finding that a CE value of approximately 300 µS/cm is associated with a 50 percent

Table 3. Regression models were developed to estimate the EC at which WVSCI scores would fall below two criteria

n=222	WVSCI Criterion > 60.6		p = WVSCI <		
Curve Type	Model	R^2	0.25	0.50	0.75
Power	$y=0.0255x^{0.4328}$	0.7221	195	969	2472
Polynomial (2°)	$y=0.00000002x^2+0.0003x+0.2$	0.6748	128	907	1620
Linear	y=0.0003x+0.1991	0.6735	170	1003	1836
Logarithmic	y=0.1517 ln(x)-0.5251	0.5853	166	860	4471
Exponential	$y=0.223e^{0.0007x}$	0.5607	163	1153	1733

n=222	WVSCI Criterion > 68.0		p = WVSCI <		
Curve Type	Model	R^2	0.25	0.50	0.75
Power	$y=0.0321x^{0.4605}$	0.8983	86	389	937
Logarithmic	y=0.2026 ln(x)-0.6545	0.8941	87	298	1025
Polynomial (2°)	y=-0.0000002x2+0.0008x+0.2	0.8599	21	368	810
Linear	y=0.0004x+0.3502	0.7609	<50	375	1000
Exponential	$y=0.3404e^{0.0007x}$	0.5534	<50	549	1129

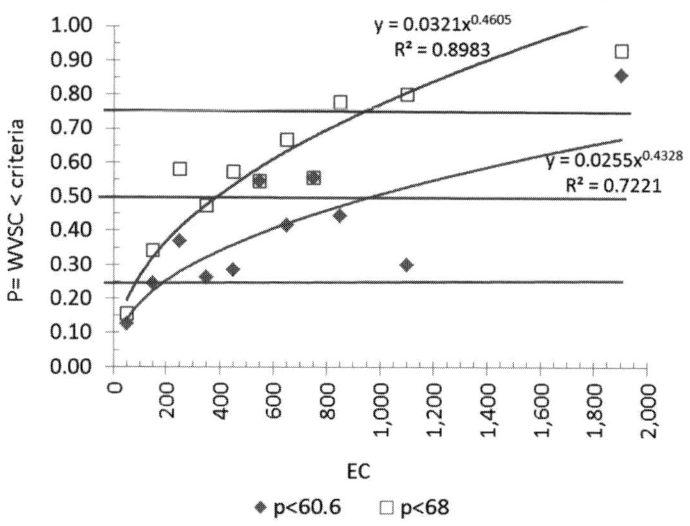

Figure 5. The probability of achieving a WVSCI score below the impairment criteria at a given EC value. The brown, green, and purple horizontal lines represent probabilities of 0.25, 0.50 and 0.75, respectively.

probability of finding a WVSCI score greater than 68. It is also consistent with previous analyses in finding that there is a very high probability of finding WVSCI scores on either side of 68 at a very wide range of EC values. This again illustrates that no level of EC is suited to predict the divide between WVSCI impairment.

Logistic Regression

Logistic regression offers an alternative approach for estimating probability in binomial populations. Figure 6 summarizes the results indicating the probabilities of exceeding a WVSCI score of 68 at different EC values. The vertical line labeled 0.79 (red) represents the average EC value of WVDEP's reference (undisturbed) sites, 68.3. At

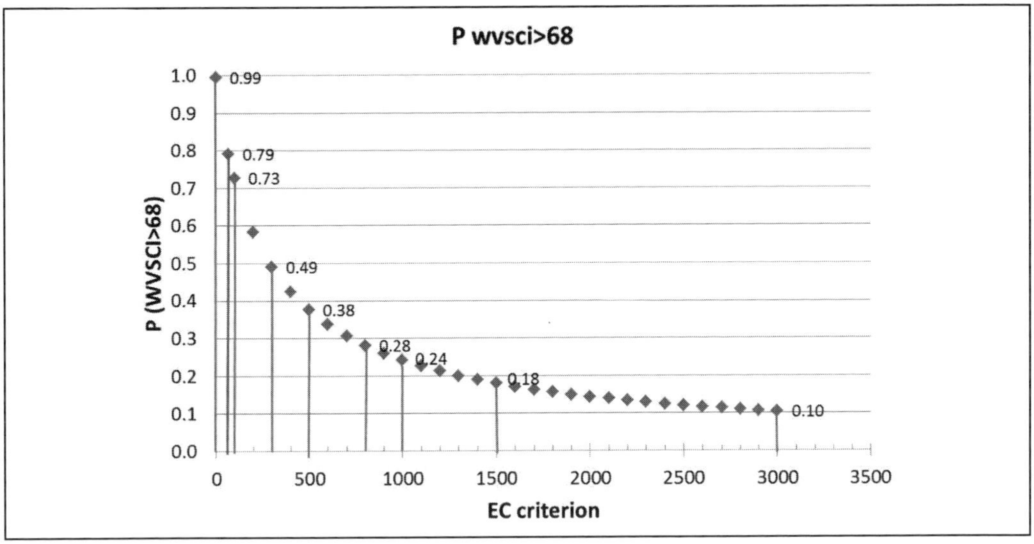

Figure 6. Logistic regression indicating the probability that WVSCI will exceed 68) at a given EC value. The vertical red line is the mean EC value of WVDEP's reference sites.

this EC, about 21 percent of sites would not meet the WVSCI criterion of 68.

This analysis is again consistent with previous analyses in finding that an EC value of approximately 300 µS/cm is associated with a 49 percent probability of finding a WVSCI score greater than 68. It is also consistent with previous analysis in finding that there is a very probability of finding WVSCI scores on either side of 68 at a very wide range of values for EC. This again illustrates that no level of EC is suited to predict the divide between WVSCI impairment.

CONCLUSIONS

The first objective of this study was to identify whether a correlation exists between the WVSCI and EC. We found a statistically significant relationship between EC and WVSCI. The second objective was to identify the best model for predicting WVSCI as a function EC: a log-linear model was the most successful model form. The third objective was to identify the precision with which EC predicts WVSCI. We found that there is very little meaningful predictability for individual observations.

It is important to note that the narrow confidence intervals presented by Bernhardt et al. (2012) only indicate the variability to be expected if the model was refit to a similar but independent data set, and not the ability of the model to predict individual observations. The population confidence intervals illustrate that a WVSCI=68 can be attained at a very wide range of ECs (perhaps encompassing the whole range of the ECs found in the data). This conclusion is supported by either simple inspection of the data or by the prediction confidence bounds for the log-linear model or through the variety of other analysis techniques presented herein.

No evidence was found that an EC value less than 300 µS/cm is a reliable predictor of WVSCI values that meet the West Virginia narrative criteria for stream impairment. This implies that factors other than EC play a significant role in determining WVSCI ranking and that efforts to lower a stream's EC alone will not reliably improve its stream condition.

ACKOWLEDGMENT

This study was sponsored by the Appalachian Research Initiative for Environmental Science (ARIES). The views, opinions and recommendations expressed herein are solely those of the authors and do not imply any endorsement by ARIES employees, other ARIES-affiliated researchers or industrial members. Information about ARIES can be found at http://www.energy.vt.edu/ARIES.

REFERENCES

Bernhardt, E.S., Lutz, B.D., King, R.S., Fay, J.P., Carter, C.E., Helton, A.M., Campagna, D., and Amos, J. 2012. How many mountains can we mine? Assessing the regional degradation of Central Appalachian rivers by surface coal mining. *Environ. Sci. Technol.* 46: 8115–8122.

Draper, N.R. and Smith, H. 1981. *Applied Regression Analysis,* 2nd ed. New York: Wiley.

Elphick, J.R., Davies, M., Gilron, G., Canaria, E.C., Lo, B., and Bailey, H.C. 2011. An aquatic toxicological evaluation of sulfate: The case for considering hardness as a modifying factor in setting water quality guidelines. *Environmental Toxicology and Chemistry*. 30(1):247–253.

Gerritsen, J., Burton, J., and Barbour, M.T. 2000. *A Stream Condition Index for West Virginia Wadeable Streams*. Prepared for U.S. EPA Region 3 Environmental Services Division and U.S. EPA Office of Science and Technology Office of Water. Available online at http://www.dep.wv.gov/wwe/watershed/bio_fish/documents/wvsci.pdf.

Petty, J.T., Fulton, J.B., Strager, M.P., Merovich, G.T., Stiles, J.M. and Ziemkiewicz, P.F. 2010. Landscape indicators and thresholds of stream ecological impairment in an intensively mined Appalachian watershed. *J. N. Am. Benthol. Soc.* 29(4):1292–1309.

Ostle, B. 1963. *Statistics in Research*, 2nd ed. Ames: Iowa State University Press.

U.S. EPA (Environmental Protection Agency). 2010. A Field-based Aquatic Life Benchmark for Conductivity in Central Appalachian Streams. Office of Research and Development, National Center for Environmental Assessment, Washington, DC. EPA/600/R-10/023A.

West Virginia Department of Environmental Protection (WVDEP). 2012. *Benthic Macroinvertebrate and Fish Information*. http://www.dep.wv.gov/WWE/watershed/bio_fish/Pages/Bio_Fish.aspx.

Evaluation of Current Measures of Aquatic Biological Integrity in the Central Appalachian Coalfields: Efficacy and Implications

Sara Sweeten
Department of Fisheries and Wildlife Conservation, Virginia Tech, Blacksburg, Virginia

Jerry Sweeten
Department of Biology, Manchester College, North Manchester, Indiana

John Craynon
Virginia Center for Coal and Energy Research, Virginia Tech, Blacksburg, Virginia

W. Mark Ford
U.S. Geological Survey, Virginia Cooperative Fish and Wildlife Research Unit, Department of Fisheries and Wildlife Conservation, Virginia Tech, Blacksburg, Virginia

Stephen Schoenholtz
Department of Forest Resources and Environmental Conservation, Virginia Tech, Blacksburg, Virginia

ABSTRACT

Within the central Appalachian Coalfields, efforts to minimize and mitigate impacts from surface coal mining have been complicated by incongruent restoration goals and poorly validated post-mine monitoring protocols that are unable to quantify ecosystem change relative to system recovery. At present, assessment techniques range from simple single species examinations to complex multi-taxa/life history decision matrices believed capable of documenting cumulative and additive cross-system impacts. Additionally, some effort has been placed in using abiotic measures or narrow biotic components as system surrogates to comment on stream health of mined streams. These metrics may not accurately indicate biological change particularly when extended out to predict impacts and assess recovery. Benthic macroinvertebrates are valuable indicator species. However, aquatic ecosystem function may not always be able to be inferred from macroinvertebrate assemblages alone, and full assessment of stream health may require use of a suite of taxa such as periphyton, macroinvertebrates, fish, and/or salamanders.

INTRODUCTION

Central Appalachia has high levels of terrestrial and aquatic biodiversity and species richness, with many endemic species (Morse et al. 1993). The New River catchment alone is estimated to have 23 endemic plant and animal species (Hocutt et al. 1978). The aquatic systems in this region contain some of the highest levels of biodiversity in North America, including ten percent of global salamander and freshwater mussel diversity (Stein et al. 2000; Bernhardt and Palmer 2011). However, many of these species are listed as threatened or endangered (Bernhardt and Palmer 2011). For example, 46 species of freshwater mussels once inhabited the Powell River in Virginia and Tennessee (Ortmann 1918). Currently 36 species, seven of which are listed as federally endangered, remain (Ahlstedt et al. 2005). In the United States, the biggest threat to biodiversity is loss/degradation of habitat (Wilcove et al. 1998).

Most stream habitats in the eastern half of the United States, including those in central Appalachia, have been significantly altered by past or current anthropogenic disturbances (Karr 1991). From the time of the European settlement of Central Appalachia in the 1700s through the mid-1900s, Central Appalachian streams were commonly used as unchecked drainages for residential wastewater, mining operation byproducts, and industrial discharges. The arrival of the railroads in the late 1800s brought large scale clearing and degradation of the land from exploitative practices such as commercial logging and coal mining that essentially impacted all of central Appalachian forests (Hibbart 1990; Nowacki and Abrams 1997).

In the 1960s and 1970s, researchers, policy makers, and industry became more aware of the degraded nature of not only central Appalachian streams but streams nationwide, which, along with a shift in society's view of environmental degradation, brought widespread, sweeping legislation and policy reform. The Federal legislation most influential to current coal mining practices are the Clean Water Act (CWA) of 1972, and the Surface Mining Control and Reclamation Act (SMCRA) of 1977 (Craynon et al. 2012). The CWA sets Federal regulatory standards for the identification, reporting, monitoring, and mitigation of degraded waters including those impacted by mining operations. However, under SMCRA, states were given primacy to set their own regulatory standards for surface mining as long as they are "no less effective than" Federal standards set by the CWA (Craynon et al. 2012). Reclamation standards of post-mine lands are also enforced under SMCRA (Starnes and Gasper 1995). Since the enactment of these legislative measures, much scientific focus has been placed on monitoring efforts to regulate mining activities, environmental impacts, and reclamation of post-mined lands (Starnes and Gasper 1995). However, data is not currently sufficient to predict the spatial impact of mining on stream conditions, nor is there a good understanding of the scope of subsequent changes in biological structure and function in aquatic ecosystems in watersheds where mining occurs (Simmons et al. 2008). Fully understanding the impacts of coal mining to central Appalachian streams is critical to develop proper reclamation/mitigation techniques.

COAL MINING IMPACTS

Currently, the extraction of Appalachian coal by mountaintop mining is common as it is cost-effective and allows for mining of shallow coal seams that cannot safely be mined by underground methods (Craynon et al. 2012). Although mountaintop mining accounts for 37% of coal production in Appalachia (National Mining Association 2012), it is the most controversial because of its disproportionately large environmental footprint. Mountaintop coal extraction uses explosives to blast rock layers (overburden) that sit above relatively shallow coal seams (Griffith et al. 2012). This process causes overburden to swell and take up much more volume as haul-able rock pieces than the original compacted geologic material. The amount of overburden produced is too costly to haul off the mine site for disposal, so overburden typically is pushed into nearby valleys. This process exposes vast quantities of unweathered geological materials that when exposed to atmospheric conditions and surface run-off often become acid-forming and/or leach heavy metals along with high levels of salts (Griffith et al. 2012).

Although it is clear that small headwater streams are buried during the mining process altering physical stream habitat, water chemistry, and aquatic biota, the longitudinal spatial scale of indirect disturbance is more challenging to empirically describe. The United States Environmental Protection Agency (EPA) estimates that in central Appalachia, ten percent of the region's land area (500,000 ha) has been affected by mountaintop mining, and over 2,000 km of headwater stream habitat buried by valley fills (EPA 2011). Coal mining is considered a point source pollutant by regulatory agencies such as the EPA and the Virginia Department of Environmental Quality (VADEQ). However, the larger sizes of many surface mines along with the hydrologic changes that occur often require nonpoint source management strategies (Simmons et al. 2008).

Along with direct physical stream loss from valley fills, typically watershed hydrology, stream chemistry, and biological communities also are altered from pre-mined conditions (Starnes and

Gasper 1995; Pond et al. 2008; Griffith et al. 2012). Physical alteration of streams resulting from the mining process such as dredging, burial of headwaters, and dam construction clearly alter stream hydrology. Changes from surface coal mining on upland terrestrial processes such as clearing of vegetation, increased impervious surfaces, and changes in topography, can also alter stream hydrology (Simmons et al. 2008; Merriam et al. 2011).

IMPACTS OF OTHER LAND USES

Many of these hydrological changes are not unique to coal mining land use, and are often seen in streams with urban development, and to a lesser extent agricultural and forestry, landuses as well. For example, research in both urban and mining watersheds suggests that as watersheds become increasingly altered, vegetation is replaced by impervious surfaces such as parking lots or coal processing areas, the flow regime (hydrology) of a stream system responds more quickly to rain events becoming "flashy," and results in more frequent and intense stream flood events (Starnes and Gasper 1995; Price et al. 2006). These high flow events can lead to channelization of the stream along with eroded, unstable stream banks (Booth et al. 2004). Instream habitat is buried by sediment as terrestrial erosion continues. If erosion is not controlled, terrestrial processes can be significantly altered by the loss of soil and nutrients (Baker 1985). These dramatic terrestrial changes alter physical stream habitat, and directly and indirectly influence water chemistry. All of these factors have had both acute and chronic effects on specific aquatic biota in mining-influenced streams (Northingham et al. 2011; Pond 2012).

Stream water changes are in part a function of upland landuse alterations caused by practices such as mining, logging, and urbanization within the watershed. In central Appalachia these terrestrial alterations degrade water quality conditions ranging from elevated sediment rates, total dissolved solid (TDS) levels, acid mine drainage (AMD), and heavy metal concentrations in mined watersheds to increases in nutrients and organic matter from improperly discharged wastewater from residential areas (Simmons et al. 2008). The largest (by volume) and most pervasive stream pollutant in the U.S. is stream sediment, which can impact stream biota in two ways: streambed sedimentation and suspended sediment in the water column (Sweeten and McCreedy 2002).

Streambed sedimentation is one of the most prevalent headwater stream pollutants in central Appalachia (Angradi 1999), and occurs when heavier soil particles settle thereby clogging, covering, and embedding in-stream habitat (Harding et al. 1998). Many aquatic species rely on interstitial spaces between rocks and rocky surfaces for breeding, feeding, and predator avoidance. These rock surfaces also provide important attachment substrate for algae, benthic macroinvertebrates, and freshwater mussels. As interstitial spaces fill and streambed substrate covered, many aquatic taxa including benthic macroinvertebrates, freshwater mussels, algae, salamanders, and fishes are negatively affected. Suspended sediment from smaller silt- and clay-sized particles can remain suspended in the water column for extended periods, reducing visibility and light penetration (Matter and Ney 1981; Hartman et al. 2005). Primary producers such as algae and sight-feeding larval fishes such as brook trout (*Salvelinus fontinalis*) found in the headwaters of central Appalachia have been shown to be particularly sensitive to suspended sediments (Marschall and Crowder 1996; Sweeten and McCreedy 2002). In some areas of the central Appalachians, such as northern West Virginia and south-central Pennsylvania, streams have been heavily impacted by AMD from mine runoff and/or discharge that cause stream pH far below natural conditions. AMD is estimated to effect over 4,000 km of streams in Appalachia, and currently it is the single largest non-point source pollutant in Pennsylvania (Reinhardt 1999).

Much is still unknown about many of the terrestrial and aquatic ecosystem dynamics and interactions caused by dramatic landscape alterations from coal mining practices (Stout and Wallace 2005; Simmons et al. 2008). Measuring landuse impacts on stream health is further complicated by the additive effects caused by multiple landuse practices such as agriculture, forest management, and residential areas (Welsh et al. 2005; Morgan et al. 2012). Studies of aquatic macroinvertebrate communities in central Appalachia found that mining stream impacts along with improper wastewater discharge from residential areas had an additive effect on species richness and diversity (Merovich and Petty 2007; Passmore and Pond 2009; Merriam et al. 2011).

CLEAN WATER ACT REQUIREMENTS

In order to regulate and mitigate many of these negative impacts to streams in coal mined watersheds, Section 303(d) of the Clean Water Act requires each state to identify water bodies with degraded water quality, and to set geographically-based total maximum daily load (TMDL) standards. These TMDL standards are designed to best allocate loadings from multiple pollutant sources so as to not exceed the load capacity of the aquatic system (Barbour et al. 1999). The purpose of this process is to assess stream quality, establish pollution reduction controls, and monitor remediation outcomes. Under SMCRA, each state may establish their own regulatory and TMDL standards as long as they are "no less effective than" Federal standards under the CWA (Craynon et al. 2012). If a state determines a waterbody to be degraded, chemically or biologically or both, the stream must be reported to the EPA and placed on the impaired waters list (303(d)).

In Virginia, the VADEQ is responsible for ensuring the compliance of mining operations under the CWA. In order to determine a stream's impairment level, the VADEQ regularly monitors water chemistry parameters, and developed the Virginia Stream Condition Index (VA-SCI), a multi-metric benthic macroinvertebrate assessment protocol to identify biological impairment of streams (Burton and Gerristsen 2003). This protocol uses eight different metrics that evaluate different levels of pollutant sensitivity of macroinvertebrates, such as total taxa numbers, percent pollutant intolerant taxa, and percent pollutant tolerant taxa, to produce an overall stream condition score ranging from 0 (devoid of life) to 100 (optimal stream life). Based on comparison to local reference stream conditions, the VA-SCI places any stream scoring below 60 on the EPA impaired waters list (CWA 303(d)).

Once a stream is listed as impaired, the VADEQ uses a three-step TMDL process to improve water quality conditions. Individual stream characteristics and specific impairments such as sedimentation, conductivity, and low pH, are used to set TMDL standards, and an appropriate improvement implementation plan is developed and implemented (MapTech 2008). However, monitoring using the TMDL approach does not always provide sufficient guidance for the development of best management practices to effectively improve water quality and biotic communities. Guidance is particularly lacking in systems heavily impacted by nonpoint source pollution such as surface coal mines (Barbour et al. 1999). Endpoints for removing a stream from the impaired waters list also are established by the VADEQ, varying on a stream-by-stream basis. Often these endpoints are not clearly outlined in TMDL implementation plans (see Map Tech 2004); therefore achieving an unimpaired status can be complicated and problematic. In order to properly manage water resources using TMDL standards, there must be a full understanding of the relationships among water chemistry, effluent sources, non-chemical stressors, chemical stressors, and control practices.

BIOLOGICAL INTEGRITY

Unfortunately, such TMDL standards often lack clear and consistent linkages to bioassessment results and/or biological integrity (Barbour et al. 1999). For example, in laboratory toxicology testing of water with elevated TDS, hardness, and alkalinity levels from two mining-impacted streams in West Virginia were acutely toxic to freshwater daphnia (*Ceriodaphnia dubia*). However, in field studies, these two streams had relatively high richness of macroinvertebrates in the orders considered to be very sensitive to poor water quality conditions: Ephemeroptera, Plecoptera, and Trichoptera (EPT) (Merricks et al. 2007). Without these linkages, TMDL standards may not be accurate. An overestimation of a stream's loading capacity can result in biological degradation, whereas an underestimation may result in time and financial resources wasted on needlessly managing non-impactful chemical parameters. Inherent stream variability also may contribute to the complexity of interactions and linkages that must be understood to develop accurate TMDL standards.

As linear ecosystems, streams exhibit high levels of natural spatial and temporal variability influenced by both stochastic and anthropogenic instream, atmospheric, and watershed changes (Cummins 1975). Appalachian headwater streams are estimated to account for up to 75% of stream mileage in this region (Peterman and Semlitsch 2009). These headwater streams are closely linked to proximal terrestrial processes, have little to no photosynthetic production, and are diverse in benthic macroinvertebrate communities and generally dominated by shredder and collector taxa (Cummins 1975). As stream order increases, the direct influence of adjacent terrestrial habitat lessens and the influence of cumulative discharge from the multiple upstream drainages becomes more significant (Cummins 1975). This river continuum has been well-accepted by the scientific community to explain general changes in biological assemblages and physical habitat along the stream gradient (Vannote et al. 1980). For example, in an evaluation of central Appalachian rivers, Kerans and Karr (1994) recommend that shredder macroinvertebrate abundance should not be used in assessment of higher order streams since natural abundance of this group is not great enough to accurately detect water impairments. However, shredder abundance may be an important metric in headwaters where shredders are more abundant and shown to be an accurate indicator of riparian alterations (Kerans and Karr 1994).

While the stream continuum theory helps to explain and minimize some of the natural variability that occurs in streams, many factors, such as local precipitation events, temporal scales, site-specific conditions, and transient dynamics (or the often unexplained variation in ecosystem behavior from its long-term trends with or without variation in external conditions) should be considered. These natural processes all create difficulty in separating natural, "background" variability from anthropogenic impacts on aquatic systems (Hastings 2004; Hastings 2010; Johnson and Hering 2010). For example, in a survey of rivers throughout central Appalachia, Kerans and Karr (1994) found some metrics such as proportion of chironomids, shredders, and detritivores showed a closer relationship to the temporal scale than to location, indicating that differences in macroinvertebrate assemblages may not always be the result of site quality (Kerans and Karr 1994). In order to best distinguish natural stream variations from anthropogenic impacts, it is important to have long-term stream condition data sets that analyze a variety of biological responses, since stressors may be revealed at various levels of biological organization as well as in physical stream habitat and water quality changes (Adams et. al 2002; Griffith et al. 2012).

Without historical data to provide insight into pre-disturbance conditions (and thereby provide an end goal), less-disturbed, i.e., unmined, reference watersheds are generally used as a surrogate (Younos et al. 2007). Finding a "truly undisturbed watershed" in central Appalachia is difficult as very few, if any, exist, considering the widespread and pervasive nature of documented and undocumented past mining, timbering, and other cultural activities since European settlement (Simmons et al. 2008). Measuring the deviation of abundance and assemblage of taxa between mining impacted and reference streams may be useful in measuring impairment (Merovich and Petty 2007). Comparisons to reference streams also are used often by regulatory agencies (EPA and VADEQ) to identify and monitor stream degradation (Barbour et al. 1999). Ideally reference watersheds should have similar topography, watershed size, geology, and soil types to impaired stream watersheds.

However, scientific standards for the selection of reference watersheds have not been established, and reference site choices are often left to "best professional judgment" (Younos et al. 2007). In central Appalachia, availability of optimal reference sites is limited, forcing water resource managers to select fewer reference sites or choose reference sites with different characteristics from the study watershed (Younos et al. 2007). For example, in the development of the statewide VA-SCI metrics that encompasses six level III ecoregions, a total of 62 reference sites were used; however, only five of these sites were in the central Appalachian ecoregion 69 (Burton and Gerritsen 2003). In a comparative study of mining impacts on headwater streams in southwest Virginia, Matter and Ney (1981) had to choose a 400% larger reference watershed that was located almost 50 km away from the study watersheds. However, even sub-optimal reference watersheds still offer the only insight into the establishment of water quality standards (Younos et al. 2007).

In response to national water quality concerns and monitoring approaches, the EPA Office of Water has advanced the Watershed Protection Approach (WPA) to consolidate and reorient state and Federal agencies that manage and regulate water resources by using watershed level management units. Past management focus has often been placed on small, highly degraded stream sections or individual pollutant sources. Often these efforts only provide a limited view of the total extent of degradation on a watershed with multiple mines or land uses (Barbour et al. 1999; Freund and Petty 2007). Ideally, the watershed approach provides an integrated and comprehensive evaluation of ecological conditions (integrity) in coal mining areas in order to best and most effectively protect aquatic resources (Barbour et al. 1999).

In central Appalachia, mining impacts are often assessed and mitigated on an individual mine basis. Using the watershed approach helps to better manage the longitudinal, additive effects of multiple land uses such as mining and residential wastewater on stream health (Freund and Petty 2007). Common in mine-impacted watersheds, altered stream chemistry from AMD and/or high conductivity can isolate biological communities in upstream reaches reducing genetic diversity and recolonization ability. Linking upstream and downstream biological communities is therefore particularly important for the protection of the many endemic species found in central Appalachia. These linkages can most effectively be created and maintained when managing water resources at the watershed level (Freund and Petty 2007). The watershed approach allows for a cost-effective, realistic management planning where focus can be placed on restoration efforts for larger sections of less impacted streams instead of wasting resources on small, highly degraded streams that may never recover (Freund and Petty 2007).

Although there is an increased effort to remediate stressors within watersheds using the watershed approach, full understanding of how to measure stream conditions in the context of post-mining recovery continues to be a highly debated topic among scientists, regulators, and mining companies (Kerans and Karr 1994; Harding et al. 1998; Allan 2004; Milman and Short 2008; Simmons et al. 2008). The EPA and many states have developed detailed techniques for assessing streams over a large spatial scale, such as a region or state (i.e., EPA Rapid Bioassessment Protocol, VA-SCI and the West Virginia Stream Condition Index or WV-SCI). For example, the VADEQ uses the same protocol and scoring metrics (VA-SCI) on streams impacted by coal mining in the Appalachian Plateau region of southwest Virginia as are used for agricultural or urban-impacted streams in the Piedmont and Coastal Plain.

DISCUSSION

Often to assess recovery and restoration, focus is placed on one stream condition parameter such as water chemistry or benthic macroinvertebrates over a relatively short time period (Kerans and Karr 1994). This approach often provides limited insight into the full extent of recovery and restoration success. Due to the natural temporal and spatial variability of stream ecosystems and the impacts of multiple anthropogenic stressors in central Appalachia on stream ecosystems, parameter choice can be somewhat problematic (Adam et al. 2002; Carlisle et al. 2009). The questions that remain are relatively simple to frame, but more difficult to answer: What is the level of aquatic ecosystem degradation? What spatial remediation is required to reach a valid benchmark of improvement? What factors should these benchmarks include?

Biological sampling "directly measures the condition of the resource at risk, detects problems that other methods may miss or underestimate, and provides a systematic process for measuring progress resulting from the implementation of water quality programs" (EPA 1990). Although not a substitute for chemical and toxicological monitoring, biological assemblages seem to show responses to the wide range of chemical conditions in mining impacted streams, even revealing additive effects of pollutants that individually do not exceed regulatory standards (Karr 1991; Freund and Petty 2007). Aquatic biological communities are sensitive to a range of environmental factors and watershed conditions across both long- and short-term temporal scales (Karr 1981). In central Appalachian coalfields, agency monitoring programs as well as research efforts often have used single taxon assessment protocols that usually focus on benthic macroinvertebrates such as the VA-SCI and WVSCI (Stout and Wallace 2005; Santos and Stevenson 2011; Hartman et al. 2005; Pond 2008; Passmore and Pond 2009; Merricks et al. 2007). Instead of focusing on a single group of organisms to monitor biological degradation, the integration of multiple taxa into bioassessment protocols has been shown to better measure both aquatic impairment and restoration status (Adams et al. 2002; Freund and Petty 2007).

In east Tennessee, Adams et al. (2002) found that using a suite of taxa such as periphyton, macroinvertebrates, and fish, improved the ability to identify the status of ecological recovery of disturbed aquatic systems. Biotic recovery often occurs on various time and spatial scales. Recolonization rates of biota may be affected by factors such as physiochemical stream conditions, taxa dispersal method, and pollution tolerance levels. Therefore, using a single taxon or trophic level for stream bioassessment may not always reflect a holistic ecosystem recovery. Incorporating bio-indicators from multiple levels of biological organizations and trophic levels helps to ensure a full view of recovery processes (Adams et al. 2002).

Although often considered to be too expensive and time-intensive, biomonitoring protocols have been established by state agencies such as the Ohio EPA and Federal agencies such as the EPA. For example, the Ohio EPA biomonitoring protocols incorporate benthic macroinvertebrates, fish, and salamanders that identify stream permanence as well as biological degradation (Ohio EPA 2012). The EPA's Rapid Bioassessment Protocol (Barbour et al. 1999) outlines empirical, yet cost-effective protocols for biomonitoring multiple stream taxa including periphyton, benthic macroinvertebrates, and fish. In order to select biomonitoring taxa that will accurately and cost-effectively detect aquatic degradation in a given region, it is important to fully understand each taxa's strengths and weaknesses for indicating environmental changes.

In central Appalachia, periphyton (algae) has been used infrequently in conjunction with water quality and/or macroinvertebrate surveys of mining impacted streams (Merricks et al. 2007). Primary producers such as periphyton can be directly affected by alterations in physical and chemical parameters, and often show significant responses when compared to other taxa at lower concentrations of stressors, such as herbicides (Stevenson 1998; Stevenson and Pan 1999; Barbour et al. 1999). Due to their short lifespan and quick reproductive rates, periphyton can be excellent indicators for examination of short-term impacts and recovery of degraded water quality. Algae are easy, quick, and cost-effective to sample and require little field training, and sampling is minimally invasive to the stream. However, laboratory analysis can be time consuming and requires specialized knowledge/training, although well-established, standard community assessment protocols do exist (Barbour et al. 1999). Periphyton biomass and benthic macroinvertebrate assemblages have been shown to have close linkages in response to sedimentation and nutrient loading in large streams (Molinos and Donohue 2010) as well as to AMD impacts (Meegan and Perry 1996).

In central Appalachia, benthic macroinvertebrates have been relied on heavily for bioassessment for a variety of reasons. First, benthic macroinvertebrates have been shown to serve well as a site-specific, short-term biological indicator. Many trophic- and pollution-tolerance levels are represented in this group, and sampling is relatively easy and inexpensive. Invertebrates are also present in headwater streams that have limited fish communities (Barbour et al. 1999). Although bioassessment using benthic macroinvertebrates has many advantages, large variations in seasonal assemblages, sampling location within a watershed, and analysis of data often causes conflicting and unclear results (Kerans and Karr 1994; Fulk et al. 2003). Seasonal variability of macroinvertebrate assemblages has been debated with little resolution. Fulk et al. (2003) and Timpano et al. (2011) found seasonal variability between spring and fall macroinvertebrate scores when using the WV-SCI and VA-SCI assessment protocols, respectively. However, in the VA-SCI development and field validations, no seasonal differences in invertebrate scores were shown (Burton and Gerritsen 2003). If macroinvertebrate assemblages show significant seasonal variations, the VA-SCI protocol scoring should be adjusted in order to account for seasonality differences (Timpano et al. 2011). Currently the VA-SCI score has no such allowances for temporal variabilities and scores below 60 are considered impaired regardless of season (Timpano et al. 2011).

Sampling location selection at both the regional and local levels has also been shown to produce variability in macroinvertebrate community structures. The VADEQ found that reference streams in the coalfields of southwestern Virginia tended to have lower scores using the VA-SCI assessment method as compared to other ecoregions in Virginia. This may be due to natural differences in macroinvertebrate assemblages of central Appalachian streams relative to those of other Virginia ecoregions (Burton and Gerritsen 2003). The Ohio EPA recommends designing quantitative biological criteria based on individual ecoregion characteristics (Yoder 1991). The VA-SCI attempts to parsimoniously fit metrics that will work for all six ecoregions in Virginia. However, this approach may sacrifice some accuracy and resolution on the ecoregion level in order to encompass the whole state with one bioassessment protocol.

Macroinvertebrate assemblages may vary at different stream order and permanence levels within a watershed. In the southern Appalachians, Feminella (1996) found little variation in community structure between intermittent first order streams and permanent lower order streams. Santos and Stevenson's (2011) results from a Massachusetts study were similar to Feminella's, whereas Brown and Brussock (1991) found lower macroinvertebrate diversity in intermittent headwater streams in the Ozarks. Although these studies were conducted in different regions of the United States, they do indicate possible variability in macroinvertebrate assemblages along stream permanence gradients. Moreover, watershed size has been shown to influence macroinvertebrate assemblages and abundances as well. In central Appalachia, macroinvertebrate species richness and abundance, particularly the number of Ephemeroptera, Plecoptera, and Trichoptera (EPT) taxa, increase rapidly along the stream gradient until watershed size of about 100 acres, where the rate of increase becomes constant (Stout and Wallace 2005). The percentage of collector macroinvertebrates also increased along this gradient whereas percent shredders decreased (Stout and Wallace 2005).

Without resolution to the scope/scale of these natural temporal and spatial influences on macroinvertebrate assemblages, it is extremely difficult to properly account for natural variation within biomonitoring protocols such as the VA-SCI and WV-SCI. Without a full understanding of these natural variations, there may be inaccuracies in assessment of macroinvertebrate abundance and diversity. Analysis of macroinvertebrate surveys varies greatly among agencies and biologists, ranging from traditional EPT evaluation to multiple-metric analysis of community function and structure (Blocksom and Winters 2006). Taxonomic identification level of macroinvertebrates has not been standardized among regulatory agencies, researchers, and managers. Hilsenhoff (1987) created a simple index of biotic diversity by assigning pollution tolerance values to macroinvertebrates at the genus-level and dividing the sum of the tolerance values by the total number of individuals identified (around 100). In order to address time and fiscal constraints faced by many management agencies, Hilsenhoff (1988) used 120 Wisconsin streams to develop a family-level version of the index that reduced assessment time by 50%. However, the family-level index lost some resolution compared to genus-level evaluation. Although Pond et al. (2008) did not find significant differences in bioassessment results between identification to the genus-level and identification to the family-level, genus-level indices were recommended for measuring stream conditions in mined watersheds. These identification level differences can make direct comparison of bioassessment results analyzing regional and temporal trends in macroinvertebrate assemblages and coal mining impacts difficult.

Due to their high mobility and relative longevity, fish can serve as good long-term indicators of chronic stressors and reflect broad-habitat conditions of central Appalachian streams (Matter and Ney 1981; Angermeier and Karr 1986; Fulk et al. 2003; Freund and Petty 2007). Fish account for almost half of the endangered vertebrate species in the United States (Warren and Burr 1994) and represent a wide range of trophic levels and life histories (Angermeier and Karr 1986). A study of mine-impacted streams in West Virginia that compared indicator characteristics of fish and benthic macroinvertebrates found fish communities to be less responsive than macroinvertebrates to degraded water quality, suggesting that fish may not be a good indicator for bioassessment of local conditions (Freund and Petty 2007).

However, Birge et al. (2000) observed that fish were good biosensors for heavy metal levels in Kentucky streams, and may more accurately reflect watershed conditions than macroinvertebrates. In addition, the life history and environmental requirements of most fish species are well understood, allowing for better understanding of community structure and function (Fausch et al. 1990). Most stakeholders, policy makers and land managers are more familiar and knowledgeable about fish (especially game species) than macroinvertebrates, thereby making fish assessment results more relevant and relatable (Karr 1981). Compared to other taxa, sampling of fish communities is cost-effective, and requires relatively little training or laboratory time (Karr 1981; Fausch et al. 1990). However, as with many taxa, sampling can be quite labor intensive and identification requires extensive knowledge. Fish communities are often influenced by stream permanence and by both diel and seasonal variations (Karr 1981). Fish assemblages may also be limited due to physical or chemical barriers (such as AMD) lower in the watershed.

Amphibians, particularly salamanders, are an important component to both terrestrial and aquatic ecosystems in central Appalachia (Davic and Welsh 2004). Long-lived stream salamander species have also been recommended for use as an indicator for aquatic bioassessment protocols due to their natural high abundances in undisturbed areas, sensitivity to stream and watershed degradation, and role as a keystone faunal group (Welsh and Ollivier 1998; Welsh and Droege 2001; Southerland et al. 2004). In headwater streams absent of fish, salamanders are the dominant, most abundant vertebrate predator. Stream salamanders provide a significant intermediate role in trophic cycling, feeding on small prey such as benthic macroinvertebrates and, in turn, are an important prey item for larger vertebrates (Petranka et al. 1993; Davic and Welsh 2004; OHEPA 2012). Central Appalachian riparian areas typically have relatively stable salamander populations (compared to macroinvertebrates), with densities of up to $1.4/m^2$ (Kleeberger 1984; Hairston and Wiley 1993; Welsh and Ollivier

1998). Life history and behavior vary greatly among salamander species with some larval forms having long aquatic phases of up to 48 months, whereas other species have very short- or even no aquatic larval phase (OHEPA 2012). Due to such variation in life history, the Ohio EPA uses larval salamander surveys as a method for accurately and cost-effectively indicating stream permanence and headwater habitat quality. Evidence of reproduction (eggs and larvae) of salamander species with larval phases exceeding 12 months has been found to indicate perennial stream flow, while presence of species with larval phases of less than 12 months indicates intermittent stream flow (Ohio EPA 2012).

Salamander species that move between aquatic and terrestrial environments provide an important link between upland and stream ecosystems (Fisher et al. 1998). Stream salamanders are considered to be sensitive to aquatic stressors and environmental degradation, and low stream salamander abundance has been shown to be closely linked to terrestrial watershed degradations (Petranka et al. 1993; Willson and Dorcas 2003). Stream salamander abundances have been shown to be negatively impacted, even to the point of local extirpation, by decreases in stream conditions such as streambed sedimentation, temperature, pH, and suspended sediment, as from watershed land uses such as mining, urbanization, and timber harvesting (Gore 1983; Willson and Dorcas 2003; Welsh et al. 2005; Moseley et al. 2008). This gradient of response to anthropogenic impacts is critical for an indicator taxa group. If a group of organisms is highly sensitive to environmental degradation they will be quickly extirpated from the area, and may only indicate mild water quality impairments (Welsh and Ollivier 1998).

Taxa that are extremely tolerant of impaired stream conditions should also not be selected for use in biomonitoring as they will not show a response until the aquatic ecosystem is heavily degraded. Movement of many Appalachian stream salamander species has been shown to be less than 100m (Pauley et al. 2000), and although stream salamanders are able to cross barriers such as dry streambed sections and waterfalls, mobility may be limited by riparian and upper watershed habitat fragmentation (Resetarits 1997; Williams et al. 2002; Willson and Dorcas 2003; OHEPA 2012). Stream salamanders may also be useful to indicate recovery endpoints as their low mobility reduces the likelihood of increasing abundances being a reflection of immigration, and helps to confirm that local, long-term stream and watershed conditions are suitable for a positive rate of salamander survival and reproduction (Welsh and Ollivier 1998; Welsh et al. 2005).

CONCLUSION

Traditionally, research and monitoring of aquatic ecosystems has focused on the effects of land use practice disturbances. However, the ability to distinguish natural stream variability from the additive effects of anthropogenic influences is currently limited. Furthermore, there is little known about the multiple temporal and spatial dynamics of the biological recovery of a degraded aquatic system (Harding et al. 1998; Adams et al. 2002; Niemi et al. 2004; Hartman et al. 2005; Milman and Short 2008; Simmons et al. 2008; Northingham et al. 2011). Currently, even identifying and quantifying aquatic ecosystem structure and function of undisturbed systems can be very difficult (Simmons et al. 2008). Fiscal and personnel constraints are often limiting factors to widespread or more comprehensive sampling. It is, however, imperative to establish scientifically-backed bench marks of success for watershed remediation efforts and/or coal mine reclamation. Biological assessment using a suite of taxa may provide the most extensive view of aquatic changes including degradation and recovery. In order to develop optimal bioassessment protocols, it is essential to first gain a better understanding the response of each taxa to water conditions as well as the relationship dynamics among taxa. Investigation of salamanders as bio-indicators for streams in mined watersheds should be a priority, since little is known about their indicator abilities in central Appalachia.

ACKOWLEDGMENT

This study was sponsored by the Appalachian Research Initiative for Environmental Science (ARIES). The views, opinions and recommendations expressed herein are solely those of the authors and do not imply any endorsement by ARIES employees, other ARIES-affiliated researchers or industrial members. Information about ARIES can be found at http://www.energy.vt.edu/ARIES.

REFERENCES

Adams, S.M., W.R. Hill, M.J. Peterson, M.G. Ryon, J.G. Smith and A.J. Stewart. 2002. Assessing recovery in a stream ecosystem: Applying multiple chemical and biological endpoints. *Ecological Applications*. 12:1510–1527.

Ahlstedt, S.A., M.T. Fagg, R.S. Butler, and J.F. Connell. 2005. *Long-term trend information for freshwater mussel populations at twelve fixed-station monitoring sites in the Clinch and Powell rivers of eastern Tennessee and southwestern Virginia, 1979–2004.* Unpublished Report, U.S. Fish and Wildlife Service, Cookeville, Tennessee.

Allan, J.D. 2004. Landscapes and riverscapes: The influence of land use on stream ecosystems. *Annual Review of Ecology Evolution and Systematics*. 15:257–284.

Angermeier, P.L. and J.R. Karr. 1986. Applying an index of biotic integrity based on stream-fish communities: Considerations in sampling and interpretation. *North American Journal of Fisheries Management*. 6:418–429.

Angradi, T.R. 1999. Fine sediment and macroinvertebrate assemblages in Appalachian streams: A field experiment with biomonitoring applications. *Journal of the North American Benthological Society*. 18:49–66.

Baker, D.B. 1985. Regional water quality impacts of intensive row-crop agriculture: A Lake Erie Basin case study. *Journal of Soil and Water Conservation*. 40:125–132.

Barbour, M.T., J. Gerritsen, B.D. Snyder, and J.B. Stribling. 1999. *Rapid bioassessment protocols for use in streams and wadeable rivers: periphyton, benthic macroinvertebrates and fish, Second Ed.* EPA 841-B-99-002. U.S. Environmental Protection Agency; Office of Water; Washington, D.C. 339 pages.

Bernhardt, E.S. and M.A. Palmer. 2011. The environmental costs of mountaintop mining valley fill operations for aquatic ecosystems of the Central Appalachians. Evaluation Studies Review. *Annual New York Academy of Science*. 1223:39–57.

Birge, W.J., D.J. Price, J.R. Shaw, J.A. Spromberg, A.J. Wigginton and C. Hogstrandt. 2000. Metal body burden and biological sensors as ecological indicators. *Environmental Toxicology and Chemistry*. 4:1199–1212.

Blocksom, K.A. and L. Winters. 2006. *The evaluation of methods for creating defensible, repeatable, objective and accurate tolerance values for aquatic taxa.* U.S. Environmental Protection Agency, Office of Research and Development, Cincinnati, OH. 52 pages.

Booth, D.B., J.R. Karr, S. Schauman, C.P. Konrad, S.A. Morley, M.G. Larson and S.J. Burges. 2004. *Journal of the American Water Resources Association*. 40:1351–1364.

Brussock, P.P. and A.V. Brown. 1991. Riffle-pool geomorphology disrupts longitudinal patterns of stream benthos. *Hydrobiologia*. 220:109–117.

Burton, J. and J. Gerritsen. 2003. *A stream condition index for Virginia non-coastal streams.* Tetra Tech, Inc. Virginia Department of Environmental Quality, Richmond, Virginia. 168 pages.

Carlisle, D.M., J. Falcone and M.R. Meador. 2009. Predicting the biological condition of streams: use of geospatial indicators of natural and anthropogenic characteristics of watersheds. *Environmental Monitoring Assessment*. 151:143–160.

Craynon, J.R., E.A. Sarver and D.P. Robertson. 2012. Could a public ecology approach help resolve the mountaintop mining controversy? *Resources Policy*. http://dx.doi.org/10.1016/j.resourpol.2012.08.001.

Cummins, K.W. 1975. *The ecology of running waters theory and practice*. Kellogg Biological Station. Hickory Corners, MI. 16 pages.

Davic, R.D. and H.H. Welsh. 2004. On the ecological roles of salamanders. *Annual Review of Ecology, Evolution, and Systematics*. 35:405–434.

Fausch, K.D., J. Lyons, J.R. Karr, and P.L. Angermeier. 1990. Fish Communities as indicators of environmental degradation. *American Fisheries Society Symposium*. 8:123–144.

Feminella, J.W. 1996. Comparison of benthic macroinvertebrates assemblages in small streams along a gradient of flow permanence. *Journal of North American Benthological Society*. 15:651–669.

Fisher, S.G., N.B. Grimm, E. Marti, R.M. Holmes and J.B. Jones, Jr. 1998. Material spiraling in stream corridors: a telescoping ecosystem model. *Ecosystems.* 1:19–34.

Freund, J.B. and J.T. Petty. 2007. Response of fish and macroinvertebrate bioassessment indices to water chemistry in a mined Appalachian watershed. *Environmental Management.* 39:707–720.

Fulk, F., B. Autrey, J. Hutchens, J. Gerritsen, J. Burton, C. Cresswell and B. Jessup. 2003. *Ecological assessment of streams in the coal mining region of West Virginia using data collected by the U.S. EPA and environmental consulting firms.* U.S. Environmental Protection Agency. National Exposure Research Laboratory. Cincinnati, OH. 152 pages.

Gore, J.A. 1983. The distribution of Desmognathine larvae (Amphibia: Plethodontidae) in coal surface mine impacted streams of the Cumberland Plateau, USA. *Journal of Freshwater Ecology.* 2:13–23.

Griffith, M.B., S.B. Norton, L.C. Alexander, A.I. Pollard and S.D. LeDuc. 2012. The effects of mountaintop mines and valley fills on the physicochemical quality of stream ecosystems in the central Appalachians: a review. *Science of the Total Environment.* 417–418:1–12.

Hairston, N.G. and R.H. Wiley. 1993. No decline in salamander populations: A twenty-year study in the Southern Appalachians. *Brimleyana.* 18:59–64.

Harding, J.S., E.F. Benfield, P.V. Bolstad, G.S. Helfman and E.B.D. Jones. 1998. Stream Biodiversity: the ghost of land use past. *Proceedings of the National Academy of Science, USA.* 95:14843–14947.

Hartman, K.J., M.D. Kaller, J.W. Howell and J.A. Sweka. 2005. How much does valley fill influence headwater streams? *Hydrobiologia.* 552:91–102.

Hastings, A. 2004. Transients: the key to long-term ecological understanding? *Trends in Ecology and Evolution.* 19:39–45.

Hastings, A. 2010. Timescales, dynamics, and ecological understanding. *Ecology.* 91:3471–3480.

Hibbard, W.R. 1990. *Virginia coal: An abridged history.* Virginia Center for Coal & Energy Research. Blacksburg, VA. 160 pages.

Hilsenhoff, W.L. 1987. An improved biotic index of organic stream pollution. *The Great Lakes Entomologist.* 20:31–39.

Hilsenhoff, W.L. 1988. Rapid field assessment of organic pollution with a family-level biotic index. *Journal of North American Benthological Society.* 7:65–68.

Hocutt, C.H., R.F. Denoncourt and J.R. Stauffer, Jr. 1978. Fishes of the Greenbrier River, West Virginia, with drainage history of the Central Appalachians. *Journal of Biogeography.* 5:59–80.

Johnson, R.K. and D. Herring. 2010. Spatial congruency of benthic diatom, invertebrate, macrophyte, and fish assemblages in European streams. *Ecological Applications.* 20:978–992.

Karr, J.R. 1981. Assessment of biotic integrity using fish communities. *Fisheries.* 6:21–27.

Karr, J.R. 1991. Biological integrity: A long-neglected aspect of water resource management. *Ecological Applications.* 1:66–84.

Kerans, B.L. and J.R. Karr. 1994. A benthic index of biotic integrity (B-IBI) for rivers of the Tennessee Valley. *Ecological Applications.* 4:768–785.

Kleeberger, S.R. 1984. A test of competition in two sympatric populations of Desmognathine salamanders. *Ecology.* 65:1846–1856.

MapTech, Inc. 2004. *General standard TMDL development for Dump's Creek Russell County, Virginia.* Prepared for the Virginia Department of Environmental Quality and Virginia Department of Mines, Minerals, and Energy, Blacksburg, VA. 184 pages.

MapTech, Inc. 2008. *Dumps Creek implementation plan summary.* Prepared for Virginia Department of Environmental Quality, Blacksburg, VA. 28 pages.

Marschall, E.A. and L.B. Crowde. 1996. Assessing population responses to multiple anthropogenic effects: A case study with brook trout. *Ecological Applications.* 6:152–167.

Matter, W.J. and J.J. Ney. 1981. The impact of surface mine reclamation on headwater streams in southwest Virginia. *Hydrobiologia.* 78:63–71.

Meegan, S.K. and S.A. Perry. 1996. Periphyton communities in headwater streams of different water chemistry in the Central Appalachian Mountains. *Journal of Freshwater Ecology.* 11:247–256.

Merovich, G.T. and J.T. Petty. 2007. Interactive effects of multiple stressors and restoration priorities in a mined Appalachian watershed. *Hydrobiologica.* 575:13–31.

Merriam, E.R., J.T. Petty, G.T. Merovich, J.B. Fulton and M.P. Strager. 2011. Additive effects of mining and residential development on stream conditions in a central Appalachian watershed. *Journal of the North American Benthological Society.* 30:399–418.

Merricks, T.C., D.S. Cherry, C.E. Zipper, R.J. Currie and T.W. Valenti. 2007. Coal-mine hollow fill and settling pond influences on headwater streams in southern West Virginia, USA. *Environmental Monitoring Assessment*. 129:359–378.

Milman, A. and A. Short. 2008. Incorporating resilience into sustainability indicators: An example for the urban water sector. *Global Environmental Change*. 18:758–767.

Molinos, J.G. and I. Donohue. 2010. Interactions among temporal patterns determine the effect of multiple stressors. *Ecological Applications*. 20:1794–1800.

Morgan, R.P., K.M. Kline, M.J. Kline, S.F. Cushman, M.T. Sell, R.E. Weitzell and J.B. Churchill. 2012. Stream conductivity: Relationships to landuse, chloride, and fishes in Maryland streams. *North American Journal of Fisheries Management*. 32:941–952.

Morse, J.C., B.P. Stark and W.P. McCafferty. 1993. Southern Appalachian streams at risk: implications for mayflies, stoneflies, caddisflies, and other aquatic biota. *Aquatic Conservation*. 3:293–303.

Moseley, K.R., W.M. Ford, J.W. Edwards and T.M. Schuler. 2008. Long-term partial cutting impacts on *Desmognathus* salamander abundance in West Virginia headwater streams. *Forest Ecology and Management*. 254:300–307.

National Mining Association. 2012. U.S. coal production by state, region and method of mining—2011.

National Mining Association Factsheet. http:// http:// www.nma.org/pdf/c_production_mines_state _type.pdf. Accessed January 15, 2013.

Niemi, G.J. and M.E. McDonald. 2004. Application of ecological indicators. *Annual Review of Ecology, Evolution, and Systematics*. 35:89–111.

Northingham, R.M., E.F. Benfeld, S.H. Schoenholtz, A.J. Timpano, J.R. Webster and C. Zipper. 2011. An assessment of structural attributes and ecosystem function in restored Virginia coalfield streams. *Hydrobiologia*. 671:51–63.

Nowacki, G.J. and M.D. Abrams. 1997. Radial-growth averaging criteria for reconstruction disturbance histories from presettlement-origin oaks. *Ecological Monographs*. 67:225–249.

Ohio EPA. 2012. *Field evaluation manual for Ohio's primary headwater habitat streams*. Version 3.0. Ohio EPA Division of Surface Water, Columbus, Ohio. 117 pp.

Ortmann, A.E. 1918. The nayades (freshwater mussels) of the upper Tennessee drainage. *Proceedings of the American Philosophical Society* 57:521–626.

Passmore, M. and G. Pond. 2009. *Evaluating appropriate existing and designated uses of Straight Creek (Lee County, VA) using current macroinvertebrate, habitat, and water quality data*. United States Environmental Protection Agency, Region III, Wheeling, WV. 39 pages.

Pauley, T.K., J.C. Mitchell, R.R. Buech and J. Moriarty. 2000. Ecology and management of riparian habitats for amphibiands and reptiles. In *Riparian Management in Forests of the Continental Eastern United States*, ed. E.S. Verry, J.W. Hornbeck, C.A. Dolloff. pp. 169–91. Boca Raton, FL. Lewis. 402 pages.

Peterman, W.E. and R.D. Semlitsch. 2009. Efficacy of riparian buffers in mitigating local population declines and the effects of even-aged timber harvest on larval salamanders. *Forest Ecology and Management*. 257:8–14.

Petranka, J.W., M.E. Eldridge and K.E. Haley. 1993. Effects of timbering harvesting on southern Appalachian salamanders. *Conservation Biology*. 7:363–370.

Pond, G.J., M.E. Passmore, F.A. Borsuk, L. Reynolds and C.J. Rose. 2008. Downstream effects of mountaintop coal mining: comparing biological conditions using family- and genus-level macroinvertebrate bioassessment tools. *Journal of the North American Benthological Society*. 27:717–737.

Pond, G.J. 2012. Biodiversity loss in Appalachian headwater streams (Kentucky, USA): Plecoptera and Trichoptera communities. *Hydrobiologia*. 679:97–117.

Price, S.J., M.E. Dorcas, A.L. Gallant, R.W. Klaver and J.D. Willson. 2006. Three decades of urbanization: Estimating the impact of land-cover change on stream salamander populations. *Biological Conservation*. 133:436–441.

Reinhardt, C.H. 1999. Acid mine drainage in Pennsylvania streams. "Ironing out" the problem. *Restoration and Reclamation Review*. 5:1–10.

Resetarits, W.J., Jr. 1997. Differences in an ensemble of streamside salamanders (Plethodontidae) above and below a barrier to brook trout. *Amphibia-Reptilia*. 18:15–25.

Santos, A.N. and R.D. Stevenson. 2011. Comparison of macroinvertebrate diversity and community structure among perennial and non-perennial headwater streams. *Northeastern Naturalist.* 18:7–26.

Simmons, J.A., W.S. Currie, K.N. Eshleman, K. Kuers, S. Monteleone, T.L. Negley, B.R. Pohlad and C.L. Thomas. 2008. Forest to reclaimed mine land use change leads to altered ecosystem structure and function. *Ecological Applications.* 18:104–118.

Southerland M.T., R.E. Jung, D.P. Baxter, I.C. Chellman, G. Mercurio and J.H. Vølstad. 2004. Stream salamanders as indicators of stream quality in Maryland, USA. *Applied Herpetology.* 2:23–46.

Starnes, L.B. and D.C. Gasper. 1995. Effects of surface mining on aquatic resources in North America (Revised). *Fisheries.* 20:20–23.

Stein, B.A., L.S. Kutner, and J.S. Adams. 2000. *Precious heritage: the status of biodiversity in the United States.* Oxford, U.K.: Oxford University Press. 399 pages.

Stevenson, R.J. 1998. Diatom indicators of stream and wetland stressors in a risk management framework. *Environmental Monitoring and Assessment.* 51:107–118.

Stevenson, R.J. and Y. Pan. 1999. Assessing ecological conditions in rivers and streams with diatoms. Pages 11–40 in E.F. Stoermer and J.P. Smol, editors. *The diatoms: Applications to the Environmental and Earth Sciences.* Cambridge University Press, Cambridge, UK.

Stout, B. and J.B. Wallace. 2005. *A survey of eight major aquatic insect orders associated with small headwater streams subject to valley fills from mountaintop mining. Mountaintop mining/valley fills in Appalachia: Final Programmatic Environmental Impact Statement.* U.S. Environmental Protection Agency, Washington, D.C. 13 pages.

Sweeten, J. and C.D. McCreedy. 2002. *Suspended sediment: An environmental indicator for warmwater streams.* Technical report for Indiana Department of Environmental Management, Office of Water Management, Indianapolis, IN ARN 98–175.

Timpano, A., S. Schoenholtz, C. Zipper and D. Soucek. 2011. *Levels of dissolved solids associated with aquatic life effects on headwater streams of Virginia's Central Appalachia Coalfield Region.* Virginia Tech, Water Resources Research Center, Blacksburg, VA. 82 pages.

U.S. EPA (Environmental Protection Agency). 1990. *Biological criteria: National program guidance for surface waters.* U.S. Environmental Protection Agency, Office of Water Regulations and Standards, Washington, D.C. EPA 440-5-90-004. 61 pages.

U.S. EPA (Environmental Protection Agency). 2011. *The effects of mountaintop mines and valley fills on aquatic ecosystems of the Central Appalachian coalfields.* Office of Research and Development, National Center for Environmental Assessment, Washington, D.C. 136 pages.

Vannote, R.L., G.W. Minshall, K.W. Cummins, J.R. Sedell and C.E. Cushing. 1980. The river continuum concept. *Canadian Journal of Fisheries and Aquatic Sciences.* 48:2477–2480.

Warren, M.L., Jr., and B.M. Burr. 1994. Status of freshwater fishes of the US: Overview of an imperiled fauna. *Fisheries.* 19:6–18.

Welsh, H.H. and S. Droege. 2001. A case for using plethodontid salamanders for monitoring biodiversity and ecosystem integrity of North American forests. *Conservation Biology* 15:558–569.

Welsh, H.H. and L.M. Ollivier. 1998. Stream amphibians as indicators of ecosystem stress: A case study from California's redwoods. *Ecological Applications.* 8:1118–1132.

Welsh, H.H., G.R. Hodgson and A.J. Lind. 2005. Ecogeography of herpetofauna of a northern California watershed: Linking species patterns to landscape processes. *Ecography.* 28:521–536.

Wilcove, D.S., D. Rothstein, J. Dubow, A. Phillips, and E. Losos. 1998. Quantifying threats to imperiled species in the United States. *BioScience.* 48:607–615.

Williams, L.R., D.L. Crosswhite and M.G. Williams. 2002. Short-term effect of riparian distribution on *Desmognathus brimleyorum* at a natural spring in Oklahoma. *Southwest Naturalist.* 47:611–613.

Willson, J.D. and M.E. Dorcas. 2003. Effects of habitat disturbance on stream salamanders: Implications for buffer zones and watershed management. *Conservation Biology.* 17:763–771.

Yoder, C.O. 1991. *Answering some concerns about biological criteria based on experiences in Ohio.* Biological Criteria: Research and Regulation, Proceedings of a National Conference, U.S. EPA, Office of Water, Washington, D.C. 8 pages.

Younos, T., G. Yagow, L. Shabman, L. Smock, E. Smith, H. Kator and J. Walker. 2007. *Protocols for reference watershed approach.* Virginia Department of Environmental Quality, Academic Advisory Committee. 7 pages.

5.2
Landforms and Geomorphic Reclamation

Sustainable Landform Construction on Coal Mining Operations That Require Valley Fills 396

Methodology for Geomorphic Landform Design of Valley Fills in Appalachia Surface Mine Reclamation 397

Investigating Regional Design Parameters in the Design of West Virginia Valley Fills to Support Application of Geomorphic Landform Design Principles 405

Slope Stability Analysis of Geomorphic Landform Profiles Versus Approximate Original Contour Applied to Valley Fill Designs 415

Sustainable Landform Construction on Coal Mining Operations That Require Valley Fills

Richard Warner
Biological Systems and Agricultural Engineering, University of Kentucky, Lexington, Kentucky

Carl Zipper
Crop and Soil Environmental Sciences, Virginia Tech, Blacksburg, Virginia

ABSTRACT

Construction of mine spoil fills, including fills used for excess spoil disposal, is essential to surface coal mining operations in the Appalachian coalfields. In Applachian terrain, valley fills are often the only viable means for excess spoil disposal. In recent years, the mining industry has been confronted with new regulations concerning water quality discharged from mining operations. Total dissolved solids (TDS) and valley fills have been a primary focus, and related issues have affected mine permitting.

We are working with mining firms to develop, demonstrate, and assess effectiveness of proposed "best management practices (BMPs)" and "adaptive mining methods (AMMs)" for construction of coal-mine valley fills and associated mined landforms that will restore environmental quality in the aftermath of coal removal. Specific goals are to build mine spoil fills that will substantially reduce TDS levels in water discharges relative to fills constructed using conventional techniques; produce surface-water flows that more closely approximate pre-mining hydrology than occur in response to conventional reclamation techniques; to restore forested post-mining land uses and ecosystem processes on those mined landforms; and to achieve the above while controlling costs to an extent necessary for a profitable mining operation.

Specific BMPs and AMMs which are being addressed in this applied research program encompass: (1) identification of high-conductivity or high-Se producing strata, (2) identification of low-conductivity strata that has the engineering characteristics to produce a low-permeable isolation barrier, (3) isolation techniques linked with mining equipment, (4) underdrain development with low-conductivity durable rock, (5) reduce entry of fines into the underdrain, (6) valley fill low-permeable spoil cover, (7) landform to convey runoff and reduce infiltration, (8) Forest Reclamation Approach and (9) stream creation with form and function.

Specific valley fill monitoring equipment will be identified. Additionally, data acquisition frequency and data analysis/usage will be discussed.

Currently, we are working to design and construct mine spoil fills that are intended to achieve the above environmental restoration goals on three mine sites. This presentation will review progress and accomplishments to date, and will describe in general terms the mine spoil fill designs being developed in collaboration with industry cooperators.

ACKOWLEDGMENT

This study was sponsored by the Appalachian Research Initiative for Environmental Science (ARIES). The views, opinions and recommendations expressed herein are solely those of the authors and do not imply any endorsement by ARIES employees, other ARIES-affiliated researchers or industrial members. Information about ARIES can be found at http://www.energy.vt.edu/ARIES.

Methodology for Geomorphic Landform Design of Valley Fills in Appalachia Surface Mine Reclamation

Alison Sears
West Virginia University, Morgantown, West Virginia

Christopher Bise
West Virginia University, Morgantown, West Virginia

John Quaranta
West Virginia University, Morgantown, West Virginia

Leslie Hopkinson
West Virginia University, Morgantown, West Virginia

ABSTRACT

Fluvial geomorphic principals are being investigated as a reclamation alternative to the conventional valley fill process in the surface mining regions of Appalachia. The purported benefits are that landscapes appear natural, possess geotechnical stability, have inherent stability against erosion, support vegetative re-growth, and offer financial benefits of lower initial costs and reduced long-term maintenance. This design alternative has received recognition for its successes in semi-arid regions, specifically New Mexico.

The purpose of this research was to perform a geomorphic design on a permitted valley fill and compare design outputs with the variance method. The objective was to evaluate alternative reclamation approaches and determine if an effective and implementable valley fill could be designed geomorphic landform design principles as applied to mountainous terrain. An industry partner assisted by providing design details regarding their permitted project site located in southern West Virginia. Data, such as stream length, drainage density, ridge to head-of-channel distance, annual precipitation, stream flow, slope stability, and approximate original contour were evaluated to compare the two designs.

INTRODUCTION

Appalachian surface mines are reclaimed to the approximate original contour (AOC) according to the Surface Mining Control and Reclamation Act of 1977. In West Virginia, the AOC guidelines are also promulgated by West Virginia Surface Mining Reclamation Regulations, Code of State Regulations §38, which require slope profile configurations constructed by the traditional backfilling technique. It is also required that grading of disturbed areas have a final profile that, in effect, closely resembles the general surface configuration of the land prior to mining. The post mining slope design is intended to ensure slope stability, control drainage, complement the drainage pattern of the surrounding terrain, and prevent stream sedimentation. These requirements are comprehensive, covering the drainage pattern of the surrounding terrain, highwalls, valley fill, and spoil piles. Special circumstances and permit variances can be approved by regulatory agencies for areas that will not comply with regulations but will provide specific desirable land uses. In addition, the West Virginia Department of Environmental Protection and the US Environmental Protection Agency implement the Clean Water Act of 1972 through the National Pollution Discharge and Elimination

System in order to provide requirements for drainage and sediment control for the quality of the discharged runoff on the disturbed area.

The conventional valley fill design was intended to ensure slope stability, control drainage, complement the drainage pattern of the surrounding terrain, and prevent stream sedimentation. The design consisted of: slope shapes exhibiting uniform benches; planar slopes having unvarying contours; and, drainage ditches located along the perimeter and/or center of the fill. To date, the AOC guidelines have provided help in the reduction in environmental degradation of mountain streams and an increase in the stability of slopes on the reclamation sites. In West Virginia, the reclamation vegetation efforts using select grasses and hardwoods have proven somewhat effective in concealing the planar slope profiles and surface drainage structures (Skousen and Venable 2008; Groninger et al. 2007). The effectiveness of post-mine land use implemented by the mining industry has been successful to an extent and, as tree canopy matures, the slopes appear natural, to some degree (Burger 2011).

The work discussed in this paper re-evaluates traditional valley fill design, incorporating landforming into the process. The purpose of this research was to perform a geomorphic design on a permitted valley fill and compare design outputs with the conventional method. An alternative reclamation design was created and evaluated to determine if an effective and implementable valley fill could be designed using Carlson's Natural Regrade® software following the Geofluv™ method as applied to mountainous terrain in the central Appalachian region of West Virginia.

BACKGROUND: GEOMORPHIC LANDFORM PRINCIPLES

Under natural conditions, landforms develop a balance between erosive and resistance forces, resulting in a system in equilibrium with low erosion rates. The fluvial geomorphic landform design approach attempts to design landforms in this steady-state condition, considering long-term climatic conditions, soil types, slopes, and vegetation types (Toy and Chuse 2005; Bugosh 2009).

The geomorphic landform design procedure builds a drainage network using a reference landform approach. With this procedure, a reference watershed must be identified and characterized. Typically, the reference watershed is the location where the geomorphic design is going to be implemented to properly mimic the pre-disturbed topography. The information that is necessary to inform successful design (Toy and Chuse 2005; Eckels and Bugosh 2010) includes the main channel slope and landform profile shape, drainage density and area, and channel characteristics.

Main channel slope especially represents the watershed slope. As the main channel slope increases, the stream power and erosion potential increase (Toy and Chuse 2005). Longitudinal profile shapes of the landforms must also be considered as the concave shape differs among headwater and downstream locations. In mountainous terrain the nature of slope profiles develop into compound surface profiles. These profiles exhibit steep convex slopes at the head of the valley and progressively transition into a concave form gradually tapering to a uniform profile. The development of natural streams on these complex profiles is affected by fluvial influence stream cutting, surficial erosion, and rill to gulley erosion. To minimize these erosive forces, the drainage density for each designed stream must meet the target drainage density requirements.

Drainage density is a measure of the average stream channel spacing and results from flow interactions with sediment and soil, vegetation, topography, and weather variables (Bugosh 2004; Toy and Chuse 2005). For a given reference landform, the drainage density describes the drainage network that can be supported without significant aggradation or erosion (Bugosh 2004). The default target drainage density in the settings is 80 ft/ac with a 20% variance but would be changed in order to be site specific.

The fluvial geomorphic design approach assumes a dendritic pattern. Natural channels vary in characteristics depending on its location in the watershed. Headwater streams are often

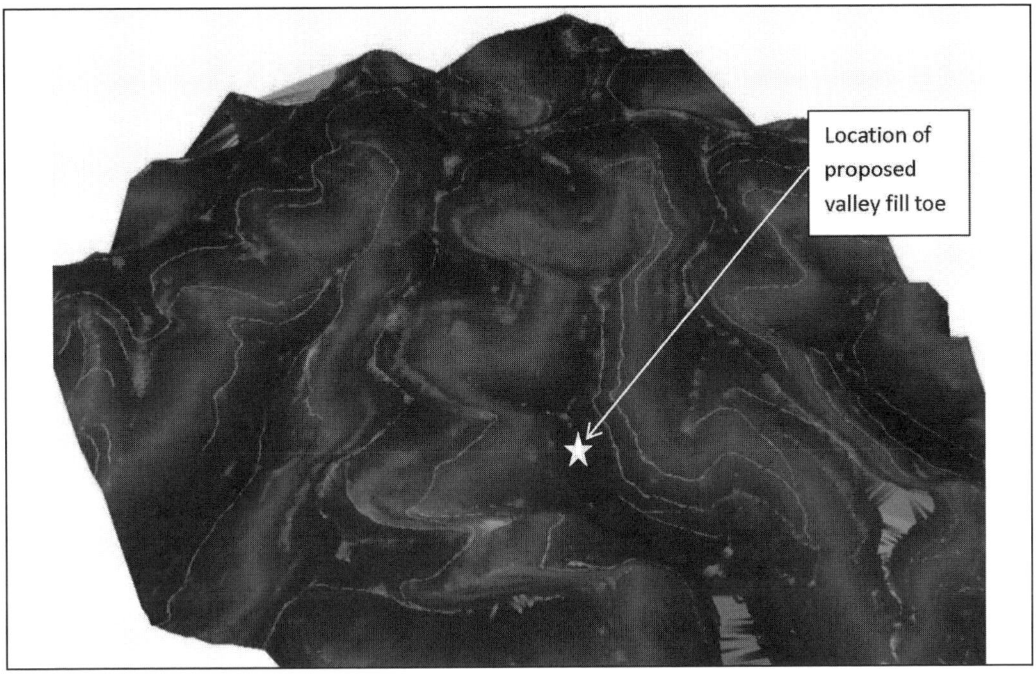

Figure 1. Pre-mined original topography of study area

steep (>4% slope), characterized as "A" channels as defined by the Rosgen classification system (Rosgen 1994), and relatively straight (sinuosity = 1.0–1.2), and down-stream channels have a lower gradient (<4% slope), C-channels, and increased sinuosity (>1.2; Rosgen 1994).

The stream characteristics that must be considered when designing systems that will properly manage both flow and sediment discharge are bankfull width, width to depth ratio, meander belt width, A-channel reach length and sinuosity (Eckels and Bugosh 2010). A-channel reach length is defined as the distance of one-half of a meander length in steep channels. Ridge to head of channel distance defines the length required to form concentrated flow, while also defining the channel head location in reference to the watershed boundary.

METHODS

The study site is a mountaintop removal surface mine site located among steep, rugged terrain in southern West Virginia (Figure 1). Approximately 75 million cubic yards of overburden and coal will be removed during the mining process, resulting in the site as shown in Figure 2.

A valley fill design was being implemented at the sample site. The post-mined land use of the area was determined to be pastureland and, therefore, a variance was obtained to leave the land relatively flat at an elevation of approximately 1,700 ft. The valley fill covered approximately 340 ac and holds over 90 million cubic yards of fill material since the original 75 million cubic yards of material removed had a bulking factor of 1.2 applied. Runoff from the property drained to Surface Water Runoff Analysis (SWROA) ditches that outlined the boundary and out National Pollution Discharge and Elimination System (NPDES) points into surrounding stream channels.

The sample site valley fill design contained numerous features that were typical for many conventional valley fill designs. These include a traditional rock core underdrain that was formed from the natural gravity segregation of the durable rock material as it was placed using end-dumping methods as well as a benched valley fill face. The

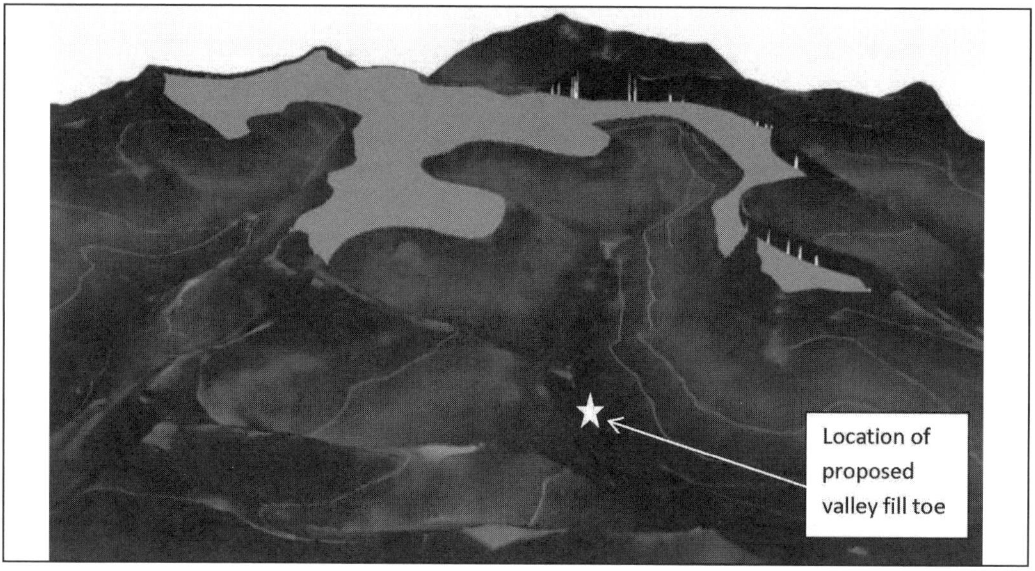

Figure 2. Post-mined and pre-reclamation topography of study area

valley fill would be highest in elevation along the top of the benched valley fill area. The fill would then slope down one to two percent toward the north as well as to both sides. Therefore, all drainage would be directed away from the valley fill face and into the pre-constructed SWROA ditches. The face of the valley fill, like many traditional fills, was constructed in benches as to provide stability to the structure. These benches were 20 ft in width and slope at 3% to 5%. The benches are constructed every 50 ft in vertical elevation (Figure 3).

Geomorphic Landform Design Approach

The design tool Carlson® Natural Regrade® with GeoFluv™ was used to apply the geomorphic landform design approach to the valley fill study site described in the previous section. The pre-mined terrain is characterized by a system of steep-sloped ridges and valleys (Figure 1).

The sample site boundary was divided into six subwatersheds to perform analysis over the entire disturbed surface mine area (Figure 4). Each subwatershed was designed to drain to existing NPDES permit points to maintain a hydrologic balance. There were a total of six subwatersheds, which ranged in size from approximately 4 ac to 241 ac. The geomorphic design principles were then applied to each subwatershed individually to create balanced, stable landforms.

For example, the design of the main, or largest, subwatershed was characterized by a main channel with 12 tributaries, forming a dendritic drainage pattern, which is typical for unconsolidated materials (Figure 4) (Toy and Chuse 2005; Eckels and Bugosh 2010). The created channel pattern mimicked the original drainage pattern of the pre-mined site, but included many more channels. Ideally, the topography, including the number of stream channels, would mimic the pre-disturbed condition. However, due to the lack of pre-mined data, the default drainage density and ridge to head-of-channel distance were used which resulted in more stream channels than the topography prior to mining. The ridge to head-of-channel distance is the length from the ridge to the point where flow transitions from sheet flow to channelized flow. The default value for ridge to head-of-channel distance is 80 ft.

More tributary channels were added to the main channel closer to its mouth in order to satisfy the target drainage density, while less tributary channels were added at the head of the main channel to meet the drainage density

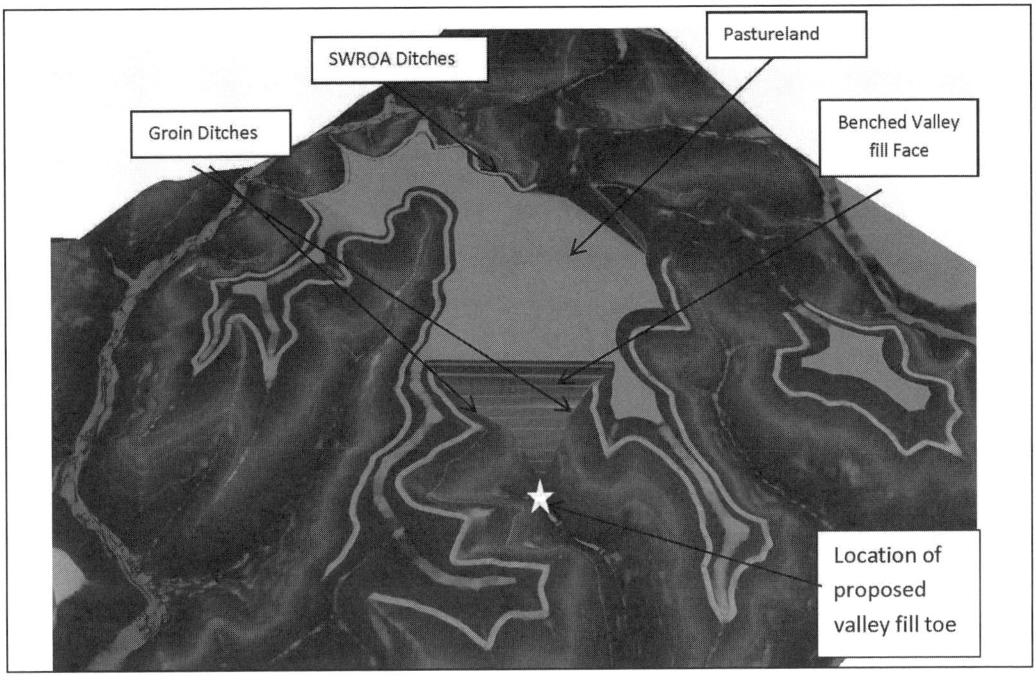

Figure 3. Variance valley fill on sample site, post-reclamation

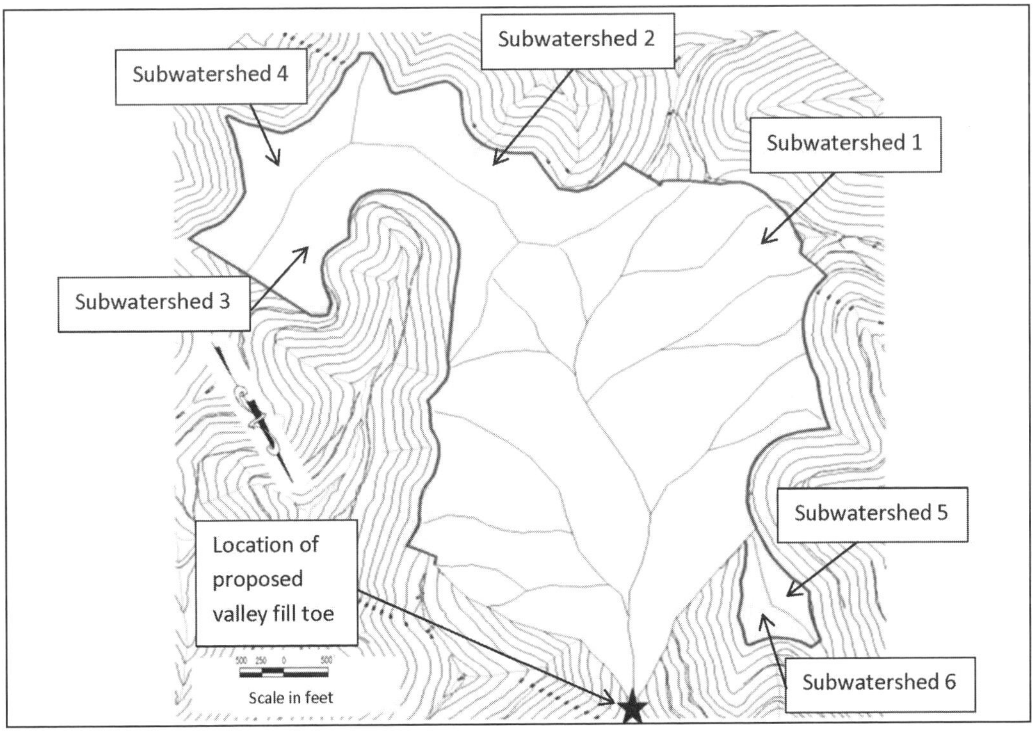

Figure 4. Stream channels created in dendritic pattern in subwatershed 1

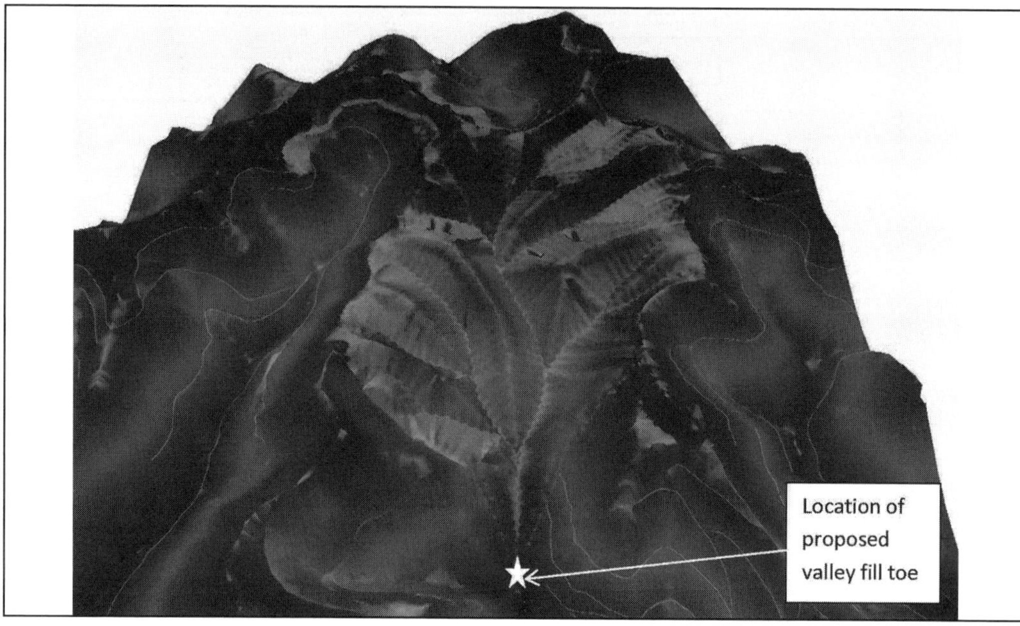

Figure 5. Completed three-dimensional geomorphic landform valley fill design

requirement. The same design principles were applied to the remaining subwatersheds. Once all of the subwatershed designs were created using these principles, they were combined so that the valley fill design was complete. All of the data for each analyzed subwatershed was compiled so that a proper analysis of the total design was completed.

A majority of the default settings in Natural Regrade® were used in order to create the geomorphic landform design for the sample site due to the lack of on-site data that could have been collected from the pre-disturbed reference watershed. Ideally, the pre-disturbed land would have been analyzed to obtain reference data so the created design could accurately mimic the pre-mined topography. However, this was not possible at the study site due to mining currently occurring at the sample site.

Natural Regrade® provides data for rainfall throughout the U.S. even if field data from the site is not obtainable. The two of the global settings that were drastically changed for the design were that the 2 yr, 1 hr rainfall changed from 0.6 in to 1.41 in and the 50 yr, 6 hr rainfall changed from 2.0 in to 4.03 in. The default rainfall values are based on semi-arid regions of the United States where precipitation is lower as opposed to the higher precipitation region of the United States in which the sample site is located, based on type II storm and precipitation data.

RESULTS AND DISCUSSION

Comparison of Traditional and Alternative Designs

Features of the completed geomorphic landform valley fill design include a natural, aesthetically pleasing appearance that includes ridges, valleys, and stream channels that are mature landforms. These erosive equilibrium landforms mimic the function of the natural landscape that would have naturally evolved over time. The created stream channels, designed and located within the valley fill as Rosgen type "A" channels, control drainage and create stable hydrologic equilibrium over the valley fill (Rosgen 1994). The geomorphic landform design also required fewer artificial elements and supported long-term erosion stability due to the mature landforms that were created (Martin-Duque et al. 2009). This also led to potentially reduced maintenance and cost because of

Table 1. Geomorphic landform design data of sample site

	Subwatershed						
	1	2	3	4	5	6	Total Watershed
Area (ac)	241	32	36	21	4	6	340
Elevation range of topography (ft)	1150–1995	1720–1995	1650–1995	1720–1993	1780–1897	1720–1900	1150–1995
Original stream length (ft)	8630	1180	NA*	460	NA	NA	10270
Number of created channels	13	1	1	2	1	1	19
Range of created stream length (ft)	814–5290	3159	3551	580–1505	487	510	487–5290
Total created stream length	21902	3159	3551	2085	487	510	31694
Original stream length saved (ft)	1315	NA	NA	NA	NA	NA	1315
Drainage density (ft/ac)†	90.62	99.35	98.76	98.42	116.78	90.09	
Rosgen Channel type	A, C	A, C	A, C	A, C	A, C	A, C	A, C

*NA = not applicable.
†Target drainage density range: 80–120 ft/ac.

decreased erosion of the landforms. Figure 5 illustrates three-dimensional views of the completed geomorphic landform valley fill designs that demonstrate all of the features described above.

The overall geomorphic reclamation design of the sample site saved a total of 1300 ft of original stream length at the toe of the valley fill (Table 1). Using geomorphic principles, almost 33,000 ft or six and a quarter miles of stream length (type A and type C channels as defined by Rosgen, 1994) was created on the property (Table 1). Figure 4 shows the location of all the created stream channels.

The variance valley fill design contained a benched valley fill face with the rest being flat pastureland. Table 2 shows that this design filled in almost 21,000 feet of original stream channel length and in return created no stream channels.

Table 2. Variance design data of sample site

	Watershed Variance Valley Fill
Area (ac)	339.9
Elevation range of topography	1693
Original stream length (ft)	20,540
Number of created channels	NA*
Range of created stream length (ft)	NA
Total created stream length	NA
Original stream length saved (ft)	NA
Drainage density (ft/ac)	NA
Rosgen Channel type	NA

*NA = not applicable.

Infiltration

Geomorphic landform design methods could minimize infiltration by reducing the amount of time surface water runoff is in contact with the surface of the valley fill. The decrease in infiltration could increase the long-term stability of the structure by reducing internal erosion, suffusion and potential for pore water pressure buildup. Fluvial geomorphic design provides precipitation with quick and easy routes into channels. Therefore, if the runoff contact time was minimized, the time the water had to collect contaminants that were within the fill would be reduced. Acid mine drainage and leaching of selenium into groundwater and streams are concerns that have received recent emphasis in the mining industry and are proposed to be reduced using geomorphic landform reclamation methods.

Slope Stability

Slope stability analysis should not be overlooked in surface mine reclamation because slope failure can have significant ecological impacts (Kenney and Lau 1985). Geomorphic landform designs created using principles do not analyze the stability of the slopes. It is important that slope structures remain durable and do not fail when constructed with surface mine overburden for several reasons.

If toxic mineralogy exists within the spoil, slope stability becomes an especially critical analysis that should be performed during the design planning process. It is also incredibly important to perform slope stability analysis to ensure the durability of the structure in Appalachia due to its steep terrain. Therefore, all created landforms, whether created using conventional or geomorphic principles, should be analyzed properly in order to limit the steepness of slopes and insure the stability of the landforms because slope failures can have a significant impact on the health and safety of downstream communities as well as the operators constructing the structures.

CONCLUSIONS

State and Federal regulations directing mine reclamation using the conventional approach have resulted in geotechnically stable designs of valley fills constructed using waste rock overburden. Environmental concerns at mountaintop mining sites abound because of the loss of headwater stream length and increased flooding risk. One promising technique to lessen the impacts involves fluvial geomorphic landform design applied to the waste rock fill and slope profiles.

The geomorphic valley fill design allowed 1300 feet of original stream length to remain undisturbed compared to the traditional valley fill design and almost 33,000 ft or six and a quarter miles of stream length (type A and type C channels as defined by Rosgen 1994) was created on the property (Table 1). In four of the six subwatersheds, only one channel was needed to satisfy the drainage density requirement due to the watershed area and the length the main channel.

These landform designs add variability and aid in establishing a site with the potential of long-term hydrologic balance. The geomorphic landform reclamation approach has potential to extend beyond current industry practices and will improve environmental impacts, flood control, water quality, and human safety.

Future application of this innovative reclamation technique has potential to protect the public and environment during reclamation.

REFERENCES

Bugosh, N. 2004. "Computerizing the fluvial geomorphic approach to land reclamation," *National Meeting of the American Society of Mining and Reclamation and The 25th West Virginia Surface Mine Drainage Task Force*, April 18–24, 240–258.

Bugosh, N. 2009. "A summary of some land surface and water quality monitoring results for constructed GeoFluv landforms. Revitalizing the Environment: Proven Solutions and Innovative Approaches," *National Meeting of the American Society of Mining and Reclamation*, May 30–June 5, 2009, Billings, MT, 153–175.

Burger, J.A. 2011. "Sustainable mined land reclamation in the Eastern U.S. coalfields: a case for an ecosystem reclamation approach," *National Meeting of the American Society of Mining and Reclamation*, June 11–16, 2011, Bismarck, ND.

Eckels, R., and Bugosh N. 2010. "Natural approach to mined land reclamation," *FIG Congress*, 2010.

Kenney, T.C., and Lau, D. 1985. "Internal stability of granular filters," *Canadian Geotechnical Journal*, Vol. 22, pp. 215–225.

Martin-Duque J F., Sanz M.A., Bodoque J.M., Lucia A., Martin-Moreno C., 2009, "Restoring earth surface processes through landform design. A 13-year monitoring of a geomorphic reclamation model for quarries on slopes," *Earth Surface Processes and Landforms*. 35: 531–548.

Groninger, J., Skousen, J., Ange,l P., Barton, C., Burger, J., Zipper, C. 2007. "Mine reclamation practices to enhance forest development through natural succession," *The Appalachian Regional Reforestation Initiative (ARRI) Rorest Reclamation Advisory, July, 2007.*

Rosgen D., 1996, "Applied river morphology," *Wildland Hydrology*, Pagosa Springs, CO.

Rosgen, D.L. 1994. A classification of natural rivers. *Catena*, 22 (3): 169–199.

Sawatsky, L., and Beckstead, G. 1996. "Geomorphic approach for design of sustainable drainage systems for mineland reclamation," *International Journal of Mining, Reclamation and Environment.* 10(3): 127–129.

Skousen, J.G., and Venable, C.L. 2008. "Establishing native plants on newly-constructed and older-reclaimed sites along West Virginia Highways," *Land Degradation & Development*, 2007.

Toy, T.J., and Chuse, W.R. 2005. "Topographic reconstruction: a geomorphic approach," *Ecological Engineering*.24: 29–35.

Investigating Regional Design Parameters in the Design of West Virginia Valley Fills to Support Application of Geomorphic Landform Design Principles

Catlin Buckley
West Virginia University, Morgantown, West Virginia

Leslie Hopkinson
West Virginia University, Morgantown, West Virginia

John Quaranta
West Virginia University, Morgantown, West Virginia

Ben Mack
West Virginia Water Research Institute, West Virginia University, Morgantown, West Virginia

Paul Ziemkiewicz
West Virginia Water Research Institute, West Virginia University, Morgantown, West Virginia

ABSTRACT

Fluvial geomorphic landform design has the potential to improve water quality while restoring productive stream channels in the reclaimed landscape. The technique is difficult to apply in the southern West Virginia coal fields in part due to the absence of unaltered landform datato serve as reference design values. This research examined the application of geomorphic landform design principles to valley fills. The objectives of this research were to quantify mature landform features in an undisturbed watershed in southern West Virginia and compare these characteristics to default parameters utilized in a current design tool. Reference landform characteristics were quantified in the Whetstone watershed located in the Panther Wildlife Management Area in southern West Virginia. A topographic survey was completed to quantify ridge to head of channel distance, channel slope, and hillslope profile. Channel grain size distributions and cross sectional geometry were quantified in both head of channel and watershed outlet locations. Findings suggest that the slope at the head of channel ranges between 35 and 50 percent, with the slope at the mouth remaining at 8 percent. Drainage density was calculated as 5.3 km^{-1}, and sinuosity remained close to one at 1.03. These design parameters substantially differ from design inputs of current design tools and will be utilized to calibrate future conceptual valley fill designs.

INTRODUCTION

Approximately 2,000 km of headwater streams were lost by 2002 due to surfacing mining disturbance in the central Appalachian region (USEPA 2011). Typically, the horizontally bedded seams are removed sequentially as overburden is placed both on the pit floor and in external, valley fill dumps. Conventional valley fills under West Virginia regulations, are designed to meet minimum design requirements to achieve geotechnical stability and to control surface runoff. State regulations (WVDEP 1993) require:

- A long-term static factor of safety of 1.5;
- 2:1 slopes with minimum 20-ft wide benches installed within every 50 vertical feet;

- Internal drainage provided by a vertical rock chimney (minimum width of 16 ft); and,
- Surface drainage for a 100-yr, 24-hr precipitation event.

The resulting surfaces often have planar slope profiles which contrast with the surrounding landscape, and their increasing size has resulted in an increasing loss of headwater streams. Studies have shown that streams below valley fills often have elevated dissolved ion concentrations resulting from water contact with the overburden (Hartman et al. 2005; Pond et al. 2008; Petty et al. 2010). Additionally, research has documented that surface mining and reclamation increase stormflow response compared to the undisturbed condition (Bonta et al. 1997; Messinger 2003; Messinger and Paybins 2003; Negley and Eshleman 2006), and selenium leaching from spoil related to coal mining is of increasing concern (e.g., Ziemkiewicz et al. 2011).

Fluvial geomorphic landform design has the potential to improve water quality while restoring productive stream channels in the reclaimed landscape. Under natural conditions, landforms develop a balance between erosive and resistance forces, resulting in a system in dynamic equilibrium with low erosion rates. The fluvial geomorphic landform design approach attempts to design landforms in this steady-state condition, considering long-term climatic conditions, soil types, slopes, and vegetation types (Toy and Chuse 2005; Bugosh 2009). Relative to traditional reclaimed landforms, fluvial geomorphic landform design appears natural, reduces long-term maintenance, requires fewer artificial elements, and supports long-term stability (Martin-Duque et al. 2009).

This design approach has been used with success (e.g., Toy and Chuse 2005; Measles and Bugosh 2007; Martin-Moreno et al. 2008; Bugosh 2009; Robson et al. 2009; Marin-Duque et al. 2009) but has not been utilized in Appalachian surface mining reclamation. The complexity of mature landform design in steep terrain presents challenges. In addition, current regulations do not support the utilization of the of the design technique (Michael et al. 2010).

Geomorphic landform design uses a reference landform approach which requires predevelopment, geomorphic data. The data needed for design are similar to those needed for stream classification systems (e.g., Schumm and Mosley 1977; Rosgen 1994, 1996; Montgomery and Buffington 1997) and stream assessments (e.g., Kaufmann and Robison 1998; VANR 2004):

- main channel slope;
- drainage density;
- longitudinal profile shape;
- channel characteristics (bankfull width, width to depth ratio, sinuosity, meander belt width, "A" channel length); and,
- ridge to head of channel distance.

Limited geomorphic data are available in West Virginia, especially in the southern coal fields (e.g., Wiley et al. 2001). This region has a history of surface mining and logging, often requiring changes of the steep terrain for site access, which has rendered limited unaltered land profiles.

The overall goal of this research was to quantify geomorphic features in an undisturbed watershed in southern West Virginia that will inform geomorphic landform design for valley fills in Central Appalachia. This research quantified geomorphic characteristics in the Whetstone watershed located in the Panther Wildlife Management Area (WMA). These characteristics were then compared to design inputs used in a recent alternative valley fill design developed by Sears et al. (2012).

METHODS

The Whetstone Branch watershed located in the Panther Wildlife Management Area in McDowell County located near the southern border of West Virginia was the experimental watershed for this research (Figure 1). The study site was identified using aerial photography, topographic maps, and communication with area officials. The Panther WMA site is managed by the West Virginia Division of Natural Resources and since the

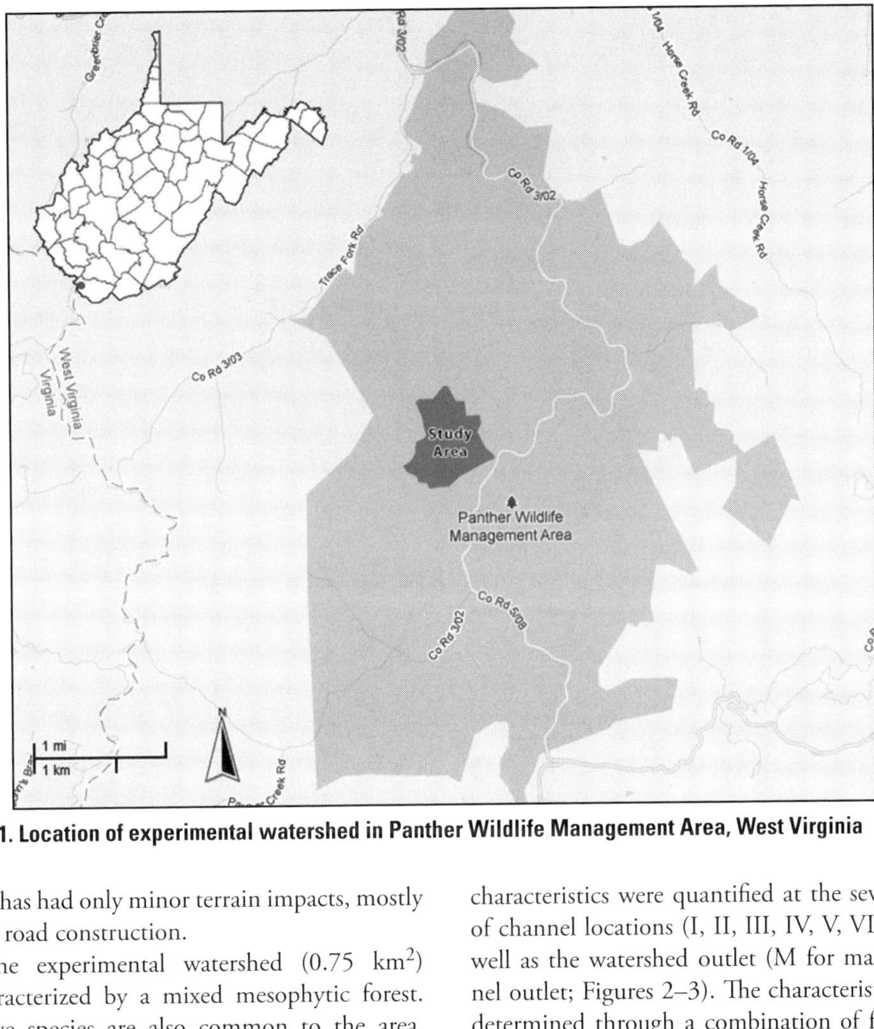

Figure 1. Location of experimental watershed in Panther Wildlife Management Area, West Virginia

1920s has had only minor terrain impacts, mostly due to road construction.

The experimental watershed (0.75 km^2) is characterized by a mixed mesophytic forest. Invasive species are also common to the area, including *Elaeagnus umbellate* (autumn olive), *Ailanthus altissima* (tree of heaven), and *Rosa multiflora* (multiflora rose). The Whetstone Branch watershed consists mainly of an extremely steep and stony soil (Pineville-Berks), with a small portion fine sandy loam (Yeager) located around the mouth of the stream. The study area receives an average of 100–122 cm of precipitation annually with a strong seasonal pattern (USDA-NRCS, NWCC 2012).

The Whetstone Branch watershed includes nine major unnamed tributaries. Seven of these branches were selected for study based on accessibility (Figure 2). Field data collection was completed June–July 2012. Geomorphic characteristics were quantified at the seven head of channel locations (I, II, III, IV, V, VI, VII) as well as the watershed outlet (M for main channel outlet; Figures 2–3). The characteristics were determined through a combination of field surveys and existing GIS data as described in the following sections.

Data Collection and Analysis

Field data needed to quantify grain size distribution, hillslopes, ridge to head of channel distance, channel slope, and cross-sectional geometry were collected in head of channel and watershed outlet locations (Figure 2). A field survey was completed using a Topcon FC-100 and Hyperlite+ receivers (Topcon, Paramus, New Jersey)using a 0.6 m horizontal error and a 1.5 m vertical accuracy; this error represented the minimum allowable error to complete measurements within the dense vegetation cover. Study reaches were surveyed to

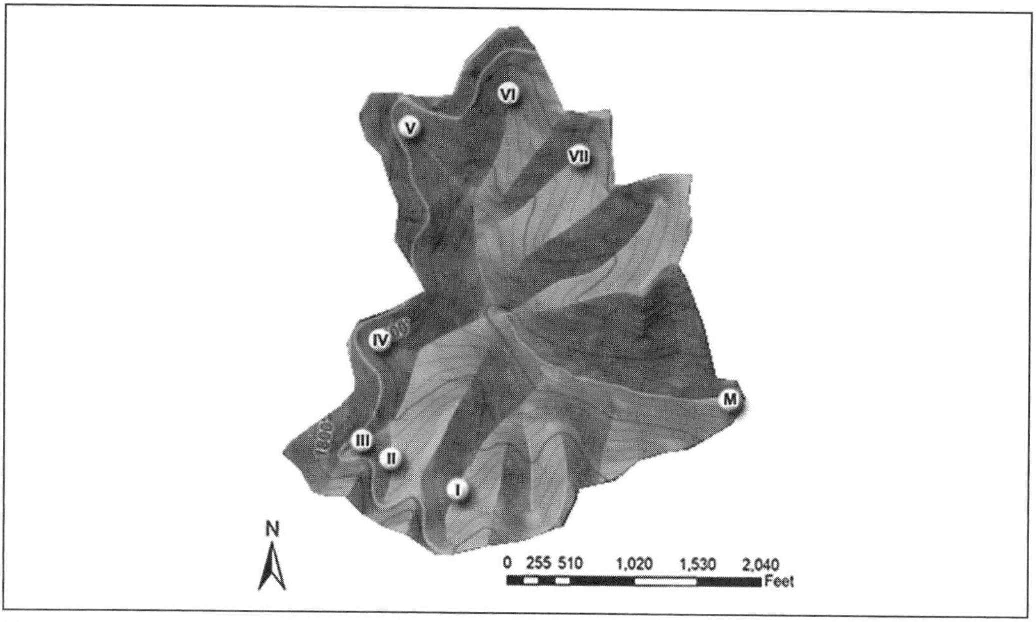

Figure 2. Head of channels surveyed in experimental watershed

Figure 3. Experimental field sites for the head of channel sampling stations (I, II, III, IV, V, VI, VII) and the watershed outlet (M)

quantify slopes, sinuosity, ridges, and channel head locations. The location of the watershed ridge and head of channel locations were identified and recorded as points; these data were used to calculate ridge to head of channel distance. Roads that altered the natural topography were also recorded. A minimum of five points were taken downslope from the start of channel to identify the channel slope and sinuosity (i.e., channel length/valley length). Bank slopes were determined through points taken a minimum of 7.5 m from the start of channel on either side of the channel. A clinometer was used to verify slope measurements. Channel dimensions were measured at the head of each channel as well as the mouth of the watershed. These sections were taken by placing an adjustable measuring rod horizontally and perpendicular to the stream; the distance from the rod to the streambed were measured and recorded at 0.3 m intervals.

Particle size distributions of bed material were quantified at the eight sampling locations (Figure 2) using modified Wolman (1954) pebble count (Harrelson et al. 1994). Bank materials were also observed and recorded. Riparian trees, shrubs, and herbaceous plants were characterized at each head of channel location through sketching sections, highlighting plant types and observations of plant cover. Additionally, a percentage of each type of cover (trees, shrubs, low lying plants) was estimated based on observation.

ArcMap was used in conjunction with digital spatial datasets for elevation (U.S. Geological Survey, WV SAMB), hydrology (U.S. Geological Survey, WV SAMB), and soils (NRCS). The field measurements were downloaded into a GIS desktop application and georeferenced with the field data. GIS was used to verify slope and sinuosity measurements. Slope and aspect maps were created and drainage density (i.e., valley length/watershed area) was calculated. Ridge to head of channel distances were calculated using survey data.

RESULTS AND DISCUSSION

Stream Pattern and Profile

Sinuosity, a measure of channel curvature, was calculated as nearly one when using both survey data (average sinuosity = 1.01) and GIS data (average sinuosity = 1.05) (Table 1). Channels with a sinuosity greater than 1.3 are considered meandering (FISRWG 1998); therefore, no meandering channels were observed in the steep, headwater watershed. The sinuosity measurements calculated with field measurements were slightly smaller than those derived from GIS. This is expected because the survey only accounted for a small stretch at the beginning of the stream (where slopes are greater) while the GIS measurements represented the entire branch.

The ridge to head of channel distance represents the distance required to form channelized flow and is essential to understand watershed runoff processes (Hancock and Evans 2006). The head of channel was determined by identifying the location where soil began to give way to gravel and there was an apparent change in slope. An apparent v-notch began to form at the head of each channel as well. The distance from the ridge to the head of channel remained relatively constant with one outlier (Table 1). The average distance recorded from the ridge to the head of channel was 121 m. When neglecting the outlier, the average was 114 m. All points were influenced by a road that ran along the slope of the site. In the sites where the outlier was measured, the road was further from the ridge.

For the seven headwater tributary locations, channel slope was greater than 16%. At the watershed outlet, the main channel had a slope of 8%, characteristic of a non-meandering stream.

Channel Material and Hillslope

Median particle size (D50) ranged from 19 to 34 mm for all headwater locations (I-VII, Table 2), representing gravel bed channels. The

Table 1. Ridge to head of channel distance, sinuosity, and channel slope for each field site

Site	Ridge to Channel Head Distance (m)	Sinuosity (from survey)	Sinuosity (from GIS)	Channel Slope (%)
I	112	1.05	1.08	16
II	113	1.01	1.12	18
III	163	1.00	1.05	21
IV	108	1.00	1.02	27
V	106	1.00	1.00	42
VI	136	1.01	1.06	34
VII	110	1.01	1.02	36
M	NA*	1.01	1.03	8

* NA = not applicable.

median particle size for the watershed outlet was also in the gravel size range (D50=20 mm). The head of channel bed material was colluvial according to the Montgomery-Buffington classification (Montgomery and Buffington 1993); it originated from hillslope debris and was formed by gravity.

Banks primarily consisted of sand and tended to have slopes from 9–25%. The heads of the channels tended to start out broad (1.8–3 m) and narrowed as they traveled down the slope (Table 3). Channel slopes were also very steep, reaching as high as 42% grade (Table 1). The steep valley slopes are also presented in Figure 4. Much of the watershed has greater than a 50% incline, with very few areas less than 30% (Figure 4). The complexity of the watershed arrangement is apparent through the aspect distribution; the watershed had 40%, 30%, 20%, and 10% of south (south, southwest, southeast), north (north, northwest, northeast), east, and west facing slopes, respectively (Figure 5).

Table 2. Gran size distributions for each field site

Site	D16 (mm)	D50 (mm)	D84 (mm)
I	9.1	31	72
II	9.4	21	59
III	11	33	66
IV	9.4	22	62
V	8.7	19	51
VI	8.3	27	76
VII	8.4	34	63
M	10	20	32

Comparison and Analysis of Design Parameters

Sears et al. (2012) recently designed an alternative valley fill for a site under construction in southern West Virginia. The design applied the geomorphic landform technique and used the design tool Carlson Natural Regrade with GeoFluv™. Default design parameters that were not specific to West Virginia were utilized in the design process (Table 4).

The measured values quantified in this research varied significantly from the default settings. All observed channels were characterized as colluvial as described by the Montgomery and Buffington (1993) classification system. All channel slopes were greater than 4% for this study and all measured sinuosity values were near one. The measured ridge to head of channel distances were nearly seven times greater than the value utilized in the Sears et al. (2012) design. The default drainage area was less than the measured value; however Sears et al. (2012) allowed a 20% error (6–9 km^{-1}). The experimental watershed value (5.3 km^{-1}) did not fall far outside this range (Table 4).

Table 3. Channel width, bank material, and bank slope for each field site

		Left Bank		Right Bank	
Site	Channel Width	Slope	Texture	Slope	Texture
I	SC	VS	Sand/Silt	VS	Sand/Silt
II	B	S	Sand	S	Sand
III	B	S	Sand	S	Sand
IV	B	H	Sand	H	Sand
V	B	VS	Sand	VS	Sand
VI	B	H	Sand	VS	Sand
VII	N	S	Sand	S	Sand
M	B	ES	Sand	S	Sand

*SC is semi-confined (0.6–1.2m), B is broad (1.8–3m), N is Narrow (1.2–1.8m), VS is very steep (16–25%), S is steep (9–15%), H is hilly (4–8%), and ES is extremely steep (>25%); notation adapted from (VANR, 2004).

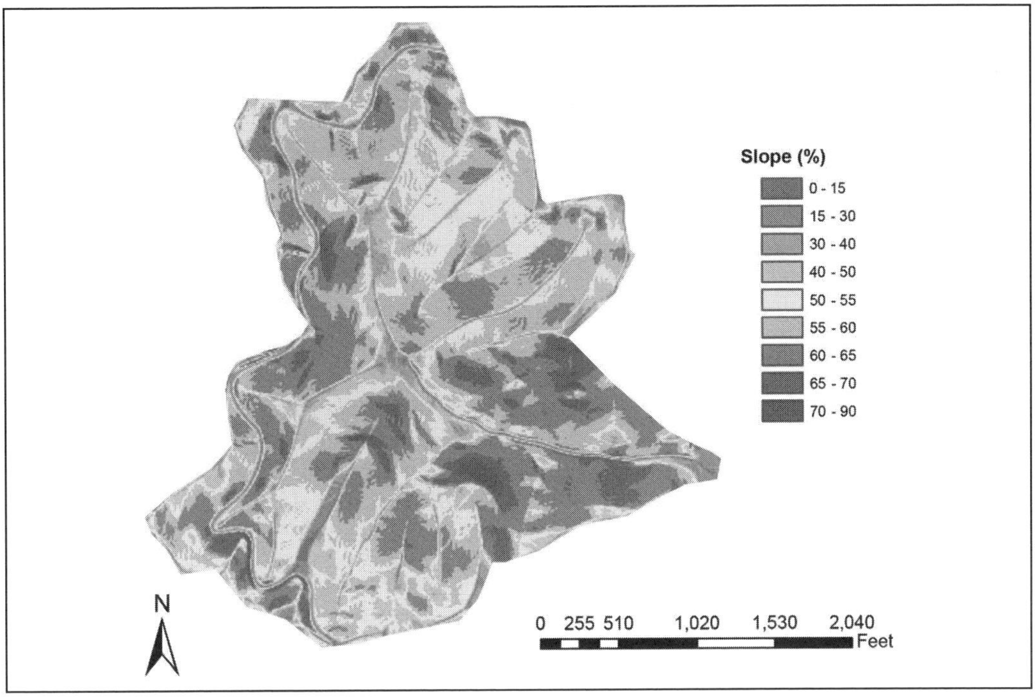

Figure 4. Slope map of Whetstone Branch

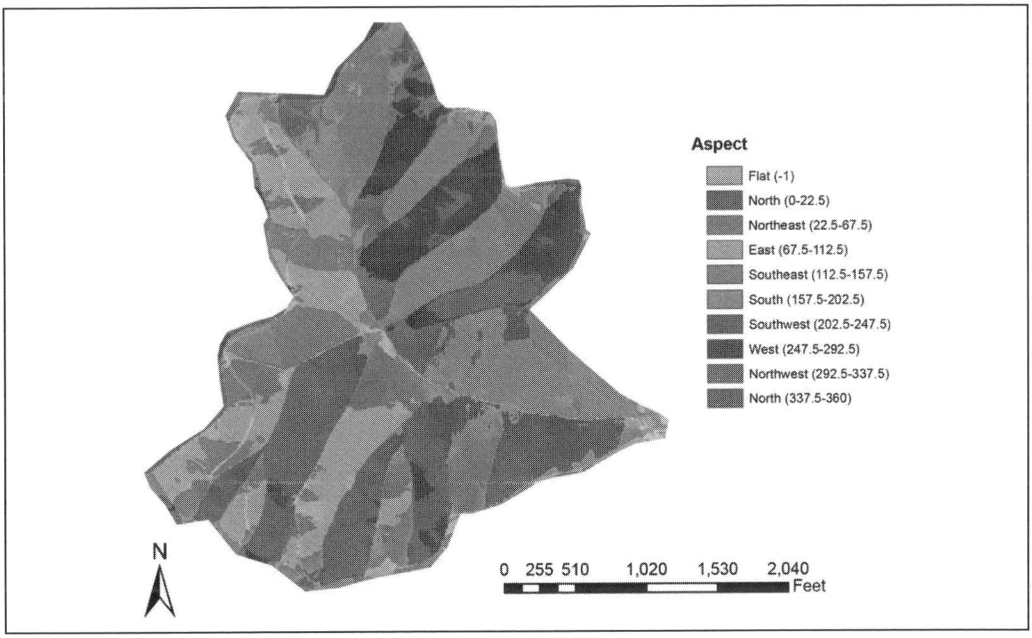

Figure 5. Aspect map of Whetstone Branch

Table 4. Comparison of default design parameters to measurements taken from experimental watershed

	Default	Measured
Max distance from ridgeline to channel's head, m (ft)	24.4 (80.00)	162.8 (534.12)
Slope at mouth of main valley channel, %	2	8
Drainage density, km^{-1} (ft/ac)	7.5 (100)	5.3 (70)
Upstream slope, %	12	28*
Downstream slope, %	2	8
Sinuosity (>–4%)	1.15	1.03*
Sinuosity (<–4%)	1.48	NA†

* Represents an average value.
† NA = not applicable.

These reference landform design values are critical to design a system with low erosion rates. Systems designed with a lower than optimum drainage density will likely promote sediment deposition, and systems designed with a greater than optimum drainage density will likely promote erosion, leading to instability. Difference between default and measured parameters noted in this study were somewhat expected. The default design parameters incorporated into the design software were based on semi-arid regions. The geomorphic characteristics in southern West Virginia are a result of the steep slopes, consolidated soil, vegetation, and climatic influences of the region. These characteristics need to be considered for future designs.

CONCLUSIONS

Because the geomorphic landform approach utilizes a reference landform design method, region specific design parameters are crucial to inform design. This research quantified the complex, steep terrain in southern West Virginia. Results from this study indicate differences between design parameters specific to southern West Virginia and default design parameters utilized in current design tools (i.e., Carlson Natural Regrade with GeoFluv™). The ridge to head of channel distance, slopes, and drainage density vary from the default values. These region-specific characteristics need to be considered for future designs.

Future work will quantify geomorphic characteristics in additional watersheds in the mining region of southern West Virginia. In addition, surveys of reclaimed sites of varying ages will also provide insight into generating successful designs. Designs will then be created using region specific design values and the differences in each design will be quantified. Ultimately, the research will provide the coal industry and regulators with data to advance watershed reclamation in Central Appalachia.

ACKNOWLEDGMENTS

The authors would like to thank Dustin Deaton for support in the completion of this work. The project described in this publication was supported by the Grant/Cooperative Agreement Number G11AP20114 from the United States Geological Survey. Its contents are solely the responsibility of the authors and do not necessarily represent the official views of the USGS.

REFERENCES

Bonta, J.V., Amerman, C.R., Harlukowicz, T.J., and Dick W.A. 1997. Impact of coal surface mining on three Ohio watersheds-Surface water hydrology. *Journal of the American Water Resources Association.* 33(4): 907–917.

Bugosh N. 2009. A summary of some land surface and water quality monitoring results for constructed GeoFluv landforms.In *Revitalizing the Environment: Proven Solutions and Innovative Approaches*. Billings, MT: ASMR.

FISRWG. 1998. *Stream Corridor Restoration: Principles, Processes, and Practices.* Federal Interagency Stream Restoration Working Group.

Hancock, G.R., and Evans, K.G. 2006. Channel head location and characteristics using digital elevation models. *Earth Surface Processes and Landforms,* 31(1): 809–824.

Harrelson, C.C., Rawlins, C.L., and Potyondy, J.P. 1994. *Stream channel reference sites: an illustrated guide to field technique.*Rocky Mountain Research Station: United States Department of Agriculture.

Hartman, K.J., Kaller, M.D., Howell, J.W., and Sweka, J.A. 2005. How much do valley fills influence headwater streams? *Hydrobiologia.* 532(1): 91–102.

Kaufmann, P.R., and Robison, E.G. 1998. Physical habitat characterization. In *Surface waters: Field operations and methods for measuring the ecological condition of wadeable streams.* Washington, DC: U.S. Environmental Protection Agency.

Martin-Duque, J.F., Sanz, M.A., Bodoque, J.M., Lucia, A., and Martin-Moreno, C. 2009. Restoring earth surface processes through landform design. A 13-year monitoring of a geomorphic reclamation model for quarries on slopes. *Earth Surface Processes and Landforms.* 35(1): 531–548.

Martin-Moreno, C., Martin-Duque, J.F., Nicolau, J.M., Sanchez, L., Ruiz, R., Sanz, M., Lucia, A., and Zapico, I. 2008. A geomorphic approach for the ecological restoration of kaolin mines at the Upper Tagus Natural Park (Spain). *6th European Conference on Ecological Restoration,* Ghent, Belgium.

Measles, D., and Bugosh, N. 2007. Making and building a fluvial geomorphic reclamation design at an active dragline mine using the GeoFluv™ design method. In *30 Years of SMCRA and Beyond.* Gillette, WY: ASMR.

Messinger, T. 2003. Comparison of storm response of streams in small, unmined and valley-filled watersheds, 1999–2001, Ballard Fork, West Virginia. *Rep. No. US Geological Survey Water-Resources Investigations Report, 02-4303,* USGS.

Messinger, T., and Paybins, K.S. 2003. Relations between precipitation and daily and monthly mean flows in gaged, unmined and valley-filled watersheds, Ballard Fork, West Virginia, 1999–2001. *Rep. No. US Geological Survey Water-Resources Investigations Report, 03-4113,* USGS.

Michael, P.R., Superfesky, M.J., and Uranowski, L.J. 2010. Challenges to applying the geomorphic and stream reclamation methodologies to mountaintop mining and excess spoil fill construction in steep-slope topography (e.g., Central Appalachia). In *Pittsburgh, PA Bridging Reclamation, Science and the Community.* Lexington, KY: ASMR.

Montgomery, D.R., and Buffington, J.M. 1993. Channel Classification, Prediction of Channel Response, and Assessment of Channel Condition. *Rep. No. TFW-SH10-93-002.* Olympia, WA: Washington Dept. of Natural Resources.

Montgomery, D.R., and Buffington J.M. 1997. Channel-reach Morphology in Mountain Drainage Basins. *Geological Society of America Bulletin.* 109(5): 596–611.

Negley, T.L., and Eshleman, K.N. 2006. Comparison of stormflow responses of surfaced-mined and forested watersheds in the Appalachian Mountains, USA. *Hydrol. Processes.* 20(1): 3467–3483.

NRCS-NWCC. NRCS National Water and Climate Center. September 13, 2012. <http://www.wcc.nrcs.usda.gov/climate/clim-data.html>.

Petty, J.T., Fulton, J.B., Strager, M.P., Merovich, G.T., Stiles, J.M., and Ziemkiewicz, P.F. 2010. Landscapeindicators and thresholds of stream ecological impairment in an intensively mined Appalachian watershed. *J. N. Am.. Benthol. Soc.* 29(4):1292–1309.

Pond, G.J., Passmore, M.E., Borsuk, F.A., Reynolds, L., and Rose, C.J. 2008. Downstream effects of mountaintop coal mining: comparing biological conditions using family and genus-level macroinvertebrate bioassessment tools. *J. N. Am. Benthol. Soc.* 27(3): 717–737.

Robson, M., Spotts, R., Wade, R., and Erickson, W. 2009. A case history: Limestone quarry reclamation using fluvial geomorphic design techniques. In *Revitalizing the Environment: Proven Solutions and Innovative Approaches.* Billings, MT: ASMR.

Rosgen, D.L. 1996. *Applied River Morphology.* Pagosa Springs, CO: Wildland Hydrology.

Rosgen, D.L. 1994. A classification of natural rivers. *Catena,* 22(3): 169–199.

Schumm, S.A., and Mosley, M.P. 1987. *Experimental Fluvial Geomorphology.* New York: John Wiley & Sons.

Sears, A., Russell, H., Hopkinson, L., and Quaranta, J.D. 2012. Incorporating geomorphic landform approaches for valley fill designs in West Virginia. *27th International Conference on Solid Waste Technology and Management,* Philadelphia, PA.

Toy, T.J., and Chuse, W.R. 2005.Topographic reconstruction: a geomorphic approach. *Ecological Engineering.* 24(1): 29–35.

USEPA. 2011. *The Effects of Mountaintop Mines and Valley Fills on Aquatic Ecosystems of the Central*

Appalachian Coalfields. 2011 Final Report. EPA/600/R-09/138F. Washington, DC: United States Environmental Protection Agency.

VANR. 2004. *Vermont stream geomorphic assessment protocol handbooks.* Waterbury, VT: Vermont Agency of Natural Resources.

Wiley, J.B., Evaldi, R.D., Eychaner, J.H., and Chambers, D.B. 2001. Reconnaissance of stream geomorphology, low streamflow, and stream temperature in the moutaintop coal-mining region, southern West Virginia, 1999–2000. *Rep. No. Water-Resources Investigations Report 01-4092.* USGS.

Wolman, M.G. 1954. A method of sampling coarse river-bed material. *Transactions, American Geophysical Union.* 35(6): 951–956.

WVDEP. 1993. *Technical handbook: Standards and specifications for erosion and sediment control excess spoil disposal haulages.* West Virginia Department of Environmental Protection, Division of Mining and Reclamation.

Ziemkiewicz, P.F., O'Neal, M., and Lovett, R.J. 2011. Selenium leaching kinetics and in situ control. *Mine Water Environment.* 30(1): 141–150.

Slope Stability Analysis of Geomorphic Landform Profiles Versus Approximate Original Contour Applied to Valley Fill Designs

Harold Russell
Civil and Environmental Engineering, West Virginia University, Morgantown, West Virginia

John D. Quaranta
Civil and Environmental Engineering, West Virginia University, Morgantown, West Virginia

ABSTRACT

Valley fills in steep Appalachian terrain offers challenges for reclamation efforts. This research was intended to analyze and compare a valley fill Approximate Original Contour (AOC) design with a geomorphic landform design alternative. Four separate slope stability numerical modeling assessments were performed in the modules available in GeoStudio™ for comparison. Each of the two profiles was analyzed for an initially unsaturated and an initially saturated durable rock underdrain condition. These analysis scenarios will be referred to as "cumulative analysis" since they were a product of results from three modules of the modeling software. Through this rigorous analysis, it was found that slope angles, phreatic elevations, and areas of increased pore pressures greatly affected the factors of safety that were yielded. The cumulative analysis for the geomorphic valley fill alternative design produced the highest factors of safety by a difference of 2.24 for the most extreme case. All geomorphic cases produced factors of safety over 2.0. The most likely reason for these high factors of safety is that the geomorphic design typically has shallow slopes and drains well. Geomorphic landform design can be utilized to reduce infiltration volumes by shortening runoff travel distances, increasing runoff water removal from a design site. A completed design should retain less water than the modeled results show because of vegetative cover and quick surface runoff. The highest factor of safety produced by the valley fill design was 1.67. The failure locations were sought out to find the lowest factors of safety for the earthen structures.

INTRODUCTION

Concerns of detrimental environmental impacts originating from mountaintop mining and valley fill construction are of constant debate, resulting in numerous lawsuits (e.g., Hasselman 2002, Davis and Duffy 2009) and scientific studies throughout Central Appalachia (e.g., Hartman et al. 2005; Pond et al. 2008; Ferrari et al. 2009). State and Federal regulations have been promulgated to control environmental impacts associated with mountaintop mining and valley fill construction through the Surface Mining Control and Reclamation Act (SMCRA) and the Clean Water Act (CWA) (WVDEP 1999).

Empirical factor of safety analysis methods such as Bishop's Method, General Limit Equilibrium, and Simplified Bishop's Method are conventionally used methods of analysis for slope design certification. Computer modeling software is utilized to emulate realistic slope scenarios and analyze them. Designers utilizing computer modeling software have the ability to change various input parameters affecting slope stability quickly to understand the implications and severity of the effects of each parameter with regards to specific slope geometry. These tools allow engineers to assess risk more adequately by analyzing a greater magnitude of variability in input parameters. The

Table 1. Material input values

Input Parameters	Blasted Fill	Blocky Core Drain	Foundation Rock
γ (kN/m^3)	18.39	19.64	18.39
c (kPa)	0	9.58	0
ϕ	27.7°	30°	27.7°
Elastic Modulus, kPa	29868	29868	1000000
Unit Weight, γ (kN/m^3)	18.39	18.39	26.48
Poisson's Ratio	0.34	0.34	0.38

research presented in this paper addresses the likelihood of a valley fill failure in steep sloped Appalachian terrain compared to a geomorphic landform design alternative with a factor of safety approach acquired by general limit equilibrium computational techniques.

The objectives of this research was to sample and identify a surface mine spoil in southern West Virginia and to characterize its associated physical properties as well as its strength aspects for appropriate input parameters to key in into a slope stability software analysis tool in order to compare stability results of a valley fill and a geomorphic landform design with regard to a factor of safety computed by the general limit equilibrium method. Laboratory values were used along with the surface mine permit documentation of the proposed slope geometry to validate, assess, and calibrate the models. A deterministic analysis method was then used along with the general limit equilibrium computational method in order to perform a stability assessment with regard to a factor of safety. The factor of safety output data was addressed to develop insights into stability assessment of the generated landform model compared to the valley fill design.

The approach taken for this study involved considering the input parameter design limitations of an overburden material retrieved from a surface mine site in the Appalachian region. The shear strength of the sampled unweathered sandstone overburden material was determined via digital direct shear testing. The geotechnical values that were found via laboratory testing according to the American Society for Testing and Materials (ASTM) were input into GeoSlope™ software for numerical finite element modeling analysis.

The modules utilized for the analysis were SIGMA/W, SEEP/W, and SLOPE/W, to more realistically model the two landform designs and determine the stability of the structures using general limit equilibrium analysis as well as in situ stress calculations involving hydraulic parameters generated with a ten years of infiltration with a set hydraulic conductivity.

Input Parameters

This data input information has been accumulated from geotechnical laboratory testing on an unweathered sandstone overburden. This overburden material is the fill material in both slope models presented. The input parameters necessary for the modeling were the dry unit weight (γ_d), the cohesion (c), and the friction angle (ϕ). The values for "blocky core drain," "blasted fill," and "foundation rock" in Table 1 were determined via ASTM laboratory testing and WVDEP mine permit information. The in situ stress calculations were determined by the input parameters elastic modulus (E) and Poisson's ratio (ν). Poisson's ratio was determined from typical values for sand (Das 2006). The elastic modulus was identified to be the lowest value for sandstone to seek out the most critical failure scenario. A summary of the values are shown in Table 1.

The valley fill model consisted of three materials. The identified materials in Table 1 are "Fill," which is the laboratory tested unweathered sandstone, "Blocky Core Drain," which refers to the 10 ft thick gravity segregated durable rock underdrain, and "Foundation Rock," which is simply the undisturbed bedrock strata. Excluding hydrologic considerations, the models included the materials at a two-dimensional homogeneous condition. The materials are shown in Figure 3.

The first material, named "Fill," was the top layer of the valley fill comprised of blasted overburden. It was modeled under a saturated/unsaturated condition, which required a function for hydraulic conductivity and an additional function for the material's water content. Hydraulic conductivity (m/s) was plotted versus pore water pressure (kPa) by inputting a saturated hydraulic conductivity and using the Van Genuchten estimation method within SEEP/W. A maximum suction was set at 0.01 kPa and a minimum set at 1,000 kPa with 20 data points to produce a function. Hydraulic conductivity was chosen from a range values found from previous work done with waste rock. These ranged from an unsaturated hydraulic conductivity of 1×10^{-5} m/s (Abdelghani 2009) to a saturated hydraulic conductivity of 1×10^{-7} m/s (Aubertin et al. 1996). These previously published values were chosen instead of the values determined from laboratory testing because they represented more realistic values. Laboratory testing calculates hydraulic conductivity based on optimum water content and compaction values, which are difficult to obtain in the field. Previously published values for hydraulic conductivity of waste rock provided a more practical value to be used in numerical modeling. This function was also compared to the function produced by Fredlund et al. (1998) and found to be quite similar. The water content function was produced in a similar method. A saturated volumetric water content was input and the same parameters as the conductivity function were used to create a function. Gravel was chosen as the material type for this function. Saturated volumetric water content was taken to be 44%, the value used by Fredlund et al. (1998).

The bottom layer of the valley fill, the "Foundation Rock" layer, was modeled in a saturated only condition with a very low hydraulic conductivity (2×10^{12} m/s) to simulate an impermeable rock layer. Between the "Fill" layer and "Foundation Rock" layer, a 10 ft "Blocky Core Drain" material was placed. This layer was modeled under a saturated only condition with a constant hydraulic conductivity value of 0.1 m/s. This hydraulic conductivity was taken from the Das (2010) average value for gravel.

VALLEY FILL LANDFORM DESIGN BY APPROXIMATE ORIGINAL CONTOUR

Geometric Input

The valley fill under inspection is located at the same area as the geomorphic design. The geometry of the slope was taken from the contour information shown in Figure 1. The fill is referred to as "Fill 2." Figure 1 shows the contour information attained from the WVDEP permit file to formulate the geometry for the valley fill profile.

The dimensions shown in Figure 1 are in U.S. customary units. This file was taken from WVDEP permit File#S500809. The models that were created used metric dimensions and metric laboratory soil property values for consistency within the scope of this research.

Numerical Modeling

The modeling process undertaken for the valley fill under inspection encompassed two scenarios. The first scenario run was for the durable rock underdrain to be initially dry, referred to as unsaturated. The second condition run was for the underdrain to be at an initially saturated state. The objective of the stability analysis performed for this valley fill profile geometry was intended to imitate more realistic in-field conditions. SEEP/W was first used to produce infiltration results over a ten year period. The results from SEEP/W illustrated areas of storage, areas of increased hydraulic head, and elevated pore pressures. The results from this analysis were utilized as parent inputs to a SIGMA/W analysis. The SIGMA/W analysis, via finite element modeling techniques, calculated in situ stresses within the fill structure. The in situ stresses produced had spatially variable areas of increased total hydraulic head, as well as spatially variable areas of elevated pore pressures. Finally, the SIGMA/W results were used as parent input parameters for the finite element SLOPE/W analysis. The SLOPE/W analysis was run to produce factors of safety at several critical areas. The critical areas that were selected for failure plane entry and exit points were chosen for general crest, toe,

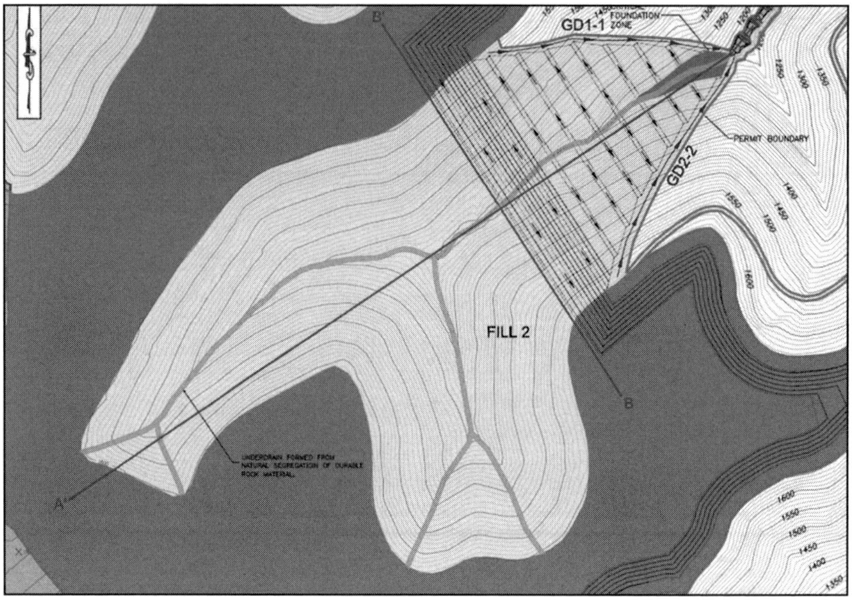

Figure 1. Valley fill plan view (WVDEP, 2007)

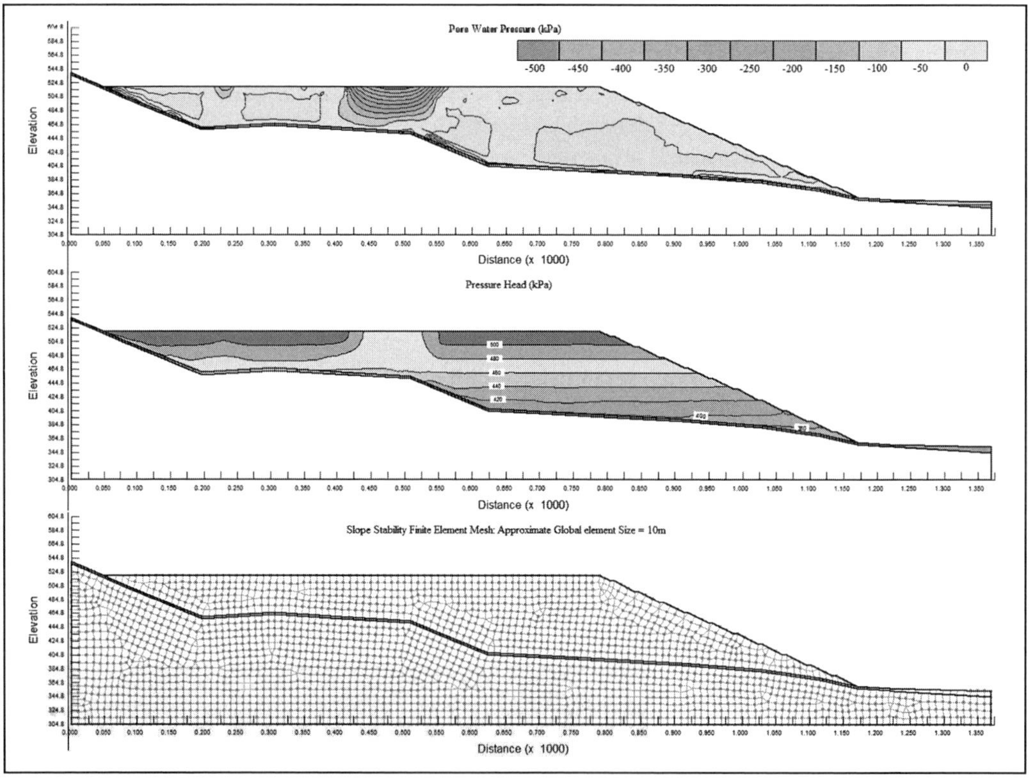

Figure 2. Valley fill diagrams of results from a cumulative analysis of SEEP/W, SIGMA/W, and SLOPE/W from GEO-SLOPE, 2007

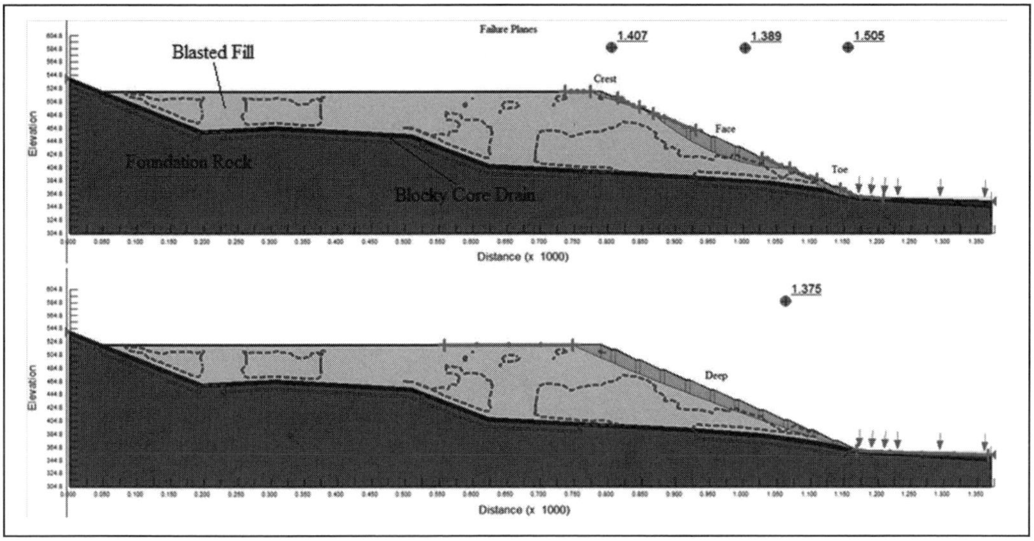

Figure 3. Failure entry and exit locations for saturated underdrain—Deterministic analysis results (GEO-SLOPE, 2007)

face, and deep foundation failure scenarios. Slope profile with cumulative analysis results shown in the following Figure 2, Figure 3, and Table 2.

GEOMORPHIC LANDFORM DESIGN ALTERNATIVE

Geomorphic design uses concepts derived from natural processes. Geomorphic landforms are designed with the intent of emulating naturally occurring landforms whose erosion potential is minimized. The minimized erosion potential is a function of drainage density. Drainage density is the ratio of the relationship between watershed area per channel length.

Geometric Input

The geomorphic landform design shown in Figure 4 was produced with GeoFluv® software. SLOPE/W was used for the slope stability analysis. The contours by which the profile dimensions were retrieved are shown in Figure 4 (left). The geometric dimensions were retrieved from a centerline cut at the identical dimensions as the valley fill centerline section A-A' in Figure 1 for direct comparison.

The objective of the stability analysis performed for this geomorphic design valley fill alternative profile geometry was intended to imitate more realistic in-field conditions than utilizing SLOPE/W alone. The modeling process undertaken for the geomorphic design valley fill alternative under inspection in this section was for two scenarios. The first scenario run was for the durable rock underdrain to be initially unsaturated. The second condition run was for the underdrain to initially be at a fully saturated state. The cumulative analysis procedure for the geomorphic design was identical to the analysis run on the valley fill design. SEEP/W was first used to produce infiltration results over a 10 year period. The results from SEEP/W illustrated areas of ground water storage, areas of increased hydraulic head, and elevated pore pressures. The results from this analysis were utilized as parent inputs to a SIGMA/W analysis. The SIGMA/W analysis, via finite element modeling techniques, calculated *insitu* stresses within the fill structure. The *insitu* stresses varied from previous piezometric head conditions by having spatially variable areas of increased total hydraulic head, as well as spatially variable areas of elevated pore pressures. Finally, the SIGMA/W results were used as parent input parameters for the finite element SLOPE/W analysis. The SLOPE/W analysis was

Table 2. Slope angles, hydraulic condition, and factors of safety

Failure Plane	Hydraulic Condition	$\beta° = \tan^{-1}(y/x)$	Critical Deterministic Factor of Safety
Cumulative Analysis: Valley Fill Design–Saturated			
Crest	Saturated	26.57	1.41
Toe	Saturated	26.57	1.51
Face	Saturated	26.57	1.39
Deep	Saturated	26.57	1.38
Cumulative Analysis: Valley Fill Design–Unsaturated			
Crest	Unsaturated	26.57	1.67
Toe	Unsaturated	26.57	1.25
Face	Unsaturated	26.57	1.37
Deep	Unsaturated	26.57	1.30
Cumulative Analysis: Geomorphic Valley Fill Alternative–Saturated			
Crest	Saturated	11.43	2.04
Toe	Saturated	17.63	2.14
Face	Saturated	15.47	2.42
Cumulative Analysis: Geomorphic Valley Fill Alternative–Unsaturated			
Crest	Unsaturated	11.43	2.31
Toe	Unsaturated	3.84	3.49
Face	Unsaturated	7.34	2.15

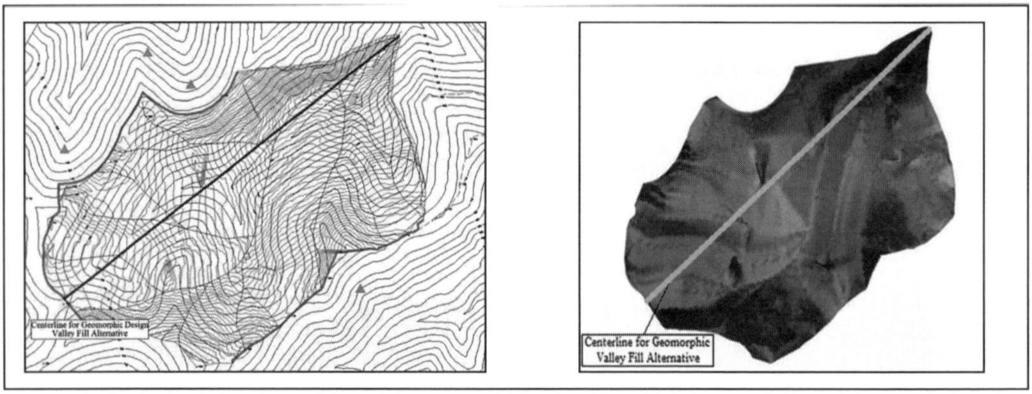

Figure 4. Geomorphic design contours superimposed on original ground contours (left); geomorphic design surface generated from a triangulated irregular network (TIN) (right)

run to produce factors of safety at several critical areas. The critical areas that were selected for failure plane entry and exit points were chosen by considering areas where pore pressures were highest. Other failure plane entry and exit locations were selected by keeping in mind general limit equilibrium concepts of driving forces and resisting forces (or moments). The failure plane entry and exit location selection also involved keeping in mind the sometimes significant effects that steep slope angles can have on the stability of an earthen structure. The slope profile with cumulative analysis results are shown in Figure 5 and Table 2.

Numerical Modeling

Figure 5 shows output from (top to bottom) SLOPE/W, SIGMA/W, SEEP/W, and finally the finite element mesh with a global element size of 10 m. SEEP/W analysis was performed first

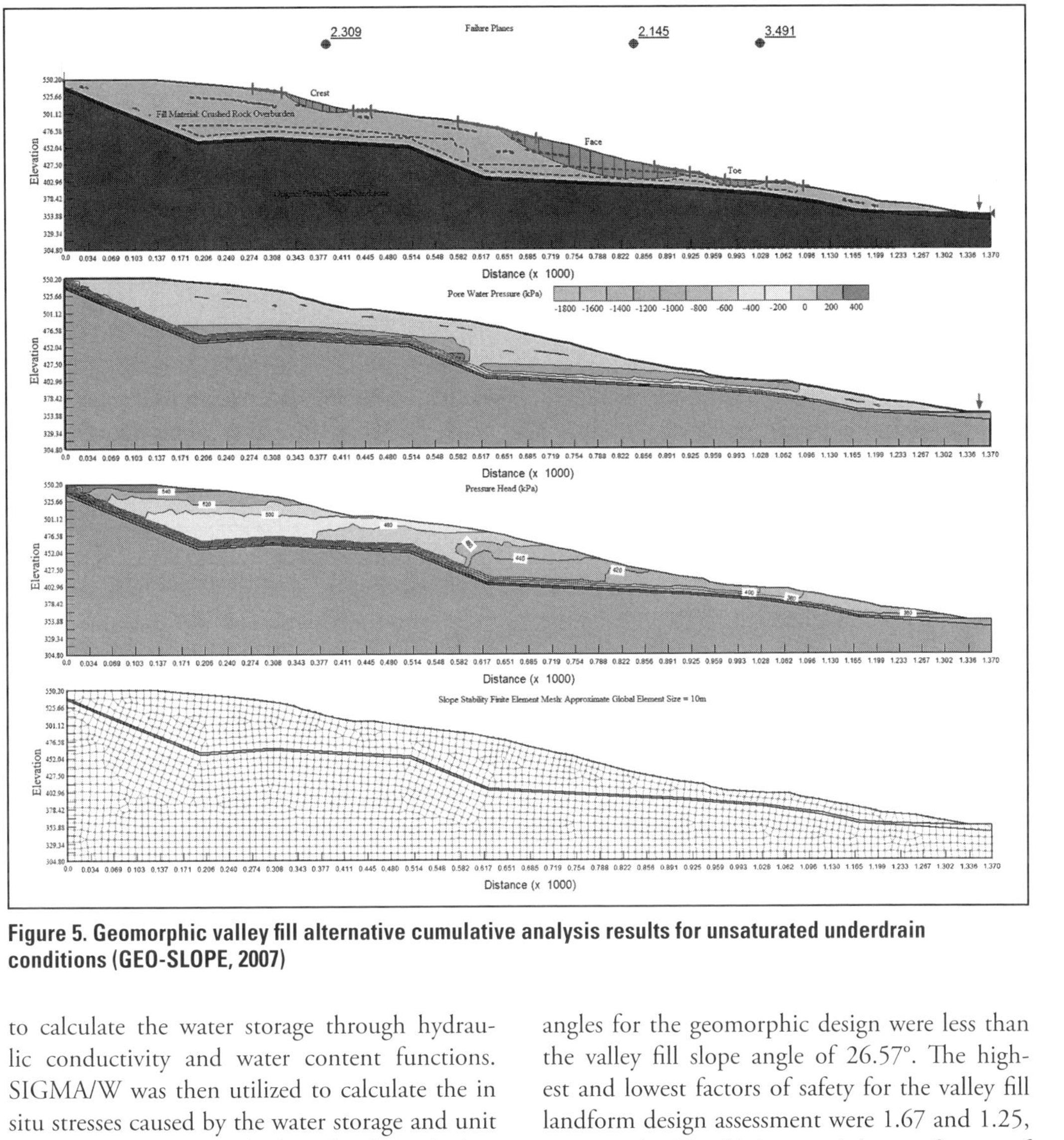

Figure 5. Geomorphic valley fill alternative cumulative analysis results for unsaturated underdrain conditions (GEO-SLOPE, 2007)

to calculate the water storage through hydraulic conductivity and water content functions. SIGMA/W was then utilized to calculate the in situ stresses caused by the water storage and unit weight. SLOPE/W was lastly utilized to calculate factors of safety at three critical locations.

RESULTS

The slope angle remained constant at a planar condition for the valley fill at 26.57°. The slope angle is defined by the equation $\beta° = \tan^{-1}(y/x)$, where y is the "rise" of the slope and x is the "run." The boundaries defined for the elevation coordinates identified for the slope angle calculations depended on the critical failure plane entry and exit points for each failure scenario. All slope angles for the geomorphic design were less than the valley fill slope angle of 26.57°. The highest and lowest factors of safety for the valley fill landform design assessment were 1.67 and 1.25, respectively. The highest and lowest factors of safety for the geomorphic landform design assessment were 3.49 and 2.04, respectively.

CONCLUSIONS

Four separate assessments were performed in the modules available in GeoStudio™ for comparison. Through this rigorous analysis, it was found that slope angles, phreatic elevations, and areas of increased pore pressures greatly affected the factors of safety that were yielded.

The height of the phreatic surface had a profound influence on the factors of safety. When an area was saturated, the factor of safety decreased. If the piezometric line was not elevated to the area of the selected failure plane, then the factor of safety changed very little. Additionally, steeper slope angles decreased factors of safety. Two initial saturation conditions were modeled for the cumulative analysis. The saturation conditions were applied in SEEP/W, and SIGMA/W computed in situ stresses to be input into SLOPE/W for a factor of safety computation. The initial saturation condition of the gravity segregated durable rock underdrain was significant. The initial saturation altered the volume of water retained within the structure, and ultimately altered the factors of safety that resulted. Table 2 shows the factors of safety for both the saturated and unsaturated initial hydraulic condition. The factors of safety vary from one hydraulic condition to the next, but do not necessarily increase or decrease accordingly. The result is an effect of the varying areas of increased pore pressure. The result of the SEEP/W analysis produced outputs that accumulated water storage within each fill in different areas, which resulted in varying factors of safety. Both cumulative analyses that were run for the geomorphic design and the valley fill proved that the initial condition could vary the factor of safety. The change in the factor of safety was not always in favor of either condition from a structural standpoint. The significance of the initial saturation condition was that the water storage areas within the fill changed and altered the factors of safety.

It is important to note that the fills may be built in lifts or by end dumping. Sometimes when contractors build slopes, they cut into the original ground, and may use compaction equipment to help integrate the components of the structure. These considerations would likely alter the factors of safety by changing the areas of water accumulation and transport within the earthen structure. Additionally, spatial variability of the material was not taken into account in these models. Each defined material region was considered isotropic within its boundary conditions. Actual field conditions of a slope structure would have a great deal of spatial variability both in compaction and in initial water content affecting soil strength and phreatic surface elevations.

The cumulative analysis for the geomorphic valley fill alternative design yielded the highest factors of safety by a difference of 2.24 in the most extreme case. All geomorphic cases produced factors of safety over 2.0. The most likely reason for these high factors of safety is that the geomorphic design has shallow slopes and drains well. Geomorphic landform design can be utilized to reduce infiltration volumes by shortening runoff travel distances, increasing runoff water removal from a design site. A completed design should retain less water than the modeled results show because of vegetative cover and quick surface runoff. The failure locations were sought out to find the lowest factors of safety for the structure.

The valley fill design was typical with its bench cuts and planar slopes. Regulations require that slope factors of safety must remain above 1.3 for back stacked material (WVDEP 1999). For fill faces, WVDEP suggests a static factor of safety of 1.5 (WVDEP 1993). The analysis performed showed that the design could withstand in situ loads and slope angle under all conditions analyzed. Elevated pore pressures tended to result at the toe of the slope, and there was an associated decrease in the factor of safety. The most critical scenario was a totally saturated toe at the unsaturated underdrain initial condition, which yielded a factor of safety of 1.25. The only assessed scenarios that produced a factor of safety above the regulatory requirement was the saturated underdrain toe scenario, with a factor of safety of 1.51, and for the unsaturated underdrain crest scenario, with a factor of safety of 1.67. Overall, the valley fill retained its structural integrity, although the highest computed factor of safety did not surpass the lowest geomorphic factor of safety. None of the factors of safety even under the most critical circumstances tested yielded factors of safety under 1.25.

The SEEP/W analysis yielded results that implied that the structure drained well for the valley fill design. There were small areas of water storage accumulation and elevated pore water pressures, but nothing which caused the factor of safety to drop below 1.0 for the cumulative analyses.

Geomorphic design decreases erosion potential and therefore decreases maintenance demands. The results shown imply that the proposed valley fill design would be adequate if it remained sufficiently drained. If particle transport can occur and alter toe pore pressures, or if the structure does not drain well, it is possible that slopes may not reach the regulatory factor of safety requirement of 1.5.

ACKNOWLEDGMENTS

The Wheeling Jesuit University–National Technology Transfer Center is the project sponsor for all work in the project and is under contract through the US Center for Disease Control (CDC). The authors wish to thank Mr. J. Davitt McAteer, Program Director, for support in completing this work.

REFERENCES

Abdelghani, F.B., Simon, R., Aubertin, M., Molson, J. & Therrien, R. (2009). "Use of the hydrogeosphere code to simulate water flow and contaminants transport through mining wastes disposed in a symmetric open pit within fractured rock." *Proc. Tailings and Mine Waste*. Taylor and Francis Group. Vail, CO. 601–611.

Aubertin, M., Bussierre, B. & Chapuis, R. (1996). "Hydraulic conductivity of homogenized tailings from hard rock mines." *Can. Geotech. J.*, 33: 470–482.

Das, Braja M. (2006). *Principles of Geotechnical Engineering* (7th ed.). Canada: Nelson Education Ltd.

Davis E.D., Duffy R.J. 2009. King coal vs. reclamation: Federal regulation of mountaintop removal mining in Appalachia. *Administration and Society*, 41(6), 674692.

Ferrari, J.R., Lookingbill, T.R., McCormick, B., Townsend, P.A. 2009. Surface mining and reclamation effects on flood response of watersheds in the central Appalachian Plateau region. *Water Resources Research*, 45, W04407.

Fredlund, M.D., Wilson, G.W. & Fredlund, D.G. (1998). "Estimation of hydraulic properties of an unsaturated soil using a knowledge-based system." *Proc., Second Inter. Conf. on Unsat. Soils.* Beijing, China.

GEO-SLOPE International Ltd. (2007). *Stability Modeling with SLOPE/W 2007 Version.* Calgary, Alberta, Canada.

H. Gercek. April 22, 2006. "Poisson's ratio values for rocks." Elsevier. *International Journal of Rock Mechanics & Mining Sciences* 44 (2007) 1–13.

Hartman, K.J., Kaller M.D., Howell, J.W., Sweka J.A. 2005. How much do valley fills influence headwater streams? *Hydrobiologia*, 532, 91–102.

Hasselman, M. 2002. Bragg v. W.Va. Coal Ass'n and the unfortunate limitation of citizen suits against the state in cooperative federalism regimes. *Ecol. Law Quart.* 29: 205–229.

National Oceanic and Atmospheric Administration. (2012). "Precipitation Frequency Data Server (PFDS)." *Hydrometeorological Design Studies Center*, http://hdsc.nws.noaa.gov (Feb. 10, 2012).

Pond, G.J., Passmore, M.E., Borsuk, F.A., Reynolds, L., Rose, C.J. 2008. Downstream effects of mountaintop coal mining: comparing biological conditions using family- and genus-level macroinvertebrate bioassessment tools. *J. N. Am.. Benthol. Soc.*, 27(3), 717–737.

West Virginia Department of Environmental Protection, Division of Mining and Reclamation. (1993). Technical Handbook.

West Virginia Department of Environmental Protection, Division of Mining and Reclamation. (1999). Permit Handbook: AOC and excess spoil disposal.

West Virginia Department of Environmental Protection (WVDEP), Permit ID S500809. Pine Creek No. 2 Surface Mine (July 12, 2007).

6.1
Selenium in the Environment

Subsurface Sequestration of Leached Waste-Rock Selenium in Reactive Ferrihydrite Beds 426

Selenium Concentrations and Solid-State Chemical Speciation in Black Shale Associated with a West Virginia Coal Seam 438

Selenium Treatment for Mine Water Discharge Compliance 443

Strategies for Remediation of Selenium-Laden Waters 450

Analysis of Low-Level Selenium in Coal Mine Discharges 451

Subsurface Sequestration of Leached Waste-Rock Selenium in Reactive Ferrihydrite Beds

P.F. Ziemkiewicz
West Virginia Water Research Institute, West Virginia University, Morgantown, West Virginia

ABSTRACT

Thirty lysimeters were installed at a large surface mine in southern West Virginia in Spring 2010 to test for transport rate of selenium in mine-run interburden as well as selenium removal by adsorption to layered ferric oxyhydroxide (ferrihydrite). Each 488 × 732 cm cell contained about 55 tons (122 to 183 cm thickness) of interburden, overlaying various thicknesses (0.6, 5.7, 22.9 and 45.7 cm, plus a zero-amendment control cell) of fine-grained ferrihydrite from a limestone-treated AMD wetland. The mass of ferrihydrite was 0 to 2.2% weight percent of the spoil. The control and each amendment were replicated six times, for estimation of uncertainty in reproducibility of results. Rainfall was collected on and infiltrated through these cells to create a leachate that drained to individual central collection tanks. Sampling was performed 46 times at approximate two week intervals from 2010 to 2012. Basal ferrihydrite layers in the three highest amendment categories successfully removed up to 76.9% selenium (in comparison to the unamended piles) from leachate by an adsorption mechanism. Results were somewhat variable for replicates. The higher amendment concentrations were demonstrably more effective, while the very thin ferrihydrite-treated piles showed no significant reduction from unamended piles. Selenium estimates were sufficiently precise to discriminate mean values among treatments. Work now focuses on continuing the time series to assess long-term persistence of Se fluxes, examine the ultimate capacity of the ferrihydrite beds, and to geochemically model the adsorption and transport processes.

INTRODUCTION

It has been widely observed that selenium (Se) fate and transport in coal overburden materials may be influenced by mining and reclamation operations (Dreher and Finkelman 1992). Predicting the fate and transport of Se mobilized from within coal mine spoil requires an understanding of the amount, form, and redox state of the element present; the rate and mechanism(s) of solubility; and the tendency for adsorption within the spoil mass post-dissolution (Naftz and Rice 1989). Se concentrations in southern West Virginia coal overburden lithologies range from 0.84 mg/kg in sandstone to 4.10 mg/kg in organic shale, with high Se concentrations covarying with high sulfur levels (Vesper et al. 2008). The proportion of mobile Se was estimated by Roy (2005) using extreme sequential extraction methods to accelerate and maximize its dissolution rate. She estimated that roughly 50% of total Se in overburden is non-extractable while another 25% is bound in non-mobile organic form. Most of the remaining 25% potentially mobile Se was inferred in this study to occur either within sulfide minerals or as adsorbed ions. Similar estimates were obtained for mobile Se (35% and 38% of total Se) using extraction by sonication of two similar organic shales from southern West Virginia (Pumure et al. 2010). All these sequential extraction estimates should be considered upper limits on field Se solubility, obtained under more intense weathering conditions than found in nature. Emerson et al. (2009) determined the more-reactive fine fraction (<2 mm) as 34% for southern West Virginia coal mine spoil that had been exposed to weathering for one to two years.

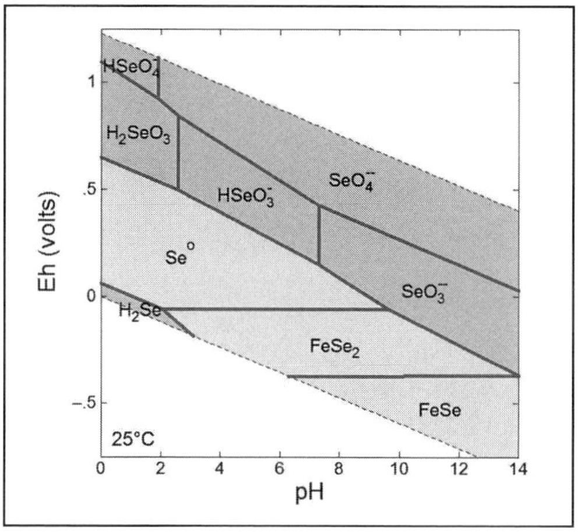

Figure 1. Eh-pH diagram for soluble Se species (blue) and insoluble solid phases containing Se (tan); calculations are for 10^{-6}M total Se and 10^{-3}M total Fe at 25°C

Diehl et al. (2005) determined Se in pyrite in southern Appalachian coals by a variety of microanalytical techniques. Whole-coal values for Se ranged from 25 to 247 mg/kg, while 670 mg/kg Se was found in pyrite alone, roughly 2 to 20% of the Se in the coal. This Se percentage for pyrite correlates well with the fraction found in sulfides by Roy (2005) (8% in the pyrite fraction of coal, 13% in shale).

Thus the most likely primary solid-phase occurrence of Se is in sulfide minerals, either in pyrite or ferroselite ($FeSe_2$), its Se-bearing analog. An Eh-pH diagram for Se species (Figure 1) shows areas of thermodynamic stability for elemental zero-valent Se (Se°) as well as more reduced iron sulfidic. The principal soluble species are selenate and selenite, obtained by oxidation of these source solid phases. The oxidation of selenite to selenate, the more mobile aqueous species, requires higher oxidation potential than for sulfide dissolution. Vesper et al. (2008) showed that the proportion of selenate to selenite only gradually increased downstream of source coal mines. Selenite is generally the most abundant Se species reported in various power plant ash materials (Huggins et al. 2007), although Hyun et al. (2006) also found significant selenate. Zhang and Sparks (1992) observed that while both aqueous phases are amenable to adsorption on crystalline iron oxides, selenite is less rapidly adsorbed, but more tightly held, than selenate.

Ferrihydrite or similar iron oxyhydroxides, a common neutralization product of treated metal-rich mine drainage, is thought to be a principal adsorbent for sequestering Se dissolved from coal waste rock. Ferrihydrite removed 95% of selenite at pH 4 and 80% at pH 9 (Merrill et al. 1985). Su and Suarez (2001) reached similar conclusions for amorphous iron oxide and its more crystalline phase goethite (α-FeOOH), although the amorphous phase showed less capacity for adsorption. Zhang and Sparks (1990) examined the reaction rates of selenate and selenite sorption on goethite and concluded that selenite formed an inner sphere complex, whereas selenate formed a less strongly-bound, but more rapidly formed, outer sphere complex. Ziemkiewicz et al. (2011) added ferrihydrite to Se-bearing mine spoil in batch reactors and found that roughly 70% of available Se was immobilized by adsorption compared to untreated controls.

While laboratory leaching results provide insight into weathering and leaching mechanisms, the more salient issue to the mining and regulatory sectors is the concentration of total Se that will discharge from operating mines over time. Se discharge dynamics under field conditions are poorly understood, largely due to the slow nature of field weathering processes, the low concentrations of Se involved, and the relatively short time for which Se has been monitored in mined areas. Se was not considered a pollutant in Appalachian coal mining until about 2003, after which several large surface mines were required to sample and report Se discharged via their regulated water outlets. Ziemkiewicz and Lovett (2012) developed a chronological sequence relating the age of various coal mine spoil units to selenium concentrations in outlets monitored over an eight year period. They reported that selenium concentrations ranged from 25 to 100 μg/L in young spoil to less than 5 μg/L in spoil that was more than about 25 years old, indicating a relatively rapid selenium

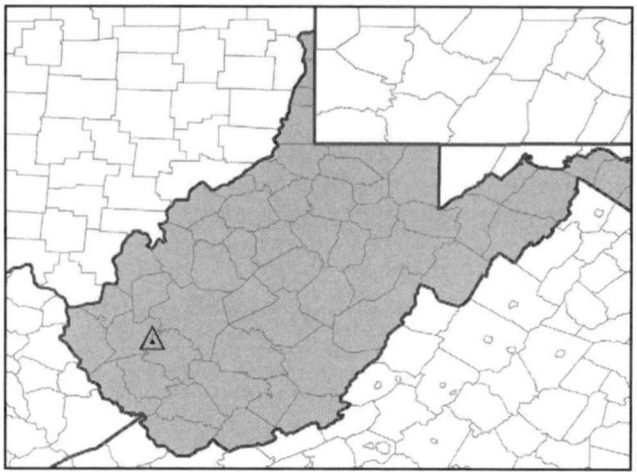

Figure 2. Location map of the study site, Lincoln County, WV

removal rate. Nonetheless, there has been insufficient opportunity to develop a regional sense for how long or at what concentrations Se will discharge from mined spoil materials via leachate. Due to close rock-water interaction, and the variation of selenium concentrations within spoil units, the regulated discharges integrate selenium rich sources like organic shales, diluted by water from selenium-poor rock units and surface runoff. So it is pivotal to find techniques to isolate selenium-rich rock units and then minimize the leaching of selenium from them.

Another complicating factor at field scales of leaching is the compositional heterogeneity of spoil materials and how this might influence its Se leaching characteristics. Large-scale surface coal mining in the dissected Appalachian plateau involves multiple coal seams, overburden and interburden sequences, as well as a mixture of sedimentary lithologies. However, spoil created and emplaced by mining methods is generally well-mixed within individual sequences. Roughly half the overburden/interburden at a modern mine is moved by cast-blasting and the other half by dragline or shovel/truck haulage. Excavated spoil is generally replaced by backfill along an advancing sloped face from 10 to 50 m long. Such handling methods result in a high degree of overburden blending. Weathering begins immediately after disaggregation of the rock by mining. Spoil age refers to the length of time since a rock unit was mined. While each lithology within a sequence will behave differently with regard to Se mobility, spoil units are highly mixed composites of the overburden lithologies. Segregation of the Se-rich rock units opens the possibility of treatment directed at specific lithologies to strategically limit Se release.

Finally, the well-known heterogeneous nature of fluid flow systems that develop in coal-mine spoil adds another aspect of variability. Double-porosity characteristics, flow through conduit-like networks, and linear-flow characteristics have been observed and documented in the Appalachians (Maher and Donovan 1997; Hawkins 2004). In field trials, the vagaries of different fluid flowpath networks may cause differences in leaching intensity to be observed between spoil piles. In summary, for reasons of inherent variability in spoil composition and in the resulting fluid flow systems through spoil piles, field-scale experiments must incorporate adequate replication to verify the significance of chemical fluxes observed at the discharge.

PURPOSE AND OBJECTIVES

The purpose of this study is to provide some initial estimates of selenium mobility in the presence of varying amounts of absorbent materials, as well as to test for the variability of spoil-pile chemistry and hydrology at field leaching scale, rather than laboratory scale. This particular investigation focuses strictly on Se leaching behavior over a period of three years of initial exposure of fresh waste-rock spoil to infiltration and weathering. The study area is a large (6,000 ha) surface mine in Lincoln County, WV (Figure 2). This is an area where surface mining has been practiced for about 30 years and is ongoing. Discharges from the mine enter the Little Coal and Mud River basins.

METHODS

Experimental Array and Design

Thirty pan-type lysimeters of large size (488 × 732 cm; Figure 3) were constructed on a reclaimed section of the mine in spring 2010. The individual lysimeter cells were constructed in a continuous

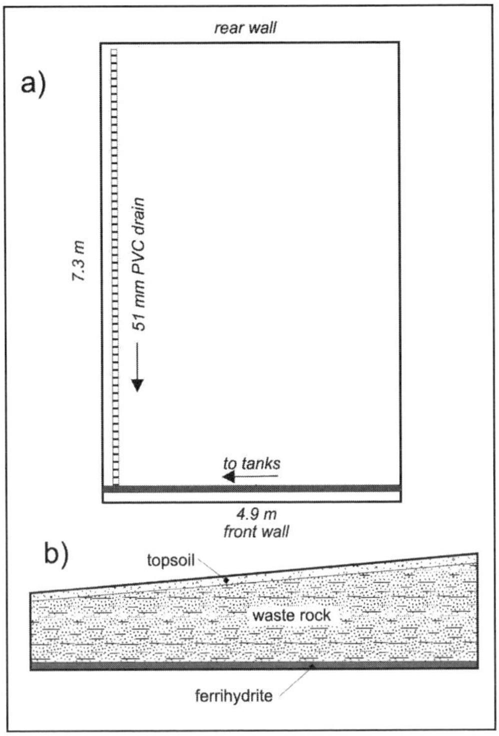

Figure 3. Plan view (a) and cross section view (b) of cell construction

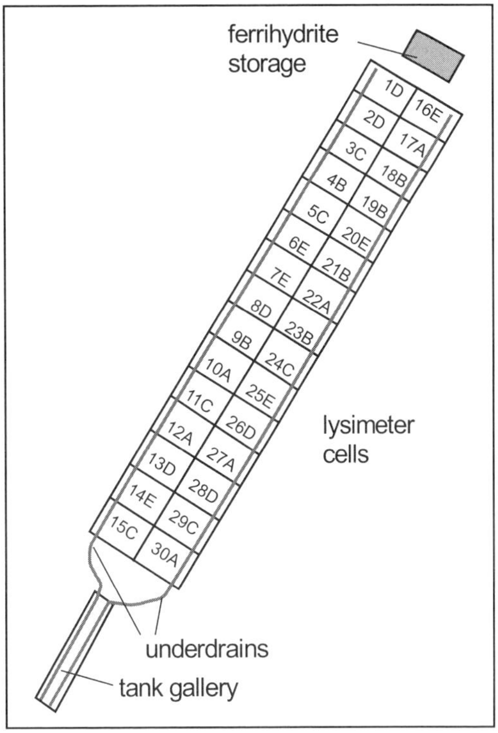

Figure 4. Layout of lysimeter array, with labels showing ID numbers of individual cells

array (Figure 4) of two rows of 15 cells each, separated by a 183 cm high plywood inner wall. The outer walls of both rows were 122 cm high; thus the side partitions between cells and the soil surface were sloped gently from inner to outer walls. The cells were lined along the base and sides with 4-mil polyethylene and filled with (from bottom to top) a thin ferrihydrite layer, shale waste rock, and a ca. 30 cm topsoil layer of weathered brown sandstone (Figure 3). The base of the liner was gently sloped roughly parallel to the soil surface, inducing basal gravity drainage from the inner to outer wall of each cell (Figure 3).

Treatments consisted of five ferrihydrite levels with six replications each. Treatments and replications were completely randomized. Each randomized experimental unit was assigned a number from 1 to 30, with a suffix to identify material composition. Additionally, the slope of the lysimeter array in each row (cells 1 to 15 and 16 to 30) was about 1:20 from north to south, allowing gravity drainage of collected lysimeter leachate in that direction. Each cell contained a two-inch perforated-PVC lateral underdrain in the ferrihydrite layer, located along the down-gradient side wall and sloped from inner to outer wall, where it was collected in a non-perforated 1.9 cm ABS pipe bundled in an array with pipes from all 15 cells in that row. These pipe bundles directed flow continuously under gravity to an array of thirty 1,250 L polyethylene tanks, each receiving water from a single lysimeter (Figure 4).

Precipitation onto the soil of each cell either infiltrated or ran off onto native soil around the array, allowing no runoff between or onto adjacent lysimeters. Perennial grass/legume cover on the surface of each cell induced some evapotranspiration. The difference between surface infiltration and evapotranspiration constitutes recharge into each cell, which seeps deeply into the overburden soils and collects on the bottom liner, flowing into the diversion pipes and each cell's respective sample tank. The sample tank for each cell thus contained a time-integrated volume of leachate fluids collected during multiple recharge events between sampling dates.

The average depth of all material in each cell was from 122 183 cm (Figure 3), with an average volume of 21.7 m^3. The total mass was approximately 55 tons for an average field-moisture bulk density of 2.53 g/cm^3.

Material Composition

Each cell contained approximately the same volume of waste rock, measured by counting excavator buckets during construction, but various amounts of ferrihydrite, measured as a uniform thickness at the base of the spoil, immediately over the liner. The waste rock was sampled during construction in three 19 L pails and analyzed chemically. The waste rock was freshly-mined organic shale that forms the interval between the Stockton and Coalburg coal seams in the Pennsylvanian Kanawha Formation. The rock was homogeneous in appearance, stockpiled in a single day of mining for use in this project. The ferrihydrite was obtained from a large wetland near Pittsburgh, PA, in which limestone-treated AMD sludge had settled over a period of 30+ years, producing a reaction product with negligible aluminum and of high uniformity. The layer thickness and weight/weight ratio of ferrihydrite to spoil for each treatment (A-E) are as follows.

A—no (0% w/w) ferrihydrite (6 lysimeters) ferrihydrite)
B—0.6 cm (0.2%) ferrihydrite (6 lysimeters)
C—5.7 cm (1.5%) ferrihydrite (6 lysimeters)
D—22.9 cm (6%) ferrihydrite (6 lysimeters)
E—45.7 cm (12%) ferrihydrite (6 lysimeters)

Therefore, the array includes 6 identical replicates of each treatment, to allow for assessment of sampling error and uncertainty in material variation/hydrologic conditions.

Water Sampling and Chemical Analysis

Sampling was performed biweekly beginning on May 14, 2010, through June 1, 2012. On each date, the fluid volumes in the collection tanks were measured, samples collected, and each tank drained. Field pH, temperature, and conductivity were measured and the filtered samples were immediately taken to a state-certified analytical laboratory (REI Consultants, Beaver, WV) and analyzed for total Se, SO_4, alkalinity to pH 4.5 endpoint, acidity, Fe, Al, Ca, Mg, Mn, Na, and Cl. The water in the tanks and collection drains was found to be frozen on December 10, 2010 (week 30). Sampling resumed on April 8, 2011 (week 48). There were several sampling events in fall 2010 when some tanks were dry due to lack of infiltration and no samples could be collected. Also, when tanks and lines thawed in April to May 2011, there were some tanks/lines which took longer to re-establish flow than others. Therefore, the number of replicates for each treatment was usually, but not always, six. During exceptionally dry periods, most if not all tanks were dry. Thus on some dates, as few as one sample per replicate set was collected, affecting precision estimates.

RESULTS

Time Series of Results

Figure 5 shows time series of selenium for the five treatments. Each plot shows a set of six different replicate time series for total selenium concentration in one treatment class, with each replicate obtained from a different lysimeter. The ferrihydrite addition increases from top (zero ferrihydrite) to bottom (maximum ferrihydrite, 45.7 cm). This plot shows general coherence in concentration between individual replicate sets, although the confidence intervals indicate variance within treatments as well. In particular, the first and second replicates in all treatments but especially C, D, and E—the circles and triangle symbols in these plots—were notably and in some cases markedly higher than the other replicates in their sets. It is conceivable that this could represent systematic natural variability in the refuse; all these cells are in the same portion of the array, from cells 1–12 (Figure 4) and were the first installed. It may also be that there was some difference in the installation of these lysimeters, either in the composition of the refuse used for

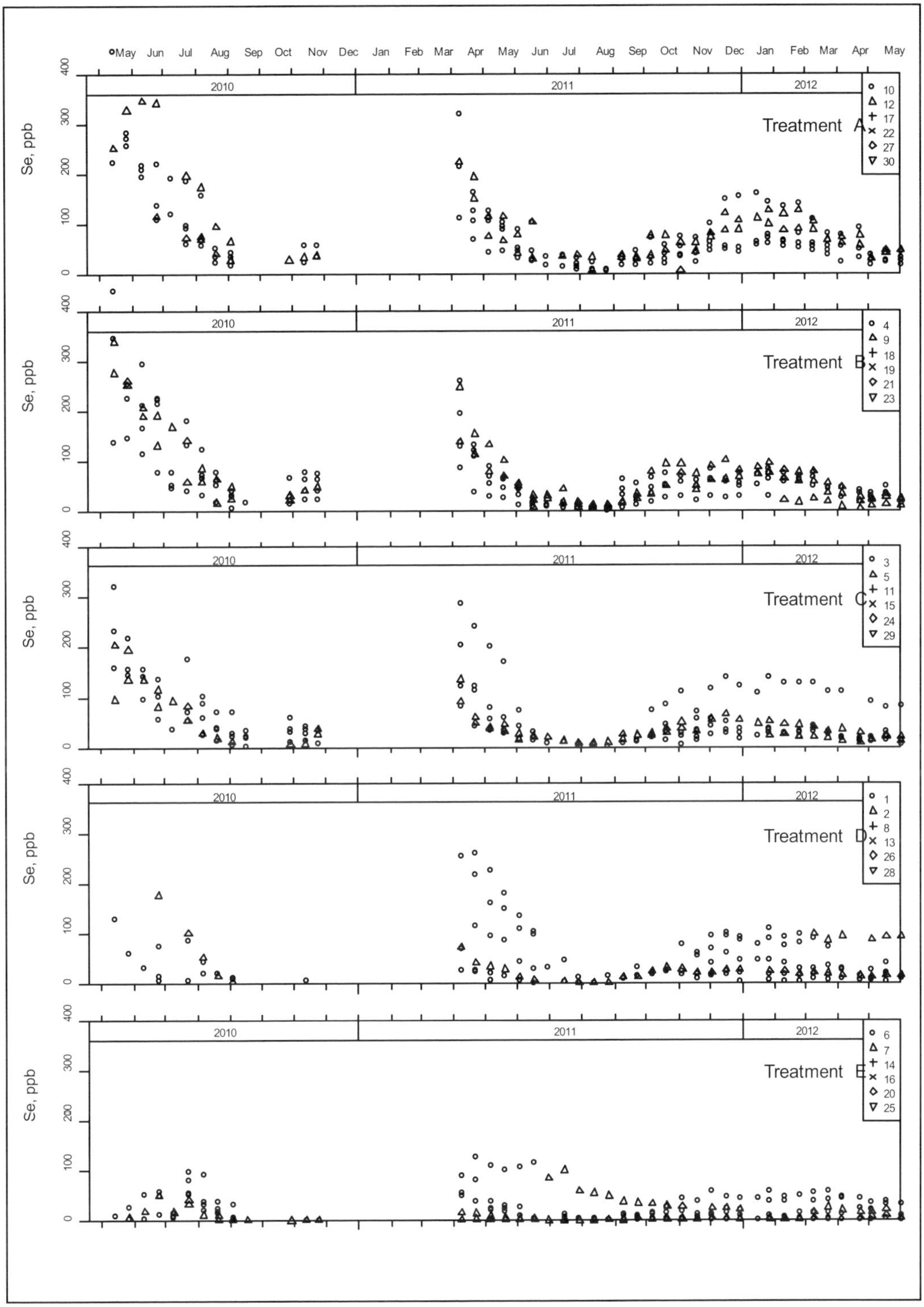

Figure 5. Raw time series of Se concentrations arranged by treatment. Each plot shows replicates for different treatments.

those cells, in the compaction method, or in the technique used for leveling the PVC drain line within the ferrihydrite layer. Excepting this variability, there is generally minor variance between replicates within treatments, and variance within treatments seems greatest at times of highest Se concentration, which occurred in spring of both 2010 and 2011. Nonetheless, the estimates were reasonably precise. The ratio of the standard deviation of the mean/mean was: 0.06, 0.07, 0.08, 0.12, and 0.08 for treatments A, B, C, D and E respectively.

There was a gap in sampling between late November 2010 and early April 2011, caused by freezing of line drainage between the lysimeters and the tanks. During winter 2011–12, no freeze-up occurred. As a result, in addition to the winter 2010–11 data gap, several sampling events in fall 2010 and spring 2011 had fewer than six replicates per treatment, as some drain lines thawed before others. Also, there were some periods in 2010 when some lysimeters produced no water during dry conditions.

Statistical Analysis of Mean Concentrations

Figure 6 shows the mean concentrations of the up to 6 replicates within each treatment (X symbols) and the 95% confidence limits on these concentrations calculated from sample standard errors and Student's t values (based on n–1 degrees of freedom) for up to six replicates. The fitted line in all plots represents the means of Treatment A replicates (no ferrihydrite). Some error brackets are significantly larger than others, either due to more variability within replicates for that date (larger standard error) or due to fewer samples than six due to freezing or dry conditions (smaller sample size and larger t statistic). Treating the zero-ferrihydrite mean as an unbiased estimate, the confidence range for the means of other treatments exceeding the zero-treatment mean curve decreases from Treatment B (0.6 cm ferrihydrite) to Treatment E (45.7 cm ferrihydrite). For treatment E, there are extensive periods when the confidence range for Se lies completely below the untreated mean Se curve. It is clear from these results that the mean Se concentrations decline in comparison to the Treatment A (control) values as the mass of ferrihydrite increases.

Se Fluxes

Figure 7 shows time series of Se fluxes calculated as the product of average discharge volume between sampling dates and the Se concentration at the end of the sampling period. The units are in g total Se/day. There appears to be no significant difference in flux between Treatments A and B, and in fact Treatment B was higher in Se flux in the initial three months of leaching, becoming more similar in 2011 and 2012 sampling. There is substantial decrease in Se loads for Treatments C, D, and E compared to both the Treatment A and B values. To a higher degree than the mean concentration plots, the fluxes demonstrate that Se fluxes are significantly lower in the Treatment C, D, and E lysimeters than in the untreated or lightly-treated spoil piles.

Figure 8 compares cumulative Se fluxes for the five treatments, again averaged for all replicates. Treatments A and B are, again, the highest in Se loads (112.8 and 149.8 g Se, respectively), with Treatment B being the highest based on elevated concentrations in the first three months of sampling. In later sampling, the A and B piles show no significant difference in their mean fluxes. Treatment C (76.8 g Se) is slightly lower in flux than treatment A and much less than B. Treatments D and E are both substantially lower (45.4 and 26.1 g, respectively). The mean Se flux for Treatment E at the end of the sampling period in 2012 was 76.9% lower than that of the untreated lysimeters.

DISCUSSION

Sequestration Mechanism

These results show conclusively that there was a very large reduction in Se flux for the lysimeters employing higher masses/thicknesses of ferrihydrite compared to the untreated or lightly-treated cells. We attribute the reduction to adsorption of selenite and possibly selenate ions, although no redox estimates were obtained due to the batch method of sampling. There was not a proportional difference between the flux reduction of the

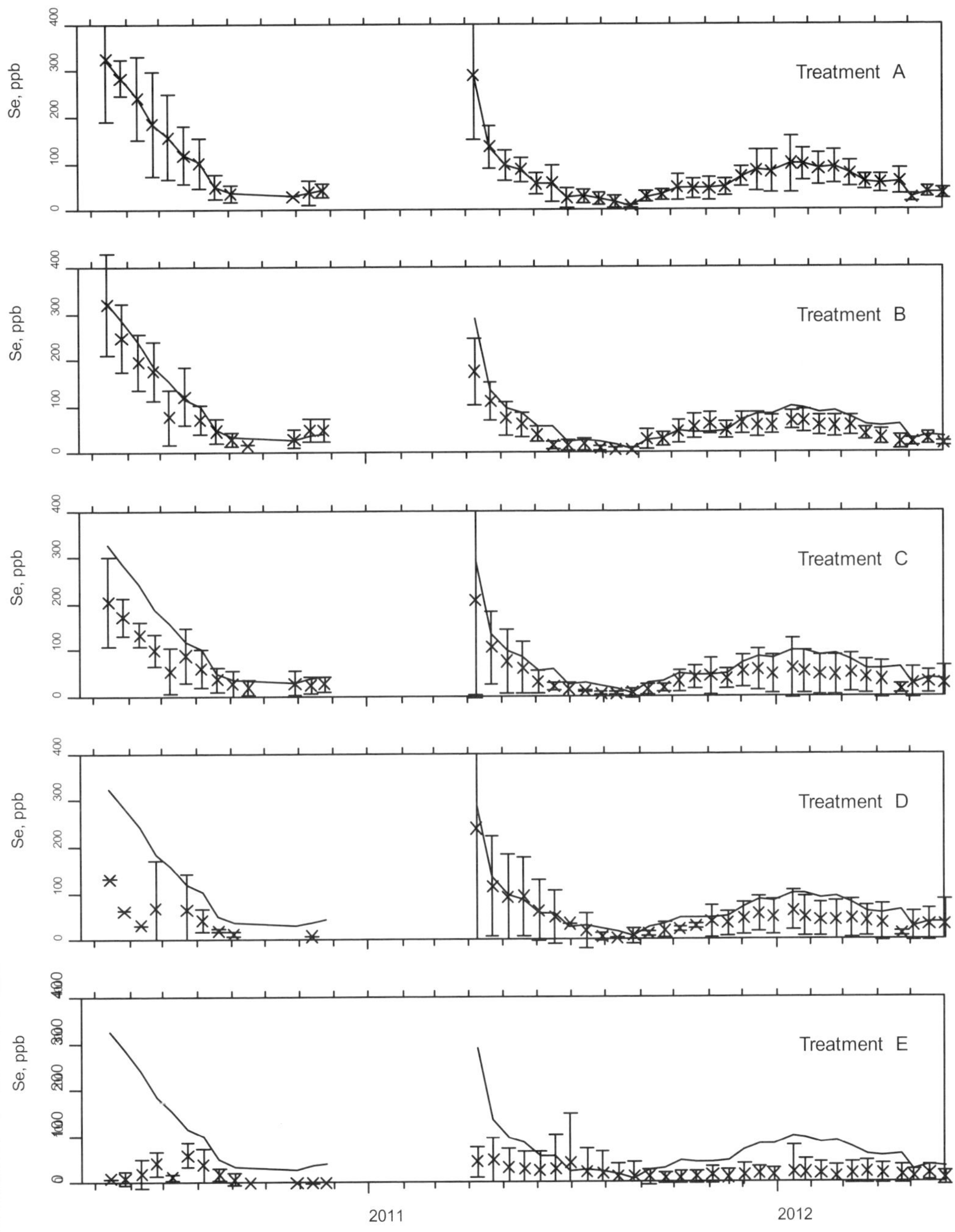

Figure 6. Time series of mean (X's) and confidence intervals (95%) for Se concentrations of all replicates within treatment classes (n=6 or less)

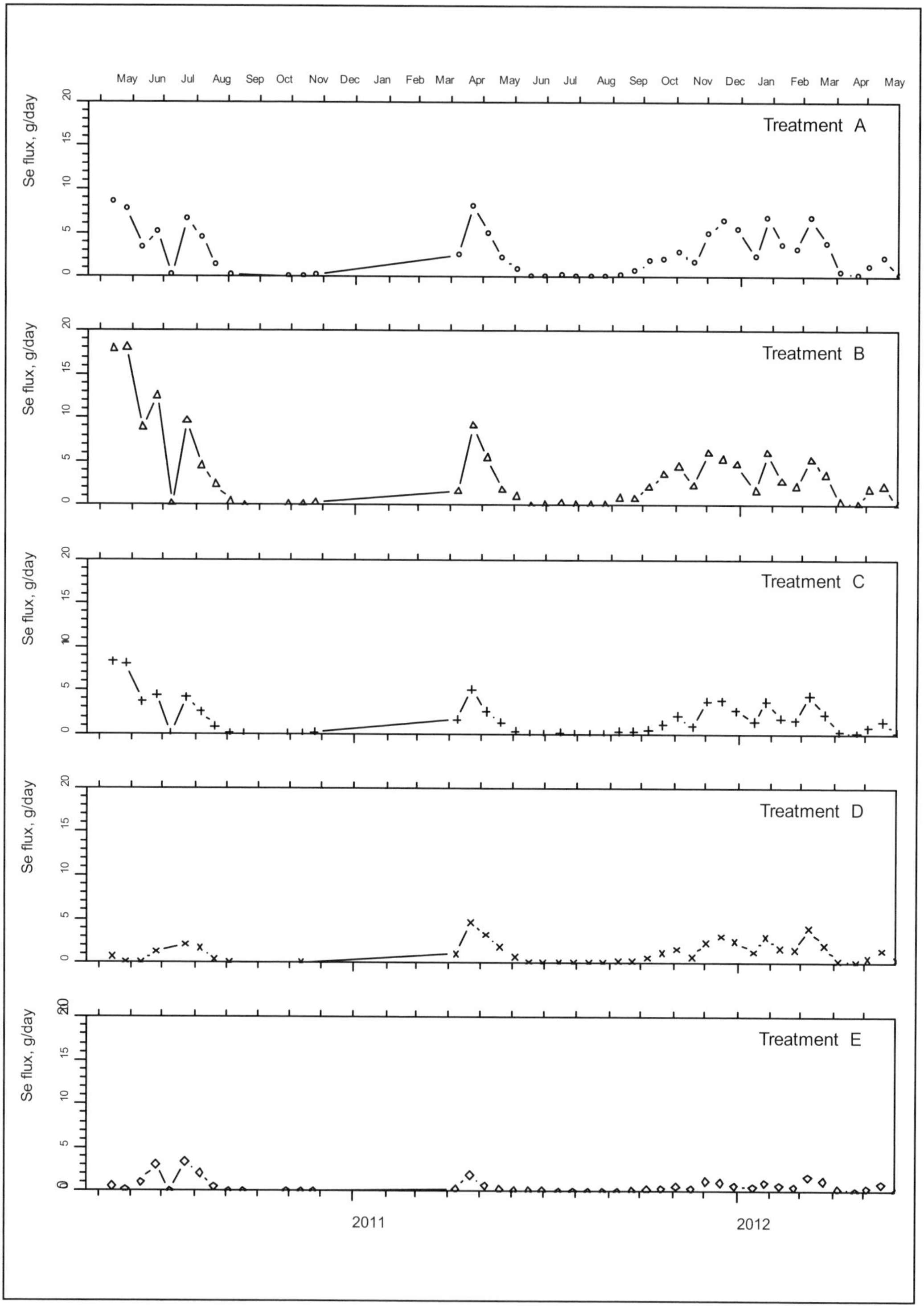

Figure 7. Time series of mean Se flux (g/day) per cell for each treatment class (n=6 or less)

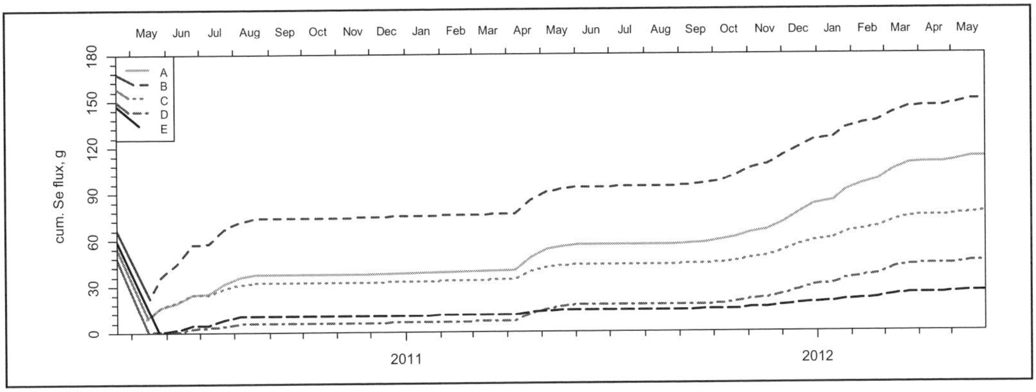

Figure 8. Cumulative Se fluxes for the five ferrihydrite treatments

22.9 cm (59.8% reduction) and 45.7 cm (76.9% reduction) ferrihydrite layers. The use of thin ferrihydrite layers appeared to have little adsorption effect, and in fact early fluxes from the Treatment B piles were much higher than the controls.

Replication and Variability

The variability of fluxes in all treatment replicates was not negligible, but within a limited range. There were minor outliers in Se concentration in both treatments C and D. However, at most sampling events for which a full number of replicates could be sampled, the standard errors were reasonably low. The largest sampling errors occurred when one or more of the replicates could not be sampled due to frozen plumbing or dry conditions. In general, the replication provides convincing evidence for reproducibility of both the Se-generating potential of untreated spoil and for its sequestration by adsorption using increasing masses of ferrihydrite.

The most surprising outcome was the elevated flux in the Treatment B piles compared to the control. Closer examination of these data indicates the leachate chemistry was similar between treatments, but more than twice the flow was collected in the B lysimeters than the A set. The reason for this difference is not understood, as the flows were much more similar for these two sets for the duration of the experiment. Sampling error of some type in flow measurement is a possibility.

Se Mobility Observations

In all treatments, strong seasonality was observed in both Se concentrations and in flow rates. The "stair-step" appearance of Figure 8 underscores the episodic nature of Se fluxes. It may be speculated that, during winter conditions, infiltration into the spoil is at minimum rate but air ingress is continuous and encourages oxidation even while discharge of water is minimal. The "slug" of elevated Se concentrations in spring may therefore represent flushing of over-winter reaction products. Strong Se flushes were observed both in early 2010 and 2011. In 2012, the flush occurred but at lower concentration than in the first two years. During drier seasons, both flow and concentration of Se drop to very low levels. The seasonality of Se discharge may offer water management options for addressing periods of high concentration.

Practical Implications

The success of this sequestration trial suggests that an inexpensive waste material (ferric iron-rich mine drainage treatment sludge) could be an effective spoil amendment that could reduce and control dissolved Se outbreaks at mine scale. Organic-rich shale tends to be among the most sulfur- and selenium-rich overburden lithologies. It is often found stratigraphically above coal seams, which, in combination with its diagnostic color, makes it amenable to selective handling during the mining process. Placing layers of ferrihydrite within and at the base of selectively-handled cells of organic shale may be a practical method for immobilizing a significant proportion of the Se flux for an entire surface mine. Scaling these results to much larger spoil masses will require testing and optimization of the technique. Nonetheless, a simple variant of this method yielded excellent sequestration in this investigation in cells for which the mass of ferrihydrite was sufficient. While many of the mean concentrations observed in Treatment E were higher than the 5 μg/L discharge limit (maximum 59.6 μg/L; mean 19.7 μg/L), these values are four to six times lower than in unamended spoil, and after July 2011, all values remained below 20 μg/L. While monitoring continues at this site, it may be hypothesized that many of the higher concentrations occurred in the first year or so of leaching. Given that such unamended spoil generates such extreme Se concentrations in leachate, the amenability of surface mine waters to Se management without active treatment may depend on such special handling techniques to control Se concentrations.

CONCLUSIONS

This investigation was an experiment at field scale to test the effectiveness of a single basal ferrihydrite layer at reducing Se mobility in leachate derived from organic-rich shale spoil. Thirty 488 × 732 cm lysimeters of identical construction, composition, and design, except for various thicknesses of ferrihydrite added, were monitored for 2.4 years on a biweekly basis. Key results include that:

- Time series of Se concentration and flux are highest during late winter and spring months, attributed both to spring flush of overwinter reaction products and to elevated recharge. Se fluxes are strongly seasonal, and winter/spring measurements of flux are critical to long-term accurate estimation.
- In a trial of four different layer thicknesses of ferrihydrite and one control (no ferrihydrite), the three thickest ferrihydrite amendments showed significant reduction in Se fluxes from the control, while the thinnest layer showed no reduction. The two thickest layers (22.9 and 45.7 cm) showed 59.4% and 76.9%, respectively, cumulative reduction in Se flux.
- All results were couched in terms of mean results for six replicates in each treatment class. The uncertainty due to variability in materials, construction, and hydrology of individual lysimeters was quantitatively assessed and considered within acceptable limits. The reported confidence intervals indicate that differences found in mean Se concentration and flux are considered to have a high probability of significance.

ACKNOWLEDGMENTS

This study was sponsored by the Appalachian Research Initiative for Environmental Science (ARIES). The views, opinions and recommendations expressed herein are solely those of the authors and do not imply any endorsement by ARIES employees, other ARIES-affiliated researchers or industrial members. Information about ARIES can be found at http://www.energy.vt.edu/ARIES.

This work was also supported by funding from the US Office of Surface Mining to the senior author. The cooperation of Patriot Coal Co. in constructing the lysimeters and providing site access was fundamental to the success of the project. Sampling assistance and laboratory chemical analyses were provided by REI Consultants, Beaver, WV.

The author thanks J.J. Donovan for his contribution to this paper.

REFERENCES

Diehl, S.F., Goldhaber, M.B., Koenig, A.E., Tuttle, M.L.W., and Ruppert, L.F. 2005. Concentration of arsenic, selenium, and other trace elements in pyrite in Appalachian coals of Alabama and Kentucky. Proceedings National Meeting of the Am. Soc. for Mining and Reclamation, p. 283–301.

Dreher, G.B., and Finkelman, R.B., 1992. Selenium mobilization in a surface coal mine, Powder River Basin, Wyoming, USA. *Environmental Geology and Water Science*, 19: 155–167.

Emerson, P., Skousen, J.G. and Ziemkiewicz, P.F. 2009. Survival and growth of hardwoods in brown versus gray sandstone on a surface mine in West Virginia. *J. Environ. Qual.* 38:1821–1829.

Hawkins, J.W., 2004. Predictability of Surface Mine Spoil Hydrologic Properties in the Appalachian Plateau. *Ground Water* 42: 119–125.

Huggins, F.E., Senior, C.L., Chu, P., Ladwig, K., and Huffman, G.P. 2007. Selenium and arsenic speciation in fly ash from full-scale coal-burning utility plants. *Environ. Sci. Technol.* 41:3284–3289.

Hyun, S., Burns, P.E., Murarka, I., and Lee, L.S. 2006. Selenium (IV) and (VI) sorption by solid surrounding fly ash management facilities. *Vadose Zone J.* 5:1110–1118.

Maher, T.P., and Donovan, J.J., 1997. Double-flow behavior observed in well tests of an extremely heterogeneous mine-spoil aquife. *Engineering Geology*, 5: 83–99 (17).

Merrill, D.T., P.M. Maroney, and Parker, D.S. 1985. Trace element removal by coprecipitation with amorphous iron oxyhydroxide: engineering evaluation. Electric Power Research Institute Coal Combustion Systems Division, Report EPRI, CS 4087, Palo Alto, CA.

Mullennex, R.H. 2005. Stratigraphic distribution of selenium in upper Kanawha-lower Allegheny formation strata at a location in southern West Virginia. The 23rd Annual International Pittsburgh Coal Conference, Pittsburgh, PA, Poster.

Naftz, D.L., and Rice, J., 1989. Geochemical processes controlling selenium in ground water after mining, Powder River Basin, Wyoming, USA. *Applied Geochemistry*, 4: 565–576.

Pumure, I., Renton, J.J., and Smart, R.B. 2010. Ultrasonic extraction of arsenic and selenium from rocks associated with mountaintop removal/valley fills coal mining: estimation of bioaccessible concentrations. *Chemosphere* 78:1295–1300.

Roy, M. 2005. A detailed sequential extraction study of selenium in coal and coal-associated strata from a coal mine in West Virginia. M.S. Thesis, Department of Geology and Geography, West Virginia University, Morgantown, WV, USA. 103 pp.

Su, C., and Suarez, D.L. 2001. Selenate and selenite sorption on iron oxides: an infrared and electrophoretic study. *Soil Sci. Soc. Am. J.* 64:101–111.

Vesper, D.J., Roy, M., and Rhoads, C.J. 2008. Selenium distribution and mode of occurrence in the Kanawha formation, southern West Virginia, U.S.A., *Int. J. Coal Geol.* 73:237–249.

Zhang, P., and Sparks, D.L. 1990. Kinetics of selenate and selenite adsorption/desorption at the goethite/water interface. *Environ. Sci. Technol.*, 24:1848–1856.

Ziemkiewicz, P.F., O'Neal, M. and Lovett, R.J. 2011. Selenium leaching kinetics and in-situ control. *Mine Water and the Environ* 30: 141–150.

Ziemkiewicz, P.F. and Lovett, R.J., 2012. Determination of natural attenuation of selenium at a coal-mining complex in West Virginia. Proceedings, 9th International Conference on Acid rock drainage, Ottawa, Ontario Canada. 20–26 May 2012.

Selenium Concentrations and Solid-State Chemical Speciation in Black Shale Associated with a West Virginia Coal Seam

Jason M. Unrine
Department of Plant and Soil Sciences, University of Kentucky, Lexington, Kentucky

Blanche Collin
Department of Plant and Soil Sciences, University of Kentucky, Lexington, Kentucky

Richard C. Warner
Department of Biosystems and Agricultural Engineering, University of Kentucky, Lexington, Kentucky

ABSTRACT

We determined concentrations of Se in strata from a West Virginia coal mine using inductively coupled plasma mass spectrometry (ICP-MS). Select samples with elevated Se concentrations were then analyzed using synchrotron-based scanning X-ray fluorescence spectroscopy (SXRF) and micro X-ray absorption near edge spectroscopy (μXANES). Our results indicate that elevated Se concentrations are contained within narrow bands of strata being found mostly in grey or black shale, fireclay and coal. Selenium was found either in elemental form or as PbSe in all of the samples that were analyzed. This suggests that waste handling techniques that slow the oxidation of these Se-containing minerals should reduce Se concentrations in mine effluents.

INTRODUCTION

Selenium has emerged as a contaminant of concern within the Appalachian region of the United States due to its potential for release from coal mining sites as well as its potential to elicit adverse population level effects in oviparous vertebrates including birds, fish and amphibians.[1] Weathering of organic matter-rich shales has been identified as a potential source of Se in coal mine effluents. These shales comprise a major component of coal mining waste. Relatively few studies have investigated the chemical speciation of Se in coal mining waste, particularly within the Appalachian region. A previous study utilized synchrotron X-ray techniques to investigate Se speciation in hard rock mine waste and found that Se was primarily associated with pyritic minerals.[2] In order to better understand the factors that influence weathering and release of Se from coal mine waste, we conducted an investigation to identify the Se-bearing strata in a core from a West Virginia coal mine and determine the chemical speciation of Se in those strata using X-ray absorption spectroscopy (XAS).

METHODS AND MATERIALS

Sample Collection and Processing

A 93 m core sample was obtained from a West Virginia coal mine. The core was divided into 93, 1 m sections and each section was crushed and sieved using a 200 μm mesh nylon sieve for ICP-MS analysis. We also sectioned 1–2 cm diameter pieces of the core that were retained after sieving using a diamond waffering blade to 1 mm for SXRF and μXANES analysis.

ICP-MS Analysis

We digested samples weighing approximately 0.1 g in sealed 55 mL Teflon bombs containing 10 mL of trace-metal grade nitric acid. The samples were heated using a microwave digestion system to 180 °C and held at that temperature for 10 min[3]. The samples were then diluted appropriately with ultra-pure (18.2 MΩ) water prior

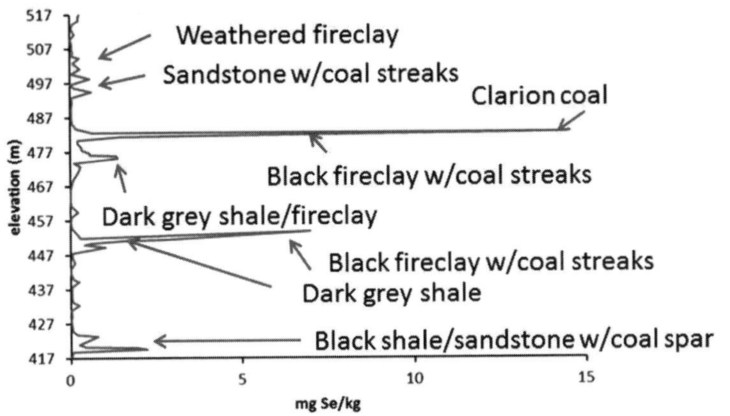

Figure 1. Se concentrations in a core from a West Virginia coal mine

to analysis. The samples were analyzed using an ICP-MS equipped with and octopole reaction cell pressurized with H_2 to remove Ar based interferences. We mixed the samples online with 5% octanol at a ratio of 1:1 to normalize the carbon content of the samples and standards. We used a standard reference shale provided by the U.S. Geologic survey (Green River shale, SGR-1b) as a control sample. This material has a reference Se concentration of 3.5 mg/kg. Our method resulted in a measured value of 3.54 ±0.09 (mean ±SD, n = 4). Method detection limits were extremely low and averaged 40 µg Se kg^{-1} for a 100 mg sample.

Synchrotron Analyses

We selected four samples with the highest Se concentrations to analyze using a synchrotron based X-ray microprobe at beam line X26A, National Synchrotron Light Source, Brookhaven National Laboratory. The thin sections were mounted in the X-ray beam, which was focused to a spot size of about 5 × 7 μm^2. A 4 element silicon drift detector was used to measure X-ray fluorescence of Se and other elements contained within the sample. The samples stage was rastered in the beam to produce 2 dimensional elemental images of the samples. Foci of elevated Se concentrations were then further analyzed using µXANES. To determine the Se speciation, the µXANES spectra were compared to a standard Se (0) foil as well as selenomethionine (Se –II), lead selenide (Se –II), sodium selenate (Se +VI) and sodium selenite (Se +IV). Spectra for standards were collected in transmission mode using a pin diode detector while spectra for samples were collected in fluorescence mode using the silicon drift detector. The spectra were normalized and processed using Athena.[4]

RESULTS AND DISCUSSION

The core sample contained a variety of rock types including sandstone, shale, coal and fireclay. Of these rock types, the highest Se concentrations were found within the coal, dark grey/black shale, and fireclay containing strata (Figure 1). Se concentrations in all rock types ranged from below the detection limit in a few sandstone strata to 14.2 mg/kg in clarion coal. Sandstone typically had the lowest Se concentrations but slightly elevated concentrations were observed in sandstone when coal streaks or spar were present.

In the SXRF images we observed localized areas of elevated Se concentrations that were several tens of µm in diameter, indicating that the Se was concentrated into relatively few mineral grains (Figure 2). Because the Se concentrations within these grains were at percent concentrations (data not shown), we were able to use the microfocused X-ray beam to obtain spectral data.

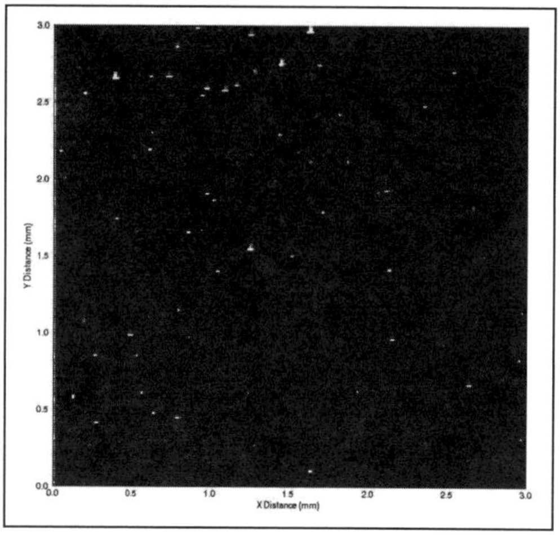

Figure 2. X-ray fluorescence micrograph showing the distribution of selenium in dark gray shale; each pixel is approx. 10×10 μm^2 brighter colors indicate higher concentrations of Se

Good correspondence between Fe and S suggested the presence of pyrite in the sample; however, Se containing structures did not show good correspondence with Fe or S, suggesting that Se was not substituted for S in pyrite (Figure 2). On the other hand, Pb and Se showed a nearly identical spatial distribution in many of the samples (Figure 3). Based on this correspondence between Pb and Se, we hypothesized that PbSe was present within the samples and included PbSe as a standard in the μXANES analysis. We analyzed multiple foci of elevated Se concentrations from four different shale samples and one coal sample. In all cases we found excellent agreement between the spectra and spectra from either elemental Se or PbSe standards (Figure 4).

Taken together, our results indicate that the majority of Se within this core was found within

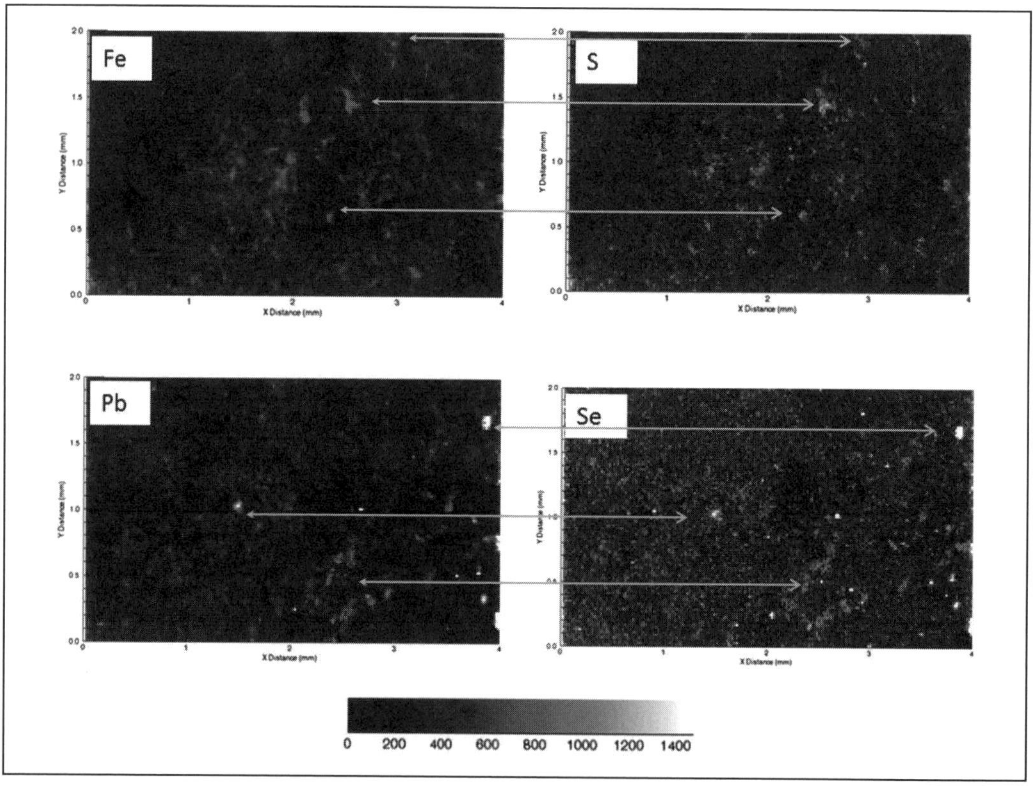

Figure 3. Distribution of Fe, S, Se and Pb in a section of dark gray shale; brighter colors indicate higher concentrations of the indicated element. Green arrows show examples of structures containing both Pb and Se or both Fe and S.

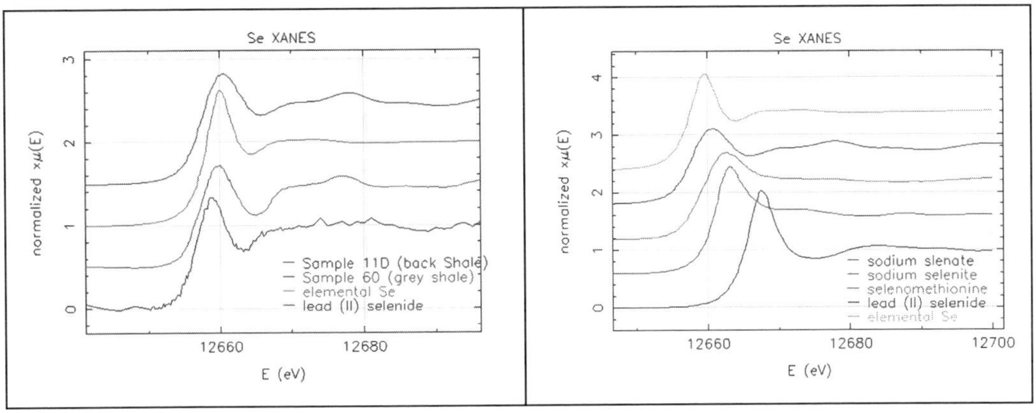

Figure 4. Se K-edge X-ray absorption near-edge spectroscopy (XANES) example spectra (left) and model compounds (right)

relatively narrow bands of strata. Strata containing dark, presumably organic-rich, shales as well as coal and fireclay contained the highest concentrations of Se. Selenium was present primarily as elemental Se or PbSe in these strata and was present in micron or sub-micron sized grains. These phases could have been formed in aquatic sediments (shale) or soils (fireclay) during the time of deposition or perhaps as a result of diagenesis. It is well known that elemental Se particle scan be formed in sediments as a result of microbial activity[5] and may have been preserved in the shale. Lead selenide (clausthalite) has been previously thought to occur in coal.[6] Our results confirm this finding and also extend its occurrence to fireclay and organic-rich shale. This is in contrast to a previous synchrotron experiment that found Se substituted into sulfide minerals.[2] Selenium in coal mine effluents is primarily found in the +IV or +VI oxidation states.[7] Based on the present study, we hypothesize that oxidation is the rate limiting reaction for Se release from mine waste. Identification of procedures that limit the oxidative dissolution of Se in mine waste may provide a cost-effective means of Se source reduction. Future studies should focus on factors that influence weathering of elemental selenium and metal selenides under conditions relevant to different mine waste handling procedures.

ACKOWLEDGMENT

This study was sponsored by the Appalachian Research Initiative for Environmental Science (ARIES). The views, opinions and recommendations expressed herein are solely those of the authors and do not imply any endorsement by ARIES employees, other ARIES-affiliated researchers or industrial members. Information about ARIES can be found at http://www.energy.vt.edu/ARIES.

REFERENCES

1. Young, T.; Adams, W.; Besser, J.; Finley, K.; Hopkins, W.; Jolley, D.; McNaughton, E.; Presser, T.; Shaw, P. Unrine, J., What you need to know about selenium. In *Ecological Assessment of Selenium in the Aquatic Environment*, P.M. Chapman, W.J.A., M.L. Brooks, C.G. Delos, N.L. Samuel, W.A. Maher, H.M. Ohlendorf, T.S. Presser, D.P. Shaw, Ed. SETAC Press: Pensacola, FL, 2010.
2. Strawn, D.; Doner, H.; Zavarin, M.; McHugo, S., Microscale investigation into the geochemistry of arsenic, selenium, and iron in soil developed in pyritic shale materials. *Geoderma* 2002, *108*, (3–4), 237–257.
3. EPA *Method 3052: Microwave assisted acid digestion of siliceous and organically based matrices*; United States Environmental Protection Agency: Washington, DC USA, 1996.

4. Ravel, B.; Newville, M., Athena, Artemis, Hephaestus: data analysis for X-ray absorption spectroscopy using IFEFFIT. *J Synchrotron Rad* 2005, *12*, 537–541.
5. Zhang, Y.Q.; Moore, J.N., Selenium fractionation and speciation in a wetland system. *Environmental Science & Technology* 1996, *30*, (8), 2613–2619.
6. Hower, J.C.; Robertson, J.D., Clausthalite in coal. *Int. J. Coal Geol.* 2003, *53*, (4), 219–225.
7. Pumure, I.; Renton, J.J.; Smart, R.B., Accelerated aqueous leaching of selenium and arsenic from coal associated rock samples with selenium speciation using ultrasound extraction. *Environ. Geol.* 2009, *56*, (5), 985–991.

Selenium Treatment for Mine Water Discharge Compliance

Timothy Harrison
CH2M HILL, Cincinnati, Ohio

Kar Munirathinam
CH2M HILL, Pittsburgh, Pennsylvania

Ken Martins
CH2M HILL, Santa Ana, California

James Bays
CH2M HILL, Tampa, Florida

Kristen Jenkins
CH2M HILL, Atlanta, Georgia

Sarah Dorminy
CH2M HILL, Denver, Colorado

ABSTRACT

The presence of selenium in fossil fuels and some groundwater formations has caused concern regarding reproduction effects as a result of accumulation in fish tissue. The growing requirement to treat wastewater to remove selenium has become a compliance challenge for many industries, including oil and gas, agriculture, power, and mining. Mining operations can release selenium into the environment if it is present in the rock surrounding coal seams. Surface runoff, in the form of precipitation, groundwater intrusion, or mine operations, transfers selenium from rock into the aquatic environment. As a result, surface coal mines, especially in the eastern U.S., are coming under increased scrutiny to control release of selenium into aquatic streams. This paper discusses the approaches the surface mining industry is taking to comply with the limits of the National Pollutant Discharge Elimination System. Most selenium in water is in the most oxidized form (selenate), which is soluble and difficult to treat. CH2M HILL's experience with active and passive treatment technologies has shown success in pilot- and full-scale applications.

INTRODUCTION

Selenium is widely distributed throughout most soils and natural waters. As regulatory limits become increasingly more stringent, the removal of selenium presents a significant challenge. As with any constituent present in water at low concentrations, it is difficult to achieve low effluent limits consistently with current treatment technologies. Management and treatment of selenium to low limits is technically challenging and expensive. This paper reviews treatment technologies that CH2M HILL has employed in the mining industry, including active biological treatment using an upflow fluidized bed reactor (FBR), passive biological using a biochemical reactor (BCR), ion exchange (IX), and zero-valent iron (ZVI). The review of technologies includes ability to remove selenium to low levels, and discussion of pretreatment or post-treatment requirements.

CORE TREATMENT TECHNOLOGIES REVIEW

Ion Exchange

Ion exchange, or IX, removes ionic-form selenium by exchanging selenate for similarly charged ions on a resin backbone structure. IX resins typically contact wastewater in fixed beds until the resin is saturated. Pretreatment requirements for IX can include removal of total suspended solids (TSS) or any scale-forming or precipitating materials, such as iron or manganese, to avoid fouling or plugging of the media. IX can cause notable pH shifts, requiring pH control as a possible post-treatment step. Sulfate, bicarbonate, and nitrate also may be removed, exerting greater demand on the IX resin, depending on the resin used. IX systems are easily automated and can be run with little operator attention. Since sulfate, and some bicarbonate and nitrate, are exchanged for chloride, chloride in the effluent will increase in proportion to the other anions removed. The regeneration process produces a spent regenerant (brine) that requires disposal of selenate following treatment. The IX media is regenerated in situ and lasts as long as three to five years, depending on water characteristics.

Key developments in the use of IX in mining applications include bench and pilot-scale tests and full-scale operation at a surface coal mine in West Virginia. Pilot testing of IX reduced selenate in runoff water from the mine from an average of 7.3 micrograms per liter (μg/L) and maximum of 12 μg/L to less than 1 μg/L over a 4-week period. Pilot testing also showed that IX can be effective for selenium (as selenate) removal, with sulfate concentrations up to 250 milligrams per liter (mg/L). The mine water did not contain selenite, so the technology was not assessed for selenite removal. A small full-scale system, with influent selenium of 7 to 14 μg/L and sulfate as high as 690 mg/L met discharge permit limits (4.7 μg/L monthly average) for a system treating 50 gallons per minute (gpm) for an initial short period of less than 1 month.

Pilot- and full-scale systems thus far have demonstrated that IX is a feasible treatment technology for selenium removal from the surface mining water. The ion exchange columns must be regenerated, and a disposal method for the regenerant containing concentrated selenium as soluble selenate must be determined.

Zero-Valent Iron

The chemical oxidation-reduction mechanisms with ZVI or elemental iron in steel wool reduce the oxidized forms of selenium (selenate or selenite) to elemental selenium. ZVI media can be in powder, granular, or fibrous forms. ZVI acts as a reducing agent in the redox reaction. The iron acts as both a catalyst and electron donor and provides a source of both ferrous and ferric iron that can provide further reduction and adsorption of selenium. The result is green rust, a complex ferrous ferrihydroxide co-precipitate. The formation of green rust is a function of pH and equilibrium concentrations of ferrous and ferric iron. It is the form of ZVI required to reduce selenate abiotically to selenite, and selenite to elemental selenium.

When water containing selenate and selenite comes into contact with ZVI, the ferrous and ferric iron products of ZVI will either directly reduce selenate to selenite and then to elemental selenium, or reduce selenate to selenite and adsorb the selenite to the ferrihydrite solids formed during the redox reaction with ZVI.

The elemental form of selenium is insoluble and will exist as nanoparticles embedded in the iron solids matrix. Any soluble iron from the ZVI will require removal. This could require oxidation or pH adjustment, followed by solids removal by sedimentation or filtration. The following equations summarize the overall reactions of ZVI with selenate and selenite:

$$3Fe^0\,(\text{solid}) + SeO_4^{2-} + 8H^+ \rightarrow 3Fe^{2+} + Se^0 + 4H_2O$$

$$2Fe^0\,(\text{solid}) + SeO_3^{2-} + 6H^+ \rightarrow 2Fe^{2+} + Se^0 + 3H_2O$$

Other oxyanions (e.g., carbonate, sulfate, nitrates, and phosphates) can contribute to oxidation of ZVI. They may also help to form green rust, an essential component for reduction of selenium. ZVI treatment systems typically include tanks or filter vessels in parallel that hold elemental iron. Elemental iron can come in several forms, including filings, steel wool, or iron-impregnated foam. The tanks generally are covered to promote anoxic conditions for selenate/selenite removal. Tanks may be in series, creating several treatment trains. The effluent from the reactor tanks can be aerated to oxidize the iron in an aeration tank before discharge to the outfall.

Passivation of ZVI has been reported in previous work. An oxide coating is believed to form on the iron, inhibiting the reactivity and increasing the doses required to meet treatment goals. Several studies, however, did not encounter this. Observed bench- and pilot-scale studies for surface mining application indicate that effective ZVI steel wool systems require the feed water pH to be reduced to a range of 6.0 to 6.5 to maintain an active ZVI media surface.

Pretreatment requirements for ZVI processes can include removal of TSS and pH adjustment, typically to 6.0 to 6.5. The ZVI process will release significant quantities of soluble iron in solution. Following ZVI treatment, the soluble iron usually must be removed, typically through air oxidation or pH adjustment, followed by solids removal by sedimentation or filtration. Waste iron rust media from the ZVI reactor and precipitated iron sludge from the precipitation process must be dewatered and disposed of.

Key advancements for selenium removal in mining applications include use of steel wool in pilot- and multiple full-scale systems at a surface coal mine site in West Virginia. A ZVI pilot study using steel wool was conducted to evaluate selenium removal from coal mine runoff. The pilot consisted of five parallel treatment trains, each with two tanks in series, configured or operated differently. The inlet selenium varied from about 15 to 29 μg/L and was primarily in selenate form. The effluent average total selenium concentration of 4.7 μg/L was observed with influent pH adjustment in the range of 5.5 to 6.5. Full-scale operation (up to 160 gpm) has shown that runoff water can reduce influent selenium (as selenate or selenite) from about 19 to less than 4.7 μg/L monthly average, or 8.2 μg/L daily maximum, continuously for a 4-month period.

ZVI, as applied to selenium treatment, is a developing technology. More engineering for the full-scale design needs to be performed for reliable long-term operation at high flows (>800 gpm). It has established a significant foothold at one Appalachian surface coal mine, with multiple full-scale installations for low flows. With proper operator attention, ZVI can be a reliable and cost-effective option at certain flow rates.

Biological Treatment

Active Biological

Selenium removal by biological treatment uses certain heterotrophic facultative bacteria to reduce selenate and selenite in wastewater to elemental selenium, in addition to reducing nitrate to nitrogen gas. Elemental selenium can then be removed as an insoluble solid, along with bacterial solids. Under anoxic/anaerobic conditions, heterotrophic facultative bacteria use selenates and selenites, as well as nitrates, chlorates, perchlorates, and sulfates, as electron acceptors for respiration. The bacteria prefer dissolved oxygen as an electron acceptor, followed by nitrites and nitrates, selenite and selenate, chlorates and perchlorates, then sulfites and sulfates. At present, two active anoxic/anaerobic biological (attached growth) water treatment technologies are feasible and proven for consistent selenium removal: a downflow granular activated carbon (GAC) biological gravity filter reactor, and an upflow biological FBR.

Downflow GAC biological filter reactor. This water treatment system is a downflow attached-growth granular activated carbon filter that uses heterotrophic bacteria. The first and second generation reactors typically are operated as parallel trains with two reactors in series, and designed as either single-stage or two-stage, depending on nitrate and selenium loadings. The third generation reactor, put on the market in 2011, has deeper carbon bed and headspace to increase hydraulic loading rate, with parallel reactors in a single stage. Microbes are seeded in the bioreactor, where they persist by attached growth. Organic carbon (electron donor) with nutrients (phosphorus and micronutrients) is used to create an ion-reducing environment for selenate and selenite reduction. The reactors are pulsed to reduce entrainment of gas (primarily nitrogen and carbon dioxide) within the bed of granular activated carbon. They also require backwashing to prevent impediment of flow through the filter by excess influent TSS and biomass growth. The system is generally operated with oxidation reduction potential (ORP) ranging from –250 to –350 millivolts (mV). The required minimum operating temperature is 50°F (10°C), although the system may perform at a lower temperature of 46°F (8°C).

Pretreatment is required to remove influent solids, if backwashing is required more frequently than the vendor design frequency of once or twice a month. It also is required to prevent limitations on hydraulic throughput capacity. Depending on discharge requirements, an aeration tank will be required as a post-treatment to increase the reactor effluent dissolved oxygen from zero (–250 to –350 mV ORP) to greater than 3 to 5 mg/L.

Advancements of this technology for mining applications include a pilot test at a coal mining site in Canada. The system included one filter bioreactor train of two cells operating in series and containing a mixture of microbes and granular activated carbon to support microbial growth. The influent water was heated to maintain a minimum operating temperature of 50°F (10°C) in the reactors, and a proprietary nutrient solution was fed as the electron donor. The influent selenium varied from about 150 to 550 µg/L with an average value of 350 µg/L, primarily in selenate form. The effluent average filtered selenium concentrations ranged from 2.0 to 10 µg/L. Total selenium concentrations ranged from 3.5 to 20.0 µg/L for three different operating conditions. The backwash water contained high selenium concentration and required further treatment for selenium before disposal.

FBR system. In the FBR system, wastewater passes through a granular solid media at velocities high enough to suspend the media and to cause it to behave as a fluid. Fluidization keeps the media with attached biomass in suspension and expanded in depth to provide good contact for water treatment. Microbes cultured in the FBR media beds create an ion-reducing environment (anoxic/anaerobic condition) using readily degradable organic carbon and nutrients for selenate and selenite reduction. Sloughed biomass, including reduced particulate selenium, is removed from the FBR effluent using a liquid/solids separation unit process (e.g., ballasted sand clarifier, membrane filter, or sand filter). In addition, unused electron donor or residual dissolved biochemical oxygen demand (BOD) from the FBR can require oxidation (removal), depending upon the BOD discharge requirements. Given typically low BOD concentrations, this is most efficiently done in an aerobic environment with another attached growth system (such as an aerobic moving bed bioreactor). Low dissolved oxygen after liquid-solids separation will require aeration to increase the dissolved oxygen concentration above 3.0 mg/L. The aerobic moving bed bioreactor will provide both BOD polishing and dissolved oxygen increase for the final effluent. The FBR system is designed as single-stage or two-stage, depending on the nitrate and selenium loadings. The typical operating ORP range is 150 to 200 mV. The required minimum operating temperature is 50°F (10°C), although the system may perform at a slightly lower temperature of 46°F (8°C).

Fluidization allows FBR systems to be more forgiving than downflow or upflow filter-based systems, for some level of influent suspended and colloidal solids. However, for excess solids of a particular nature, pretreatment can be required to reduce the in-bed cleaning frequency in the FBR. Observed total selenium concentrations in the FBR effluent typically are high, since the reduced elemental selenium particles exit in the FBR effluent along with the sloughed off biomass. From three different pilot studies treating mining water, for influent selenium concentrations ranging from 20 to 550 µg/L, an effluent filtered selenium concentration range of 3.5 to 10 µg/L was observed. However, the FBR effluent total selenium (primarily reduced elemental or particulate selenium) concentrations ranged from 10 to 500 µg/L. Hence, a liquid-solids

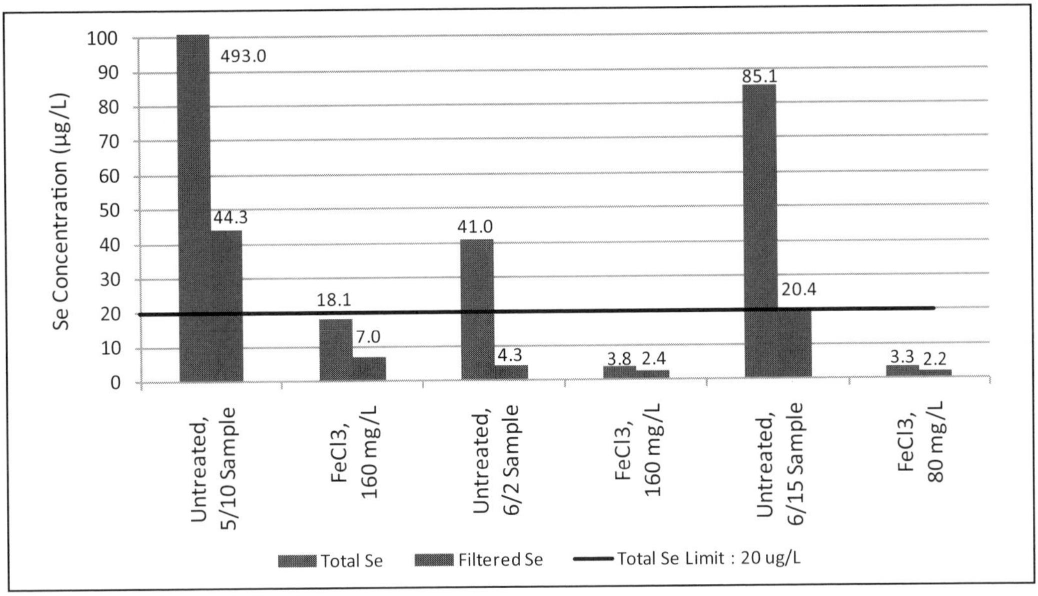

Figure 1. Jar test results for FBR effluent selenium removal

separation system is necessary to remove biomass and elemental selenium from the FBR effluent. Generally, as selenium concentrations increase, the amount of reduced selenium particles increases proportionally, thereby requiring greater efficiencies in the liquid-solids separation technology. Jar tests were performed to simulate high-rate clarification using ferric chloride as the coagulant and an emulsion polymer as the flocculant. Figure 1 shows the results from the three sets of jar tests performed to demonstrate removal of biosolids containing particulate selenium.

Several full-scale treatment systems have been installed to remove nitrates and perchlorates from contaminated groundwater. Two FBR-based full-scale treatment plants that will remove selenium from mining water are under construction and are scheduled to be operational in early 2013 and mid-2014.

Passive Biological

Passive biological treatment systems employ naturally occurring biological, chemical, and physical processes. They typically are more land-intensive than active treatment systems, but they can be less expensive to operate and manage because of lower energy or chemical inputs. In passive systems, oxidized forms of selenium (selenite and selenate) can be reduced to elemental selenium and selenides through microbial reduction. This is followed by sequestration of the product in soil and sediments. Labile organic carbon released from the substrate serves as an electron donor. Dissolved oxygen and nitrate are common electron acceptors that must first be removed before selenium.

Passive treatment systems designed for vertical or horizontal subsurface flow through a reducing organic substrate can achieve microbial and chemical reduction of selenium with relatively shorter residence times and smaller areas than can constructed wetlands. Termed biochemical reactors (BCRs), these passive systems have been used to treat a variety of mine-impacted waters, but their incorporation into passive systems for selenium reduction is new. Substrates consisting of wood chips, saw dust, mushroom compost, horse manure, field hay, yard waste, and limestone have been used in varying proportions.

The geochemistry of BCRs is relied on in a staged approach for selenium removal. Sulfate-reducing BCRs precipitate trace metals with biogenic sulfide (i.e., chemical reduction of selenate to elemental selenium), whereas selenate-reducing

Table 1. Barrel study media compositions (CH2M HILL. March 2011. Owner Confidential)

Tank	Peat Moss	Organic Peat	Composted Manure	Sawdust	Wood Chips	Hay	Limestone Chips
A	100%	—	—	—	—	—	—
B	20%	20%	—	20%	20%	15%	5%
C	—	—	15%	47%	16%	16%	6%
D	—	—	23%	30%	20%	20%	7%

BCRs remove selenium as elemental precipitates. Because selenium compounds are more readily reduced than are sulfur compounds, and because reduced sulfur compounds can act as a chemical reductant, any sulfides precipitated in the BCR provide additional reducing capacity within the substrate. As additional selenium-bearing water passes through the reduced substrate, the benefit of biological reduction and potential for chemical reduction is realized.

Volatilization and adsorption processes contribute to the removal of selenium in passive systems. Volatilization of selenium through bacterial, fungal, or algal-mediated methylation of selenium has been shown to reduce selenium in wetlands significantly through conversion to organic forms, such as dimethyl selenide. Adsorption of selenite to iron, aluminum, or manganese oxy-hydroxides in soil or sediments and also to organic matter is physical removal that occurs readily in passive systems.

In general practice, pretreatment for a BCR required reduction of TSS to low concentrations (< 20 mg/L) in order to minimize the potential for excessive accumulation of solids within the BCR. The influent pH should be adjusted if necessary to an average of 6.5 to 8.5 to sustain a normal range of microbial activity. Because the BCR contains organic media, byproducts, such as BOD, sulfide, and reduced nitrogen, are generated that require treatment before discharge. Varying across different projects, post-BCR treatment has included aerated and non-aerated ponds, surface flow constructed wetlands, and single or combined subsurface flow gravel beds. Dynamic influent flow to passive systems treating surface flow has been found to have adverse effects on system treatment performance. For this reason, flow equalization ahead of passive systems is required to reduce the maximum range of inflow rates.

Pilot studies conducted for coal mining companies in southern Appalachia have consistently indicated that total selenium can be reduced to levels below the state regulatory standard of 4.7 μg/L monthly average and 8.2 μg/L daily average. One BCR pilot system was constructed and sampled at an outlet from a reclaimed coal mine valley fill drain in southern Appalachia. Four media compositions were tested (Table 1). A media mix of 20 percent wood chips, 20 percent sawdust, 15 percent field hay, 20 percent peat moss, 20 percent organic peat, and 15 percent limestone chips appeared to balance high treatment performance with moderate production of byproducts. Figure 2 shows total selenium removal by each media composition for the various high rate treatment systems tested.

SUMMARY

Ion exchange has been evaluated for mining applications using bench and pilot tests. IX was able to reduce selenium concentrations of 12 to 14 μg/L to 1.0 to 4.7 μg/L in the presence of sulfate at concentrations varying from 250 to 690 mg/L. Competing anions such as sulfate, nitrate, and carbonate should be taken into consideration when designing a system. Pretreatment includes reduction of TSS concentrations and removal of scale-forming ions. The IX process produces brine that must be disposed of after treatment for selenium. During the exchange process, chlorides are released from the resin, which increases the level of chlorides in the effluent.

ZVI has been evaluated for mine water treatment in bench, pilot, and full-scale systems. The systems treated influent selenium concentrations in the range of 15 to 29 μg/L to an effluent

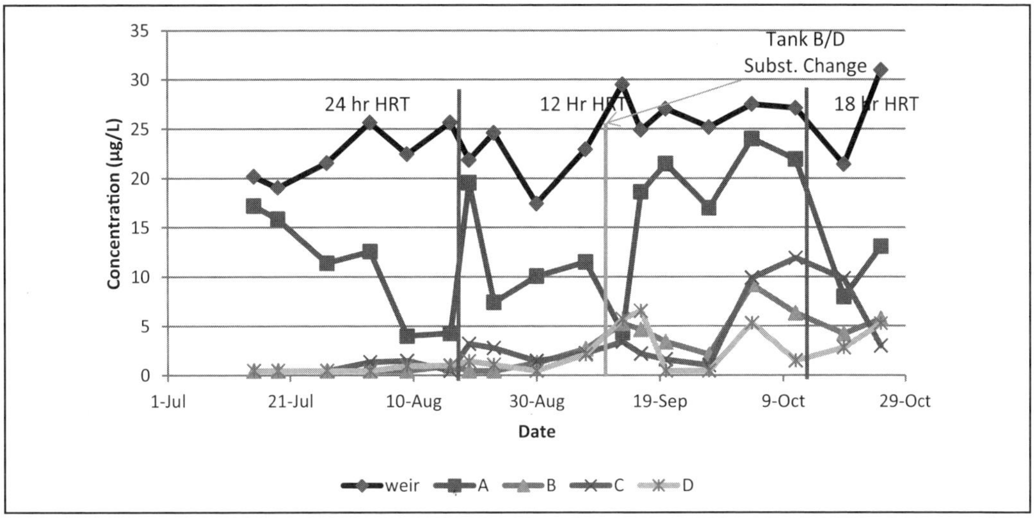

Figure 2. Total selenium removal by media type. Note: The vertical blue lines in the figure separate the data between different HRTs.

average level below 4.7 µg/L. The systems were column-based using steel wool as the media. Pretreatment requirements include removal of TSS and adjustment of pH to the target range. Post-treatment includes removal of residual ferrous iron through aeration, pH adjustment and settling, as well as pH adjustment. Nitrate, dissolved oxygen, and other electron donors in the influent will consume ZVI.

The downflow GAC biological filter reactor and FBR systems were evaluated at the pilot scale for active biological treatment. A full-scale FBR system is under construction in Southern Appalachia. Influent selenium concentrations for both systems were as high as 550 µg/L, with effluent filtered concentrations in the range of 3.5 to 10 µg/L. However, for the FBR system, liquid-solids separation is required following FBR.

BCRs have been tested at the pilot scale for mining wastewater, and full-scale systems have been installed. Flows treated by full-scale systems constructed since 2010 have ranged from 75 to 1,500 gpm at valley fill drains in southern Appalachia. In general, a flow range of about 200 gpm is typical of BCRs. Pilot studies have shown that hydraulic application rate designed to achieve a nominal media residence time of 1.0 day is sufficient to reduce selenium concentrations. Both pilot and full-scale systems have demonstrated that compliance with a selenium limit of 4.7 µg/L is achievable with BCR influent concentrations ranging from 10 to 30 µg/L.

Strategies for Remediation of Selenium-Laden Waters

Matt Huddleston
Cardno ENTRIX, Clemson, South Carolina

John H. Rodgers Jr.
Clemson University, Clemson, South Carolina

James Castle
Clemson University, Clemson, South Carolina

Michael Spacil
DiamondV, Greenville, South Carolina

ABSTRACT

Selenium can be a constituent of ecological concern in waters associated with energy production, such as coal ash basins, flue gas desulfurization (FGD) effluents, and produced waters. An effective and reliable approach is needed to treat Se in these waters to meet stringent water discharge limits established under the National Pollutant Discharge Elimination System (NPDES) and the Clean Water Act. Constructed treatment wetlands (CTWs) have been developed that target biogeochemical mechanisms for removal of selenium and other constituents (e.g., arsenic). This paper will focus on Se removal using CTW systems by presenting case studies on FGD effluents and produced waters. For example, in pilot scale CTWs for FGD, Se was decreased from approximately 4 mg/L (influent) less than 1 mg/L (effluent). In subsequent studies focused on enhancing Se removal from produced waters using organic amendments to the wetlands, inflow Se concentrations of approximately 50 µg/L were decreased to a range of 3.4 to 9.8 µg/L, depending on the experimental treatment. This work illustrates that CTWs can reliably remove Seto <5 µg/L to meet stringent discharge limits.

Analysis of Low-Level Selenium in Coal Mine Discharges

M. Harris
ToxStrategies, Inc., Katy, Texas

ABSTRACT

The identification of selenium (Se) as a contaminant of concern has prompted the development of strict regulatory levels for its concentration in coal mine discharges. The current West Virginia DEP water quality standard (WQS) for Se is 5 µg/L (47CSR2) and it appears that USEPA is going to propose a new WQS for Se in early 2013. Fish and birds have proven to be sensitive to low level exposure to Se. Importantly, Se has a very narrow range of what is beneficial and what is detrimental for biota. Aquatic organisms require 0.5 µg/g of Se in their diet, whereas concentrations of Se that are only an order of magnitude greater have been shown to be toxic. Se is found as a natural component of metal sulfide ores and coal, and is a by-product of the coal mining process. Thus, given the narrow range between beneficial and toxic levels, the concentration of Se in mine water tailings and other industrial effluents must be accurately quantitated to demonstrate compliance with applicable regulatory criteria and to determine if remediation is needed to protect biota.

The West Virginia WQS for Se is very close to the method detection limit (MDL) for most Se analytical methods (by 40CFR Part 136). As a result, the analysis of low-level Se is influenced by a number of factors, and the accurate quantitation of low-level Se in mine water is dependent on carefully following strict sampling and analytical protocols.

One factor to consider in the analysis for Se is the possibility of contamination introduced during the sample collection process. In this context, "contamination" can be defined as the unintended addition of Se into a sample from sources other than the sample. Unclean sampling equipment, unclean sample bottles, soil, dust, unclean preservatives, and poor sampling technique are common sources of contamination.

Once a sample has been collected and transferred to the laboratory, it must be satisfactorily analyzed to accurately determine the actual concentration of Se in the sample. The erroneous data results in expensive treatment where none is required, or worse, inadvertent discharges of Se into surface waters that are undetected and untreated.

Sensitivity is also an important factor in choosing which analytical technique is best suited to the analysis of Se. Commonly used techniques for the analysis of metals do not possess the required sensitivity (i.e., they are not capable of detecting Se at or below the 5 µg/L concentration level). Typical analytical techniques that are used to detect Se at the 5 µg/L level generally possess the required sensitivity, but each technique has specific challenges that must be addressed to ensure that the results are accurate.

The final consideration in the analysis of Se is speciation. Different molecular forms of Se (Se(IV), Se(VI), Se(0), etc.) can influence how well the analytical technique works for a given sample. The different Se species in the sample can have an effect on which digestion and analytical techniques are appropriate, and must be understood to determine the best approach for an analysis.

The analysis of Se in complex matrices often encountered in mine discharges can present numerous challenges. The additional consideration of a very low action level requires that strict sampling, analytical, and quality assurance protocols be followed to ensure accurate results are available for decision-making. Expensive and unnecessary remediation efforts, potential fines and liability, or the unintentional release of Se into the environment are possibilities if the appropriate protocols are not followed. A thorough understanding of the site-specific chemistry of Se and strict attention to sample collection and analytical details is vital to the production of accurate analytical data to support an effective Se management program. This presentation will address these issues.

ACKNOWLEDGMENT

The author thanks R.J. Vitale, and G. Kirkpatrick for their contributions to this paper.

6.2
Stream Restoration

Natural Streams: Biological Effects of Natural Channel Design for Aquatic Habitat Restoration **454**

Hydrologic Assessment of the Guy Cove Stream Restoration Project **463**

Guy Cove Stream Restoration Project: Recreating a Headwater Stream System on a Head-of-Hollow Fill **464**

Leaf Litter Breakdown as a Functional Assessment Tool for Appalachian Coal Mine Stream Restorations **465**

Substrate and Seeding Treatment Effects on Native Tree Establishment and Herbaceous Cover on an Appalachian Coal Mine: First Growing Season Results **467**

Natural Streams: Biological Effects of Natural Channel Design for Aquatic Habitat Restoration

Joshua J. Noble
Wallace & Pancher, Inc., Boardman, Ohio

ABSTRACT

Natural channel design (NCD) has been effectively used for a wide range of projects to satisfy Federal and state permitting conditions for resource mitigation, as well as for stream stabilization and enhancement projects conducted for the general benefit of the stream, riparian corridor, and greater watershed. Hydraulic engineering approaches to stream mitigation focused on the linear-footage aspect of such projects. Streams were often designed as rip-rap lined trapezoidal channels with a cross-sectional area appropriate to contain and convey a specific flood event. The primary flaw with such a technique is that it places little emphasis on the fundamental physical functions of a natural stream, with the exception of water conveyance, and it ignores the ecological functions and values of a natural stream. The intent of the NCD technique is to mitigate streams by replicating the innate condition of a reference reach. This includes a design that accounts for the stream's natural dimension, pattern, profile, substrate particle distribution, habitat heterogeneity, hydraulic source(s) for the stream system, and overall watershed condition. By considering the holistic physical environment of an aquatic system and its companion terrestrial habitat, stream mitigation projects can be designed in a manner that provides the greatest chance for the successful re-establishment of a biological community. Regulatory agencies require stream mitigation projects to exhibit a measureable "ecological lift." Properly designed and constructed, NCD projects provide the greatest opportunity to achieve the mandated lift and can result in stable, low-maintenance mitigation projects that may need little to no long-term investment.

INTRODUCTION

Throughout the history of human technological and societal development, a constant by-product has been degradation of surface waterways. Streams and rivers have long been the convenient recipients of wastes and excesses in many forms. The manipulation of land and exploitation of land resources, such as timber and minerals, altered the landscape and degraded the natural control measures (i.e., forests and wetlands) that attenuate excess rain water and bind up organic chemicals that might otherwise compromise the physical stability or chemical balance of surface waters (Karr 1991).

In recent decades, society has become conscious of the long-term effects of our actions to develop the world around us. In response, steps have been taken to make consideration of the natural environment a point of focus when projects are conceived (Bernhardt et al. 2005). Regulations have been enacted at every level of government, from Federal regulations, such as the Water Pollution Control Act of 1972 (i.e., the Clean Water Act), to municipal ordinances protecting stream corridors.

As this awareness of the fragility of streams has grown, so have the schools of thought on effective and reliable methods for restoring degraded stream systems. The discipline of stream design has grown to include several approaches, often at odds with each other, which emphasize different properties of a stream's function. These design methods are divided into two broad categories: empirical methods and analytical approaches (Shields & Copeland 2006). A potential causative factor for why stream mitigation has failed to (1) create physically stable streams, (2) support a biological community, and (3) achieve a specific hydraulic capacity is that the over-riding design stresses one performance factor over the others (Rosgen 1996).

The methods of stream design commonly referred to as Natural Channel Design (NCD) are empirical approaches that focus on relationships (i.e., ratios) of hydraulic geometry variables recorded at a reference reach and then applied to the project stream. These empirical approaches are applicable primarily for meandering streams with mobile bed materials, and are accepted as the appropriate approaches for habitat restoration. Analytical approaches are process-based and are more mathematically complex. These methods base channel design on stream flow and sediment continuity. A supply reach (i.e., reference reach) is a necessity for analytical approaches as well. Yet, the critical data gathered from the supply reach is the relationship of sediment supply to channel form (Shields & Copeland 2006). It is not the purpose of this paper to pursue a comparative argument in favor of either empirical or analytical methods. This paper will focus on the empirical methods of NCD as they are accepted as being better suited for habitat restoration, as stated by Shields & Copeland (2006).

The predominant use for NCD projects is to satisfy mitigation requirements for impacts incurred on streams. By employing NCD, the intent of the design is to create new or improve degraded habitats such that the biological community within the stream and riparian corridor exhibit a targeted "ecological lift," demonstrated by an increase in species richness and/or population densities (Miller et al. 2010). Ecological lift can be defined as a net positive change in the biological communities or populations within a targeted area. This may be observed, for example, as an increase in the number of individuals of a pollution-intolerant species, the reduction in the populations of pollution-tolerant species, or an increase in the diversity of taxa (e.g., macroinvertebrates, fishes, amphibians, and reptiles) within the target area. There is currently no standard for what defines ecological lift, as it typically tailored to each project's specific pre-construction condition and anticipated outcome.

A pervasive optimism exists in the NCD-community, and it is generally assumed that channel stability and habitat improvements designed into NCD projects guarantee that the ecological lift will occur. This is referred to as the "Field of Dreams" hypothesis (i.e., "if you build it, they will come") (Miller et al. 2010; Palmer et al. 1997). But despite the best efforts of many designers/builders of NCD projects, some projects have fallen short of achieving the proposed lift, even in the presence of a stable, habitat-rich stream. Conversely, there are multitudes of projects where the designs have been simplistic and limited, yet have shown significant improvements in biological productivity and have exceeded the proposed ecological lift (Miller et al. 2010). These inconsistencies lead critics of NCD to further voice their opposition and argue that NCD projects should not be accepted as viable options for stream mitigation and habitat restoration projects. However, the apparent failures of NCD projects to achieve ecological lift are not attributable to a poor methodology, but rather a result of misguided expectations of the ecological lift that could occur, or could if ecological lift was possible. It may also be attributed to selecting improper metrics to evaluate the project.

This paper discusses the process and design considerations of NCD, as well as the applicability and benefits of NCD to stream mitigation, and investigates the biological and ecological responses to NCD projects. It is the author's intent to provide information on NCD as a valuable tool that can be used creatively to benefit the owner of a project and the environment. The term "mitigation" will be used to represent mitigation, restoration, and enhancement projects for streams.

HYDRAULIC ENGINEERING OF STREAMS

Traditional approaches to stream design, referred to as " hydraulic engineering," focus on the linear-footage aspect of mitigation projects. "Streams" were and still are designed as rip-rap lined trapezoidal channels with a cross-section appropriate to contain a targeted storm event. This design approach places little if any emphasis on channel planform or the fundamental physical stream functions beyond water conveyance, and ignores the stream's biological and ecological functions and values (Shields & Copeland 2006).

Hydraulic engineering assumes uniform flow, which does not account for mobile boundaries of alluvial streams. Streams designed using hydraulic engineering have uniform slopes and minimal sinuosity, as the primary design goal is to move water. Uniform channel morphology is drastically different from that of natural streams, which have innate variability in channel dimensions (Shields & Copeland 2006).

Streams designed using hydraulic engineering are typically oversized as they are designed to convey a high-volume flow (often the 100-year storm flow). Flow energy is reduced in an oversized channel, and the stream cannot convey its normal sediment load during average daily flows or even in storms with volumes less than the design storm discharge. Over time, the stream will build a channel with floodplains within the deposited sediment. This is desirable from a natural prospective, but is problematic for hydraulic engineers.

NATURAL CHANNEL DESIGN

In contrast to hydraulic engineering, NCD assumes that overbank flooding will occur. Rather than designing the channel to convey the 100-year storm discharge, NCD relies on floodplains and vegetated buffers to provide attenuation of floodwaters and sediment, and to provide stability to streambanks.

Natural Channel Design does not undervalue the importance of stream flow. The evaluation of flow and hydraulic discharge is critical to NCD. But, NCD does not view the "catastrophic flood" as the channel-forming event. It is the more frequent bankfull discharge event that dictates channel form. Although significant channel and bank erosion may occur during extreme floods, it is modest flows that entrain and transport a larger quantity of sediment over time as lesser flood events occur more frequently (Rosgen 1996).

The term "bankfull" is defined as the incipient point of flooding, or the bank elevation where a naturally-formed stream channel is full and first begins to overflow onto the floodplain. The accepted definition of bankfull was offered by Dunne and Leopold (1978): "The bankfull stage corresponds to the discharge at which channel maintenance is the most effective, that is, the discharge at which moving sediment, forming or removing bars, forming or changing bends and meanders, and generally doing work results in the average morphologic characteristics of channels." The term bankfull discharge is synonymous with "effective discharge" and "dominant discharge." The bankfull flow discharge is equated typically with the 1.5-year storm event, but that may differ based on regional or watershed conditions (Rosgen 1996).

Stream morphology is dictated by the interaction of channel-forming forces (e.g., stream flow, sediment transport) and resisting forces (e.g., channel bed material, channel bank material, rooted vegetation; a.k.a. "boundary conditions"). Channel form may be driven by the pattern of alluvial deposits (e.g., channel forming force), or by the presence of a controlling feature, such as bedrock (e.g., resisting force) (Rosgen 1996). From the standpoint of channel design in practice, it is critical that a design have proper planform and dimensioning to prevent channel failure. Streams utilize meanders to manage flow and velocity and if these variables are not given due consideration, the product channel is likely to experience some failure.

Natural Channel Design methods are considered empirical approaches (Shields & Copeland 2006). The term "NCD" refers to multiple stream design approaches. The United States Department of Agriculture (USDA) recognizes six NCD methods: Threshold Channel, Alluvial Channel, Two-Stage Channel, Rosgen Geomorphic Channel, Channel Alignment and Variability, and Treatment Technique (USDA 2007).

The intent of NCD approaches is to design stable streams that neither aggrade, degrade, or significantly migrate laterally or horizontally. All approaches, with the exception of the Treatment Technique, consider the stream's sediment load and capacity to mobilize that load. But the emphasis on sediment is not equal across the methods. The more holistic approaches include an evaluation and design that accounts for the stream's natural dimension, pattern, profile, and substrate particle distribution, and attempt to replicate those conditions in a manner that reflects a reference reach located within a watershed that is similar in topography, flow regime, and boundary conditions. Aquatic habitat variability, density, and spacing, quantity of organic materials (e.g., woody debris and boulders), and biological recovery potential are weighted equally with the physical parameters of the design (Rosgen 1996).

Below, the six NCD methods are summarized as presented in USDA 2007.

Threshold Channel:

- Achieves channel stability by having immobile channel boundaries.
- Flows are not sufficient to entrain the boundary particles.
- Sediment transport capacity greatly exceeds the inflowing sediment load.

Alluvial Channel:

- Utilized for streams with mobile boundary materials.
- Assumes a continual exchange of incoming and outgoing boundary material.
- Channel designs require sediment continuity and performance assessment for a range of flows.

Two-Stage Channel:

- Originally developed for use in modifying agricultural ditches.
- Often uses designs in the trapezoidal configuration.
- Incorporates benches that provide flood plains.
- Attempts to restore natural alluvial channel processes.

Rosgen Geomorphic Channel:

- Focuses designs on reestablishing channel dimension, pattern, and profile.
- Relies on a reference reach to guide design.
- Places equal emphasis on channel stability, biological integrity and habitat quality.

Alignment and Variability:

- Used for channel planform design (i.e., Rosgen Geomorphic Channel).
- Accounts for natural variability in channel width and planform patterns.

Treatment Technique:

- Utilizes treatments at specific points within a stream; "spot treatment."
- Analyzes each treatment for the location where it is to be constructed (as opposed to the larger-scale analysis used for the other techniques described above).
- Examples of treatment: adding rock to streambanks, adding Large Woody Debris (LWD), installing grade control structures (a.k.a. "bioengineering").
- May be applied to an existing channel alignment or to an entirely new channel.

Despite defining NCD methods as six separate approaches, stream designs may incorporate multiple NCD approaches. Of these six methods, the Rosgen Geomorphic Channel and Treatment Technique methods are the most widely used and most well-known, and are commonly used in combination.

These methods incorporate natural components and materials that perform physical and ecological functions. This is true particularly for the Rosgen Geomorphic Channel design approach. Some examples of this include:

- Pools: reduce flow energy; provide preferred habitat.
- LWD: streambank erosion protection (e.g., Rosgen's toe-wood structure); food source for macroinvertebrates; cover for juvenile fishes.
- Boulders: diverge flow creating backpool eddys; eddys provide resting areas for fishes, collection areas for floating food particles.
- Converging Rock Clusters: channel bed grade control; create riffles that provide preferred habitat; riffles oxygenate water.

APPLICATION OF NCD TO STREAM MITIGATION

In earlier years of stream mitigation, a knowledge gap existed between the disciplines of hydraulic engineering and ecological science. Although fluvial geomorphology has been accepted by ecologists since the 1950s, the two camps rarely worked cooperatively on stream projects (Shields & Copeland 2006). Stream mitigation projects were intended to compensate for an exact linear footage of stream impact. However, regulatory agencies and advocates became concerned about the lack of biological production from mitigation projects designed by hydraulic engineers (Karr 1991). This re-focused the evaluation of stream mitigation projects to ecological lift as well as channel stability.

This shift lead to NCD becoming the preferred approach for mitigation designs (Lave 2009), and has been effectively used for a range of resource mitigation projects to satisfy permit conditions or for projects conducted for general ecological benefit.

Because the various NCD approaches are applicable at any scale, NCD is applicable to the range of mitigation requirements that may be necessary for activities in the energy industry.

Presently, the energy industry is proficient at limiting direct impacts to surface waters. The industry operates at an accelerated pace, which does not allot months for regulatory agencies to issue waterway encroachment permits. Designers of energy projects have become diligent in the practice of impact avoidance.

But in cases where resource impacts are unavoidable, mitigation may be required. In these cases, it is important for the mitigation project to be successful so that a company is not bound to a mitigation project for years to come. When mitigation is necessary, permit conditions require the mitigation monitoring for a minimum of five years post-construction. This period may be extended if the project fails to meet minimum requirements for measuring success. For mitigation projects in designated high-quality watersheds, the length of the monitoring period may be set at ten years. Generally speaking, energy companies do not want to be saddled to lengthy mitigation monitoring. The knowledge of this potential has contributed to the high rate of resource impact avoidance in the energy industry.

Resource impact avoidance is readily accomplished when designing projects with small footprints (e.g., well pads). However, projects with larger extents, such as pipelines or coal processing surface facilities, have a more challenging task when trying to avoid impacts. A pipeline corridor, for example, is likely to encounter numerous resources. This would lead to smaller individual impacts, but may have a cumulative effect that would trigger the requirement for mitigation.

In the case of point-location mitigation, continuity of habitat and channel conditions from the impacted reach to the unaltered reaches is critical to prevent isolating aquatic species populations and destabilizing the reach. The mitigated reach must be consistent with the undisturbed condition upstream and downstream. The mitigation effort must be designed and constructed in a manner that, for example, does not form nick points or significant gradient shifts in the stream profile, does not create homogenous habitat, does not create improper velocity gradients, and does not introduce streambed material that is different than the adjacent reaches.

Beyond the benefits of NCD to the channel and stream ecology, NCD methods benefit the entity that is ultimately responsible for the work and/or is financing the work. These projects provide the designer with opportunities to develop creative solutions for the design that utilize materials occurring at the project site, thereby reducing costs. For example, if trees must be removed before work begins, the downed trees can be re-used to build in-stream and bank structures. This eliminates costs for purchasing and hauling materials to and from the site. Along similar lines, small trees and shrubs growing at the site can be transplanted and live whips can be cut and planted, thereby saving money on plant materials. Cost-saving measures like these are not exclusive to NCD. However, designers of NCD projects are more likely to consider re-use of materials already present at a project due to the emphasis on ecology.

In general, NCD projects are more cost-effective because they require less maintenance. Yet, there are NCD projects that are not designed with the full execution of an NCD technique. Theoretically, the mitigation would be designed with proper consideration of stream flows, sediment capacity and competence, and boundary conditions (Selego et al. 2012). Yet designs may be developed without completing full analyses of the reference reach and target stream. This is commonly due to the designer having limited time and/or funding with which to complete the design. As such, some designs have inherent flaws that are not realized until the project fails. Critics of NCD point to these failures as evidence that it is not a reliable method for channel design (Smith & Prestegaard 2005; Juracek & Fitzpatrick 2003). Realistically, it is not the method that is unreliable, but more so the inadequate execution of the methodology. Any project, whether using NCD or hydraulic engineering or another method, will be compromised or require maintenance if it is not designed properly.

BIOLOGICAL RESPONSE TO NCD

Regulatory agencies require stream mitigation projects to exhibit an ecological lift expressed by increases in biological productivity. Therefore, stream mitigation projects are designed with consideration of maximizing the potential to achieve that lift. Despite skepticism about the effectiveness of NCD for designing stable streams (see Smith & Prestegaard 2005; Lave 2009), the success of mitigation projects is predicated on the recovery of the biological community. This has directed a shift in attitudes toward the use of NCD with an emphasis on habitat heterogeneity.

By emphasizing habitat, a project has a greater chance of successfully increasing biological density and diversity, when compared to hydraulic-based designs (Miller et al. 2012). Using native, on-site materials or incorporating regionally-appropriate organic materials (e.g., sandstone vs. limestone; deciduous wood vs. coniferous wood) will more accurately re-establish or replicate native conditions within a stream, which favors repopulation by aquatic species. This also applies to restoration of the riparian corridor. Properly-designed and constructed NCD projects provide greater opportunities to achieve an ecological lift (Selego et al. 2012).

It is important to quantify the density of stream facets (e.g., riffle, run, pool, glide) and assess the distribution of rock, woody debris, and other abiotic factors of a reference reach so the mitigation design replicates the distribution of these factors (Taniguchi and Tokeshi 2004). Some designers of NCD projects default to using only rock for stream structures, erring toward armoring the stream against erosion. However, if this condition does not reflect the reference reach, the macroinvertebrate community in the mitigation reach may have difficulty even returning to the pre-construction condition (Miller et al. 2012).

Large woody debris has been shown to have a correlation to significant increases in biomass and taxa richness in streams, whereas designs that heavily utilize large rocks for structural and habitat improvements do not show significant biomass and richness improvements (Miller et al. 2010). Additionally, LWD provides a structural and organic component to a lotic system that may be critical habitat for multiple phases of aquatic organisms' life histories. Species that have preferences for LWD will repopulate reaches that lack sufficient LWD. However, the populations are likely to be transient in nature, only residing in less suitable habitat for short periods (Noble 2002).

Ideally, ecological lift is best expressed by the return of formerly extirpated sensitive taxa or increases in taxa/species that are intolerant of pollution. Such taxa would include Ephemeroptera/Plecoptera/Trichoptera macroinvertebrates (EPT taxa; e.g., mayflies, stoneflies, caddisflies), non-game fishes, and amphibians (e.g., salamanders). In the past, evaluations of the success of mitigation projects were focused on monitoring game fish populations, largely due to the financial value of a thriving fishery (Larson et al. 2001; Thompson 2006). However, more recent studies have underscored the importance of the macroinvertebrate community's role in maintaining the function of an aquatic ecosystem (Baxter et al. 2005), which facilitated a shift of focus to monitoring of macroinvertebrates. Additionally, macroinvertebrates are a more applicable for analyses of ecological lift in first-order streams as few fishes will reside in the uppermost headwater reaches of streams at the top of a watershed.

Miller et al. (2010) conducted an evaluation of stream mitigation strategies in ten replicated studies. Restoration strategies included boulder additions, LWD additions, and channel reconfigurations. Miller's team targeted the evaluation metrics of macroinvertebrate taxa richness (e.g., total taxa present) and macroinvertebrate density (e.g., total individuals). Their evaluation determined that significant, positive effects occurred in macroinvertebrate communities when a combination of restoration strategies was used to increase habitat heterogeneity. The addition of LWD and boulders returned increases in taxa richness and density to varying degrees. Richness and density increases associated with channel reconfiguration were not significant. Cumulatively, the addition of LWD showed the largest increases in richness and density. The addition of LWD had a consistent improvement of habitat heterogeneity by affecting facet transition patterns (i.e., created more riffle-pools, or reduced riffle-pool spacing).

But despite the general trends in increases of taxa richness and density observed by Miller et al. (2010), they acknowledge that there was no correlation between the two metrics. An increase in one metric was not necessarily accompanied by an increase in the other. Taxa richness was a more significant metric in general, whereas macroinvertebrate density was largely inconclusive.

Taxa richness and macroinvertebrate density are commonly used to evaluate mitigation projects. But using both to evaluate a project can provide conflicting results. Evaluations of mitigation projects across a range of habitat-improvement strategies propose that macroinvertebrate richness may improve steadily, although it may not be an immediate response (Miller et al. 2010).

However, despite the ecological lift envisioned when designing a stream mitigation project, there is no certainty that ecological lift will be realized. The uncertainty of the success can be attributed to numerous factors. The majority of stream projects target a stream reach, rather than employ watershed-level mitigation (Lake 2007). As such, although the mitigation reach may be stable and provide ample habitat, there may be external influences from the surrounding watershed that can limit recovery potential. These may include land use, water chemistry, point-source/non-point source pollution, among other limiting factors (Miller et al. 2010).

A common problem with stream mitigation projects being "successful" is that the success criteria are often not appropriate for the specific project. Regardless of the evaluation metrics being used, it is too easy, particularly for biologists and ecologists, to be optimistic when predicting post-mitigation recovery. This may result in projects being seen as failing, when they may in fact be successful at achieving a level of ecological lift (Miller at al. 2010). It is imperative that designers anticipate a realistic outcome in terms of ecological lift and thoroughly contemplate the evaluation metrics in order to apply the correct metrics for the nature of the project.

SUMMARY

The rapid growth of the energy industry in America has revitalized the scrutiny to which energy companies are subjected. Impacts to natural resources are a premier topic in the industry, public, and government. The exploration, extraction, and transportation of energy sources will have certain unavoidable environmental impacts. It is vital that the energy industry take measures to ensure that impacts to aquatic resources from their activities are mitigated properly and successfully to provide an acceptable off-set to the impacts.

In terms of mitigation for impacts to stream resources, NCD methods provide the best prospect for achieving the ecological lift required by regulatory agencies. These methodologies, the Rosgen Geomorphic Channel method in particular, place equal emphasis on channel hydraulic geometry, planform, sediment capacity and competence, and habitat. This builds a foundation for successful recovery of biological communities within the mitigation area.

It is important, however, that the process of NCD be implemented properly. The design approaches have been developed over many years, and neglecting to consider the details of the methodologies will compromise the project. It is also critical that the measures by which the project's success is judged are carefully selected, and that those responsible for conducting monitoring understand the purpose and applicability of those measures. Yet agencies and designer must establish standard guidelines for the ecological lift requirement. Natural systems are intrinsically variable and their response to any activity varies and therefore cannot be assumed.

ACKNOWLEDGMENTS

The argument for the biological effects of NCD is based largely on the in-depth study and work published by Miller et al. (2010). As this paper is intended to serve as a platform for discussion, specific data, graphs, and other supporting information from Miller et al. (2010) are not included. Refer to the referenced paper for detailed results from said study.

REFERENCES

Baxter, C.V., K.D. Fausch, and W.C. Sanders. 2005. Tangled webs: reciprocal flows of invertebrate prey link streams and riparian zones. *Freshwater Biology* 50:201–220.

Bernhardt, E.S., E.B. Sudduth, M.A. Palmer, J.D. Allan, J.L. Meyer, G. Alexander, J. Follastad-Shah, B. Hassett, R. Jenkinson, R. Lave, J. Rumps and L. Pagano. 2007. Restoring rivers one reach at a time: results from a survey of U.S. river restoration practitioners. *Restoration Ecology* 15:482–493.

Dunne, T. and L.B. Leopold. 1978. *Water in Environmental Planning.* San Francisco: W.H. Freeman & Company.

Juracek, K.E. and F.A. Fitzpatrick. 2003. Limitations and implications of stream classification. *Journal of the American Water Resources Association* 39(3):659–670.

Karr, J.R. 1991. Biological integrity: a long-neglected aspect of water resource management. *Ecological Applications* 1(1):66–84.

Lake, P.S., N. Bond, and P. Reich. 2007. Linking ecological theory with stream restoration. *Freshwater Biology* 52:597–615.

Lave, Rebecca, 2009. The controversy over natural channel design: substantive explanations and potential avenues for resolution. *Journal of the American Water Resources Association.* 45(6):1519–1532.

Larson, M.G., D.B. Booth, and S.A. Morley. 2001. Effectiveness of large woody debris in stream rehabilitation projects in urban basins. *Ecological Engineering* 18:211–226.

Miller, S.W., P. Budy, and J.C. Schmidt. 2010. Quantifying macroinvertebrate responses to in-stream habitat restoration. *Restoration Ecology* 18(1):8–19.

National Resource Conservation Service. 2007. *Part 654, National Engineering Handbook; Stream Restoration Design.* United States Department of Agriculture. http://directives.sc.egov.usda.gov/viewerFS.aspx?hid=21433.

Noble, J.J. 2002. *Distribution and life history characteristics of fish species-at-risk in the Great Lakes basin.* Youngstown: Youngstown State University.

Palmer, M.A., R.F. Ambrose, and N.L. Poff. 1997. Ecological theory and community restoration ecology. *Restoration Ecology* 5:291–300.

Selego, S.M., C.L. Rose, G.T. Merovich, S.A. Welsh, and J.T. Anderson. 2007. Community-Third-Order Tributary of the Potomac River, USA. *International Journal of Ecology* 2012:1–9.

Shields, F.D. and Copeland, R.R. 2006. *Empirical and analytical approaches to stream channel design.* Proceedings, Eighth Federal Interagency Sedimentation Conference. Advisory Committee on Water Information, Subcommittee on Sedimentation. Washington D.C.

Smith, S.M. and K.L. Prestegaard, 2005. Hydraulic performance of a morphology-based stream channel design. *Water Resources Research* 41(W11413).

Rosgen, D. 1996. *Applied River Morphology.* Pagosa Springs: Wildland Hydrology.

Taniguchi, H., and M. Tokeshi. 2004. Effects of habitat complexity on benthic assemblages in a variable environment. *Freshwater Biology* 49:1164–1178.

Thompson, D.M. 2006. Did the pre-1980 use of in-stream structures improve streams? A reanalysis of historical data. *Ecological Applications* 16:784–796.

Hydrologic Assessment of the Guy Cove Stream Restoration Project

Travis P. Maupin
Biosystems and Agricultural Engineering Department, University of Kentucky, Lexington, Kentucky

Carmen T. Agouridis
Biosystems and Agricultural Engineering Department, University of Kentucky, Lexington, Kentucky

Richard C. Warner
Biosystems and Agricultural Engineering Department, University of Kentucky, Lexington, Kentucky

ABSTRACT

The surface mining method of mountaintop removal is a common method used to extract coal seams in the mountainous terrain of the Central Appalachian coal fields. The excess spoil or overburden produced from this practice is placed in adjacent valleys; this produces valley fills (VF). These VFs bury headwater stream systems resulting in alterations to the hydrologic characteristics of down-gradient stream reaches. Traditional reclamation practices create highly compacted VF's which resulted in low infiltration rates, increased discharge, and inhibited the reforestation potential of the mine site. Recent studies have shown that forest productivity can be increased when restoration efforts follow the Forestry Reclamation Approach (FRA). This approach has been shown to increase infiltration, root penetration, and gas exchange which in turn have decreased surface runoff and erosion potential. Research has also indicated that the hydrological characteristics on test cells built to FRA standards, for a variety of spoil types, were trending towards those of a forested watershed. It has been suggested by the U.S. Environmental Protection Agency (USEPA) that limiting the amount of water allowed the infiltrate completely through the VF could possibly improve water quality. The Guy Cove stream restoration project was the first to demonstrate that a stream could be constructed atop a retrofitted VF. The project incorporated natural channel design techniques in conjunction with the FRA in an effort to restore the hydrologic, water quality, and ecological characteristics of the watershed.

The hydrologic characteristics (discharge volume, peak discharge, discharge duration, time to peak, lag time, and response time), which resulted from the restoration project, were evaluated for 2010 and 2011. The characteristics were compared to a stream from an unmined, forested watershed and to a stream from a mined but unrestored, traditionally reclaimed, watershed. Preliminary results indicate that for storm events, the restored stream on the crown is performing similarly to the unmined, forested watershed for the hydrologic parameters evaluated. When considering base flow conditions, such as the number of days of flow at each reach, significant differences emerged for the section of stream located at the crown and face junction. This location was the only site without a baseflow contribution, and as such, it had the least number of days with flow, particularly during the summer and fall months.

ACKNOWLEDGMENT

The authors thank Christopher D. Barton for his contribution to this paper.

Guy Cove Stream Restoration Project: Recreating a Headwater Stream System on a Head-of-Hollow Fill

Carmen T. Agouridis
Biosystems and Agricultural Engineering Department, University of Kentucky, Lexington, Kentucky

Richard C. Warner
Biosystems and Agricultural Engineering Department, University of Kentucky, Lexington, Kentucky

ABSTRACT

A growing body of research has illustrated how crucial first-order streams are to the overall health and function of aquatic ecosystems. Mountaintop mining poses a significant environmental risk to aquatic and terrestrial life through the direct conversion of headwater streams to uplands, and through the associated degradation of downstream receiving waters, particularly with regards to electrical conductivity and sedimentation. In an effort to replace lost headwater streams in both form and function, improve water quality, and restore watershed function, the University of Kentucky created a headwater stream system on a retrofitted valley fill, which is located in Breathitt County, Kentucky. The design was largely built on natural channel design (NCD) principles as well as the Forestry Reclamation Approach (FRA), which encourages a non-compacted spoil medium to promote tree growth, in an effort to address concerns related to water quantity and quality as well as habitat development. The major components of the design included (1) modifications to the crown geometry, (2) compaction of the crown to control infiltration, (3) utilization of natural channel design techniques, (4) use of the FRA to promote tree growth, (5) creation of ephemeral channels and vernal ponds, and (6) implementation of a novel bioreactor-wetland treatment system to improve water quality.

A total of 3,335 ft of intermittent and perennial channel, 1,555 ft of ephemeral channel, and 0.6 ac of wetland and vernal pools were constructed plus 30,000 trees were planted on 40 ac from mid-2008 to early-2009. To date, four years of geomorphic and water quality data and three years of streamflow and habitat data have been collected. Geomorphic surveys indicated that the stream reaches are laterally and vertically stable. Elk browsing in the riparian area is an issue; the effect is seen in tree growth but not tree survival. However, in the uplands where FRA was employed, the trees are growing rapidly. The average electrical conductivity levels, for the downstream portion of the stream along the crown, continue to drop. In 2012, the average value was 595 $\mu S\ cm^{-1}$ which is close to the 500 $\mu S\ cm^{-1}$ threshold recommended by the U.S. EPA. Salamanders and aquatic invertebrates began to recolonize the project site soon after construction. Preliminary macroinvertebrate data are encouraging with increases in EPT taxa noted.

ACKNOWLEDGMENT

The authors thank Christopher D. Barton and Hwa-seong Jin for their contributions to this paper.

Leaf Litter Breakdown as a Functional Assessment Tool for Appalachian Coal Mine Stream Restorations

Robert J. Krenz III
Department of Forest Resources and Environmental Conservation, Virginia Water Resources Research Center, Virginia Tech, Blacksburg, Virginia

Stephen H. Schoenholtz
Department of Forest Resources and Environmental Conservation, Virginia Tech, Blacksburg, Virginia

Carl E. Zipper
Department of Crop & Soil Environmental Sciences, Virginia Tech, Blacksburg, Virginia

ABSTRACT

Stream restoration is a standard means of compensatory mitigation, a requirement of the Clean Water Act [section 404], intended to produce ecological benefits to offset those lost due to mining impacts. Although assessment of stream restorations has historically focused on structural measures of stream condition such as physicochemical and biotic metrics, recent guidance from USEPA requires direct measurement of both structure *and* function to assess the efficacy of stream restoration efforts. Functional metrics are derived from serial measurements representing dynamic processes occurring through time rather than at discreet points, and are expressed as rates. Leaf litter breakdown provides a relative index of an important energetic function over fairly long intervals (typically ≤1 yr), and does not require exceedingly expensive equipment or technical expertise. To address the needs of regulators, industry, and stream restoration professionals, we are measuring breakdown rates of *Quercus alba* leaves for eight first and second order mining-impacted stream restorations in the coalfield region of southwestern Virginia, evaluating those rates through comparisons to four minimally impacted reference streams, and determining relationships of physicochemical factors with those rates. Coarse-mesh and fine-mesh bags were loaded with 6.5 g (±0.01 g) of *Q. alba* leaves, and 288 of each type were deployed in study streams in December 2010. Bags were retrieved in triplicate monthly through April 2011, and bi-monthly thereafter through October 2011. After washing, mean percent ash-free dry mass remaining was determined, and an exponential decay model was used to determine breakdown rates (day^{-1} and degree-day^{-1}). Mean coarse-mesh breakdown in forested reference streams (0.023 ±0.003 d^{-1}) was nearly twice as fast as that measured in mined streams (0.012 ± 0.005 d^{-1}). Similarly, mean fine-mesh breakdown in mined streams (0.007 ±0.003 d^{-1}) was approximately half the rate measured in forested reference streams (0.013 ±0.003 d^{-1}). Median breakdown rates were significantly different between stream type for both coarse-mesh [$z = 2.63$; $p = 0.0085$] and fine-mesh [$z = 2.12$; $p = 0.0338$]. Comparison of breakdown rates from fine-mesh, designed to exclude benthic macroinvertebrates, and coarse-mesh bags will provide insight into the influence of the benthic macroinvertebrate assemblage on leaf litter breakdown relative to that of microbial and physicochemical factors alone. Additionally, we anticipate that relating leaf litter breakdown with structural factors will help guide future stream reconstruction efforts and aid development of functional assessment protocols. This study is being augmented with an additional season of observations with final retrieval projected for September 2012.

ACKOWLEDGMENT

This study was sponsored by the Appalachian Research Initiative for Environmental Science (ARIES). The views, opinions and recommendations expressed herein are solely those of the authors and do not imply any endorsement by ARIES employees, other ARIES-affiliated researchers or industrial members. Information about ARIES can be found at http://www.energy.vt.edu/ARIES.

Substrate and Seeding Treatment Effects on Native Tree Establishment and Herbaceous Cover on an Appalachian Coal Mine: First Growing Season Results

S.C. Koropchak
Crop and Soil Environmental Sciences, Virginia Tech, Blacksburg, Virginia

C.E. Zipper
Crop and Soil Environmental Sciences, Virginia Tech, Blacksburg, Virginia

D.M. Evans
Crop and Soil Environmental Sciences, Virginia Tech, Blacksburg, Virginia

ABSTRACT

This study deals with the effects of various substrate treatments and seeding regimes on the survival and growth of native tree species, as well as the establishment of native herbaceous vegetation from the local species pool on a reclaimed surface coal mine in southern WV. Four substrates were placed by the Apogee Coal Company in winter 2011/2012: weathered sandstone (brown), unweathered sandstone/shale mix (gray), mixture of weathered and unweathered rock (mixed), and native topsoil, a mixture of the soil solum and unconsolidated soil parent material (soil). The experimental area occupies approximately 16 acres, with each substrate replicated 4 times. Each substrate plot was split into two subplots; one seeded with a tree-compatible seeding mixture and one left unseeded. Trees were planted in March 2012, tallied and measured for initial height in June 2012, and measured for height and survival in late October 2012. Herbaceous groundcover and species richness were measured in late July. After one growing season, mean percent survival and growth of planted trees differed among tree species and seeding treatments. There were no interactive effects. There were no differences in tree survival or growth among substrate treatments ($P > 0.10$). Of planted tree species, survival was highest for hawthorn and black cherry (~85%) and lowest for Eastern white pines (25.3%) and shagbark hickory (24.3%). Hawthorn (10.2 cm) and black cherry (11.3 cm) height increased more than any other tree species, and white pines height decreased more than any other species (14.2 cm) ($p<0.0001$). Unseeded treatments had higher survival (70.4%) than seeded treatments (56.4%). Unseeded treatments showed a mean height increase of 1 cm, whereas seeded treatments had a mean height decrease of 5.7 cm. Gray sandstone and soil treatments had the highest overall herbaceous richness ($p=0.0007$) and soil treatments had the highest volunteer richness ($p=0.0032$). Mixed substrate treatments had the lowest total and volunteer species richness of all treatments. Seeded treatments had less bare ground and higher mean species richness than unseeded plots ($p<0.0001$). After one growing season, planted trees' survival and growth vary among species and with seeding practice, and herbaceous vegetation cover varies with seeding practice and substrate treatment. Further development of the plant communities may increase differences among soil and seeding treatment effects in future years.

ACKNOWLEDGMENT

The authors thank J.A. Burger for his contribution to this paper.

7.1 Total Dissolved Solids and Environmental Impacts 2

Conductivity Sensor Accuracy and Temporal Stability: A Laboratory and Field Study 470

Screening-Level Specific Conductivity Testing of Drill Core Strata 471

Field Data and Regulatory Aquatic Life Criteria 472

Chronic Toxicity Testing in Mining Influenced Streams of West Virginia 474

Conductivity Sensor Accuracy and Temporal Stability: A Laboratory and Field Study

Travis P. Maupin
Biosystems and Agricultural Engineering Department, University of Kentucky, Lexington, Kentucky

Carmen T. Agouridis
Biosystems and Agricultural Engineering Department, University of Kentucky, Lexington, Kentucky

Richard C. Warner
Biosystems and Agricultural Engineering Department, University of Kentucky, Lexington, Kentucky

ABSTRACT

Water quality characteristics fluctuate in response to changes in environmental factors such as precipitation, land use, time of day (diurnal), and season or climate. In order to adequately account for these variations in water quality, continuous water quality monitoring sensors are needed. Such high frequency in-situ monitoring is best suited for capturing cyclical trends associated with seasons or diurnal fluctuations as well as rapid changes associated with storm events. But which continuous monitoring sensor to select becomes an important question. Studies on soil moisture sensors have shown large variations in performance. Is this the case for water quality sensors, and in particular specific conductivity sensors? In the Appalachian region, specific conductivity ($EC_{25°C}$) is of most interest because elevated conductivity levels have been linked to declines in EPT taxa. Because of this, the U.S. Environmental Protection Agency (USEPA) issued guidance stating that waters discharged from mine sites in this region should have $EC_{25°C}$ levels less than 300–500 µS cm^{-1}. This guidance has implications not only for mining but also for stream restoration projects, which are conducted in the region especially if done for mitigation. Thus, being able to accurately determine the $EC_{25°C}$ levels has significant implications, especially as $EC_{25°C}$ levels approach this designated threshold. As such, four commercially available continuous recording conductivity sensors were compared in in the laboratory as well as in the field in stream reaches with low to high $EC_{25°C}$ levels. Laboratory results showed that (1) sensor over- and under-estimation typically ranged from 10–150 µS cm^{-1}, and (2) at least one sensor within each sensor type performed quite differently than the others of the same type. Field results indicated that fouling, even within days, significantly interfered with performance at high conductivity levels. The mixed chemistry of high conductivity waters means that sensor ruggedness is a critical factor.

ACKOWLEDGMENTS

This study was sponsored by the Appalachian Research Initiative for Environmental Science (ARIES). The views, opinions and recommendations expressed herein are solely those of the authors and do not imply any endorsement by ARIES employees, other ARIES-affiliated researchers or industrial members. Information about ARIES can be found at http://www.energy.vt.edu/ARIES.

The authors thank Christopher D. Barton for his contribution to this paper.

Screening-Level Specific Conductivity Testing of Drill Core Strata

Tessa Monday
Mining Engineering Department, University of Kentucky, Lexington, Kentucky

Richard C. Warner
Biosystems and Agricultural Engineering Department, University of Kentucky, Lexington, Kentucky

ABSTRACT

New federal restrictions on the specific conductivity of water discharge have inhibited the permitting of hollow fills in the Kentucky and West Virginia states of the Appalachian coal field. The objective of this applied research was to develop/modify and evaluate a screening-level testing protocol that may assist mines in determining the propensity of selected geologic strata to generate low, medium or high levels of specific conductivity. The method of overburden field leach testing utilizes geologic exploration cores to perform a comparative analysis of potential generation of specific conductivity. The test was modified from a USGS field leach test that simulates the reactions that occur when materials are leached by water. The overburden field leach test is inexpensive, easy to perform without the need for reagents other than distilled water and hydrogen peroxide and assessments can be rapidly completed. The results of this screening-level testing procedure, based on five cores from three mines in Kentucky and West Virginia, have indicated the potential capacity of identifying geologic strata that requires isolation from oxidation processes. This test can assist in selective mine spoil placement to reduce the specific conductivity of water discharge. Furthermore, strata has been identified that can be used for French underdrains and for isolating materials if rock durability and permeability engineering specifications are respectively satisfied. There are preliminary indications that test results can be regionally extrapolated to gain knowledge of strata—conductivity relationships. Leachate samples can be brought to a lab for further chemical analysis to identify which constituents are contributing to lessened water quality. This test can assist mining companies in obtaining the necessary permits for hollow fills, provide guidance for mine planning and operations associated with material handling and in meeting specific conductivity restrictions on water quality.

ACKOWLEDGMENT

This study was sponsored by the Appalachian Research Initiative for Environmental Science (ARIES). The views, opinions and recommendations expressed herein are solely those of the authors and do not imply any endorsement by ARIES employees, other ARIES-affiliated researchers or industrial members. Information about ARIES can be found at http://www.energy.vt.edu/ARIES.

Field Data and Regulatory Aquatic Life Criteria

Robert Gensemer
GEI Consultants, Inc., Washington, D.C.

Steve Canton
GEI Consultants, Inc., Washington, D.C.

Shaun Roark
GEI Consultants, Inc., Washington, D.C.

Karen Bennett
Hunton & Williams LLP, Washington, D.C.

Brooks Smith
Hunton & Williams LLP, Washington, D.C.

ABSTRACT

The current practice for deriving numeric water quality criteria for protection of aquatic life is to use results from standard laboratory toxicity tests with a range of surrogate test species using defined protocols to calculate both acute and chronic criteria. Derivation of these criteria are based on specific data quality objectives and requirements intended to result in broad levels of protection for aquatic communities exposed to pollutants, while recognizing that aquatic ecosystems can tolerate some stress and occasional adverse effects, and, therefore, protection of all species at all times and places is not deemed necessary.

Such criteria derivation protocols have been used for decades to develop water quality standards, effluent limitations in discharge permits for a wide range of chemical pollutants under EPA's National Pollutant Discharge Elimination System (NPDES), and for other uses, such as 303(d) impairment listings and ecological risk assessments. As a matter of both practice and precedent, the development of such standards and limitations typically accounts for additional factors such as social, legal, economic and hydrological considerations.

Recently, however, new approaches have been proposed to derive numeric chemical criteria on the basis of empirical relationships between benthic macroinvertebrate community structure and pollutant concentrations in synoptic surveys. The most recent examples of this approach by EPA include derivation of numeric nutrient criteria in Florida and a conductivity benchmark in central Appalachian streams chemically dominated by sulfate and bicarbonate salts. While use of field data is attractive because it tends to better reflect reality (i.e., what is actually happening in the water environment as opposed to in a particular lab test), identifying causality, minimizing the influence of confounding factors, and acknowledging inherent weaknesses using data from samples collected for other purposes, makes derivation of such numeric criteria or benchmarks extremely challenging. These impediments to devising sound and defensible criteria become all the more difficult when placed in the context of regulatory programs where the criteria are applied with compliance and enforcement consequences.

Using the conductivity benchmark as a case study, this presentation explores the difficulties of using synoptic field survey data in deriving regulatory aquatic life criteria. Such difficulties include statistical methods and models used, lack of established causality, and approaches for identifying or minimizing the influence of confounding environmental factors. We conclude that, in this case, increases in conductivity (or the ions measured by conductivity) have not yet been shown to be causally related to changes in benthic macroinvertebrate community composition. Furthermore, the inability of such benchmarks to account for confounding factors such as benthic habitat characteristics, stream flow, and other anthropogenic influences, as well as basic issues with sampling design, prevent rigorous application of a numeric aquatic life benchmark based on conductivity.

A critical flaw in the conductivity benchmark is that the macroinvertebrate data used to calculate the benchmark were not collected and enumerated in a manner appropriate for the assumptions of the model used to develop the benchmark. Specifically, US EPA developed the conductivity benchmark by repurposing a regional dataset of benthic invertebrate community samples collected by the West Virginia Department of Environmental Protection (WVDEP). The WVDEP data were originally collected to calculate the West Virginia Stream Condition Index (WVSCI). Identification and enumeration of invertebrates in samples collected for the WVSCI was done using fixed-count, random grid-based subsamples of 200 ±40 individuals from kicknet or dipnet samples representing a one square meter riffle area.

Using the WVDEP data, US EPA estimated a value called the "extirpation concentration" (XC95) for each genus by calculating the 95th percentile of the sampling density-weighted cumulative distribution function of conductivity values where the genus was present. The final conductivity benchmark value was then determined as the 5th percentile of the ranked genus XC95 values.

We evaluated the effect of fixed-count subsample size on the final benchmark value by recalculating the conductivity benchmark multiple times with successively reduced fixed-count subsample size. We used resampling to simulate subsampling and evaluate the influence of subsampling on the inference of absence of taxa. We also determined how the variation in presence/absence conclusion in turn influences the XC95 calculation.

Our results demonstrate that as the subsample size is reduced, the variability of the XC95 estimate increases, suggesting that erroneous presence/absence determination due to random sampling variation in each subsample results in uncertainty in the estimated XC95 for less common taxa. Ultimately, our results demonstrate that the final HC05, or conductivity benchmark, is biased downward due to the effect of using a small fixed-count.

In addition, in the process of replicating the model used to calculate the conductivity benchmark, we found that arbitrary and/or subjective decisions made in the mathematical model substantially affect the final benchmark value, such as decisions on "bin" size for their weighting process and the minimum number of sites a genus had to be present at.

The technical problems associated with derivation of the conductivity benchmark raise a fundamental question about whether the benchmark approach is appropriate or defensible for application in various Clean Water Act programs. This paper will contrast the EPA's benchmark approach against the robust data requirements of EPA's 1985 National Guidelines for Deriving Aquatic Water Quality Criteria in an attempt to answer the question, is there an appropriate role for a "benchmark" based on field data in the regulatory context? And, if so, what safeguards needs to be in place to ensure that the benchmark is defensible and can be used effectively—within its own limitations—for other Clean Water Act purposes?

Chronic Toxicity Testing in Mining Influenced Streams of West Virginia

Mindy Armstead
Integrated Science and Technology Department, Biology Department and Marshall University, Huntington, West Virginia

ABSTRACT

Whole effluent toxicity (WET) tests have become a common tool in the evaluation of effluent for discharge acceptability. In this study, four years of toxicity data from 119 sampling locations are analyzed to determine relationships with ions and conductivity as indicators of toxicity. WVSCI scores are also examined to evaluate correlations between stream scores, conductivity, and IC25 endpoints from toxicity results. Conductivity was not an indicator of toxicity in the range of conductivities tested. Streams dominated by mining effluent sometimes exhibited toxicity to *Ceriodaphniadubia*; however, toxicity was not found to be related to ionic concentration in the range tested. Although mortality and reproductive impairment were often demonstrated in the mining effluent dominated streams, there were no relationships established between survival and reproductive endpoints and the ionic concentrations. Benthic macroinvertebrate communities in the streams sampled indicated some level of impairment. Only a weak relationship was demonstrated between habitat assessment scores and WVSCI scores. No apparent relationship between conductivity and WVSCI was observed.

ACKOWLEDGMENT

This study was sponsored by the Appalachian Research Initiative for Environmental Science (ARIES). The views, opinions and recommendations expressed herein are solely those of the authors and do not imply any endorsement by ARIES employees, other ARIES-affiliated researchers or industrial members. Information about ARIES can be found at http://www.energy.vt.edu/ARIES.

The author thanks Leah B. Creathers for her contribution to this paper.

7.2
Environment, Coal Mining, and Sustainability

Surface Coal Mine Blasting Optimization and Mitigation of Environmental Impacts: Mine to Fill 476

Production, Cost, and Environmental Impact of Loading Equipment in Surface Coal Mining in Appalachia 490

NOx Emission of Equipment and Blasting Agents in Surface Coal Mining 501

Micro-Hydroelectric Power Generation Utilizing Mine Drainage 512

Minepool Geothermal in Pennsylvania 520

Sustainability in Surface Coal Mine Operations: The Importance of Good Geotechnical Design 529

Ethics in Energy Sustainability 530

Surface Coal Mine Blasting Optimization and Mitigation of Environmental Impacts: Mine to Fill

Catherine E. Johnson
University of Kentucky, Lexington, Kentucky

Braden T. Lusk
University of Kentucky, Lexington, Kentucky

Joshua M. Hoffman
University of Kentucky, Lexington, Kentucky

John T. Rathbun
University of Kentucky, Lexington, Kentucky

ABSTRACT

Environmental impacts from blasting are traditionally related to the control of vibration, airblast levels and flyrock near mine operations. None or very little work either experimental or theoretical has been performed regarding environmental impacts from blasting with reference to surface water and air quality. Other aspects could be considered, resulting from the fragmentation of the blasted material and its final use as fill material. This paper discusses the optimization of mine blasting activity in surface coal mines based on the premise of mitigating environmental impacts different from those traditionally studied. This mine to fill analysis is conceptualized through applying classical fragmentation analysis.

A conceptual downstream cost analysis of the blasting activity is introduced to evaluate the impacts of the environmental improvements in the blast in the final mining cost. Fragmentation is a key element in the analysis due to the nature of the use of overburden material throughout the mine processes for reclamation and fill purposes. Environmental impacts of the earth fills in coal mines can be directly related to the properties of material used for the earth fills. The final result of this study is a concept for specifying material for specific use, placement and handling in surface coal mines. A matrix of specifications is introduced and will allow for future optimization of overburden handling with consideration of blasting parameters once finalized specifications are determined.

INTRODUCTION

In the past environmental effects of blasting have predominantly been evaluated by measuring its direct effects. The effects most easily measured traditionally were ground vibrations and air overpressures. Blast optimization therefore sought to minimize ground vibrations and air overpressures while maintaining needed fragmentation and production rates at an economically feasible level. An optimized blast design is one that will break or move rock to the required fragment size for secondary equipment, to efficiently handle it, while minimizing secondary components such as cost and ground vibration effects. To that end, the optimization of blasting is not achieved in the same way at every site.

In addition to these optimization regimes one is proposed that considers indirect environmental effects of blasting. These environmental effects include conductivity, selenium content, Total Suspended Solids (TSS), Total Dissolved Solids (TDS), reforestation, vegetation growth, wildlife and wildlife habitat reintroduction in water runoff from fills created with blasted overburden material. Just as designs to optimize fragmentation and vibration can run counter to one another so can optimization for the different environmental effects. The materials created from blasting events need to be evaluated to determine their short and long term environmental impacts of the fill areas into which the blasted materials are placed. Means of reducing targeted impacts can be achieved through redesigning blasts to produce rock materials with particle sizes, densities, distributions, and compositions that minimize negative environmental impacts. Optimized fill material will change throughout the construction of the fill and this temporal component must be considered as well.

These material particle sizes, densities, distributions, and compositions vary depending on the different environmental impacts that are to be mitigated and the materials in question. Creating these different material specifications can be aided through blasting practices. This process potentially means moving away from the most economical method of removing overburden to one that weighs environmental impacts. This is the cost/benefit associated with the optimization of blasting to minimize the impacts on a given environmental concern. Mining operations must be optimized through their entire operational cycles. There are real costs associated with environmental considerations such as lengthy permitting process, regulatory fines, reclamation, and high capital water treatment facilities. Increased costs associated with processes early on in the cycle, such as blasting, could offset long term remediation costs.

Optimizing blasts to produce materials with required sizes and distributions can be aided by mathematical models which predict particle size and distribution. These models have typically been utilized to design blasts which produce fragment sizes that are best for secondary crushing or handling. It is shown in later sections that these models can be applied to typical blasts from the Appalachian Region. Provided the final product specifications of the rock material are known, blasts can be designed using this optimization process.

TRADITIONAL ENVIRONMENTAL IMPACTS OF BLASTING

Traditionally environmental effects of blasting have been limited to areas of vibration, air overpressure, and flyrock. These effects have been targeted for research and improvement due to the immediacy of their effects, the regulations which govern their limits, and the relative ease by which they are measured. Early work in these areas was conducted by the United States Bureau of Mines. These efforts led to the development of field feasible methods of measuring ground vibrations (Lee, Thoenen and Windes 1936) and air overpressure (Ireland 1942). Over time better methods and tools for measuring ground vibrations and air overpressure were researched (Blair, 1954, Stachura 1981). Eventually, as reliable measurements were made possible, attention was turned to monitor the effects that measurable vibrations had on nearby structures (Edwards 1960, Duvall 1962, Northwood 1963, Nicholls 1971, Medearis 1977, Siskind 1980a, Siskind 1980b, Clark 1983, Stagg 1984, Siskind 1985, Shaw 1989, Siskind 1993, Siskind 2000, Adhikari 2005). Much of this work aided in the promulgation of legislation which regulated mine blasting and defined ground vibration and air overpressure limits. Legislation such as Surface Mining Control and Reclamation Act of 1977 created a regulatory environment that encouraged improvement.

While vibration and damage monitoring work continues today, it has been joined by the companion focus of altering aspects of a blast with the intention of controlling ground vibrations and air overpressures. This research led to theories and methods to reduce ground vibration and air overpressure (Duval 1963, Siskind 1985, Stachura 1986, Kopp 1986, Siskind 1989). Although much of the theory for ground vibration waveform tailoring was developed in the 1980s it has been revisited in recent years due to the introduction of the economically viable electronic detonator.

Flyrock incidents have occurred for as long as there has been blasting. Most research looks into causes and offers up best practices for the prevention of flyrock (Ash 1975, Watson and Roth 1978, Fletcher and D'Andrea 1986, Bajpayee 2003, Bajpayee 2004, Little 2007). This was made imperative as regulations defined flyrock and contained provisions for flyrock control and avoidance (Dick 1979).

SECONDARY ENVIRONMENTAL IMPACTS OF BLASTING

Contemporary environmental concerns that have come under more scrutiny are those of long term environmental impacts. These are focused around water quality, air quality, and resulting downstream habitat effects. Although these are often a result of mining practices as a whole, solutions could be found in the manner in which blasting occurs.

Much of the work investigating gas emissions from blasting has taken place on the laboratory scale. These experiments emulated differential conditions in the field such as explosive contamination and confinement to study their effects on gaseous product production (Chaiken, Cook, and Ruhe 1974, Persson and Persson 1980, Mainiero 1997). Field work on the subject has proven less than conclusive given the physical challenges associated with monitoring gas emissions from entire blasts (Schettler and Brashear 1996, Onederra et al. 2012).

Lashgari et al. (2012) discuss work related to NOx emissions from both blasting and diesel equipment at a specific mine site. Highest NOx emissions come from Haul trucks (44.69%), clearly a truck that is not at full capacity due to a poor distribution of particle sizes will increase the number of trucks and reduce the air quality. Furthermore, blasting is calculated to contribute to the next highest level of NOx emissions at a mine site, increasing explosive quantity or strength to produce different particle sizes will also increase the level of NOx produced at the mine site. Shovel productivity in terms of swings per hour could be reduced if the required size distribution is found to be too great for the bucket size, increasing both cost and air quality. A balance between all environmental effects needs to be established; simply improving one area can have a negative effect on another.

This paper focuses on the potential environmental impacts from materials that have been blasted based on their fragmentation sizing, distribution, and placement. The fills in which the blasted materials are placed are the areas from which environmental concerns arise. These environmental effects include conductivity, selenium content, TSS, TDS present in runoff waters, as well as reforestation, vegetation growth, wildlife and wildlife habitat reintroduction. While these effects will vary with different strata and rock materials (Williams 1994) they will also vary based on the fragment size, and distribution. These latter two qualities can be altered through different blasting practices. Work can be done to develop better ways to construct fills using material with properties that has been historically produced from surface coal mines. This however takes blasting to be a "black box" from which the same material will always be produced. Instead blasting must be considered in this work as well for its ability to produce materials better suited for fill construction.

COMPREHENSIVE EFFECTS OF BLASTING PRACTICES

The cycle associated with mining and reclamation is: Drill, Blast, Dig, Move, Place, Wait, Monitor, React. This overall cyclical operation must be optimized to allow for continued operation and adaptation during increased regulatory scrutiny on the back end of the cycle. These alterations must be evaluated economically as well as environmentally. The effects changes in blasting have on overall operational costs cover everything from drilling and blasting to final placement and reclamation of the material. The concept of mine to mill is not new to the mining industries that process blasted materials (Kanchibotla and Valery 2010). Considering an overall process for blasted overburden is very similar to mine to mill and could be termed "mine to fill." What is proposed in this paper takes into account environmental impacts of these alterations and the associated economics with remediating those impacts. Fragmentation is the first step in this cycle and therefore the first place to make alterations with the intention of improving downstream effects.

Hissem (2013) advises a Lean Six Sigma cost and waste elimination model for an aggregate quarry that demonstrates a relationship between the degree of rock fragmentation and cost per ton. Poor fragmentation from blasting will simply increase costs through hauling, crushing or grinding. There is an optimum that can be found to satisfy all aspects of the quarry cycle for both cost and productivity with just a few changes to personnel attitude and pit design. Drilling is the start of the cost structure; it directly affects the blast, which in turn affects the cost. In a similar way to how the initial blast can directly affect the final cost to the mine, it also has a direct effect to the final environmental issues of the mine. Rather than the pit and plant being seen as separate entities, they should be considered one smooth transition with each affecting the next step, meaning improvements from the start will affect the end result. A similar model and idea can be applied to the 'environmental cost.'

Optimized mine fill for environmental purposes should comprise of a series of rock filters at the base before the main bulk fill is placed. This practice is well established and allows for reduced environmental effects from water runoff at the fill sites. Filters are best formed from medium grain grey sandstone, generally found at surface coal sites (Warner, personal email communication). These materials are particularly robust and allow for less outfall issues such as TDS, conductivity, and Selenium content. Other material found in Appalachian surface coal mines would be used in the main fill material above the filter and drains thus isolating them (Figure 1).

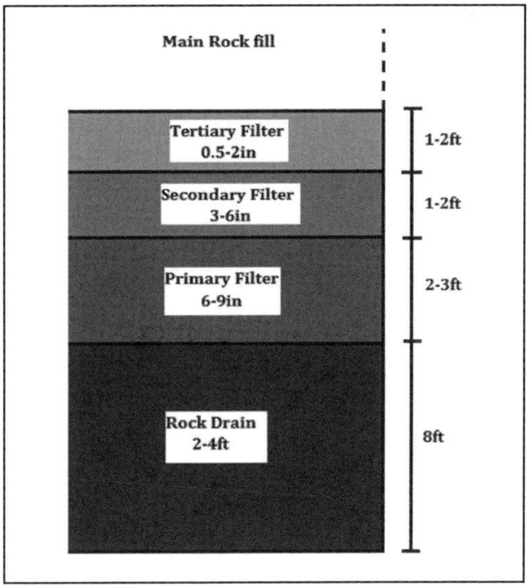

Figure 1. Proposed filter dimensions for fill (Warner, personal email communication)

Through a series of blast designs which target the size ranges for the rock drain, primary, secondary and tertiary filters, it is possible to produce enough material to successfully create the proposed filter design. Only the initial blasts would have to be specifically designed to produce the material desired for the filters, not every blast at the mine site. Changes in blast design have been shown to allow for shifting of the distribution curve for fragmentation. These changes would suggest that blast design could be altered to create more material of a desired size fraction. Considerable adjustment to the blast design may be necessary to achieve such drastically different materials. In the initial stages of fill construction, considerable effort and attention would be required in the blasting phase to achieve the "specified" material. Some additional processing may be necessary, but could be as simple as a grizzly to allow for fine material to pass and not be utilized for drain and filter construction. If a sufficient filter is created for the fill without the use of mechanical separation, solely adapted blast design, it is possible to produce a cost effective, environmentally improved reclaimed mine site.

INTRODUCTION TO FRAGMENTATION MODELLING AND ASSESSMENT

Blasting is used throughout the surface coal mining industry for effective overburden removal. Ideal fragmentation sizes have long been created through adaptations to blast design in order to optimize mine productivity and for commercial

Table 1. Rock factor parameters (Lilly, 1983)

Rock Mass Description (RMD)	Powdery/Friable	10
	Blocky	20
	Totally Massive	50
Joint Plane Spacing (JPS)	Close (<0.1m)	10
	Intermediate (0.1 to 1m)	20
	Wise (>0.1m)	50
Joint Plane Orientation (JPO)	Horizontal	10
	Dip Out of Face	20
	Strike Normal to Face	30
	Dip into Face	40
Specific Gravity Influence (SGI)	SG in tonnes/m^3	25(SG-50)
Hardness Factor (HF)	Elastic Modulus(GPa)/3	1–10

use. Fragmentation has a central influence over mine productivity and cost; MacKenzie (1966) described optimum fragmentation to be "that blasting practice which gives the degree of fragmentation necessary to obtain the lowest unit cost of the combined operations of drilling, loading, hauling and crushing." Overburden is effectively mine waste and has been removed in the most cost effective manner and is often used as fill material in mine reclamation. Typically, blast optimization for overburden has been guided towards optimizing loader and haul truck productivity. Loader productivity is a direct function of the range of rock sizes a loader bucket has to juggle in each filling; both large rocks too big for a loader bucket, and small rocks that simply fall out of the bucket while in a swing motion will impede productivity, consequently increasing the number of bucket loads per blast and therefore costs. Having a medium and uniform size range may consequently increase productivity and reduce overall operation costs. Fill material is usually taken from the muck pile to an overburden waste dump to be used in a fill at a later date for rehabilitation purposes or directly to a fill. Typically no grading or sorting of particle size is carried out in these dumps, even though for stability and water contamination purposes, separate layers of sized material are beneficial. Fine particles are unwanted in fills due to their large surface area and susceptibility to the transport of contaminants such as selenium, TDS or conductivity.

A number of models have been established to predict size distributions from specific blast designs. Predictions are made either through empirical or mechanistic modeling. Empirical modeling assumes finer fragmentation from higher energy input whilst mechanistic modeling tracks the physics of detonation and the energy transfer for specific blast layouts. Quite clearly the mechanistic model requires more data so is more difficult to apply on a day to day basis, therefore, empirical models are most often used for the prediction of fragmentation size. One of the first empirical fragmentation models with respect to blasting is the Kuz-Ram model. The model was established by Cunningham in 1983 as an easy method of estimating fragmentation based on geometric parameters of the drilling and blast design. A mean size is calculated using blast parameters (Kuznetsov 1973) and the fragmentation spread is based on an adapted version of the Rosin-Rammler equation. Four equations are required for the model:

Adapted Kuznetsov equation:

$$x_m = AK^{-0.8} Q^{\frac{1}{6}} \left(\frac{115}{RWS}\right)^{\frac{19}{20}} \quad (1)$$

where x_m = mean particle size (cm); A = rock factor; K = powder factor (kg/m^3); Q = mass of explosive in the hole (kg); RWS = weight strength relative to ANFO, 115 being the RWS for TNT.

Rock Factor, A:

$$A = 0.06(RMD + JPS + JPO + SGI + HF) \quad (2)$$

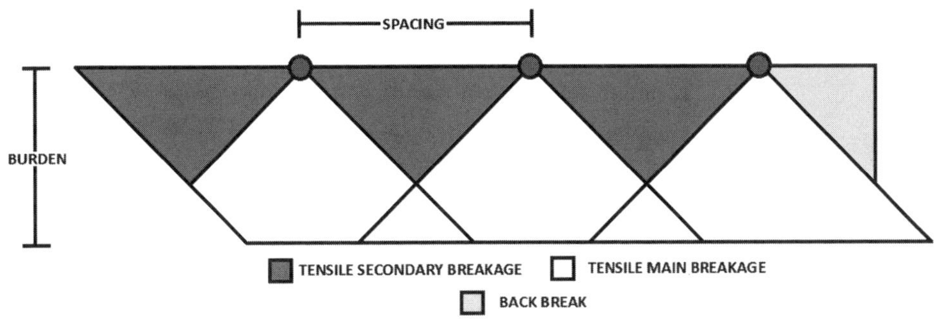

Figure 2. Breakage mechanisms identified by Rollins (1990)

Adapted Rosin-Rammler equation:

$$R_x = \exp\left[-0.693\left(\frac{x}{x_m}\right)^n\right] \qquad (3)$$

where R_x = mass fraction remained on screen opening x; n = uniformity index, between 0.7 and 2.

Uniformity equation:

$$n = \left(2.2 - \frac{14B}{d}\right)\sqrt{\left(\frac{1 + S/B}{2}\right)}\left(1 - \frac{W}{B}\right)\left(abs\left(\frac{BCL - CCL}{L}\right) + 0.1\right)^{0.1}\frac{L}{H} \qquad (4)$$

where B = burden (m); S = spacing (m); d = hole diameter (mm); W = standard deviation of drilling precision (m); L = charge length (m); BCL = bottom charge length (m); CCL = column charge length (m); H = bench height (m) (Cunningham 1983).

The model is simple to use due to the ease of access to data for all parameters involved and does not involve a lengthy image analysis process. Despite this, it has been scrutinized for its lack of inclusion of all parameters involved with rock fragmentation and an underestimation of the prediction of fines (fragments less than 50 mm). Lownds (1983) identified that the uniformity parameter is not influenced by mechanical properties of the rock or characteristics of the explosive. Fragmentation prediction by Cunningham (1983) assumes that a single distribution of pre-existing discontinuities is present within the rock mass and that tensile failure is the main mechanism of failure. A tensile stress field is created around each blasthole extending radial cracks as it initiates. The interaction between radial cracks from different blastholes and the free face create rock fragments.

Rollins (1990) acknowledged that explosive properties and mechanical rock properties are not fully considered within the original Kuz-Ram model, only the drilling pattern and general rock mass factors based on an original model by Lilly (1986) (Table 1). He developed a separate model that takes into account different tensile breakage mechanisms that occur due to explosive interaction with the rock. Breakage mechanisms identified include the main breakage area, secondary breakage and a back break area, shown in Figure 2. The boundary of the main breakage area is found from the magnitudes of the strain from the tensile stress waves and the tensile strength of the rock. The secondary breakage area is determined by tensile stress wave reflections at discontinuities and interactions from adjacent boreholes. The back break area is that beyond the main breakage area where reflected tensile waves no longer break the rock. A rock volume for the main breakage area is calculated for each hole and individual mean fragmentation sizes and a fragmentation distribution for each borehole are calculated using Equations 1 and 3, respectively; an average is taken of all boreholes calculated. In order to calculate the breaking area more in depth analysis of the specific explosive used and individual rock parameters around each blast hole are required. Comparing this model to field data showed excellent results for the fine

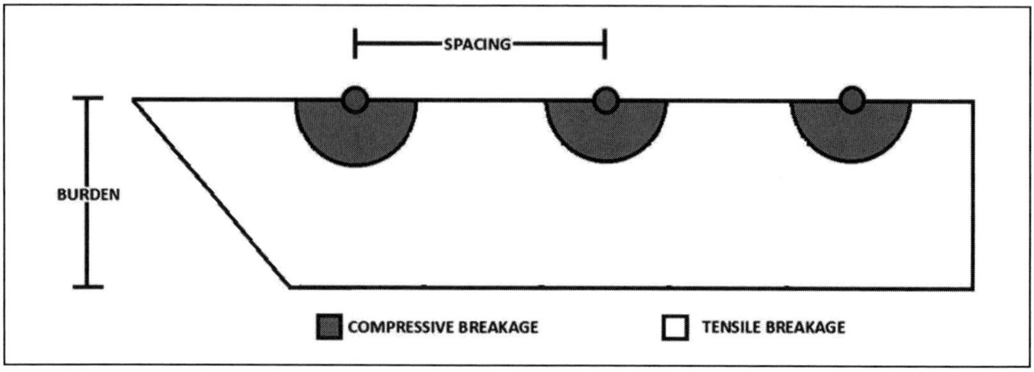

Figure 3. Compressive and tensile stress rock breakage

end of the spectrum, far more accurate than using the total volume of rock as in the original model by Cunningham (1983), however, the model is based on a small scale, ideal, test (Rollins 1990).

The Rollins model still only takes into account tensile breakage; in reality, there are two very different mechanisms that cause rock fragmentation, both tensile failure, and compressive shear failure (Figure 3). Compressive shear failure occurs close to the blasthole and creates much smaller particles. In hard rock blasting, fragmentation models such as the Kuz-Ram give a good result since the effects of compressive shear failure are minimal. Softer rocks, however, have a significant amount of compressive shear failure and therefore this parameter should be taken into account in the prediction of fragmentation size, this also explains the underestimation of fines in the original Kuz-Ram model. Djordjevic (1999) developed a two component model for cases where fragmentation of the given rock volume occurs due to two different mechanisms. The d_{50} size calculated from equation 1 is used for the coarse distribution of sizes. The mean size of fines, typically under 50 mm, are considered separately. The model gives a better estimation to the full spectrum of fragmentation sizes. The two separate d_{50} sizes are input into a modified version of Equation 3 to give:

$$R_x = 1 - (1 - F_c) \cdot \exp\left[-0.693\left(\frac{x}{a}\right)^b\right] - F_c \cdot \exp\left[-0.693\left(\frac{x}{c}\right)^d\right] \quad (5)$$

where a and c = mean fragment sizes in the tensile and compressive failure regions, respectively; b and d = uniformity coefficients in the tensile and compressive failure regions, respectively and F_c = the total mass of rock failed by shear compressive strength, given by:

$$F_c = \frac{M_o}{M} \quad (6)$$

where M_0 = mass of rock failed in compression and M = total mass of rock per blast hole.

Kanchibotla et al. (1999) hypothesize that fines are produced by the crushing action of the rock adjacent to the blast holes. A cylindrical volume, similar to that shown in Figure 3 as compressive breakage, is determined by calculating the point at which radial stress around the blast hole exceeds the dynamic compressive strength of the rock. The distribution of the coarse fraction (>1 mm) and the fine fraction of the rock are considered separately. Predictions using this model have been used on several mine sites, the resulting size distributions on the fines zone are far superior to those from the original Kuz-Ram model (Kanchibotla et al. 1999). Ouchterlony (2004) incorporates a third parameter to account for the upper limit cut-off of block sizes from the original model giving correlation coefficients with sieved data of at least 0.997.

DISCUSSION

The Cunningham model has typically been popular for fragmentation distribution prediction,

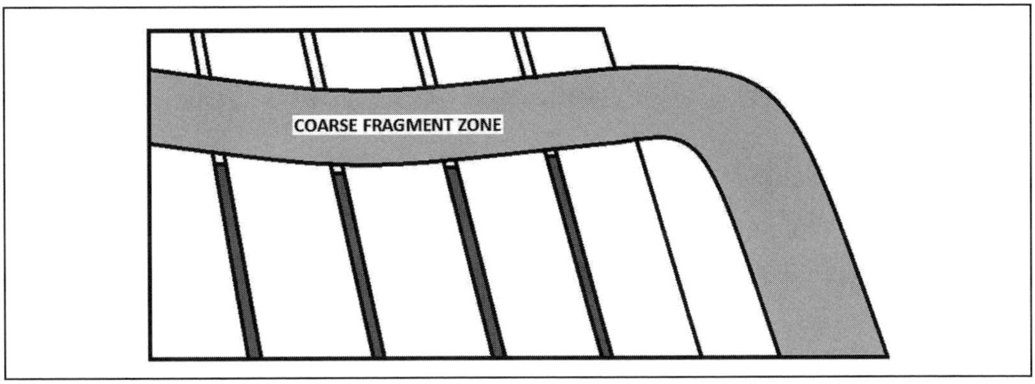

Figure 4. Coarse fragmentation zone from a blast (Kanchibotla et al. 1999)

despite its known underestimation of fines (Kojovic 1995, Comeau 1996, Rollins 1990, Djordjevic 1999, Kanchibotla et al. 1999). This is because blasting engineers are typically interested in the +250mm size due to its affects in the loading and hauling operations, a crucial part of the overall mine cost. Accurate estimation of fines developed by the Djordjevic (1999), Kanchibotla et al. (1999) and Ouchterlony (2004) models are only necessary in operations where proportion of fines affects the price of the final product, such as iron ore, so the addition of even more parameters to an already extensive, predictive model is not often required. Nevertheless, fines could be a big factor in fill design since they will inevitably transport more waterborne contaminants. Therefore knowing the exact proportions of fines and their specific size could be crucial in mitigating these environmental effects.

Fragmentation distribution can be analyzed through photographic image analysis. A number of computer models have been developed including, but not limited to, GoldSize, (Kleine 1997) BLASTFRAG (Exedaktylos 1989) and WipFrag. Kanchibotla et al. (1999), discuss the systematic bias involved with image analysis due to the coarse fragmentation zone on the surface of the muck pile caused by the stemming zone where no explosive charge is present (Figure 4). Fragmentation in this zone is due to gas heave from the underlying explosive charge and secondary breakage through crushing and grinding during movement. To avoid sampling bias, images need to be taken at regular intervals throughout the digging process, possibly hindering the shovel and truck productivity. An image based software system needs to be able to distinguish between different particles, usually by identifying particle boundary pixels in a grey level scale. It is easy to confuse shadows with particle separation and the fine portion of the scale is near impossible to predict. Due to this, pre-processing work is often required to outline separate rock particles and identify areas of fines. Fines are not fully taken into account through image analysis; first, they are not always present on the surface due to settling, wind or rain, and secondly, individual fragments are too small to segregate due to image resolution limitations. Since fines are not often of high interest when looking at fragmentation distribution, they are often eliminated from the final size distribution in image analysis, but this is the specific users' choice.

When looking at Figure 4 and the consideration for the coarse fragmentation limitation for image analysis, it is apparent that the coarse fragmentation zone is not taken into account for the Kuz-Ram model. It has been discussed that the column charge (red in Figure 4) causes two types of failure, compressive and tensile, where the tensile fraction is considered in the original model. Attempts have been made to include the fines fraction into the model through two and three component models by Djordjevic (1999) and Ouchterlony (2004), respectively. However, no consideration for the upper end of the size

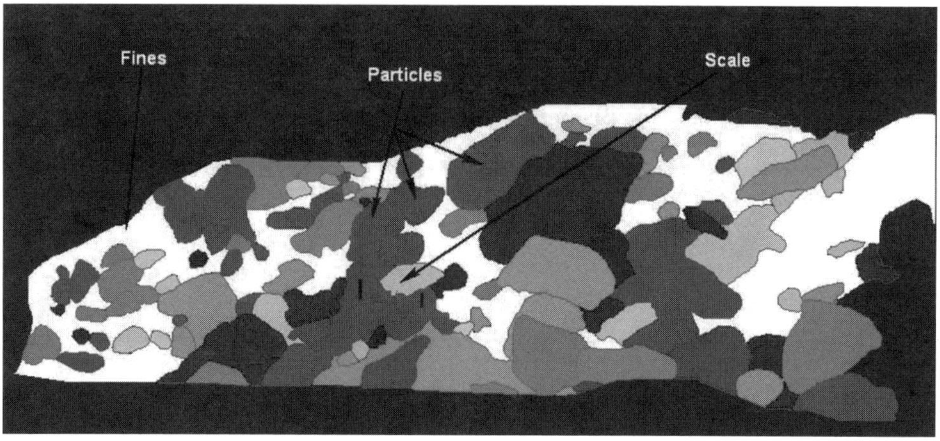

Figure 5. Digital pre-processing work (Lee and Lusk 2012)

distribution has been considered. When looking at blast optimization specifically for fill material, both extents of the scale are important and need to be considered within the models used.

A lot of research has been carried out on blast fragmentation to optimize overall mine costs, yet with increased legislation with respect to air and water quality in recent years more needs to be known with respect to the costs associated with optimizing size for mitigating these environmental effects. A comparison of ANFO versus 100% emulsion performance with equal drilling and blasting costs was carried out by Lee and Lusk (2012). Photographic fragmentation analysis, quantitative dig rates from time studies and qualitative rates about the digability of the blast from video recordings were utilized to evaluate the performance of the blast. Keeping the drill and blast cost per bank cubic yard the same for the two blasts ($0.46) produced a cost per hole of $170.76 for ANFO and $233.23 for emulsion. Results from the time study showed that the emulsion shot had a slightly lower average bucket time than the ANFO shot, of 31.98 seconds compared to 32.64 seconds. It was concluded that this was an insignificant time difference in terms of productivity and cost, but a lower standard deviation showed more consistency in digging. Digital photographs were taken of the muck pile throughout the shovel excavation to get a representative sample of the whole pile, two lasers were used for scale. WipFrag was used to analyze the fragmentation distribution after pre-processing was carried out to represent the rock particles. Fines, shown in white (Figure 5) were not taken into consideration in the WipFrag analysis.

A comparison of the average size distribution for the ANFO and emulsion shots is shown in Figure 6. The emulsion shot results in a smaller mean particle size than ANFO, which will become important in calculating costs for the full mine cycle and bucket load efficiency. The ANFO shot contained more blast holes for the same volume compared to the emulsion shot, 206 to 158, respectively. The higher amount of fines for ANFO, 45% passing 0.1in compared to 23% for emulsion can be attributed to the facts that there is a higher prevalence of holes in the ANFO shot and that fines are produced around each blast hole due to compressive stress.

Costs for the full mine cycle are required to gauge an idea of how varying the fragmentation size distribution of a blast muck pile will affect the overall mining costs. Ozdemir et al. (2007) developed an equation to estimate the bucket loading time (T_L) based on a CAT 330C LME:

$$T_L = 0.0103 * P_{50} + 3.60 \quad (7)$$

The lower mean particle size for emulsion of 9 inches compared to 11 inches for ANFO

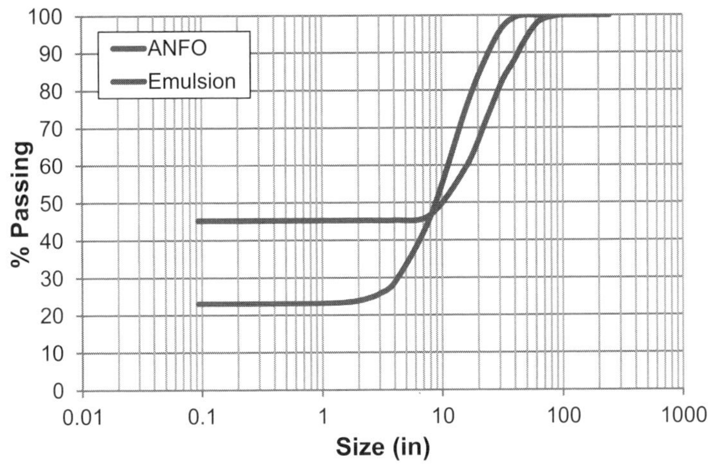

Figure 6. Particle size distribution comparison ANFO–Emulsion (Lee and Lusk 2012)

outputs a difference of 0.53 seconds in loading times and 49 buckets per hour, consequently increasing productivity; probably due to the increased powder factor for the emulsion shot (Lee and Lusk, 2012). The lower mean particle size will decrease cost with respect to bucket loading times, but the smaller particle size yields greater surface area for the transport of contaminants in water, such as TDS will be increased. There is clearly a balance between operational cost and the mitigation of environmental effects that has to be established. Calculating the final cost from mine to fill is possible through time studies and simple calculations.

The Kuz-Ram model (Cunningham 1983) can be utilized by using the same blast parameters collected by Lee and Lusk (2012). Input parameters for Equations 1–4 are shown in Table 2.

Figures 7 and 8 show comparisons of the calculated size distribution using the Kuz-Ram model with the average WipFrag photo analyzed for sized distribution for ANFO and emulsion.

Figures 6 and 7 clearly show the lack of inclusion of fines in photographic analysis; despite this the main curve of the graph has a relatively good fit between the calculated and analyzed distributions for both ANFO and emulsion. If the calculated and analyzed size distributions were compared to actual sieve data, it is hypothesized that the fines end of the curve will be between that already shown. This is due to the calculated models' underestimation of fines, and the fact that the photographic image analysis model has not considered the fines end of the spectrum at all. Mean particle sizes are shown in Table 3, the close proximity of values gives a good

Table 2. Kuz-Ram parameters

		ANFO	Emulsion	
Diameter	d	7.875	7.875	in
		200	200	mm
Burden	B	17	20	ft
		5.182	6.097	m
Spacing	S	18	20	ft
		5.487	6.097	m
Column height	L	33	33	ft
		10.06	10.06	m
Charge weight	Q	394.9	580.7	lb
		179.5	263.95	kg
Powder factor	k	1.06	1.19	lb/yd^3
		0.63	0.707	kg/m^3
RWS		100	107	% ANFO

Table 3. Comparison of mean particle size for calculated and analyzed size distributions

Explosive	d_{50} Calculated (in)	d_{50} Analyzed (in)
ANFO	12	11
Emulsion	10.5	9

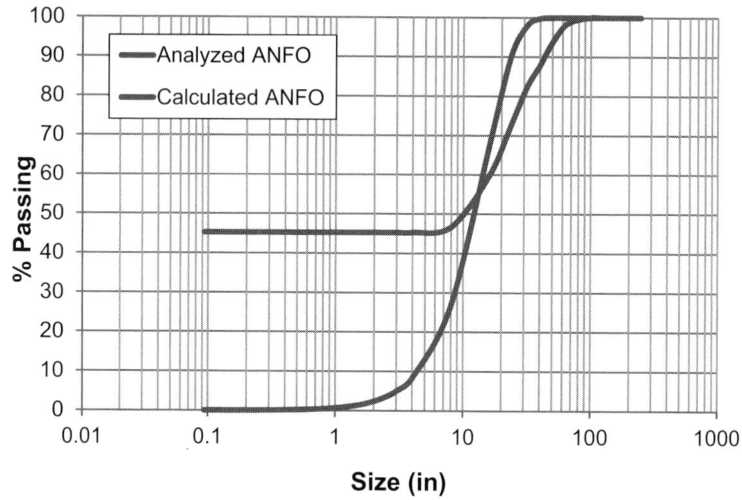

Figure 7. Comparison of calculated and analyzed size distribution for ANFO shot

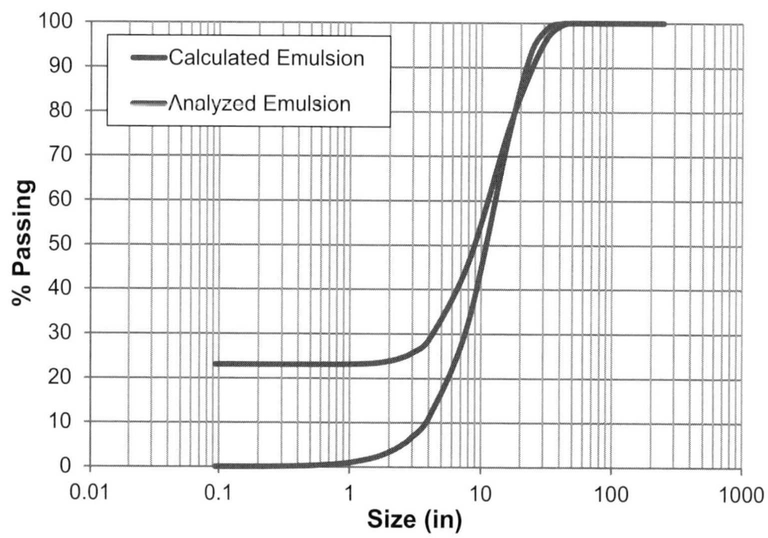

Figure 8. Comparison of calculated and analyzed size distribution for emulsion shot

indication that the Kuz-Ram model can estimate a reasonable average based solely on blast design parameters. The uniform size distribution of the Rosin-Rammler distribution and cost analysis based on time studies governed at a specific mine site can therefore give a good estimation as to the cost involved from mine to fill through obtaining different optimum size distributions.

CONCLUSION

Further research through the ARIES project can potentially produce the input parameters similar to that in Table 4 for optimum particle sizes to minimize water quality effects from fill material. The shape and size distribution can be altered though blast design parameters and a reasonable distribution can be calculated using the Kuz-Ram

Table 4. Recommended values for specific use in mining

	Property	Fill	Filter	Stability	Weathering Protection (erosion control)
Blast induced properties	Shape				
	Size Distribution				
	Density				
	Porosity				
	Resistance to weathering				

model (Cunningham 1983). Ideally, a number of tables specific for each layer in the drain and filter system for grey sandstone are required. Cost analysis on the ability to adapt a blast design to produce the quantities required for a single layer in the filter system should be carried out and studies as to whether enough of the specified material is available at one site.

It has been proven that optimizing specific particle sizes for final use is possible though known blast design parameters and explosive type. Parameters from Table 4 to minimize environmental effects with respect to water quality, could however have a negative effect on other environmental factors, such as air quality (Lashgari et al. 2012). An optimum balance between cost, production and environmental issues should be found, similar to that discussed for productivity at an aggregate quarry by Hissem (2013).

Further research into the specific sizes required for minimizing water contamination in order to determine the actual cost and negative effects with respect to other environmental effects such as air quality is required before it can be determined whether the benefits will outweigh the possible increase in costs. Surface coal mines can use the concept outlines in this paper to optimize operations through mine to fill.

ACKOWLEDGMENT

This study was sponsored by the Appalachian Research Initiative for Environmental Science (ARIES). The views, opinions and recommendations expressed herein are solely those of the authors and do not imply any endorsement by ARIES employees, other ARIES-affiliated researchers or industrial members. Information about ARIES can be found at http://www.energy.vt.edu/ARIES. Their financial contribution is gratefully acknowledged.

REFERENCES

Adhikari, G.R., Venkatesh, H.S., Theresraj, A.I. Roy, S., Balachander, R. Jain, N.K., and Gupta, R.N., *Role of Blast Design Parameters on Ground Vibration and Correlation of Vibration Levels to Blasting Damage to Surface Structures*. National Institute of Rock Mechanics, September, 2005.

Bajpayee, T.S., Bhatt, S.K. "Fatal Accidents due to Flyrock and Lack of Blasting Area Security and Working Practices in Mining," Pittsburgh, Pennsylvania: National Institute for Occupational Safety and Health, 2003.

Bajpayee, T.S., Verakis, H.C., Lobb, T.E., "An Analysis and Prevention of Flyrock Accidents in Surface Blasting Operations," Center for Disease Control, 2004.

Blair, B.E., and Duvall, W.I., *Evaluation of Gages for Measuring Displacement, Velocity, and Acceleration of Seismic Pulses*. United States Department of the Interior—Bureau of Mines, Report of Investigations 5073, Washington, D.C., April 1942.

Chaiken, R.F., Cook, E. B, Ruhe, T.C., "Toxic Fumes from Explosives: Ammonium Nitrate-Fuel Oil Mixtures," 1974.

Clark, D., Larson, B., and Lande, G., *Vibration: Its Effect & Measurement Techniques at or Near Dwellings*. VME-Nitro Consult, Inc. Louisville, Kentucky 1983.

Cunningham, C.V.B., *The Kuz-Ram Fragmentation Model—20 Years On*. Briighton Conference Proceedings 2005, African Explosives Limited, Modderfontein, South Africa, 2005: 201–210.

Dick, R.A., "A Review of the Federal Surface Coal Mine Blasting Regulations," Proceedings of the Annual Conference on Explosives and Blasting Technique, 1979.

Ash, R.L., "Good Blasting Practice: Need for Uniform Standards," Proceedings of the Annual Conference on Explosives and Blasting Technique, 1975.

Djordjevic, N., A *Two-Component Model of Blast Fragmentation.* The AusIMM Proceedings, November, 2 1999: 9–13.

Duvall, W.I., and Fogelson, D.E., *Review of Criteria for Estimating Damage to Residences from Blasting Vibrations.* United States Department of the Interior—Bureau of Mines, Report of Investigations 5968, Washington, D.C., 1962.

Edwards, A.T., Northwood, T.D., *Experimental Studies of the Effects of Blasting on Structures.* The Engineer: Technical Contributors Section, September 30, 1960: 538–546.

Exedaktylos, G.E., C. Haycocks, Y. Zhou, "*Computer-Aided Blast Fragmentation Prediction,*" Proceedings of the Annual Conference on Explosives and Blasting Technique, 1989.

Fletcher, L.R., D.V. D'Andrea, "Control of Flyrock in Blasting," Proceedings of the Annual Conference on Explosives and Blasting Technique, 1986.

Hissem, D. W, "How to Improve Safety, Create Value, and Maximize Profit in the 'New Economy,' Proceedings of the Annual Conference on Explosives and Blasting Technique, 2013.

Ireland, A. T., *Design of Air-Blast Meter and Calibrating Equipment.* United States Department of the Interior—Bureau of Mines, Technical Paper 635, Washington, D.C., 1942.

Kanchibotla, S.S., W., Valery, "*Mine to Mill Process Integration and Optimization—Benefits and Challenges,*" Proceedings of the Annual Conference on Explosives and Blasting Technique, 2010.

Kanchibotla, S. S, Valery, W, and Morrell, S. Modelling Fines in Blasting Fragmentation and its Impact on Crushing and Grinding. Explo '99, Kalgoorlie, WA, 7–11 November, 1999: 137–144.

Kleine, T.H., A.R. Cameron, "A Blast Fragmentation Measurement and Prediction System for Blast Optimization," Proceedings of the Annual Conference on Explosives and Blasting Technique, 1997.

Kopp, J.W., Siskind, D.E., *Effects of Millisecond-Delay Intervals on Vibration and Airblast From Surface Coal Mine Blasting.* United States Department of the Interior—Bureau of Mines, Report of Investigations 9026, Washington, D.C., 1986.

Kuznetsov, V.M., The Mean Diameter of the Fragments Formed by Blasting, Institute of Mining, Siberian Branch, Academy of Sciences of the USSR, Novosibirsk, October 17, 1972.

Lashgari, A., Johnson., C.E., Kecojevic, V., Lusk, B.T., Hoffman, J.M., *NOx Emission of Equipment and Blasting Agents in Surface Coal Mining.* Appalachian Research Initiative for Environmental Science, 2012.

Lee, F.W., Thoenen, J.R., and Windes, S.L., *Earth Vibrations Caused by Quarry Blasting,* Report of Investigations, Department of the Interior—Bureau of Mines, November 1936.

Lee, N. and Lusk, B.T., ANFO vs. 100% Emulsion: Quantitative Assessment of Performance with Constant Drilling and Blasting Costs, Blasting and Fragmentation, Vol. 6 No. 1, November, 2012: 1–22.

Lilly, P. A, An Emperical Method of Assessing Rock Mass Blastability, The AusIMM/IE Aust Newman Combined Group, Large Open Pit Mining Conference, University of Queensland, October 1986: 89–92.

Little, T.N., "Flyrock Risk" Carine, Australia: TNL Consultants, 2007.

MacKenzie, A.S., Cong. J., *Cost of Explosives—Do You Evaluate it Properly?* May 1966: 32–41.

Madearis, K., The Development of Rational Damage Criteria for Low-Rise Structures Subjected to Blasting Vibrations. Kenneth Medearis Associates, Research, Engineering, Computer Consultants, Fort Collins, Colorado, June 22–24, 1977.

Mainiero, R.J. 1997. A Technique for Measuring Toxic Gases Produced by Blasting Agents. Proceedings of the Twenty Third Annual Conference on Explosives and Blasting Technique, Las Vegas.

Nicholls, H.R., Johnson, C.F., and Duvall, W.I., *Blasting Vibrations and Their Effects on Structures.* United States Department of the Interior—Bureau of Mines, Bulletin 656, Washington, D.C., 1971.

Northwood, T.D., Crawford, R.., and Edwards, A.T., Blasting Vibrations and Building Damage. The Engineer: Technical Contributors Section, May 31, 1963: 973–978.

Onederra, I, Bailey, V, Cavanough, G, Torrance, A., "Understanding main Causes of Nitrogen Oxide fumes in Surface Blasting," Institute of Materials, Mineral and Mining, 2012.

Ouchterlony, F., The Swebrec Function: Linking Fragmentation by Blasting and Crushing. Institute of Materials, Minerals and Mining & Australian Institute of Mining and Metallurgy, Stockholm, Sweden, 2005: Vol 114, A29–A44.

Ozdemir, K., Kahriman, A., Ozer, U., and Turner, G., Relation Between Excavator Bucket Loading

Time and Particle Size Distribution of Shot Rock. Proceedings of the Annual Conference on Explosives and Blasting Technique, Vol 1, ISEE, 2007.

Persson, G, Persson, P.A., "Estimation of Toxic Fume Characteristics of Explosives from Steel Tube Blasting," Propellants and Explosives, Vol. 5., 1980.

Rollins, R. R and Wang, S., Fragmentation prediction in bench blasting The 3rd International Symp. on Rock fragmentation by blasting, Brisbane, Australia, August 26–31, 1990: 195–199.

Schettler, L, Brashear., S., "Effect of Water on ANFO/Emulsion Blends in Surface Mine Blasting," Proceedings of the Twenty-Second Annual Conference on Explosives and Blasting Technique, Orlando, FL, February 4–8, 1996.

Shaw, D.E., *Evaluation of Potential Blasting-Induced Damage to the Meason House.* 1989.

Siskind, D.E., Crum, S.V., Otterness, R.E., and Kopp, J.W., *Comparative Study of Blasting Vibrations From Indiana Surface Coal Mines.* United States Department of the Interior—Bureau of Mines, Report of Investigations 9226, Washington, D.C., 1989.

Siskind, D.E., Crum, S.V., and Pils, M.N., *Blast Vibrations and Other Potential Causes of Damage in Homes Near a Large Surface Coal Mine in Indiana.* United States Department of the Interior—Bureau of Mines, Report of Investigations 9455, Washington, D.C. 1993.

Siskind, D.E., Stachura, V.J., Stagg, M.S., and Kopp, J.W., *Structure Response and Damage Produced by Airblast From Surface Mining.* United States Department of the Interior—Bureau of Mines, Report of Investigations 8508, Washington, D.C., 1981.

Siskind, D.E., Stachura, V.J., Nutting, M.J., *The Generation of Low-Frequency Long-Duration Vibrations from Surface Mine Blasting at Blandford, IN.* United States Department of the Interior—Bureau of Mines. November 15, 1985.

Siskind, D.E., Stagg, M.S., *Assesment of Blasting Vibration Impacts from Quarry Blasting in Dade County.* D.E. Siskind & Associates, LLC. May 10, 2000.

Siskind, D.E., Stagg, M.S., *Blast Vibration Measurements Near and On Structure Foundations.* United States Department of the Interior—Bureau of Mines, Report of Investigations 8969, Washington, D.C., 1985.

Stachura, V.J., Siskind, D.E., Engler, A.J., *Airblast Instrumentation and Measurement Techniques for Surface Mine Blasting.* United States Department of the Interior—Bureau of Mines, Report of Investigations 8508, Washington, D.C., 1981.

Stachura, V.J., Fletcher, L.R., and Peltier, M.A., *Delayed Blasting Tests to Improve Highwall Stability—A Final Report.* United States Department of the Interior—Bureau of Mines, Report of Investigations 9008, Washington, D.C., 1986.

Stagg, M.S., Siskind, D.E., Stevens, M.G., and Dowding, C.H., *Effects of Repeated Blasting on a Wood-Frame House.* United States Department of the Interior—Bureau of Mines, Report of Investigations 8896, Washington, D.C., 1986.

Watson, R.W., J. Roth, "A Review of Blasting Accidents in Surface Mining," Proceedings of the Annual Conference on Explosives and Blasting Technique, 1978.

Williams, M.A., "Mining to Reduce Future Environmental Liabilities," Fifth High-Tech Seminar Blasting Technology Instrumentation and Explosives Applications, New Orleans, Louisiana, 1994

Production, Cost, and Environmental Impact of Loading Equipment in Surface Coal Mining in Appalachia

Ali Lashgari
West Virginia University, Morgantown, West Virginia

Vladislav Kecojevic
West Virginia University, Morgantown, West Virginia

ABSTRACT

This paper presents the results of research related to development of a software tool for the determination of the production, cost, and environmental impact of loading equipment in surface coal mining in Appalachia. Loading equipment includes rope shovels, front hydraulic shovels, backhoe shovels, and front-end wheel loaders. The design and procedures used in this research include three interrelated modules for each type of loading equipment: (i) production, (ii) ownership and operating costs, and (iii) environmental impact of loading operations. The latter includes particulate matter (PM_{10}), total suspended particulate matter (TSP), carbon dioxide (CO_2), carbon monoxide (CO), nitrogen oxides (NO_x), sulfur oxides (SO_x), volatile organic compounds (VOCs), and sound pressure level. This research is a part of a broader industrial project related to the development of software systems for the selection of productive, cost-efficient, and eco-friendly mining systems.

INTRODUCTION

Geologic formations in the Appalachian region consist of overburden and coal seams of varying thicknesses interspersed between multiple layers of interburden. Overburden and interburden must be removed before the coal is extracted. Basic surface mining unit operations include drilling, blasting, loading, hauling, and dumping. Loading equipment includes rope shovels, hydraulic front shovels, backhoe shovels, and front-end wheel loaders. The main goal in the selection of loading equipment is to maximize production rate while minimizing overall cost and environmental impact.

Surface coal mining in the Appalachian region presents many environmental challenges. Major environmental disturbances in loading operations are air pollutants and sound exposure. Air pollutants include particulate matter (PM_{10}), total suspended particulate matter (TSP), car bon dioxide (CO_2), carbon monoxide (CO), nitrogen oxides (NO_x), sulfur oxides (SO_x) and volatile organic compounds (VOCs). The mining industry is compelled by public pressure and stringent governmental regulations to reduce its environmental impact. Future surface coal mining operation will need to substantially decrease environmental disturbances during overburden removal and coal extraction to achieve "low impact" mining by incorporating new practices and design features.

To date, different approaches and software packages have been developed to determine the production and cost of loading equipment in surface mining. The Fleet Production and Cost Analysis Program (FPC) is a software package developed by Caterpillar that assists in predicting fleet productivity and associated costs within given operational conditions. FPC is a very useful tool for comparing the various types of loader-truck systems based on production and cost parameters. A similar software package, the Haulage Analysis Tool (HAT), was developed by Komatsu. The

TALPAC, by Runge, is another software package commonly used by mining engineers to estimate loader and truck requirements, productivity, and overall costs for surface mining operations. This software uses the stochastic Monte Carlo method to simulate loading and hauling operation over a specific haul profile. It also provides sensitivity analysis over a wide range of production factors. However, all of these software tools lack any consideration of the environmental impact of loading and hauling equipment.

A number of studies have been conducted to analyze emissions from mining and construction equipment. Lewis et al. (2009) described governmental regulations that limit emission of air pollutants, identified construction equipment emission sources, and compared their data with data obtained from other sources. Sharrard et al. (2007) conducted research on the environmental and energy implications in the construction industry and concluded that equipment fuel consumption is almost twice the level indicated in various governmental reports; and therefore, that the impact of air emission is 30% greater for particulate matter and almost twice the levels for NO_x and VOCs. Kean et al. (2000) conducted a study to determine emissions of NO_x and PM_{10} for off-road diesel equipment based on diesel fuel consumption. Gautam et al. (2002) used an in-field testing method to determine emission factors for diesel powered off-road engines, including excavators, front-end loaders, dozers, and street sweepers. Bogunovic and Kecojevic (2009) conducted research to determine surface mining equipment CO_2 emissions. Lewis (2009) estimated fuel consumption, exhaust, and dust emissions of excavators, track loaders, wheel loaders, backhoes, dozers, off-road trucks, and motor graders. Frey et al. (2010) used a portable emission monitoring system to gather data from excavators, backhoes, dozers, track type loaders, wheel loaders, graders, generators and off-road trucks. Dallmann and Harley (2010) conducted research to determine exhaust emissions for NO_x and fine particulate matter ($PM_{2.5}$) from mobile sources using a fuel-based methodology. Kecojevic and Komljenovic (2010) determined the quantity of CO_2 emitted by haul trucks and associated costs that may arise from potential CO_2 legislation.

Organiscak and Reed (2004) described the average and instantaneous peak dust levels 100 feet from haul roads. The authors also published the results of research related to the evaluation of safe following distance for equipment in order to avoid overexposure to respirable dust from lead trucks (Reed and Organiscak 2005).

Overexposure to sound is an important health hazard. According to Kovalchik et al. (2009), many health hazards associated with mining operations have improved, with the exception of hearing loss. Excessive sound levels are detrimental to mine workers. Bolt, Baranek, and Newman, Inc. (1971) established empirically-based relationships of heavy equipment sound exposure as a function of horsepower. In 1982, the Federal Highway Administration (FHWA) (Bowlby and Cohn 1982) published a standardized construction sound model called Highway Construction Noise Model (HICNOM). More recently, a number of models have been developed for the prediction of sound exposure in construction projects, such as CadnaA, SoundPLAN, and the Environmental Noise Model (FHWA 2006). In these models, equipment sound data is expressed as a sound pressure level at a reference distance.

The objective of the present study was to develop a software tool that can be used for the assessment of production, cost, and environmental impact of various loading equipment in surface coal mining in the Appalachian region. This research is a portion of a broader project on development of a software system for the selection of productive, cost-effective, and eco-friendly mining systems, a project that is sponsored by the Appalachian Research Initiatives for Environmental Sciences (ARIES). The text that follows provides a description of the methodology used in this research, results, and concluding remarks.

METHODOLOGY

To achieve the objectives of this research, three interrelated modules were developed in the

Microsoft Visual Studio.NET programming environment. Module I determines production parameters, specifically hourly and annual production rates. Module II considers different components of ownership and operating costs associated with loading equipment, and determines total operating costs in dollars per hour, dollars per year, and dollars per cubic yard. Module III determines the environmental impact of loading equipment, including dust and exhaust emissions and sound pressure level. Emissions are presented in pounds per hour, pounds per year, and pounds per cubic yard.

Various sources and equations given by Caterpillar (2010), Komatsu (2009), Hartman (1992), Kennedy (1990), InfoMine (2010), Runge (1998), and P&H Mine Pro (2003) are used and incorporated into Modules I and II in order to determine production rates and cost.

Several estimation methods can be used to determine the exhaust emissions of mining equipment. These methods include: direct measurement or sampling, mass balance, fuel analysis, and emission factors (NPI 2008). Direct measurement uses continuous emission monitoring or sampling data from the working area. In the mass balance method, the difference between amount of a specific substance in output and input of the system is used to calculate the emission of a specific pollutant. The fuel analysis method is based on chemical or physical properties of substance (e.g., ideal gas law) and the application of a set of mathematical relationships.

The emission factors method is commonly used for determination of exhaust emissions. This method is based on multiplication of the activity rate (in units of weight, volume, distance, or duration per unit of time) by the appropriate emission factors. These factors are expressed as the emission rate when a unit of equipment is operated in an average manner. Emission factors are usually mass-based and stated in units of mass per unit of fuel consumed, mass per unit of activity (e.g., pound per horsepower hour), or mass per mile traveled. The latter method is applied to Module III of this study.

Exhaust emission of loading equipment (CO, NO_x, SO_x, VOC, and CO_2) is determined as follows:

$$E_i = EF_i \times HFC \times H \tag{1}$$

where E_i is annual emission of the substance i (lb/year), EF_i is emission factor of substance i (lb/gallon), HFC is hourly diesel fuel consumption (gallon/hr), and H is the number of operating hours per year (hr/year). Values of emission factors are adopted from NPI (2002) and EPA (1985).

Dust emission is categorized according to the size range of the component particles: TSP and PM_{10}. The TSP is the mass loading of airborne particles determined gravimetrically by a high volume air sampler. The PM_{10} refers to the mass loading of airborne particles that pass through a size selective inlet with a 50% efficiency cut-off at 10 μm aerodynamic diameter (NPI, 1999). In other words, TSP is the total of all particles suspended in the air from loading operation. The PM_{10} refers to the subset of TSP, including particles smaller than 10 μm in diameter. In this study, dust emission is determined as follows:

$$E_i = A \times EF_i \times (1 - CE_i/100) \tag{2}$$

where E_i is emission rate of pollutant i (lb/year), A is production rate (tons/year), EF_i is uncontrolled emission factor of pollutant i (lb/ton), CE_i is overall control efficiency of pollutant i (%), and TSP and PM_{10} are pollutants i.

According to the Australian National Pollution Inventory Agency (NPI 2012), dust emission factors for loading equipment working on overburden material can be determined as follows:

$$EF_i = K_i \times 0.0035 \times \frac{\left(\frac{U}{7.22}\right)^{1.3}}{\left(\frac{M}{2}\right)^{1.4}} \tag{3}$$

where U is mean wind speed (ft/sec), and M is moisture content (% by weight). The factor K_i for TSP and PM_{10} is equal to 0.74 and 0.35, respectively.

The TSP emission factor for digging and loading equipment working on coal extraction can be determined as follows:

$$EF_{TSP} = \frac{1.28}{M^{1.2}} \quad (4)$$

The PM_{10} emission factor is determined as follows:

$$EF_{PM_{10}} = \frac{0.0985}{M^{0.9}} \quad (5)$$

In the absence of on-site specific data on wind speed and moisture content, the default values for emission factors derived from NPI (2012) can be used. Based on this reference, TSP and PM_{10} emission factors for loading operation in overburden material are 0.0125 and 0.006, and for coal extraction, these factors are 0.0145 and 0.007 lb/ton, respectively.

The sound pressure level is the level of sound at a measuring point. Therefore, the sound produced by equipment should be described by specifying the measurement distance along with sound pressure level. Sound pressure level (L_p) can be expressed as follows (Mollenhauer and Tschoeke 2010):

$$L_p = 20 \times \log\left(\frac{p}{P_0}\right) \quad (6)$$

where p is sound pressure (Pa), and P_0 is the reference sound pressure ($P_0 = 2\times10^{-5}$ Pa).

An alternative way to describe sound produced by a machine is the sound power level (L_w):

$$L_w = 10 \times \log\left(\frac{W}{W_0}\right) \quad (7)$$

where W is sound power emitted by the source (watts) and W_0 is reference sound power level ($W_0 = 10^{-12}$ watts).

Therefore, the relation between sound pressure level and sound power level can be written as follows:

$$L_w = L_p + 10 \times \log\left(\frac{A}{A_0}\right) \quad (8)$$

where A_o is reference surface which is 1 square meter and A represents the area of measuring surface, which is determined as follows:

$$A = 2\pi \times r^2 \quad (9)$$

where r is distance from the sound source.

Both sound power level and sound pressure level are defined on a logarithmic scale, called the decibel (dB). Decibels are a useful way of handling very small or very large scalar values, defined as follows:

$$dB = 10 \times \log_{10}\left(\frac{\text{Quantity measured}}{\text{Reference level}}\right) \quad (10)$$

It should be noted that decibels defined for sound power and sound pressure level are completely different, because reference level for sound pressure level is $P_0 = 2\times10^{-5}$ Pa, while reference level for sound power level is $W_0 = 10^{-12}$ watts. It is a means for comparing two sounds, which can be defined by comparing the sound level with a reference sound.

Based on a report by the British Standards Institution (BSI, 2009), the permissible sound power level of shovels is calculated as follows:

$$L_w = \begin{cases} 103 & P \le 55\text{ kW } (\le 73.76\text{ hp}) \\ 84 + 11 \times \log P & P > 55\text{ kW } (> 73.76\text{ hp}) \end{cases} \quad (11)$$

and for front-end wheel loader:

$$L_w = \begin{cases} 101 & P \le 55\text{ kW } (\le 73.76\text{ hp}) \\ 82 + 11 \times \log P & P > 55\text{ kW } (> 73.76\text{ hp}) \end{cases} \quad (12)$$

where P is engine power.

Therefore, the sound pressure limit measuring at 50 ft from shovel will be:

$$L_p = \begin{cases} 71.36 & P \le 55\text{ kW } (\le 73.76\text{ hp}) \\ 52.36 + 11 \times \log P & P > 55\text{ kW } (> 73.76\text{ hp}) \end{cases} \quad (13)$$

And, for front-end wheel loader:

$$L_p = \begin{cases} 69.36 & P \leq 55\ \text{kW}\ (\leq 73.76\ \text{hp}) \\ 50.36 + 11 \times \log P & P > 55\ \text{kW}\ (> 73.76\ \text{hp}) \end{cases} \qquad (14)$$

It should be noted that exact value of sound pressure level for mining equipment should be determined by on-site measurements since its level varies widely (FHWA 2006). Values presented in above equations only represent maximum level of allowed sound pressure for different equipment based on the BSI (2009).

RESULTS

Production, cost, and environmental modules developed for the hydraulic front shovel are shown in Figure 1.

In the production module, it is necessary to input values for the following factors: bucket volume; engine power; fill factor; material density; shovel cycle time; availability; operating efficiency; number of operating hours per year in order to determine volume of the material in the bucket; bucket payload; and hourly and annual production rates.

The cost module requires input values for these parameters: purchase cost; ownership period; depreciation rate; interest, insurance and tax rates; cost per gallon of fuel; operator rate; and maintenance factor. The following parameters are calculated: residual value; value to be recovered through work; capital cost; interest, insurance and tax costs; total ownership cost per hour and year; maintenance, wear parts, and labor cost; fuel cost; total operating cost per hour and year; and the total cost in dollars per cubic yard, hour and year. It should be noted that two options are available for the determination of fuel consumption. The first one allows determination of fuel consumption based upon on-site site measurements, while the second option calculates the fuel consumption based upon specific engine load factor.

The environmental module requires input values for these factors: emission factors of pollutants as follows: total suspended particular matter, airborne particles, sulfur oxides, nitrogen oxides, carbon monoxide, volatile organic compounds, and overall control efficiency factors. Output values in pounds per hour, year, and cubic yard include PM_{10}, TSP, CO_2, CO, NO_x, SO_x and VOCs. The sound pressure value is given in dB(A).

Figures 2 and 3 show production, cost, and environmental modules developed for the backhoe shovel and front-end wheel loader, respectively. The same procedure described for the hydraulic front shovel is applied to the backhoe shovel and front-end wheel loader. However, the latter requires inclusion of the cost of tires.

Figure 4 shows production, cost, and environmental modules developed for the rope shovel. The production module is the same as the one used for hydraulic shovel. Cost module includes the cost of electricity rather than diesel fuel, while environmental impact includes only PM_{10}, TSP, and sound pressure value.

The next phase of the research should include: (i) field measurements of environmental parameters, (ii) calibration, validation and sensitivity analysis of the model, (iii) development of the software help function and a User's Manual, and (iv) training for mining personnel regarding usage and data manipulation within software application.

CONCLUSION

The objective of this study was to develop a software tool for the assessment of production, cost, and environmental impact of various loading equipment in surface coal mining in the Appalachian region. Three interrelated modules (production, cost, and environmental impact) were developed in the Microsoft Visual Studio. NET programming environment. The research presented here may be used by mining professionals to help determine hourly and annual production rates, ownership, and operating costs associated with loading equipment, and environmental impact related to loading equipment such as dust and exhaust emissions and sound pressure level.

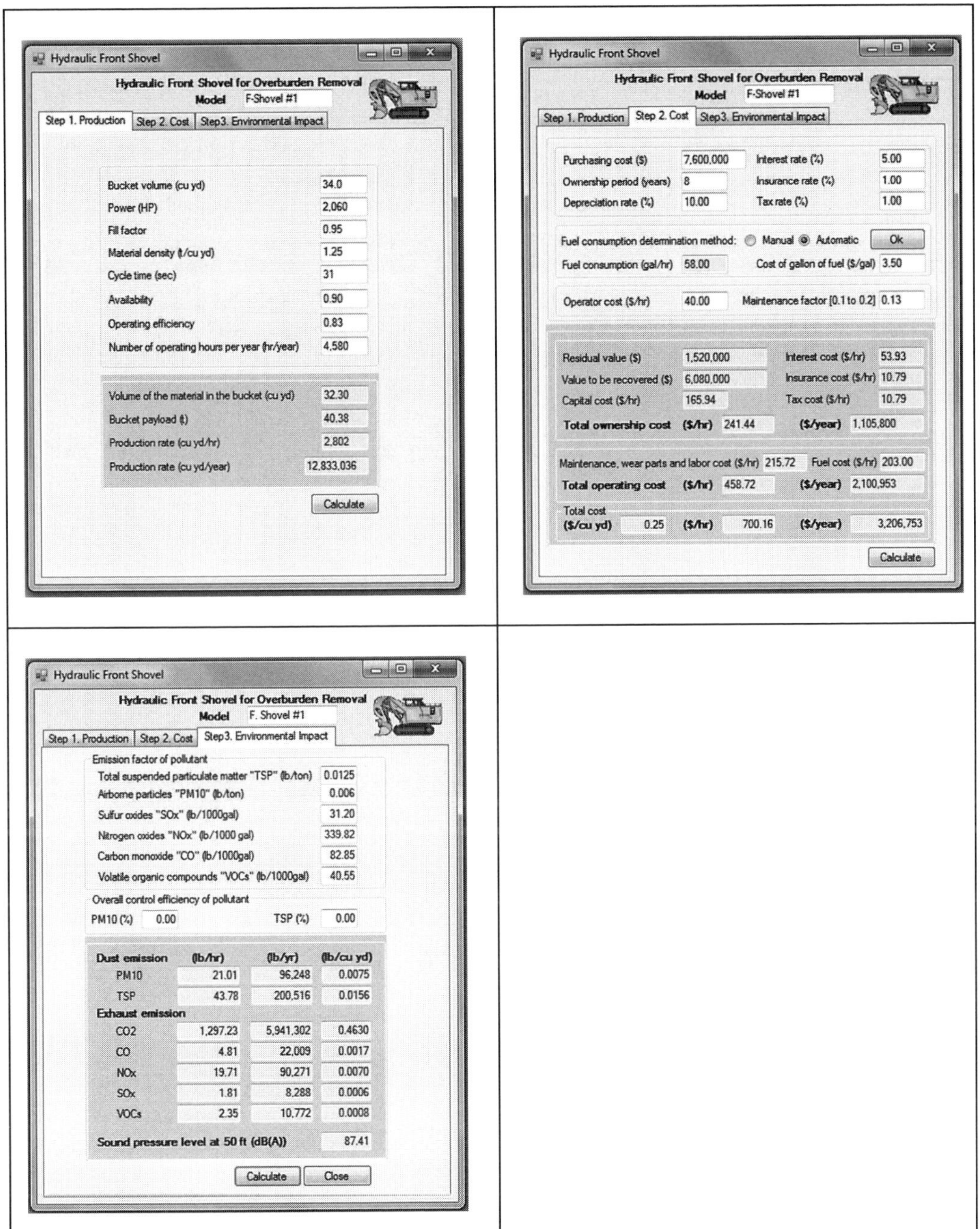

Figure 1. Production, cost, and environmental impact modules for hydraulic front shovel

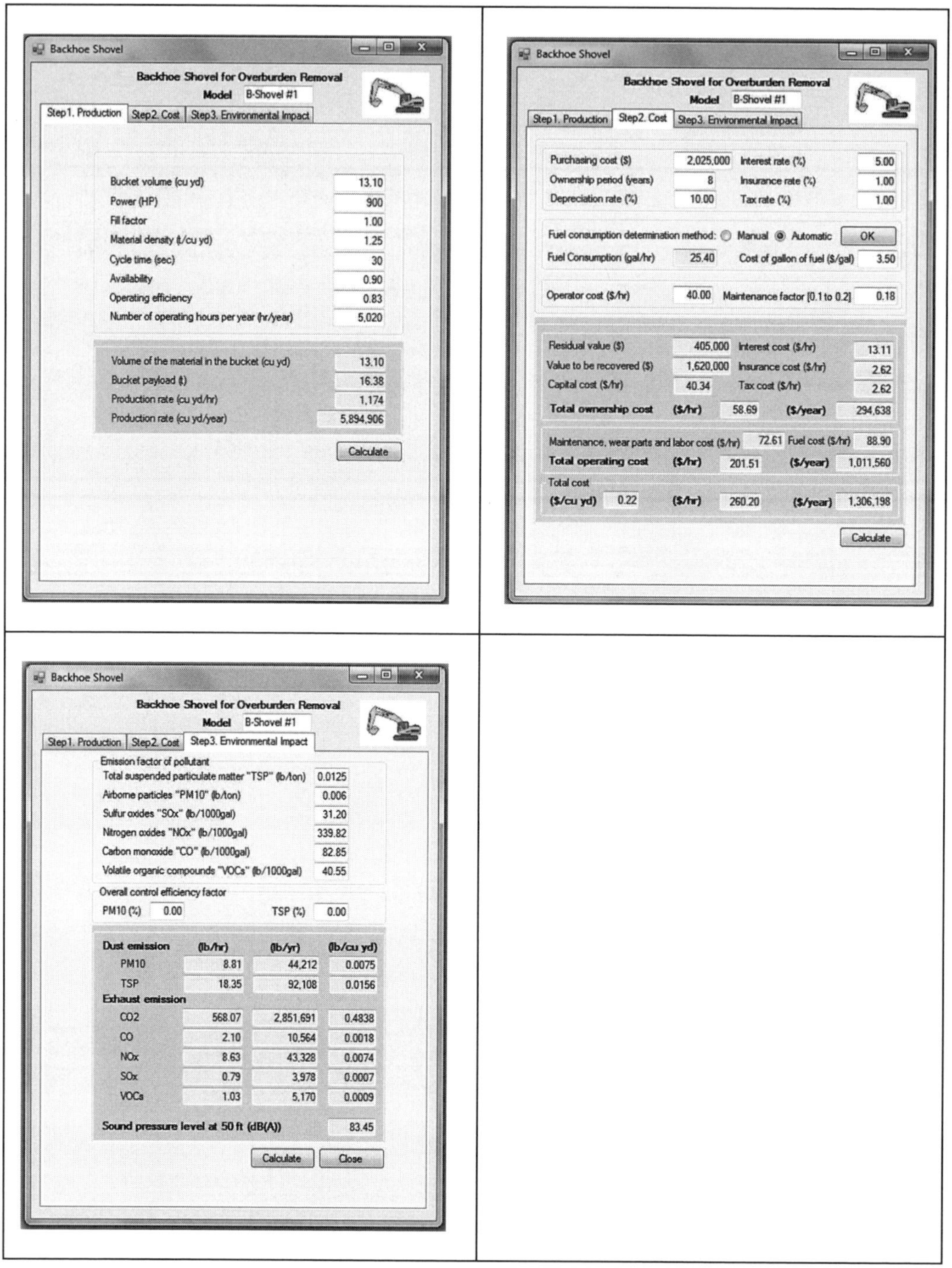

Figure 2. Production, cost, and environmental impact modules for backhoe shovel

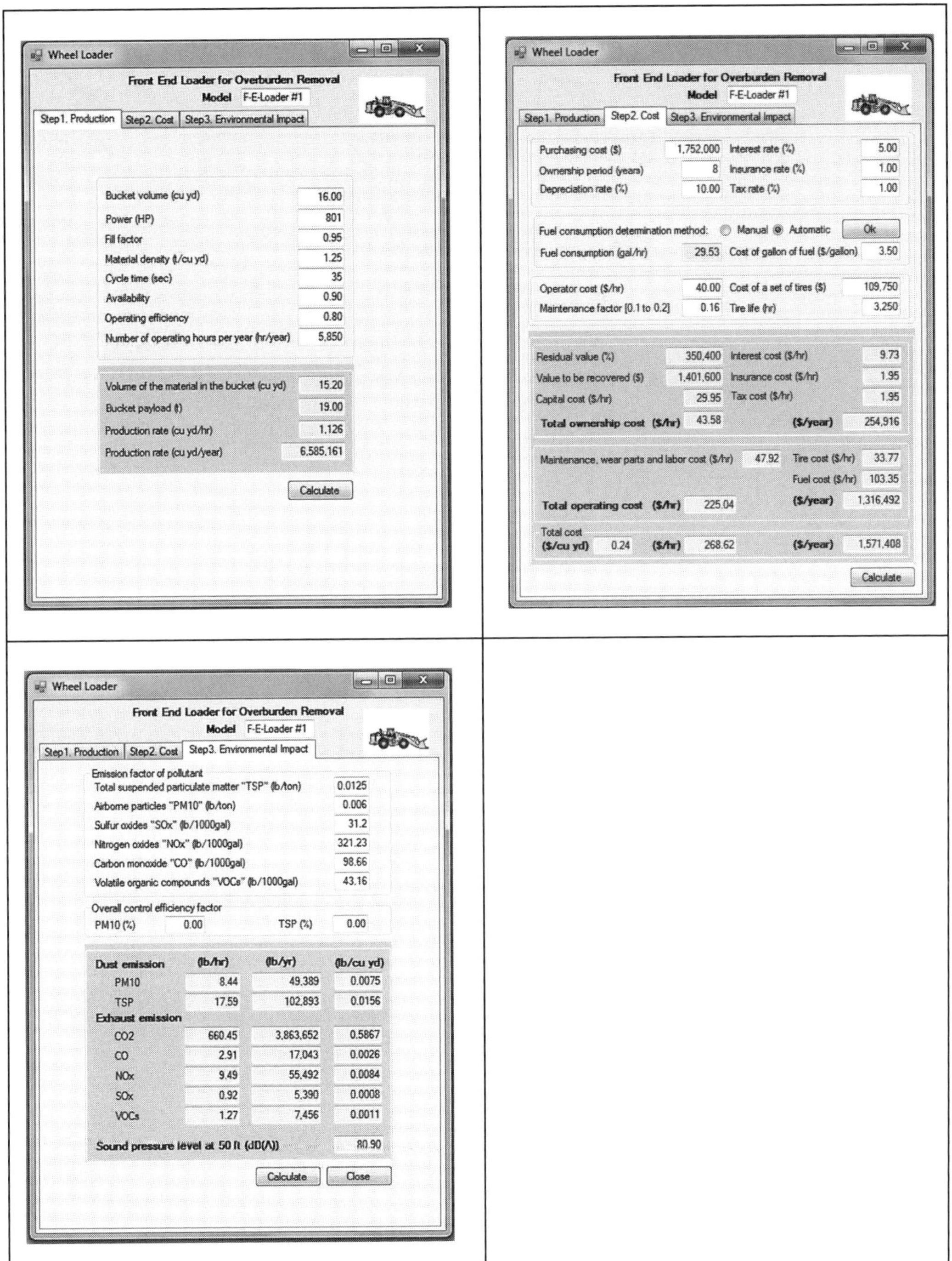

Figure 3. Production, cost, and environmental impact modules for front-end wheel loader

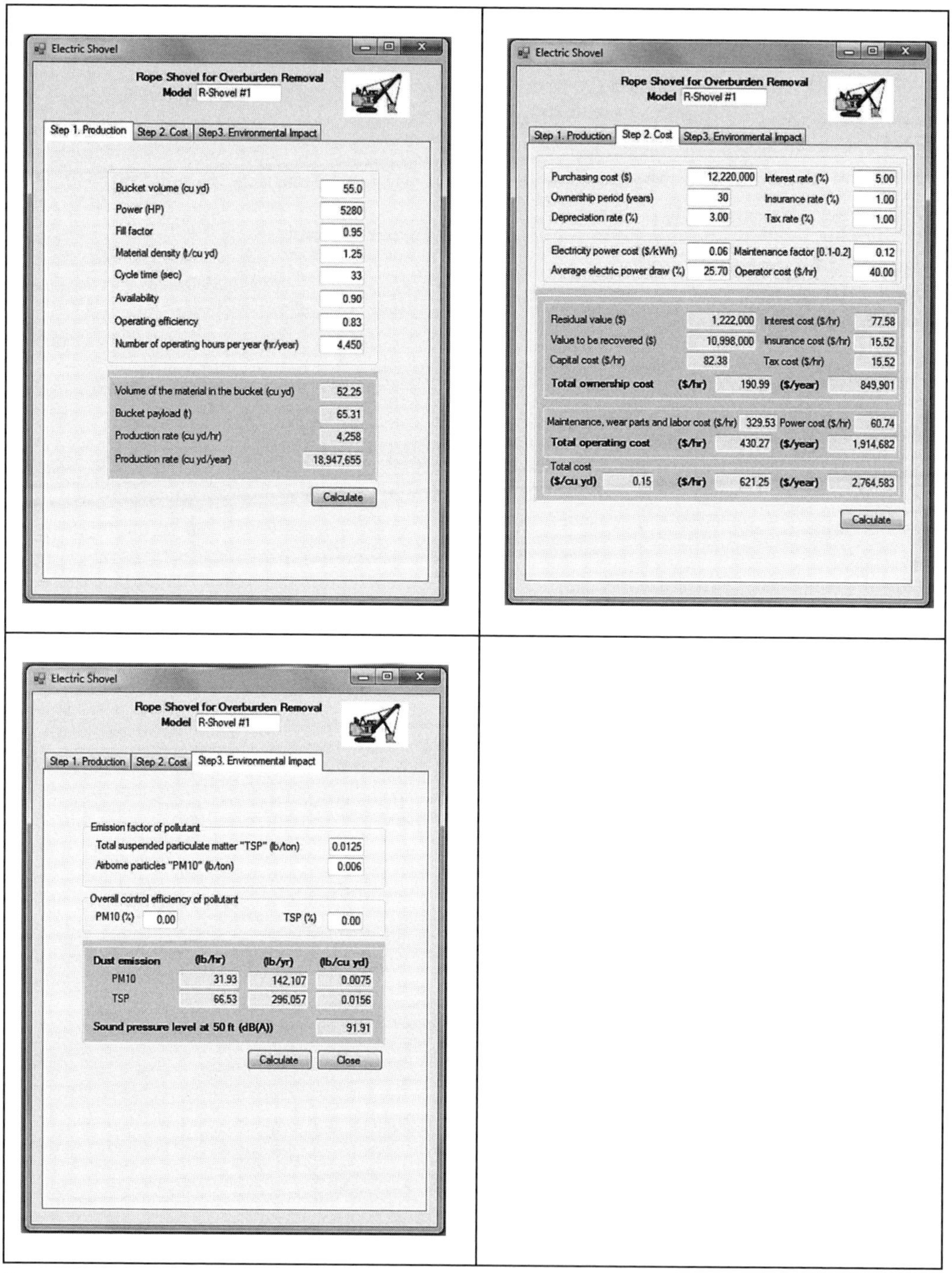

Figure 4. Production, cost, and environmental impact modules for cable (electric) shovel

ACKOWLEDGMENT

This study was sponsored by the Appalachian Research Initiative for Environmental Science (ARIES). The views, opinions and recommendations expressed herein are solely those of the authors and do not imply any endorsement by ARIES employees, other ARIES-affiliated researchers or industrial members. Information about ARIES can be found at http://www.energy.vt.edu/ARIES. Their financial contribution is gratefully acknowledged.

REFERENCES

Bogunovic, D. and Kecojevic, V. 2009. Equipment CO_2 emission in surface coal mining. International Journal of Mining and Mineral Engineering 1, 2: 172–180.

Bolt, Beranek and Newman Inc. 1971. Noise from Construction Equipment and Operations, Building Equipment, and Home Appliances. U.S. Environmental Protection Agency. Office of Noise Abatement and Control, Washington, D.C. 20460.

Bowlby, W. and Cohn, F.L. 1982. Highway Construction Noise—Environmental Assessment and Abatement. Vol. 1—Executive Summary and Simplified Prediction Methods. Vanderbilt University, Nashville, TN.

British Standard Institution. 2009. Code of Practice for Noise and Vibration Control on Construction and Open Sites. Part 1: Noise. (BS 5228–1:2009).

Caterpillar. 2010. Caterpillar Performance Handbook. Edition 40, Caterpillar, Inc. Peoria, IL.

Dallmann, R.T. and Harley, A.R. 2010. Evaluation of Mobile Source Emission Trends in the United States. Journal of Geophysical Research Atmospheres. 115(D14305), doi:10.1029/2010JD013862.

Environmental Protection Agency (EPA). 1985. Compilation of Air Pollutant Emission Factors. Vol.2: Mobile Sources. U.S. Environmental Protection Agency, Office of Transportation and Air Quality, Ann Arbor, MI.

Federal Highway Administration (FHWA). 2006. Highway Construction Noise Handbook. U.S. Federal Highway Administration, Office of Natural and Human Environment, Washington, D.C. 20590.

Frey, H.C., Rasdorf, W. and Lewis, P. 2010. Comprehensive Field Study of Fuel Use and Emissions of Nonroad Diesel Construction Equipment. Journal of the Transportation Research Board, Transportation Research Board of the National Academies, Washington, D.C., 2158:69–76.

Hartman, H.L. and Mutmansky, J. 2002. Introductory Mining Engineering, 2nd edition, John Wiley and Sons, Hoboken, NJ, pp. 570.

Gautam, M., Carder, K.D., Clark, N.N. and Lyons, W.D. 2002. Testing for Exhaust Emissions of Diesel Powered Off-Road Engines. APB Contract 98–317. Prepared by West Virginia University for the California Air Resources Board and the California EPA, Sacramento, CA.

InfoMine. 2010. Mine and Mill Equipment Costs. Spokane Valley, WA: InfoMine USA Inc. pp. [SU-2]-[SU-43].

Kean, J.A., Sawyer, F.R. and Harley, A.R. 2000. A Fuel-Based Assessment of Off-Road Diesel Engine Emissions. Journal of the Air & Waste Management Association. 50(11):1929–1939.

Kecojevic, V. and Komljenovic, D. 2010. Haul Truck Fuel Consumption and CO_2 Emission Under Various Engine Load Conditions. SME Mining Engineering. 62(12):47–52.

Kennedy, B.A. (ed.). 1990. Surface Mining Handbook, Society for Mining, Metallurgy, and Exploration, Littleton, CO, pp. 1,194.

Komatsu. 2009. Specifications and Application Handbook. Edition 30, Komatsu America Corp., Peoria, IL.

Kovalchik, G.P., Duda, T.F. and Harper, G.S. 2009. Technique for Estimating the Sound Power Level Radiated by Pneumatic Rock Drills and the Evaluation of a CSIR Prototype Rock Drill with Engineering Noise Controls. Journal of the Southern African Institute of Mining and Metallurgy. 109(295–299).

Lewis, M.P. 2009. Estimating Fuel Use and Emission Rates of Non-Road Diesel Construction Equipment Performing Representative Duty Cycles. PhD thesis, North Carolina State University, http://repository.lib.ncsu.edu/ir/bitstream/1840.16/5008/1/etd.pdf .

Lewis, M.P., Rasdorf, W., Frey, H.C., Pang, S., and Kim, K. 2009. Requirements and Incentives for Reducing Construction Vehicle Emissions and Comparison of Nonroad Diesel Engine Emissions Data Sources. ASCE, Journal of Construction Engineering and Management. 135(5):341–351.

Mollenhauer. K. and Tschoeke, H. 2010. Handbook of Diesel Engines. Springer-Verlag, 1st ed., Berlin Heidelberg.

National Pollutant Inventory (NPI). 1999. Emission Estimation Technique Manual for Mining and Processing of Non-Metallic Minerals. Available online: http://www.npi.gov.au.

National Pollutant Inventory (NPI). 2002. Emission Estimation Technique Manual for Combustion Engines, Version 2.2. Available online: http://www.npi.gov.au.

National Pollutant Inventory (NPI). 2008. Emission Estimation Technique Manual for Combustion Engines, Version 3.0. Available online: http://www.npi.gov.au.

National Pollutant Inventory (NPI). 2012. Emission Estimation Technique Manual for Mining Version 3.1. Available online: http://www.npi.gov.au.

Organiscak, A.J. and Reed, W.M.R. 2004. Characteristics of Fugitive Dust Generated from Unpaved Mine Haulage Roads. International Journal of Surface Mining, Reclamation and Environment. 18(4):236–252.

P&H Mine Pro. 2003. Peak Performance Practices: Excavator Selection. Harnischfeger Corporation.

Runge, I., 1998. Mining Economics and Strategy, Society for Mining, Metallurgy, and Exploration, Littleton, CO.

Reed, W.M.R. and Organiscak, A.J. 2005. The Evaluation of Dust Exposure to Truck Drivers Following the Lead Haul Truck. SME Annual Meeting. Salt Lake City, UT.

Sharrard, L., Aurora, M.H.S. and Roth, M. 2008. Environmental Implications of Construction Site Energy Use and Electricity Generation. ASCE, Journal of Construction Engineering and Management. 133(11):846–854.

NOx Emission of Equipment and Blasting Agents in Surface Coal Mining

Ali Lashgari
West Virginia University, Morgantown, West Virginia

Catherine Johnson
University of Kentucky, Lexington, Kentucky

Vladislav Kecojevic
West Virginia University, Morgantown, West Virginia

Braden Lusk
University of Kentucky, Lexington, Kentucky

Joshua M. Hoffman
University of Kentucky, Lexington, Kentucky

ABSTRACT

Surface coal mining industry in the Appalachian region consumes a significant amount of energy in the form of diesel fuel and blasting agents. Emission of oxides of nitrogen (NOx) from these sources represents an environmental challenge to the mining industry. This paper presents the results of research work related to determination of NOx emission of mining equipment and blasting agents, which is part of a broader industrial project conducted by the Appalachian Research Initiative for Environmental Research (ARIES). Data for this project are collected from an operating surface coal mine in West Virginia. The research work presented here may be used by mining professionals to aid in quantifying NOx emission and determine strategies for reducing its overall environmental impact.

INTRODUCTION

The primary substances emitted from combustion engines include: volatile organic compounds (VOC), carbon monoxide (CO), carbon dioxide (CO_2), oxides of nitrogen (NOx), particulate matter (PM), and sulfur dioxide (SO_2). The NOx is a group of highly reactive gasses such as nitric oxide (NO) and nitrogen dioxide (NO_2). They are toxic chemicals which help form acid rain and hamper the growth of plants. They also contribute to the creation of ground-level ozone. In combination with some other substances, NOx can cause major human respiratory problems, and may lead to shortness of breath, irritated nose and eyes, nausea and fluid forming in lungs (Konac, 2004). High amounts of NOx cause visual impairment, reduced oxygen intake, a larger buildup of fluids in lungs, swelling of the throat, rapid burning spasms and even death (Baukal, 1998).

In order to avoid potential human and environmental problems, the U.S. Environmental Protection Agency (EPA) established standards for NO_2 emission in 1971. A primary standard (to protect health) and secondary standard (to protect the public welfare) were set at 0.053 parts per million (ppm), averaged annually (EPA, 2012a). In 2010, EPA established an additional primary standard at 0.100 ppm, averaged over one hour (EPA, 2012a).

Surface coal mining unit operations include drilling, blasting, loading and haulage. Various mining equipment and blasting agents are used in these operations, and they are also a dominant source of NOx emission. The objective of this research was to determine NOx emission of mining equipment and blasting agents for a specific surface coal mining operation in Appalachian region. The text that follows provides a background to NOx emission from mining equipment and blasting agents, a description of the methodology used in this study, results, analysis of obtained results and concluding remarks.

BACKGROUND TO NOx EMISSION FROM MINING EQUIPMENT

There are several estimation methods that can be used to determine the amount of NOx emission. These methods include: direct measurement or sampling, mass balance, fuel analysis, and emission factors (NPI, 2008). Direct measurement uses continuous emission monitoring or sampling data from the working area. In the mass balance method, the difference between the amount of a specific substance in output and input of the system is used to calculate the emission of specific pollutants. The fuel analysis method is based on chemical or physical properties of substance (e.g., ideal gas law) and application of the set of mathematical relationships. The emission factors method is commonly used for estimating exhaust emissions. This method is based on multiplication of the activity rate (in units of weight, volume, distance or duration per unit of time) by the appropriate emission factors. Emission factors are expressed as the average emission rate when a unit of equipment is operated in an average manner (i.e., load). These factors are usually mass-based and stated in units of mass per unit of fuel consumed, mass per unit of activity (e.g., pound per horsepower hour), or mass per mile traveled.

Several studies have been conducted to determine exhaust emission for construction and mining equipment. Gautam et al. (2002) conducted the study to determine the exhaust emission for off-road equipment, focusing on front-end loaders, street sweepers, excavators and dozers. Emission factors have been established based on the field data. Frey et al. (2010) published results of a comprehensive field study on fuel consumption and emissions of non-road diesel construction equipment. Kecojevic and Komljenovic (2010, 2011) studied impact of engine load factors on fuel consumption and CO_2 emission of haul trucks and bulldozers, while Bogunovic and Kecojevic (2009) presented the results of research work related to determination of surface mining equipment CO_2 emission in an operating surface coal mine in the southern United States. Lewis (2009) conducted research related to emission of backhoes, dozers, excavators, motor graders, off-road trucks, track loaders and wheel loaders. The author estimated that in 2008 the annual NOx emission of this equipment was approximately 56% of total NOx emitted by all equipment used in construction and mining projects. Kean et al. (2000) presented a method for a fuel-based determination of NOx emission for off-road diesel engines at regional and national levels.

The NONROAD model developed by EPA (2004) provides an equation for determination of equipment emission as follows:

$$E = N \times HP \times LF \times EF \times A \quad (1)$$

where E is emission amount (lb/year), N is engine population for a given base year (year of 2000 for mining equipment), HP is engine power (hp), LF is load factor (%), EF is emission factor (lb/hp-h) and A is activity rate (h/year).

Engine populations in NONROAD model are classified by fuel type, application and power level. In the case that engine activity rate and engine load factor are not known, default input values suggested by NONROAD model can be used (EPA, 2005).

EPA presents emission factors for heavy duty diesel powered mining and construction equipment in lb/hr, lb/kWh, lb/hph and lb/gallon (EPA, 1985). The National Pollutant Inventory (NPI) Agency in Australia has also established the same values for exhaust emission factors of surface mining equipment (NPI, 2002).

BACKGROUND TO NOX EMISSIONS FROM BLASTING

NOx emissions from blasting are released rapidly in relatively high concentrations that can dissipate out of the mine area in toxic plumes of gas. Due to the extensive use of blasting agents used in mining and increasing constraints with respect to their environmental impacts it is important to have a gage on the quantity of toxic fumes generated.

As an oxygen balanced explosive, such as Ammonium Nitrate/Fuel Oil (ANFO), detonates, a large quantity of expanding hot gases is produced. In an ideal situation only hot steam, carbon dioxide and nitrogen are produced.

$$3NH_4NO_3 + CH_2 \rightarrow 3N_2 + CO_2 + 7H_2O \quad (2)$$

In reality, efficient blasting practice is only possible with sufficient confinement and priming. Furthermore, the explosive products are often contaminated with water and drill cuttings, all of which affect the explosives quality and chemical kinetics. This results in non-ideal explosive reactions and production of toxic gases: nitrogen dioxide, nitric oxide and carbon monoxide (NO_2, NO and CO), as well as a reduction in energy imparted onto the rock mass. The concentrations Immediately Dangerous to Life or Health (IDLH) for NO_2, NO and CO are 20, 100 and 1,200 ppm, respectively (NIOSH, 1994). NOx consists of multiple species of nitrogen and oxygen (N and O) in varying combinations, including N_2O, NO, NO_2, N_2O_2 N_2O_3, N_2O_4 and N_2O_5. NO_2 is the primary component of the NOx series but for the purpose of this research NOx will refer to the sum of pounds of NO and NO_2. Explosives are often contaminated with ground or rain water, leading to increased levels of NOx gases through the breakdown of ammonium nitrate into nitrate and ammonia in solution. The desensitized solution will not fully decompose when detonated, producing non-ideal products. Under and over fueling an explosive mixture results in positive and negative oxygen balances, respectively. A positive oxygen balance will produce more NOx gases, whilst a negative oxygen balance will produce more CO.

The effect of over fueling ANFO is:

$$2NH_4NO_3 + CH_2 \rightarrow 2N_2 + CO + 5H_2O \quad (3)$$

The effect of under fueling ANFO is:

$$5NH_4NO_3 + CH_2 \rightarrow 4N_2 + 2NO + CO_2 + 7H_2O \quad (4)$$

Nitric oxide is unstable in air and reacts with oxygen to produce the more toxic nitrogen dioxide, which can be seen after a blast by the generation of an orange/brown cloud (Onederra, 2012):

$$2NO + O_2 \rightarrow 2NO_2 \quad (5)$$

For every 2.2 lbs of ammonium nitrate that is diverted along a reaction path similar to that above, over 30 gallons of NOx will be generated. In a worst case scenario of no added fuel, theoretically 2.2 lbs of ammonium nitrate can generate about 160 gallons of NOx (AEISG, 2011). This calculation shows that only a small fraction of the explosive reacting in the wrong way can produce significant volumes of NOx gases. Even in a perfectly oxygen balanced explosive mixture; NOx could be produced if other contaminants, such as drill cuttings, prevent ammonium nitrate fully decomposing through to nitrogen, by slowing down the velocity of detonation (VoD), causing deflagration instead of detonation.

Toxic fumes are not always considered within environmental licenses but annual emissions of NOx should be estimated at some mines for tax purposes. These estimates are also recommended to prepare for increased scrutiny for emissions during the permit process. Estimates are calculated by multiplying the total annual amount of explosive consumed by a chemical product factor. EPA suggests 17 pounds of NOx per ton of ANFO (lbs/ton); however no conversion factor is available for emulsion. This conversion factor is based on the chemical composition of the explosive and does not take into account other properties such as rate of detonation, explosive composition, priming or confinement. Since an ideal chemical composition will not produce any NOx, it is unknown what chemical composition is used to calculate the product factor; in addition, the number has not been updated since 1980 so is subject to high uncertainty (EPA, 2011).

Factors affecting the emissions from explosives detonation are difficult to measure accurately in the field and almost impossible to duplicate in lab scale tests. Currently almost all tests have been carried in laboratory test chambers that differ significantly from the actual environment where surface mine blasting takes place.

An ideal way of measuring toxic fumes produced from a blast would be to sample the atmosphere after a typical blast takes place. This is, however, impractical as the fumes will not be confined for a long enough period of time to be accurately sampled and there is no such thing as 'typical blast' (Rogers et al., 1997; Abata et al., 1978). In addition, research has shown that degree of confinement of an explosive charge and the material being blasted have significant impact on fume production, therefore measurement of fumes at one mine location may not give a good indication of fumes at a different blast location, pattern or material.

A facility was established at the Pittsburgh Research Center's Experimental Mine which has the ability to detonate large, confined charges in a controlled volume (Mainiero, 1997). A blasting agent fumes test is described involving a 27 inch length of 4 inch schedule 80 seamless steel pipe or 4 inch diameter galvanized sheet metal pipe for confinement. A velocity probe was taped to the inner surface before it was loaded with 10 lbs of explosives charge. Initiation was achieved using a 3 inch diameter, 1 inch thick cast pentolite booster initiated by a number 8 instantaneous electric blasting cap. Fans are equipped within the chamber to mix the atmosphere and fumes samples are taken through 0.25 inch Teflon tubes for analysis. This was the first small scale test that had detonation velocities similar to those measured in the field of 4,000 m/s.

Mainiero (1997) carried out a number of tests at the experimental mine looking at NOx emissions as a function of fuel oil content in ANFO. Results were compared to previous tests from Chaiken et al. (1974) and Persson (1980) and predicted values for an ideal blast by Chaiken et al. (1974). As expected the lowest levels of NOx occur at an optimum ANFO ratio of 94/6. Comparison to other data however had little correlation and it was concluded that a standard technique, through testing or computer modeling, needed to be established.

Exploratory laboratory-scale studies were carried out by The National Institute for Occupational Safety and Health (NIOSH) to determine factors that contribute to the production of nitrogen oxides associated with the non-ideal detonation of ANFO in the Pittsburgh Research Center's Experimental Mine (Rowland et al., 2000). Degree of confinement, water contamination, oxygen balance, aluminum content and the addition of rock dust were tested at an optimum fuel oil content of 6%. The addition of water, confinement and rock dust were found to significantly increase NOx production, as does a decreased oxygen balance but aluminum, often added to ANFO to increase velocity and output energy, had no effect.

Rowland et al. (2001) published a second paper regarding NOx emissions from emulsion and ANFO/emulsion blends when exposed to water for prolonged periods of time. Tests were again completed at the Pittsburgh Research Center's Experimental Mine. Blends of 70/30 and 50/50 ANFO/emulsion and emulsion were tested after being exposed to water for 1 day, 1 week, 1 month and 2 months in both steel and galvanized sheet metal. For comparison, explosives were also tested with no exposure to water. Results show that 70/30 blends exposed to water for longer than a week failed to detonate and those confined with galvanized sheet metal failed after exposure for just one day. It has been noted that blends with 30 percent emulsion are not sufficient to make it waterproof and 50 percent emulsion can only be exposed for short periods of time. Results produced by Rowland (2001) agree with Schettler and Brashear (1996) that 50/50 bends are water resistant for short exposures but not for exposures of one week or more. Since it is known that emulsion blends should not be exposed to water for long periods of time but are only suitable for loading into wet holes,

NOx emissions from other contaminants such as drill cuttings, priming, and confinement should be further investigated to get a full understanding of the emissions from emulsion blends in order to produce a feasible product factor for surface mine operations. NOx emissions for emulsion peaked at an exposure to water for 1 week, so prolonged contamination with water will not continue to increase NOx emissions.

Sapko (2002) conducted an exploratory lab scale study into factors that contribute to NOx production, identifying that the addition of drill cuttings, loss of fuel oil in ANFO from wicking, ammonium nitrate dissolution with water, degree of confinement, ANFO density, and critical diameter all contribute to increased NOx levels. Experiments to determine the effectiveness of additives to reduce NOx production were also conducted, concluding that aluminum powder, coal dust, urea, and excess fuel oil in ANFO all reduced NOx production. ANFO, emulsion and emulsion blends were all tested.

Table 1 shows a summary of the highest and lowest product factors in lbs/ton taken from the current literature on NOx emissions from blasting both ANFO and emulsion, along with the contaminant tested. Product factors were converted from liters per kilogram (l/kg) to lbs/ton for NO and NO_2 separately using Avogadro's law and their individual molecular weights.

Product factors for ANFO range from 6.4 lbs/ton to 112.9 lbs/ton and from 8.6 lbs/ton to 24.6 lbs/ton for emulsion. The highest product factors for all authors come from under fueled mixtures of ANFO or very heavy confinement. High NOx emissions from emulsion are from water exposure for 1 week and the addition of drill cuttings, it should, however, be noted that fewer contaminants have been tested with respect to emulsion and the range is based on fewer sources.

METHODOLOGY

Data for this project are collected from an operating surface coal mine in West Virginia. The mine has been active since the early 1970s. The geologic formations in mine consist of sandstone overburden, with some shale streaks, five coal seams of varying thicknesses interspersed between layers of interburden. The mine produces approximately 2.8 million tons of coal and about 42 million cubic yards of overburden per year. The operation utilizes diverse mining equipment: dragline, cable shovel, drills, bulldozers, hydraulic shovel, graders, haul trucks, front-end loaders, water trucks and various auxiliary equipment.

Data on fuel consumption and number of working hours for all mining equipment were obtained from the mine. These data are classified for eight drills, 14 bulldozers, a hydraulic shovel, three graders, 32 haul trucks, nine front-end loaders as well as two water trucks. A dragline and cable shovel are also used for overburden removal but they are omitted from further consideration since they consume electricity in the process.

For the purpose of this study, the Emission Factors method is applied, and the NOx emission for mining equipment has been determined as follows:

$$E_{NOx} = EF \times HFC \times A \tag{6}$$

where E_{NOx} is annual NOx emission of the equipment (lb/year), EF is emission factor (lb/gallon), HFC is hourly diesel fuel consumption (gallon/h), and A is number of operating hours per year (h/year). Values of emission factors were adopted from NPI (2002) and EPA (1985).

Explosives usage was also available for this surface coal mine. The mine uses both ANFO and emulsion in quantities of 12,263 tons and 13,035 tons, respectively. Minimum and maximum product factors were taken from Table 1 for both ANFO and emulsion in order to calculate a range of NOx emissions in pounds for that working year. A total emission from explosives is then compared to total emissions for diesel equipment at the mine.

RESULTS AND ANALYSIS

Figures 1, 2 and 3 show a total number of operating hours of the mining equipment, annual and hourly fuel consumption, respectively. Annual fuel consumption of the equipment is more than 5.3 million gallons, and haul trucks consume

Table 1. Range of product factors (lbs NOx/ton) for ANFO and emulsion

	ANFO		Emulsion	
Author	Lowest Product Factor (lbs/ton)	Highest Product Factor (lbs/ton)	Lowest Product Factor (lbs/ton)	Highest Product Factor (lbs/ton)
Chaiken (1974)	—	42.2 Under fueled	—	—
Mainerio (1997)	7.4 6% fuel oil	15.7 Under fueled	—	—
Rowland (2000)	6.4 No water	28.7 Under fueled	—	—
Rowland (2001)	—	—	8.6 No water	15.2 Water:1 week
Sapko (2002)	42.4 Confinement: Steel pipe	112.9 Confinement: Galvanized sheet metal	14.1 Confinement: Steel pipe	24.6 Additive: Drill cuttings

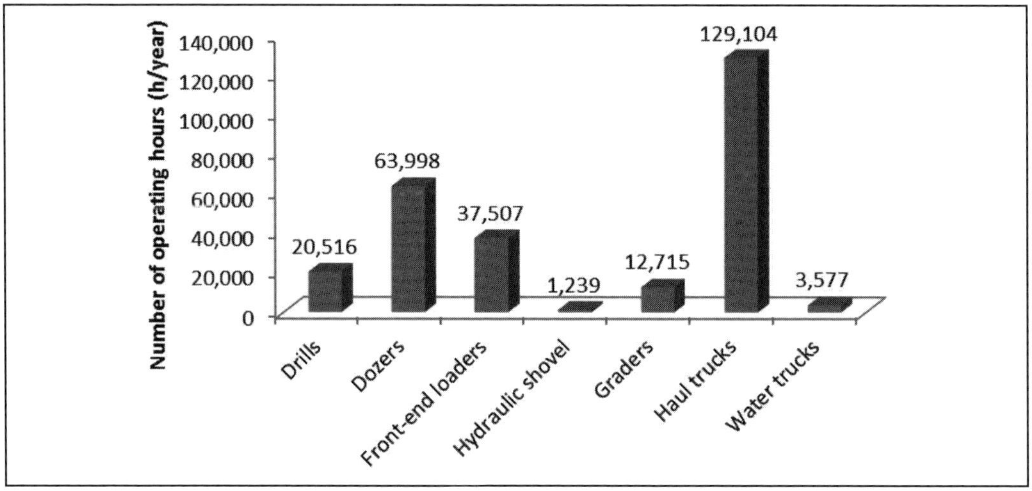

Figure 1. Total number of operating hours of the mining equipment

around 57% of all fuel in the mine, followed by bulldozers and front end loaders with 17% and 15%, respectively.

Hourly NOx emission is shown in Figure 4. It can be seen that hourly NOx emission ranged from 1.05 lbs/h for hydraulic shovel to 206.94 lbs/h for 32 haul trucks. The total hourly NOx emission of diesel equipment in the mine is determined to be 373.23 lbs/h.

Minimum product factors for ANFO and emulsion were found to be 6.4 lbs/ton and 4 lbs/ton, respectively and maximum values of 112.9 lbs/ton and 24.6 lbs/ton, respectively (Table 2). These values were multiplied by the total pounds of explosive for the year to produce a minimum NOx emission of 190,304 lbs (95 tons) and a maximum emission of 1,828,791 lbs (914 tons). This is a huge range due to the complexity of factors that contribute to the production of NOx from blasting. The upper limit is based on very heavy confinement of galvanized sheet metal for ANFO and long exposure to water for emulsion, confinement of galvanized sheet metal would not be found on a mine site and emulsion should not be exposed to water for prolonged periods of time but this maximum value gives an indication to the possible levels of NOx that can be produced from blasting. Using

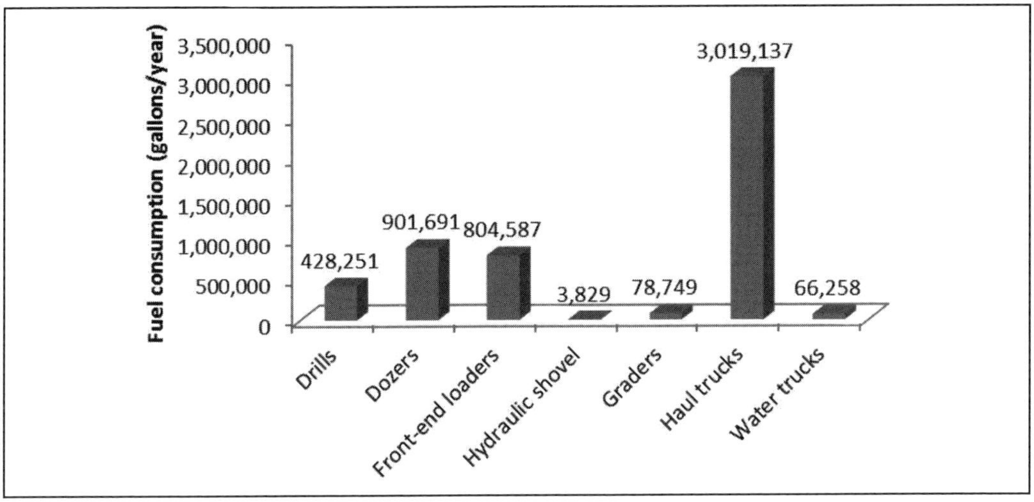

Figure 2. Annual fuel consumption of the mining equipment

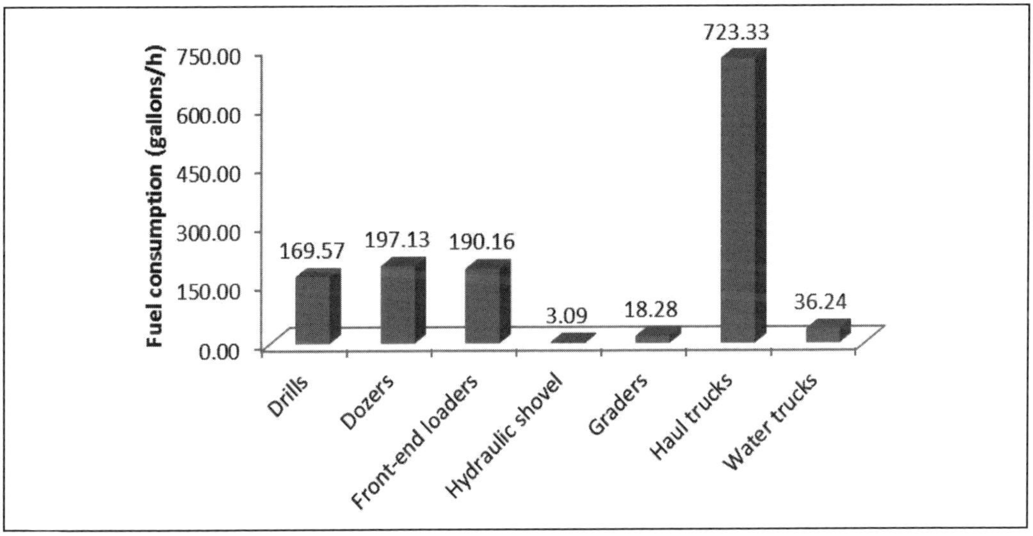

Figure 3. Hourly fuel consumption of the mining equipment

the EPA standard product factor of 17 lbs/ton for ANFO gives a total NOx emission of 430,059 lbs (215 tons).

An annual NOx emission for mining equipment and blasting is shown in Figure 5. The total annual NOx emission of mining equipment is 1.5 million lbs (751 tons). It can be noted that the haul trucks generated the largest amount of NOx, i.e., 863,775 lbs (431.9 tons), which is more than all other equipment combined. Dividing cumulative annual NOx emission of the equipment with total annual fuel consumption shows that an average of 0.283 lb of NOx has been emitted per each gallon of diesel fuel used by the equipment. The NOx emission from blasting is 430,059 lbs (215 tons).

Figure 6 shows the pictorial representation of contribution of specific mining equipment and blasting as cumulative annual NOx emission in the mine. It can be seen that haul trucks are represented by 44.69%, followed by blasting (22.25%), front-end loaders (13.37%), dozers

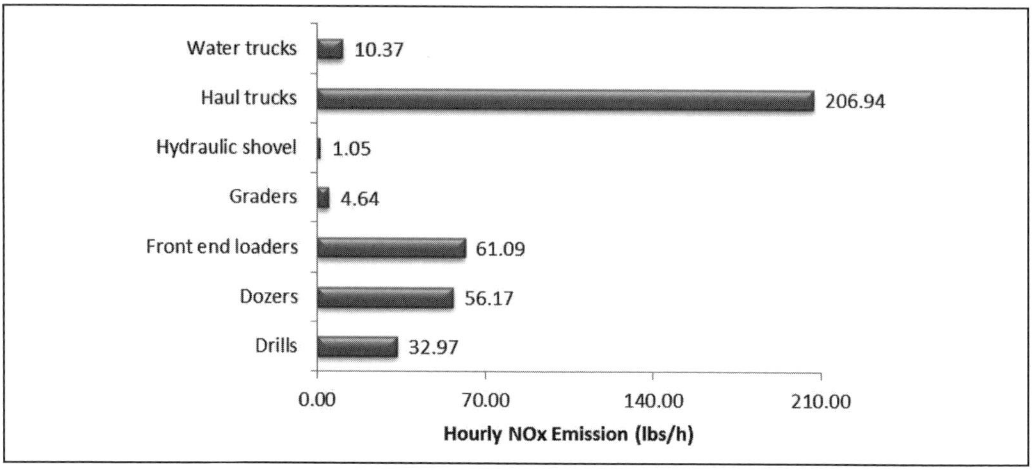

Figure 4. Hourly NOx emission of the mining equipment

Table 2. Range of NOx emissions from explosives

	Minimum		Maximum		EPA
	ANFO	Emulsion	ANFO	Emulsion	
Product Factor (lbs/ton)	6.4	4	112.9	24.6	17
lbs NOx	190,304		1,828,791		430,059
tons NOx	95		914		215*

* Value based on the total ANFO and emulsion tons, since no standard product factor for emulsion is known.

(13.29%), drills (4.31%), graders (1.03%), water trucks (0.98%), and hydraulic shovel (0.07%).

CONCLUSIONS

The objective of this research was to determine NOx emission of mining equipment and blasting agents for a specific surface coal mining operation in West Virginia. Results show that a total of 966 tons of NOx per year was emitted in the mine. Haul trucks and blasting contributed to the total emission by 44.69% and 22.25%, respectively.

The emission factors method was used to determine the NOx emission for mining equipment and it was a straight forward process. An extensive review of literature into NOx emissions from explosives demonstrated that there is a complex set of factors contributing to increased levels of NOx produced from blasting. These include but are not limited to, under fueling ANFO, excessive confinement and contamination from water or drill cuttings. It has also been established that emissions will differ for each blast design and rock type making predicting or calculating precise NOx emissions very difficult. A range of NOx emissions for a year at a mine is West Virginia was calculated to be from 95 to 914 tons, based on converted product factors established from published literature for ANFO and emulsion. The high end of the range is greater than emissions for all diesel equipment used at the mine; however the value is based on a worst case scenario for each blast over the year. A standard product factor is available for ANFO of 17 lbs/ton but it has not taken into account contamination or confinement, it is based solely on chemical composition. Despite this, when used to calculate NOx emissions for the mine, the result of 215 tons is in the middle of the range calculated from literature, making it a more feasible estimation then either the minimum or maximum values. There is less research into emulsion and no standard product factor for NOx is available, even though it is one of the most frequently used explosive types. It has been well established that both confinement

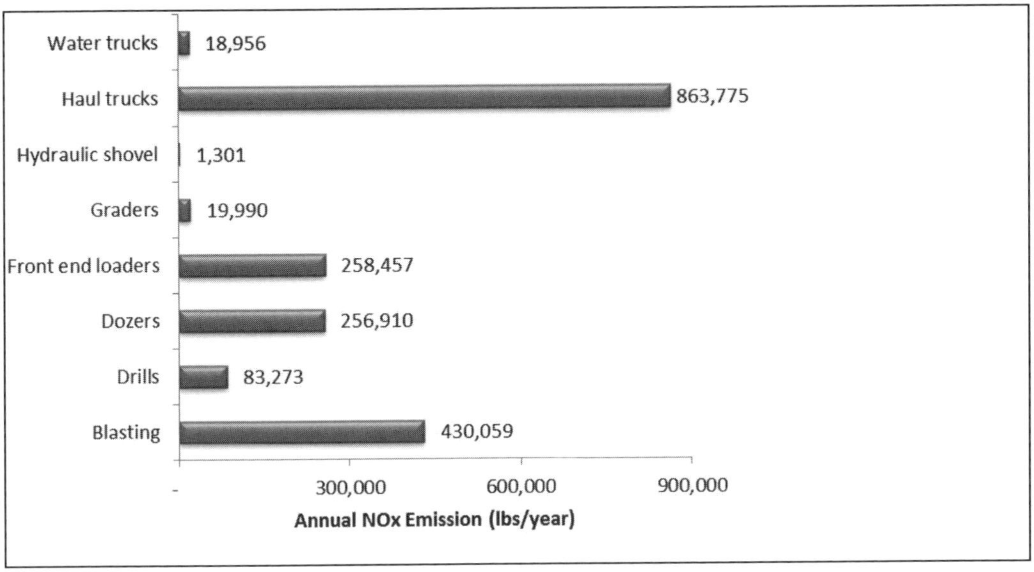

Figure 5. Annual NOx emission for mining equipment and blasting

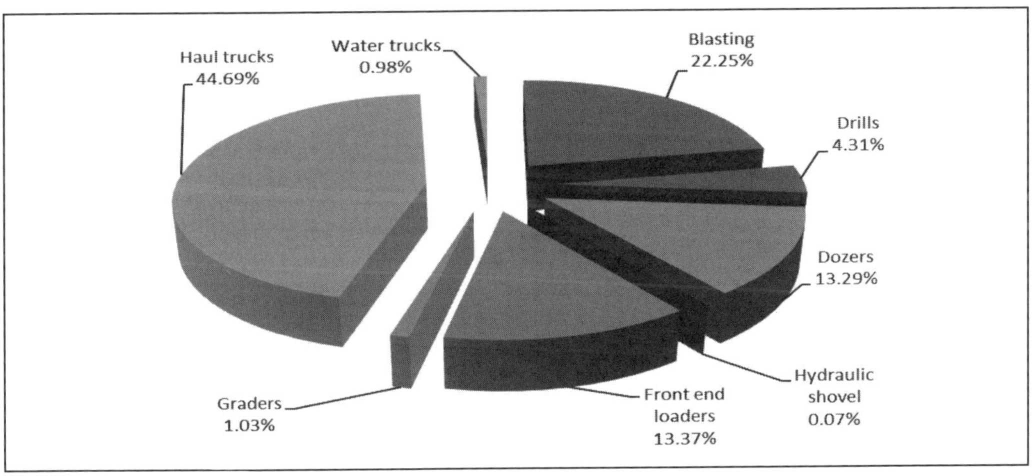

Figure 6. Contribution of mining equipment and blasting in a cumulative annual NOx emission

and contamination from drill cuttings and water have a significant impact on the volume of NOx produced from a blast, however little is known with regards to which explosives react with specific ground conditions and contamination from them. Matching explosives product to ground conditions could be crucial in controlling and accurately quantifying NOx emissions. Further work is identified towards NOx emissions from explosives with respect to contamination from drill cuttings and varying rock types.

ACKOWLEDGMENT

This study was sponsored by the Appalachian Research Initiative for Environmental Science (ARIES). The views, opinions and recommendations expressed herein are solely those of the authors and do not imply any endorsement by ARIES employees, other ARIES-affiliated researchers or industrial members. Information about ARIES can be found at http://www.energy.vt.edu/ARIES. Their financial contribution is gratefully acknowledged.

REFERENCES

Abata, D.L., Johnson, J.H., Bunting, B.G. and Robb, J. 1978. Monitoring mine air blast pollutants from six explosives tested in an underground mine. USBM OFR 44–78.

Australian Explosives Industry and Safety Group Inc. (AEISG). 2011. Code of practice: prevention and management of blast generated NOx gases in surface blasting. AEISG, Inc.

Baukal, C.E. 1998. Oxygen-enhanced combustion; CRC Press: Boca Raton.

Bogunovic, D. and Kecojevic, V. 2009. Equipment CO_2 emission in surface coal mining. International Journal of Mining and Mineral Engineering 1, 2: 172–180.

Chaiken, R.F., Cook, E.B. and Ruhe, T.C. 1974. Toxic fumes from explosives: ammonium nitrate-fuel oil mixtures.

Environmental Protection Agency (EPA). 1985. Compilation of air pollutant emission factors. Volume II: Mobile sources, U.S. Environmental Protection Agency, Office of Transportation and Air Quality, Ann Arbor, MI.

Environmental Protection Agency (EPA). 2005. User's guide for the final NONROAD 2005 model. EPA-420-R-05-013, U.S. Environmental Protection Agency, Office of Transportation and Air Quality, Ann Arbor, MI.

Environmental Protection Agency (EPA). 2011. Technology transfer network clearinghouse for inventories & emissions factors. Chapter 13.3 Explosives Detonation. U.S. Environmental Protection Agency.

Environmental Protection Agency (EPA). 2012. Nitrogen dioxides, Office of Transportation and Air Quality. Washington DC, Available at: http://www.epa.gov/airquality/nitrogenoxides/.

Environmental Protection Agency (EPA). 2004. Median life, annual activity, and load factor values for non-road engine emissions modeling. EPA420-P-04-005, NR-005c, Washington, D.C.

Frey, H.C., Rasdorf, W. and Lewis, P. 2010. Comprehensive field study of fuel use and emissions of nonroad diesel construction equipment, in transportation research record. Journal of the Transportation Board, No. 2158, Transportation Research Board of the National Academies, Washington, D.C., pp. 69–76.

Gautam, M., Carder, D., Clark, N. and Lyons, D. 2002. Testing for exhaust emissions of diesel powered off-road engines. APB Contract 98-317. Prepared by West Virginia University for the California Air Resources Board and the California EPA, Sacramento.

Kean, J.A, Sawyer, F.R.and Harley, A.R. 2000. A fuel-based assessment of off-road diesel engine emissions. Journal of the Air & Waste Management Association, 50:11, 1929–1939.

Kecojevic, V.and Komljenovic, D. 2010. Haul truck fuel consumption and CO_2 emission under various engine load conditions. SME Mining Engineering 62, 12: 47–52.

Kecojevic, V.and Komljenovic, D. 2011. Impact of bulldozer's engine load factor on fuel consumption, CO_2 emission and cost. American Journal of Environmental Sciences 7, 2: 125–131.

Konac, H. 2004. Environmental issues and sustainable development in OIC countries. Journal of Economic Cooperation 25, 4: 1–60.

Lewis, M. 2009. Estimating fuel use and emission rates of non-road diesel construction equipment performing representative duty cycles. PhD thesis, North Carolina State University, http://gradworks.umi.com/33/57/3357743.

Mainiero, R.J. 1997. A technique for measuring toxic gases produced by blasting agents. Proceedings of the 23rd Conference on Explosives and Blasting Technique, Las Vegas.

National Pollutant Invention (NPI). 2002. Emission estimation technique manual for combustion engines, Version 2.2, Commonwealth of Australia. Available online: http://www.npi.gov.au.

National Pollutant Invention (NPI). 2008. Emission estimation technique manual for combustion engines, Version 3.0, Commonwealth of Australia. Available online: http://www.npi.gov.au.

National Institute for Occupational Safety and Health (NIOSH). 1994. Pocket guide to chemical hazards. Cincinnati. Department of Health and Human Services, Public Health Service, Centers for Disease Control and Prevention, DHHS. Publication 94-116. pp. 54–55.

Onederra, I, Bailey, V, Cavanough, G, Torrance, A. 2012. Understanding main causes of nitrogen oxide fumes in surface blasting. Maney. Institute of Materials, Mineral and Mining.

Persson, G. and Persson, P.A. 1980. Estimation of toxic fume characteristics of explosives from steel tube blasting. Propellants and Explosives, Vol 5.

Rogers, S.J. 1977. Analysis of toxic fumes generated during shot firing. USBM OFR 44–79.

Rowland, J.H. and Mainiero, R. 2000. Factors affecting ANFO fumes production. Proceedings of the 26th Annual Conference on Explosives and blasting technique, (ed. I.S.O.E. Engineers), 163–174, Cleveland, OH, International Society of Explosives Engineers.

Rowland, J., Mainero, R. and Hurd, D. 2001. Factors affecting fumes production of an emulsion and ANFO/emulsion blends. Proceedings of the 27th Annual Conference on Explosives and blasting technique, (ed. S.O.E. Engineers), 133–142.

Sapko, M., Rowland, J., Mainiero, R. and Zlochower, I. 2002. Chemical and physical factors that influence NOx production during blasting—exploratory study. Proceedings of the 28th Conference on Explosives and blasting technique, 317–330, Las Vegas, NV, International Society of Explosives Engineers.

Schettler, L. and Brashear, S. 1996. Effect of water on ANFO/emulsion blends in surface mine blasting. Proceedings of the 22nd Annual Conference on Explosives and Blasting Technique, Orlando, FL, February 4–8.

Micro-Hydroelectric Power Generation Utilizing Mine Drainage

B.J. Page
BioMost, Inc., Mars, Pennsylvania

T.P. Danehy
BioMost, Inc., Mars, Pennsylvania

C.A. Neely
BioMost, Inc., Mars, Pennsylvania

S.L. Busler
BioMost, Inc., Mars, Pennsylvania

C.F. Denholm
BioMost, Inc., Mars, Pennsylvania

R.M. Mahony
BioMost, Inc., Mars, Pennsylvania

M.H. Dunn
BioMost, Inc., Mars, Pennsylvania

ABSTRACT

With a combined flow rate ranging from about 63 to 126 lps (1,000 to 2,000 gpm), the treated effluent and seasonal untreated overflow at a conventional mine water treatment plant are being used to generate electricity. Two micro-hydroelectric turbines, individually rated at 20 kW, were installed in 2012 at the Antrim Acid Mine Drainage Treatment Plant. The Babb Creek Watershed Association has operated the plant since 2001 and the onsite generation of electricity will defray operation and maintenance costs in order to help assure sustainable treatment. This is the first project in Pennsylvania to use treated and untreated acidic mine drainage (AMD) to generate electricity with impulse (Turgo) turbines. In addition to furnishing the power needs of selected treatment components at the plant, the intention is to connect to the local electric utility, or, "the grid." The complex terrain coupled with the utilization of both the treated effluent and seasonal untreated overflow created some unique design and construction issues. This is a case study on the design, permitting, construction, and startup of a micro-hydroelectric power facility located in the bituminous coal fields of north-central Pennsylvania within the Susquehanna River Basin.

INTRODUCTION

During the 20th century, medium-volatile bituminous coal was extensively mined in the North Central Coalfields of Tioga County in northern Pennsylvania. As a result, metal-laden, acidic mine drainage (AMD) from some of the historic surface and underground mining operations degraded the receiving streams. Over the past few decades, the Babb Creek Watershed Association (BCWA) has been able to successfully treat or mitigate many of these discharges (Barr 2004). The Antrim Acid Mine Drainage Treatment Plant (Plant), operated by the BCWA, is a conventional

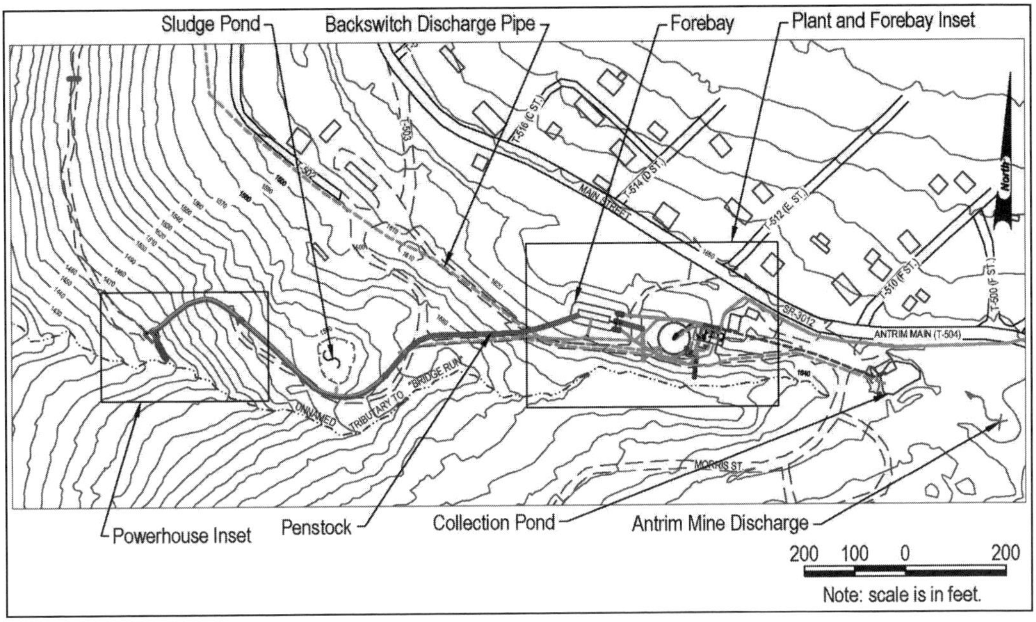

Figure 1. Site diagram (3.0-m [10-ft.] contour interval)

Table 1. BI-16, BI-14, 009 water quality (median values)

Location (collection period)	Flow cfs (gpm)	pH s.u.	Alkalinity mg/l	Acidity mg/l	Iron mg/l	Manganese mg/l	Aluminum mg/l	Sulfate mg/l
BI-16 (n=16) (1993–2008)	126 (2,000)	3.00	00	208	06.3	08.3	17.5	414
BI-14 (n=14) (1992–2008)	5.7 (90)	3.10	00	248	25.7	23.5	16.4	720
009 (n=197) (1996–2006)	114 (1,800)	8.85	66	000	00.3	03.6	00.8	675

BI-16 antrim discharge; BI-14 backswitch discharge; 009 antrim acid mine drainage treatment plant discharge; total metals.
Source: Babb Creek Watershed Association.

facility utilizing waste lime as an alkaline reagent to treat some of the most pollutive AMD in the watershed. The Plant is located within the town of Antrim, Duncan Township, Tioga County (Figure 1). The Pennsylvania Department of Environmental Protection (PADEP) oversees the operation and maintenance of the Plant.

Two sources of AMD are conveyed to a collection pond, which feeds the Plant. Both discharges issue from abandoned underground mines. A drainage ditch conveys discharge BI-14, which is located within the town of Antrim, and a 6-inch gravity pipeline conveys discharge BI-16 (Backswitch Discharge), which is located 1.6 km (1 mile) from the Plant. The combined flow rate ranges between 32–252 lps (500–4,000 gpm) and averages 95 lps (1,500 gpm). The Backswitch Discharge has considerable less flow, 5.7 lps (90 gpm), but the water is more significantly degraded. Table 1 provides the median water quality data for both of the discharges (BI 16 and BI-14) and the treatment Plant effluent (009). From the collection pond, the combined discharges are conveyed to and through the Plant via a 30-cm (12-in.) PVC pipe (Main Line). A 30-cm (12-in.) valve within the Main Line controls the flow to the Plant. Waste lime is pulverized by two ball mills to create lime slurry that is injected into the Main Line. The raw water and slurry are mixed in a static mixer, before being conveyed to an 18-m (60-ft.) clarifier. The treated water is decanted from the clarifier and is then conveyed via a 41-cm (16-in.) PVC pipe to a 1.8-m (6-ft.) diameter outlet structure, which in turn discharges to an unnamed tributary to Bridge Run. During seasonal high-flow periods, the Plant is unable to treat the entirety of the AMD and the excess untreated water discharges from the collection pond directly into the unnamed tributary. During these periods, additional lime slurry is injected into the Plant effluent in order to provide excess alkalinity to treat the by-passed raw water in-stream.

In 2008, due to an economic downturn, the sustainability of the trust fund that supplied the funding for the operation and maintenance for the Plant was in question. To offset operation and maintenance costs and potentially generate revenue, BCWA working with the PADEP considered building a micro-hydropower facility. BCWA then applied for and was awarded a $430,000 Energy Harvest Grant in 2009 to fund the project. BioMost, Inc., was selected to design and to install the micro-hydropower facility.

Design considerations for the Antrim Micro-Hydropower Project (AMHP) included: (1) utilization of both the treated AMD and the periodic raw water overflow, (2) maximum utilization of the available change in elevation at the site, (3) protection of vital parts of the turbines from potentially aggressive water, (4) management of potential scaling issues in the penstock, and (5) installation of facilities to enable the sale of excess electricity to the local utility.

FEDERAL ENERGY REGULATORY COMMISSION APPROVAL

In addition to state and local permits and notifications relating to earth disturbance and building construction, one of the design criteria for the AMHP was for the facility to be connected to the local utility company (grid) in order to sell the excess energy generated by the turbines. Connecting to the grid, however, is considered to be affecting foreign and interstate commerce; therefore, the AMHP falls under the jurisdiction of the Federal Energy Regulatory Commission (FERC) (16 USC 817, 2012). To obtain FERC approval, there are three licensing tracks and two exemption tracks (exemptions considered licenses) (FERC 2004). If a project does not connect to the grid, the project does not fall under the jurisdiction of FERC, but a Declaration of Intent (DI) must be completed and approved to confirm that the project does not need to obtain a license (18 CFR 24.1, 2012).

The three licensing tracks are Traditional, Integrated and Alternative. All licensing tracks have a time-line that, on average, are approximately two years or longer, which extended beyond the Energy Harvest Grant period. The two exemption tracks are the 5MW and Conduit. The time-lines for the exemptions are similar to the regular licensing tracks, but with elimination of certain sections. The time-lines of the Conduit and 5MW Exemption tracks are less than one year and about a year, respectively (FERC 2004). The Conduit Exemption was not applicable for the AMHP, and, even with the shortened time-line for the 5MW Exemption, the time needed to obtain the approval and construct the AMHP exceeded the grant period.

To ensure that the project was going to be built within the required grant period, the decision was made for the AMHP to be an off-grid facility until FERC approval is received. The off-grid configuration required completion of a DI, with an average approval time-line of about two to three months. This allowed for a longer construction period for the AMHP. In the design change, select motors in the Plant that run continuously were to be segregated from the grid and powered solely by electricity generated by the AMHP.

CONSTRUCTION

In the context of this paper, the construction phase, which was completed in the spring of 2012, is divided into different sections, which are not chronologically ordered. The different sections are the Plant, Forebay, Penstock and Powerhouse.

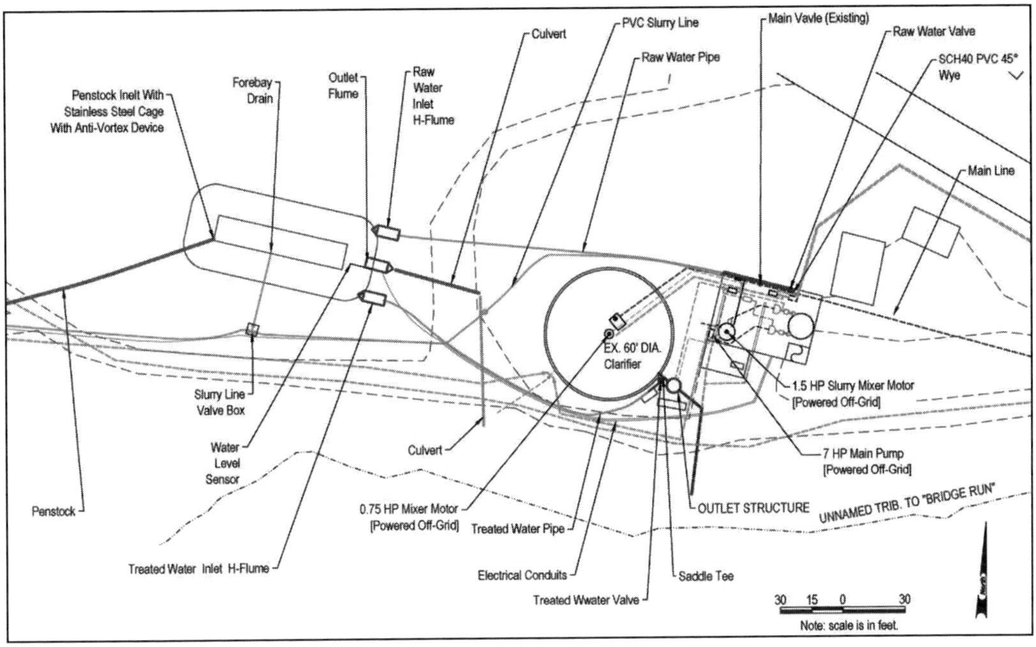

Figure 2. Plant and forebay inset

Plant

To utilize the maximum flow rate, as feasible, two pipes were extended from existing facilities to direct both the treated and untreated water to the forebay.

Treated Water

To redirect the treated water to the forebay, a connection to the existing pipeline between the clarifier and outlet structure was installed (Figure 2). The added plumbing consists of a 41-cm × 30-cm (16-in. × 12-in.) saddle tee, 30-cm (12-in.) knife-gate valve (Treated Water Valve), and 30-cm (12-in.) pipe (portions SCH40 PVC and DR32.5 HDPE). The Treated Water Valve in the open position directs the water to the forebay and in the closed position the water is directed to the outlet structure for discharge through the pre-existing effluent piping. The added plumbing was also installed in a manner to maintain the water level within the outlet structure as a small portion of the treated water is continuously transferred by means of a 7 hp pump (Main Water Pump) to the Plant for use as process water.

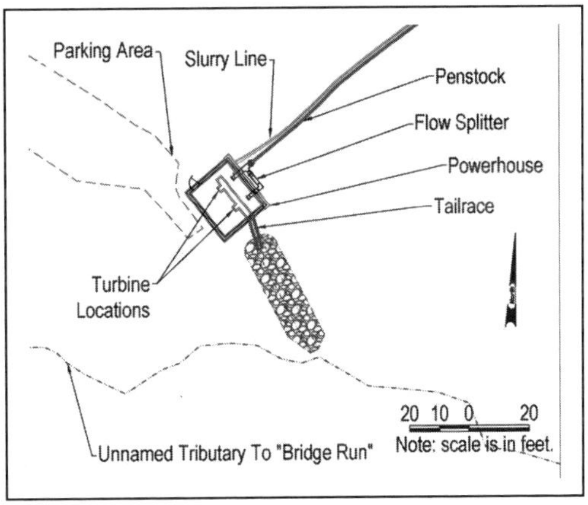

Figure 3. Powerhouse inset

Raw Water

In order to utilize the majority, if not all, of the excess raw AMD during high-flow periods, a 30-cm (12-in.) 45° wye was connected to the existing Main Line above the valve that controls the flow to the Plant. A 30-cm (12-in.) stainless steel gate valve (Raw Water Valve) was installed in the piping (Raw Water Pipe) that was extended from the wye to the forebay. The configuration of the added plumbing allowed excess raw water to be directed to the forebay without affecting the functionality of the Plant. Portions of the Raw Water Pipe are SCH40 PVC and DR32.5 HDPE. When the Raw Water Valve is closed, overflow water will discharge from collection pond spillway.

Forebay

A forebay was installed in order to utilize both the treated water and the periodically bypassed raw water. The forebay for the AMHP is a 12-m × 28-m (40-ft. × 91-ft.) pond, 2.1-m (7-ft) in depth, which stabilizes and controls the flow of water before entering the penstock. The Raw Water Pipe discharges in the northeast corner and the Treated Water Pipe discharges in the southwest corner of the forebay. The outlet of the forebay is located between the inlets from the Plant. To measure the flow entering the penstock, the forebay inlets and the outlet include 0.6-m (2.0-ft.) H-flumes with 117-cm × 122-cm × 305-cm (46-in. × 48-in. × 120-in.) concrete approach channels. Any flow in the forebay outlet can be subtracted from the combined inlet flows in order to calculate the flow rate into the penstock. The inlet for the penstock in the southeast corner of the forebay is protected with a trash rack/anti-vortex device. The custom-fabricated, 304 stainless steel, trash rack/anti-vortex device is a cubic cage topped with a 2.67-mm × 305-mm × 457-mm (0.105-in. × 12-in. × 18-in.) 304 stainless steel plate designed to prevent objects >15 cm (6 in.) from entering the penstock and to prevent the penstock from drawing air via a vortex.

Penstock and Other Pipe/Conduits

Penstock

A penstock over 335 m (1,100 ft.) in length was installed in order to utilize the 49-m (160-ft.) change in elevation. The penstock conveys water, under pressure, from the forebay to the turbines. For discussion purposes, the penstock is divided into five sections: inlet, upper, middle, lower and flow splitter. The inlet section of SDR35 PVC pipe extends from the forebay to the beginning of the HDPE pipe. The SDR35 PVC inlet section connects to the DR26 upper section of the penstock with a 46-cm (18-in.) flexible coupler. Flange connections join the upper section with the middle and lower sections, DR17 HDPE and DR11 HDPE, respectively. A flange connection is also used to join the lower section of the penstock to the DR11 HDPE flow splitter immediately upgradient of the powerhouse (Figure 3).

Electrical

A 10-cm (4.0-in.) electrical conduit was used to house the four, 4/0-gauge, aluminum wires that extend from the powerhouse to the Plant. The four wires convey electricity generated by the turbines to the Plant. Three #4 and one #2 THHN copper wires from the Plant convey high-voltage power from the grid to the powerhouse for the lights, controls and other electrical components. A 2.5-cm (1.0-in.) electrical conduit houses shielded, twisted-pair communication wire from the water level sensor control box in the Plant to the water level remote display in the powerhouse. Electrical risers with above-grade junction boxes, installed along the conduits, allowed the wire to be pulled through the sections. A separate 2.5-cm (1-in.) electrical conduit houses coaxial communication wire that runs from the water level sensor to the water level sensor control box located in the control room of the Plant.

Slurry Line (Acid Recirculation Line/ Forebay Drain)

The slurry line was installed to serve the following three main functions: (1) to enable supplemental water treatment in the tailrace at the powerhouse; (2) to enable acid-washing of the penstock; and (3) to enable the forebay to be drained to the sludge pond. For simplicity, this pipe and related connections are referred to as the slurry line. There are 15-cm (6-in.) PVC cleanouts with threaded caps located approximately every 91 m (300 ft.) along the length of the slurry line.

The primary function of the pressure-rated, 15-cm (6-in.), SCH40 PVC slurry line is to allow lime slurry from the Plant to be dispensed into the combined untreated and treated AMD in order to provide supplemental alkalinity after power generation in the tailrace prior to entering the receiving stream. The utilization of the lime slurry for supplemental treatment was selected, in part, due to the pre-existing availability from the Plant. Within the powerhouse the slurry line is extended to the tailrace by installation of a 7-cm (3-in.) PVC pipe with ball valve. Note that if lime slurry were added at the forebay, the non-reactive material (grit) in the lime slurry may have accumulated in the forebay and been conveyed through the penstock which would have resulted in damage to the turbines.

The slurry line can also be used to drain the forebay or to acid-wash the penstock. A 15-cm (6-in.) perforated riser and solid lateral, connected to the slurry line by a 15-cm (6-in.) knife gate valve, allows the water from the forebay to drain to either an existing sludge pond via a 7-cm (3-in.) PVC ball valve or to the tailrace. If needed to remove scale, acid can be circulated within the slurry line and penstock via a 5-cm (2-in.) PVC pipe with ball valve installed in the powerhouse.

Powerhouse

The powerhouse is the building the where turbines, electrical equipment and other components are located. The features discussed in this section are the penstock flow splitter, powerhouse penstock components, turbines, overspeed controllers and tailrace.

Penstock Flow Splitter

The penstock flow splitter (PFS) is comprised of 46-cm & 30-cm (18-in. & 12-in.) HDPE piping that splits the flow to the two turbines. The main section of the penstock is connected to the PFS immediately upgradient of the powerhouse with a flange connection. The PFS includes an 46-cm × 46-cm × 30-cm (18-in. × 18-in. × 12-in.) reducing tee with a 46-cm × 30-cm (18-in. × 12-in.) reducer on the primary leg. The 30-cm (12-in.) HDPE pipes extend through a 0.6-m (2-ft.) thick, steel reinforced concrete wall. A 45° bend in each leg of the PFS, cast into the concrete wall, helps the wall to act as a thrust block for the PFS.

Figure 4. A portion of the powerhouse penstock components

Powerhouse Penstock Components

Each turbine has the same components (Figure 4). From the PFS, there is a 30-cm × 20-cm (12-in. × 8-in.) reducer followed by a 20-cm (8-in.) butterfly-type shutoff valve. This valve is for shutoff purposes only and not intended to control the quantity of flow to the turbine. A 20-cm (8-in.) expansion joint is located between the butterfly valve and the Turbine Flow Splitter (TFS). A TFS is installed downstream of the expansion joint and directs the flow to the two spear valves. A pressure gage with a range of 0.0–0.6 MPa (0–87 psi) is located on the TFS.

Turbines

The impulse, Turgo-type, turbines (Figure 5) installed are designed to operate at 1,800 rpm and generate 240v at 60 Hz using approximately 16 to 57 lps (250 to 900 gpm). Due to the possible aggressive nature of the water, turbine components in contact with water were coated with high phosphorus, electroless nickel (EN) plating. The EN plating is approximately 2 to 3 mils thick and provides a high level of corrosion and abrasion resistance. Each turbine has dual adjustable spear valve nozzles. These spear valve-type nozzles provide precise flow control and allow the operator to adjust the amount of generated power provided to the off-grid components in the Plant. The water flows through the nozzles directed at the runners causing the runners to rotate. The

Figure 5. Installed turbines

runners are attached to the generator shaft, which when the runner is turning, the generator produces electricity. Each turbine has a three-phase 138V/240V 20 kW (appx. 27 hp) generator.

Overspeed Controllers

Each turbine has an overspeed control that uses a 420 VME-D-2 Bender Relay that activates mechanical contacts connected to the dump loads (electric heaters submerged in the tailrace). If the rpms for a turbine runner increases, the voltage and frequency (Hertz) increases, which causes the dump loads to engage. The engaged dump loads increase the electrical demand, which slows the generator. The relay, programmed to automatically check the voltage and frequency, disengages the dump loads once the voltage and frequency fall within the specified range. The overspeed control protects the turbine from spinning too fast and prevents over-generation of electricity. Excessive voltage can damage the electric motors in the Plant.

Tailrace

The tailrace is used to convey the water from the turbines to the discharge location. The AMHP tailrace is comprised of three sections: the trough, the culvert pipe and an outlet channel. The trough is located in the powerhouse floor and is cast-in-place reinforced concrete. The 46-cm (18-in.) HDPE Type-S (dual-walled) culvert pipe conveys the water from the powerhouse to the outlet channel. The outlet channel is an open channel lined with 46-cm × 10-cm (18-in. × 4-in.) riprap, approximately R-5, that conveys the water to the receiving stream.

DISCUSSION

Currently, the electricity generated is being used to operate selected motors within the Plant. In the off-grid configuration, only 21 lps (330 gpm) are needed to produce the required 6.9 kW (9.25 hp). At the current usage, the AMHP has reduced the operating costs of the Plant by about $5,000 a year. As the discharge flow averages 95 lps (1,500 gpm), however, significant electricity generation (over 200,000 kWh per year) is not being realized; therefore, an application for a FERC license via the 5MW Exception track, is in preparation. Once licensed, operating at full capacity and selling electricity to the local utilities, the payback period is projected to be approximately ten years.

With basic maintenance, the electrical components (turbines, overspeed controllers, etc.) as well as the mechanical components (runner, valves, etc.) that come into contact with the treated water, have a projected design life of approximately 20 years. The forebay, penstock, powerhouse, tailrace, etc., are considered permanent infrastructure.

CONCLUSION

In summary, the AMHP, as one of the first of its kind, provided design and construction challenges; however, the project serves as an example of turning an environmental liability into an environmental and economic asset. Needless to say, the general principals used in the design and the installation of micro- (up to 100 kW) and pico- (under 5 kW) hydropower plants are similar with power generation potential being dependent on AMD flow rates and available hydraulic head. This project demonstrates that hydroelectric power generation utilizing AMD is possible and this technology can be applied throughout the Appalachian coalfields.

REFERENCES

"Declaration of Intention." *Code of Federal Regulations* Title 18 Pt. 24.1. 2012.

Barr, J.P. 2004. *Restored! A History of the Reclamation of the Babb Creek Watershed from Acid Mine Drainage Pollution*. Pennsylvania: Babb Creek Watershed Association Inc.

"Projects not affecting navigable water: necessity for Federal license, permit or right-of-way: unauthorized activities," Title 16 *U.S. Code*, Pts. 817. 2012.

U.S. Federal Energy Regulatory Commission. 2004. *Handbook for Hydroelectric Project Licensing and 5MW Exemptions from Licensing*. Washington, DC.

Minepool Geothermal in Pennsylvania

Michael C. Korb
Pennsylvania Department of Environmental Protection, Bureau of Abandoned Mine Reclamation, Wilkes-Barre, Pennsylvania

ABSTRACT

In 2009, Marywood University, in Scranton, Pennsylvania, received a grant from the Pennsylvania Department of Environmental Protection (PADEP) to construct, test, and monitor a geothermal energy system capable of exchanging heat with a minepool that exists beneath the University's campus. Marywood installed the geothermal system in 2010 to serve a portion of the cooling needs of Marywood's new Center for Architectural Studies. The system has been expanded to be utilized in Phase II of the Center's construction and is adequately cooling the entire facility.

Coal underlies about one third of Pennsylvania, with anthracite coal in the north-eastern and north-central portions of the state and bituminous coal in the central and western portions. There are between 10,000 and 15,000 abandoned underground coal mines and numerous abandoned underground metal and non-metal mines in Pennsylvania, and many are flooded. With more than three million people living in and around cities located over known minepools in locations such as Pittsburgh, Scranton, Wilkes-Barre, Johnstown, and Hazleton; the possibilities for use of geothermal heat from minepools seem to be unlimited. Yet, only a few facilities are using this apparently great energy source. Pennsylvania and other states are looking for sustainable options for minewater treatment, and this may be a potential resource.

This somewhat unscientific paper discusses the Marywood minepool geothermal project, other installations in Pennsylvania, the background, benefits, and potential problems with the use of minewater from abandoned mines for geothermal heating, and attempts to discover why so few are using this process.

INTRODUCTION

Marvine Colliery of the Delaware and Hudson Coal Company, in Scranton, was opened in 1872, and coal was mined underground until 1954. In 1915, 940 men working at the Marvine Colliery produced 391,000 tons of anthracite coal. Some 400 feet above them, Marywood College opened on September 8, 1915, with a class of thirty-four women. Four years later, as the first Marywood class of seventeen graduated, a concrete and steel colliery building—Marvine II—was under construction at the bottom of the hill from the Marywood Campus.

Operations ceased at Marvine in 1963, and its dewatering pumps were shut down. This allowed the mine workings to flood to the elevation of a gravity outlet, the Old Forge Borehole, ten miles southwest, which was drilled in 1961 to prevent rising abandoned minepools from flooding area basements and low-lying neighborhoods.

Marywood University was the first college for women in Northeastern Pennsylvania and the first Catholic college for women in the state. It has grown from a one-building liberal arts college into a coeducational university consisting of five colleges. Marywood's 115-acre campus enrolls over 3,400 students in undergraduate and graduate degree programs.

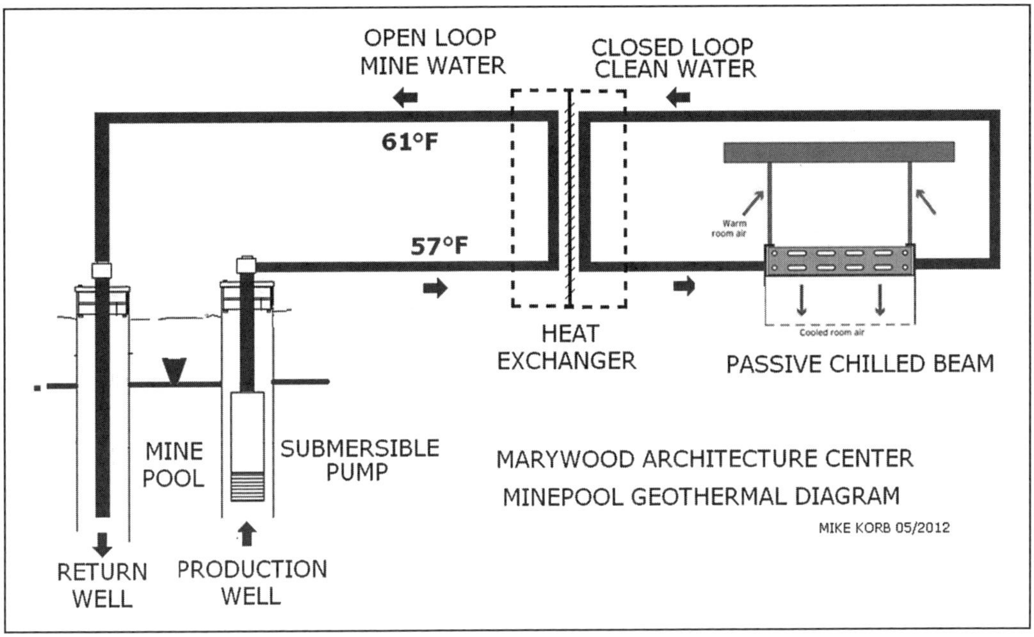

Figure 1. Marywood geothermal system

MINEPOOL GEOTHERMAL AT MARYWOOD UNIVERSITY

In 2009, Marywood received a grant from PADEP to construct, test, and monitor a geothermal energy system capable of transferring heat to the minewater pool under the university's campus. Marywood installed the system in 2010 to serve the cooling needs of its new Center for Architectural Studies.

The Marywood Geothermal system includes one production well and one recharge well (Figure 1). The production well houses a submersible pump, and the recharge well is simply a cased well that extends to a mine seam that serves as a path for the water to be discharged back into the minepool. The campus is located on a hill above the Marvine Mine pool, and the boreholes are some 400 feet deep. Depth to the minepool is 340 feet.

Water from the production well is pumped to and through a plate heat exchanger in which heat is transferred to a fresh water loop. The heat exchanger separates the geothermal loop from the building loop and generates water 2–3°F warmer than the water extracted from the mine. The fresh water reports to a passive chilled beam cooling system that utilizes the ~60°F water to cool the architecture studios, providing cooling without the use of mechanical cooling or forced air. While chilled beams are common in Europe, they are still considered an emerging technology by many in the U.S. 100% of the cooling for both design studio floors of the renovated gymnasium—the School of Architecture—is being provided by the chilled beams. 2012 was the first full cooling season with the completed building (Figure 2).

Several potential problems will merit watching at Marywood, besides the economics. Water quality may change because of de-stratification of the minepool–pumping may draw mineralized water from deeper, poorly circulated strata, causing mixing with the uppermost layers, which are the mine discharge waters. Iron scaling and corrosion may occur over time, causing gradual loss of conductivity in the heat exchanger and increased pumping pressures.

An open geothermal system utilizing a minepool such as at Marywood has a huge volume of water for heat exchange, not just a finite amount of recirculating fluid as may be used in

Figure 2. Marywood Center for Architectural Studies and Marywood Production Well (Photos–Michael Korb 2012)

a closed loop or in open loops utilizing wells. The Marywood project utilizes the large existing underground pool of water that has resulted from the abandoned anthracite mines in Northeastern Pennsylvania.

MINEPOOLS IN PENNSYLVANIA

Information on the full extent of the minepool water resources in Pennsylvania and comprehensive inventories of minepools or minepool complexes do not exist. However there are over 1,000 identified water discharges from abandoned coal mines in the state—252 in the anthracite region and 793 in the bituminous—all discharges from minepools. Work is ongoing on estimating minepool resources in the Pittsburgh seam and in the anthracite region, and apparently there are several billion gallons of 45°–60°F water in these mines.

The minepool reservoir water temperatures in the anthracite region vary between 41°F and 65°F. Measured temperatures of the Old Forge borehole, where the water from the Marvine minepool discharges, vary from 58°F to 61°F. See Figure 3.

The water chemistry at Marywood hardly resembles what people picture as mine drainage.

Six independent power generators in the anthracite region have used abandoned mine drainage (AMD) as cooling water. These all treat the minewater on-site before using it. Depending on the water quality and the application, minepool geothermal may or may not require treatment. Based on these experiences, long-term pumping tests may be appropriate to evaluate potential changes in water quality for design of treatment strategy or pumping strategy to avoid treatment. Groundwater modeling may be useful to evaluate the area of influence of pumping.

The public often perceives minewater quality as "acid mine drainage"—uniformly contaminated and very acidic. Many of the mine outfalls in Pennsylvania and especially in the anthracite region are not particularly acidic, typically more than pH 5, and many are alkaline, whereas Wilkes-Barre rain sometimes approaches pH 4. People tend to be reluctant to use minewater for anything if they think the water is bad.

One of the reviewers of this paper expressed concerns about some heavy metals reported in mine water being dangerous to human health. Minewater samples taken from the Marywood borehole exhibit corrosiveness and hardness but few parameters that are in excess of maximum drinking water standards, as shown in Figure 4.

KINGSTON COMMUNITY CENTER

I have been able to find only two coal-minepool geothermal installations operating in Pennsylvania—Marywood in Scranton and the Kingston Recreation Center near Wilkes-Barre.

In 1979, Kingston Borough built a 17,000-sq.ft. recreation center. The U.S. Department of

MARYWOOD BOREHOLE 2010	
pH	6.65
conductivity	1140 μmohs/cm
alkalinity	202 mg $CaCO_3$/L
sulfate	300 mg/L
Fe	0.43 mg/L
Mn	0.27 mg/L

OLD FORGE BOREHOLE 2012	
pH	6.4
conductivity	1090 μmohs/cm
alkalinity	99 mg $CaCO_3$/L
sulfate	317 mg/L
Fe	16.2 mg/L
Mn	1.9 mg/L

Figure 3. Marywood and Old Forge borehole comparison; Source (Marywood): B.F. Environmental Consultants Inc.; Source (Old Forge): LRCA

Energy (DOE) was actively promoting the potential of utilizing minepools as heat sinks for minepool geothermal heating and cooling systems and provided a grant to install a system in the center. Jim Palumbo, one of the project's engineers and current president of QUAD3 Group, recounts, "It just seemed natural to use the resources at hand (the minepool) to save money for the recreation center, and the grant sure helped to make the decision." The Kingston installation has been the site to visit for engineers working on minepool systems around the world. The Center opened in 1981, and the system has been working successfully, albeit with some repairs, for more than 30 years.

The 8"ϕ production and return wells have 102 ft casings through alluvium and extend another 90 ft through rock into the Abbot seam, the first mined seam. The production well was constructed as a sanitary well, sealing off zones that could result in the formation of poor water quality, such as coal seams and bedrock units high in iron and manganese. The water is returned directly to the well to minimize aeration and subsequent oxidation of iron in the water. Static water levels after drilling were 7 ft with no pumping drawdown. See Figure 5.

A submersible pump in the production well pumps 90 GPM to the heat exchanger. Minewater flows along one side of the plate exchanger with clean water flowing on the other side. The clean water loop transmits cooling to the water/air

MARYWOOD BOREHOLE WATER ANALYSIS			
PARAMETER	CONTAMINANT LEVEL	VALUE	UNITS
pH	6.5 - 8.5	6.65	pH units
Conductivity	no standard	1140	μmohs/cm
Alkalinity	no standard	202	mg $CaCO_3$/L
Total Hardness	no standard	720	mg $CaCO_3$/L
Ca Hardness	no standard	560	mg $CaCO_3$/L
Chloride	< 250 mg/L	132	mg/L
Sulfate	< 500 mg/L	300	mg/L
Nitrate+Nitrite	< 10 mg N/L	1.4	mg/L
Orthophosphate	no standard	nr	mg/L
Turbidity	< 1 NTU	8.7	NTU
Turbidity (filtered-0.45μ)	< 1 NTU	--	NTU
Total Dissolved Solids	< 500 mg/L	> 750	mg/L
TON (Odor Number)	< 3	1	unitless
Saturation Index	Rec. ± 0.5	-1.0	unitless
Copper	< 1.3 mg/L	< 0.05	mg/L
Lead	< 0.015 mg/L	--	mg/L
Iron (total)	< 0.3 mg/L	0.43	mg/L
Iron (dissolved)	< 0.3 mg/L	--	mg/L
Manganese (total)	< 0.05 mg/L	0.27	mg/L
Manganese (dissolved)	< 0.05 mg/L	--	mg/L
Zinc	< 5 mg/L	0.19	mg/L
Tannin Levels	no standard	--	mg/L
Sodium	no standard	94.1	mg/L
Potassium	no standard	8.16	mg/L

Source: B.F. Environmental Consultants Inc.[1]

Figure 4. Marywood borehole water analysis

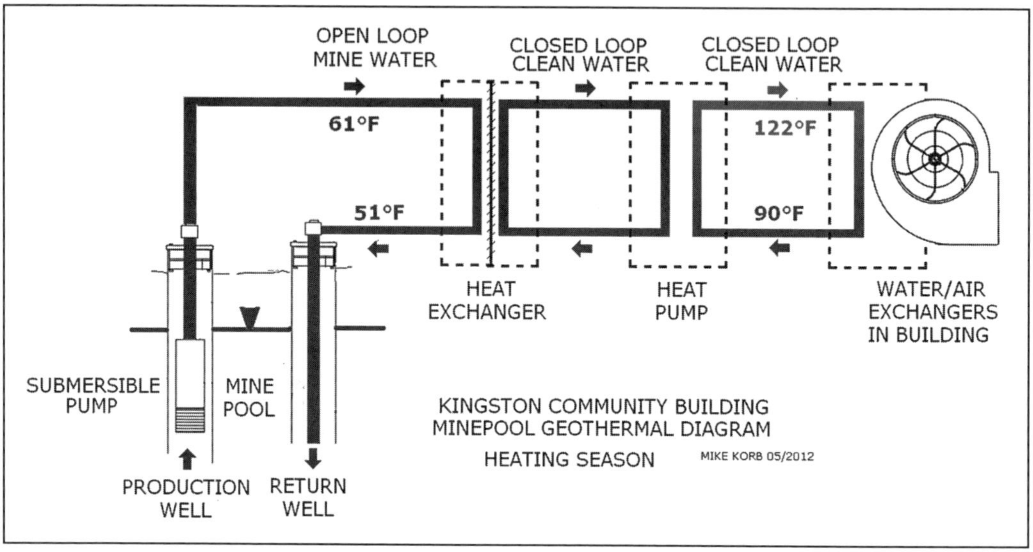

Figure 5. Kingston Community Building geothermal system

heat exchangers during the cooling season, and heat to a closed loop that passes through a heat pump during the heating season. Around 2000, the original well pipes deteriorated, the wells collapsed and the heat exchanger had to be replaced. Plastic piping was installed, and the system now undergoes maintenance every three months. The production well has been replaced once and the pumps three or four times in thirty years.

The building seems comfortable during both heating and cooling seasons, even in the gymnasium section. The Kingston Community Center may have the nation's oldest continuously operating minepool geothermal system.

The annual cost of heating and hot water, and a comparison with other heating systems that might have been installed in 1980, estimated using February 2012 Northeast Pennsylvania fuel costs, are as follows (Table 1).

PAST MINEPOOL GEOTHERMAL PROJECTS

A number of past projects are worthy of mention. A *Popular Science* article reported in 1981 that the Radio Shack in the Wyoming (north of Kingston) Midway Shopping Center was being heated and cooled by minewater. For some twenty years until the store closed, the 3,200 sq.ft. store used a commercial heat pump system utilizing the minepool as its heat sink. The author expressed that "using the flooded coal mines as a heat sink with heat-pump technology might turn what once was a disaster (the Knox Mine disaster caused flooding of the underground mines) into a bonanza for Northeast Pennsylvania."

Also in Kingston, First Hospital Wyoming Valley–Nesbitt Memorial Medical Center utilized a geothermal heat pump system for cooling and for some of its heating for 25 years starting in the early 1980s. This system had early problems with its production well collapsing because of subsidence, and with sediment buildup in the heat exchanger and pump abrasion due to silty sediment, likely from backflushed mine subsidence support material. However, after those problems, the system operated well for some 20 years, with only regular cleanouts of "coal dust."

In Carbondale Township, Lackawanna County, Pennsylvania, Adams Cable Service also installed a minepool geothermal system in the early 1980s. The system operated for about ten years, but was removed because of excessive buildup of iron scale in the heat exchanger and piping. The system had a 500-gallon surge tank for the minewater, installed to keep the pump from running continuously, and the tank allowed aeration and oxidation of the water.

Table 1. Kingston Community Center estimated heating costs (Source: IMWA; updated by Michael Korb 2012)

Fuel oil	$40,000	500%
Electric resistance	$24,000	300%
Gas	$14,000	175%
Minepool heat pump	$8,000	—

John Wesley A.M.E. Zion Church is a historic church in the Hill District of Pittsburgh. In 2006, the PADEP Bureau of Abandoned Mine Reclamation (BAMR) installed piping to prevent flooding of the church by water draining from a 100-acre abandoned mine under the city. In 2008, grants from PADEP and the Foundation for Pennsylvania Watersheds (FPW) provided funds to construct a geothermal circuit to capture the energy from the mine drainage to heat and cool the church, and to provide for a potential development next to the church. The untreated minewater passes through an underground vault with a plate heat exchanger, and safely drains into the stormwater and sanitary sewer drainage system beneath the street. The project operated with no problems for about a year, when the church stopped being used due to some restoration needs. Opportunity still remains for use of this system, with potential development next door to the church, but these options remain in limbo.

I am sure there are more people living over minepools that have put a well down to harvest the energy. There are "urban legends" about a doctor's office in Kingston, a fire hall in Mahanoy City or Minersville, a knitting mill in the Wyoming Valley in the 1960s, and others. The internet makes finding today's "news" easier, but not so much for the past. The International Mine Water Association 1982 Conference Proceedings were just scanned in 2012, so I would not have been able to access that good source if I had attempted this paper in 2011.

POTENTIAL MINEPOOL GEOTHERMAL PROJECTS

Several potential projects are at different early stages. In Johnstown, an ambitious project to treat mine drainage from the minepool that underlies 4,969 acres and contains 3.89 billion gallons of water is under consideration. This project would be a model in resource-recovery by using the value of the energy to pay for treatment of the Inclined Plane Discharge. The geothermal energy might be used to create a geothermal heating district to heat and cool downtown Johnstown. Near Shamokin, a "feasibility study" to treat AMD in Quaker Run, a tributary of Shamokin Creek, was done. The study also explored options for passive and/or active treatment and the potential beneficial reuse of the treated water. A number of beneficial reuses were identified, including low-temperature geothermal. Both of these projects were also funded with grants from PADEP and FPW.

Austin Burke, of the Greater Scranton Chamber of Commerce, has boosted projects for use of systems like the one at Marywood, which would use the constant-temperature water to help control manufacturing processes or for computer room climate control, and has facilitated an assessment of utilizing minewater as a method to reduce energy costs in manufacturing processes for Quadrant EPP, Scranton, which produces engineered polymer products. The assessment was funded by a government-supported nonprofit, Ben Franklin Technology Partners.

The assessment suggests minepool cooling has potential for significant reduction in energy demand for the company's plant expansion, replacing the need for mechanical refrigeration. The work included a detailed review of current and planned future cooling loads, review and preliminary design of the pumping and cooling systems, and a review of mine grids to assess the feasibility of using the minepool. The project looked promising, with an adequate anticipated return on investment, but the drawbacks to the project, such as timing, capital costs, and caution, will cause the project not to proceed.

The company is involved in an overall technology upgrade as well as expansion, and permitting considerations, deed/title searches, and system specifications for a somewhat 'unknown' technology might mean that minepool utilization would delay the overall project. The projected capital and operating costs of the geothermal system are much less than the costs for conventional cooling towers, but the cooling system has to work, and many companies would be cautious of pioneering an "untested" technology.

CONCLUSIONS

The number of such minepool systems in place is incredibly small, and none approach the volumes or with the mission-critical status that this system would be expected to operate at. Validation of the technology has ramifications for the entire manufacturing community in Northeast Pennsylvania, as well as in other coal regions.

It appears that the author of the 1981 *Popular Science* article was being overly optimistic in his hope for utilization of minepool geothermal systems in Pennsylvania. The possibilities for use of geothermal energy from minepools seem to be unlimited, yet only a few facilities are looking at, while even less are using, this apparently great energy source.

Why Aren't We Using Minepool Geothermal in Pennsylvania?

There are a number of opinions, predominantly that people don't know enough about minepools, legal and regulatory questions, geothermal technology and economics, and the risks involved in working with minewater and minepools.

Knowledge About Minepools

There is a lack of information about the extent and locations of minepools, even in government agencies. Ohio, Indiana, West Virginia, and Pennsylvania are all at various stages of surveying, outlining, and defining their minepools.

People are not really aware there are minepools beneath their feet. The public's attention is drawn when something bad happens. Minepools were part of national news as we watched newspapers and television as the Quecreek miners were rescued from beneath Somerset County, Pennsylvania after they mined into an ill-defined minepool.

Legal and Regulatory Questions

There needs to be a clarification and better understanding of water ownership rights and obligations, as well as the permissions and permitting required for drilling, withdrawal and injection. Marywood addressed this issue by approaching the mineral rights owner who had no objection to the use of the water in the way Marywood proposed. Pennsylvania's groundwater law is based on the reasonable use or "American Rule," which provides that a landowner may withdraw groundwater from beneath the property for "natural and ordinary" usage, whereas extraction for use off-site is "unreasonable" and "unlawful." In light of the benefits associated with the withdrawal, treatment, and use of the minepool water, PADEP continues to explore resolutions to minepool ownership issues.

These issues are also complicated by the potential liability of affecting the abandoned mine drainage discharge. Pennsylvania did some work on this in its Mining and Reclamation Advisory Board Orphan Mine Discharge Task Force Action Plan several years ago. Recent interest in utilization of mine drainage in hydraulic fracturing of gas wells in the state likely will increase the clarification of these issues; however, the PADEP recently released "White Paper on Utilization of Mine Influenced Water for Natural Gas Extraction Activities" gives little guidance or resolution.

Lack of Knowledge About Heat Pump Technology

A problem in comprehending this technology is that it is difficult for a layman (or me for that matter) to understand how heat extracted from 55° water can heat anything. I was aware of geothermal heat pumps (GHPs) when I replaced the air-source heat pump in my home (not over a minepool) five years ago but couldn't find a contractor that had expertise in the field. More HVAC contactors are informed and involved today, but it still isn't a large pool to choose from, and HVAC contractors who are involved complain that architects "are the ones holding back GHP because they don't understand the calculations." The people that installed the equipment in the 1980s were certainly "thinking outside the box."

Lack of Knowledge About Minepool Geothermal Economics

It appears that the economics of minepool geothermal are good, and that usage of minepool water as a constant temperature bath for computer applications or manufacturing processes has the best chance for success. Exploitation of the method for heating has been limited in the past due to relatively low conventional energy costs, and is still affected by current low gas pricing. A DOE project to install minepool geothermal systems at University of Scranton, the Scranton Army Ammunition Plant, and at a public housing project in Scranton in the 1980s was deemed not feasible because of pumping costs.

I have not been able to duplicate many of the published cost and comparative calculations on geothermal heat pumps. They all seem to make some assumptions that I can't identify. Much of the published information indicates that the investment involved is not great, but that does not seem to be the case. Ball State University is installing a geothermal system estimated to cost $83 million which will save $2 million annually. Also, in my discussions with some who have rejected minewater geothermal proposals, they cited the high estimated capital cost of the heat pump equipment.

I recently used an on-line calculator to look at a system for my 2,400-sq.ft. home. Estimated cost of installation of a ground-source heat pump and ground loops, with no ductwork, was "in the ballpark of" $23,250 (about $10,700 for the ground loop system and $12,500 for the heat pump). That would result in a cost to me of $15,190 "including the recent 30% federal tax credit" and a $1,085 PPL rebate. That seems like a lot of money compared to my air-source heat pump which cost $5,578.45 five years ago.

Grants have been important in the small amount of promotion the technology has had. Likely a grant which would build a functioning system and monitor and report the results would be the best way of promoting minepool geothermal. The Commonwealth of Pennsylvania offers some fourteen different grant or loan programs that apply to geothermal systems; non-profits and government supported agencies also provide funding and expertise.

However, a process that has been functioning for 30 years really shouldn't need a grant if the process is a good, economical one. Depending upon grants may limit the possibilities, and may promote unsustainable technology. If we are unable to do anything without a grant, our efforts are directed into getting grants rather than solving problems. Most grants have no requirement for follow-up monitoring or reports of their results. It's too bad that the past 30 years of operation of the Kingston system wasn't monitored and studied—maybe a paper some other day?

But of course, the technology should stand on its own economic merit, and there hasn't been much building in the areas over minepools. GSHPs can be difficult and costly to install in retrofit applications. The Pennsylvania coal regions are not areas where large "green" housing developments are going up and never have been. There has been some discussion about utilizing the old steam heat systems in Scranton and Wilkes-Barre in district heating, but it has met with underenthusiastic response. Industrial development has declined for the past couple years, but the only real growth has been in industrial parks, mostly located on reclaimed mine lands. Local community development agencies have boosted minepool geothermal, but to no avail. With no indication that any business or group stands to make a lot on minepool geothermal energy, it may be that public grants will be the only way for this technology to progress.

Risks Involved

Like minepools, very little common knowledge about mines and mining exists. "My granddad worked in the mines" is the extent most people know except for the annual commemoration of the disasters; the news coverage of miners trapped in Chile, or some-such.

However, the risks and uncertainties of drilling into a mine, hitting the mine opening with the hole, potential failure and subsidence of the boring, water quality, avoiding pollution of the minewater, and collapse of mine workings caused by drilling, are certainly concerns for a developer looking at minepool geothermal. Water quality, buildup of yellow-boy, and corrosiveness of some minewaters are all concerns to be addressed when a minepool is being used. Access to this information isn't readily available.

Reputation of Mining

Nevertheless, the #1 opinion on why we don't have more systems involving minepools is just simply that people, especially people in the coal regions, just don't like coal and mining.

It's still coal country, even where there's hardly any coal mined anymore. The old timers still mine coal in the bars and can point to where they used to work under this street or that, but most people don't know anything about the mines except for the gob piles that the off-roaders kick up dust on each weekend, the orange-bottomed "cricks," the dumps in the "strippins," and the fire that's burning underground. People still are from the "patch"—one of the remnants of double-block company houses near the mine in William Penn or Brady's Bend. The onion domes of the eastern-rite churches are still in town, but the breakers and beehive ovens are fallen or torn down. The Fell Tavern, where anthracite was first burned, is now a parking lot. Those who worked in the mines, whose livelihood depended on the hard, often dirty work, didn't dislike coal mining like the public does today.

Robert F. Kennedy, Jr., president of the environmental advocacy group Waterkeeper Alliance, in a May protest to a proposal to export coal from West Coast ports, said, "Coal will undermine everything that you love...I've seen what it's done to small Appalachian towns. Its ruined democracy, corrupted politicians and literally drove people out of town. Do not let it corrupt this community. You are at the front lines of this battle...Anybody who touches coal gets poisoned by it. You don't just get sick. It poisons democracy, it poisons communities, it poisons values."

SUMMARY

If you Google "Michael C. Korb" you get about 2,320 results, and if you Google "minepool geothermal" you get about 306 results. I know I'm a legend in my own mind, but it ought to be a little closer than that! More people should be more interested in inquiring about minepool geothermal than about Mike Korb. The public, architects, contractors, and developers all need more information about minepool availability, about minewater, about legal and regulatory questions, about heat pump technology and the economics of minepool geothermal; and there has to be a better understanding of the risks involved with utilizing minepools, and the tools that can be used to minimize these risks. We have to start building houses, commercial buildings, and industry in the coal regions. Perhaps utilizing the old mines as a "green" energy source will help put the pride of the coal region's heritage back into the community.

To paraphrase from a presentation by Terry Ackman and George Watzlaf: "Water from a mine is a terrible thing to waste."

REFERENCES

Brian Oram, B.F. Environmental Consultants Inc., Laboratory Results–Marywood—8 Hour Test and Purging Evaluation, unpublished, 2010.

Bernie McGurl, personal communication from McGurl (Lackawanna River Corridor Association) to Michael Korb (PADEP BAMR), 2012.

J.F. Schubert and M.J. McDaniel, Argonne National Lab, Using Mine Water for Heating and Cooling–IMWA Proceedings 1982, http://www.imwa.info/docs/imwa_1982/IMWA1982_Schubert_063.pdf, ©International Mine Water Association 2012.

Sustainability in Surface Coal Mine Operations: The Importance of Good Geotechnical Design

Z. Agioutantis
Department of Mineral Resources Engineering, Technical University of Crete, and Virginia Tech, VCCER, Blacksburg, Virginia

C. Steiakakis
Geotechpedia Team

ABSTRACT

Sustainable operation of large open pit coal mines requires substantial geotechnical involvement starting from the initial investigation to the mine design and finally to the mine operation phase. Typically, geotechnical involvement is limited in the initial steps of mine investigation and mine planning. During this phase, slopes are usually designed based on best practice or prior experience, without significant geotechnical involvement. During mine operations, geotechnical involvement is kept at a minimum, used only when stability problems occur, at which point the experts are called in to resolve the issues. Certainly this approach can disrupt mine operations either with slope failures or abandonment of unstable areas. In the case of extended events, mine operations can be jeopardized with significant impacts to mine life as well as to nearby communities.

This presentation will show some of the ties between the technical aspects of mine design with emphasis on the geotechnical component on one side and sustainability of operations as well as the public image on the other side. A case study will be presented in which the geotechnical involvement has significantly improved production in unstable or even moving slopes and also ensured sustainability of operations.

Ethics in Energy Sustainability

Peter H. Denton
Red River College, Winnipeg, Manitoba

ABSTRACT

Efforts to achieve energy sustainability in domestic, continental or global contexts require much more than the solution of technical problems relating to the discovery, production and distribution of resources. These efforts also require a thorough understanding of the environmental, social and cultural systems involved. If the decision-making process behind these efforts does not reflect such an understanding, public support will be short-lived, fragmented and inherently unstable. In turn, this will undermine the necessary policy decisions in the direction of sustainability that require significant changes in social and cultural patterns of energy consumption.

Index

Note: *f.* indicates figure; *t.* indicates table.

A

Acid mine drainage, in micro-hydroelectric power generation (Pennsylvania), 512–519
Antrim Acid Mine Drainage Treatment Plant, 512–513
  acid mine drainage in micro-hydroelectric power generation, 512–519
Apogee Coal Company, 467
Appalachia
  central, and inventory of bacterial water quality impairments, 214–228
  central, defined, 173, 216–217
  chronic disease in coal-dependent communities (research evaluation), 172–188
  comparison of mortality rates between coal mining and non-coal mining counties (focus on W. Va.), 189–201
  effect of TDS on benthic macroinvertebrates in coal mine streams, 371
  electrical conductivity in streams flowing from headwater valley fills (Va.), 286
  evaluation of current measures of aquatic biological integrity in coalfields, 381–394
  identification of toxic agents, potential exposure routes, and possible associated diseases in coal mining communities (literature search), 229–243
  metastudy of research used by EPA et al. to evaluate impact of coal mining discharge on watersheds, 2
  Mine Land Stewardship Initiative, 20
  modeling in comparison of Approximate Original Contour (AOC) design with geomorphic landform design alternative for valley fill, 415–423
  population and exposure characteristics of coal mining and non-coal mining counties (W. Va.), 202–213
  prediction of TDS release from mine spoils, 275–285
  prediction of TDS release from overburden in coal fields, 255–264
  problems with point source discharge regulations in protection of streams, 4
  public ecology studies of, 99–110
  and regulatory, market, and energy policy challenges to coal industry, 124
  scenario analysis in prediction of aquatic conditions in intensively mined watershed (W. Va.), 5–19
  SDPS model in sizing of barrier pillars for stream protection, 319–335
  study of mining industry and community poverty, 86–98
  "value of information" approach in assessing modeling of landscape-to-water quality relationships, 3
  See also Kentucky; Valley fills; Virginia; West Virginia
Appalachian Regional Commission (ARC), 86–98
  and definition of central Appalachia, 173, 216–217
Appalachian Research Initiative for Environmental Science (ARIES), 486–487, 491, 501
  and TDS research, 275, 284
Appalachian Wildlife Foundation, 20

Armstrong Dock and Preparation Facility (Centertown, Ky.), 33–34
  and Portland cement in stabilization of fine coal waste, 32–41

B

Babb Creek Watershed Association (Pennsylvania), 512–513
  acid mine drainage in micro-hydroelectric power generation, 512–519
Benthic macroinvertebrates
  and evaluation of effects of TDS in mine effluent, 348–361
  flaw in electrical conductivity testing as determinant of health of, 472–473
  response to TDS in coal mine streams (Appalachia), 371
  suggested inadequacy as sole indicator of aquatic biological integrity in Appalachian coalfields, 381–394
  and suggested inapplicability of TDS/conductivity as metric for stream health (Va.), 362–370
Biochemical reactors (BCRs), in removal of selenium from mine water discharge, 443–449
Blasting
  assessment of secondary environmental effects, 478
  comprehensive assessment of environmental effects (mine to fill), 478–479
  and environmental effects, 476–477, 487–489
  fragmentation modeling and assessment, 479–487
  NOx emission of blasting agents in surface coal mining (W. Va. study), 501–511
  traditional assessment of environmental effects, 477–478

C

Carlson Natural Regrade with GeoFluv software, 400, 410, 412, 419
China, and modeling of well designs for sustainable degasification of coal bed methane (Qinshui Basin), 153–169
Clean Water Act, 4, 450, 454, 465, 473
Coal bed methane (CBM)
  modeling of CO_2 injection in production of (Russell County, Va.), 140–152
  modeling of well designs for sustainable degasification of (Qinshui Basin, China), 153–169
  Russell County, Va., 140–141
Coal gasification. *See* Underground coal gasification
Coal mining
  and chronic disease in coal-dependent central Appalachian communities (research evaluation), 172–188
  comparison of mortality rates between coal mining and non-coal mining Appalachian counties (focus on W. Va.), 189–201
  effects of substrate treatments and seeding regimes on survival of native tree species and herbaceous ground cover at reclaimed surface mine, 467
  electrical conductivity in streams flowing from headwater valley fills (Va.), 286
  and evaluation of current measures of aquatic biological integrity (Appalachia), 381–394
  geotechnical engineering in sustainability of surface mining operations, 529
  geothermal heat exchange using minepool of closed coal mine (Scranton), 520–528
  identification by literature search of toxic agents, potential exposure routes, and possible associated diseases in coal mining communities (Appalachia), 229–243
  and inventory of bacterial water quality impairments (central Appalachia), 214–228
  leaf litter breakdown in assessment of coal mine stream restorations (Appalachia), 465
  metastudy of research used by EPA et al. to evaluate impact of discharge on watersheds (Appalachia), 2

population and exposure characteristics of coal mining and non-coal mining counties (W. Va.), 202–213
prediction of TDS release from mine spoils (Appalachia), 275–285
prediction of TDS release from overburden in Appalachian coal fields, 255–264
regulatory, market, and energy policy challenges to industry in Appalachia, 124
and response of benthic macroinvertebrates to TDS in streams influenced by (Appalachia), 371
scenario analysis in prediction of aquatic conditions in intensively mined watershed (Coal River, W. Va.), 5–19
software for determination of production, cost, and environmental impact of loading equipment in surface mining (Appalachia), 490–500
surface coal mine blasting optimization and mitigation of environmental impacts (mine to fill), 476–489
treatment technologies for removal of selenium from mine water discharge, 443–449
voluntary performance standards for ecosystem conservation (Mine Land Stewardship Initiative), 20
West Virginia water quality standards, and analysis of low-level selenium in coal mine discharges, 451–452
See also Longwall mining; Mountain-top mining; Room and pillar mining; Streams; Underground coal gasification; Valley fills

Coal preparation
dewatering of slurries with pilot-scale plate filter press, 42–50
DriJet electronic coal sorter in reduction of environmental impacts, 22–31
leaching tests in determination that co-disposal of coarse and fine wastes could retard acid generation and mobility of trace elements, 60–76
persistence of diesel range organic compound residuals in process waters, 51–59
Portland cement in stabilization of fine coal waste, 32–41
See also Underground coal gasification

Coal River watershed, West Virginia, 5, 10
scenario analysis in prediction of aquatic conditions in intensively mined watershed, 5–19

Coal Seam Group, Southeast Regional Carbon Sequestration Partnership (SECARB), 140

Community well-being
boundary crossings, defined, 99, 101
edgework, defined, 99, 101
modeling determinants of local community acceptance of energy resources and materials mining, 111–120
public ecology in Appalachia, 101–110
public ecology, defined, 99, 100–101
and relationship between ethics and technology, 121
study of mining industry and community poverty (Appalachia), 86–98
See also Environmental considerations; Public ecology; Public health; Sustainability/sustainable development

Conductivity. *See* Electrical conductivity

Constructed treatment wetlands, in removal of selenium from flue gas desulfurization effluents, 450

D

Diabetes mellitus (Appalachia), 172–188

Diesel
diesel range organics (DROs), and persistence of compound residuals in coal process waters, 51–59
NOx emission of diesel-fueled equipment (W. Va. study), 501–511

Discrete choice modeling, 114

DriJet electronic dual-energy X-ray coal sorter, 22–31

Dumps Creek watershed, Va., 365–366
and suggested inapplicability of TDS/conductivity as metric for stream health, 362–370

E

Electrical conductivity
- comparative laboratory and field study of four commercially available sensors (specific conductivity), 470
- flaw in conductivity testing as determinant of benthic macroinvertebrate health, 472–473
- as indicator of stream impairment, evaluation of, 372–380
- overburden field leach test utilizing drill cores to determine specific conductivity of geologic strata, 471
- in streams flowing from valley fill (Va.), 286
- suggested inapplicability of TDS/conductivity as metric for stream health (Va.), 362–370
- whole effluent toxicity (WET) tests in study of ions and conductivity as indicators of toxicity, 474

Electronic dual-energy X-ray coal sorter (DriJet), 22–31

Energy sustainability, and ethics, 530

Environmental considerations
- algorithm for site-specific adjustments of horizontal displacements due to sloping terrain (Surface Deformation Prediction System), 308–318
- DriJet electronic coal sorter in reduction of environmental impacts, 22–31
- evaluation of current measures of aquatic biological integrity in Appalachian coalfields, 381–394
- evaluation of effects of TDS in mine effluent on benthic macroinvertebrates, 348–361
- event tree analysis in risk assessment of subsidence for safety of mines and nearby surface water, 288–296
- gas migration/stray gas problem (Marcellus shale-gas play, Pa.), 137–139
- NOx emission of blasting agents in surface coal mining (W. Va. study), 501–511
- SDPS in assessment of factors affecting springs undermined by longwall mining, 336–346
- SDPS model in sizing of barrier pillars for Appalachian stream protection, 319–335
- software for determination of production, cost, and environmental impact of loading equipment in surface mining (Appalachia), 490–500
- subsurface subsidence model in prediction of longwall mining influences on overburden hydrological system, 297–307
- suggested inapplicability of TDS/conductivity as metric for stream health (Va.), 362–370
- surface coal mine blasting optimization and mitigation of environmental impacts (mine to fill), 476–489
- of underground coal gasification, 125, 127–129, 134–136
- *See also* Community well-being; Public health; Selenium; US Environmental Protection Agency

Environmental Protection Agency (EPA). *See* US Environmental Protection Agency

Ethics
- and technology, 121
- and understanding of environmental, social, and cultural factors in energy sustainability, 530

Event tree analysis, 291
- in risk assessment of subsidence for safety of mines and nearby surface water, 288–296

F

Fleet Production and Cost Analysis Program (FPC), 490

Flue gas desulfurization, and constructed treatment wetlands in removal of selenium from effluents, 450

Fluidized bed reactors (FBRs), in removal of selenium from mine water discharge, 443–449

Forestry Reclamation Approach (FRA), 463, 464

G

GAC. *See* Granular activated carbon
GeoFluv software, 419–420. *See also* Carlson Natural Regrade with GeoFluv software
Geomorphic landform design, 397, 398–399
  as alternative approach to valley fill construction, in comparison with Approximate Original Contour (AOC) design (Appalachia), 415–423
  application to valley fill design, with quantification of mature landform data, 405–414
  applied to creation of more natural valley fills, 397–404
  purported benefits of, 397
GeoSlope software, 416–419
GeoStudio software, 421
Geotechnical engineering, in sustainability of surface coal mine operations, 529
Geothermal heat exchange
  minepools as resource for, 520
  using minepool of closed coal mine (Scranton, Pa.), 520–528
Granular activated carbon (GAC), in removal of selenium from mine water discharge, 443–449
Gross Happiness Index, 121
Groundwater
  consideration of, in event tree analysis of subsidence for safety of mines and nearby surface water, 288–296
  modeling in comparison of groundwater seepage quantity and quality in conventional and geomorphic valley fill design, 246–254
  See also Valley fills
Guy Cove stream restoration project (Kentucky), 463, 464

H

Haulage Analysis Tool (HAT), 490
Hydraulic engineering, 456

IC 8741, 288–296
Ion exchange (IX), in removal of selenium from mine water discharge, 443–449

K

Kentucky
  Guy Cove stream restoration project, 463, 464
  valley fills, and leach test utilizing drill cores to determine specific conductivity of geologic strata and reduce conductivity in water discharge, 471

L

Longwall mining
  event tree analysis in risk assessment of subsidence for safety of mines and nearby surface water, 288–296
  SDPS in assessment of factors affecting springs undermined by longwall mining (Pa.), 336–346
  and SDPS model in sizing of barrier pillars for Appalachian stream protection, 319–335
  and subsurface subsidence model in prediction of influences on overburden hydrological system, 297–307

M

Marcellus shale-gas play (Pennsylvania), 137
  gas migration/stray gas problem, 137–139
Marywood University (Scranton, Pa.), 520
  geothermal heat exchange using minepool of closed coal mine, 520–528
Mine Land Stewardship Initiative, 20
Mineral Separation Technologies, Inc., 27
Mining
  algorithm for site-specific adjustments of horizontal displacements due to sloping terrain (Surface Deformation Prediction System), 308–318
  and community poverty, study of assumed link, 86–98
  evaluation of effects of TDS in mine effluent on benthic macroinvertebrates, 348–361
  modeling determinants of local community acceptance, 111–120
  and suggested inapplicability of TDS/conductivity as metric for stream health in Dumps Creek watershed (Va.), 362–370

See also Coal mining; Community well-being; Longwall mining; Room-and-pillar mining; Sustainability/sustainable development
Mixed multinomial logit (ML) model, 116
MNL model. *See* Multinomial logit model
MNP model. *See* Multinomial probit model
Modeling
algorithm for site-specific adjustments of horizontal displacements due to sloping terrain (Surface Deformation Prediction System, or SDPS), 308–318
blasting fragmentation and assessment, 476–489
in comparison of Approximate Original Contour (AOC) design with geomorphic landform design alternative for valley fill (Appalachia), 415–423
in comparison of groundwater seepage quantity and quality in conventional and geomorphic valley fill design, 246–254
of CO_2 injection in production of coal bed methane (Russell County, Va.), 140–152
of determinants of local community acceptance of energy resource mining, 111–120
digital elevation, terrain, and surface models, 309
discrete choice modeling, 114
mixed multinomial logit (ML) model, 116
multinomial logit (MNL) model, 114–115
multinomial probit (MNP) model, 115–116
SDPS in assessment of factors affecting springs undermined by longwall mining, 336–346
SDPS in sizing of barrier pillars for Appalachian stream protection, 319–335
subsurface subsidence model in prediction of longwall mining influences on overburden hydrological system, 297–307
of well designs for sustainable degasification of coal bed methane (Qinshui Basin, China), 153–169
Mountain-top mining (MTM)
and chronic disease in coal-dependent central Appalachian communities (research evaluation), 172–188
and evaluation of current measures of aquatic biological integrity in Appalachian coalfields, 381–394
finding of no statistical relationship to poverty (Appalachia), 86–98
and ionic composition of valley fill, 265–274
and valley fills, 246–247
See also Streams; Valley fills
Multinomial logit (MNL) model, 114–115
Multinomial probit (MNP) model, 115–116

N

National Pollutant Discharge Elimination System (NPDES), 443, 450, 472
Natural channel design, 454–458, 460–462
biological response to, 459–460
compared with hydraulic engineering, 456
and Forestry Reclamation Approach in Guy Cove stream restoration project, 463, 464
and stream mitigation, 458–459
"Natural resource curse," 86–98
NOx, and emission from blasting agents and diesel equipment in surface coal mining (W. Va. study), 501–511
NPDES. *See* National Pollutant Discharge Elimination System

P

Panther Wildlife Management Area (West Virginia), 405, 406–407
quantification of mature landform data for enhancement of geomorphic landform design of valley fills, 405–414
Pennsylvania
acid mine drainage in micro-hydroelectric power generation (Antrim Acid Mine Drainage Treatment Plant), 512–519
Department of Environmental Protection (PADEP), 137–139, 513, 520

geothermal heat exchange using mine-pool of closed coal mine (Scranton), 520–528
Marcellus shale-gas play, and gas migration/stray gas problem, 137–139
SDPS in assessment of factors affecting springs undermined by longwall mining, 336–346
Plate filter presses
in dewatering of coal slurries, 42–50
plate-and-frame press filter in association with Portland cement in stabilization of fine coal waste, 32–41
Public ecology
in Appalachia, 99–110
defined, 99
See also Community well-being; Sustainability/sustainable development
Public health
chronic disease in coal-dependent central Appalachian communities (research evaluation), 172–188
comparison of mortality rates between coal mining and non-coal mining Appalachian counties (focus on W. Va.), 189–201
identification by literature search of toxic agents, potential exposure routes, and possible associated diseases in coal mining communities (Appalachia), 229–243
inventory of bacterial water quality impairments (central Appalachia), 214–228
population and exposure characteristics of coal mining and non-coal mining counties (W. Va.), 202–213
See also Community well-being

Q

Quantitative Risk Assessment, 291

R

Reagents (diesel range organic compound residuals), persistence in coal process waters, 51–59
Risk assessment, and event tree analysis of subsidence for safety of mines and nearby surface water, 288–296
Room-and-pillar mining
event tree analysis in risk assessment of subsidence for safety of mines and nearby surface water, 288–296
and SDPS model in sizing of barrier pillars for Appalachian stream protection, 319–335
Russell County, Va., 140–141
modeling of CO_2 injection in production of coal bed methane, 140–152
and suggested inapplicability of TDS/conductivity as metric for stream health in Dumps Creek watershed, 362–370

S

Scenario analysis. *See* Watershed Futures Planner
Scranton, Pennsylvania
geothermal heat exchange using mine-pool of closed coal mine (Marywood University), 520–528
Marywood University, 520
SDPS. *See* Surface Deformation Prediction System
SECARB. *See* Southeast Regional Carbon Sequestration Partnership
Selenium
analysis of low-level presence in coal mine discharges (W. Va.), 451–452
concentrations and solid-state chemical speciation in black shale associated with West Virginia coal seam, 438–442
in flue gas desulfurization effluents, and constructed treatment wetlands in removal of, 450
transport rate of leachate and removal by adsorption to layered ferrihydrite (W. Va.), 426–437
treatment technologies for removal from mine water discharge, 443–449
Southeast Regional Carbon Sequestration Partnership (SECARB), 140

Spectrometry and spectroscopy, in determination of selenium concentrations and solid-state chemical speciation in black shale associated with West Virginia coal seam, 438–442
Springs, and assessment of factors affecting longwall undermining of (Pa.), 336–346
Streams
  electrical conductivity in streams flowing from valley fill (Va.), 286
  evaluation of electrical conductivity as indicator of stream impairment, 372–380
  Forestry Reclamation Approach to restoration, 463, 464
  and hydraulic engineering, 456
  leaf litter breakdown in assessment of coal mine stream restorations (Appalachia), 465
  natural channel design and biological effects, 454–462
  problems with point source discharge regulations in protection of, 4
  SDPS model in sizing of barrier pillars for protection of (Appalachia), 319–335
  suggested inapplicability of TDS/conductivity as metric for stream health in Dumps Creek watershed (Va.), 362–370
Subsurface subsidence model, 297–307
Surface Deformation Prediction System (SDPS), 308
  and algorithm for site-specific adjustments of horizontal displacements due to sloping terrain, 308–318
  in assessment of factors affecting springs undermined by longwall mining, 336–346
  in sizing of barrier pillars for Appalachian stream protection, 319–335
Surface Mine Control and Reclamation Act of 1977, 276
Sustainability/sustainable development, 78
  definitions, 78–80
  ethics, and understanding of environmental, social, and cultural factors in energy sustainability, 530
  geotechnical engineering in sustainability of surface coal mining operations, 529
  implementation, 80–81, 83–85
  modeling of well designs for sustainable degasification of coal bed methane (Qinshui Basin, China), 153–169
  operational, 79–80
  organizational, 79
  project, 80
  SD Action Plan, 81, 82–83
  SD Checklist, 81–82, 83
  See also Community well-being; Public ecology

T

TALPAC, 490–491
Total dissolved solids (TDS)
  and conductivity as metric for stream health, and suggested inapplicability to Dumps Creek watershed (Va.), 362–370
  development of valley fill construction techniqes to reduce levels of, 396
  effect on benthic macroinvertebrates in coal mine streams (Appalachia), 371
  evaluation of effects of mine effluent on benthic macroinvertebrates, 348–361
  ionic composition and natural attenuation in valley fill, 265–274
  prediction of release from mine spoils (Appalachia), 275–285
  prediction of release from overburden in Appalachian coal fields, 255–264

U

Underground coal gasification, 125–127
  environmental considerations, 125, 127–129, 134–136
  operational parameters, 125, 129–136
United Nations Conference on Happiness, 121
US Environmental Protection Agency (EPA)
  and flaw in electrical conductivity testing as determinant of benthic macroinvertebrate health, 472–473
  metastudy of research used by EPA et al. to evaluate impact of coal mining discharge on watersheds (Appalachia), 2
US Geological Survey, 471

V

Valley fills
- and application of geomorphic landform design, with quantification of mature landform data, 405–414
- best management practices and adaptive mining methods for construction of, 396
- comparison of Approximate Original Contour (AOC) design with geomorphic landform design alternative (Appalachia), 415–423
- defined and described, 246–247, 265
- electrical conductivity in streams flowing from (Va.), 286
- geomorphic landform design in construction of more natural and stable versions, 397–404
- ionic composition and natural attenuation of TDS in, 265–274
- leach test utilizing drill cores to determine specific conductivity of geologic strata and reduce conductivity in water discharge, 471
- modeling in comparison of groundwater seepage quantity and quality in conventional and geomorphic fill design, 246–254
- and natural channel design and Forestry Reclamation Approach in stream restoration, 463, 464
- *See also* Streams

Virginia
- and coal bed methane (Russell County), 140–141
- Department of Mines, Minerals and Energy, 286
- electrical conductivity in streams flowing from headwater valley fills, 286
- modeling of CO_2 injection in production of coal bed methane (Russell County), 140–152
- suggested inapplicability of TDS/conductivity as metric for stream health in Dumps Creek watershed, 362–370

Virginia Stream Condition Index (VSCI), 362–370

W

Water quality
- discharge limits to protect streams, 4
- electrical conductivity as measure of, and comparative study of four commercially available sensors, 470
- electrical conductivity in streams flowing from headwater valley fills (Va.), 286
- flaw in electrical conductivity testing as determinant of benthic macroinvertebrate health, 472–473
- inventory of bacterial water quality impairments (central Appalachia), 214–228
- leach test utilizing drill cores to determine specific conductivity of geologic strata and reduce conductivity in water discharge, 471
- metastudy of research used by EPA et al. to evaluate impact of mining discharge (Appalachia), 2
- modeling in comparison of groundwater seepage quantity and quality in conventional and geomorphic valley fill design, 246–254
- and prediction of TDS release from overburden in Appalachian coal fields, 255–264
- transport rate of leached selenium and removal by adsorption to layered ferrihydrite (W. Va.), 426–437
- treatment technologies for removal of selenium from mine water discharge, 443–449
- "value of information" approach in assessing modeling of landscape-to-water quality relationships, 3
- West Virginia standards, and analysis of low-level selenium in coal mine discharges, 451–452
- whole effluent toxicity (WET) tests in study of ions and conductivity as indicators of toxicity, 474

Watershed Futures Planner (software), 5–19

Watersheds
- metastudy of research used by EPA et al. to evaluate impact of mining discharge (Appalachia), 2

problems with point source discharge regulations in protection of streams, 4
scenario analysis in prediction of aquatic conditions in intensively mined watershed (Coal River, W. Va.), 5–19
suggested inapplicability of TDS/conductivity as metric for stream health in Dumps Creek watershed, 362–370
valley fills, and modeling in comparison of groundwater seepage quantity and quality in conventional and geomorphic fill design, 246–254
"value of information" approach in assessing modeling of landscape-to-water quality relationships, 3

West Virginia
Coal River watershed, 5, 10
comparison of mortality rates between coal mining and non-coal mining counties (including Appalachian counties outside W. Va.), 189–201
modeling in comparison of groundwater seepage quantity and quality in conventional and geomorphic valley fill design, 246–254
NOx emission of blasting agents in a surface coal mine, 501–511
Office of Miners' Safety and Training, 202
population and exposure characteristics of coal mining and non-coal mining counties, 202–213
quantification of mature landform data for enhancement of geomorphic landform design of valley fills, 405–414
scenario analysis in prediction of aquatic conditions in intensively mined watershed (Coal River), 5–19
selenium concentrations and solid-state chemical speciation in black shale associated with coal seam, 438–442
transport rate of leached selenium and removal by adsorption to layered ferrihydrite, 426–437
valley fills, and leach test utilizing drill cores to determine specific conductivity of geologic strata and reduce conductivity in water discharge, 471
water quality standards, and analysis of low-level selenium in coal mine discharges, 451–452
whole effluent toxicity (WET) tests in study of ions and conductivity as indicators of toxicity, 474

West Virginia Stream Condition Index (WVSCI), 4, 474
and evaluation of electrical conductivity as indicator of stream impairment, 372–380, 472–473

Whetstone watershed, West Virginia, 405, 406–407
quantification of mature landform data for enhancement of geomorphic landform design of valley fills, 405–414

Whole effluent toxicity (WET) tests, 474

World Happiness Report (United Nations), 121

Z

Zero-valent iron (ZVI), in removal of selenium from mine water discharge, 443–449